Geography of the Physical Environment

The *Geography of the Physical Environment* book series provides a platform for scientific contributions in the field of Physical Geography and its subdisciplines. It publishes a broad portfolio of scientific books covering case studies, theoretical and applied approaches as well as novel developments and techniques in the field. The scope is not limited to a certain spatial scale and can cover local and regional to continental and global facets. Books with strong regional focus should be well illustrated including significant maps and meaningful figures to be potentially used as field guides and standard references for the respective area.

The series appeals to scientists and students in the field of geography as well as regional scientists, landscape planners, policy makers, and everyone interested in wide-ranging aspects of modern Physical Geography. Peer-reviewed research monographs, edited volumes, advance and undergraduate level textbooks, and conference proceedings covering the major topics in Physical Geography are included in the series. Submissions to the Book Series are also invited on the theme 'The Physical Geography of…', with a relevant subtitle of the author's/editor's choice. Please contact the Publisher for further information and to receive a Book Proposal Form.

U. C. Sharma · M. Datta ·
Vikas Sharma

Soils in the Hindu Kush Himalayas

Management for Agricultural Land Use

Springer

U. C. Sharma
Division of Natural Resources
Management
Indian Council of Agricultural
Research
Jammu, Jammu & Kashmir, India

M. Datta
Government Agricultural College
Indian Council of Agricultural
Research and Principal
Kolkatta, West Bengal, India

Vikas Sharma
Division of Soil Science
and Agricultural Chemistry
Shere-Kashmir University
of Agricultural
Jammu, India

ISSN 2366-8865 ISSN 2366-8873 (electronic)
Geography of the Physical Environment
https://doi.org/10.1007/978-3-031-11458-8

This Springer imprint is published by the registered company Springer Nature Switzerland AG
The registered company address is: Gewerbestrasse 11, 6330 Cham, Switzerland
(www.springernature.com/mycopy)

Foreword by I. P. Abrol

Mountains cover more than 22% of the earth's land area and are home to around one billion people, or about 12% of the world's population. Producing crops and livestock in mountain environments can present many difficult challenges. Compared to lowlands, highland farmers are faced with shorter growing seasons, hostile climatic conditions, complex and variable soils, high gradients, and greater distance to the markets. The soils are usually less favorable, and the weather is more prone to extreme events that can lead to sudden disasters. Hunger and malnutrition are widespread in the Hindu Kush Himalayan (HKH) countries. In the HKH region, more than 30% of people face food insecurity and malnutrition. Eradication of hunger and all forms of malnutrition is one of the main Sustainable Development Goals (SDGs) and a core mandate of the Zero Hunger Goal (SDG2) of the United Nations.

Mountain regions, despite their peculiarity, environmental sensitivity, and inaccessibility, are quite significant for global sustainability as nearly 70% of the world's population depends on the mountains for the supply of freshwater, sources of renewable energy, and rich biodiversity. Indigenous communities who have adapted to living in mountain regions over thousands of years have evolved a diversity of cultures that comprise traditional resource development and environmental conservation practices. Such communities have also developed a variety of farming and food systems that can be adapted to cope with environmental changes, and a range of institutions that have immense relevance and practical significance in ensuring food and livelihood security for a large part of the world. However, mountains have long been marginalized from the point of view of the sustainable development of natural resources and the wellbeing of their inhabitants.

The Hindu Kush Himalayan region currently faces diverse challenges, such as hunger and malnutrition, climate change, deforestation, environmental degradation and desertification, loss of biodiversity, population growth, and migration. These challenges disproportionally and more severely affect the inhabitants. There are various reasons for this, including harsh climatic conditions, weak infrastructure, and limited economic opportunities, due to the inaccessibility, fragility, and seasonality of mountain areas. Regional and global changes, including climate change, land use change, deforestation, and overgrazing, are affecting mountains in an unprecedented manner. While the global average of food insecure people is one in eight, one in two food insecure people live in rural mountain areas. Many mountain peoples in the Himalayas are farmers who live on subsistence agriculture. Over generations, living in their particular high-risk environments, they have developed solutions and techniques, indigenous practices, knowledge, and sustainable soil management approaches that shape and protect ecosystems that ultimately provide water for at least half the world's population.

Himalayan soils have long performed a host of essential ecosystem services that helped to ensure food security and nutrition. Soils play a key role in providing food and nutritional security. Soils help people to mitigate and adapt to climate change by playing a dominant role in carbon sequestration and water management. However, over-exploitation of soil resources has rendered them vulnerable to erosion and degradation. With the continuous increase in population, there will be more pressure on limited land resources. Understanding and judicious management of soil resources are, therefore, necessary to fulfill future needs of the human and livestock population. While mountain agriculture has made some headway, identifying innovative responsive solutions that target mountains directly in relation to sustainable development remains a distant dream.

This publication by Sharma et al. outlines how understanding HKH soils and their judicious management would transform mountain agriculture and could better contribute to food security and better nutrition in the region. The objectives of this publication were to identify constraints, gaps, and opportunities in mountain agriculture to tap underutilized areas and resources for zero hunger and exchange of good experience and practices related to mountain agriculture. Conventional approaches in mountain agriculture have not been able to ensure food security. The volume offers fairly good coverage of physiography and resources, soil geology, soil microbes, nature and properties of soils, and land use and management, for livelihood security in the HKH region. I wish to congratulate the authors for their hard work in the consolidation of the important available information on, soils and their management in the HKH region, in one place for easy access and use to improve food security.

New Delhi, India Dr. I. P. Abrol
February 2020

I. P. Abrol, Director, Centre for Advancement of Sustainable Agriculture and Former Deputy Director General, Division of Natural Resources Management, Indian Council of Agricultural Research

Foreword by K. M. Bujarbaruah

I am extremely happy to have received a request from one of my closest friends cum colleagues, Dr. Uttam Chand Sharma to write the foreword for their book "Soils: Their Management for Agricultural Land Use in Hindu Kush-Himalayas". I have known Dr. Sharma as a developer of simulation models for natural resource management, particularly for hill ecosystems using the available data and his personal and team research experience. He learnt this art of natural resource management much before the advent of Artificial Intelligence applications in agriculture. Similarly, the other two authors, Dr. Mrinmoy Dutta and Vikas Sharma, are also equally known to me for their contribution to their expert area domain which is Soil. I, therefore, considered the request of Dr. U. C. Sharma as an opportunity to push forward the book to the soil scientists, students, farmers, policymakers, and all other stakeholders.

We are all increasingly aware that the world is going to demand around 60 percent more food by 2050 than the current level of production that too in the face of shrinking natural resources and challenges ranging from climate change to various other stresses, both biotic and abiotic. This realization has encouraged crop scientists globally to devise newer methods and methodologies; tools, techniques, and technologies right from soilless agriculture to zero gravity to vertical to precision agriculture bringing in innovations in the traditional methods or innovating and applying newer sciences like Artificial Intelligence together with RS and GPS tools. While we wait for perfection and wider use of these upcoming technologies, our continued effort to produce more has to have its base on Soil—the main food production medium. Although our survival depends on this vital resource called Soil, it is often seen that its study including the study materials is comparatively less which is why perhaps Leonardo Da Vinci had once said

"we know more about the movement of celestial bodies than about the soil underfoot." To me particularly, our soil is just like our mothers. All the mothers while carrying their babies need extra care in the form of that additional dose of protein, vitamins, calcium, etc. which we provide. Similarly, a plot of land pregnant with a crop of wheat or any other crop needs that care, that management so that it can yield a bumper harvest like a very healthy baby. This art of managing the soil backed adequately by scientific knowledge is the core issue for our food security and I am happy that Dr. U. C. Sharma and his co-authors have tried to enlighten us all as to how to expertly and sustainably manage the soils of Hindu Kush Himalayas, known for enriching the food basket of the country with its diversified food products.

The Hindu Kush Himalayan (HKH) region is one of the largest and most assorted mountain settings in the world embracing 4.3 million km^2 of land with several parallel mountain ranges, such as the Karakoram, the Himalayas, the Hindu Kush, and the Tibetan Plateau—all comprising of diverse landscapes of mountains, plateaus, river valleys, and adjoining foothills. The region is famous for geo-hydrological, biological, cultural, and aesthetic values with a hidden treasure of floral and faunal resources. Due to its topographical advantages ranging from foothills to alpine zones, a wide range of crops are grown in the region. On the livestock front, the HKH mountains have all sorts of animals found in the plains with some special livestock like mithuns and yaks, unique to this region only. The climatic conditions range from tropical to high alpine zones with a principal vertical vegetation regime representing tropical and subtropical rainforests; temperate broadleaf, deciduous, or mixed forests; temperate coniferous forests; alpine moist and dry scrub; meadows; and desert steppe. The HKH region is inhabited by more than 210 million people representing diverse ethnic and socio-cultural groups. In addition, 1.3 billion people living in the downstream areas depend on the ecosystem goods and services flowing from the HKH region.

The past few decades have witnessed unprecedented changes in the patterns of resource use and developmental activities in the HKH region under the influence of fast population growth, change in lifestyle, additional requirement of food and essential goods and services, and socio-economic transformation of the societies. These, coupled with a rapidly changing climate, pose serious threats to the sustainability of the ecosystems, especially at higher altitudes, which are ecologically fragile and extremely sensitive. With the exception of a few local and regional studies, there have been no serious efforts to enhance scientific understanding of the significance of ecosystem interfaces and the value of the ecosystem services of the HKH region. The fundamental characteristic of mountain agriculture in the mountain system is its extreme internal variability and complexity, with a multiplicity of highly localized micro-ecosystems providing the habitats for many unique crop varieties and animal species.

The most important factor responsible for agriculture development and its sustainability in HKH is the medium of plant growth itself, that is, "Soil". Understanding the nature of the soil and its management is very important to raise crop productivity and production and, improve the quality of life of the Himalayan people. Soil management is the application of operations, practices, and treatments to protect soil and enhance its performance. It includes soil conservation, soil amendment, proper handling, and management of its biota—the bacteria, fungi, actinomycetes, etc. for ensuring its optimal and sustainable health. The focus of agriculture in the Himalayan region is slowly shifting from traditional cereal crops to high-value cash crops farming such as fruits and vegetables. This transformation from subsistence systems to commercial agriculture poses new challenges for improving and maintaining the productivity and quality of soil. Judicious soil management will play a crucial role in maintaining an optimum level of soil fertility to accommodate a new set of crops and cropping systems. Soils have been and still are the primary medium for large-scale agricultural production upon which human civilizations have depended for sustenance. Efforts to reduce the poverty of people in remote mountainous areas by responding to the growing demand for mountain agricultural products often call for more intensive use of natural resources, in particular soil and water. Communities living in mountainous areas have developed unique and sustainable production systems adapted to their local environments. Successful soil management in mountainous areas necessitates that these locally adapted production and processing techniques are both retained and improved upon injecting state-of-the-art modern technologies without harming the prevailing ecological balance and ecosystem. I am further happy that the authors have kept these indicators in view while penning down this book and also incorporated additional basic information on soils required for developing appropriate strategies to get maximum from limited arable soil resources of the HKH region for food and nutritional security.

I compliment the authors once again and hope that the publication will fulfill the purpose for which it has been written.

Guwahati, India
February 2020

Prof. K. M. Bujarbaruah

K. M. Bujarbaruah, Former Deputy Director General, Indian Council of Agricultural Research & Vice Chancellor, Assam Agricultural University

Foreword by Saroj Kumar Sanyal

Soils are the foundation of agriculture, supporting crops, livestock, and forests necessary for feeding the ever-increasing population of the World, a population expected to be about nine billion by 2030. More than half of these people live in cities and depend on farmers to produce their food from limited soil resources. The mountain soils are particularly susceptible to climate change, deforestation, unsustainable farming practices, and resource extraction methods that affect their fertility and trigger land degradation, desertification, landslides, floods, and droughts. Mountain soils are vulnerable to these changes and so are mountain peoples, their livelihoods, and food security. The mountain peoples have learned to survive at high altitudes, developing resilient cropping systems and solutions to adapt to their difficult environments. Besides agriculture, the peoples are dependent for their livelihoods on diversification, practicing forestry, pastoralism, and fishery at the family level. Regional and global changes, including climate change, land use change, deforestation, and overgrazing, are affecting mountains adversely.

Human activities have greatly accelerated the degradation processes. This led to widespread soil erosion and, in many areas, the natural redevelopment of soils has remained insufficient to restore the productive capacity. Today, human impact on mountain soils continues, with land use activities, resource exploration, and tourism contributing to, and often accelerating, these degradation processes. Soil is a limited resource that has a crucial role in sustaining ecosystem services and human life, and ensuring environmental stability and agricultural productivity. It is essential for life on Earth as it provides nutrients, water, and minerals to plants and trees and is home to an infinite array of insects, bacteria, and small animals.

Mountain peoples have to cope with high runoff rates from steep slopes and associated soil erosion and landslides. They also face challenges of highly differentiated climatic conditions due to altitudinal differences, within short distances, and they are vulnerable to climate change. Crop growth is slower due to the lower temperatures at high altitudes, so many farmers have only one crop per year. Also, soils in mountain areas are often degraded due to leaching of nutrients and erosion by water on steep slopes and erosion by wind on exposed areas. Moreover, mountain peoples often live in remote areas that are not easily accessible.

Most studies have excluded the Himalayan region because of its extreme and complex topography and the lack of adequate data. For robust agricultural land use practices to enhance production to sustain the ever-increasing human and livestock population as well as sustainable ecosystem services, there is a strong need to conserve the resource base and optimize inputs. Soil would play a crucial role in our future endeavors in this regard. Judicious soil management is necessary to achieve the objectives of maintaining good soil health for optimum crop production and ecosystem services. There is an urgent need to close the knowledge gap by establishing monitoring schemes for soil physical, chemical, and biological properties and developing basin-wide scenarios which also take water demand and socio-economic development into account.

The present publication "Soils: Their Management for Agricultural Land Use in Hindu Kush-Himalayas" by Drs. U. C. Sharma, M. Datta, and Vikas Sharma intends to raise awareness of the importance of mountain soils in providing critical ecosystem services and the need for their sustainable management. The book will go a long way in fulfilling the aspirations of researchers, students, and field workers alike, who want the information in one place. The authors have successfully accomplished the task of bringing related information on Himalayan soils through appropriate chapterization.

Kolkatta, India
January 2020

Dr. Saroj Kumar Sanyal
Former Vice Chancellor, BCKV
Mohanpur, W. Bengal, India

Preface

The Himalayas represents the youngest but complex mountain system on the earth, which is still evolving. The temporal and spatial variations caused by diversity in geological processes or orogenesis have resulted in a significant difference in climate and physiography and consequently in the distribution pattern of biotic elements. Despite the rich endowment of agro-climatic and geographical diversity, the mountain region has lagged behind in economic development. Nonetheless, there is ample potential for development with an abundance of natural resources. Soils in the valleys are rich in soil organic matter but are mostly degraded. The organic matter content is low to medium in arable lands on steep slopes because of accelerated soil erosion. Due to variations in altitude, vegetation, and parent material, a lot of variation in soils are found, both horizontally and vertically. The Himalayas are very important in terms of provisioning the ecosystem services that have thus far sustained a huge population of people and high levels of biodiversity. The survival of these ecosystems and wildlife sustained are now threatened by human activities like timber harvesting, intensive grazing by livestock and agricultural expansion into forestlands, and above all climate change. Though HKH has rich soils and a variety of climates, crop productivity has remained relatively low. The major reason was that mountain agriculture remained, by and large, organic by default because new crop production technologies have eluded the region. This was partly due to the reason that this region was paid less attention by the concerned authorties because of rugged terrain and less availability of arable land and partly due to perpetual poverty of the farmers and their low access to costly inputs. Only sustenance type of agriculture was followed, and modern farming machinery could not be used due to undulating topography and extreme gradient. Since the economy of HKH countries mostly depends on agriculture as more than 70% of people, directly or indirectly, depend on this profession, raising crop productivity and diversification of agriculture would ensure a stable economy of the region as well as improve the quality of life of people. Soil is a medium of crop growth, and its good health and stability are important for optimum crop productivity and overall production. Soil erosion and degradation, soil acidity, alkalinity, as well as fragmented land holdings, less availability of arable land, high gradient, and low absorption of modern technologies have been found to be the main constraints in crop production in the Himalayas. A great range of diversity in micro-environments, soils, and farming systems, as well as in crops in each mountain region existed in the Himalayas. Nevertheless, striking similarities were also found when the various regions were compared. There are many common components in the traditional and originally sustainable farming systems that are now coming under increased pressure due to population growth, their changing needs, and lifestyle.

Since we have worked for long in the Himalayas and looked into the plight of farming communities in mountains, the idea came to mind that a comprehensive book on "basic natural resource and medium of crop production", that is the "The Soil", should be there to understand "what we have" and "what we need to have". So, the book "*Soils: Their Management for Crop Production in Hindu Kush-Himalayas*" was planned and written. The book has seven chapters, besides "Introduction". In all the chapters, very interesting material, relevant to soil formation, such as natural resources, climate, geological features, soil biota, soil classification, and management for optimum crop production, has been given in a simple and succinct manner. The readers would find it interesting and quite useful. This would serve as a practical guide for scientists, extension workers, state/regional governments, and farmers for ecological soil management that provides background information as well as details of soil-improving practices. This book is meant to give the reader a holistic appreciation of the importance of soil health and to suggest ecologically sound practices that help to develop and maintain healthy soils as well as build this resource for better and optimum crop production. By maintaining adequate levels of organic matter in soil, you have less reason to use as much commercial fertilizer, lime, and pesticides as many farmers now do.

We have more than three decades of experience working on Himalayan soils, both in Eastern Himalayas and Western Himalayas. While Dr. U. C. Sharma worked for more than three decades in Eastern and Western Himalayas, Dr. M. Datta worked in Eastern Himalayas and Dr. Vikas Sharma is working in Western Himalayas. Writing a book on the Himalayas was really a Himalayan task but all your good wishes and our experience of working in the mountains helped us in achieving this goal. We hope this book will be of use not only to scientists but also to extension workers, policymakers, and farmers as well as all those interested in more productive and sustainable mountain agriculture.

We leave up to the readers to decide how far we have been successful in writing this book and its usefulness. Suggestions for improvement are most welcome. In the end, we genuinely thank Springer Publishers for providing life to our efforts by publishing the book.

Jammu, India
March 2022

Dr. U. C. Sharma
Dr. M. Datta
Dr. Vikas Sharma

Acknowledgements

While it is difficult to acknowledge individually all the authors, whose work has been quoted in the book, we are really grateful to them for getting immense help from their publications and bringing the book to present form.

We are indebted to the Food and Agricultural Organization (FAO) of the United Nations for using their number of publications in shaping this book. Without this, the information chain on various aspects of soils of the Hindu Kush Himalayas could not have been completed.

Established in December 1983, eight countries of the Hindu Kush Himalayan region (Afghanistan, Bangladesh, Bhutan, China, India, Myanmar, Nepal, and Pakistan) realized the importance of mountains and established the International Centre for Integrated Mountain Development (ICIMOD). With an 8-country charter, ICIMOD serves as a regional intergovernmental center for cooperation on the sustainable development of the Hindu Kush Himalayan region. ICIMOD believes in a future in which the region's mountain people can experience enhanced livelihoods, equity, and social and environmental security, in which they can adapt to environmental, socio-economic and climate change, and in which generations of mountain and downstream populations can enjoy the benefits and opportunities afforded by the region's natural endowment. The mission of the ICIMOD is to enable sustainable and resilient mountain development for improved and equitable livelihoods through knowledge and regional cooperation. It aims to serve the region through information and knowledge generation and sharing to find solutions to critical mountain problems. It connects science with policy and practice. We are extremely grateful and very special thanks are due to ICIMOD and a number of highly talented scientists working in the institute for taking immense help from their work and using a few of their tables and figures, for the interest of the readers. As an organization, the ICIMOD has done pioneering and wonderful work on many aspects of the Himalayas, including soils, and without citing their work, this publication would have been incomplete.

The National Bureau of Soil Survey and Land Use Planning (Indian Council of Agricultural Research), Nagpur, Maharashtra, India, has been engaged in research, classification of soils, and land use planning in India since its inception in 1976. We gratefully acknowledge taking liberal help from the work done by talented scientists of the Institute. We are also indebted to ICAR Research Complex for Northeast Hill Region for using a few of their photos and other information generated by this institute. We want to place on record our appreciation and gratitude to all authors and organizations whose work has been quoted in this book. Last, but not the least, we are extremely grateful to the Indian Council of Agricultural Research (ICAR), which made us capable of writing scientific research papers and books by gaining experience during serving this organization for a long period.

Sincerely and gratefully, we want to place our "sincere thanks" to all on record.

Dr. U. C. Sharma
Dr. M. Datta
Dr. Vikas Sharma

Contents

About the Authors

Dr. U. C. Sharma obtained his Ph.D. degree in Soil Science from Punjab Agricultural University, Ludhiana; Graduation course in Cropping Systems from International Rice Research Institute; and The Philippines and Advance course in Research Management from National Academy of Agricultural Research Management, Hyderabad, India. He is specialized in Soil Fertility and Chemistry, Integrated Water Resources Management, Catchment Management, Soil Hydrology, Environment, and Modeling water regimes. He has served the Indian Council of Agricultural Research as Principal Scientist, Joint Director, Director, and National Coordinator (National Agriculture Technology Project, A World Bank Project). Dr. Sharma has more than 350 research papers/publications in National and International Journals of repute. He has experience working in both, Western Himalayas in Jammu and Kashmir (12 years) and Eastern Himalayas (20 years). He won Phosphate and Potash Institute of Canada and Fertilizer Association of India Award for maximization of crop yields and Fakhruddin Ali Ahmed Award for Natural Resources Conservation and Management. Dr. Sharma is a fellow of many societies, including National Academy of Agricultural Sciences, India, Indian Society of Agronomy, Indian Association of Hill Farming, and Indian Potato Association. He has visited about 60 countries to see the agricultural activities in other countries of the World.

Dr. Sharma was Vice President, International Commission on Water Quality (IAHS) from 2003 to 2011. He was ex-officio Advisor, Northeastern Council, Ministry of Home Affairs; Consultant, Northeastern Development Finance Corporation, Guwahati, and Chairman, Research Advisory Council for "Community Natural Resources

Management Project for Northeastern states of India" (International Fund for Agricultural Development, IFAD). He was a member of Research Advisory Panel for Hills and Mountains, a member of Research Advisory committees of several Research Institutes, and a member of many Review and Evaluation committees of Research Institutes. Dr. Sharma is an office bearer and member of many International and National professional societies, and a reviewer of many National and International Research Journals.

Dr. M. Datta was born on January 1, 1954, and had passed M.Sc. (Ag) in 1974 and Ph.D. (Ag) in Soil Science & Agricultural Chemistry in 1989 from the University of Calcutta, West Bengal, India. Dr. Datta had started working in University of Calcutta as a Research Scholar and thereafter in All India Soil Survey and Land Use Planning, Govt. of India. Dr. Datta was Selected as Scientist in 1977 through Agricultural Research Examination conducted by ASRB, ICAR. Dr. Datta has 38 years of research experience in Soil management and allied Activities as well as Administrative experience. He worked as Principal Scientist and Joint Director in Indian Council of Agricultural Research for 40 years and Principal, Government of Tripura College of Agriculture for 3 years. He has experience in Agri Education, Research, Extension, and Administration. Dr. Datta is a Fellow of Indian Association of Hill Farming (FIAHF), has obtained D. N. Borthakur Award for outstanding Contribution in Farming System Research and Bharat Excellence Award by Friendship Forum of India. Dr. Datta had contributed as a Member/Chairman of various committees in Agri-Allied Activities of ICAR and Govt. of Tripura. Dr. Datta also has Teaching Experience for a couple of years and had worked as a Guide for M.Sc. (Ag) students at College of Agriculture and Technology, Kanpur, Co-Guide in Ph.D. in Fisheries in CIFE (Mumbai), Co-Guide in Agronomy at Palli Shiksha Niketan Biswabharati, Santiniketan. Dr. Datta had special attainments in Technology development in Farming System Research and Agroforestry models, Transfer of Technology in Soil Health Card and Agro-Meteorology Services, Maintenance of Crop

Local Germplasm, release of 25 Crop Varieties followed by notification by CVRC. Dr. Datta had visited Morocco in connection with Pulse Production Technology of ICARDA. Dr. Datta was involved in Research Projects sanctioned by National Rainfed Authority, IWMP, Department of Earth Sciences, NABARD, Authority of Organic Farming, NATP, NAIP, DBT, NICRA, Tribal Sub-Plan, Bill Gates Foundation, and ICARDA, and published 85 research papers in National and International Journals, 16 books/research bulletins, 23 book chapters, and a number of popular articles/folders.

Dr. Vikas Sharma is a Professor and Head, Division of Soil Science and Agricultural Chemistry at the Sher-e-Kashmir University of Agricultural Sciences and Technology of Jammu, India. He obtained his doctorate from the Punjab Agricultural University, Ludhiana, India, and Post-doc from University of Virginia, USA. His Research interests include soil and water conservation, land degradation, nutrient dynamics, and soil health. Most of his work pertains to soils in Indian Himalayas. He has worked on improving the moisture retention in the moisture deficit *kandi* areas of Jammu through both in situ and ex situ management of water, soil erosion risk mitigation, carbon pools, and nutrient dynamics in the erosion-prone foothill Himalayas. He is also involved in teaching and advising students' research. He has over 50 research articles and 15 book chapters to his credit.

At present, Dr. Vikas Sharma is teaching Soil Science to Undergraduate and Postgraduate students at the University as well as managing the Department of Soils and guiding research programs of the University. He has guided M.Sc. and Ph.D. students in their research work to complete their respective degree programs. Dr. Sharma has traveled widely throughout the globe to present his research work on soils. He is a member of many professional societies and won many awards for his work on soil conservation in the mountain regions.

Abbreviations

ADB	Asian Development Bank
AFCC	Air Force Combat Control Centre
AFS	Agroforestry Systems
AHDC	Autonomous Hills District Council
AICC	Agriculture Information Communication Centre
BCCSAP	Bangladesh Climate Change Strategy and Action Plan
BSh	Hot Semi-arid Climate
BSk	Cold Semi-arid Climate
BWh	Hot Desert Climate
BWk	Cold Desert Climate
CBD	Convention on Biological Diversity
CEPF	Critical Ecosystem Partnership Fund
CHT	Chittagong Hill Tracts
CITES	Convention on International Trade in Endangered Species
CSO	Central Statistics Organisation
DST	Department of Science and Technology, GOI
EDW	Elevation Dependent Warming
EH	Eastern Himalayas
EIHR	Eastern Indian Himalayan Region
FAO	Food and Agricultural Organization of the United Nations
FSI	Forest Survey of India
FSMP	Forestry Sector Master Plan
GBH	Global Biodiversity Hotspots
GLOF	Glacier Lake Outburst Floods
GOI	Government of India
GSP	Growing Season Precipitation
GST	Growing Season Temperature
HIMAP	Hindu Kush Himalayan Monitoring and Assessment Programme
HIPP	Himalayan Interdisciplinary Paleoclimate Project
HKH	Hindu Kush Himalayas
HKKH	Hindu Kush Karakoram Himalaya
IBIS	Indus Basin Irrigation System
ICAR	Indian Council of Agricultural Research
ICARDA	International Centre for Agricultural Research in the Dry Areas

ICIMOD	International Centre for Integrated Mountain Development
IFAD	International Fund for Agricultural Development
ILEIA	Information on Low External Input and Sustainable Agriculture
IPCC	Intergovernmental Panel on Climate Change
ITZS	Indus–Tsangpo Suture Zone
IWMP	Integrated Watershed Management Programme
JHUM	Shifting Cultivation
JICA	Japan International Cooperation Agency
LIC	Lithogenic Inorganic Carbon
LULC	Land Use Land Cover
MAP	Mean Annual Precipitation
MAT	Mean Annual Temperature
MBT	Main Boundary Thrust
MCT	Main Central Thrust
MEA	Millennium Ecosystem Assessment
MFT	Main Frontal Thrust
MHR	Main Himalayan Ranges
MOAF	Ministry of Agriculture and Forests, Bhutan
MWD	Mean Weight Diameter
MWHC	Maximum Water Holding Capacity
NAT	North Almora Thrust
NBSSLUP	National Bureau of Soil Survey and Land Use Planning
NCA	National Commission on Agriculture
NEC	North Eastern Council
NFT/S	Nitrogen Fixing Trees and Shrubs
NLWR	Nonlimiting Water Range
NOAA	National Oceanic and Atmospheric Administration
NRSA	National Remote Sensing Agency
NSSC	National Soil Services Center
NTFP	Non-Timber Forest product
NWFP	North-West Frontier Province
PIC	Pedogenic Inorganic Carbon
QTP	Qinghai–Tibet Plateau
QUINTAL (q)	100 kg
REID	Research Extension and Irrigation Division
SAGs	Sustainable Development Goals
SALT	Sloping Agricultural Land Technology
SBD	Soil Bulk Density
SIC	Soil Inorganic Carbon
SLM	Sustainable Land Management
SOC	Soil Organic Carbon
SSP	Soil Survey of Pakistan
SSR	Solar Surface Radiation
TP	Third Pole
TSS	Tibetan Tethys Series
UIB	Upper India Basin

UNCCA	United Nations Convention to Combat Desertification
UNDP	United Nations Development Programme
UNEP-WCMC	United Nations Environment Programme World Conservation Monitoring Centre
USGS	United States Geological Survey
WIHR	Western India Himalayan Region
WWF	World Wildlife Fund
ZSI	Zoological Survey of India

Introduction

1

Abstract

The name "Himalayas" means *hima* ("snow") and *alaya* ("abode") in the *Sanskrit* language. This range of mountains was earlier known as "Himalaya" in singular form, but now the range is mentioned as Himalayas (plural form). The Himalayas have many peaks above 8000 m but the highest peak, Mount Everest, has an elevation of 8,848 m above mean sea level. The Hindu Kush is an 800-km-long mountain range that stretches near the Afghan–Pakistan border, from central Afghanistan to northern Pakistan. It forms the western section of the Hindu Kush Himalayan Region (HKH). The HKH region, which is also known as the greater Himalayan region or "the roof of the World," extends from 15.95° to 39.31° N latitudes and 60.85° to 105.04° E longitudes. The countries included in the Hindu Kush Himalayas are Afghanistan, Bangladesh, Bhutan, China, India, Myanmar, Nepal, and Pakistan. The Hindu Kush Himalayan (HKH) region extends 3,500 km over all or part of eight countries from Afghanistan in the west to Myanmar in the east. The region spans over 3.441 million km^2, which is about 2.9% of the global land area and approximately 18% of the global mountain area. Because of the great amount of tectonic motion still occurring at the site, the Himalayas have a proportionally high number of earthquakes and tremors. The Himalayas can be divided into four major tectonic zones according to their altitude and geographical position. These zones are (i) the Shiwaliks or sub-Himalayas, with an elevation varying from 250 to 800 m; (ii) the Lesser Himalaya; (iii) the Higher Himalayan Crystallines; and (iv) the Tethyan Himalaya. The Higher Himalayan crystalline is extremely rugged with perennial snow and glacier-capped peaks with an elevation of 3000 m to more than 8000 m. The Tethyan Himalayan Range has developed into a cold and dry desert, with sparse vegetation and naked rocks. The climate of thc HKH Rcgion varies from subtropical to alpine, which is conducive to a wide range of crops, forest species, range and grasslands, pastures, flora, and fauna. The Himalayan region holds significant importance in terms of biological (species) richness, biodiversity, socio-cultural diversity, and wealth. The region is one of 34 worldwide "biological hotspots" (i.e. a natural environment with a high biodiversity containing a large number of endangered endemic species). The unique floral and faunal wealth of the Himalayas is undergoing structural and compositional changes due to climate change. The Himalayan range alone has a total snow and ice cover of 35,110 km^2 containing 3,735 km^3 of eternal snow and ice. Increasing demands on ecosystem goods and services from the mountains are putting pressure on the natural

U. C. Sharma et al., *Soils in the Hindu Kush Himalayas*, Geography of the Physical Environment,
https://doi.org/10.1007/978-3-031-11458-8_1

resources that they contain. These demands stemming from a burgeoning human population and haphazard infrastructural development, combined with unsustainable use, poor management, and limited investment in conservation, have led to habitat degradation, biodiversity loss, and decreased agricultural productivity.

Keywords

Future action plan · Hindu-Kush Himalayas · Major problems of HKH · Sub-divisions of Himalayas · Tibetan Tethys

1.1 General Description

Mountains in the world occupy about 24% of the area of the land (about 2.5 billion ha) and inhabit about 0.9 billion of the global population (GTOS 2008). About 65 million years ago, Powerful global plate-tectonic forces have changed the Earth's crust to form a number of Eurasian mountain ranges, which include the Himalayas also. Schild (2008) and Schild and Sharma (2011) mentioned that the mountains have ecological, aesthetic, and socio-economic importance for the people who depend for their day-to-day needs from mountain resources and also for about 40% of the world population residing downstream and depending on mountain resources for important commodities; which include water, hydropower, medicinal plants, forest produce, timber, and fuel wood. The studies conducted in mountains show that out of 34 global biodiversity hotspots, about 18 hotspots have been found in the mountains environment. Other than this, mountains have one-third of global terrestrial biodiversity (ICIMOD 2009). Mountains are useful in studying changes and variability in climate over time as well as assessing climate-related impacts (Nogués-Bravo et al. 2007, 2008). Green revolution and economic development in the world have not affected mountain communities to a large extent. Though endowed with rich natural resources, the mountains have been deprived of proper development and improving the quality of life of the people. Different farming communities and ethnic groups of the Hindu Kush Himalayas (HKH) have remained poor and marginalized despite some good approaches by the respective governments. Kandel et al. (2021) concluded that research in ecosystem services needs a local partnership between various stakeholders from various disciplines with a greater emphasis on socio-political aspects, together with the bio-physical sciences, for a thorough understanding, since ecosystem services is a multidisciplinary concept. It is essential that ecosystem services research include the local context and issues to ignore the supremacy of global perspectives and the downgrading of local issues.

Since times immemorial, many pilgrims used to visit various shrines in the Himalayas, and, in fact, the highest peak is believed to be the abode of Lord Shiva, the Hindu God. They gave the Sanskrit name Himalaya (*Hima* means snow + *alaya* means abode) for this great mountain network (Fig. 1.1). During the same time, this mountain system has been a challenge and attracted several mountaineers from all over the world. Earlier, the Himalayan range of mountains was spelt in singular form "Himalaya"; however, at present, the range is spelt in plural form, that is, "Himalayas." The Himalayas stand like a wall between the northern border of India and the southern border of China and it is almost impossible to cross over from north to south or vice versa. Though the Himalayas have a number of peaks having altitudes above 8000 m but the highest peak, Mount Everest, has an elevation of 8,848 m above msl. The countries included in the present study of the Hindu Kush Himalayan (HKH) Region are Afghanistan, Bangladesh, Bhutan, China, India, Myanmar, Nepal, and Pakistan, as listed by the International Centre for Integrated Mountain Development (ICIMOD). HKH region and adjacent mountain areas include Afghanistan; the Chittagong Hill Tracts in Bangladesh; Bhutan; China (parts of Yunnan, Diqing, Nujiang, Dali, Sichuan, Gansu, the whole of Tibet Autonomous Region, and Qinghai); India —the 11 Northern mountain states of India; Myanmar—the states of Kachin, Chin, Shan, and

Fig. 1.1 Magnificent Himalayas

Rakkhain; and almost entire Nepal and northern areas of Pakistan, including 12 districts of Baluchistan. The Himalayas have an area of 3.441 km^2 with a population of about 253 million. The Himalayas can be characterized on the basis of its very high peaks, big-sized glaciers, undulating topography deeply cut by erosion, deep gorges, complex geologic features, and a number of altitudinal zones that exhibit various ecological associations of flora, fauna, biodiversity, and climate.

Because the Himalayas served as a natural wall between the people of India and that of China and Mongolia for millions of years, there was little contact between the people on both sides. Colonel Sir George Everest, a British surveyor, was based in India during the early-to-mid-nineteenth century and so, the Himalayan peak was named after him as "Mount Everest". In Nepal, people call Mt. Everest as "*Samgarmatha*" which in English means "Goddess of the Universe" or "Forehead of the Sky." The Hindu Kush Himalayas give ecosystem services to the communities residing in the immediate vicinity and downstream and livelihood security to a population of around 253 million inhibiting the region. The Himalayas have more than fifty peaks above 7,200 m in elevation, of which 10 peaks are above 8000 m high. The Everest has two partners, that is, Tibet (China) in the north and Nepal in the south. Figure 1.2 shows a satellite image of the Himalayas.

The Himalayan ranges lie between the Indus river from Nanga Parbat with an altitude of 8125 m on the western side to the Yarlung Tsangpo-Brahmaputra near Namche Barwa with an altitude of 7,756 m on the eastern side. The directional position of Himalayan peaks is rising from west–northwest to east–southeast, forming an arc with a total length of about 2,500 km and encompassing the political–administrative regions of Afghanistan, Pakistan, India, Nepal, Bhutan, and China (Ives and Messerli 1989). As defined by ICIMOD (Hindu kush Himalayan Region or HKH Region), however, if the mountain area of Myanmar and Bangladesh (Chittagong Hill Tract) is included, the total length stands at about 3500 km. In the world, it is one of the geologically youngest and highest mountain ranges with many peaks above 7000 m. The width of the Himalayas varies between 250 and 300 km along its length. Wadia (1931) and Valdiya (1980) explained that the mountain range is, more or less, arc-shaped and convex southwards with syntaxial bends at the western and eastern ends. Karki et al. (2011) explained that the HKH region, which is also known as the greater Himalayan region or "the roof of the World," extends from 15.95° to 39.31° N latitudes and 60.85° to 105.04° E longitudes.

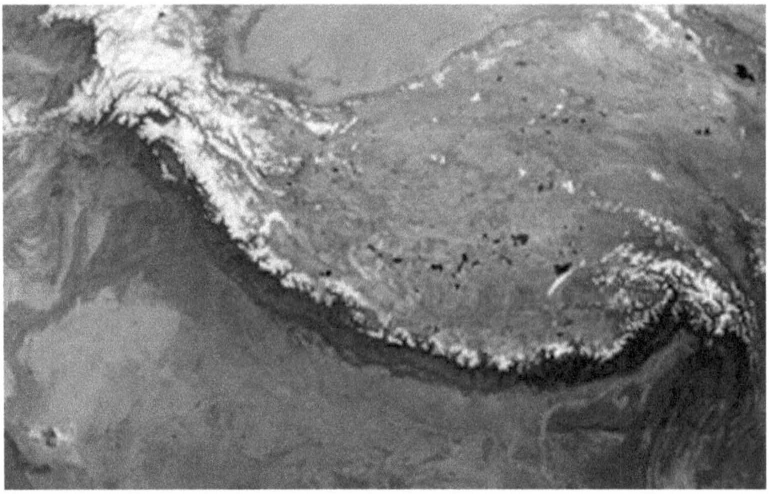

Fig. 1.2 Image showing the arc of the Himalayas (Adapted from NASA)

They further elaborated that it is a geomorphologically unique region because the robust Himalayas act as the geological buffer protecting the lush green south slopes from the dry and arid winds of the north thus supporting a unique mosaic of several biota and eco-regions.

Another mountain range forming part of the Hindu Kush is an 800 km long mountain range that extends from the central Afghanistan to northern Pakistan areas (Kohn 2006; Searle 2013). Hindu Kush mountain combined with the Himalayas is known as Hindu Kush Himalayan (HKH) region. The Hindu Kush forms the western section of the Hindu Kush Himalayan Region. The valley of the Amu Darya is divided by the Hindu Kush to the north from the Indus River valley to the south. The Hindu Kush range has also a number of high snowcapped peaks, with the highest point in the Hindu Kush being Tirich Mir at 7,708 m in the Chitral District of Khyber Pakhtunkhwa, Pakistan. In the north, the eastern end of the Hindu Kush merges with the Karakoram Range (Heuberger 2004). It connects with the Spin Ghar Range near the Kabul River (Bloom and Blair 2009) at the southern end. The communities settled in the Hindu Kush have quite a good number of ancient monasteries, important trade networks, and travelers between Central Asia and South Asia (Neelis 2010). The global atmospheric circulation that influences two major monsoon and the Westerly weather systems is significantly influenced by HKH mountain ranges, where the system plays a key role.

1.2 Settlements in HKH

The information about the behavior and magnitude of settlements of ancient communities in the Himalayas is also scant and scattered. This has made it difficult to re-arrange the existing knowledge and whatever could be done is primarily derived from Sanskrit epic literature taken from oral stories and traditions. The literature from the Britain mentions about three patterns of settlements in the Indian Himalayas. It is described that the settlements in the western Himalayas took place from 1500 BC onward by a band of nomadic warriors called as *Khas,* who were part of the succession of the insurgence of Aryan into India from the northwest and central Asia. The *Khas* are believed to be strong and they pushed the indigenous dwellers to the other areas, downgrading their social status and treating them as inferior people. The reports of the study of Karan (1987) indicate that there were around five mountain cultural regions in the Himalayas. He pointed out that the population of these regions has trebled from 11 million to over 33 million within a span of 80 years, between 1901 and 1981. It needs to be mentioned here that the HKH region has a great potential of provisioning the requirements for livelihood security for about 50 million mountain people of the region and more than 450 million people residing in plains and basins of the Indus, Ganges, Brahmaputra, and upper and middle Jiang (Sichuan Basin) which inhabits a significant proportion of the population. If we look into the modern development and progress, the region presents a discouraging scenario as it is still underdeveloped, and poverty, malnutrition, and unabated population growth are common issues. The Himalayas also have a wide range of phytogeographic regions.

1.3 Hindu Kush Himalayan Region

The Hindu Kush Himalayas are very unique mountain ranges surrounded by some important reference points such as Karakoram on the northwest, on the north by the Tibetan Plateau, and on the south by the Indo-Gangetic Plain (Sharma and Pratap 1994). Length-wise, India holds the number one position as it has about a 2400 km stretch of the Himalayas, running from the state of Jammu and Kashmir in the northwest to the state of Arunachal Pradesh in the southeast. Two countries in the HKH region (Nepal and Bhutan) are fully covered by the Himalayas. The homo-sapiens were under the awe of natural objects such as mountains, oceans, rivers, and forests from the pre-historic age. The people were small in number, primitive in technology, and their needs were extremely limited. Slowly, their number increased and needs multiplied. Though endowed with abundant natural resources, the HKH region has remained economically backward due to its indiscriminate use and mismanagement. Socio-economic constraints, such as shifting cultivation, the land tenure system, the small size of land holdings, unabated deforestation, free-range grazing, and undulating terrain, have affected the sediment yield and quantity and quality of available water (Sharma 1998, 2003). The conducive or non-conducive conditions prevailing in the Himalayas have shaped the cultures of its inhabitants. Some Himalayan peaks are religiously important and sacred in Hinduism and Buddhism.

1.3.1 Important Features of the HKH Region

According to some estimates, the total estimated area of the HKH Region has been put at 34,41,719 km^2 (Sharma and Pratap 1994). This part of the world has many mountain ranges and a number of famous peaks with altitudes of more than 8000 m. The HKH countries such as Afghanistan has mountain ranges of Hindu Kush and Pamir; Bhutan is in Himalayan range; China has Ganhdise Shan, Hengduan Shan, Himalaya, Karakoram, Kulun Shan, Nyainqentanglha Shan, Pamir, Qiantang Plateau, Qilian Shan, Tanggula, and Tien Shan ranges; India has Hinmalayan and Karakoram ranges; Hengduan Shan range is in Myanmar; Himalayan range is in Nepal; and Himalaya, Hindu Kush, Karakoram, and Pamir ranges are in Pakistan (Sharma and Pratap 1994). Some of the highest mountain peaks of the World fall in HKH countries and Nepal has the distinction of having the majority of the highest peaks. It encompasses Annapurna, Manaslu, Dhaulagiri, Chu Oyu, Makalu, Lhotse, Kangchenjunga, and Everest, with an elevation of 8091 m, 8156 m, 8167 m, 8201 m, 8462 m, 8516 m, 8586 m, and 8848 m above mean sea level, respectively. Nanga Parbat falls in India and Pakistan and has an elevation of 8125 m. Kangchenjunga peak also falls in India.

1.3.2 Sub-divisions of Himalayas

On the basis of altitude and geographical position, the scholars have tried to divide the Himalayas into four major tectonic zones, namely (i) Sub-Himalayas (also called Shiwaliks hills), with an elevation varying from 250 to 850 m; (ii) the Lesser Himalaya; (iii) the Higher Himalayan Crystallines; and (iv) the Tethyan Himalaya. The hills in the Shiwaliks rise above the

Indo-Gangetic Plains along the Main Frontal Thrust. These hills consist of Tertiary and Quaternary sediments and give rise to undulating topography in the area. The Lesser Himalayan Range is very rugged and difficult, with an elevation of about 2500 m or more. The Pir Panjal Range rises to heights greater than 3500 m at some places and strongly stands between Kashmir valley and Kishtwar and Poonch areas of Jammu Province. The Precambrian and Cambrian sequences of the Damtha, Tejam, Jaunsar-Garhwal, and Mussoorie Groups are found in Lesser Himalayas in the Garhwal region of the western Himalayas. Along the Main Central Thurst, the Higher Himalayan Crystallines are separated from the Lesser Himalaya. The Higher Himalayan crystalline is extremely complex and rugged. The higher Himalayas have perennial snow and glacier-capped peaks with an elevation of 3000 m to more than 8000 m. The Mount Everest having an elevation of 8848 m and Mount Kanchanjanga with an elevation of 8598 m lie in this zone. The Chail Group consists of a metamorphosed sequence of green schist, phyllites, phyllitic quartzite, psammitic schists, orthoquartzites, arkose, chlorite schist limestones, and metabasic rocks amphibolites. Along the South Tibetan Detachment System, the Higher Himalayan Crystallines pass into the Tethyan Himalayan ranges in the north direction. Developing into a cold and dry desert, the Tethyan Himalayan Range has extremely poor vegetation cover and naked rocks are found frequently. The Himalayan mountains have four principal tectonic zones, from south to north, such as the Sub-Himalaya, the Lesser Himalaya, the Higher Himalayan Crystallines, and the Tethyan Himalayas (Ahmed 2011). The Sub-Himalaya, also known as the Shiwalik Range (250–800 m high), rises above the Indo-Gangetic Plains along the Main Frontal Thrust; separating it from the overlying Lesser Himalaya, the Sub-Himalaya consists predominantly of Tertiary and Quaternary sediments and is bounded to the north by northward dipping Main Boundary Thrust.

1.4 Characteristics of Himalayan Sub-zones

1.4.1 Sub-Himalaya (Shiwaliks)

Shiwaliks are an important group of hills on the southern side of the Himalayas. In some areas, it is also known as *Kandi* area, characterized by very low and erratic rainfall and high temperatures during summers. The scarcity of water is a major feature of this zone. The water table in *Kandi* area is very low. It results in low crop productivity and shortage of food grains, jeopardizing the livelihood security in this area. On the southern side of the Sub-Himalayas borders, it is the Indo-Gangetic plain region which runs along the Himalayan Frontal Fault. The sub-Himalayas consist of thick Tertiary mollassic deposits, which came into existence because of accumulating fluvial deposits on the southern front of the emerging Himalayas. Southern parts of Nepal and India are mainly the Shiwalik hills. The youngest sediments on the top of hills consist of conglomerates, sandstones, and mudstones in the lower portions. The Shiwaliks hills in Nepal are almost similar in constitution to the Shiwaliks of Pakistan and that of northern India (Burbank et al., 1996).

1.4.2 Lesser Himalayas

The major constituents of Lesser Himalaya are mostly the unfossil ferous sedimentary as well as metasedimentary rocks such as shale, sandstone, conglomerate, slate, phyllite, schist, quartzite, limestone, and dolomite and situated between the Sub-Himalayas and Higher Himalayas separated by the Main Boundary Thrust and the Main Central Thrust, respectively (Bollinger et al. 2004). The total width of lesser Himalayas varies from 60 to 80 km at different locations. The rocks in this stretch vary in age from Precambrian to Miocene. The northern boundary of the Shiwaliks is marked by the Main Boundary Thrust and the low-grade metasedimentary rocks of the Lesser Himalaya overlie on this. The argillo-arenaceous rocks dominate the lower half of the succession, whereas both carbonate and siliciclastic rocks constitute the upper half (Le Forte 1975, Stöcklin 1980). The Lesser Himalaya is folded into a vast post-metamorphic anticlinal structure known as the Kunchha-Gorkha anticlinorium and the southern side of the anticlinorium is weakly metamorphosed, whereas the northern side is highly metamorphosed, as reported by Pêcher (1977).

1.4.3 Higher Himalayas

The higher Himalayas extend from the main central thrust to Tibetan-Tethys Zone in China. Consisting of about 10 km-thick succession of the crystalline rocks, called Himal Group, it forms the major part of this zone. This sequence has been divided into four main units, known as Kyanite–Sillimanite gneiss, pyroxenic marble and gneiss, banded gneiss, and augen gneiss in the ascending order as mentioned by Bordet et al. (1972). The High Himalayan Crystalline units (HHC) are mainly composed of kyanite to sillimanite-grade gneisses (Bollinger et al. 2004), intruded by High Himalayan leucogranites at structurally higher levels (Upreti Bishan 1999). The Higher Himalayan Crystallines are separated from the Lesser Himalaya along the Main Central Thrust. The Higher Himalayan crystalline is extremely complex and difficult to transverse, with perennial snow and glacier-capped peaks. The elevation varies from 3000 m to more than 8000 m. A magnificent view of the higher Himalayas is shown in Fig. 1.3.

1.4.4 Tibetan-Tethys

The Tibetan-Tethys Himalayas start from the Higher Himalayan Zone and extends to the north in Tibet. Having a width of about 40 km, on an average, this zone is composed of fossiliferous sedimentary rocks. The area north of the Annapurna and Manaslu ranges in central Nepal consists of metasediments that overlie the Higher Himalayan zone. It has undergone very little metamorphism except at its base where it is close to the Higher Himalayan crystalline rocks. The Himalayan Province ends along the Indus–

Fig. 1.3 Higher Himalayas engulfed by clouds

Table 1.1 Area share of different countries falling under the Eastern HKH region

Country	Area	% of area of Eastern Himalaya
Bangladesh	The Chittagong Hill Tract	2.53
Bhutan	Entire country	7.58
China	Southeastern states of the country	6.24
India	In India, the states of Arunachal Pradesh, Assam, Meghalaya, Nagaland, Mizoram, Manipur, Sikkim, Tripura, and the northern part of W. Bengal	50.58
Myanmar	In Myanmar, mainly Chin and Kachin states fall in this zone	17.45
Nepal	Almost the entire country except the Terai area	15.62

Tsangpo Suture zone of Ladakh–Kohistan range where the Indian plate collided with the Asian (Tibetan) plate around 55 million years ago. During the process, oceanic and continental arcs developed between the Indian and the Eurasian plate, and these are represented by the Kohistan–Ladakh Arc with sequences of rocks varying in composition from ultramafic (melange rocks) through mafic (gabbro, basalt) to acidic (granite–granodiorite–diorite) and their volcanic equivalents (Ahmad et al. 1998). Tibetan-Tethys rocks consist of a thick and nearly continuous lower Paleozoic to lower Tertiary marine sedimentary succession and are deposited in a part of the Indian passive continental margin (Liu and Einsele 1994).

1.4.5 The Eastern Himalayas

The Eastern Himalayan territory has an area of 52.419 million ha. It extends from Eastern Nepal to Yunnan in China. Country-wise percent area falling in the eastern Himalayas is given in Table 1.1. The region has a great significance because many countries share their borders in the zone and has historical importance due to its uniqueness of having different cultures, geopolitical and environmental conditions, and tectonic orogeny of its encompassing Himalayan system (Figs. 1.4.

Fig. 1.4 View of Eastern Himalayas

1.4.6 Indian Himalayas

Out of about 3500 kmlength of HKH region, the Himalaya in Indian territory is spread over a length of about 2,500 km and a width of 220–300 km. The Indian states of Jammu and Kashmir, Himachal Pradesh, Uttrakhand, Arunachal Pradesh, Assam, Meghalaya, Manipur Mizoram, Nagaland, Sikkim, and Tripura are covered fully, while only the northern part of the state of West Bengal comes under Himalayan mountain ranges. The Indian Himalayas have an area of approximately 591 thousand km^2 with a population of about 82 million in 2021. Physiographically, the Indian Himalayan region is subdivided into three subregions; Western Himalaya (Jammu and Kashmir, Himachal Pradesh states), Central Himalaya (Uttarakhand state), and Eastern Himalaya (Assam, Arunachal Pradesh, Manipur, Meghalaya, Mizoram, Nagaland, Sikkim, Tripura, and northern fringe of West Bengal state).

1.5 Ecosystem Services

1.5.1 Agriculture

The Himalayas are providing ecosystem services to different adjacent areas in Asia and to the world, one can say, for regulating hydrological processes, ecosystem stability, and helping in the conservation of biodiversity in the region. Studies have shown that the Hindu Kush Himalayas is a dynamic area on earth with a very rich and unique biodiversity (Pei 1995; Guangwei 2002). The flora and fauna are unique and in plenty, as is evident from many studies undertaken in this regard, and it has a high level of endemism (Myers et al. 2000) and quite a good number of critical eco-regions of global significance (Olson et al. 2001; Olson and Dinerstein 2002).Agriculture is the major profession of the dwellers of the HKH region (Fig. 1.5).

About 39% of the HKH is comprised of grassland, 20% forest, 15% shrub land, 5%

Fig. 1.5 View of agriculture fields in the lower Himalayas (Photo by Dr. A. Pattnayak)

agricultural land, and 21% of other types of land use such as barren land, rock outcrops, built-up areas, snow cover, and waterbodies. However, the survival of the mountain ecosystems is now threatened by various drivers of change like timber harvesting, intensive grazing by livestock, conversion of forest land for agriculture, and the climate change. The landscapes and communities in the HKH region are being simultaneously affected by rapid environmental and socio-economic changes. Schild (2008) reported that the welfare of about 1.3 billion people downstream is closely linked to the state of natural resources of the HKH region.

The dependence of the population on the Himalayas is not hidden from anybody as the local and basin dwellers are getting all the basic needs of life like food, fiber, water, shelter, and other ecosystem services from these mountains and the vast treasure of amenities these possess. Soils are the most important component of the ecosystem, but healthy soils are also dependent on human civilization. Due to the unique and intimate relationship of human wellbeing with soil, it is very important that detrimental exploitation and misuse and abuse of soils should not be encouraged. This necessitates that we must have a complete understanding of the soils with regard to their origin and evolution as well as their distribution, properties, and behavior. Such information is not only important in agriculture but many other professionals, such as construction engineers, geologists, forest people, ecologists, and planners would require this knowledge for land use planning and soil management, etc.

The soils are the basic medium of plant growth and the very existence of humanity is dependent on good soils. Many civilizations of the world have perished due to the misuse and degradation of this important resource. As of the year 2021, human density in the HKH region is about 59 persons km^{-2} and about 24 person ha^{-1} of cultivated land. The demand for the land is increasing due to constant population increase, and the gap between demand and supply is widening due to an increase in demand and a decrease in crop productivity. The fragmentation of land holdings has further aggravated the misuse of soil either through its over-exploitation and/or mismanagement. There has been large-scale conversion of forest land in mountains to an open cropped area by the local inhabitants and migrated settlers. When land under forest or grassland vegetation is brought under cultivation, the nutrient status of soil changes quickly. Therefore, it is necessary to characterize the supply of nutrients in the soils to have a better understanding of the rates of macro and micronutrients supply to crops in these soils. It has been suggested that many of the problems in minimizing nitrogen losses are related to inadequate knowledge of the amounts, forms, and patterns of change of mineral nitrogen in agricultural soils. The major part of N is lost from soils through leaching, volatilization, nitrate reduction, denitrification, and runoff water. Low N use efficiency is another factor that is attributable to mismanagement and the main reason for environmental pollution too. The changes in the soil's physical and biochemical properties associated with land use change need to be further researched to generate more information for better understanding. A view of the Himalaya mountain is shown in Fig. 1.6.

1.5.2 Ice and Water

The water provision is one of the most important services provided by HKH mountains to civilization, including people living within the mountain areas as well as inhabitants of river basins. The HKH mountains are called "water towers" (Viviroli et al. 2007) because these feed ten major rivers in the area, the water from which is used by almost one-fifth population of the world for daily household needs and irrigation purposes. Gurung et al. (2011) mentioned that snow cover in the HKH region is about 760,000 km^2. The region encompasses a large number of freshwater lakes and aquifers providing water and adding to the scenic beauty of the ecosystem encouraging tourism. Other than the supply of water, the Himalayan mountains are a storehouse of biodiversity including food, fiber, timber,

Fig. 1.6 View of the Himalaya mountain range (Photo by Dr. A. Pattnayak)

firewood, fruits, vegetables, grasses, and aromatic and medicinal plants. The large and varied types of the gene pool in the HKH region are vital for science and humanity (Karki and Gurung 2012). It has been observed that the rapid rate of melting of glaciers from the HKH region has increased the water flow in the rivers, which may subsequently decline. This has changed the river courses in some places. This may damage agriculture and can have severe consequences for environmental and economic activities, including substantial negative impacts on food security and human health (IPCC 2008). Karki and Gurung (2012) opined that the shift in the hydrological regime may significantly change rainfall patterns, thereby affecting river flow and nutrient cycles along the river basin, leading to adverse impacts on ecosystem and agricultural productivity as well as the gene pool including from agro-biodiversity and wild relatives of food crops in the natural systems.

One can say that adaptation, mitigation, and resilience have remained very close to the people of the HKH region and the landscape they inhabit. The communities of the region have their own rich cultural identity, which is frequently exhibited through their dress code, festivals, traditional dances, and routine working behavior. Schild (2008) was of the view that because of having the largest snow and ice in the world outside the polar regions, the HKH region has been referred to as the "third pole." He further elaborated that extending from the northern fringe of the Indian subcontinent, from the Indus River in the northwest to the Brahmaputra River in the east, the Himalayas directly or indirectly support the livelihood of over 300 million people. Xu et al. (2009) expressed that the Himalayas influence flow of thousands of rivers and rivulets that converge into the three main river systems in the region: the Ganges, the Brahmaputra, and the Indus. The annual mean discharge and basin area of the major rivers of the HKH Region are given in Table 1.2 (IUCN/IWMI, WRI 2003) and a waterbody in Fig. 1.7.

1.6 Water Availability

The Himalayas are the storehouse of water benefitting both the local inhabitants and people living downstream. The Himalayas are also known as "the water towers of the world." There are abundant glaciers in the region (Figs. 1.8) which capture and store snows during the winter and release meltwater slowly throughout the year, thereby making a constant supply to the rivers and streams and helping in irrigating

Table 1.2 The ten major river basins of the Himalayan region

River	Annual mean discharge (m^3 s^{-1})	Basin area (km^2)
Amu Darya	1,376	534,739
Brahmaputra	21,261	651,335
Ganges	12,037	1,016,124
Indus	5,533	1,081,718
Irrawaddy	8,024	413,710
Mekong	9,001	805,604
Salween	1,494	271,914
Tarim	1,262	1,152,448
Yangtze	28,811	1,722,193
Yellow	1,438	944,970
Total	90,237	8,594,755

Source IUCN/IWMI, WRI (2003)

Fig. 1.7 Ice and water in the Himalayas (Photo: courtesy Dr. S. U. Bhat)

agricultural fields, hydropower generation, and ecosystem stability. Snowmelt is the main source of water in the rivers and its darkening by deposited light-absorbing particles (LAP) has the capacity to enhance snowmelt and shift the snowmelt-out date (Strom et al. 2021). Many rivers of the region, important ones being Amu Darya, Indus, Ganges, Brahmaputra, Irrawaddy, Salween, Mekong-Lancang, Yangzi, Huang, and Tarim, depend for their water intake on the Himalayas. About 1.5 billion people of the world, residing within river heads and basins, are being benefitted from the water supply from these rivers fed by the Himalayas. However, the climate change would be the most important sustainability issue in the region. Melting of glaciers and snow loss admonish mountain people who depend on snowmelt for domestic water. Wood et al. (2020) found that snowmelt contributes between 34 and 90% of water depending on the season (dry, pre-monsoon season, and monsoon), the average being 65%. About two-thirds of domestic water during the dry season is at risk in Khumbu valley of Nepal and the people are vulnerable to climate change causing glaciers-melt as a consequence of global warming. Mountain people are very vulnerable to the impacts of climate change (Salerno et al. 2008),

Fig. 1.8 The Himalayan view showing snowbound mountains and glaciers

and it is feared that even if global warming is kept below 1.5 °C, about 33% of the glaciers in the HKH region and more than 50% of those in the eastern Himalaya will be lost by the end of this century (Bolch et al. 2019a, b). Analysis of surface waters by Balestrini et al. (2014) concluded from a study that the upper Khumbu Valley, Nepal, showed relatively low ionic concentrations and that surface water composition did not entirely match precipitation composition, indicating another source also affected the local water resources. Mountain communities may rely much on seasonal melting, and annual recharge of glaciers, snow, and ice since regional precipitation decreases with increasing altitude (Asahi 2010). Almost a similar study from Langtang Basin showed that meltwater controlled the hydrograph except in the monsoon season (Wilson et al. 2016).

A proper and focused attention is desired to deal with the resultant impacts of climate change. Both mitigation and adaptation strategies have to be worked out and shared with people. Studies have shown that the temperatures in the Himalayas are increasing at a faster rate than that of the average of the world. Climate change is also affecting rainfall pattern in the region and monsoon rains have been affected to the extent that their frequency and intensity have been disturbed. An increase in temperature has led to glacial melting, thereby causing unpredictable water release from glaciers to the rivers. This water supply from glaciers will be more in the beginning but will constantly decrease with the retreating of glaciers. This may be very detrimental to the livelihoods of the people in the days to come as sufficient water will not be available for irrigation and domestic use. If the present rates of glacier melt continue, small glaciers at lower elevations may disappear by the year 2100, and the glaciers at higher elevations will reduce in size (ICIMOD 2010). The glacier at Gaumukh–Gangotri, which is supplying about 70% of water in the river Ganges, is diminishing by about 37 m per year. More than one billion people in mountains and downstream areas will have to face a dark and uncertain future if glaciers continue to shrink at the current rate.

Water is the essence of life. The present increase in GHG emissions will have a direct effect on the hydrological cycle and impact the supply and demand balance (Bates et al., 2008). The dependent communities in the HKH region are likely to be deprived of a better future and decent living. Water regimes are an important component of larger ecosystems which support life and all socio-economic processes. The provision of freshwater is therefore an ecosystem service which, when disrupted, threatens both the health of ecological systems and human wellbeing, which are interlinked and have complex interactions. These changes will in turn create other higher order effects (Chalecki and Gleick 1999). It has been predicted that there will be a net negative impact on water availability in the HKH region and on the sustainability of the freshwater ecosystem. Climate change is the major factor affecting the water resources. Besides, there are other drivers of change that will affect the water regimes in the long run. The geosphere, biosphere, and cryosphere ecosystems in the HKH region have been extremely affected by the climate change due to GHG emissions. The impact has been accelerating further during the last few decades due to a rapid increase in anthropogenic activities such as modernization, industrialization, and urbanization, along with energy demands (Ningombam et al. 2021). Mention may be made of high and uncertain population growth, industrialization, development of roads and infrastructure, urbanization, deforestation, soil erosion from hill slopes, land and environmental degradation, lifestyle of the people in the region, and unresponsible attitude toward common property and goals. All these factors of change are interlinked and so sprout complex management problems for natural resources.

Important research requirements and policies to mitigate climatic impacts are as follows:

1. Not only to strengthen but recast policies, more resilient and with a capacity to mitigate climate change and its impacts.
2. Evaluation of ecosystem services from an economic point of view.
3. Studies on biodiversity as impacted by climate and land use change as well as their interactions.
4. A comprehensive database of species.
5. Appropriate documentation of indigenous wisdom and practices on adaptation to climate change.
6. Detailed survey and inventory of distribution of plants and animals.
7. Evaluation of exotic and invasive species and linkages to native species.
8. Socio-economic survey with regard to land tenure, land allotment methodology, and conditions imposed.
9. Livelihood security.
10. Capacity building.

1.7 Economy

The economies of many HKH countries are linked to the Himalayas for agriculture, water for domestic use and irrigation, hydropower, and tourism, besides several other forest commodities. Countries like India, Nepal, Bhutan, China, Pakistan, Myanmar, and Afghanistan heavily depend on the mountain ecosystem for goods and services. A World Bank report (2009) indicated that in Nepal, agriculture has remained a key economic factor, contributing about 34% of GDP and Asian Development Bank (2004) document showed that this sector employs 93% of the workforce in 2004. Over 70% of the hydropower potential of India lies in the Himalayan states of the country (GOI 2012). Xizang Province of China also relies on the Himalayas for hydropower, tourism, and agriculture for sustaining its economy.

It is a known fact that about 53 million people that inhabit the Himalayan region sustain largely on subsistence farming. Pratap (2011) reported that the cultivation is mainly confined to marginal rainfed and some irrigated farmlands occupying 15.8% of the total area of the Himalayas, which is approximately 53.8 million ha. About 15.2% area is under permanent snow cover and rocky mountains, serving as a perennial source of water to the hill people as well as to the rest of the population, directly or indirectly depending on Himalayan ecosystem services. The rest of the Himalayan landscape includes forests, rangelands, pastures, and wastelands. The tribal communities which used to be self-sufficient in food grains about six decades ago became food deficient over time due to the increase in population, non-availability of sufficient land to produce as much as required, and also government policies that were not conducive to optimum production of crops. For example, in the northeastern Himalayas of India, the tribal communities used to produce the food grains to suffice their needs but with the passage of time, these families started facing food shortages. Besides the often quoted factor of spurt in population, some government policies were responsible for low productivity and production. In a bid to provide sufficient ration to the rural people, the government started providing food grains (mostly rice and wheat) through the public distribution system. If these food grains had been supplied on landing cost, there might not have been any problem with the production system but subsidizing the supply of food grains to a drastic low level (about 5% of the actual cost) discouraged the farmers to produce from their own land as the production costs were about ten times higher than at what rate they were getting from the government. Getting food grains from the public distribution system became the first priority of the families, and the option of producing from their own land was relegated to low priority. The food grain production may get a further jolt and decline with the inclination of the cultivators toward cash and horticultural crops. In the eastern Himalayas, rice is the staple food crop occupying about 70% of the cropped area under food crops. Shifting cultivation is prevalent in this region where a diverse mixture of crops is grown together. The practice has become uneconomical over time. Whereas wheat is rarely grown in the eastern Himalayas, it is the major crop of the western Himalayan region. Besides wheat crop, rice, maize, millets, barley, potato, pulses, oilseeds, fruit, and vegetable crops are also widely grown. A massive decline in the area under rice and maize has been observed in the recent past but the area under wheat has remained stagnant. A change toward fruit and horticultural crops (about 16% of cropland) may be the reason for the decline in area under rice and maize crops. The proportion of area under fruit crops is about 18% in the western Himalayas as compared to about 7% in the eastern region. It must be kept in mind that with the diversification of farming in the mountains, many future generational problems of ecological and social nature, such as equity, gender, and ethnicity, may crop up. The issues of unsustainability are raising their heads in the mountain ecosystems. Farming systems, particularly farm families with a small holding in the region, are in transition as indicated by the studies undertaken in the Central Himalayas. Communities and associated institutions are experiencing a situation of deteriorating conditions in the agriculture sector, in which the family farm economy, ecological environment, and agro-biodiversity components are being adversely affected.

The respective governments of the HKH region must enact such policies with regard to research and extension relevant to small farms, subsidies on farm inputs (not on finished products like rice, wheat, millets, pulses, etc.) and credit policies, land use, wildlife conservation, use of forest products, etc., which may address these issues on balanced and equity basis. It has been noticed that until the recent past, most of the governmental policies have benefitted the haves and have-nots have been ignored. We may say that intention of the government was good but the implementation of the policies went on the wrong track. The farmers with large land holdings being influential remained at the forefront in taking the benefit of subsidies offered for raising orchards, floriculture, new varietal seeds, chemical fertilizers, medicinal plant cultivation, farm implements, irrigation, fish farming, etc. Interestingly, this happened at the cost of small and marginal farmers who were deprived of these benefits. The population scenario in HKH countries is given in Table 1.3. Bhutan and Nepal have 100% of their geographical area in the Himalayas, while Bangladesh has about 11.3% of its area in the region.

We may infer that the scope of creation of economic opportunities in the HKH region is, though, vast, but with the limitation of riders. An exodus of people from the region to outside states or towns in search of jobs is sufficient proof that the agencies concerned have not been able to do their job up to the mark. The natural and social environment of the Himalayas is so

Table 1.3 Population (million) in the HKH area of different countries

Country	Exact population in different years	Approximate population during 2021
Afghanistan	22.85 (2016–17)	24.25
Bangladesh	1.60 (2011)	1.90
Bhutan	0.78	0.82
China	32.51 (2015)	34.89
India	80.56 (2011)	95.96
Myanmar	11.18 (2014)	12.41
Nepal	26.49 (2011)	30.31
Pakistan	47.43 (2017)	52.70
Total	223.86	253.24

unique that a comprehensive understanding of the constraining issues as well as potentials associated with the development of the region is a prerequisite to evolve clearly defined strategies and programs. Of all the issues, the basic issue of soil health needs immediate attention to realize its full potential. This would help to conceive, prepare, and implement development projects that can result in natural resources conservation and watershed-based sustainable livelihoods. Project interventions for the fragile environments of the Himalayas must be capable of withstanding potentially damaging events, besides being resilient enough to recover within a short span of time.

1.8 Vegetation Cover

Broadly, the HKH region can be divided into two landforms, viz., (i) the region with extremely difficult terrain, sudden altitudinal differences, and having all types of vegetation like different forests species, crops, and rangelands and (ii) high plateaus with relatively friendly terrain where, generally, grasslands are found. The grass and rangelands are extensible in the region. Land use in the HKH region is given in Table 1.4.

Variable temperature regime and constant low temperature for the most part of the year are major constraints in optimum agricultural production in the HKH region due to the short growing period and limited choice of crops. Schild (2008) mentioned that changes in the river systems and their basins have affected the economic condition of millions of people directly in the region. The other factors causing low productivity of crops are rocky mountains with steep slopes; excessive runoff and coarse-textured soils; shallow, sandy, and gravelly soils; and low WHC. People in large areas are vulnerable to river basin changes as these changes impact entire river basins in HKH (ICIMOD 2008). Like many other areas of the world, climate change has emerged as the most important driver of change in the HKH region. The other reason for the change can be attributed to population increase, globalization, risks involved in farming, and mass-scale migration. A general view of land use in the lower eastern Himalayas is presented in Fig. 1.9. Fast population increase, unplanned infrastructural development, unsustainable land use, poor management practices, and misuse of resources have further pressurized the ecosystem for goods and services (Chettri et al. 2008; Xu et al. 2008; GOI 2009; Sharma et al. 2009; Tsering et al. 2010). The extensive changes in ecosystems may affect their natural processes and reduce their capacity to provide services in the future. However, some exceptions may be there.

Table 1.4 Approximate land use in the HKH countries

Country	Area within HKH (000' ha)	Protected area (%)	Area under Agriculture (%)	Area under Forest (%)	Area under Grassland, scrubland, and others (%)
Afghanistan	39,156	6.7	24.2	8.9	60.2
Bangladesh	1,554	4.1	17.5	31.6	50.9
Bhutan	3,839	23.0	7.5	64.8	5.6
China	239,510	63.5	28.7	9.5	57.9
India	40,470	15.4	24.7	34.6	25.3
Myanmar	32,365	7.4	19.7	44.3	28.6
Nepal	14,752	16.9	46.6	28.1	8.4
Pakistan	47,904	3.9	17.7	4.5	73.9
Total	419,550	39.7	26.6	14.2	19.5

Fig. 1.9 Land use and land cover in the HKH region (Eastern Himalayan Region of India) (Photo by Dr. U. C. Sharma)

1.9 Major Problems of Mountain Agriculture

Having very high-altitude peaks and ridges, the HKH remains snowbound throughout the year and experiences extremely cold weather during winters and very low temperatures during summers. Under such a situation, the agriculture operations are done on steep slopes with cropping systems conducive to local physical features and prevailing weather conditions. Agriculture, per se, includes not only agricultural crops but also horticultural plantations, livestock rearing, agroforestry, and fish farming. On visiting hill slopes, one can find beautiful terraced fields where mostly rice is grown. Agricultural activities are mostly confined to the summer/rainy season and, by and large, fields remain fallow during winters. During this season, the availability of fodder is scarce and the livestock often depends on free-range grazing for their feed. The free-range grazing discourages growing crops during winters as there is no fencing, and the free livestock trespasses agricultural fields and can also graze on crops grown. In many areas in Eastern Indian Himalayas, the headman issues an order in December mentioning that the cultivators can let loose their livestock for free-range grazing until next February or so. The main feature of the mountain farms is their small size, which is uneconomic to cultivate. Though there are many government institutions in the HKH region, besides non-governmental organizations, engaged in the development of various disciplines of agriculture and improvement of living of mountain communities, the impact has been far from satisfactory. This may also be due to the complex nature of the mountain agriculture. The agrarian economy plays a major role in the livelihood security of the mountain people, and, therefore, top priority needs to be given to increase the productivity of various sectors like farming, livestock, agroforestry, pisciculture, forestry, indigenous crafts, etc. for ensuring enduring sustainability. The efforts of the respective government in this regard would help in enhancing the livelihood security of mountain people in the HKH region in due course of time.

Food and water security in the Himalayan region is greatly impacted by the climate change and variability. This may happen mostly because of inadequate water storage facilities. It is a well-known fact that agriculture in the HKH region is mostly rainfed and any changes in rainfall pattern, amount, and frequency can affect the agriculture production leading to food insecurity. In the HKH region, a good harvest is dependent on the timely arrival and well-distributed monsoon rains. In the past, inadequate and erratic monsoon has caused low crop production and famine-like situations, whereas heavy rainfall has resulted in floods, landslides, and soil loss due to erosion. Challinor et al. (2005) reported that besides an adequate amount of water, crop productivity depends on many biophysical processes and variables and their complex interactions. Similar views were expressed by Sheehy et al. (2006) who mentioned that the relationships between crop yield and biophysical processes and variables are controlled by several factors such as uncertainties in the interrelationship among variables, lack of data from realistic controlled field studies (IRRI 2011), as well as study scale, among others. Rising CO_2 and favorable temperature and humidity may affect the crop yield in the HKH region in the future due to pest proliferation (Gornall et al. 2010). There is a need to generate more scientific data on climate change impacts on agriculture in the region because at present the requisite data are extremely limited to understand the physical processes and variables and make final conclusions.

Like many other regions, the availability of arable land for cultivation in the HKH region is going to be a big bottleneck in the face of fast and unbridled population growth. As per the available data, about 0.14 ha of land is available per capita in the region, which is likely to come down to 0.10 ha by the year 2025. More than 75% of farms in the region are below 2.0 ha, which do not produce enough food for a family even during good harvest years because of conducive weather and choicest inputs. Some male members of the household have to move to towns to seed alternate sources of income to impart better facilities to the family and better education to the children. As a result, cultivation activities of a large proportion of farms are carried out by women who are already burdened with other daily chores. Due to low cash availability, the cultivators cannot use required inputs like quality seed, fertilizers, and pest control chemicals. The use of machinery to make the agricultural operation easy is not possible in the hilly terrain. Proper assessment of resources and dependable database for strategic planning is crucial for the upliftment of agriculture in the region. Crops intensification is, perhaps, the only key for enhancing agricultural productivity and production in the region. Loss of soil fertility may prove to be a hindrance and pose a challenge in this direction. The inputs level available and the growing conditions are the other challenges to be met on the way to increase agricultural production. Diversification can reduce some of the problems of low soil fertility, pests, and pathological diseases as well as can eliminate or decrease the effect of fluctuation in market prices, thereby ensuring assured income from the land. Diversification, however, is constrained by the marginal land holdings and market demand for new crops as well as uncertainty about forward linkages for the marketing of produce. In the HKH region, there is also a problem of wild and stray animals, which may require fencing of the crop fields for safety. The Himalayan mountains are rich in biodiversity; however, there has been gross genetic erosion in the region during the last five decades. This issue needs to have the full attention of the concerned people. This is a worrying matter that has still been either ignored or intentionally put under the carpet. There has been a reduction of 50–55% in maize and wheat germplasm during the last three decades.

In the Himalayas, the exploitation and depletion of natural resources base are much faster than these resources are being replenished. This must be viewed as a matter of concern by the ones who are at the helm of affairs with regard to resource conservation. In many cases, the possible reason for this is not exactly known. The increase in population has multiplied the demands of the people with an increase in feeding mouths and change in lifestyle of the people. There has been growing social and economic disparity as well as deterioration in environmental quality. Human interferences and routine activities in the HKH region have led to land and environmental degradation. Besides, there are some basic problems that need immediate attention of the concerned, such as the following:

- The fast growth of the population.
- Depletion of biodiversity in the HKH region.
- The slow decline in crop productivity.
- Generation of technologies to cope with changes in precipitation pattern, hydrological behavior, and inclement weather conditions.
- Change in land use pattern.
- More people opting for urban life.
- Below-poverty population.
- Absence of adequate livelihood options.
- Mismanagement of natural resources.

Increasing slopes with an increase in altitude enhance the possibility of an increase in natural hazards and associated calamities. Many scholars (Eckholm 1979; Ives and Messerli 1989; Jodha 1990) have opined that expansion of agriculture on marginal land and declining crop productivity are major factors responsible for unsustainability in the Himalayas. It is also believed that a visible preference for industrial sector development could have negatively affected the agriculture sector in the HKH region. Although migration of rural people from villages to cities is not a new trend, with an increase in population and consequently increase in demands, the mountain people are compelled to leave farming and search for alternative jobs in urban areas. This is due to the reason that mountain agriculture is no longer ensuring food and livelihood security for the inhabitants. The food grains which always used to be a domestic commodity in the past has now become a market commodity. It means that the farmers are not producing sufficient food grains to feed them throughout the year, not to say to satisfy the demands of the inhabitants who are in other professions.

1.10 The Outlook

Though there may be many problems and shortcomings faced by the people of the HKH region, one thing is clear that mountain people have developed many indigenous techniques to find just solutions to waive-off emerging issues. Not only the scholars but the respective governments have also taken cognizance of such sustainable indigenous technologies developed by the local/indigenous/tribal communities. Many of the problems faced by the people are mostly due to socio-economic reasons and the natural setting of the HKH region. Perhaps, no single answer or solution can be offered to solve these issues overnight; however, continuous concerted efforts in the right direction by the governments, non-governmental organizations, and most importantly dweller's own interest and participation can make a difference in improving a lot of people of the region. The menace of poverty, the ever-growing population, and the prevailing status of women in the society are some of the ticklish issues hard to tackle with. Many of these social indicators need improvement through regional, national, and international policies and their judicious implementation. This would help in resource conservation and help in saving the region from land and environmental degradation. Some governments have started many schemes like social forestry, crop insurance, subsidies on inputs, and compensation for losses of livestock and crops due to fire incidences, floods, and droughts. Besides, the government in some countries has initiated to focus on ecotourism and religious tourism to enhance the income of local mountain people.

1.10.1 Areas to Be Focused

There is a need to focus on seven broad thematic areas as follows:

(i) Increase in population has necessitated to ensure sustainable and enduring security for food.
(ii) Conservation of the resource base for the benefit of the future generations.
(iii) Ensuring water supply for domestic and irrigational use.
(iv) Environmental conservation and security.
(v) Livelihood resource base conservation and development.
(vi) Capacity building.
(vii) Ensuring the quality of life for the dwellers.

The following technologies are important to achieve the above objectives:

- The authorities should be concerned about the policy initiatives and framework for the implementation of requisite strategies. The people's participation is significant to achieve the necessary goals.
- Easy access to cost-effective technologies.
- Development of water and land resources conservation technologies, relevant to undulating and hostile mountain terrain.
- Development and dissemination of technologies for food security.
- Watershed approach for catchment management and protection.
- Identification of technological gaps across the HKH region which require immediate redressal.
- Arrangement for Solid Waste Management (SWM).
- Creation of infrastructural facilities and civic amenities.
- Awareness programs through public demonstrations and replication of technologies.
- Ensure projects with low environmental impact but high returns.
- Active participation of women and youths.

There is an apprehension among many inhabitants about declining ecosystem conditions in HKH for the last five decades. Many fruitful debates and discussions have been held and arguments and solutions offered but judicious implementation of various programs on the Himalayas has always remained under the clouds of uncertainty. The northern plains, adjacent to the Himalayas, have also been affected and it is likely that there could be a deteriorating environmental and socio-economic scenario if the situation will not be handled properly. The major causes of the collapse of the environment in the HKH region are gross human interferences, socio-economic and political reasons, flooding in valleys, deforestation, unabated growth of population, landslides, and up to some extent, poverty. This may be due to the theory of the "Himalayan Environmental Degradation" put forward by Ives and Messerli (1989). The following be concluded:

1. No clear policy on population growth has triggered a fast increase in population.
2. Everybody knows what happens when there is a sudden spurt in population; more pressure on natural resources due to higher demand for food and shelter, as well as cash requirements for daily needs and children's education.
3. Coming up with new settlements and indigenous settlers resort to deforestation as this activity makes their living easy. The fragile mountain ecosystems are under huge pressure.
4. Deforestation, intensive use of land, and misuse of natural resources have resulted in soil erosion, soil and environment degradation, and a decline in ecosystem services in the HKH region.
5. Depletion of resource base has been responsible for the continuance of poverty, endangering livelihood security in the region.

We cannot put all the blame on the people for the deterioration of the situation in mountains. With a fast population increase and increased demand from the ecosystem, anybody may be compelled to think on those lines. The need is to enact new policies by the government and their proper implementation to discourage deforestation by the dwellers, find alternate avenues of income sources, and improve the public distribution system.

1.11 Action Plan

Wester et al. (2019) called the HKH region as the pulse of the planet, and that for its sustainability and welfare of the communities living therein, immediate actions are needed in the right direction to revive its dynamic attributes within the normal ranges of activity relative to its development. For this, the Sustainable Development Goals have to be given top priority by the respective governments of different countries falling under HKH. Gioli et al. (2019) mentioned that poverty rates in the mountainous areas of the HKH are higher by about 33% than the national average of 25% in the countries of the region. They further elaborated that the development in mountain areas depends on "mountain specificities," such as remoteness, feebleness, and ostracism, which restrict development while biological diversity, natural niches, and conformation mechanisms offer opportunities for the development of mountain communities. About 30% of the population of the region face food insecurity and about 50% suffer from malnutrition, women and children being more vulnerable (Rasul et al 2019). Migration to cities outside the region is another social problem of the region and required to be solved on a priority basis (Siddiqui et al. 2019).

Trends in GHG emissions indicate that temperatures are likely to increase above 5 °C by 2100 in the HKH region (Krishnan et al. 2019). Two-thirds of the region's glaciers will melt by 2100 if the pace of emission continues as of today (Bolch et al 2019a, b). They mentioned that in the extended Hindu Kush Himalaya (HKH) region, the cryosphere is a key freshwater resource, and if this source is disturbed by climate change, the industry, agriculture, and hydroelectric power generation will suffer considerably. The HKH region has also experienced a rapid increase in air pollution in recent years, with ill consequences for the environment and the people (Saikawa et al. 2019).

The HKH Call to Action

In order to make HKH Region self-reliant in all spheres, including food, health care, natural resources conservation, and management as well as gender equity, Wester et al. (2019) came out with "The *HKH Call to Action"* which proposed six immediate actions to be taken up jointly by the eight HKH countries. The broad outline of these actions is summarized as follows:

1. Promote and strengthen cooperation at all levels and take actions at all levels to sustain the HKH as a global asset.
2. Arrange and prioritize the distinctiveness of HKH mountain people in decision-making.
3. Take restrict global warming to 1.58 C by 2100 by taking appropriate measures, and achieve carbon–neutral societies in the HKH by 2035.
4. Actions to achieve the Sustainable Development Goals and leaving no one behind.
5. Take decisive action to enhance ecosystem resilience for perpetual availability of ecosystem services by eliminating biodiversity loss and land degradation.
6. Promote regional data and information sharing and science and knowledge cooperation.

References

Ahmad T, Thakur VC, Islam R, Khanna P, Mukherjee PK (1998) Geochemistry and geodynamic implications of magmatic rocks from the Trans-Himalayan arc. Geochemical J 32:383–404

Ahmed T (2011) Geology of the Himalayan Mountain Range, with special reference to the western Himalaya. Geologi 63:142–147

Asahi K (2010) Equilibrium-line altitudes of the present and last glacial maximum in the eastern Nepal Himalayas and their implications for SW monsoon climate. Quat Int 212:26–34. https://doi.org/10.1016/j.quaint.2008.08.004

Balestrini R, Polesello S, Sacchi E (2014) Chemistry and isotopic composition of precipitation and surface waters in Khumbu valley (Nepal Himalaya): N dynamics of high elevation basins. Sci Total Environ 485–486:681–692. https://doi.org/10.1016/j.scitotenv.2014.03.096

Bates BC, Kundzewicz ZW, Wu S, Palutikof JP (2008) Climate change and water. Technical paper for Intergovernmental Panel on Climate Change. Geneva, Switzerland

Bloom JM, Blair SS (2009) The grove encyclopedia of Islamic art and architecture. Oxford University Press, pp. 389–390. ISBN 978-0-19-530991-1

Bolch T, Shea JM, Liu S, Azam FM, Gao Y et al (2019a) Status and change of the cryosphere in the extended Hindu Kush Himalaya region. In: Wester P, Mishra A, Mukherji A, Shrestha A (ed) The Hindu Kush Himalaya assessment. Springer, Cham

Bolch T, Shea JM, Liu S, Azam FM, Gao Y, Gruber S, Immerzeel WW, Kulkarni A, Li H, Tahir AA, Zhang G, Zhang Y (2019b) Status and change of the cryosphere in the extended Hindu Kush Himalaya region. In: Wester P, Mishra A, Mukerji A, Shrestha AB (ed) The Hindu Kush Himalaya assessment: mountains, climate change, sustainability and people. Springer Nature, Cham, Switzerland, pp 209–255

Bollinger L, Avouac JP, Beyssac O, Catlos EJ, Harrison TM, Grove M, Goffé B, Sapkota S (2004) Thermal structure and exhumation history of the lesser Himalaya in central Nepal. Tectonics 23:19P. TC 5015. https://doi.org/10.1029/2003TC001564

Bordet P, Colchen M, Le Fort P (1972) Some features of the geology of the Annapurna range Nepal Himalaya. Himalayan Geol 2:537–563

Burbank DW, Beck RA, Mulder T (1996) The Himalayan foreland basin. In: Yin A, Harrison TM (eds) The tectonic evolution of Asia. Cambridge University Press, Cambridge, pp 149–188

Chalecki EL, Gleick PH (1999) A framework of ordered climate effects on water resources: a comprehensive bibliography. J Am Water Resour 35:1657–1665

Challinor AJ, Wheeler TR, Slingo JM, Hemming D (2005) Quantification of physical and biological uncertainty in the simulation of the yield of a tropical crop using present-day and doubled CO_2 climates. Philos Trans R Soc B360:2085–2094. https://doi.org/10.1098/rstb.2005.1740

Chettri N, Shakya B, Sharma E (2008b) Biodiversity conservation in the Kangchenjunga landscape. ICIMOD, Kathmandu, Nepal

Eckholm E (1979) Planting for the future: forestry for human needs. World Watch Paper 26. World Watch Institute, Washington, DC

Gioli G, Thapa G, Khan F, Dasgupta P, Nathan D, Chhetri N, Adhikari L, Mohanty SK, Aurino E, Mapstone Scott L (2019) Understanding and tackling poverty and vulnerability in mountain livelihoods in the Hindu Kush Himalaya. In: Wester P, Mishra A, Mukerji A, Shrestha AB (ed) The Hindu Kush Himalaya assessment: mountains, climate change, sustainability and people. Springer Nature, Cham, Switzerland, pp 421–455

Gornall JR, Betts E, Burke R, Clark J, Willett CK, Wiltshire A (2010) Implications of climate change for agricultural productivity in the early twenty-first century. Philos Trans R Soc b: Biol Sci 365:2973–2989

Government of India (GoI) (2009) Governance for sustaining Himalayan ecosystem: guideline and best practices. Government of India, Ministry of Environment and Forests, New Delhi, India

Government of India (GOI) (2012) Status of hydroelectric potential development. http://www.cea.nic.in/reports/hydro/he_potentialstatus_region.pdf

GTOS (Global Terrestrial Observing System) (2008) Why focus on mountain regions? www.fao.org/gtos/tems/mod_mou.jsp. Accessed 03 Sept 2009 (Accessed 15 Apr 2018)

Guangwei C (2002) Biodiversity in the Eastern Himalayas: conservation through dialogue. Summary reports of the workshops on biodiversity conservation in the Hindu Kush-Himalayan Ecoregion. ICIMOD, Kathmandu, Nepal

Gurung DR, Amarnath G, Khun SA, Shrestha B, Kulkarni AV (ed) (2011) Snow-cover mapping and monitoring in the Hindu Kush-Himalayas. ICIMOD, Kathmandu, Nepal

Heuberger S (2004) The Karakoram-Kohistan suture zone in NW Pakistan – Hindu Kush Mountain range. vdf Hochschulverlag AG, pp 25–26. ISBN 978-3-7281-2965-9

Hua (2009) The Himalayas: water storage under threat. ICIMOD Newsl 56(Winter 2009):3–5

ICIMOD (2008) Food security in the Hindu Kush Himalayan region: a position paper. International Centre for Integrated Mountain Development (ICIMOD), Kathmandu, Nepal

ICIMOD (2009) Mountain biodiversity and climate change. ICIMOD, Kathmandu, Nepal. http://books.icimod.org/index.php/search/publication/613. Accessed 15 Apr 2018

ICIMOD's comments on a turbulent debate. http://www.icimod.org/?page=737. Accessed 28 Feb 2010

International Rice Research Institute (IRRI) (2011) Coping with climate change. http://beta.irri.org/news/images/stories/ricetoday/63/feature_coping%20with%20climate%20change.pdf. Accessed 14 Mar 2019

IPCC (2008) Climate change and water. Technical paper VI: 200. Intergovernmental Panel and Climate Change, Geneva, Switzerland

Ives JD, Messerli B (1989) The Himalayan dilemma - reconciling development and conservation. Routledge, London

IWMI (International Water Management Institute) (2003) Framework for the comprehensive assessment of water management in agriculture. Colombo, Sri Lanka

Jodha NS (1990) Mountain perspective and sustainability: a framework for development strategies. In: Paper presented at international symposium on strategies for sustainable mountain agriculture, 10–14 September 1990. International Centre for Integrated Mountain Development, Kathmandu, Nepal

Kandel P, Chettri N, Chaudhary S, Sharma P, Uddin K (2021) Ecosystem services research trends in the water tower of Asia: A bibliometric analysis from the Hindu Kush Himalaya. Ecological Indicators, 121: 107152, https://doi.org/10.1016/j.ecolind.2020.107152

Karan PP (1987) Population characteristics of the Himalayan region. Mt Res Dev 7:271–274

Karki R, Gurung A (2012) An overview of climate change and its impact on agriculture: a review from least developing country, Nepal. Int J Ecosyst 2(2):19–24. https://doi.org/10.5923/j.ije.20120202.03

Karki M, Sharma S, Mahat T, Aksha S, Tulashar A (2011) From Rio 1992 to Rio 2012 and beyond, draft regional report – Hindu Kush Himalayas: sustainable mountain development, prepared for the Lucerne World Mountain conference, 10–12 October 2011. Swiss Agency for Development and Cooperation (SDC)

Kohn GC (2006) Dictionary of wars. Infobase Publishing, p 10. ISBN 978-1-4381-2916-7

Krishnan R, Shrestha AB, Ren G, Rajbhandari R, Saeed S, Sanjay J, Syed MA, Vellore R, Xu Y, You Q, Ren Y (2019) Unravelling climate change in the Hindu Kush Himalaya: rapid warming in the mountains and increasing extremes. In: Wester P, Mishra A, Mukerji A, Shrestha AB (eds) The Hindu Kush Himalaya assessment: mountains, climate change, sustainability and people. Springer Nature, Cham, Switzerland, pp 57–97

Le Fort P (1975) Himalayas, the collided range: present knowledge of the continental arc. Am J Sci 275A:1–44

Liu G, Einsele G (1994) Sedimentary history of the Tethyan basin in the Tibetan Himalaya. Geol Rundsch 83:32–61. Bibcode:1994GeoRu..83...32L. https://doi.org/10.1007/BF00211893

Myers N, Mittermeier RA, Mittermeier CG, Fonseca GAB, Kent J (2000) Biodiversity hotspots for conservation priorities. Nature 40:853–858

Neelis J (2010). Early Buddhist transmission and trade networks: mobility and exchange within and beyond the northwestern borderlands of South Asia. BRILL Academic, pp 114–115, 144, 160–163, 170–176, 249–250. ISBN 90-04-18159-8

Ningombam SS, Dumka UC, Mugil SK, Kuniyal JC, Hooda RK, Gautam AS, Tiwari S (2021) Impacts of aerosol loading in the Hindu Kush Himalayan region based On MERRA-2 reanalysis data. Atmosphere 12:12901317. https://doi.org/10.3390/atmos12101290

Nogués-Bravo D, Araújo MB, Martinez-Rica JP, Errea MP (2007) Exposure of global mountain systems to climate change during the 21st Century. Glob Environ Change 17:420–428

Nogués-Bravo D, Araújo MB, Lasanta T, López-Moreno JI (2008) Climate change in Mediterranean mountains during the XXIst century. Ambio 37:380–385

Olson DM, Dinerstein E (2002) The global 200; priority ecoregions for global conservation. Ann Mo Bot Gard 89:199–224

Olson DM, Dinerstein E, Wikramanayake ED, Burgess ND, Powell GNB, Underwood EC, D'Amico JA, Itoua I, Strand HE, Morrison JC, Loucks CJ, Allnutt TF, Ricketts TH, Kura Y, Lamoreux JF, Wettengel WW, Hedao P, Kassem KR (2001) Terrestrial ecoregions of the world: a new map of life on Earth. Bioscience 51:933–938

Pêcher A (1977) Geology of the Nepal Himalaya: deformation and petrography in the Main Central Thrust Zone. Ecologieetgeologie De L'himalaya Science De La Terre. 268:301–318

Pei S (1995) Banking on biodiversity: report on the regional consultations on biodiversity assessment in the Hindu Kush Himalaya. ICIMOD, Kathmandu, Nepal

Pratap T (2011) Hill agriculture: challenges and opportunities. Indian J Agric Econ 66:33–52

Rasul G, Saboor A, Tiwari PC, Hussain A, Ghosh N, Chettri GB (2019) Food and nutrition security in the Hindu Kush Himalaya: Unique challenges and niche opportunities. In: Wester P, Mishra A, Mukerji A, Shrestha AB, (eds) The Hindu Kush Himalaya assessment: mountains, climate change, sustainability and people. Springer Nature, Cham, Switzerland, pp 301–338

Saikawa E, Panday A, Kang S, Gautam R, Zusman E, Cong Z, Somanathan E, Adhikary B (2019) Air pollution in the Hindu Kush Himalaya. In Wester P, Mishra A, Mukerji A, Shrestha AB (eds) The Hindu Kush Himalaya assessment: mountains, climate change, sustainability and people. Springer Nature, Cham, Switzerland, pp 339–387

Salerno F, Buraschi E, Bruccoleri G, Tartari G, Smiraglia C (2008) Glacier surface-area changes in Sagarmatha national park, Nepal, in the second half of the 20th century, by comparison of historical maps. J Glaciol 54:738–752. https://doi.org/10.3189/002214308786570926

Schild A (2008) The case of the Hindu Kush-Himalayas: ICIMOD's position on climate change and mountain systems. Mt Res Dev 28:328–331

Schild A, Sharma E (2011) Sustainable mountain development revisited. Mt Res Dev 31:237–241

Searle M (2013) Colliding continents: a geological exploration of the Himalaya, Karakoram, and Tibet. Oxford University Press, p 157. ISBN 978-0-19-165248-6

Sharma UC (1998) Water management by the tribes of north-eastern region – the traditional wisdom and future strategies. Fert News 43:43–49

Sharma P, Pratap T (1994) Population, poverty, and development issues in the Hindu-Kush Himalayas, development of poor mountain areas, ICIMOD, Kathmandu, Nepal

Sharma UC, Vikas Sharma (2003) Mathematical model for predicting soil erosion by flowing water in ungauged watersheds. In: Boer DH, Froehlich W, Mizuyama T, Pietroniro A (eds) Erosion prediction in ungauged basins: integrating methods and techniques. IAHS Publications 279. IAHS Press, Wallingford, U. K., pp 79–83

Sharma E, Chettri, N , Tsering K, Shrestha AB, Jing F, Mool P, Eriksson M (2009) Climate change impacts and vulnerability in the eastern Himalayas. ICIMOD, Kathmandu, Nepal

Sheehy JE, Mitchell PL, Ferrer AB (2006) Decline in rice grain yields with temperature: models and correlations can give different estimates. Field Crop Res 98:151–156

Siddiqui T, Bhagat RB, Banerjee S, Liu C, Sijapati B, Memon R, Thinley P, Ito M, Nemat O, Arif GM (2019) Migration in the Hindu Kush Himalaya: Drivers, consequences and governance. In: Wester P, Mishra A, Mukerji A, Shrestha AB (eds) The Hindu Kush Himalaya assessment: mountains, climate change, sustainability and people. Springer Nature, Cham, Switzerland, pp 517–544

Stöcklin J (1980) Geology of Nepal and its regional frame. J Geol Soc Lond 137:1–34. https://doi.org/10.1144/gsjgs.137.1.0001

Strom J, Svensson J, Honkanan H, Asmi E et al (2021) Shift in seasonal snowpack melt-out date due to light-1 absorbing particles at a high-altitude site in Central Himalaya. Atmos Chem Phys. https://doi.org/10.5194/acp-2021-158

Tse-ring K, Sharma E, Chettri N, Shrestha A (eds) (2010) Climate change vulnerability of mountain ecosystems in the eastern Himalayas – synthesis report. ICIMOD, Kathmandu, Nepal

Upreti Bishan N (1999) An overview of the stratigraphy and tectonics of the Nepal Himalaya (PDF). J Asian Earth Sci 17:577–606

Valdiya KS (1980) Geology of Kumaun lesser Himalaya. Wadia Institute of Himalayan Geology, Dehradun, 291p

Viviroli D, Dürr HH, Messerli B, Meybeck M, Weingartner R (2007) Mountains of the World, water towers for humanity: typology, mapping, and global significance. Water Resour Res 43(7):(13p) W07447. https://doi.org/10.1029/2006WR005653

Wadia DN (1931) The syntaxis of Northwest Himalaya: its rocks, tectonics and orogeny. Rec Geol Surv India 65:189–220

Wester P, Mishra A, Mukherji A, Shrestha AB (eds) (2019) The Hindu Kush Himalaya assessment: mountains, climate change, sustainability and people. Springer Nature, Cham, Switzerland

Wilson AM, Williams MW, Kayastha RB, Racoviteanu A (2016) Use of a hydrologic mixing model to examine the roles of meltwater, precipitation and groundwater in the Langtang river basin, Nepal. Ann Glaciol 57:155–168. https://doi.org/10.3189/2016AoG71A067

Wood LR, Neumann K, Nicholson KN, Bird BW, Dowling CB, Sharma S (2020) Melting Himalayan glaciers threaten domestic water resources in the Mount Everest Region, Nepal. Front Earth Sci 8:128. https://doi.org/10.3389/feart.2020.00128

Xu J, Yang Y, Li Z, Tashi N, Sharma R, Fang J (2008) Understanding land use, livelihoods, and health transitions among Tibetan nomads: a case from Gangga Township, Dingri County, Tibetan Autonomous Region of China. EcoHealth 5(2):104–114

Xu J, Grumbine RE, Shrestha A, Eriksson M, Yang X, Wang Y, Wilkes A (2009) The melting Himalayas: cascading effects of climate change on water, biodiversity, and livelihoods. Conserv Biol 23:520–530

Climate and Biodiversity

2

Abstract

The Hindu Kush Himalayan (HKH) region is characterized by mountainous environments, having a variety of regional climatic conditions. High-altitude regions in the HKH have the recent warming amplifications, especially during the global warming period. The rapid warming cause solid-state water (snow, ice, glacier, and permafrost) to shrink, leading to increase in meltwater, and there have been found more frequent incidences of flash floods, landslides, livestock diseases, and other disasters. Meanwhile, the HKH region is often referred to as the water towers of Asia as many high-altitude regions store their water in the form of snow and/or glacier, feeding ten major large rivers in Asia. Therefore, the impacts of climate change on water availability in these river basins have huge influences on the livelihood of a large number of population, especially in downstream regions. However, the scarcity of basic hydro-meteorological observations particularly in high-altitude regions of HKH limits rigorous analysis of climate change. The temperature in the HKH undergoes a significantly increasing trend, and the warming rate is greater than the global change. Thus, the elevation-dependent warming (EDW) is clear over the Tibetan Plateau, and the warming has occurred at a greater rate at higher elevations. The exact driving mechanisms responsible for EDW need further investigation. Because of the great range of climate variability in the HKH region, it is possible here to make only very general comments about past, present, and future climate trends and variability. The degree of warming appears to increase with elevation, as observed in Tibet and Nepal, and warming is more pronounced in winter than in summer. Changes in precipitation have been observed throughout the region, but these are highly variable. The summer monsoon dominates the climate, lasting eight months (March–October) in the Eastern Himalayas, four months (June–September) in the Central Himalayas, and two months (July–August) in the Western Himalayas. The east–west variation is based on the dominance of different weather systems, which in turn causes the monsoon to weaken from east to west. The monsoon penetrates northwards along the Brahmaputra River into the southeast Tibetan Plateau. Mountains are still facing enormous pressure from various drivers of global change, including climate change. Under the influence of climate change, mountains are likely to experience wide-ranging effects on the environment, biodiversity, and socio-economic conditions. Changes in the hydrological cycle may significantly change precipitation patterns leading to changes in river runoff and ultimately affecting hydrology and nutrient cycles along the river basins, including agricultural productivity and human

U. C. Sharma et al., *Soils in the Hindu Kush Himalayas*, Geography of the Physical Environment,
https://doi.org/10.1007/978-3-031-11458-8_2

wellbeing. Recent scientific understanding, led by the Intergovernmental Panel on Climate Change (IPCC), is that global climate change is happening and will present practical challenges for local ecosystems. These changes will impact everything from the natural landscape to human health, built infrastructure, and socio-economic conditions. The global community is currently trying to understand the nexus between climate change and mountain vulnerability, especially in the most remote and highest mountains of the world—the Hindu Kush Himalayas.

Keywords

Aerosols in Hindu-Kush Himalayas · Agrobiodiversity · Biodiversity in HKH · Climate · Ecosystem restoration

2.1 General Description

Outside the poles, the Hind Kush Himalayas, having hundreds of the world's most famous mountains, are the storehouse of glacier ice, estimated to be spread over 77,700 km^2. These mountains inhibit and provide food security to about 1.5 billion people on their slopes and valleys. The major climatic drivers influence the behavior of various topographic reliefs in the HKH region. The arrangement of different mountain ranges makes the ecosystem of this area so complex that the connections and interactions between rainfall and topography could not be well established until now and remains vaguely defined (Palazzi et al. 2013). It will be a major challenge in the region from both science and geopolitical positions to estimate present precipitation and its possible dynamics and variability in the coming years. This would be a constraint in taking necessary steps to develop judicious adaptation strategies in the future. Due to its vastness and variable interactions, considering the HKH as a single region may be difficult as a uniform approach in planning and development will not work properly. The eastern Himalayas mostly receive rainfall during the monsoon season. This is due to the southwest Indian monsoon's northward breeze from the Indian Ocean, which has a high moisture content (Wu and Zhang, 1998). Himalayan glacier mass balance has been markedly negative in recent years due to high regional warming (Solomon et al. 2007). This has been proven by numerous studies conducted in this area. Some researchers, like Schmidt and Nusser (2009) and Gardelle et al. (2012), have suggested that some glaciers in the central Karakoram have remained stable and even acquired mass. Recent estimations by Tahir et al. (2011) indicate that the snow cover in this area is increasing. The climatic conditions of the Hindu Kush Himalayan (HKH) region are diverse (You et al. 2017a, b). The seasonal variation of precipitation and temperature in the Himalayan region is shown in Fig. 2.1.

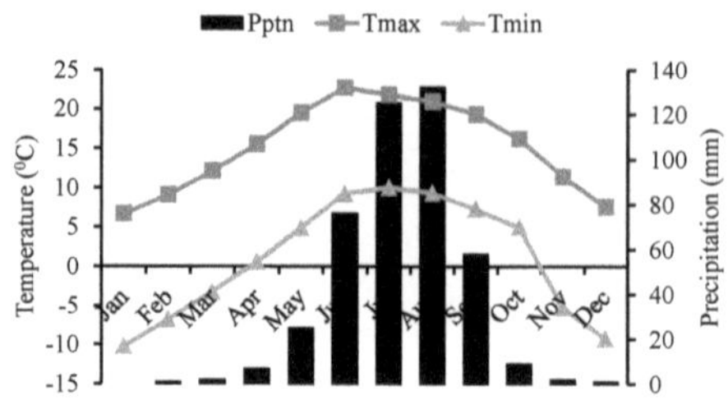

Fig. 2.1 Seasonal variation of weather variables in the Himalayan region

2.2 Climate Extremes in the Hkh Region

The areas of Hindu Kush Himalayas are remote and unapproachable compared to other areas of the world. But even most peaks and valleys have been influenced by the change in the climate. The climate change has already made the life of the people who live in valleys and slopes difficult and complex. This may crystallize in due course of time, jeopardizing the livelihood security of the people residing in the region. The mountains ecosystem is likely to be affected more even if the temperature rises by 1.5 °C by end of the century due to climate change. There is a need to reduce GHG emissions through concerted efforts of the world countries, particularly those with high emission records. At the current rate of GHG emissions, the temperature in high mountains may rise by about 3.0 °C higher by the end of the twenty-first century. And if emissions are still higher, that number could rise to over 5.5° C. During 1961–2005, temperature extremes in the HKH and Tibetan Plateau (TP) showed patterns consistent with warming, and a significant increase in temperature has been observed in the HKH region, more than the global average. Yan et al. (2016) used the term "elevation-dependent warming (EDW)" which they noticed in TP. They explained that the warming at higher elevations is more as compared to lower elevations. There is a need to enhance research efforts to unlock the exact driving forces behind the EDW in this region. It may not be easy to determine the contribution of each of the factors responsible for EDW, individually, but the greenhouse gas emission through the human interferences appears to be the most important factor (You et al. 2010).

Changes in temperature, precipitation patterns, and the frequency and intensity of extreme events are all influencing the Hindu Kush Himalayan (HKH) region as a result of climate change (Sun et al. 2017; You et al. 2017a, b; Zhan et al. 2017; Krishnan et al. 2019). The yearly mean temperature of the HKH region increased by 0.104 °C every decade from 1901 to 2014 (Ren et al. 2017). The temperature increased by 1.5 °C in the last two and a half decades (1982–2006), compared to 0.6 °C at the global level (Shrestha et al. 2012). In the Hindu Kush Himalayan (HKH) region, observed weather and expected climate change point to an increase in the spread of vector-borne diseases (VBDs) (Dhimal et al. 2021a, b). The HKH region has been reported to be warming at a higher rate than the global average. The precipitation and frequency and intensity of some extreme events have increased significantly over the last about 60 years (Dhimal et al. 2021b). Changes in temperature and precipitation have impacted climate-dependent sectors such as hydrology, agriculture, biodiversity, and human health. The Hindu Kush Himalayan (HKH) region is one of the world's most sensitive ecosystems, making it particularly vulnerable to climate change (Shrestha et al. 2012; Pandit et al. 2014; Wester et al. 2019). An evaluation of the physical and biological environment in the HKH region reveals a dramatic increase in the region's inhabitants' social and economic fragility, posing a serious threat to their physical and mental wellbeing (Wester et al. 2019).

On the other hand, the Upper Indus Basin (UIB) has shown different hydro-meteo-cryospheric processes mainly due to the rugged and difficult terrain of the HKH region and the distinctive interaction of existing regional circulations (Hasson 2016). It has been advocated that poor meteorological networks, both in high as well as low altitudes, under such rugged topographical situations cannot cater to the needs of the region. It is unable to cover vertical and/or horizontal extents fully to generate desired meteorological data. Tendencies ascertained from the observations at local sites further need to be evaluated for their field importance under such conditions. This will also offer a dominating and correct position of change, as well as a much clearer understanding of the effects of the observed contradictory climatic change on the HKH region's overall hydrological processes and regimes. Adaptive capacity, exposure, and resilience are all factors that influence the local implications of regional climate variability (Salik et al. 2015). This is more true for the security of the WEF (water, energy, and food) sectors. Climate change may pose a serious problem and a matter of concern when looking at its recent trends and movements. Falvey and Garreaud (2009) reported that such a globally averaged warming pointer is spatially heterogeneous and not necessarily synchronous between or within different regions.

2.2.1 Change in Climate

The HKH region has a great range of climatic variability, and due to this complex nature of variability, one can make only general and, up to some extent speculative, observations for the past, present, and future climate trends and the extent of change and variability. On the whole, there is a consistent trend of warming in the HKH region during the past 100 years, though, some areas have shown more warming than the global average of 0.74 °C in one century. The major feature of the trend exhibited is that the magnitude of warming has been found to increase with an increase in elevation. This has been observed in Tibet and Nepal. The warming is more conspicuous in winter than in summer. However, in some mountains, such as the Hindu Kush and the Hengduan in southwest China, summer temperatures have shown a slow but continuous decline. It is a fact that climate change has a direct effect on the hydrological cycle but, at the global level, this effect is more pronounced in the HKH region than any other place, with a large amount of water moving from mountains to the sea. Highly variable changes in precipitation have been observed in the region. In the arid plains of Pakistan and northeast India, decreasing trends in annual mean rainfall have been observed.

Sharma et al. (2009), based on the available data on precipitation and temperature trends in the HKH region, observed that, by and large, the precipitation increased from Northwest to Southeast of the region (Fig. 2.2). Tibet (China), though lying North-mid-Himalayas receives less precipitation, largely due to the reason that monsoonal winds weaken before crossing the Himalayan peaks. The Tibetan plateau can be considered as a rain shadow zone of the Himalayas. The average annual rainfall recorded over the HKH region is about 880 mm (Shrestha et al. 2015). The number of extreme rainfall events has been found to be decreasing; however, the intensity of each event is reported to be increasing, more so in the western parts of the HKH.

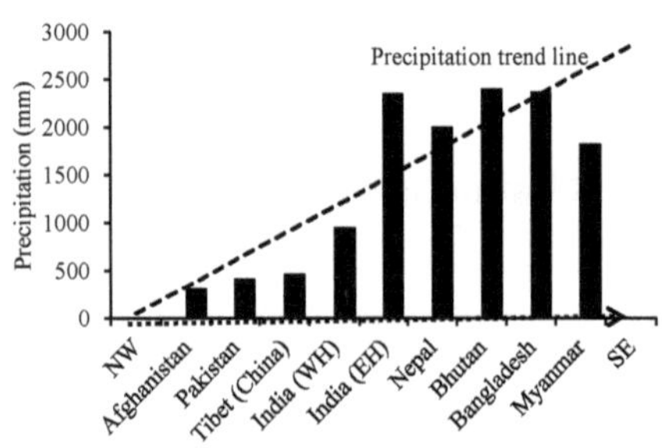

Fig. 2.2 Precipitation trend in HKH region

Much of the Himalayas are under the influence of monsoon with about 70–80% of the annual precipitation occurring during the typical monsoon months, from June to September (Bookhagen & Burbank 2010). However, in several areas, particularly in the north of the main Himalayan ranges, westerlies, the upper tropospheric waves with sources in the Mediterranean sea and the Atlantic Ocean exert a considerable influence on the seasonal pattern of precipitation (Brookhagen and Burbank 2010; Burbank et al. 2012). While much of the outer Himalayan ranges, directly exposed to monsoon thrust, receive up to 3000–4000 mm annual precipitation, the valleys in the North of the Main Himalayan Ranges (MHR) are dry receiving generally <500 mm precipitation. The precipitation of the areas in the north of MHR is characterized by the low amounts of precipitation, the difference in seasonal distribution receiving only cursory attention. Researchers have pointed out that such areas are getting more winter and spring precipitation due to the westerlies (Bhutiyani et al. 2010; Burbank et al. 2012), but the overall modification of the seasonal and spatial distribution of precipitation due to them at a regional level has drawn little attention.

The country-wise increasing trend of precipitation in HKH from NW to SE is supported by the longitudinal variations. It was observed that the data on precipitation showed an increasing trend from West to East depending on the longitudinal variation (Fig. 2.3).

This is not, however, true with the prevailing temperature, and no particular trend has been observed for the NW–SE direction (Fig. 2.4). However, the temperature has been found to be elevation dependent in the HKH region. According to the IPCC's Fourth Assessment Report (IPCC 2007a, b), anthropogenic GHG emission is responsible for about 90% of the estimated warming in the HKH region since the 1950s.

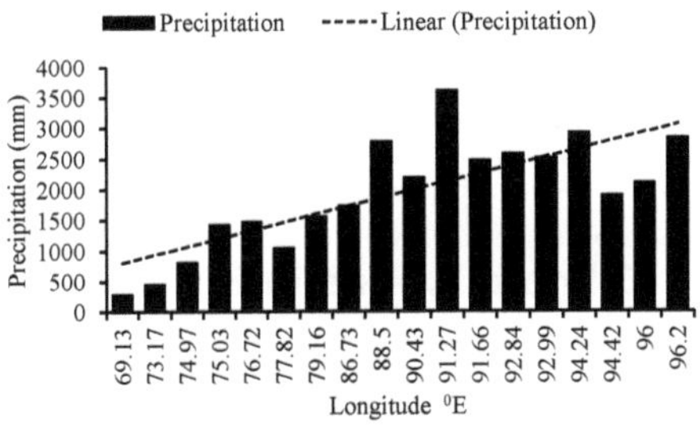

Fig. 2.3 Pattern of annual precipitation in relation to longitude along the Himalayan Arc (based on Singh et al. 2017)

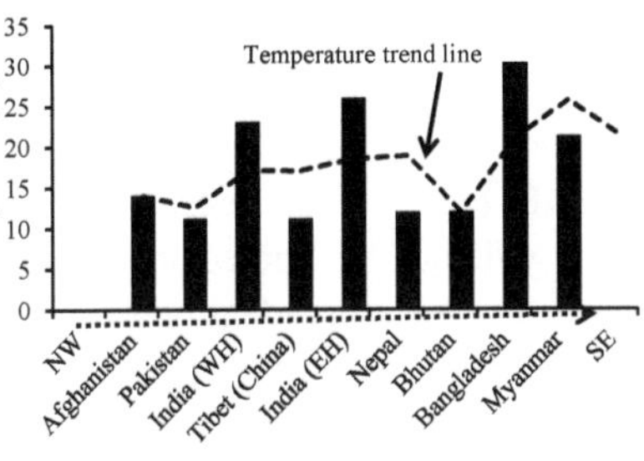

Fig. 2.4 Mean temperature in different HKH countries

2.3 Relative Humidity and Sunshine Hours

It is evident that climate affects crop growth interactively. Crop growth is significantly affected by solar radiation, temperature, rainfall, and the length of the growing season. Information on the relationship between climatic factors and the plant's ability to produce and sustain crop yield attributes is very important. Sunshine is extremely important for the growth of plants due to the reason that the solar radiation is responsible for supplying light and heat required by growing plants. The quality and quantity of the sunlight transmitted to the plants are dependent on atmospheric conditions. The sunshine influences the distribution of crops on a global basis, directly through its radiation and indirectly, through its effect on air temperature. Reduced transpiration influences the translocation of food materials and nutrients. Some ranges of humidity are helpful in encouraging the growth of fungal and bacterial diseases, which often need very high humidity to infect tissues. Humidity, atmospheric moisture, is an important factor for plant growth and development. The influence of humidity on plant growth is apparent from the sequence of plant patterns that have evolved in different agro-climatic zones of the world and the species that grow and mature during different seasons. The most visible effect of the humidity is the control of the rate of transpiration from the leaves. Atmospheric moisture has a direct impact on the regulation of stomatal opening.

There are great variations in the daily sunshine hours and relative humidity in different areas of the HKH region, depending on the altitude, amount of precipitation, and sky cloudiness (Fig. 2.5). The sunshine hours have been found to vary from 5.5 to 11.5 h, 4.9 to 7.6 h, and 4.5 to 12.8 h in Western, Central, and Eastern HKH region. The mean values for relative humidity in these zones have been found to vary from 37 to 70%, 80 to 91%, and 62 to 82%, respectively.

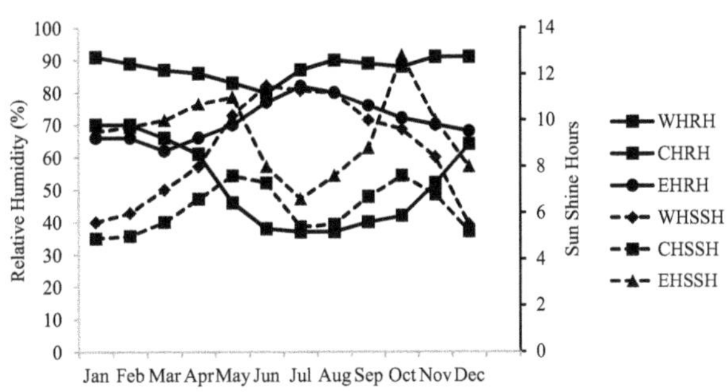

Fig. 2.5 Relative humidity (%) and sunshine hours in HKH Region (WH, CH, and EH are Western, Central, and Eastern Himalayas; RH is Relative Humidity; and SSH is Sunshine Hours)

2.4 Himalayan Glaciers

Glaciers in the HKH region have shrunk by 20–47% since 2000, mainly on the Tibetan Plateau and in the eastern reaches of the mountain range, such as the central Himalaya range and the Khumbu. The mountain ranges of Tien Shan, Pamir, Qilian Shan, Kunlun Shan, Karakoram, Qiantang Plateau, Tanggulla, Gandishi, Nianqingtangla, Hengduan, Himalayas, Hindu Kush, and Hinduradsh have glaciers area of 15,417; 12,260; 1,930; 12,260; 16,600; 3,300; 2,210; 620; 7,540; 1,620; 33,050; 3,200; and 2,700 km^2, respectively (Dyurgerov and Meier 2005).

Snowfall accounts for a significant portion of yearly precipitation, particularly at higher elevations (over 3000 m), and is a source of Himalayan glaciers (Table 2.2). The broader Himalayan region's glaciers encompass an area of more than 112,000 km^2 (Table 2.1). The melting of glaciers takes place in the summer. Snow and glacier melt account for anywhere from 2 to 50% of the average river flow. Snow and ice melt contribute around 70% of the flow of the main Ganges, Indus, Tarim, and Kabul rivers during the "shoulder seasons," before and after precipitation from the summer monsoon, according to Barnett et al. (2005). The contribution of Nepalese rivers to the annual flow in the Ganges Basin is approximately 40%. The Ganges basin alone is home to 600 million people or around 10% of the region's total human population. According to Xu et al. (2008), glacial melt is the primary supply of water in western China during the dry season.

2.5 Climate Trends in Himalayas

Mountains are likely to face different kinds of effects due to climate change on the biodiversity, environment, and socio-economic conditions, as reported by Beniston (2003). More severe weather, higher temperatures, longer droughts, heat waves, changes in local biodiversity, and reduced ground and surface water quantity and quality have been predicted by the Intergovernmental Panel on Climate Change (IPCC 2007a) in the HKH region if the present rate of GHG emissions will not be bridled in due course of time. For this, we must identify and understand the research gaps and implementation of regulatory laws with respect to GHG emissions in the Greater Himalayan region (Sharma et al. 2009). Precipitation is increasing in the Tibetan Plateau in the northeast Himalayan region (Zhao et al. 2004), as well as in eastern and central China (Xu et al. 2007). Precipitation is expected to increase by 20–30% in the western Himalayan Kunlun Shan and Tien Shan ranges (Rupa Kumar et al. 2006). Monsoon precipitation is expected to decline by up to 20% in most parts of Pakistan and southeastern Afghanistan by the end of the century. The southern and eastern

Table 2.1 Average annual rainfall and temperature in Afghanistan

Location	Average annual precipitation (mm)	No. of years	Average annual Temperature (°C)	
			Max	Min
Baghlan	251	13	24	−2
Ghazni	327	18	20	−10
Par wan	401	6	25	−4
Nangarhar	157	21	36	+5
Kabul	338	10	24	−4
Lagman	404	2	36	+5

Tibetan Plateaus, as well as the central Himalayan range, are expected to see comparable precipitation reductions. Understanding the changes in the amount of water and its seasonal availability in different regions of the Himalayas is of critical importance not only to the people who live in mountains but also to nearly one-fourth of the global population living in the 10 connected river basins (Immerzeel et al. 2010) where population density in some areas exceeds 1000 km^{-2}. Because of the sparse distribution of meteorological stations, particularly in higher mountain ranges, the spatial and temporal distribution of precipitation in the Himalayas is poorly documented. For example, contrary to the earlier understanding that precipitation in the Himalayas decreases with elevation, recent observations from Nepal show no sign of a decrease in precipitation even beyond 3000 mm (Immerzeel et al. 2014).

2.5.1 Indus River Basin

About 90% of Pakistan's agriculture depends on Indus and other associated rivers like Jehlum, Chenab, Ravi, Sutlej, and Beas. The upper Indus basin gets more rain than the lower basin and is a source of water for the whole basin. The upper Indus receives about 500 mm of rainfall in 132 rainy days, whereas the lower basin receives just below 300 mm in 84 days in a year. The share of annual rainfall during winters is 17% for the whole basin and 30% for the upper basin.

2.5.2 Mekong and Salween River Basins

Out of the total precipitation, the rainfall received during pre-monsoon, monsoon, post-monsoon, and winter season varies from 17%, 69%, 12%, and 2%, respectively. The Southern parts of these basins receive higher amounts of rainfall than other parts, where the rainfall exceeds 2,500 mm per year. The Winter months in these basins are almost dry. Rainfall intensity has not changed significantly over time. Across both the Salween and Mekong basins, temperatures have risen, especially during the winter months.

2.6 Afghanistan

Afghanistan is located in the low latitude arid and prairie zone of the Asiatic-climatic system. The climate is defined by a subtropical high-pressure zone of sinking air, which is influenced by cold, dry northeast winds from Asia's interior in the winter and warming with west and north-west breezes on the western edge of the Asiatic lowlands in the summer. Most of the precipitation in the country results from the orographic uplift of air masses. The country is known for its dry and harsh climate. However, the mountains influence the weather, causing many local variations. The average annual temperature is −10 °C during winter to 34 °C in summer. Low rainfall and high temperature during summer in Afghanistan make it as an arid nation. It has been observed that more than 50% of the area of the country receives less than 300 mm of rain, annually. The influence of the monsoon reaches up to the eastern border areas of Afghanistan and so, the rainfall in these regions is high as compared to western parts of the country. The season-wise distribution of precipitation shows that about 50% is received in winter, 30% in spring, and 20% during summer. A significant amount of precipitation falls as snow in the central mountains. Agriculture in Afghanistan is dependent on runoff from snowmelt during the spring and summer months when day temperatures are high. The complex arrangements of mountain slopes, valleys, and plains in the country result in an intricate microclimatic pattern (Table 2.1).

Factors influencing the climate of the region need to be assessed after careful study of the regional climate patterns and the effect of other factors such as local topography, altitude, and wind patterns to develop and understand a clearer picture of the local climate in several areas of the country. Factors influencing the potential for continued drought have also to be studied, particularly in light of the parameters mentioned above. In the first glance, it may be inferred that the climate of Afghanistan is an arid desert or semi-arid steppe climate. However, due to macroscale pressure fluctuations related to the Asian monsoon, this may not be the case, and the reality is substantially different. For years, some parts of the country have been battling drought. Afghanistan's climate can be described as subtropical to mid-latitude steppe in most parts, with dry subtropical deserts in others. Afghanistan has hot summers and cold winters, similar to the desert Southwest; however, winters in the highlands are harsher. Summertime high temperatures regularly exceed 38 °C, especially in the Southwest, while wintertime low temperatures can fall below −25 °C. Because of the presence of high mountains, local climates are influenced by lower temperatures at higher elevations, which affects precipitation patterns. In Afghanistan, more precipitation falls during the winter and early spring, with many locations, including Kabul, remaining gloomy, foggy, and snowy. It has been noted that flash flooding occurs frequently during periods of heavy precipitation when ephemeral stream channels are quickly filled with water. Climate-controlling factors in Afghanistan work in different ways in different sections of the country.

Due to the variability in topography, the temperatures in Afghanistan vary widely throughout the year. As a function of the environmental lapse rate, which averages roughly 6.5 °C per 1000 m of height, air temperatures decline as elevation increases. Because the high Hindu Kush mountains dominate much of Afghanistan, altitude must be taken into account when evaluating a location's potential temperature. Because of the wide range of temperatures, it is difficult to classify large areas as having homogeneous climate types. In the summer, when the Asian landmass heats up, the Afghan region receives warm, dry air from the north and northeast, resulting in little or no precipitation. Winters in Afghanistan are impacted by the Siberian high-pressure system, which sends cold, dry continental air outward in all directions. During this time, the prevailing winds are usually from the northwest or north. Storms from the western Mediterranean may travel across Afghanistan every few days as a result of this. As a result, the chance for precipitation is particularly significant in the winter and early spring. Snow is a common example of this. Orographic precipitation occurs when moist air is driven upwards and chilled isothermally on the windward side of mountain ranges. Topography has a noticeable impact on precipitation patterns across the United States. Air drops and warms on the downward side of mountain ranges, resulting in a "rainshadow" effect where the air is warmer and drier. The Leeward region of the mountain generally remains dry as compared to the windward direction.

The Hindu Kush Mountain Range in Afghanistan plays a significant role in shaping the local climate in the country. On peaks above about 3600 m, the snow remains permanently throughout the year and the passes remain closed during winters when mountains remain highly packed with snow. The precipitation, which is received on peaks in the early spring, is highly variable in amount and frequency. Areas of unpredictable rainfall encircle the region, which in certain circumstances could be categorized as subtropical steppes. Because of the average temperatures above freezing and low precipitation, the northern region near Mazar-I-Sharif is classed as a subtropical desert climate. Due to the increased frequency of low-pressure disturbances, southwest Afghanistan receives relatively high rainfall. Throughout the year, the climate in the area of Khandahar and Farah is subtropical, and it is quite dry. The Herat region, on the other hand, has a subtropical steppe climate, with precipitation being highly variable. The mountain ranges to the north of this region shield it from the northeast monsoon's effects in the winter, resulting in dry winters with temperatures below freezing. Surprisingly, the summers are scorching, with temperatures reaching 45 °C in July. Both Jalalabad and Khowst have subtropical steppe climates, according to the available data on climatic parameters. The northern plains, having an area of 10.368 million ha, encompasses Afghanistan's most fertile soils and very conducive climate (Saba 2001). In this area, the vegetation cover in Afghanistan is classified as mid-latitude grassland-steppe. In general, the climate's temperature and precipitation regimes foster a favorable soil moisture balance. Mollisols soils have been shown to predominate in areas where soil moisture and plant development are beneficial. Aridisols, on the other hand, are present when these prerequisites are not met. In moist areas (Mollisols), short grasses and herbaceous forbs are common, but on drier sites, bunch grasses and dry bushes are common (Aridisols). In this region, agriculture is a major part of the economy as a high percent of arable soils in this region have been under use for growing crops. In a way, this zone has been added to ensure food security in Afghanistan. Saba (2001) reported that only about one-half of Afghanistan's agricultural lands are currently used as farmland. However, some of the arable lands have degraded due to environmental problems such as desertification, salinization, and chemical contamination. Continuous warfare has also damaged prime agricultural lands, and the cultivators have abandoned some farmlands due to war problems (Saba 2001). Irrigation is needed to grow crops in the dry areas, though this is constrained by the non-availability of modern irrigation technology because the farmers cannot afford this due to poverty.

The Hindu Kush region in the northwest includes Badakhshan province, as well as the Wakhan corridor, which includes parts of the Pamir and Karakoram regions (approximately 4.5 million acres) (CSO 2017). The region receives the most precipitation, with around 7.5% of the land permanently covered in snow and ice. In theory, the Hindu Kush region is a major water supply, feeding important rivers like the Amu Darya (FAO 2011).

The Northern Plains of Afghanistan are famous for growing Pistachio, which is, perhaps, the most important Pistachio Belt in the world, one of the largest forest systems in the country, as well as the most extensive flatlands of the country around the foothills of the Amu Darya River (Favre and Kamal 2004). Deep valleys and mountain areas of up to 6400 m are important features of the Central Highlands in the middle of Afghanistan. They cover around 20% of or 12.5 million ha area of the country (FAO 2011). Rainfed agriculture is predominant (16.4%), while only about 7% of the land is available for irrigated agriculture (FAO 2011). The Eastern Highlands constitute the smallest region of the country, covering around 11% or 7.2 million ha area of Afghanistan. The eastern mountains are the only regions that are directly influenced by the monsoon, which get trapped at the high mountain slopes and bring rain. The rains, however, also can cause flooding and land/mud slides. The Northern region is also known as the food basket of the country because it produces a range of grain crops such as wheat, rice, barley, and maize as well as fruit crops such as peaches, almonds, pomegranates, apples, grapes, saffron, and cotton (CSO 2017). The southern region of Afghanistan has an area of about 12.5 million ha. The Southern Plateau is mainly covered by barren land, which is more than 50% of the total land. About 20% of the area is under rangelands, while about 15% of the area is sandy in texture. Agriculture is practiced in about 5% of the area, mostly irrigated (FAO 2011). In absence of proper irrigation, getting the full potential of crops will not be possible, due to a shortage of soil moisture at critical stages of plant growth. Although the southern region is the driest in the country, it still produces wheat and other crops and some fruit crops (CSO 2017).

2.7 Bangladesh

Already the most climatically vulnerable country in the world, Bangladesh is expected to become even more so due to climate change. Floods, tropical cyclones, sea water intrusion, the vulnerability of population in Sundranan delta, storm surges, and droughts are becoming more common and are likely to be more severe in the future. Climate change and variability have already affected the lives and livelihoods of people living in coastal areas and in arid and semi-arid regions of Bangladesh (Sarkar et al. 2018). IPCC (2012) indicated that the weather or climate events, even if not extreme, can still cause extreme conditions or impacts, either by crossing a critical threshold in a social, ecological, or physical system or by occurring simultaneously with other events. These changes in climate are threatening the significant achievements. Karmakar and Shrestha (2000) have made detailed studies about the recent climatic changes and variations in Bangladesh by making use of the surface climatological data on monthly and annual mean maximum temperature, minimum temperature, and monthly and annual rainfall data for the period 1961–90. Interestingly, they found that the annual mean maximum temperature over Bangladesh had a significant decreasing trend up to 1975 and a very significant increasing trend after 1975. Quite likely, Bangladesh may get warmer with increasing rains in the future. It has been projected the country may get warmer by about 0.5–2 °C between 1961 and 2030, based on a range of global climate model outcomes. It was revealed that the annual mean temperature over Bangladesh has a slightly increasing trend during the whole period of study.

Bangladesh has three main sources of rainfall: (i) the depressions from western side during winter; (ii) the early summer thunderstorms, and (iii) the summer rains from southwestern winds called as monsoon. Three seasons are generally recognized: a hot and humid summer; a rainy monsoon season; and a cool dry winter. The summer monsoon in Bangladesh begins in the southeastern portion of the nation, usually in the first week of June. According to the records, Bangladesh receives more than 70% of its yearly rainfall from June to September on average. During the monsoon season, average rainfall ranges from 1000 to 3000 mm, with peaks in the Shillong Plateau in the northeastern section of the country (Matsumoto, 1988). The monsoon's north–south oscillation is mostly responsible for Bangladesh's rainfall. Bangladesh is known for its heavy rainfall (Alamgir 2009). More receipt of rainwater during monsoon proves beneficial in growing paddy crops even without the facility of irrigation. Puddling is possible only if a lot of water is available at the initial stages of field preparation to make the soil's physical condition appropriate for transplanting paddy seedlings. For the past six decades, the average annual rainfall in Bangladesh is 2360 mm; however, the variation in rainfall at different locations of the country is large, that is, it varies from 1200–5600 mm in the northeast of Bangladesh.

It is predicted that Bangladesh will receive a 5–6% increase in rainfall by 2030 (IPCC 2007a, b, c). The estimated trend shows that the average rainfall in Chittagong during monsoon is increasing over time (Fig. 2.6). The average rainfall for three years from 1979 to 2008 is given in Fig. 2.6 and the annual Rainfall Variability Index (RVI) is given in Fig. 2.7. The linear trend of rainfall shows an increase over time.

During the southwest monsoon season, the lowest temperature in most parts of Bangladesh has been rising in all months. In July, the largest growing trend was +0.081 C/year at Dinajpur, and in September, the maximum falling trend was −0.035 C/year at Sandwip. During the southwest monsoon season, the minimum temperature has a higher growing trend in the northern half of Bangladesh (Sarkar et al. 2018).

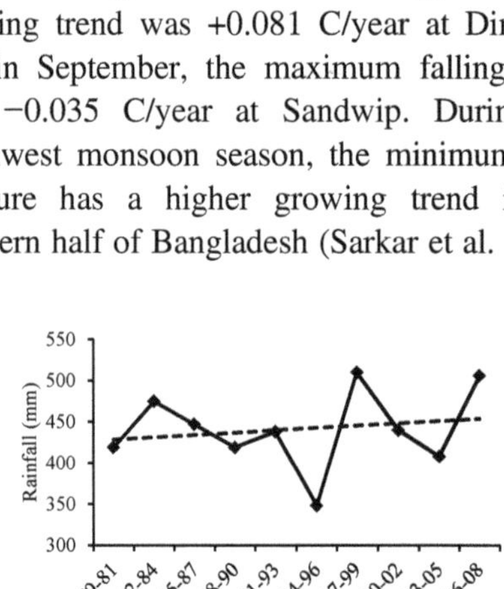

Fig. 2.6 Average (1979–2008) annual rainfall in Chittagong District in Bangladesh

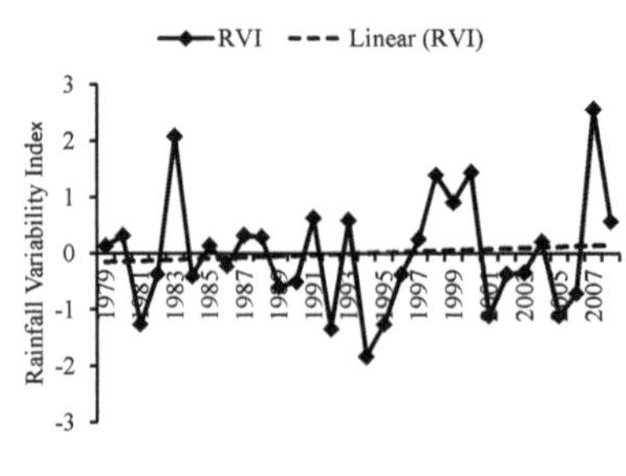

Fig. 2.7 Annual rainfall variability Index

During the southwest monsoon season, the maximum temperature in most parts of Bangladesh has been rising. The maximum growing rate is 0.094 C/year in September at Syedpur, and the maximum declining rate is −0.01 C/year in June at Srimangal. The southeastern section of the country has the highest increasing trend in mean maximum temperature in June, M. Court to Cox's Bazar in July and August, and the Syedpur–Dinajpur–Tangail region has the highest positive area in September. It has been observed that in general, maximum summer temperatures reach as high as 38 °C. January is the coldest and April is the warmest month in the country. Gentle winds blow mostly from the north and northwest in the winter, at one to three kilometers per hour in northern and central areas and three to six kilometers per hour in the coastal belt.

The mean annual temperature of Bangladesh has increased during the period of 1895–1980 at 0.31 °C over the past two decades. Chittagong Hill Tract (CHT), an outgrowth of the Himalayas in Bangladesh, has shown variation in temperature (Noorunnahar and Rahaman 2013). In CHT, the mean minimum temperature is during January (19.7 °C), and the mean maximum temperature is during 28.6 °C during the month of May. The standard deviation has been lowest of 0.445 during September and highest of 1.054 during August.

2.8 Bhutan

Bhutan is a small country situated between the Indian state of Arunachal Pradesh and Nepal and the entire country is within the Himalayas. The climate of this mountainous country is, more or less, a mix of the adjoining Eastern Indian Himalayas and Nepal. The strong influence of elevation-dependent warming (EDW) is observed in Bhutan too as average temperatures decrease with a rise in elevation, which has also been reported for Nepal (Kattel et al 2013). The effect of geographical position, elevation, slope direction, and slope angle on the spatial and temporal temperature regime in mountainous regions is complex (Dobrowski et al 2009). Because of the Himalayas' hilly and undulating landscape, it's difficult to accurately measure and assess the numerous climate characteristics (Lambert and Chitrakar 1989). This is especially true in Bhutan, where temperature variations are common. The drop in temperature with altitude, according to Blandford et al (2008) and Dobrowski et al (2009), is not only a fundamental and crucial physical attribute, but it also determines the ecosystem's appropriateness for diverse plant species. The fact that different forest kinds are found based on local temperature backs this up. Vertical traveling in the Himalayas shows that approximately up to 1000 m, broad-leaved vegetation is found, which slowly changes to pine and pine-like vegetation between 1000 and 1500 m, and still above 1600 m, deodar and similar vegetation is found.

Though Bhutan is a small country, precipitation varies significantly within Bhutan. Interestingly some areas, such as Khomachu, received a mean annual precipitation of less than 800 mm, whereas other areas, such as Phuntsholing, received more than 4000 mm, with 800 mm in the month of July alone (Dorji et al. 2018). This may be due to the fact that high Himalayan peaks create rainshadow areas where the moisture-laden clouds find it difficult to transverse. The southern and northern parts of Bhutan received the most precipitation. Winter, summer, and the entire year had the same overall regional pattern of precipitation distribution. The strong quadratic relationship between latitude and precipitation shows that there is a noticeable difference in precipitation in Bhutan along the north–south gradient. It has been discovered that latitude has the greatest impact on precipitation, whereas other factors such as longitude and slope have a lesser impact.

2.9 Tibetan Plateau (China)

Because the Tibetan steppe slopes southeast, rainfall from the southwest monsoons makes its way up to gorges in the east and south. Summer precipitation falls in a gradient from east to west and south to north, according to research. The

Tibetan steppe has a continental climate, according to Huang and Wu (1989), which is influenced by the southeastern monsoon in the summer and western air circulation patterns and high Mongol-Siberian air pressures in the winter. The Tibetan steppes are damp in the east, semi-arid in the south, and arid in the extreme west. The northern parts of the steppe are frigid and arid. Annual precipitation in the eastern portion of the steppe varies from about 600 mm to less than 60 mm in the western portion, mostly coming as wet snow and hail from June to September. Most of the pastures and grassland area receives less than 400 mm of annual precipitation. Dry winters are generally witnessed, but heavy snow events are most common and this snow covers available forage lands, with accompanying low temperatures causing an acute shortage of fodder, resulting in a difficult situation for livestock and wildlife in this zone. At an elevation of 3,658 m, the average temperature in Lhasa is −20 °C in January and 150 °C in July, with an absolute minimum of −160 °C and roughly 130 frost-free days. The average temperature in northern Tibet, at an altitude of 4,507 m, is −140 °C in January and 90 °C in July, with an absolute minimum of −410 °C and just 20 frost-free days. The average amount of hours of sunshine each day ranges from 6.8 to 8.2 h. Strong winds blow across much of the steppe, especially in the west, with wind speeds exceeding 17 m/s on 100–150 days each year.

The Tibetan Plateau (TP) has most of its area having very undulating topography. The climate change in the plateau has affected the area with spatial differences. Most of the recent studies in the area target on climate variation in the whole plateau. However, there are a few analyses about local climate change, done in totality in each region of the TP. Contrasting analysis of various regions of TP is essential and is a prerequisite to understand the climate variation and its impacts on the ecosystems of the TP. Very uneven and erratic annual distribution in precipitation in the Tibetan Plateau has been noticed. It has been reported that the mean (1961–2015) annual precipitation in the TP is about 465.5 mm. Mean monthly precipitation in the plateau is mainly received in the summer months (May to October) (Fig. 2.8). On a monthly average, precipitation is 69.8 mm during summer and 7.8 mm during winter months.

The Tibetan Plateau (TP) is often mentioned for its prominent mechanical and thermal effects at the regional and global level of circulation and climate change (Duan et al. 2012). For our understanding of climate behavior, several studies, such as data analyses (Yanai et al. 1992), theoretical studies (Wu and Zhang 1998), numerical modeling (Broccoli and Manabe, 1992), as well as rotating annulus studies (Yeh and Chang, 1974), have been conducted since the 1950s. The forcing of TP is critical in decoding the Asian summer monsoon circulation, according to numerical modeling studies (Chou, 2003). The temperature trends as a function of height are depicted in Fig. 2.9. According to Yeh and Gao (1979), the heating effect of several mountain ranges plays a significant role in hemispheric circulation patterns. TP has the most complicated terrain, with an average height of more than 4000 m above means sea level. Due to a lack of quantitative understanding, simulating the whole energy and water cycles across the TP and their impacts on global climate change is hampered by interactions between the land surface and atmosphere. As a result, research into the interaction of the atmosphere and land over the TP has grown in recent years (Xu et al. 2008; Ma et al. 2010). Nepal and the Tibetan Plateau have been experiencing increasingly more warming with elevation (Fig. 2.8) (Liu and Chen 2000).

The Tibetan plateau with an elevation of about 5000 m above msl is heated to its maximum during spring and early summer by

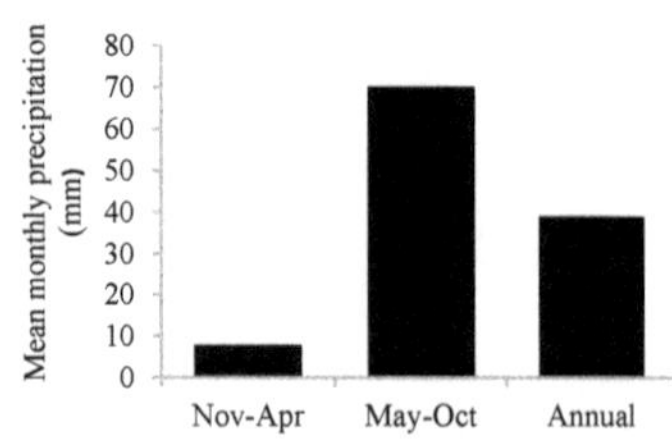

Fig. 2.8 Mean monthly precipitation trends in the TP during summers and winters

insolation which acts as a strong heat source known as Tibetan High and winds over most parts of the Indian Subcontinent as easterlies. The inter-annual fluctuations of the strength of this heat source have considerable influence on monsoon activity over the Indian subcontinent. A warm Tibet is associated with a wet monsoon and a cold Tibet with a dry monsoon. Snow cover is one of the important climatic controlling factors, which interacts with the global atmosphere by changing the energy processes due to large albedo and net radiation deficit. Massive snow cover over Eurasia during spring was observed to be followed by a longer advance period of the monsoon in India. The part played by snow and ice over the Himalayas in the radiation balance of the region is important.

2.10 India

2.10.1 Western-India Himalaya

India is a country where more than 70% of the population live in rural areas. The agriculture being the major profession of these people, their food and livelihood security depends on the farm produce. The agricultural productivity, in turn, depends mostly on intensity, pattern, and amount of annual rainfall. Climatic is the most important environmental factor affecting the production of crops anywhere on the globe. In Himachal Pradesh state, a part of the western Himalayas, major rivers like Beas, Satluj, and Ravi are mostly snowfed. The water from these rivers is used for irrigating crops. The climate change will affect these rivers' hydrology, thereby affecting the current production systems. The seasonal departure of rainfall in this part of western Himalayas was between −51 and 2%, −59 and −11%, −36 and 13%, and −83 and −18% during winter, pre-monsoon, monsoon, and post-monsoon, respectively, whereas the annual departure was between −39.0 and 8.0%(Tewari et al. 2017).

The crop production system is bound to be affected by adverse changes in precipitation, temperature, sunshine hours, and relative humidity, be it annual or seasonal basis. The Western Indian Himalayas receive less rainfall than the Eastern Indian Himalayas. In the Himalayas, rain is the primary source of water. Crops that may be produced effectively in any place are chosen based on the amount of water required by that specific crop. Besides rainfall, the temperature and character and health of the soil are the factors desired for determining which crops are to be grown. On the other hand, climatic conditions are determined by the altitudes, slope, seasons, and the presence of other specific conditions in Himalayan ranges. The rainfall analysis in the Himalayan region of Uttrakhand state of India showed that the mean annual rainfall in the frontal Himalayan ranges was highest compared to the mid and interior ranges. Seasonal rainfall distribution indicated that the southwest monsoon season contributes to 68.5% of mean annual rainfall in the mid-region, while 87.9% in the frontal ranges. Only 12.1% of the annual rainfall was distributed among the other seasons in the frontal ranges. In the interior ranges of the Himalaya, 52% of the mean annual rainfall was contributed by the monsoon season

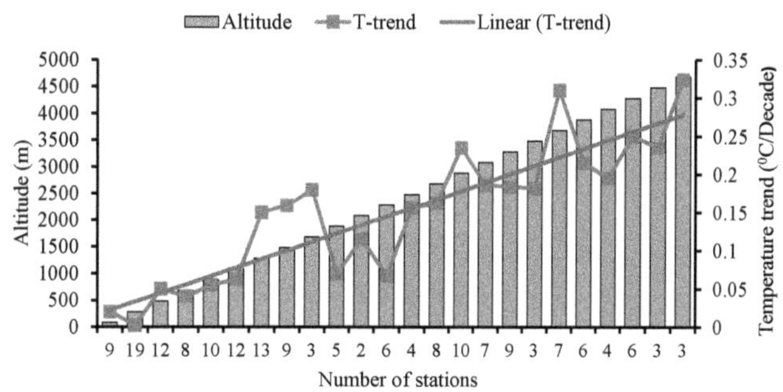

Fig. 2.9 EDW on the Tibetan Plateau (Based on Liu and Chen, 2000)

while 43.9% of the rainfall was recorded during the winter and summer seasons. The Uttrakhand state of India, situated in the western Himalayas, has very contrasting climatic conditions. Summers are hot, while winters are harsh, with strong cold waves sweeping over the region. Almost the entire year, the temperature in the Alaknanda Basin is below freezing. During the winter, the locations above 2000 m receive a lot of snow. Summers at the mid-altitudes are ideal for good health. During the summertime, the valley sections between Karnprayag and Devprayag are extremely hot, with monthly temperatures reaching 300 C or even higher. Many farming families migrate to the higher mountain areas during this period in search of summer pastures.

During the summer months from June to October, monsoon winds bring substantial rainfall, which is known as orographic rain. Heavy rain falls across the entire region, ranging in intensity from low in the valleys to very high on mountain ridges and from north-facing (leeward) to south-facing (windward) slopes. Leeward and windward phenomena occur due to high and long mountain ranges. The region is subject to frequent cloudbursts, debris flows, landslides, landslips, mass movements, and flashfloods. During the monsoon season, these disasters have a devastating impact on the entire region. The wind from western disturbances keeps the entire region cool, and cold winds blow throughout the season. Precipitation falls primarily in the form of snow at the higher elevations above 2000 m and rain in the mid-altitudes and valley region during the months of December–January. Okhimath received the most rainfall (199.4 cm), followed by Karnprayag (147.1 cm), and Srinagar received the least (92.5 cm). Valley regions receive less rainfall than high-altitude locations. However, some places located even at higher altitudes get less precipitation. For example, Joshimath, which is located at 1875 m, received only 107.5 cm annual rainfall because it is located in the leeward direction. The climatic parameters are also affected by the physiographic characteristics of a place. Low-lying river valleys, mid-altitude, highlands, alpine pastures, and snowcapped mountain peaks are the five physiographic zones that the Alakananda basin can be split into. Temperature changes can be seen from the valleys to the high mountains and from winter to summer. During the four months of winter, the places above 2000 m receive a lot of snow. In the valleys, summer is hot and humid. Mid-altitudes, on the other hand, stay pleasant during the summer and receive a lot of rain.

The Western Indian Himalayan region is characterized by high humidity during the whole year though the percentage of humidity differs depending upon the season and altitude of a place. Table 2.2 shows the status of rainfall at various locations in the Alakananda basin of the western Himalayas.

In the mid-Himalayan ranges (Fig. 2.10), 12.6 and 14.5% of rainfall was recorded during the winter and summer seasons, respectively. The rainfall variability index varied between −1.47 and 1.95, indicating temporal variations in rainfall. During the post-monsoon season, the rainfall of 1.2–4.0% of the mean annual rainfall was

Table 2.2 Rainfall at some locations of the Alakananda Basin in the western Himalayas (2008)

Parameter	Srinagar	Kamprayag	Okhimath	Pauri	Joshimath
Elevation (m)	550	883	1578	1630	1875
Yearly rainfall (mm)	925	1471	1994	1303	1075
Seasonal rainfall (%)					
Winter	16.0	10.5	8.8	14.8	15.4
Premonsoon	17.7	13.4	11.3	14.7	10.3
Monsoon	58.8	65.9	71.3	61.5	53.1
Postmonsoon	8.5	10.2	8.6	9.0	21.2

Source Forest Working Plan, Nainital Working Circle (2010)

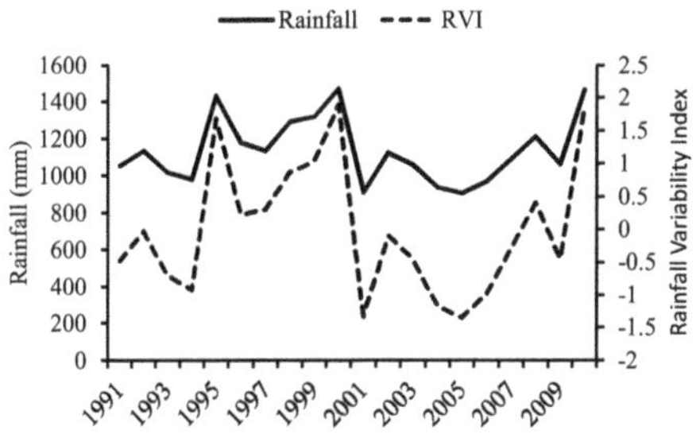

Fig. 2.10 Rainfall and rainfall variability index in mid-Himalayan Region. *Source* Aggarwal (2012)

received thereby indicating a dry period in the hill zones.

The water balance implies a comparison of rainfall and evapotranspiration with reference to the water-holding capacity of the soil as basic parameters. The general water balance equation with reference to vegetation is

$$P = Et + S + R + D$$

where P = Precipitation, Et = Evapotranspiration, S = Storage change, R = Surface runoff, and D = Deep percolation losses.

Evapotranspiration, more precisely actual evapotranspiration (AET), is most difficult to estimate accurately. Thornthwaite (1948) defined potential evapotranspiration (PET or PE) as the water loss from a large homogeneous vegetation cover area, which never suffers from lack of water. It is a function of wind, radiation, and humidity and represents atmospheric water demand or water need. The objective of water balance is to develop a soil site suitability model for different crops based on the soil and climatic requirements of crops being grown in the region. The moisture availability period or length of the growing period is the key parameter, which helps in assessing climatic suitability to attain maturity within an available length of the growing period.

The Himalayan range works as an effective barrier to the cold continental air blowing southwards into the Indian subcontinent, during winter. During the monsoon period, the Himalayas force the rain-bearing clouds up the mountains to deposit most of their moisture southwards in India, Pakistan, and Bangladesh. The Indian monsoon occurs in response to differential heating between land to the north and sea to the south. The intensity of monsoon circulation is proportional to the degree of thermal contrast between the land and the ocean. One of the most important effects of the Himalayas on southwest monsoon circulation over India is the formation of a large low-pressure area known as heat low formed over Pakistan and its adjoining areas with an elongated low pressure extending through the Gangetic plains to the North Bay of Bengal. The axis of the low-pressure area is oriented roughly from the northwest to southeast, that is, parallel to the southern periphery of the Himalayas known as the monsoon trough. Its northward and southward movements have an important bearing on the short-period fluctuations in monsoon rainfall.

2.10.2 Eastern India Himalaya

Eastern Himalaya gets more rainfall compared to the western, the extreme west being out of reach of monsoon rains. Both the intensity and duration of the monsoon current decrease as it moves to the west. During summer, the snow melts resulting in flooding of Himalayan rivers. The contribution due to snowmelt water to the annual river flow is greater in Western than in Eastern Himalayas. About 4000 km^3 of water is received as annual precipitation in the country out of which 510 km^3 is received in North Eastern Region. The average annual rainfall of the region is around 2500 mm; however, the distribution is uneven varying both in time and space. The annual precipitation received in the region comes largely from the southwest monsoon which sets in the middle of May and continues till October. On an average, Meghalaya, Arunachal Pradesh, Karbi Anglong district of Assam, and parts of Nagaland received more than 3000 mm of rainfall. The Cherrapunjee–Mawsynran range receives about 11,500 mm of rainfall annually which is the highest in the world. Manipur, Mizoram, Tripura, and parts of Nagaland and lower Assam receive below 2000 mm of rainfall while in Upper Assam districts, the rainfall varies from 2000 to 3000 mm, annually. The region receives 122.7% of the country's rainfall against 8.7% of the geographical area. A large fraction of the rainwater received is lost as runoff and in situ retention is very low (Sharma et al. 2006).

The altitudinal differences coupled with varying physiographical conditions contribute to great climatic variations in the Himalayas. The precipitation in the Eastern Himalayas varies from <1500 mm to >4000 mm. The mean yearly precipitation recorded at five stations is given in fig 2.10. The climate of Meghalaya state (Umiam station) is humid and warm. The rainfall varies from 1270 mm to more than 2500 mm. The relatively low elevation of this part is responsible for fairly high temperature from April to October. It has been reported that the probability of occurrence of a dry spell was very high (53–100%) in the first 15 weeks of the year (Satapathy 2000). The mean monthly precipitation in Manipur state (Imphal station), Meghalaya state (Umiam station), Mizoram (Aizwal station), Sikkim (Gangtok station), and Nagaland (Medziphems station) in Eastern Himalayas varied, respectively, from 8 mm in December to 435 mm in July, 13 mm in January to 588 mm in June, 15 mm in January to 358 mm in June, 19 mm in January to 540 mm in July. and 17 mm in December to 269 mm in September (Sharma et al. 2006). Rainfall of a 7-day period leaving at least 10 mm of rain after meeting the evaporational demand is designated as the start of an effective monsoon (Michael et al. 1978). Keeping in view this approach, the mean date of effective monsoon in the foothills of Eastern Himalayas was 14th April to ending August. Maximum rainfall is received in the Siang district of Arunachal Pradesh state of India which is around 6250 mm. Though there are great variations in the amount of rainfall received but on an average, the state receives 3087 mm of rainfall annually. The World's highest rainfall zone (Cherrapunjee–Mawsynram range) lies in the Meghalaya state, which receives about 11,500 mm of rainfall, annually. The climate of Northern Arunachal Pradesh state of India is alpine with temperatures touching below −7.0 ° C. The climate of central-eastern Arunachal Pradesh is bracing with temperatures varying from 5 to 28 °C and rainfall from 2000 to 5000 mm. Manipur state has a moderate climate but great variation has been recorded in hills and valley areas.

Though there are variations in temperature in Mizoram, the annual mean temperature varies from 16.5 to 29.6 °C in summer and 10.6 to 26.1 °C in winter. The temperature variation in Nagaland state was 18.3–31.7 °C in summer and 9.4–27.2 °C in winter. In Sikkim (Gangtok station), extreme variations in climatic conditions were observed. Very low temperature prevails at heights varying from 2000 to 6000 m. The areas falling above 5000 m amsl are perpetually snowbound. The average temperature in the state varies from 17.0 to 26.2 °C in summer and 9.1 to 20.2 °C in winter. Choudhury et al. (2012) investigated long-term meteorological variables (1983–2010) in mid-altitude Meghalaya, Northeast India, to find variations in rainfall trends. The overall annual rainfall trend grew at a non-significant rate of 3.72 mm/year, according to the findings.

On an average, minimum sunshine hours were recorded in July due to overcasting of the sky during monsoon season. Minimum sunshine hours are 3.0 at Imphal and 2.2 at Gangtok. Maximum average sunshine hours are 8.1 in January at Imphal and 5.9 in October/November at Gangtok. The highest average relative humidity was 89% in December at Imphal and 90% from July to September at Gangtok. Sharma et al. (2006) studied the climatic parameters in different zones of India's Eastern Himalayas (IEH) and the details are given in Table 2.3.

Sharma (1996), based on climatic features, topography, soil characteristics, and cropland use. suggested 14 agro-climatic subregions of eight Northeastern states of India, comprising IEH. This exercise was undertaken for agricultural development based on a watershed approach as well as to introduce location-specific farming systems in hills and mountain areas.

Table 2.3 Agro-climatic zones of IEH

S. No	Zone	Altitude range	Area (km^2)	Climatic characteristics
1	Alpine	>3500 m	47,068	Average rainfall varies from 750 to 1500 mm and air temperature 2 to 17 °C. The area remains covered with perpetual snow
2	Temperate subalpine	1500–3500 m	33,564	This zone receives an average rainfall of 2000 mm, annually. The temperature varies from 11 to 20 °C
3	Subtropical hills	1000–1500 m	29,021	Average temperature is 12 °C in winter and 30 °C in summer and the average rainfall is 2000 mm
4	Subtropical plains	400–1000 m	96,812	The rainfall in this zone varies from 1200 to 2000 mm and the temperature is 10 °C in winter and 28 °C in summer
5	Mild-tropical	200–400 m	26,349	This zone receives heavy rainfall varying from 2000 to 3500 mm and temperature from 9 to 26 °C
6	Mid-tropical plain	73–200 m	29,333	Temperature varies from 12 to 31 °C and rainfall from 2000 to 3000 mm

2.10.3 Weather and Agriculture in Indian Himalayas

Weather is one of the most important factors which determine the success of farming. It influences farm operations and crop production through its effect on soil, growth, and development of crops thereby affecting the productivity. Agriculture is vulnerable to climatic risks, and many techniques of agriculture have been designed to mitigate these risks. Some technologies have helped in mitigating some aspects of climatic disturbances. The climatic parameters affecting growth and productivity include rainfall, temperature, solar radiation, humidity, and rate of evaporation. Different crops and their varieties have their own optimal requirement of various climatic parameters such as rainfall, temperature, soil moisture, and light intensity.

In India, rainfed agriculture is done in about 86 million ha of land, and in terms of area and value of produce, it is at the top in the world. Cultivation of coarse grains (91%), pulses (91%), oilseeds (80%), and cotton (65%) takes place in rainfed areas. According to Rosegrant et al. (2002), rainfed lands will produce one-third of India's cereal production by 2025. Broadly speaking, with the present agricultural practices and technologies available, 30% of the area under irrigation in India produces enough food for 65–70% of the country's population whereas 70% of the cultivated area as rainfed agriculture produces only for 30–35% of the population. Looking into the problems of rainfed agriculture, the Government of India constituted a National Rainfed Area Authority in 2006 to address the rainfed agriculture-related issues. This authority shall develop a plan to improve agriculture in rainfed cultivated areas of India, based on the present status and available technologies to get optimum production from these areas. The authority will suggest a future course of action to be taken up by the government for rained cultivated areas.

2.11 Myanmar

The global geographical position plays a significant role in influencing the precipitation and average temperature of a country. This is further affected by the altitude, direction toward the sun, and nature of winds blowing, besides many other minor local factors. Climate data shows that some meteorological disturbances, such as cyclones with the high destructive force for coastal areas near Landfall sites, the passage of Western disturbances from NE India, and Easterly Waves from Thailand, have occurred. Everything has an impact on the weather in the

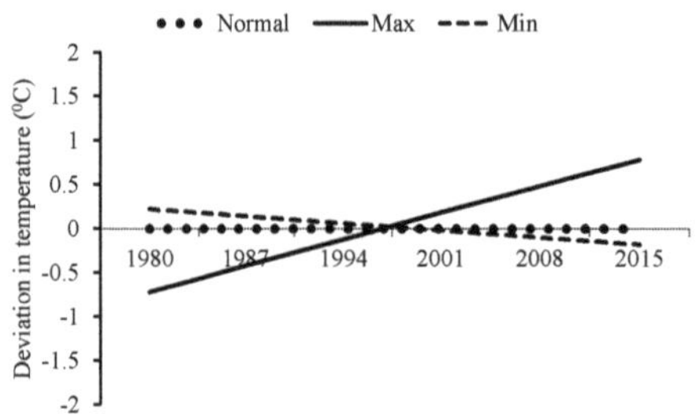

Fig. 2.11 Deviation in maximum and minimum temperature between 1980 and 2015 at Dewei, Myanmar (Aung et al. 2017)

area. Myanmar has also seen aberrant climatic circumstances, primarily since the 1980s, such as atypical cyclone landfall, late monsoon commencement, early rain withdrawal, monsoon intensity variation and unpredictable rainfall, and an increase in summer and winter temperatures (Fig. 2.11).

The overall trend of mean maximum temperature anomalies at four stations in Mandalay, Meikthila, Monywa, and Pyinmana increased throughout the latter half of ***the*** observed periods in all stations except Minbu. The anomalies in Minbu were varied, moving between positive and negative ones. Most of the station reaches the highest anomalies at about + 1.5 °C versus the normal value. The mean maximum temperature in the dry zone area started to increase since 1977 (Sein et al. 2016). The continuous positive anomalies started from 1993 ranging from a minimum of + 0.01 °C (2001) to a maximum of +1.12 °C (2010). The significant positive anomalies can be observed in 2009, 1998, 2005, 2014, 2012, and 1979 (descending order of anomalies value) and almost the years correspond with the declining phase of strong El Niño events (NOAA website 2016). During the twentieth century, temperatures have continued to increase from 2003 to 2015 except in 2007 and 2011, which experienced the La Niña events (NOAA website 2016).

The mean minimum temperature anomalies for all stations demonstrated an overall increasing trend with values becoming positive after the 1990s. However, the nature of the anomaly is not consistent among stations, and both positive and negative anomalies were found. On average, the dry zone area shows continuous positive anomalies from 2001 to 2014 with the range from + 0.01 °C (2001) to + 1.01 °C (2005). Generally, it can be expected that warm periods started after 2000. During the warm periods, the significant positive anomalies were detected in 2005 and 2010 which coincided with El Niño events.

It has been observed that during the winter season, Myanmar receives very low rainfall, but the Northern and Southern parts get a higher amount of rainfall than other parts of the country. The Southern and coastal areas receive high to very high rainfall during July and August because of strong monsoon clouds laden with sufficient water. The Central Myanmar areas receive double maxima rainfall due to the onset and withdrawal phase of the monsoon. The monthly rainfall data collected from 78 stations of the department of meteorology and hydrology (DMH) for 30 years (1981–2010) is presented in Table 2.4.

The trends of annual total precipitation illustrate the increasing trend but are not significant for all stations. The highest precipitation was found in El Niño year 2006 and 2010 and La

Table 2.4 Rainfall variability in some selected stations of Myanmar (1981–2010)

Station	Average rainfall (mm)	SD	Station	Average rainfall (mm)	SD
Putao	4057	457	Loikaw	1073	194
Taunggyi	1509	184	Hinthada	2289	439
Yangon	2686	350	Pyay	1240	238
Dawei	5342	817	Mandalay	851	197
Kalaywa	1661	264	Shwegyin	5339	903

Niña year 2011. It could be explained by the fact that the precipitation pattern in the dry zone area cannot be directly interpreted based on the links between El Niño and La Niña during the study period 1975–2015. In Myanmar, the higher elevations are subject to heavy snowfall during winters, especially in the northern mountains of the country.

Tun et al. (2017) reported that the production of agricultural crops has been negatively affected by the climate change in the dry zone of Myanmar. This has caused a decline in food crops' productivity in these areas. Interestingly, the farmers are, now, trying to adapt to the changes related to climatic parameters in this dry region of Myanmar. However, farmers' choice of adaptation strategies has been found to be influenced by many factors such as socio-economic conditions of the farmers, availability of necessary resources to adapt, and feasibility of adaptation. It also depends on the level of awareness of the farmers on climate change, perception of climate change, and other constraints and specific problems prevailing in the land use systems.

2.12 Nepal

Nepal lies between Gangetic plains on the southern side and Tibet (China) on the northern side. The country's climate is influenced by two major weather systems: summer monsoon circulation from June to September and westerly circulation from November to May. The influence of these two circulation systems varies, according to Nayava (1980), with summer precipitation being stronger in the southeast and westerly derived winter precipitation being greater in the northwest. The country is separated into five physiographic regions. These are (a) Terai, the region adjacent to India's Gangetic plains; (b) Shiwaliks, the lower hills; (c) Middle Mountains; (d) High Mountains; and (e) High Himalayas, in order of increasing altitude from south to north (Kansakar et al. 2004). Terai is a low-lying relatively plain tract, toward the north lie the Shiwalik hills and middle mountains, which are elongated east–west with a series of complicated and difficult valleys, such as Kathmandu, Pokhara, Dang, and Surkhet. During the summer monsoon, Nepal receives roughly 80% of its yearly precipitation (Shrestha et al. 2000).

It has been observed that the mean annual air temperature in the Terai plains reaches above 26 °C, while at the same time the mountain tops have a temperature below 12 °C. In Nepal, the temperature strictly follows the elevation pattern according to the north–south gradient or we can say that the air temperatures are mostly affected by the altitude of a place. As far as precipitation is concerned, three high rainfall areas are found in Nepal such as (i) the highest rainfall of about 5000 mm in Nepal is received at Lumle, situated North of Pokhara valley; (ii) Gumthang area, lying to the south of Langtang Himalayan range, just northeast of the Kathmandu, gets around 3500 mm of precipitation; and (iii) annual precipitation of more than 3500 mm is also found east of Mount Everest region in Num area. The available data on precipitation show that many high-altitude regions receive low rainfall and snow. The lowest annual precipitation of less than 200 mm is observed in and around the Mustang, Manang, and Dolpa regions located at the leeward side north of the Annapurna range.

Two major air movements influence the precipitation in Nepal. During the winter, western disturbances primarily affect the western sections of the country, resulting in snowfall in the Himalayan highlands. Though Nepal's average annual rainfall is 1,857.6 mm, the greatest annual rainfall location was found in Lumle, Kaski District, where the annual mean annual rainfall was 5,402.8 mm. The lowest rainfall site is in Mustang District's Lomanthang, with a mean annual rainfall of only 143.6 mm. When both of the country's highest and lowest rainfall areas are in the Annapurna area, the difference in annual mean rainfall between these two places is obvious. The monsoon season, post-monsoon, winter, and pre-monsoon seasons received 79.6, 4.2, 3.5, and 12.7% of the total annual rainfall, respectively, from 1976 to 2005.

Due to the increase in temperature in Nepal by 0.06 °C between 1977 and 2000, high-altitude glaciers are melting at a very fast rate, resulting in the creation of glacial lakes and the threat of catastrophic Glacial Lake Outburst Floods (GLOFs). These increases in temperature are more conspicuous at higher altitudes, particularly during winter. Due to the melting of glaciers, there is a possibility of significant changes to hydrological regimes (flows) and an increased risk of GLOFs. As glacier melt increases, the runoff will increase initially, followed by a slow but continuous decline. The trends are less certain as far as precipitation is concerned but there is evidence of an increase in intense rainfall events and flood days with more river flows. The poor farming community will be the most vulnerable to the impacts of climate change if the predictions made come true. Baidya et al. (2008) found an increasing trend in daily temperature from a study, based on the data for 36 years from 1971 to 2006 and the precipitation data for 46 years from 1961 to 2006. The study found that the days and nights are increasing warmer, with fewer cool days and cool nights.

According to Marahatta et al. (2009), yearly rainfall in Nepal varies greatly by location, ranging from less than 150 mm to more than 5,000 mm. According to Marahatta et al. (2009), the Kaski area in the western development region, the Sindhupalchowk area in the central development region, and the Sankhuwasabha area in the eastern development region have high annual precipitation above 5000 mm.

Two significant lowest annual precipitation pockets were observed over the Manang and Mustang Districts of the western development region where the mean annual precipitation is only about 145 mm. The areas with low precipitation are mostly the rainshadow regions. The mid-western and far-western development regions were found to be relatively dry. The temperature drops down with the start of the monsoon rains, though the relative humidity increases. So, May or maximum up to early June is the hottest period.

2.13 Pakistan

About 51% of Pakistan (404,195 km^2) is part of the Hindu Kush Himalayan (HKH), while around 23% of the country 's population (40 M) live in the Hindu Kush Himalayas region (Banskota and Sharma, 1995). Due to its geographical position, being located on a great landmass north of the Tropic of Cancer, between latitudes 25° and 36° N, Pakistan has a continental type of climate, showing extreme variations of temperature, both seasonally and on daily basis. True to the general pattern, places at very high altitudes and snow-covered mountains experience cold weather, more severe during winters. Higher temperatures on the Balochistan Plateau have been recorded. In the coastal areas, the climate is influenced by sea breezes and winds. In other parts of the country, typical June temperatures in the plains reach 38 °C, with highs exceeding 47 °C. During the summer, scorching breezes sweep across the plains during the day, causing trees to drop their leaves to prevent moisture loss. Dust storms and thunderstorms occur on occasion, lowering the temperature temporarily. Winters in Pakistan are bitterly cold, with minimum mean temperatures of around 4 °C in January in locations bordering the Himalayan range and sub-zero temperatures in Balochistan's northern regions. The weather of Pakistan is mainly influenced by Southwest monsoon and Western Disturbances monsoon. July to September are the wettest months of the year and more than 60% of the annual rainfall is received during this time of the year.

By 2100, Pakistan's annual mean temperature is expected to climb by 3–5 °C, while global emissions may cause a rise of 4–6 °C (Chaudhry

2017). The average annual rainfall is not projected to have a long-term trend, but it is expected to vary greatly from year to year. The largest city of Pakistan, Karachi, which is also the country's industrial center, is more humid than Islamabad but receives less rain. There is a great variation in temperature in Islamabad. During winters, the mean daily temperature remains as low as 2 °C in January and an average daily high temperature of 38 °C in June. About 500 of rainfall is received annually, half of which is received in July and August. During the spring season, hailstorms are fairly prevalent. The weather in Karachi is also more consistent than in Islamabad, which swings from an average daily low of 13 °C in the winter evenings to an average daily high of 34 °C during summer days. Although the summer temperatures are not as hot as they are in Punjab, there is still a lot of humidity. Islamabad is subjected to cold winds from the north, namely the Himalayas. On May 26, 2010, the highest temperature ever recorded in Mohenjo-daro, Sindh, Pakistan, was 53.5 °C. Not only was it the hottest temperature ever recorded in Pakistan, but it was also the warmest temperature ever measured on the Asian continent. On July 24, 2001, Islamabad received 620 mm of rain in 24 h. Due to quick runoff, most summer rains are also unavailable for crop production, according to Ahmad (1993, 1999, 2005). Rainfall during the Rabi season (October to March) ranges from 50 mm in Sindh province to >500 mm in the NWFP. The average rainfall during the Kharif season (April to September) ranges from 50 mm in portions of Balochistan to over 800 mm in the northern Punjab and NWFP. Because 92% of the country's land is semi-arid to arid, there is a severe water scarcity (Ahmed 2007).

Northern parts of the Hindu Kush mountains and Himalayas located in Pakistan generally remain snow-clad throughout the year. So, the temperature in the HKH region of Pakistan is sub-zero during winters and mild during summers. The rainfall in this region is scanty.

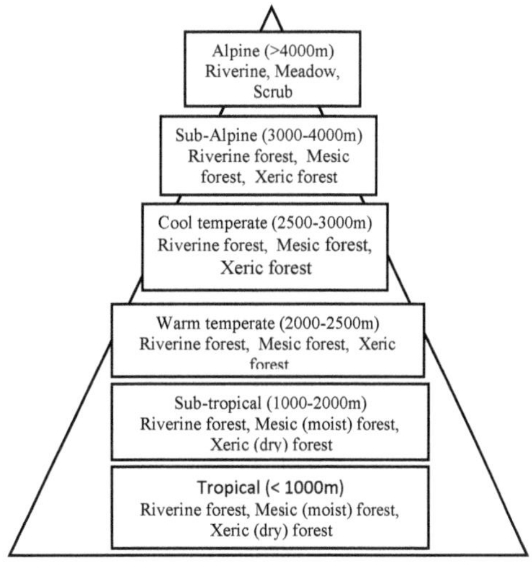

Fig. 2.12 Altitudinal divisions and major forest types found in the Eastern Himalayas

2.14 Biodiversity in Himalayas

The Himalayas are diverse in vegetation, soils, and water resources. Chettri et al. (2010), categorized vegetation types in the Himalayas into six groups as presented in Fig. 2.12.

2.14.1 Biodiversity Hotspots in the Himalayas

If we have to search for the most biodiverse region in the world, the HKH may stand at the top, though some of these mountain ranges still remain undiscovered. The abrupt change in altitude in the Himalayas from south to north, that is, from 500 m to more than 8000 m, creates a diverse ecosystem range from alluvial grasslands and subtropical broadleaf forests along the foothills to temperate broadleaf forests in the mid-hills, mixed conifer and conifer forests in the higher hills, and alpine meadows above the tree line. The total number of species of plants, mammals, reptiles, amphibians, freshwater fishes, and birds in the HKH region is 10,000,

300, 176, 105, 269, and 977, respectively. The respective endemism percent is 31.5, 4.0, 27.3, 40.0, 12.3, and 1.5 for the above taxonomic groups.

Five plant families, such as the Tetracentraceae, Hamamelidaceae, Circaesteraceae, Butomaceae, and Stachyuraceae, are endemic to the region. The largest family of flowering plants found in the hotspot of the region is the Orchidacea, with 750 species, and a large number of orchids, most of which are young endemic species, have recently been reported, signaling that further exploration will probably reveal a much higher degree of plant endemism. Though the terrain is harsh and rugged in the eastern Himalayas, there is quite a good number of vascular plants occurring at some of the highest places in the world. Some new plants, such as the cushion plant, have been discovered at altitudes of over 6100 m, while Ermania himalayensis, a high-altitude scree plant in the mustard family, was discovered at an altitude of 6300 m on the slopes of Mt. Kamet in the northwestern Himalayas. East of the Ganges–Brahmaputra lowlands, the Indo-Myanmar hotspot covers 2.373 million km^2 of tropical Asia. It used to cover the Himalayan range and the foothills of Nepal, Bhutan, and India.

2.14.2 Taxonomic Diversity

The Eastern Himalayan region is home to 45 species of mammals, 50 species of birds, 16 species of reptiles, 12 species of amphibians, 2 species of invertebrates, and 36 species of plants. All these are world-famous species.

2.14.3 Floral Diversity

Dhar et al. (2002) reported that the Eastern Himalayan (EH) region has one of the world's most rich and important alpine flora harboring about one-third of them. The Eastern Himalayan region is very rich in endemic flora and many medicinal and edible plants species are found here (Sundriyal 1999). Many local, subtropical endemic trees from the eastern Himalayan region of Nepal have been reported by Shrestha et al. (1999).

2.14.4 Faunal Diversity

Faunal diversity has been studied more extensively than floral diversity. India having diverse kinds of biogeographic zones endowed with a variety of unique ecosystems with their characteristic physiographic, climatic variations and rich biodiversity is identified as one of the 17 megadiverse countries of the world. India, with only about 2.4% of the global land area, harbors ~ 6.4% of the global diversity of faunal elements. Constituting about 30.16% of the total Indian fauna (100,762 species), about 30,377 species/subspecies of both Protozoa (372) and Animalia (30,005) have been recorded from Indian Himalayas. Out of 1,00,708 species known from India, 80,318 species are present in the landscape whereas 20,444 species are exclusively found in seascape (Chandra et al. 2016, 2017a, 2017b). To compare with the species diversity of terrestrial ecosystems, the faunal diversity of IH represents approximately 38% of the land diversity of the country. The highest faunal diversity is found in the Central Himalayas, which has about 14,183 species/subspecies. This is closely followed by West Himalaya, having 12,022 species. The Northwest Himalaya, East Himalaya, Ladakh Mountains, Tibetan Plateau, and Trans Himalaya-Sikkim have 8,731; 5,542; 1,561; 1,320; and 1,112 species, respectively.

2.14.5 Agro-Biodiversity

Agriculture is the mainstay of the inhabitants of the HKH region. Their livelihood security remains protected through agriculture and agroforestry practices. Though the practices followed vary considerably in western and eastern Himalayas, agriculture remains the main source of the economy of the whole HKH region. The eastern Himalayas (EH) can be divided broadly into five major systems, based on the agro-climatic zones and farming practices followed, and in each of these specialized systems, there are differences in crops and cropping patterns. Out of 800 species of plants used for food in India, about 300 species are found in Northeast India alone (Rao and Murti 1990). The genetic diversity of the HKH also has global significance. Nepal alone has about 2,500 species of rice and 100 types of basmati in the Western Himalayas. We can say that the variety of crops grown by farmers has a potential to serve as a genetic resource for improving crop yield because of adaptability to specific climatic conditions, responsiveness to applied nutrients, and resistance to pests and diseases. Geographically, the EH is located between the two most populated countries of the world namely China and India, both of which have a tremendous requirement for resources for economic and overall development. One can imagine the heavy load on ecosystems as the region faces pressure from migration, economic development, and population growth as well as from climate change (Beniston 2003). Assessing the impacts of climate change in the EH region is indeed a big challenge because of available data limitations, uncertainties due to the climate consequences, and the uncertainties of non-linear feedbacks between impacts. This may be important to support food security in the world where reduction in crop diversity has left the people vulnerable to climate change.

While evaluating the ecosystem, it is observed that a critical vegetation belt is formed by the Himalayan subtropical broadleaf forests. The flora and fauna of the Himalayas change with altitude and exhibits great diversity, as well as play a key role in shaping the ecosystem dynamics. The lower elevations, which also form the outer Himalayas, are covered by the Bhabar, Terai grasslands, and the Shiwalik ranges. It is a common phenomenon in the region that as one moves from the foothills to the higher elevations, the tropical forests merge into the temperate-type forests. The vegetation transforms to oak, rhododendron, dwarf hill bamboo, Alpine pastures, and then sparse or no vegetation above the snow line. The high-altitude desert plateau of Ladakh, for example, is desolate and sparsely vegetated. The Himalayan Tahr, Markhor, and Ibex are high-altitude mammals with hooves. These ungulates provide food for the region's carnivores. Leopards and bears are the main predators at mid-elevation, but leopards and wolves rule at higher elevations. The Western Himalayas are home to 11 bird species that are unique to this region. Wild sheep of three kinds can be found here: Nayan, Bharal, and Urial. Alpine meadows and green mountain slopes provide food for these sheep.

The vegetation mainly consists of shrubs, the major being *Adhatoda vasica, Jasminum multiflorum, Solanum erianthum, Murraya koenigii, Carissa opaca, Clerodendrum viscous, Cal/icarpa macrophylla, Eranthemum nervosum, Phlogacanthus thyrsiformis, Jatropha curcas, Rhus parviflora, Dodonaea viscosa, Woodfordia fruticosa,* and many others (ENVIS Newsletter, 2013). Lying completely in the Himalayas, the Temperate zone of Uttarkhand state in the western Himalayas has *Ibex dipyrena, Quercus semecarpifolia* (Kharsu Oak), *Q. dilatata* (Moru Oak), *Quercus leucotrichophora* (Banj oak), *Rhododendron arboretum* (Burans), *Myrica esculenta* (Kaphal), *Leonia ovalifolia* (Aynor), etc. Pure stands of Abides windrow (Raga), Pinus wal/ichiana (Kail), *Cedres deodar* (Deodar, Cedar), and Taxus wal/ichiana (Thuner, Himalayan Yew) at different locations give a unique and beautiful look to the ridges and slopes of the temperate zone. Generally, the subalpine and alpine vegetation zones are those areas that have an altitude above 3,000 m. The tree species found here are *Pinus wallichiana, Abies pindrow, Prunus cornuta, Acer caesium, Betula utilis* (Bhoj Patra) and *Salix* sp. Species of the genera *Cotoneaster, Rosa, Berberis, Ribes, Junipers, Rhododendron anthropogenic,* and *Rhododendron campanulas* (ENVIS Newsletter 8 (2), 2013. This constitutes the shrub components of the zone.

Of late, it has been observed that many areas in the HKH region are becoming less resilient to outside pressures. Possibly, an increase of pressure on the natural resources due to population increase and misuse of these resources due to enhanced need and greed has caused this. Gross interferences by the human population and their settlements at non-traditional locations have resulted in the loss of wildlife populations and a decline in crop productivity. Areas frequently visited by wild animals have now been occupied by the humans and even the abodes of these animals have not been spared. The severity of natural disasters has increased in the HKH region due to climate change and a breakdown of the traditional joint family system as well as systems of management. It is expected that as pathogens relocate to new habitats, biodiversity loss would result in the spread of vector-borne diseases, as well as the emergence of novel diseases due to changes in water availability and processes. The Hindu Kush Himalaya (HKH) region is also a culturally varied and vital part of the world. The HKH mountain ranges are densely populated, and its cultivated valleys are home to more than 1,000 ethnic groups, each having their own traditional practices, agriculture crops, and local traditional knowledge and wisdom. The importance of Himalayan biodiversity on human health has been undervalued and underestimated. The region has been known for millennia as the home of numerous medicinal plants, and this aspect needs to be exploited for enhancing the economy and quality of life of the inhabitants. Many of the species of medicinal plants may be lost before their useful utilization. The revival of these routes may prove beneficial in the long run. However, the interrelationships of HKH countries are important for the revival of trade among these countries through these routes.

Effects of Climate Change on Biodiversity and Health in the HKH Region.

The HKH region is known for having many biodiversity hotspots in the world. However, this biodiversity is in danger due to anthropogenic activities, such as overgrazing, land use change and exploitation, alien species, and the impact of climate change (Sharma et al. 2008). Climate change is likely to change species distribution and spread, change landscape, and will influence agriculture patterns (Romanelli et al. 2015). In Nepal, Alien species, which have an effect on health, agriculture, and livelihood in Nepal, are likely to increase in the future with climate change (Shrestha and Shrestha, 2019).

2.15 Aerosols in Hindu Kush Himalayas

Light-absorbing particles (LAPs) are found in abundance in glacier snow of the central Himalayas. These LAPs are most commonly found between younger and older snow. In these several snow layers, the LAPs have a particular vertical distribution. The deposition constant of the younger snow layers is around 50 gm-2 mm-1 snow water equivalent (SWE), whereas the deposition constant of the older snow layers is around 150 gm-2 mm-1 SWE (Svensson et al. 2021). Climate change has accelerated in recent decades as a result of increased anthropogenic activities such as development, mechanization, and expansion, as well as increased energy demands (Ningombam et al. 2021). Using long-term homogenous MERRA-2 reanalysis data, they sought to investigate aerosol optical depth (AOD) across the HKH region. The HKH region was divided into three subregions, namely the northwestern and Karakoram, the Central and southeastern Himalaya, and the Tibetan Plateau, due to its enormous landmass, fragile topology, and harsh climatic conditions.

The aerosols can be transported by air mass from the Indo-Gangetic plains to the Himalayas, which acts as a barrier (Hooda et al. 2018; Kopacz et al. 2011; Zhang et al. 2015). Having large quantities of snow, the Himalayan cryosphere gets carbonaceous aerosol deposited onto its surface (He et al. 2018; Jacobi et al. 2015; Xu et al. 2009). Gertler et al. (2016) reported that in Himalayan snow, there has been spatial and temporal variation in LAP content. In the Tibetan plateau, He et al. (2018) found black carbon concentrations ranging from 1 to 3600 ppbw. It was observed that other than long-range transported LAPs, local sources in Tibetan Plateau have also contributed LAP content in snow and ice (Li et al. 2016). Simulation of the black carbon impact on snow albedo and glacier runoff in the Hindu Kush Himalayan region has been done by Santra et al. (2019). The increase in dust particles with carbonaceous aerosols such as black carbon over the HKH region has been a matter of concern during the last few decades (Dumka et al. 2019; Sarangi et al. 2020). Wester et al. (2019) mentioned in the Hindu Kush Himalaya Assessment report about the consequences of climate change impacts on the people as well as flora and fauna of the region. The increase in black carbon aerosol could be responsible for the melting of glaciers as it results in a decrease in snow surface albedo by depositing over snow and darkening its surface (Svensson et al. 2021). There is a great contribution from anthropogenic aerosols undergoing long-range transportation through the wind over the high-altitude pristine region as the whole HKH region is located near the Indo-Gangetic plains (Dumka et al. 2015, 2020) and these aerosols could reach the Arabian Sea and central south India (Badarinath et al. 2009) as well as remote areas of the northwestern and central Himalayas (Dumka et al. 2015, Hooda et al. 2021, Sebastian et al. 2021).

2.16 Ecosystem Restoration

The world is confronted with challenges of the climate change, food and water insecurity, and the COVID-19 pandemic. Ecosystems can help the mankind in overcoming these challenges through their protection and proper management. Besides, the global degraded ecosystems and the benefits they give must also be restored (UNEP 2021). There is no denying the fact that we are a part of our environment, which we are degrading at a more than expected pace. We need both to conserve and restore ecosystems as they absorb our wastes, keep alive the economic sectors and the livelihoods of millions of people, as well as health, culture, and spirituality (IPBES 2019).

The people have misused and deteriorated the ecosystems and natural species, causing the disintegration of the very services we depend on (UNEP 2021) due to the methods we use for food production (Benton et al. 2021) as well as altering our landscapes, along with climate change, pollution, and intruding species (IPBES 2019). Our massive wealth increases come at the cost of the biosphere's health deteriorating. To maintain our existing lifestyle, we use the equivalent of 1.6 Earths (Global Footprint Network 2021), putting the fate of our economy in jeopardy (Dasgupta 2021).

There are many practices that can help in ecosystem restoration, such a; climate-smart agriculture, conservation agriculture, organic farming, sustainable land management, and bio-innovations (CPF 2021; FAO 2020). Agroforestry alone has the promise to increase food security for 1300 million people (Smith et al. 2019); decrease soil erosion by 50%; and increase soil carbon by 21%, inorganic nitrogen by 46%, and phosphorus by 11% (Muchane et al. 2020). Water quality can be improved by restoring wetlands and riverine areas as this practice will entrap pollutants and sediment from land degradation.

Under such a scenario, it will be useful to invest in ecosystem restoration as a basic element of conflict resolution, peace, improving access to resources as well as managing climate-induced security risks (UNCCD 2018; UN 2020; IRP 2019b). It would help in building trust in affected communities by combating for resolution of problems faced due to ecosystem degradation (Monga 2018; UN 2020; IRP 2019b) and reducing pressure on people to migrate (UN 2020; Barbut and Alexander 2016). An increase in access to resources would help to alleviate poverty and open the door to youths for new opportunities (Monga 2018; Wetlands International 2018). A study by the UN revealed that 50% of former contenders undergoing reintegration programs choose agriculture as a livelihood, but may face problems in access to land and resources (UNEP and UNDP 2012). Dasgupta (2021) explained that by over-using the ecosystem services, the people are misplacing the ecosystem's capacity to regulate climate, safe drinking water, crop pollination, floods, droughts, etc., and for this, they may face dire fallout. Ramsar (2018) and IRP (2019a, b) mentioned that the revival of terrestrial and aquatic ecosystems is helpful in delivering all 17 SDGs. Success in restoration requires an integrated, holistic approach, and it will need some substitutes; some of them may be negative. One hundred and fifteen governments have committed to restoring a total of around 1 billion hectares of land, as a contribution to achieving the objectives of the CBD, UNCCD, UNFCCC, or the Bonn Challenge on forest landscape restoration (Sewell et al. 2020).

Kuniyal et al. (2021) made the assessment of ecosystem restoration using the model of the Department of Forest, Government of Uttrakhand, considering three major variables, namely direct management measure (M), environmental desirability (E), and socio-economic feasibility (SE) and 22 individual variables. "M" comprised both biotic (grazing, tourism, etc.) and abiotic pressures (soil erosion). Vegetation and soil nutrient profile (N, P, K, pH, and water-holding capacity) formed the environment desirability component, while social acceptability of the local communities and stakeholders who are supposed to be the beneficiaries was the third component.

2.16.1 Approaches and Principles of Restoration

The ecosystem restoration involves various activities that would help in conserving and redressing the spoiled ecosystems (UNEP and FAO 2020). There is a need to recover the terrestrial, freshwater, or marine ecosystem that has been degraded, as well as to reverse the degradation of soils, agricultural areas, forests, and watersheds to revive their ecological functionality. The ecosystem restoration also includes maintaining ecosystem structure and other activities which may help in food production and regulating supporting services, recovering original biota, and making the ecosystem self-sustaining and resilient.

2.16.2 Principles for the UN Decade on Ecosystem Restoration

The FAO-led Task Force on Best Practices, the Society for Ecological Restoration, and IUCN's Commission on Ecosystem Management constituted an expert team who proposed nine principles of ecosystem restoration. The final wording and full descriptions of these principles will be developed collaboratively in 2021 to support effective restoration efforts in the world. The outline of these principles would be as follows: (i) to promote inclusive and participatory governance, social fairness, and equity; (ii) continuum of restorative activities; (iii) maximum possible recovery of ecosystem health and better quality of life of the people; (iv) to eliminate drivers of ecosystem degradation; (v) to integrate and incorporate available knowledge and exchange of information; (vi) the restorative process has to be viable under local conditions and acceptable under prevailing socio-ecological and socio-cultural situations, (vii) is based on ecological and socio-economic objectives and goals; (viii) to undertake monitoring, evaluation, and adaptive management throughout the project life; and (ix) maintenance of funding and scaling up of interventions.

References

Aggarwal RK (2012) Forecast verification of rainfall during monsoon. Res J Agric Environ Manag 1(1):21–24

Ahmad S (1993) Viability of agriculture resource base: a critical appraisal. In: Agricultural strategies in the 1990s: issues and options. Pakistan Association of Agricultural Social Scientists, Islamabad, pp 449–466

Ahmad S (1999) Achievements and issues of irrigation in the 20th century. In: Proceedings of the national workshop on "water resources achievements and issues in the 20th century and challenge of the next millennium", Islamabad. PCRWR/UNESCO, 28–30 June1991, pp 188–201

Ahmad S (2005) Background paper on evapotranspiration and water balance. Water Country Assistance Strategy. Pakistan Economy Getting Dry. World Bank Resident Mission, Pakistan

Ahmed S (2007) Land and water resources of Pakistan-a critical assessment. Pak Rev 46:911–937

Alamgir S (2009) Characterization and estimation of rainfall in Bangladesh based on ground radar and satellite observations. Institut National de la Recherche Scientifique INRS-ETE, Université du Québec

Aung O, Huylenbroeck GV, Stijn Speelman S (2017) Determining factors for the application of climate change adaptation strategies among farmers in Magwe District, dry zone region of Myanmar

Badarinath KVS, Kharol SK, Sharma AR (2009) Long-range transport of aerosols from agriculture crop residue burning in Indo-Gangetic Plains-a study using LIDAR, ground measurements and satellite data. J Atmos Sol Terr Phys 71:112–120

Baidya SK, Shrestha M, Sheikh MM (2008) Trends in daily climatic extremes of temperature and precipitation in Nepal. J Hydrol Meteorol 5(1):38–51

Banskota K, Sharma B (1995) Carrying capacity of Himalayan resources for mountain tourism development. MEI Discussion Paper No. 95/14. ICIMOD, Kathmandu

Barbut M, Alexander S (2016) Land degradation as a security threat amplifier: the new global frontline (Chap. 1.1). In: Chabay I, Frick M, Helgenson J (eds) Land restoration: reclaiming landscapes for a sustainable future. Academic Press, pp 3–12. https://doi.org/10.1016/B978-0-12-801231-4.00001-X

Barnett TP, Adam JC, Lettenmaier DP (2005) Potential impacts of a warming climate on water availability in a snow-dominated region. Nature 438:303–309

Beniston M (2003) Climatic change in mountain regions: a review of possible impacts. Clim Change 59:5–31

Benton TG, Bieg C, Harwatt H, Pudasaini R, Wellesley L (2021) Food system impacts on biodiversity loss: Three levers for food system transformation in support of nature. In: Energy, environment and resources programme. Chatham House, London

Bhutiyani MR, Kale VS, Pawar NJ (2010) Climate change and the precipitation variations in the Northwestern Himalaya: 1866–2006. Int J Climatol 30:535–548

Blandford TR, Humes KS, Harshburger BJ, Moore BC, Walden VP, Ye H (2008) Seasonal and synoptic variations in near-surface air temperature lapse rates in a mountainous basin. J Appl Meteorol Climatol 47:249–261

Bookhagen B, Burbank DW (2010) Toward a complete Himalayan hydrological budget: spatiotemporal distribution of snowmelt and rainfall and their impact on river discharge. J Geophys Res 115. https://doi.org/10.1029/2009JF001426

Broccoli AJ, Manabe S (1992) The effects of orography on mid-latitude Northern Hemisphere dry climates. J Clim 5:1181–1201

Burbank DW, Bookhagen B, Gabet EJ, Putkonen J (2012) Modern climate and erosion in the Himalaya. CR Geosci 344:610–626

Chandra K, Raghunathan C, Mondal T, Dash S (2016) Current status of marine faunal diversity in India. Director, Zoological Survey of India, Kolkata, pp 1–525

Chandra K, Gopi KC, Rao DV, Valarmathi K, Alfred JRB (2017a) Current status of freshwater faunal diversity in India. Director, Zoological Survey of India, Kolkata, pp 1–624

Chandra K, Sheela S, Das D (2017b) Animal discoveries 2016 new species and new records. Director, Zoological Survey of India, Kolkata, pp 1–95

Chaudhry QUZ (2017) Climate change profile of Pakistan. Asian Development Bank, 6 ADB Avenue, Mandaluyong City, 1550 Metro Manila, Philippines

Chettri N, Sharma E, Shakya B, Thapa R, Bajracharya B, Uddin K, Oli KP, Choudhury D (2010) Biodiversity in the Eastern Himalayas: Status, trends and vulnerability to climate change. Climate change impact and vulnerability in the Eastern Himalayas. Technical report 2. ICIMOD, Kathmandu

Chou C (2003) Land-sea heating contrasts in an idealized Asian summer monsoon. Climate Dyn 21:11–15

Choudhury BU, Das A, Ngachan SV, Slong A, Bordoloi LJ, Chowdhury P (2012) Trend analysis of long term weather variables in mid altitude Meghalaya, North-East India. J Agric Phys 12(1):12–22

Collaborative Partnership on Forests [CPF] (2021) Challenges and opportunities in turning the tide on deforestation. Joint Statement of the Collaborative Partnership on Forests

CSO (Central Statistics Office) (2017) Annual report 2017–18. Department of Agriculture, Cooperation & Farmers Welfare Ministry of Agriculture & Farmers Welfare Government of India Krishi Bhawan, New Delhi-110 001. www.agricoop.nic.in

Dasgupta P (2021) The economics of biodiversity: the Dasgupta review. HM Treasury, London

Dhar U (2002) Conservation implications of plant endemism in high-altitude Himalaya. Curr Sci 82:141–148

Dhimal M, Kramer IM, Phuyal P et al (2021a) Climate change and its association with the expansion of vectors and vector-borne diseases in the Hindu Kush Himalayan region: a systematic synthesis of the literature. Adv Clim Chang Res 12(3):421–429

Dhimal M, Bhandari D, Dhimal ML, Kafle N, Pyakurel P, Mahotra N, Akhtar S, Ismail T, Dhiman RC, Groneberg DA, Shrestha UB, Müller R (2021b) Impact of Climate change on health and well-being of people in Hindu Kush Himalayan region: a narrative review. Front Physiol 12:651189. https://doi.org/10.3389/fphys.2021.651189

Dobrowski SZ, Abatzoglou JT, Greenberg JA, Schladow SG (2009) How much influence does landscape-scale physiography have on air temperature in a mountain environment? Agric for Meteorol 149:1751–1758

Dorji T (2018) Long-term climate impacts on hydropower in Bhutan. ICIMOD, Kathmandu, Nepal

Duan AM, Wu G, Liu Y, Ma Y (2012) Weather and climate effects of Tibetan plateau. Adv Atmos Sci 29 (5):978–992. https://doi.org/10.1007/s00376-012-1220-y

Dumka UC, Kaskaoutis DG, Srivastava MK, Devara PCS (2015) Scattering and absorption properties of near-surface aerosol over Gangetic-Himalayan region: the role of boundary-layer dynamics and long-range transport. Atmos Chem Phys 15:1555–1572

Dumka UC, Kaskaoutis DG, Francis D, Chaboureau JP, Rashki A, Tiwari S, Singh ES, Liakakou E, Mihalopoulos N (2019) The role of the intertropical discontinuity region and the heat low in dust emission and transport over the Thar desert, India: a premonsoon case study. J Geophys Res Atmos 23(23):13197–13219

Dumka UC, Ningombam SS, Kaskaoutis DG, Madhavan BL, Song HJ, Angchuk D, Jorphail S (2020) Long-term (2008–2018) aerosol properties and radiative effect at high-altitude sites over western trans-Himalayas. Sci Total Environ 734:139354

Dyurgerov DM, Meier MF (2005) Glaciers and changing earth system: a 2004 snapshot. Institute of Arctic and Alpine Research, University of Colorado, Boulder (Colorado), p 117

ENVIS Newsletter (2013) ENVIS Centre, UEPPCB, Dehra Dun. ENVIS Newsl 9(2):20p

Falvey M, Garreaud RD (2009) Regional cooling in a warming world: recent temperature trends in the southeast Pacific and along the west coast of subtropical South America (1979–2006). J Geophys Res 114 (D04102):1–5

FAO (2011) The state of food insecurity in the world - how does international price volatility affect domestic economies and food security? Food and Agricultural Organization of the United Nations, Rome, Italy

FAO (2020) Position paper on "ecosystem restoration" of production ecosystems, in the context of the UN Decade of Ecosystem Restoration 2021–2030. COFI/2020/Inf.15.2

Favre R, Kamal GM (2004) Watershed atlas of Afghanistan, First edition – working document for planners. Kabul, Afghanistan

Gardelle J, Berthier E, Arnaud Y (2012) Slightmass gain of Karakoram glaciers in the early twenty-first century. Nat Geosci. https://doi.org/10.1038/NGEO1450.

Gertler CG, Puppala SP, Panday A, Stumm D, Shea J (2016) Black carbon and the Himalayan cryosphere: a review. Atmos Environ 125:404–417. https://doi.org/10.1016/j.atmosenv.2015.08.078

Global Footprint Network (2021) Ecological footprint. https://www.footprintnetwork.org/our-work/ecological-footprint/. Accessed 30 Nov 2021

Hasson SU (2016) Future water availability from Hindukush-Karakoram-Himalaya Upper Indus Basin under conflicting climate change scenarios. Climate 4:40–67. https://doi.org/10.3390/cli4030040

He C, Flanner MG, Chen F, Barlage M, Liou KN, Kang S, Ming J, Qian Y (2018) Black carbon-induced snow albedo reduction over the Tibetan Plateau: uncertainties from snow grain shape and aerosol–snow mixing state based on an updated SNICAR model. Atmos Chem Phys 18:11507–11527. https://doi.org/10.5194/acp-18-11507-2018

Hooda RK, Kivekäs N, O'Connor EJ, Collaud CM, Pietikäinen JP, Vakkari V et al (2018) Driving factors of aerosol properties over the foothills of central Himalayas based on 8.5 Years continuous measurements. J Geophys Res-Atmos 123:13421–13442. https://doi.org/10.1029/2018JD029744

Huang R, Wu Y (1989) The influence of ENSO on the summer climate change in China and its mechanism. Adv Atmos Sci 6(1):21–32

Immerzeel WW, Van Beek LPH, Bierkens MFP(2010) Climate change will affect the Asian water towers. Science 328:1382–1385

Immerzeel WW, Petersen L, Ragettli S, Pellicciotti F (2014) The importance of observed gradients of air temperature and precipitation for modelling run off from glacierized watershed in the Nepalese Himalayas. Water Resour Res 50:2212–2226

IPBES (2019) Summary for policymakers of the global assessment report on biodiversity and ecosystem services of the intergovernmental science-policy platform on biodiversity and ecosystem services (Díaz S, Settele J, Brondízio ES, Ngo HT et al (eds)). IPBES, Bonn. https://doi.org/10.5281/zenodo.3553579

IPCC (2007a) Climate change 2007a: the physical science basis. Contribution of working group I to the fourth assessment report of the intergovernmental panel on climate change (Solomon S, Qin D, Manning M, Chen Z, Marquis M, Averyt KB, Tignor M, Miller HL (eds)). Cambridge University Press, Cambridge and New York

IPCC (2007b) Climate change 2007b: impacts, adaptation and vulnerability. Contribution of working group II to the fourth assessment report of the intergovernmental panel on climate change (Parry ML, Canziani OF, Palutikof JP, van der Linden PJ, Hanson CE (eds)). Cambridge University Press, Cambridge

IPCC (2007c) Climate change impacts, adaptation and vulnerability. IPCC WGII: fourth assessment report. IPCC, Geneva

IPCC (2012) Managing the risks of extreme events and disasters to advance climate change adaptation. Special report of the Intergovernmental Panel on Climate Change. Cambridge University Press, p 115

IRP (International Resource Panel) (2019a) Land restoration for achieving the sustainable development goals: an international resource panel think piece (Herrick JE, Abrahamse T, Abhilash PC, Ali SH et al (ed)). United Nations Environment Programme, Nairobi

IRP (2019b) Restore landscapes to push ahead on sustainable development, says International Resource Panel. https://www.unep.org/news-and-stories/press-release/restore-landscapes-push-ahead-sustainable-development-says. Accessed 5 Sept 2019

Jacobi HW, Lim S, Ménégoz M, Ginot P, Laj P, Bonasoni P, Stocchi P, Marinoni A, Arnaud Y (2015) Black carbon in snow in the upper Himalayan Khumbu Valley, Nepal: observations and modeling of the impact on snow albedo, melting, and radiative forcing. Cryosphere 9:1685–1699. https://doi.org/10.5194/tc-9-1685-2015

Kansakar SR, Hannah DM, Gerrard J, Rees G (2004) Spatial pattern in the precipitation regime of Nepal. Int J Climatol 24:1645–1659

Karmakar S, Shrestha ML (2000) Recent climatic changes in Bangladesh, SMRC No. 4. SAARC Meteorological Research Centre, Dhaka, pp 1–43

Kattel D, Yao T, Yang K, Tian L, Yang G, Joswiak D (2013) Temperature lapse rate in complex mountain terrain on the southern slope of the central Himalayas. Theoret Appl Climatol 113:672–682

Kopacz M, Mauzerall DL, Wang J, Leibensperger EM, Henze DK, Singh K (2011) Origin and radiative forcing of black carbon transported to the Himalayas and Tibetan Plateau. Atmos Chem Phys 11:2837–2852. https://doi.org/10.5194/acp-11-2837-2011

Krishnan R, Shrestha AB, Ren GY et al (2019) Unravelling climate change in the Hindu Kush Himalaya: rapid warming in the mountains and increasing extremes. In: Wester P, Mishra A, Mukherji A et al (eds) The Hindu Kush Himalaya assessment. Springer, Berlin, pp 57–59

Lambert L, Chitrakar BD (1989) Variation of potential evapotranspiration with elevation in Nepal. Mt Res Dev 9:145–152

Li C, Bosch C, Kang S, Andersson A, Chen P, Zhang Q, Cong Z, Chen B, Qin D, Gustafsson O (2016) Sources of blackcarbon to the Himalayan-Tibetan Plateau glaciers. Nat Commun 7:12574. https://doi.org/10.1038/ncomms12574

Liu X, Chen B (2000) Climatic warming in the Tibetan Plateau during recent decades. Int J Climatol 20:1729–1742

Ma Y, Menenti M, Feddes R (2010) Parameterization of heat fluxes at heterogeneous surfaces by integrating satellite measurements with surface layer and atmospheric boundary layer observations. Adv Atmos Sci 27:328–336. https://doi.org/10.1007/s00376-009-9024-4

Marahatta S, Dangol BS, Gurung GB (2009) Temporal and spatial variability of climate change over Nepal, 1976–2005. J Hydrol Meteorol 6:44–48

Matsumoto Y (1988) Floods in Bangladesh and flood disasters in 1987. In: Muramoto Y (ed) Investigations of floods disasters caused by heavy rainfall in Bangladesh during the 1987 monsoon season. Report of the scientific research for natural disasters, No, B-62-5, pp 1–10

Michael AM, Sarma PBS, Ashok Raj PC, Chowdhary RK (1978) Resources inventory of Garo Hills. Water Technology Centre, IARI, New Delhi, India

Monga, P (2018) Landscapes of peace: using land restoration for conflict resolution. Keynote speech to the United Nations convention to combat desertification. Caux, Switzerland. https://www.unccd.int/sites/default/files/inline-files/CAUX%20Forum%20Final%20Speech.pdf. Accessed 17 July 2018

Muchane MN, Sileshi GW, Gripenberg S, Jonsson M, Pumariño L, Barrios E (2020) Agroforestry boosts soil health in the humid and sub-humid tropics: a meta-analysis. Agric Ecosyst Environ 295:106899. https://doi.org/10.1016/j.agee.2020.106899

Nayava JL (1980) Rainfall in Nepal. Himalayan Rev 51–61. Nepal Geographical Society

Ningombam SS, Dumka UC, Mugil SK, Kuniyal JC, Hooda RK, Gautam AS, Tiwari S (2021) Impacts of aerosol loading in the Hindu Kush Himalayan region based On MERRA-2 reanalysis data. Atmosphere 12:1290. https://doi.org/10.3390/atmos12101290

NOAA website (2016) The NOAA website. http://www.el-nino.com/

Noorunnahar M, Rahman MA (2013) Estimating regional trends of temperature in Bangladesh. Int J Environ Monit Anal 1:175–181. https://doi.org/10.11648/j.ijema.20130105.12

Palazzi E, von Hardenberg J, Provenzale A (2013) Precipitation in the Hindu-Kush Karakoram Himalaya: observations and future scenarios. Geophys Res 118:85–100. https://doi.org/10.1029/2012JD018697

Pandit MK, Manish K, Koh LP (2014) Dancing on the roof of the world: ecological transformation of the Himalayan landscape. Bioscience 64:980–992. https://doi.org/10.1093/biosci/biu152

Ramsar (2018) Scaling up wetland conservation, wise use and restoration to achieve the sustainable development goals. https://www.ramsar.org/document/wetlands-and-the-sdgs

Rao RR, Murti SK (1990) North-East India: a major centre for plant diversity in India. Indian J for 13 (3):214–222

Romanelli C, Capon AG, Maiero M, Campbell-Lendrum D, Butler C, Corvalan C et al (2015) Climate change, biodiversity and human health. In: WHO and CBD, connecting global priorities: biodiversity and human health - a state of knowledge review. World Health Organization, Geneva and Montreal

Rosegrant M, Ximing C, Cline S, Nakagawa N (2002) The role of rain-fed agriculture in the future of global food production. EPTD Discussion Paper 90. International Food Policy Institute, Washington D.C., USA, p 252. www.ifpri.org/divs/eptd/dp/papers/eptdp90.pdf

Rupa Kumar K, Sahai AK, Krishna Kumar K, Patwardhan SK, Mishra PK, Revadkar JV, Kamala JVK, Pant GB (2006) High resolution climate change scenario for India for the 21st Century. Curr Sci 90:334–345

Saba DS (2001) Afghanistan: environmental degradation in a fragile ecological setting. Int J Sustain Dev World Ecol 8(4):279–289. https://doi.org/10.1080/13504500109470086

Salik KM, Jahangir S, Zahdi WZ, Hasson S (2015) Climate change vulnerability and adaptation options

for the coastal communities of Pakistan, Ocean & Coastal Management, under review article # OCMA-D-14-00289
Santra S, Verma S, Fujita K, Chakraborty I, Boucher O, Takemura T, Burkhart JF, Matt F, Sharma M (2019) Simulations of black carbon (BC) aerosol impact over Hindu Kush Himalayan sites: validation, sources, and implications on glacier runoff. Atmos Chem Phys 19:2441–2460. https://doi.org/10.5194/acp-19-2441-2019
Sarangi C, Qian Y, Rittger K, Leung LR, Chand D, Bormann KJ, Painter TH (2020) Dust dominates high-altitude snow darkening and melt over high-mountain Asia. Nat Clim Chang 10:1045–1051
Sarkar H, Chowdhury AM, Karmakar S (2018) Study on the trends of rainfall and different temperatures during the Southwest monsoon season in Bangladesh. J Eng Sci 9(2):41–54
Schmidt S, Nusser M (2009) Fluctuations of Raikot glacier during the past 70 years: a case study from the Nanga Parbat massif, northern Pakistan. J Glaciol 55:949–959. https://doi.org/10.3189/002214309790794878
Sebastian M, Kanawade VP, Soni VK, Asmi E, Westervelt DM, Vakkari V, Hyvärinen AP, Pierce JR, Hooda RK (2021) New particle formation and growth to climate-relevant aerosols at a background remote site in the Western Himalaya. J Geophys Res Atmos 126(7):e2020JD033267. https://doi.org/10.1029/2020JD033267
Sein KK, Chidthaisong A, Pimple U, Htay TM, Varnakovida P (2016) Changes in temperature and precipitation and their extreme indices over dry zone area in Central Myanmar. In: Proceedings of the international conference on climate change, biodiversity and ecosystem services for the Sustainable Development Goals (SDGs): policy and practice, 27–29 June 2016, Cha-Am, Phetchaburi, Thailand
Sewell A, van der Esch E, Löwenhardt H (2020) Goals and commitments for the restoration decade: a global overview of countries' Restoration Commitments Under the Rio Conventions and other Pledges. PBL Netherlands Environmental Assessment Agency, The Hague. https://www.pbl.nl/sites/default/files/downloads/pbl-2020-goals-and-commitments-for-the-restoration-decade-3906.pdf
Sharma UC (1996) Agro-climatic zones of Northeatern Hills Region. In: Paper presented at the Second Crop Science Congress, held in 4–6 November 1996, New Delhi, India
Sharma UC, Datta M, Das Chowdhury D (2006) Weather and its variation. In: Sharma UC, Datta M, Samra JS (eds) Soils and their management in North East India. ICAR Research Complex for NEH Region, Umiam, Meghalaya, India
Sharma E, Tse-ring K. Chettri N, Shrestha A (2008) Biodiversity in the Himalayas–trends, perception and impacts of climate change. In: Proceedings of the international mountain biodiversity conference, 16–18 November 2008, Kathmandu, Nepal
Sharma E, Chettri N, Tsering K, Shrestha AB, Fang J, Mool P, Eriksson M (2009) Climate change impacts and vulnerability in the Eastern Himalayas. ICIMOD, Nepal, Kathmandu
Shrestha UB, Shrestha BB (2019) Climate change amplifies plant invasion hotspots in Nepal. Divers Distrib 25:1599–1612. https://doi.org/10.1111/ddi.12963
Shrestha AB, Wake CP, Mayewski PA, Dibb JE (1999) Maximum temperature trend in the Himalaya and its vicinity: an analysis based on temperature records from Nepal for the period 1971–94. J Clim 12:2775–2786
Shrestha AB, Wake CP, Dibb JE, Mayewski PA (2000) Precipitation fluctuations in the Nepal Himalaya and its vicinity and relationship with some large scale climatological parameters. Int J Climatol 20:317–327
Shrestha UB, Gautam S, Bawa KS (2012) Widespread climate change in the Himalayas and associated changes in local ecosystems. PloS One 7(5). Article e36741
Shrestha AB, Agrawal NK, Alfthan B, Bajracharya SR, Maréchal J, van Oort B (eds) (2015) TheHimalayan climate and Water Atlas: impact of climate change on water resources in five of Asia's major river basins. ICIMOD, GRID-Arendal and CICERO
Singh SP, Singh RD, Gumber S, Bhatt S (2017) Two principal precipitation regimes in Himalayas and their influence on tree distribution. Trop Ecol 58:679–691
Smith P, Nkem J, Calvin K, Campbell D, Cherubini F, Grassi G et al (2019) Interlinkages between desertification, land degradation, food security and greenhouse gas fluxes: Synergies, trade-offs and integrated response options. In: Shukla PR, Skea J, Calvo Buendia E, Masson-Delmotte V et al (ed) Climate change and land: an IPCC special report on climate change, desertification, land degradation, sustainable land management, food security, and greenhouse gas fluxes in terrestrial ecosystems. IPCC
Solomon S, Qin D, Manning M, Marquis M, Averyt K, Tignor MMB, Miller Jr HL, Chen Z (2007) Climate change 2007: the physical science basis. In: Contribution of working group I to the fourth assessment report of the Intergovernmental Panel on Climate Change. Cambridge University Press, 32 Avenue of the Americas, New York, NY 10013-2473, USA
Sun XB, Ren GY, Shrestha AB et al (2017) Changes in extreme temperature events over the Hindu Kush Himalaya during 1961–2015. Adv Clim Chang Res 8 (3):157–165
Sundriyal M (1999) Distribution, propagation and nutritive value of some wild edible plants in the Sikkim Himalaya. PhD thesis, H. N. B. Garhwal University, Srinagar, India
Svensson J, Ström J, Honkanen H, Asmi E, Dkhar NB, Tayal S, Sharma VP, Hooda R, Leppäranta M, Jacobi HW, Lihavainen H, Hyvärinen A (2021) Deposition of light-absorbing particles in glacier snow of the Sunderdhunga Valley, the southern forefront of

the central Himalayas. Atmos Chem Phys 21:2931–2943. https://doi.org/10.5194/acp-21-2931-2021

Tahir A, Chevallier P, Arnaud Y, Ahmad B (2011) Snow cover dynamics and hydrological regime of the Hunza River basin, Karakoram Range, Northern Pakistan. Hydrol Earth Sys Sci 15:2275–2290. https://doi.org/10.5194/hess-15-2275-2011

Tewari VP, Verma RK, von Gadow K (2017) Climate change effects in the Western Himalayan ecosystems of India: evidence and strategies. For Ecosyst 4:13–21. https://doi.org/10.1186/s40663-017-0100-4

Thornthwaite CW (1948) An approach towards rational classification of climate. Geogr Rev 38:55–94

Tun OA, Van Huylenbroeck G, Spleenman S (2017) Determining factors for the application of climate change adaptation strategies among farmers in Magwe District, dry zone region of Myanmar. Int J Clim Chang Strat Manag 9:36–55. https://doi.org/10.1108/IJCCSM-09-2015-0134

UN (2020) Climate emergency 'a danger to peace'. UN Security Council hears. UN News. https://news.un.org/en/story/2020/07/1068991. Accessed 24 July 2020

United Nations Environment Programme (2021) Becoming #GenerationRestoration: ecosystem restoration for people, nature and climate. Nairobi. ISBN: 978-92-807-3864-3

UNCCD (2018) Caux dialogue on land and security: creating landscapes of peace, 27 July 2018. https://www.unccd.int/news-events/caux-dialogue-land-and-security-creating-landscapes-peace. Accessed 30 Nov 2021

UNEP (2021) Making peace with nature: a scientific blueprint to tackle the climate, biodiversity and pollution emergencies. UNEP, Nairobi. https://www.unep.org/resources/making-peace-nature

UNEP and FAO (2020) Strategy for the UN decade on ecosystem restoration. https://www.decadeonrestoration.org/strategy. Accessed 26 Apr 2021

UNEP and United Nations Development Programme [UNDP] (2012) The role of natural resources in disarmament, demobilization and reintegration: addressing risks and seizing opportunities. UNEP, New York; UNDP, Nairobi. https://postconflict.unep.ch/publications/UNEP_UNDP_NRM_DDR.pdf

Wester P, Mishra A, Mukherji A, Shrestha AB (2019) The Hindu Kush Himalaya assessment. Mountains, climate change, sustainability and people. Springer Nature, Cham

Wetlands International (2018) Degradation of wetlands contributes to social instability and insecurity, 29 August 2018. https://www.wetlands.org/news/degradation-wetlands-contributes-social-instability-insecurity/. Accessed 26 Apr 2021

Wu G, Zhang Y (1998) Tibetan plateau forcing and the timing of the monsoon onset over south Asia and the south China sea. Mon Weather Rev 126:913–927. https://doi.org/10.1175/1520-0493

Xu J, Shrestha A, Vaidya R, Eriksson M, Hewitt K (2007) The melting Himalayas: regional challenges and local impacts of climate change on mountain ecosystems and livelihoods. ICIMOD technical paper. ICIMOD, Kathmandu

Xu Z, Gong T, Li J (2008) Decadal trend of climate in the Tibetan Plateau-regional temperature and precipitation. Hydrol Process 22:3056–3065

Xu B, Cao J, Hansen J, Yao T, Joswiak DR, Wang N, Wu G, Wang M, Zhao H, Yang W, Liu X, He J (2009) Black soot and the survival of Tibetan glaciers. Proc Natl Acad Sci USA 106:22114–22118. https://doi.org/10.1073/pnas.0910444106

Yan LB, Liu Z, Chen G (2016) Mechanisms of elevation-dependent warming over the Tibetan plateau in quadrupled CO_2 experiments. Clim Change 135:509–519

Yanai M, Li C, Song Z (1992) Seasonal heating of the Tibetan Plateau and effects of the evolution of the Asian summer monsoon. J Meteorol Soc Jpn 70:189–221

Yeh TZ, Chang CC (1974) A preliminary experimental simulation on the heating effect of the Tibetan Plateau on the general circulation over Eastern Asia in China. Sci China (d) 17:397–420

Yeh TZ, Gao YX (1979) Meteorology of the Qinghai-Xizang (Tibet) Plateau. Science Press, Beijing

You QL, Kang SC, Pepin N (2010) Relationship between temperature trend magnitude, elevation and mean temperature in the Tibetan Plateau from homogenized surface stations and reanalysis data. Glob Planet Change 71:124–133

You QL, Ren GY, Zhang YQ et al (2017a) An overview of studies of observed climate change in the Hindu Kush Himalayan (HKH) region. Adv Clim Change Res 8(3):141–147

You QL, Ren GY, Zhang YQ, Ren YY, Sun XB, Zhan YJ, Shrestha AB, Krishnan R (2017b) An overview of studies of observed climate change in the Hindu Kush Himalayan (HKH) region. Adv Clim Change Res 8:141–147

Zhan YJ, Ren GY, Shrestha AB et al (2017) Changes in extreme precipitation events over the Hindu Kush Himalayan region during 1961–2012. Adv Clim Change Res 8(3):166–175

Zhang R, Wang H, Qian Y, Rasch PJ, Easter RC, Ma PL, Singh B, Huang J, Fu Q (2015) Quantifying sources, transport, deposition, and radiative forcing of black carbon over the Himalayas and Tibetan Plateau. Atmos Chem Phys 15:6205–6223. https://doi.org/10.5194/acp-15-6205-2015

Zhao L, Ping CL, Yang DQ, Cheng GD, Ding YJ, Liu SY (2004) Change of climate and seasonally frozen ground over the past 30 years in Qinghai-Tibetan plateau, China. Glob Planet Change 43:19–31

3 Physiography and Resources

Abstract

The HKH Region can conveniently be divided into four distinct zones such as the outer southern part called "Shiwaliks", the lesser Himalayas, the higher Himalayas, and the Tibetan-Tethys. The Shiwalik Hills form the southern stretch of the Himalayas starting from the Indus River to about 2,400 km eastwards, spanning across the northern parts of Pakistan, India, and Nepal. It is 10–50 km wide with an average elevation of 1,500–2,000 m. The Lesser Himalaya lies in between the Sub-Himalayas and Higher Himalayas separated by the Main Boundary Thrust (MBT) and the Main Central Thrust (MCT), respectively. The total width ranges from 60 to 80 km. The rocks in the lesser Himalaya range in age from Precambrian to Miocene. The northernmost boundary of the Shiwaliks Group is marked by the Main Boundary Thrust (MBT), over which the low-grade metasedimentary rocks of the Lesser Himalaya overlie. The higher Himalayas extend from the main central thurst to Tibetan-Tethys Zone and consists of about 10 km-thick succession of crystalline rocks, known as the Himal Group. The Tibetan-Tethys Himalayas generally begins from the top of the Higher Himalayan Zone and extends to the north of Tibet. This zone is about 40 km wide. The Himalayas are the tallest mountain ranges in the world spread across 8 countries and can conveniently be categorized as (a) Western Himalayas (Afghanistan, Pakistan, India), (b) Central Himalayas (India, Nepal, Tibet Autonomous Region of China), (c) Eastern Himalayas (India, China, Bhutan), and (d) Far-eastern Himalayas (India, Myanmar, China, Bangladesh). The Hindu Kush Himalayas mountain ranges function as water towers, providing water to people throughout much of Asia. The glacier-fed rivers originating from the Himalayan mountain ranges surrounding the Tibetan Plateau comprise the largest river runoff from any single location in the world. The rivers shape the landscape and ecosystems and are important in terms of biodiversity. The high mountain ranges strongly influence atmospheric circulation and meteorological patterns across the region. As a result of its varied topography, the HKH is endowed with rich biodiversity and diverse ecosystems, which provide a basis for the livelihoods of the many people who live there. Its physical characteristics, however, also make the Hindu Kush Himalayas one of the most hazard-prone regions in the world. Heavy rains, steep slopes, weak geological formations, accelerated rates of erosion, and high seismicity contribute to serious flooding and mass movements of rock and sediment affecting the lives and livelihoods of millions. The rivers flowing from the Hindu Kush Himalayas provide the region with one of the

U. C. Sharma et al., *Soils in the Hindu Kush Himalayas*, Geography of the Physical Environment,
https://doi.org/10.1007/978-3-031-11458-8_3

most valuable resources, that is, freshwater. Besides huge water resources, the HKH Region is endowed with sufficient human, animal, land, soil, vegetation, rangelands, grasslands, pasture, and mineral resources.

Keywords

Agro-ecological zones · Demographic features · Natural resources of HKH · Permafrost · Physiography

3.1 General Description

The Hindu Kush Himalayas mountain ranges have world's uppermost mountains above 8000 m in height and supporting millions of people in both up and downstream areas for their needs of food, fuel, and fiber (Karki et al. 2012). The HKH region also helps in providing water for irrigation, industry, and domestic use. One thing unique to the region is that the rivers originating from the Himalayan mountain ranges form the largest river system from any single location in the world. All these rivers are fed by the Himalayan glaciers, having the highest runoff from any single source.

Mountains in HKH are in complex, diverse, and risk-prone regions (Shrestha et al. 2015) in rugged topography and fragile environment coupled with heavy rains, undulated and steep terrains, high seismicity, and excessive land erosivity thus causing severe flooding/landslides and movements of rock /sediments. The outcome is the prevalence of poverty and malnutrition among inhabitants of the region with limited access to development services (Vaidya and Sharma 2014), though the region has rich biodiversity with potential sources of various medicinal herbs and aromatic plants still remaining unexplored/unutilized, extensive grasslands and forests, workable mineral deposits, easy to harness waterpower, etc. It is striking to note that out of nearly 220 million population in the Hindu Kush Himalayan region, poverty is severe with 5% more than that of the National average of the respective HKH countries, and 31% population remains below the representative poverty mark (Gerlitz et al. 2014).

The drainage basins of Himalayan rivers cover an area of 8.6 million km^2 in the HKH region from Afghanistan in the west, through Pakistan, India, Nepal, Bangladesh, and Bhutan to Myanmar and China in the east, extending a stretch of 3,500 km over eight countries. The river basin in the HKH region varies from 271 to 1016 thousand km^2, being the longest in the Ganges covering India, Nepal, and Bangladesh and the shortest in the Salween covering Tibet, Myanmar, and Thailand. Mean discharge varies from 146 to 340,00 m^3/s in Tarim and Yangtze, respectively. HKH region shows a high vulnerability to climate warming followed by a retreat of glaciers and marked variability in summer monsoon rains producing a threat to create drought situations in many areas during dry periods. The percent irrigated cropland presently varying from 0.6 to 24.1, may undergo a steep decline causing a fall in crop production and increasing poverty threat. In addition to water stress and water quality deterioration, it is important to note that many natural hazards in the HKH region like earthquakes, windstorms, droughts, floods, etc. are often observed in the HKH region. The contamination of Arsenic, Fluoride, and Iron in groundwater is rampant. Thus, the socio-economic development approach and poverty reduction program in the HKH region may not be achieved to the desired extent causing misery to inhabitants.

The HKH region being a geologically active zone on Earth is highly seismic. The Himalayas formed through the collision of the Eurasian and Indian plates have steep, cold, and dry upper catchment areas having a heavenly picturesque sight with hot and humid lower catchments. Anthropogenic activities coupled with climatic variability have been causing a degradation in the Himalayan ecosystems. In the HKH region, particularly in high-altitude rangeland covering a huge area of 2.1 million km^2, it leaves behind serious repercussions for the safety, food security, and water security of downstream populations. Percent of National water use for Agriculture is the highest in Afghanistan

(98.19%) and lowest in China (67%). But China (25%) has high water use for industrial purposes and Afghanistan does not at all use water for any sort of Industrial setup. India also has comparatively high water use for Agriculture (86.46%), and in addition to 8% water for domestic use, Industrial use of water (5%) is 2nd highest after China.

The Himalayan region has indeed the biggest reserves of water in the form of snow and ice outside the polar region, and Agriculture, here, is utilizing, on an average, 95% of water in contrast with 70% of water use for global agriculture. It indicates that agriculture, apart from domestic and industrial demand, will be the major user of freshwater mainly used for the production of cereal crops in the region dominated by rice consumption. Considering the increase in consumption of fruits, sugar, edible oils, meat, and dairy products in HKH, increased competition for water supplies is being faced in Agriculture from industrial and domestic usage, thus making the World's most intensively irrigated continent. In some countries of the region, besides surface water, groundwater using water pumps is also under extensive use for irrigating the land for crop production. Due to high production costs and low returns, the production of food grains has become stagnant in the region with the substitution of grain crops for other crops or keeping the land fallow. Nearly 30% of land in Nepal is abandoned, and in China, 14.5 million hectares of arable land are used for urban and industrial development.

The Indian Himalayan Region (IHR) is the most frangible ecosystem which has experienced changes due to difficult situations, degradation, and fracturing (Kumar et al. 2021). In the IHR, springs are the main source of water for rural communities (Negi and Joshi 2004, Tambe et al. 2012, Niti Aayog 2018). About 50% of the people of the IHR use water from about 3 million springs for their sustenance (Daniel et al. 2021). There is a need to identify vulnerable springs due to biophysical and socio-economic reasons for their judicious management. Many springs in the mid-hills of the Hindu Kush Himalayas (HKH) are drying (Valdiya and Bartarya 1991; Tiwari 2008; Tiwari and Joshi 2011; Mahamuni and Kulkarni 2012, Chapagain et al. 2017), and so, their proper management is necessary for the survival of the population residing in the area (Scott et al. 2019). The HKH region has the largest deposit of ice and snow in the world (Barnett et al 2005; Bajracharya and Shrestha 2011; Schickhoff et al 2016).

Due to the predominance of rice–wheat cropping sequence being water-intensive crops, Indo-Gangetic plain areas and Punjab in India are facing water scarcity as well as an increase in contaminated water particularly in groundwater reserve. There may be a need to think seriously about increasing the water use efficiency and commercial pricing of water may be a necessity, thus compelling the marginal farmers to migrate to other professions besides Agriculture for their livelihood. Wastewater management systems and trans-boundary water supply with multiple countries may be thought upon to address the problems of water shortage and pollution in the near future.

Agricultural land degradation is the cause of a decline in crop and livestock production. Agricultural land degradation affecting the soils (FAO and ITPS 2015, UNEP 2021) can be due to the loss of wild species that provide pest control and crop pollination (Dainese et al. 2019). Soil erosion is caused primarily due to deforestation and cropland expansion (Prăvălie et al. 2021). Land degradation could reduce global food productivity by about 12% (Noel et al. 2015; Kopittke et al. 2019).

3.2 Sources of River Water: Glacier, Rainfall, and Groundwater

The Himalayas are the abode of rivers that roll down along the steep terrains in a crisscross heavenly snowmelt network displaying boundless natural scenic splendor in the surroundings and the ten rivers (Lutz et al. 2014). In addition to snow and glacier melt contributing 22–41% of runoff, the precipitation contributed only 27% to total runoff. As reported by Lutz and Immerzeel (2013), the rainfall contributed by 66%, 59%,

42%, and 44% to total runoff is perceptible in Upper Ganges, Upper Brahmaputra, Upper Salween, and Upper Mekong rivers, respectively. In Upper Mekong, the glacier melt has contributed only about 1% to runoff. It is observed from the study conducted by the Indian Space Research Organization (ISRO) that 75% of Himalayan glaciers are retreating at an alarming rate (Prakash 2020). These retreats will augment water flow variability and cause danger to water sustainability in the entire basins. Groundwater which may undergo a decline due to the anticipated retreat of glaciers is an essential part of HKH's hydrology.

Annual Runoff (mm) from River Basin Base flow, Rainfall runoff, Snowmelt, and Glacier melt in Indus is 62, 154, 125, and 233 but in the Ganges, 151, 718, 94, and 125; in Brahmaputra, 112, 407, 62, and 110; in the Salween, 106, 204, 150, and 40; and in Mekong, 106, 204, 150, and 4, respectively. The highest runoff both from base flow and rainfall is noted in the Ganges and the lowest in Indus. It may be pertinent to mention that the rainfall is also dominant in the upper Salween and Mekong basin areas. The highest runoff from snowmelt is observed in both Salween (2815 km long) known in China as Nu River flowing from the Tibetan plateau into the Andaman sea in Southeast Asia and Mekong (4350 km long) being a trans-boundary river in East Asia and Southeast Asia flowing from Tibet through China, Myanmar, Laos, Thailand, Cambodia, and Vietnam. There is the highest runoff from Glacier melt in Indus but the lowest, in the Mekong. Glacier melt is appreciably high in both Ganges and Brahmaputra having high-altitude basin areas in Western Himalaya in Uttarakhand, India, and Himalayas in Tibet, respectively.

The runoff dynamics as delineated in the balance sheet of river systems of the Himalayas could determine the climatic unevenness in the surrounding areas in the Southeast Asia. Himalayan Rivers make available the crucial source of water to 210 million people living in these mountains and some 1.3 billion people living in the downstream of HKH region, for drinking, agriculture, energy, navigation, or other purposes. A huge hydropower potential is existing in the water resources of the Himalayas and India accounts for 79% of the total hydropower potential in the region. But a meager fraction of this potential has been developed in India (Government of India, 2010) and in Pakistan, 11% of hydropower potential has been realized. As reported by Shrestha et al. 2015, Nepal where 25% of people live at the poverty line and massive load shedding (16 h a day) occurs has the highest untapped hydropower potential in the World. It is a matter of grave concern that in South Asia having 35% of the World's undernourishment, the population is expected to reach 2.2 billion from 1.89 billion presently with a cereal demand doubling from 246 to 476 million tones in 2025 producing a stress in water use utilization.

3.3 Land Distribution, Use, and Suitability

The global area of cultivated land has grown by a net 159 Mha since 1961 (Fig. 3.1). Irrigated area has more than doubled over the period, and the number of hectares needed to feed one person has reduced dramatically from 0.45 to 0.22 ha per person (FAO 2010).

In the area of land degradation (Fig. 3.2), 25%, 8%, and 37% are under highly, moderately, and slightly degraded, respectively.

3.4 Afghanistan

Afghanistan (Fig. 3.3), with a geographical area of 652,860 km^2, is a landlocked country, located in Asia and is surrounded by Pakistan, Iran, Turkmenistan, Uzbekistan, Tajikistan, and China. The country has a very undulating and rugged topography, with altitudes ranging from about 150 m to 8,000 m above mean sea level. About 75% of the country comprise of mountains and hills, out of which about 25% has an elevation of more than 2,500 m above msl. Rout (2008) explained that major river valleys spread out from these mountains, creating fertile valleys along which most agricultural and irrigation

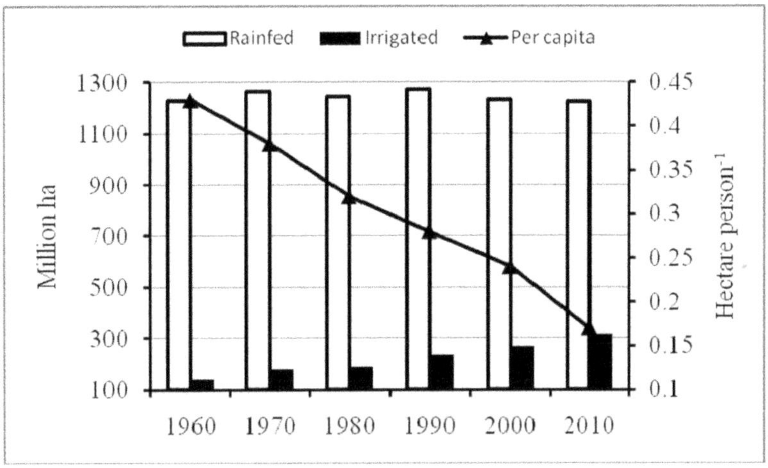

Fig. 3.1 Land under irrigated and rainfed cropping (1960–2010). *Source* FAO (2010)

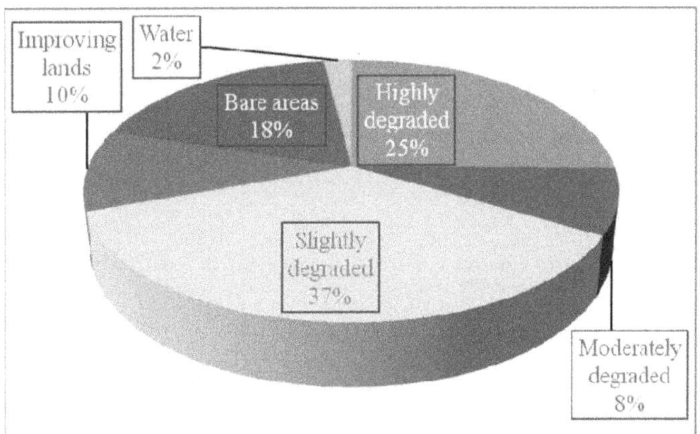

Fig. 3.2 Trends in global land degradation

development takes place. The passes through several high mountains in Afghanistan serve as a lifeline to the people as most of the trade within and outside the world is conducted through them. Historically, these passes have remained of prime importance in the history of the country.

3.4.1 Human Resources

The total population of Afghanistan was 38.4 million with a population density of 58 per km^2. The median age in Afghanistan is 17.4 years. The total land area is 652,860 km^2. In rural areas, 77% of people live and 16.7 million population is economically active being 59% of the total population. It is notable that 6.2 million people are economically active in Agriculture having 7.9 million ha as a cultivable area being 12% of the total area. As reported by FAO (2012), the life expectancy is 65.98 years. Out of the total population, half (50%) of them is having access to improved drinking water with 78% and 42% in urban and rural populations, respectively.

3.4.2 Livestock Resources

The National Livestock Census started in Afghanistan in September 2002 with the arrival of an external consultant and the Livestock Census Manager (National Livestock Census 2003). The consultant was responsible for the design of census survey forms in consultation with the Census Manager. A basic form (Level I Livestock Census) was drawn up covering

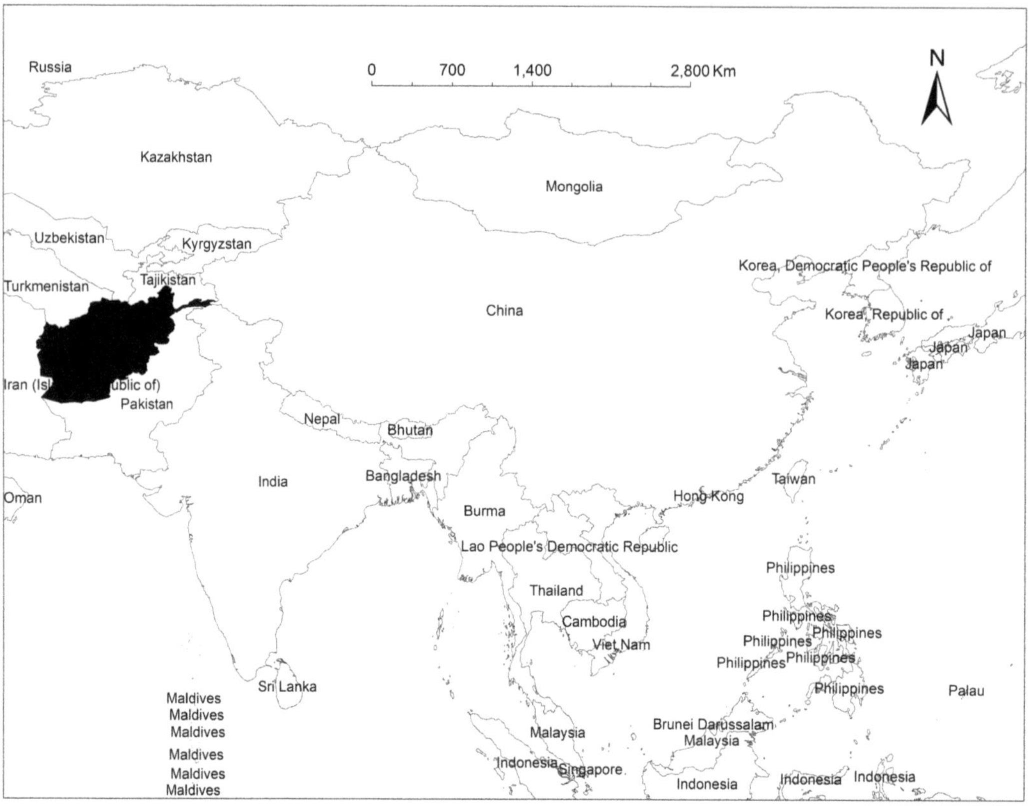

Fig. 3.3 Afghanistan

ruminants and monogastric animals including poultry to determine livestock numbers and herd breakdown by age and sex. Level I was designed to be conducted by enumerators on every village in Afghanistan, or on component communities within large villages. A second census form (Level II Survey), to be administered by the Supervisors, covered details of production and marketing systems. The Level II form was more detailed and designed to be administered to individual households in randomly selected districts, villages, and households. A further survey, the Women's Survey, was administered by female enumerators to women livestock keepers, selected randomly in selected districts.

As per livestock census 2002–03 (FAO 2008), the livestock population in Afghanistan is presented in Table 3.1. The total number of cattle, sheep, goats, donkeys, camels, and horses, was 3.7, 8.77, 7.28, 1.58, 0.17, and 0.14 million,

Table 3.1 Livestock population in Afghanistan

Livestock category	Number	Livestock/family
Cattle	3,715,409	1.22
Sheep	8,772,351	2.88
Goats	7,280,866	2.39
Donkeys	1,587,594	0.52
Camels	175,277	0.06
Horses	142,094	0.05
Chickens	12,155,846	3.99

respectively, according to this census. Out of 7.28 million goats, over 0.5 million were recorded in Hirat, Hilmand, and Nuristan.

3.4.3 River Basin

Though Afghanistan is a relatively dry country as far as annually received rainfall is concerned, the country has five major rivers, the water from which is used for irrigation. However, the full potential of these rivers has not been exploited so far. The possible reasons could be political unrest, lack of finance and resources to invest, and lack of judicious and purposeful planning. The Kabul river originates in the central region of the Hindu Kush, about 100 km west of Kabul, and has about 74.4% of its total basin area in Afghanistan. The Amu Darya river originates in the Afghanistan part of the Pamir river. From west to east, the main rivers include the Shirin Tagab, the Sarepul, the Balkh, and the Khulm rivers (Favre and Kamal 2004; Rout 2008; Uhl and Tahiri 2003). These river basins cover about 75,000 km^2 area or about 12% of the total geographical area of Afghanistan.

Total renewable water resources in Afghanistan amount to 65 km^3/year with water resource availability of 2019 m^3/year per inhabitant. But water use per inhabitant as recorded in 2017 was 1839 m^3/year. But safe drinking water is scarce with 15.4 million people (44.7% of the population) not getting access. Problems may arise in the Kabul and Eastern Helmand river basins due to the shortage of available water for various uses. There is a strong need for improving system efficiency and productivity through modernizing infrastructure, proper water distribution, and developing and improving water storage systems and their capacity (Rout 2008). Measures have to be adopted to improve structures to reduce desilting and protection against water loss. Agriculture is the major user of water to the quantum of 20 km^3/year, but both municipalities (203 million m^3/year) and Industry (170 million m^3/year) are using a low volume of water. Besides wheat, all other crops including rice (200 thousand ha) grown in Afghanistan are having low irrigated areas (Fig. 3.4).

3.5 Bangladesh

People's Republic of Bangladesh (Fig. 3.5), a country in South Asia, with about 165 million people, increased at the rate of 1.01%. Bangladesh has an area of 147,570 km^2, sharing land borders with India on three sides. About 9

Fig. 3.4 Irrigated area under major crops in Afghanistan

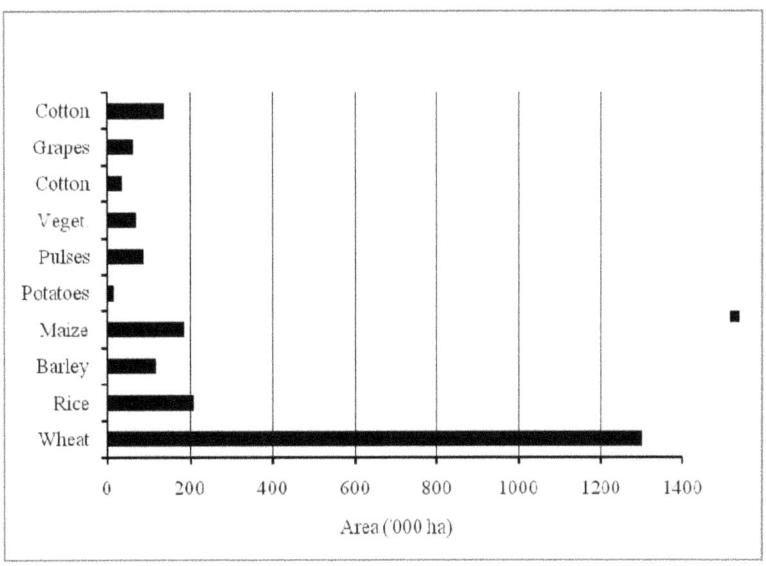

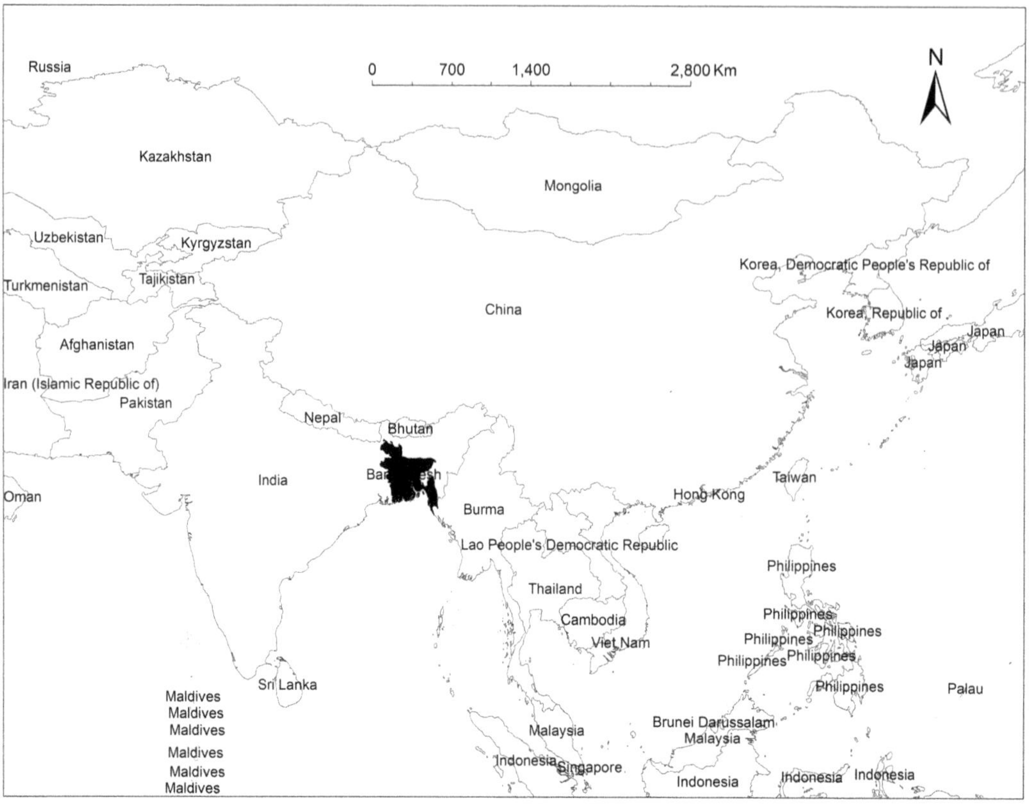

Fig. 3.5 Bangladesh

million hectares of area are cultivable land in Bangladesh. Due to favorable weather conditions in Bangladesh, three crops of rice are grown on the same piece of land at some places during the year. A revenue survey conducted by the government has revealed that the total cropped land in the country is nearly 14.1 million ha, including single, double, and triple cropping land. Due to population pressure, each person working in the agriculture sector has only 0.12 ha of cropland, on an average. Out of the total geographical area, 63% are being used for cultivation and 4.38% for rural and urban housing. About 37% of the area is the forest land and cultivable wasteland.

3.5.1 Water Resources

Bangladesh lies at the end of the Ganges, the Brahmaputra, and the Meghna basins. This country was part of India before 1947, East Pakistan from 1947 to 1971, and emerged as Bangladesh in 1971. Formed by the alluvial soil deposited by three great rivers, such as the Brahmaputra, the Ganges, and the Meghna, most of its area is low-lying floodplain. Only 8% of the catchment area of 1.72 million km^2 of these three rivers lies within Bangladesh, while the rest is in India, China, Nepal, and Bhutan. Except for the little hilly areas in the southeastern part, called the Chittagong Hill Tract (CHT), the topography of the country is almost flat. The optimum production potential is not fully harnessed due to the occurrence of frequent flood events during the monsoon season. Acute scarcity of irrigation water is experienced from March to April, which limits the cultivation of highly responsive varieties of rice, which accounts for about 36% of total rice production in the country.

3.5.2 Sources of Water in Bangladesh

Rainfall is the major source of water in Bangladesh, though heavy rains many a time cause more harm than good. A substantial part of rainwater flows as runoff and a part of it infiltrates into the soil to recharge groundwater. Khan (1993) reported that about 93% of the stream flow passing through the country originates from outside the Bangladesh. The other sources of water are surface water and groundwater. The Ganges–Brahmaputra–Meghna river system discharges a huge amount of surface water through Bangladesh, and a part enters into the ground to form groundwater.

Surface Water

Bangladesh has sufficient surface water resources, particularly during the wet season, the surface water is abundant. In the downstream area of the confluence of the Ganges and the Brahmaputra, about 795,000 million m^3 of surface water is discharged through the Ganges–Brahmaputra system. When spread over the surface, this water can cover the area of the country up to 5.52 m in depth. About one-third of Bangladesh is submerged in a normal flood and two-third area during severe floods occurrence. However, water scarcity persists in many areas of the country. The perennial waterbodies are decreasing with the use of more and more surface water.

Groundwater

Though Bangladesh receives heavy rainfall and three giant rivers pass through the country, the dependence on groundwater for irrigation and other uses has increased since the 1980s. The scarcity of surface in the country during dry season (winters) has made groundwater an essential source to supplement irrigation. According to estimates, about 79% of the water demand for irrigation, livestock, households, and industry is met from this source. The recharge from surface water, particularly rainwater, forms the major part of the groundwater. The presence of arsenic in groundwater in Bangladesh beyond the critical limit, particularly in shallow aquifers, is a matter of concern. It has created many health problems for the people of the country, and many physical deformities and other associated diseases have been reported. It is estimated that about 16% of the present population is exposed to arsenic contamination exceeding 0.05 mg/l, much more than the standard set by the government of Bangladesh.

3.6 Bhutan

Bhutan (Fig. 3.4) lies in Eastern Himalayas between 26° 41′ 52” and 28° 14′ 52” N latitude and 88° 44′ 54” and 92° 41′ 07” E longitudes (Fig. 3.6). Total area of the country is 38,686 km^2 and elevation ranges from 160 m above mean sea level to about 7000 m amsl (Yangchen et al. 2015). The general climatic conditions are subtropical in the south, temperate in the central, and alpine in the northern parts. According to the Bhutan census 2005, the total population of the country was 672,425 persons, out of which 70% live in rural areas (NSB 2014). Due to urbanization and infrastructure development, not only the area under forests but also the area under agriculture and shrub lands has decreased (Giri and Singh 2013).

3.6.1 Agro-Ecological Zones of Bhutan

Bhutan is mountainous and 49.45% area has more than 50% slope. About 71% (2.7 million ha) of the area of the country is under forest cover, while the non-forest area is 1.1 million ha. A decrease of 13% in bare and 6% in area under agriculture was observed between 2000 and 2013. However, the area under all categories increased during this period except for lakes which remained stagnant.

3.6.2 Population

Bhutan's first census was conducted in 2005 by the Population and Housing Census of Bhutan (PHCB 2005). The population of Bhutan was

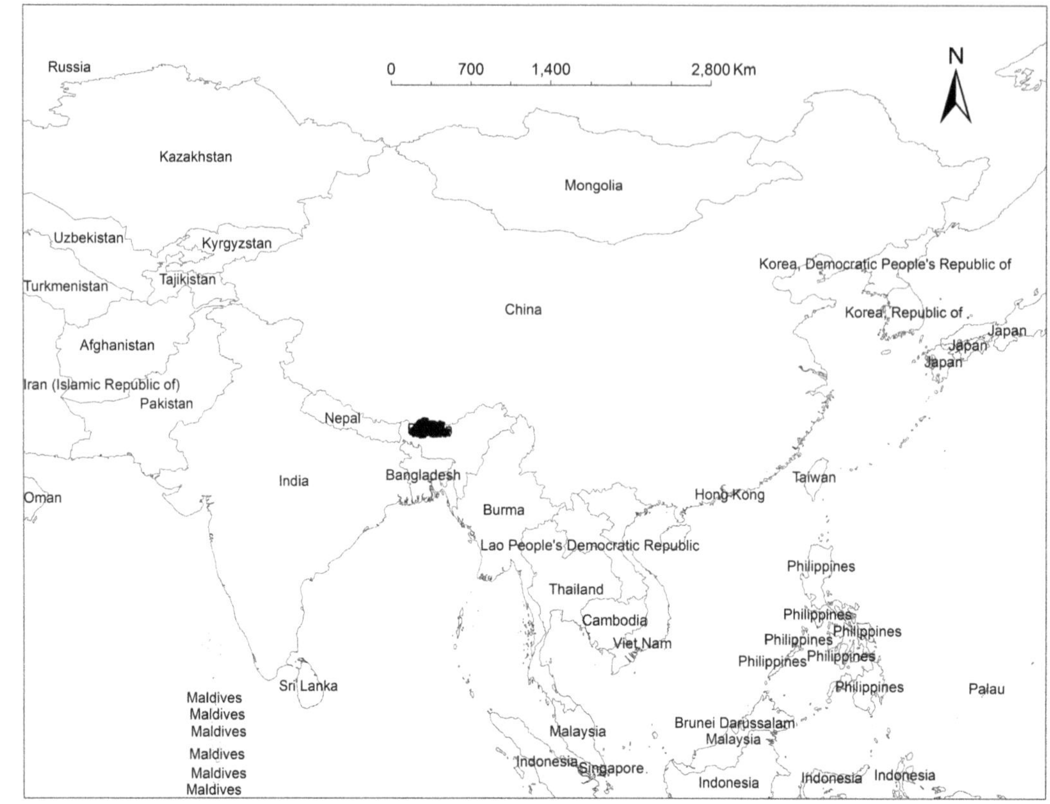

Fig. 3.6 Bhutan

0.634 million in 2005, is growing at an annual compound rate (ACGR) of 1.35%, and is expected to grow to around 0.887 million by 2030 (Fig. 3.7), an increase of 39.9% within 25 years. The youth population in the age group of 15–24 is expected to increase from 145,810 persons to 161,280 in 2030 at an ACGR of 0.404%; while the population in the school-going age of 5–14 years is expected to decline from 147,406 persons in 2005 to 140,037 persons in 2030 at an ACGR of −0.204%.

3.6.3 Water Resources

Bhutan is endowed with abundant water resources of good quality (Chhophel and Chhophel 2006) and has the following features:

- Climate dominated by monsoon bringing in the rain from June–September,
- Mean annual rainfall between 500 and 5000 mm,
- Major rivers originate from alpine zones with flow from north to south,
- Water flow is more in monsoon, and
- Scanty Information is available on ground water.

The mountainous terrain limits agricultural production, and less than 8% of the total land area is arable but 79% of the population is engaged in agriculture. Therefore, sustainable land and water management is necessary for this sector as it forms the backbone of the rural economy. The main sector for water resource management is hydropower, generating substantial revenue for the Government. Seasonal fluctuations and increased sediment transport affect the optimal

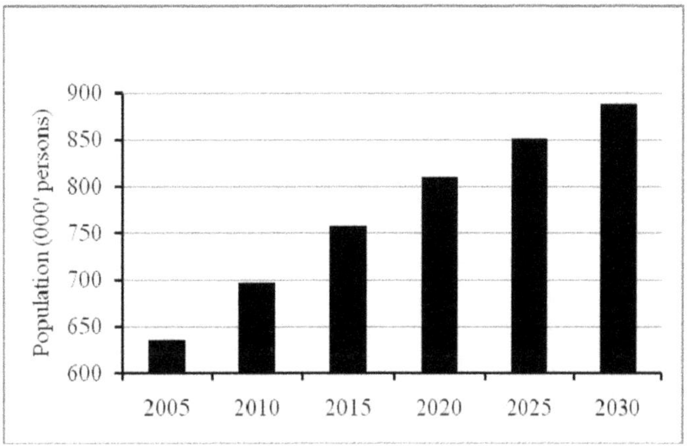

Fig. 3.7 Population projections in Bhutan

generating capacity and thus the life of hydropower plants. The Government is now moving toward Integrated Water Resources Management (IWRM) as the main paradigm for the management of its water resources. The Hydro-met Services and NEC with support from FAO are implementing a one-year project for capacity building. The project aims at developing and implementing good water governance through the development of the knowledge base, institutional setup, and planning tools for IWRM within the context of integrated watershed management.

3.6.4 Livestock Population

Among livestock, Cattle has got the largest (303 thousand nos) followed by goat and pig, etc. Dzo, a male hybrid between yak and domestic cattle, is also found in Bhutan while the female is known as Dzomo. Bhutan has also got Mithun population and poultry is 660 thousand numbers (Table 3.2).

3.6.5 Land Use

Land is one of the most important elements in agriculture and hence statistics on land and its use is of special interest to users. Based on the interpretation of panchromatic 1:50,000 SPOT satellite imagery taken in December 1989 and color SPOT images taken in June–August 1990, combined with ground thrusting exercises conducted during 1993 and 1994, the Ministry of Agriculture (MoA) published the first ever atlas of Bhutan showing the land cover and area statistics in 1997 (MoA 1997). According to that publication, the total area under agriculture including natural pastures and tseri cultivation was calculated as 777,392 acres. The second land cover mapping exercise was done by the ministry in 2009–2010 using ALOS images (AVNIR-2) and the updated maps published in 2011 (MoAF 2011). According to MoAF (2011), the total agricultural land was calculated as 277,998.5 acres, which as a percentage comes to 2.9% of the total country's land area.

Table 3.2 Animal population in Bhutan, 2016

Livestock	Population	Livestock	Population
Cattle	303,297	Equine	18,890
Mithun	577	Sheep	11,277
Yaks	40,438	Goats	39,513
Dzo	9,179	Pigs	15,324
Buffaloes	532	Poultry	660,000

Source MoEF, RGB (2016)

3.7 Tibet (China)

The Tibetan Plateau lies between the Himalayan range to the south and the Taklamakan Desert to the north. The Tibetan Plateau is also known in China as the Qinghai–Tibet Plateau (Jones et al. 2012). It has an approximate width of 1,000 km north to south and 2,500 km east to west. The Tibetan Plateau contains the headwaters of the drainage basins of most of the streams lying in the region. Thousands of glaciers on the plateau help in maintaining water flow in the rivers throughout the year. The impact of global warming on the Tibetan Plateau has created great scientific interest among the scholars to unveil the truth behind and understand mechanisms of change (Liu and Chen 2000).

To plateau has many forested gorges and ridges toward the east and southeast and controls headwaters of the Salween, Mekong, and Yangtze rivers in northwest Yunnan and western Sichuan. The Indus river originates from West Tibet near Manasarovar lake. North of the mountains runs the Gansu corridor which was the main route from China during the earlier period. The plateau is interspersed with mountain ranges and large brackish lakes and forms a high-altitude arid steppe. Permafrost is experienced over most parts of the plateau. About 45 thousand glaciers are found in the plateau, covering a huge area of 105,000 km^2. Norbu (2011) mentioned that with an area of 2.5 million km^2, the Tibetan Plateau is the largest and highest region on Earth. There are numerous grassland areas on the southern and eastern edges of the steppe which are supporting indigenous communities and their livestock. The occurrence of frost in the region for about six months in a year causes problems for the local population as well as the livestock, which face feeding difficulties. Figure 3.8 shows the geographical setting of China.

3.7.1 Livestock

In 1999, the livestock population in Tibet Autonomous Region (TAR) comprised 3.87, 1.00, 0.32, 0.24, 10.38, and 5.19 million of yaks, cattle, zos, horses, sheep, and goats, respectively (Tashi et al. 2005). There were about 5.5 million large animals in TAR. The pastoral system had the highest number of livestock, particularly, yaks, sheep, and goats. This is due to the reason that these three categories of animals do not depend on stall feeding and, in general, have free-range grazing in pastures and grasslands. Pastoral and agro-pastoral systems are very important for these animals. There is a shortage of fodders during late winter and early spring; however, adequate supplies are available during summers. Tashi et al. (2005) mentioned that energy and protein intakes are the major nutrients, limiting livestock productivity in the winter season.

3.7.2 Traditional Pastoral Livestock System

Comprising of shrublands, alpine steppes, cold deserts, semi-arid steppes alpine steppes, and alpine meadows, the rangelands of the Tibetan steppe cover a large area of Asia. Some major rivers of Asia originate from the Tibetan steppe (Dennis et al. 2006). There has been transhumant grazing by yak, sheep, and goats in this region. This is a traditional practice. During the last six decades, management of rangelands has undergone major shifts from individual to collective and also privatization of livestock with individual grazing rights. Most of the rangelands on the Tibetan plateau are above 4,000 m in elevation and some livestock owners camp permanently at higher altitudes along with their animals. The droppings of animals in pastures and grasslands maintain the soil fertility for better productivity from rangelands. These rangelands provide forage for an estimated 12 million yak and 30 million sheep and goats. Livestock supports the livelihoods of pastoralists and agro-pastoralists of the region.

Sheehy (2001) explained that the traditional forage-based, well-managed pastoral livestock-production systems, however, are showing

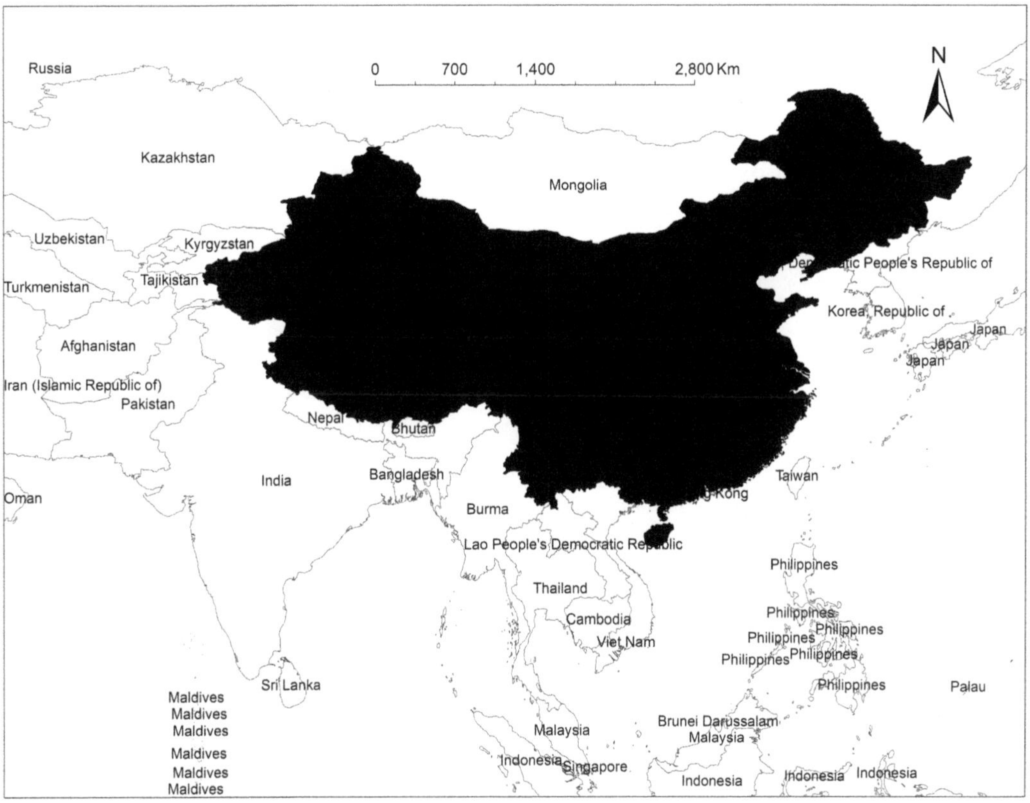

Fig. 3.8 Geographical location of China

declines in overall productivity. About a third of the rangelands on the Qinghai–Tibetan Plateau is currently considered moderately to severely degraded (Sheehy 2001), and the percentage of degraded rangelands increased from 18 to 30% of the total area between 1980 and 1990.

3.7.3 Vegetation

Vegetation types of the Tibetan steppe and the area are presented in Table 3.3.

3.7.4 Permafrost

Due to climate warming, caused as a result of anthropogenic activities, the permafrost in the Qilian Mountains in the northeastern parts of the Qinghai–Tibet Plateau has suffered a radical change, thereby affecting the aplomb of the biome, water, and GHG cycles (Qi et al. 2021). In a study conducted in the Qilian mountains, Qi et al. (2021) found that during 2004–2019, the area of permafrost decreased by 20% at an average rate of $0.074 \times 104\ km^2 \cdot yr^{-1}$. In the past about four decades, permafrost has degenerated due to climate warming and increasing human activities, which is shown by an increase in ground temperatures and active layer thickness, and a decrease in maximum frozen soil depth (Ran et al. 2018; Wang et al. 2015; Biskaborn et al. 2019; Wu et al.2015; Cao et al. 2018). Peng et al. (2020) and Way et al. (2018) observed that the change in permafrost significantly affects hydrological cycles, while Nitzbon et al. (2020) found a change in thermokarst encouraging courses. There was an escalation of biogeochemical cycles (Wang et al. 2020; Mu et al. 2020), a shift in geomorphologies (Nitzbon

Table 3.3 Vegetation types in the Tibetan steppe (Ni 2002; Chen and Fischer 1998)

Vegetation type	Area (000' ha)	% of the total area
Alpine meadow–steppe	5626	4.28
Alpine steppe	37,762	28.75
Alpine desert–steppe	8679	6.61
Alpine desert	5967	4.54
Alpine meadows	58,622	44.64
Temperate meadow–steppe	210	0.16
Temperate steppe	3833	2.92
Temperate desert–steppe	988	0.74
Temperate steppe–desert	107	0.08
Temperate desert	2084	1.59
Lowland meadow	1168	0.88
Temperate mountain meadow	6067	4.61
Tropical tussock	9	–
Tropical shrub tussock	28	0.02
Temperate tussock	1	–
Temperate shrub tussock	140	0.10
Marsh	21	0.01
Total	131,322	99.93

et al. 2020; Zhang et al. 2019), and a loss of regulation of surface hydrothermal dynamics (Guo et al. 2011; Stiegler et al. 2016) because of the change in permafrost.

3.8 India

The Himalayan region in India (Fig. 3.9) covers partially or fully twelve states of India (MoEF, GoI 2010). After Mt. Everest in Nepal, Kanchenjunga in India has the 2nd highest peak (8598 m) followed by Nanga Parbat (8,126 m), Nanda Devi (7,817 m), Namcha Barwa (7,782 m), and Kamet (7,756 m).

The greater part of Indian Himalayas, like the Shiwalik hills and some areas of lesser Himalayas, is below the snow line. However, the altitudes above 2000 m remain perpetually covered with snow and glaciers. The hills and mountains are dissected by rivers and streams, forming valleys, gorges, and peaks. The Himalayan region is unstable and fragile. It is the youngest and highest mountain system with about 30 peaks having a height of above 7600 m. Some peaks in the Himalayas are still rising. The range also exerts a major influence on monsoon and rainfall patterns.

3.8.1 Human Resources

India was having a population of 1210 million according to the 2011 census. However, it is expected to be about 1390 million in 2021. The Himalayan states of India had a population of 74.11 million in 2011, which is expected to be around 86.68 million in 2021. More than 60% of the population in India is young (<45 years), indicating a sizeable workforce. Administratively, in addition to 10 states in India, it covers a couple of states, like Assam and West Bengal, partly. It has been estimated that more than 65% of the geographical area of the Himalayas in India is under forests, which constitutes one-third of the total forest area of the country.

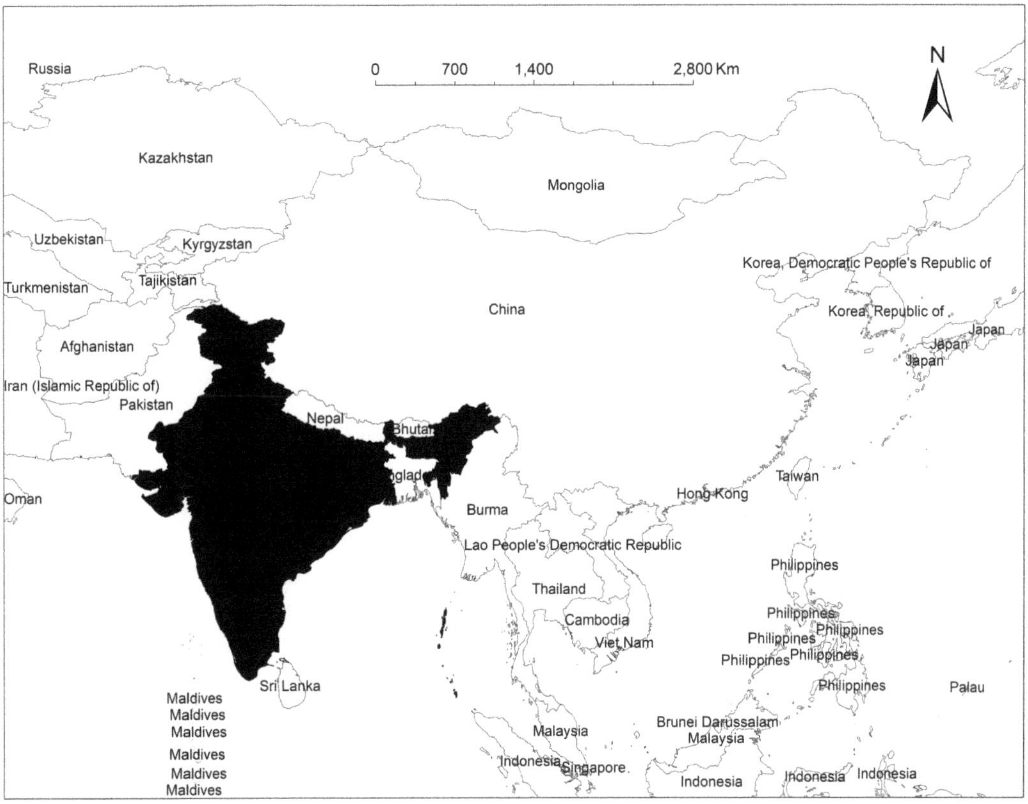

Fig. 3.9 India

3.8.2 Water Resources in India—The Regional Distribution

In order to support 18% of the World's population, India has only 4% of renewable water resources. Besides fresh inland water resources of rivers and lakes (Mall et al. 2007), other available water resources are groundwater, deep sub-surface water, glaciers, and permanent snowfields. The oceans are the largest reservoir of water on earth and the estimate is that 96% of all earth's water is present in oceans providing evaporated water to the water cycle and allowing water to move all around the globe as ocean currents. The annual precipitation is estimated at 4,000 km^3, including snow and rainfall (Table 3.4). The average annual natural runoff in rivers and aquifers in India is about 1,869 km^3.

Rivers

The three main Himalayan rivers in India are the Indus, the Ganges, and the Brahmaputra. In addition to runoff from rainfall, the snow and glacier melt contribution from the Himalayan region are the main sources of water yield in these rivers. The average intensity of mountain glacier melt varies from 3.4% for the Indus to 3.2% for the Ganga and 1.3% for the Bramhaputra.

A large number of tributaries like Alkananda, Ramganga, Kali, Yamuna, Gomti, Kosi, etc. draining 11 states in India join Ganges at various confluence points. The Indus is the westernmost river system in the subcontinent, and Jhelum, Chenab, Ravi, Beas, and Sutluj are the main tributaries. Zanskar river (Photo 3.3), a tributary of the Indus, flows entirely in Ladakh. The

Table 3.4 Water resources of India

Parameter	Km^3
Estimated annual precipitation (including snowfall)	4000
Runoff received from upper riparian countries	500
Average annual natural flow in rivers and aquifers	1869
Estimated utilizable water	1123
(i) Surface	690
(ii) Ground	433
Water demand _ utilization (for the year 2000)	634
(i) Domestic	42
(ii) Irrigation	541
(iii) Industry, energy, and other	51

Source National Water Mission under National Action Plan on Climate Change, MoWR, GOI (2010)

primary tributaries of the mighty Brahmaputra are Dhansiri river, Lohit river, Dabang river, Subansiri river, Kameng river, Manas river, Sankosh river, and Teesta river draining through the eastern Himalayas. Owing to variability in land use leading to deforestation, agricultural practices, and urbanization, hydrological catastrophes, viz., variable rainfall, runoff followed by sedimentation, and lake pollution are observed (Ramakrishnan 1998).

Glaciers

The number of glaciers and the area covered are given in Table 3.5. The contribution of snow to the runoff of major rivers in the eastern Himalayas is about 10% (Sharma 1993) but more than 60% in the western Himalayas (Vohra 1981). It is estimated that Himalayan mountains have a surface cover (10–20%) with glaciers and the melt of glaciers has kept rivers perennial. Seasonal snow cover (30–40%) is also observed in the mountains as reported by Upadhayay (1995) and Bahadur (1999). As estimated, snow and icemelt water are contributing 500–515 km^3/yeasr of water in the Himalayan streams.

Groundwater

Groundwater being extracted through dug wells and borewells is being used both for drinking and irrigation purposes in Himalayan states in India. This water is often having a high presence of Iron, Arsenic, Nitrate, and Fluoride thus making it unsafe for drinking purposes. The static fresh groundwater reserve in aquifers is estimated as 10,812 billion m^3 and the dynamic component being replenished annually is amounting to 432 billion m^3.

3.8.3 Eastern India Himalaya

Ethnically, Northeast India is linguistically and culturally very distinct from the other states of India. The North Eastern Region states have a total population of 44.9 million, which is about 3.8% of the total population of the country (2011 census). More than 68% of this population lives in the state of Assam only. The North Eastern Region is recognized as one of the world's biodiversity hotspots due to its rich natural endowments. The forest cover is about 52% of the total geographical area, and petroleum and natural gas reserves of this region constitute about one-fifth of the country's total potential.

3.8.4 The Middle Himalaya

This middle Himalaya comprises of the highlands and mid-altitudes, elevated from 1000 m to about 2800 m. The highlands (2000–2800 m) are mainly covered by the dense forests, including oak and coniferous, and snow is received in the upper reaches of the mountains. A higher percent

Table 3.5 Major glaciers in the three Northwestern Himalayan States of India

Basin	No. of glaciers	Area covered (km^2)	Ice volume (km^3)
Indus Basin			
Ravi	172	193	8.04
Chenab	1278	3059	206.30
Jhelum	133	94	3.30
Beas	277	579	36.93
Sutlej	926	635	34.96
Upper Indus	1796	8370	73.58
Shyok	2454	10,810	NA
Nubra	204	1536	NA
Gilgit	535	8240	NA
Kishenganga	222	163	NA
Total	**7997**	**33,679**	**363.10**
Ganga Basin			
Yamuna	52	144	12.20
Bhagirathi	238	755	67.02
Alaknanda	407	1229	86.38
Ghaghra	271	729	43.77
Total	**968**	**2857**	**209.37**

of population resides in this zone, being a relatively mild area where the temperatures are not too harsh during summers and winters. These summer resorts located in this zone play a significant role in the tourism development of the region.

3.8.5 Potentials of Natural Resources and Their Distribution Pattern

The Himalayan ecosystem is important to the ecological security of India, as it is a source of forest, glaciers for feeding perennial rivers, and conservation of biodiversity. This zone has more cultivable land as compared to high mountains and can sustain more number of people. It is vulnerable and susceptible to the impacts and consequences of (i) changes on account of natural causes; (ii) climate change resulting from anthropogenic emissions; and (iii) developmental paradigms of the modern society. The abundant natural resources and favorable agro-climatic conditions are not harnessed sustainably to find solutions for issues like food insecurity and malnutrition. Lack of infrastructural facilities and improper planning further aggravate this problem.

3.8.6 Forest Resources

About 42% of the total area of the Alaknanda Basin in the Himalaya is covered by forests. The basin is very rich in terms of forest and its diversity. The pine tree is a common vegetation in the mid-altitudes while at higher elevations (>1800 m), temperate coniferous forest mainly Kharsu (*Quercus semicarpifolia*). Tilonj (*Q. dilitata*), Rianj (*Q. lanuginose*), and Banj oak (*Q. leucotricophora*) are found in plenty (Sati 2006a, b). In the Jammu and Kashmir state of India, deodar (*Cedrus deodara*) is a common tree at high elevations. Except for these forest types, many other fodder plants, like bhimal and khadik, are also grown along with edges of agricultural fields. Deodar Forests (*Cedrus deodara*) are found between the heights of 1650 and 2300 m. While moving up altitudenally in the Himalayas, one can calculate the altitude of a place by looking at a sudden change in the type of vegetation at different specific heights. These forests help in increasing the soil fertility at higher altitudes through the fall of litter in abundance. The runoff from higher slopes also brings fertile soil from different slopes and deposits it in valley areas. In the Himalayas, where high slopes encourage soil loss through runoff and consequential erosion, forests serve as conservation tools. Generally, Oak and pine forests are useful for firewood, fodder, and timber. The dwellers of this region are highly dependent on forests and their produce for their livelihood. Forests have the linkages with agriculture and crop production.

3.8.7 Land Resource

In Indian Himalayas, only about 8% of the land is available for cultivation. This land is unable to sustain high population pressure and high cropping intensity. So, the management of land is a crucial and important aspect of agricultural production in the Himalayan region. Shrinking agricultural land and increasing population pressure has resulted in the migration of a large population, particularly youths and their families, outside the region in search of livelihoods. Big industrial cities are the most sought-after destinations, as getting jobs is relatively easy, particularly the labor class non-permanent jobs.

3.8.8 Livestock Rearing

Livestock plays an important role in the livelihood of mountain dwellers as it is the second main occupation after farming. A pair of bullocks is required for plowing the land as tractor cultivation is rather difficult due to the undulating topography of the hills and mountains. By and large, every household has at least one pair of bullocks for plowing the land, one cow or buffaloe for milk, as well as their followers. Further, the animals also provide manure for fields to improve soil fertility. Draft animal constitutes the composition of the domestic animal kingdom mostly used in the field of agriculture as draft power. A recent study on livestock farming reveals that the numbers of domesticated animals are decreasing (Sati and Singh 2010).

It has been found that grazing and trampling by a large population of domestic livestock have resulted in severe degradation of the natural habitat. The nomadic communities in the Himalayas are mobile and reside for half of the year in the mountains during the summer season and move to lower elevations and plain areas during winters during the rest of the year or winter season. This is mainly due to the availability of fodder in these areas during summers and winters, respectively.

3.8.9 Land and Genetic Diversity

Out of a total of 26,224 thousand ha of the geographical area of eight northeastern states, 23,238 thousand ha is the reported area (88.68%). Most of the area of the region is covered by rugged and inaccessible terrain, steep hills, high mountain ranges, undulating topography, deep valleys, and marshy and swampy lands. The area available for cultivation is limited. Due to climatic diversification, the region has a great potential for varied types of agricultural and horticultural crops. Judicious utilization of the available area for the cultivation of suitable crops and their proper management can ensure enduring food security in the region. As per the available data on land use (Anonymous 2002), about 50.6% of the total reported area is under forests, 5.3% under non-agricultural uses, 4.4% is barren land, 5.0% fallow, 8.1% area is uncultivated, and only 16.6% area is the net sown area. According to the state of forest report (1999), the area under forests is about 16,704 thousand ha, which is 63.7% of the total area of the region. The land use figures have been based on estimates as most of the hilly tracts have not been cadastrally surveyed.Arunachal Pradesh has the highest percent area under forests (93.8% for the

reported area and 62.2% of the geographical area) followed by Mizoram (61.9%), Tripura (57.7%), Nagaland (56.2%), Meghalaya (41.9%), Sikkim (34.3%), Manipur (27.2%), and Assam (25.3%). Relatively highest percent of area (12.4%) has been put to non-agricultural uses in Tripura, followed closely by Assam (11.6%) while the minimum (0.5%) is in Arunachal Pradesh. It is revealed that a substantial portion of the landmass is available in different states as barren, uncultivated land, and fallow land which can be put to agroforestry, horticulture, forestry, pastures, etc. with suitable reclamation measures and management practices.

The region as a whole is endowed with an array of native flora and economically important species. The landraces available in rice are plenty and Assam Rice Collections (ARC) are commendable (Shastry et al. 1971) in this pursuit. Some of the landraces grown in the high elevations of 2000–2500 m above msl are resembling japonica plant types. Some of the accessions are tolerant to various biotic and abiotic stress. Datta and Laskar (1992) reported that some of the local rice races in Tripura are highly tolerant to Fe toxic conditions. Local gene pool in maize is also noteworthy. The region is rich in diverse genotypes of grain legumes such as perennial types of pigeon pea, rice beans,Dolichos bean, and winged beans. The tree bean (*Parkia roxburghii*) grows almost wild in the hilly areas of Manipur and Mizoram.A number of species of *Vigna vaxillata* are also reported in this region. Indigenous cotton varieties having short staples but colored are known to exist in Meghalaya and Tripura. The region is home to many horticultural crops (Pal and Parthasarathy 2001) including fruits, vegetables, spices, condiments, medicinal plants, and ornamentals. It is recorded that the 17 species, 52 varieties, and 7 natural hybrids of citrus come from this region. Among the tropical fruits, *Citrus*, *Musa* sp., *Mangiferra,* and *Artocarpus* exhibit wide variations. Variability is also very high among the subtropical and temperate fruits. The region is rich in cucurbits and solanaceous crops. This is also a gene center of *Alocasia, Colocasia,* and *Dioscorea* occurring in great diversity. Ginger, turmeric, chillies, cinnamon, and large cardamom have wild relatives in the region. Black pepper also grows in abundance in the wild in forest areas. Many ornamentals also occur wild in this region. Out of 1300 species of orchids belonging to 163 genera reported in India, more than 600 are from the eastern Himalayas. Plantation crops like tea, coffee, rubber, coconut, and areca nut were introduced in this region long ago. Medicinal and aromatic plants also occur in plenty in this region. Over 2500 fodder grasses and 1500 fodder shrubs and trees/climbers are known to exist in this region. Over 250 fleshy mushrooms indigenous to north–east India could be collected. Out of about 500 species of bamboo grown in hot and humid tropical regions, about 136 species are available in this region. These species generally belong to *Bambusa, Dendrocalamus, and Melocana* having large commercial potentialities. Bamboo shoots are the delicious edible parts consumed largely in the region. The important forest trees of the NE region belong to evergreen/semievergreen, moist/deciduous, subtropical, temperate, and subalpine types.

3.9 Myanmar

Myanmar (Fig. 3.10), having seven states in the Hills and seven divisions in the plains (FAO Aquastat 2011), has a total area of 676,590 km^2. Myanmar with coastlines in the Andaman sea and Bay of Bengal is having an International border with India, Bangladesh, China, Lao People's Democratic Republic, and Thailand. The highest peak in Mynamar is observed to be 5,800 m above msl. The country has a cultivable area of 18.3 million ha. The total cultivable area is almost 18.3 million ha, mainly available in the Ayeyarwady river basin, and further expansion of cultivation may be possible in Upper Mynamar in the Chin, Kachin, and Shan states.

Southwest monsoon is prevalent in Mynamar with hot and humid period in the months from May to October. Relatively dry and cool period is in northwest monsoon from December to March (FAO Aquastst 2011). Out of the mean annual rainfall of 2,341 mm, the coastal area has

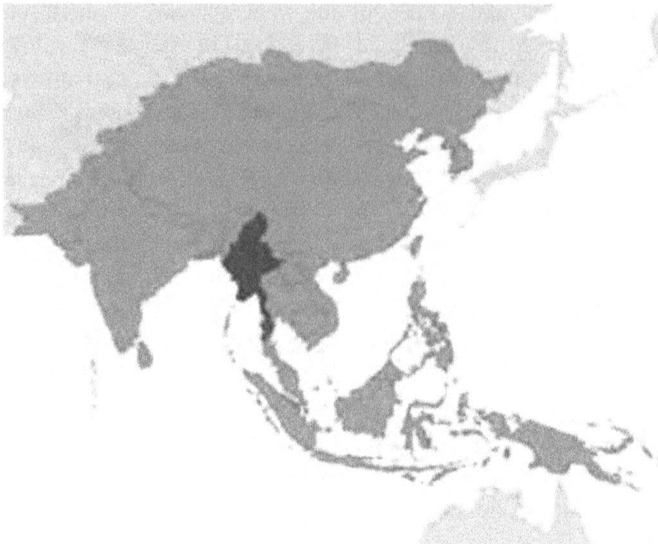

Fig. 3.10 Myanmar's position in Asia

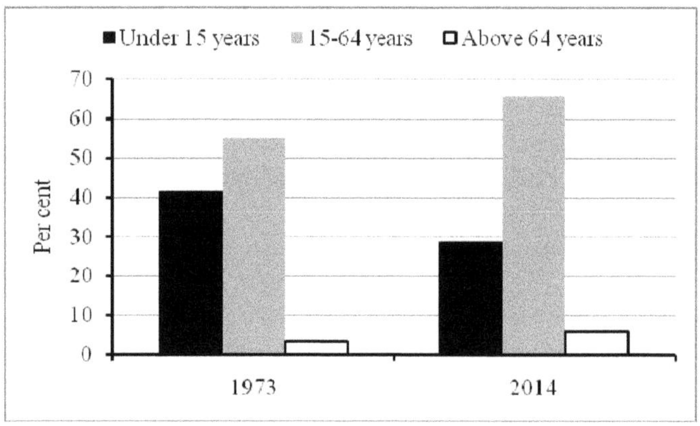

Fig. 3.11 Population of Myanmar in different age groups (%)

been receiving the maximum rainfall of 4,000–6,000 mm and the mountains/central dry zone are having low rainfall of 500–1000 mm.

3.9.1 Demographic Features

According to the census report of 2014, the population of Myanmar was 51.48 million. This has reached approximately 58.0 million in 2021 as per the annual growth rate of the country. This population constituted 48.2% males and 51.8% females. The population in different age groups of 15 years of age, 15–64 years, and 65 years was 28.6, 65.6, and 5.8%, of the total population, respectively. Trend analysis across the last three censuses shows that the proportion of these broad age groups has changed significantly since 1973 (Fig. 3.11).

3.9.2 Water Resources

The surface water resources of Myanmar are plentiful and estimated to be around 1082 km^3 draining an area of 737,800 km^2 through 10 nos. of river basin (Table 3.6). The largest catchment area is present in Ayeyarwady (Irrawaddy)–Chindwin river basin draining 58% of the area in Mynamar and the lowest river basin is in Bago

Table 3.6 Catchment areas and surface flows of major rivers in Myanmar

River basin	Catchment area (km^2)	Inflow (km^3)
1. Sittaung	34,400	41.95
2. Rivers in Rakhaing State	58,300	139.25
3. Rivers in Tanintharyi Division	40,600	130.93
4. Chindwin	115,300	141.29
5. Ayeyarwady (upper)	193,300	227.92
6. Ayeyarwady (lower)	95,000	85.80
7. Bilin River and other rivulets	8,400	31.17
8. Bago River	5,300	8.02
9. Thanlwin (in Myanmar)	158,000	257.92
10. Mekong (in Myanmar)	28,000	17.63
Total	**737,600**	**1081.88**

Source Ministry of Agriculture and Irrigation. 2011

river. Thanlwin river originated from China drains 18.4% of the territory in Mynamar and is mainly present in the Shan plateau in the east. It is noticeable that the annual inflow from rivers of other neighboring countries such as China, Thailand, and Laos People's Democratic is also observed to be of high magnitude. The urban population has access to 75% superior drinking water resources with 86% improved sanitation, but in rural areas, 69% of improved drinking water resources with 79% good sanitation are available.

In Myanmar, groundwater resources have the potential of 453.7 km^3/year. Nearly 98% of the groundwater (about 443 km^3/year) accounts for the base flow of the rivers and also surface runoff. Two natural lakes, Inle lake in Shan state and Indawgyi lake in Kachin state, cover an area of 312 km^2 and 242 km^2 and are available in Myanmar, respectively. In the constructed reservoirs, 15.46 km^3 of water from surface runoff is being stored and used to irrigate 1 million ha as reported by Naing (2005).

3.9.3 Agricultural Resources

In Myanmar, agricultural cultivation is done on about 12.4 million ha of land, which forms 18.32% of the country's total land resource of 67.657 million ha. There is some 5.7 million ha of additional arable land available but has not been put to use at present (ADB 2013). A comparison of crop production between 1990 and 2010 is shown in Table 3.7. Rice is the major crop of Myanmar and is cultivated in about 8.05 million ha or 64.3% of the total area under cultivation. The other major crops of Myanmar are beans, sesamum, and vegetables. The area under vegetables has more than doubled between 1990

Table 3.7 Growth in Food Crop Area (000' ha) and Production (000' t), 1990–2010

Crop	Area	1990	Area	2010	% increase	Over 1990
		Production		Production	Area	Production
Rice	4760	13,971	8051	33,204	69.1	137.6
Beans	433	263	2745	3029	533.9	1051.7
Sesamum seed	924	206	1570	722	69.9	250.4
Vegetables	136	163	277	371	103.6	127.6

Source Food and Agriculture Organization of the United Nations. FAOSTAT: http://faostat.fao.org/site/339/default.aspx (accessed March 2021)

and 2010. Myanmar is the seventh-ranking country in the world in the production of rice. A substantial quantity of rice is exported from Myanmar to other countries of the world. The rice production in Myanmar was 12.5, 12.2, and 12.5 million tons during 1914–15, 1915–16, and 1916–17; out of which 1.73, 1.8, and 1.6 million tons were exported during the above years, respectively. Beans and pulses have recently become major export crops of the country. Oilseeds cultivation is done on 3.3 million ha of land and is mostly concentrated in the central dry zone of the country. Vegetables and chilies are grown on 0.8 million ha of land, mainly on high lands. Other crops are maize, cotton, rubber, sugarcane, and tropical fruit crops.

There has been some change in the production of cereal in Myanmar after 2010. The production of wheat, coarse cereal, and rice is given in Table 3.8.

The irrigated area under rice in Myanmar is the maximum (1.8 M ha) followed by pulses and other annual crops (Fig. 3.12). The rice crop is mostly produced in rainfed areas.

In Myanmar, there are several factors that affect the growth of agricultural production. The draft power is one of the constraints to increased agricultural production due to the shortage of draft animals in the country. The major constraints are low prices of produce; high input costs; lack of proper facilities for marketing, credit, and irrigation; and land tenure system. In most of the cases, the land tenure gives tilling rights but not ownership of the land. Frequent natural disasters are also responsible for the low production of agricultural crops. There is a need to create a strong base for research, training, and extension services. The physical, policy, and institutional/governance constraints for agriculture development in Myanmar are, for food crops, Rural infrastructure, irrigation, electrification, Land tenure, land laws, taxes, Improved farm inputs, rural finance, extension services, and R&D (Mir 2013). In water resources above constraints are groundwater, farm level structures, soil erosion, salinity, water policy, the institutional framework for water resources management, and institutional framework for

Table 3.8 Production of cereals in Myanmar (Million tones)

Crop	2018	2019 (estimates)	Average of 5 years (2014–2018)
Paddy	26.1	25.6	26.3
Wheat	0.1	0.1	0.1
Coarse grains	2.5	2.7	2.3
Total cereals	28.7	28.5	28.8

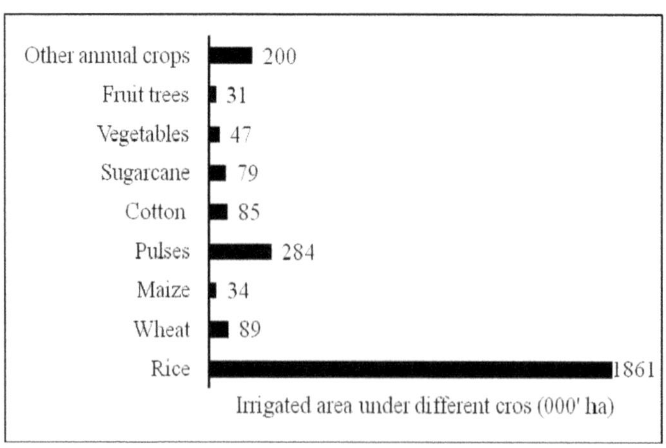

Fig. 3.12 Irrigated area under different crops in Myanmar

water resources management. These constraints with regard to the environment are Size of the country, accessibility, land degradation, Legislations and regulations, protection of forest reserves, Framework for environmental management, and human resource capacity (Mir 2013).

3.9.4 Livestock Sector

The agriculture sector (agriculture, forestry, livestock, and fishery) accounts for about 43.74% of GDP and about 66.7% of export earnings. The livestock and fisheries contributed 7.9% to the National GDP while its growth rate was at 16.5%. The livestock share of the GDP under Livestock and fisheries was 54% and the livestock share of total GDP was 4.4%, respectively. The livestock population was 11.2, 2.7, 1.4, and 1.4 million cattle, pigs, buffaloes, and goats in 2010, respectively. This population changed to 9.7, 5.8, 1.8, and 2.2 million heads of cattle, pigs, buffaloes, and goats in 2018, respectively (Myint et al. 2020). In Myanmar, being an agriculture-dependent country, draft cattle development has been given the highest priority in the livestock development program.

3.9.5 Forest Resources

Myanmar is famous for its forest resources as the country has the largest forest reserves in Southeast Asia. The forest offers a rich habitat for tropical and subtropical flora and fauna. Most of the areas on hill slopes, which are not suitable for agricultural production, have been relegated to forests. Myanmar has about 322,220 km^2 (FAO 2005) of the area under forests which constitutes 47.62% of its geographical area. The forests of Myanmar are rich habitats of biodiversity. There are 1,347 species of big trees, 741 species of small trees, 1,696 species of shrubs, 96 species of bamboo, 36 species of rattan, and 841 species of orchids found in these forests.

3.10 Nepal

Nepal (Fig. 3.13), like Afghanistan and Bhutan, is also a landlocked and mountainous country with a total land area of 147,181 km^2. The country lies between a latitude of 26° 22' to 30° 27' north and a longitude of 80° 04' to 88° 12' east. With a length of about 885 km from west to east and a width varying from 130 to 260 km from north to south, the boundary line almost gives a rectangular look. Nepal has the distinction of encompassing eight of the world's fourteen peaks, having an altitude of more than 8,000 m amsl. Nepal is also the custodian of the world's highest peak, the Mt. Everest or the Sagarmatha (Marahatta et al. 2009).

Based on ecology, Nepal has been divided into three broad ecological zones, namely *tarai*, hills, and valleys in the middle and high mountains in the north. However, physiographically, the country can be conveniently divided into five regions. These regions are the *tarai*, the Shiwalik hills, middle hills, high hills, and high mountains (Marahatta et al. 2009). About 47% of the population of Nepal lives in the *Tarai* region and 46% of the total population inhabits the hills and valleys of the middle zone. The rest 7% of the population live in the high mountains (CBS 2004). More population in the *tarai* area and middle mountains is mainly due to the reason that arable land is available in these regions for cultivation. These regions ensure livelihood security for the people if proper land management practices are followed for the cultivation of crops. The terrain and climate in high mountains are rugged and harsh, and the survival of people at extremely low temperatures and on high slopes is very difficult.

Tarai in Nepal is located in the south of the country and extends nearly 800 km from east to west and 30 to 40 km north to south with elevations ranging from 60 to 200 m above msl. Shiwalik hills are lower hills, mostly dry during the major part of the year. The elevation of the Shiwalik ranges from 200 to 1,500 m amsl. The Shiwalik hills cover about 13% of the total area

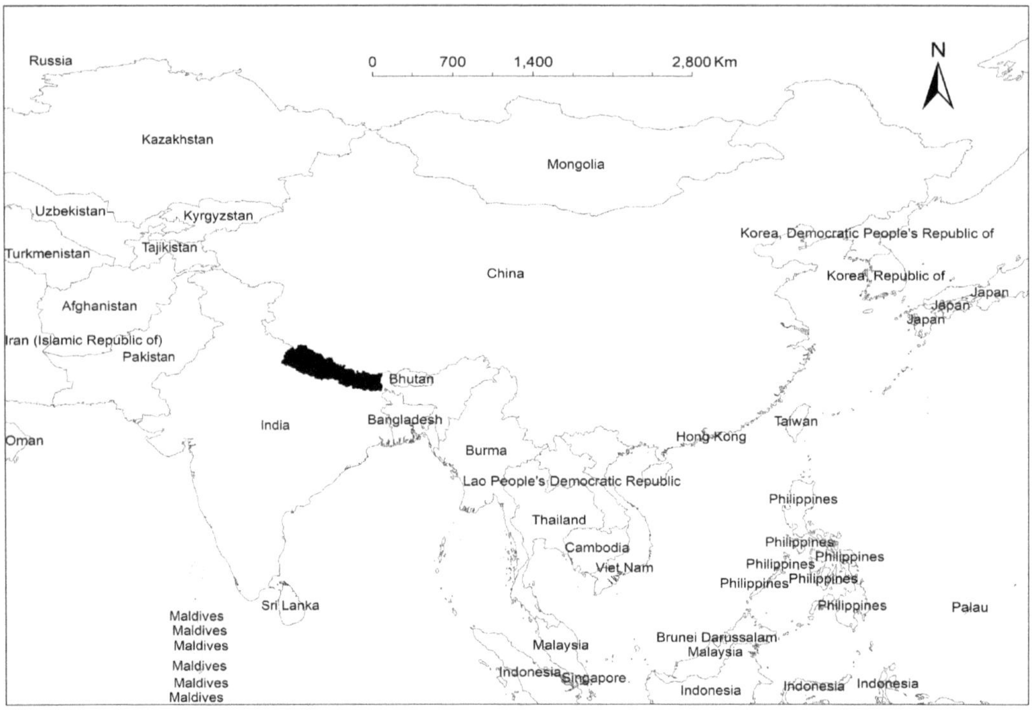

Fig. 3.13 Nepal

of the country. The elevation of the middle hills ranges from 1,000 to 2,500 m amsl and extends throughout the length of the country. The high hills region lies further north of the middle hills and has an elevation ranging from 2,200 to 4,000 m amsl. This region has a cool temperate climate. High mountains rise slowly to the north and make up the snowcapped high Himalayas. The elevation ranges from 4,000 to 8,848 m amsl.

3.10.1 Demographic Features

According to the 2011 census, the population of Nepal was about 26 million. The population is expected to have reached 29.7 million by 2021 if the present annual compound growth rate of 1.35% is taken into account. According to the 2011 census, 65.9% of the total population is literate (Hrabovszky and Miyan 1987). The current population is roughly 30 million which contributes to an increase of about 3 million people every 5 years. The population of Nepal in different age groups is given in Table 3.9.

3.10.2 Land Resources

Land is one of the major natural resources available for an agrarian country like Nepal, where more than 90% of the country's

Table 3.9 Population of Nepal in different age groups

Age group	Male	Female	Total	Percent
0–14	4.71	4.53	9.24	34.91
15–64	7.24	8.41	15.65	59.82
65+	0.70	0.71	1.41	5.27
Total	12.65	13.65	26.30	

population is dependent upon the land for food, fodder, fuel, fiber, and timber. Nepal has about 31% of arable land, though less than 18% of the total land area of Nepal is presently being cultivated (LRMP 1986). With the drastic population increase from 8.0 million in 1950 to 26.3 million in 2011, the country now needs more land to cultivate for ensuring food security. Without major changes in forest management and the use of synthetic fertilizer, soil fertility will restrict the future expansion of the cultivated area, particularly in the mountain region.

South Asian countries of the HKH region such as India, Pakistan, Bangladesh, Bhutan, and Nepal with a total land area of 441.4 mha are home to over 1760 million people (Table 3.10). In South Asia, about 19.5% of the total land area is under forests. Sri Lanka has the minimum forest cover (2.7%) and Bhutan has the highest percent of land under forests (70.2%).

3.10.3 Irrigation

Irrigation, like in other countries, is a major factor in the development of agriculture in Nepal. Irrigation is an important factor in rainfed areas for ensuring proper crop seed germination as well as during the critical stages of plant growth, for optimum crop productivity. It is the largest water use subsector, affecting the livelihood of farming communities practicing agriculture, which is the major contributor (40%) to the Gross Domestic Product (MOF 2005). The government of Nepal has given due importance to irrigation, both in its yearly and five-year plans. Irrigation potential and development in Nepal are given in Table 3.11.

It is indicated that nearly 42% of the cultivated area is irrigated in Nepal with the maximum in the Terai region and the minimum recorded in Hills. The volume of water available for groundwater extraction is estimated to be between 5.8 BCM and 12 BCM. The Nepal *Terai* is the food basket of Nepal, and if properly managed, it can serve the country on a sustainable basis.

3.10.4 Livestock

Eighty-three percent of the total population of Nepal lives in the rural areas and their principal means of livelihood are agriculture and livestock. Livestock plays a significant role in the Nepalese economy contributing around 11% to the

Table 3.10 Land use pattern (000'ha) in some HKH countries

Land use	Bangladesh	Bhutan	India	Nepal	Pakistan
Total area	14,400	4,700	328,769	14,080	79,610
Land area	13,391	4,700	297,329	13,680	77,088
Arable land	8,895	96	165,570	2,310	20,320
Forest and woodland	2,115	3,300	67,100	2,308	3,140

Source FAO (1988)

Table 3.11 Irrigation development in Nepal (WECS 2003)

Geographic region	Cultivated area ('000 ha)	Irrigable ('000 ha)	Irrigated ('000 ha)	% of cultivated area
Terai	1360	1338	889	65.3
Hills	1054	370	167	15.8
Mountain	227	60	48	21.1
Total	2642	1766	1104	41.7

Table 3.12 Length, drainage area, and estimated runoff from major rivers of Nepal

River	Length (km)	Drainage area in Nepal (km^2)	Estimated runoff from Nepal (m^3 s^{-1})
Mahakali	223	5410	247
Karnali	507	41,890	1371
Babai	190	3400	103
West Rapti	257	6500	224
Narayani	332	28,090	1409
Bagmati	163	3700	178
Sapta Koshi	513	31,940	878
Kankai	108	1330	68
Other rivers		24,921	1001
Total		147,181	5479

national Gross Domestic Product. Nepal has a cattle population of 6,530,397 (6,214,326 local and 316,071 improved) and buffaloe population of 3,174,389 (3,035,876 local and 138,513 improved). The population of sheep, goats, and pigs is 612,884, 11,225,130, and 870,197, respectively. The livestock sector has a high potential for growth in Nepal. Realizing this huge potential of the sector, the Ministry of Livestock Development (MoLD 2017) has been formed. This Ministry has been implementing various programs for the sustainable upliftment of the sector.

3.10.5 Water Resources

3.10.5.1 Status of Water Resources in Nepal

Nepal having an annual rainfall of 1530 mmhas plentiful water resources from its availability point of view with 173,000 million m^3 annual runoff. There are about 6000 rivers and streams in Nepal with a total drainage area of 194,471 km^2 (WECS, GoN 2011). These rivers originate in the Midlands or the Mahabharat Range and are fed by precipitation as well as groundwater regeneration (including springs). These rivers too are perennial but are commonly characterized by a wide seasonal fluctuation in discharge. The details about important major rivers of Nepal are mentioned in Table 3.12.

3.10.6 Snow and Glaciers

The glaciers in the High Himalayan region of Nepal are very important features that sustain water availability in the region. About 23% of Nepal's total area lies above the permanent snow line of 5000 m {Ministry of Population and Environment (MoPE 2004)}. About 10% of the total precipitation in Nepal falls as snow (UNEP 2001). There are altogether 1,025 glaciers in the Gandaki River Basin which cover an area of 2,030.15 sq. km with an estimated ice reserve of 191.39 km^3. The Karnali River Basin has 1,361 glaciers with an estimated area of 1,740.22 km^2 and an estimated ice reserve of 127.72 km^3.

3.11 Pakistan

Pakistan (Fig. 3.16) has 24°–37° N latitude and from 61° to 75° E longitude. Having the geographical area of 881,913 km^2 being 33rd largest country in the World, Pakistan is the 5th most populous country with a population of 212 million. Pakistan is composed of four provinces, viz., the Punjab, Sindh, Khyber Pakhtunkhwa, and Balochistan with a federally administered Capital territory of Islamabad. Punjab is the largest province in terms of population thus dominating other provinces in contributing to the national economy. Pakistan (Fig. 3.15) has the geographical area divided into Northern

Fig. 3.15 Pakistan

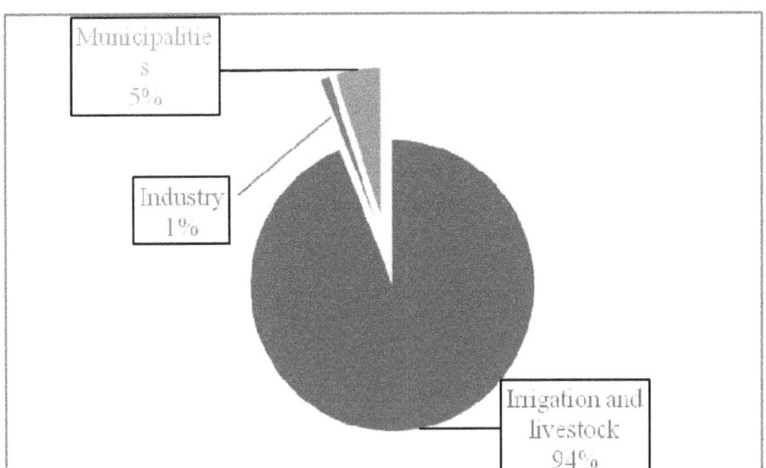

Fig. 3.16 Sector-wise water withdrawal

Highlands with Hindu Kush, Karakoram, and the Himalaya's ranges; Pothwar plateau surrounded by Indus, Jhelum, Margalla hills, and Kala Chitta range; and Indus plain in the valley of Indus river and Balochistan plateau. Pakistan is a land of plains, mountain ranges, deserts, and coastal belts. In most of the areas in Pakistan, the climate is tropical or semitropical, semi-arid or desert. In the north near mountains, the area is quite rainy and cold /frigid mountain peaks are also available in Pakistan. The Himalayas occupy the entire northern part of Pakistan extending about 320 km in area. Pakistan has more than hundred Himalayan peaks having altitudes more than

7000 m amsl. The mountain ranges of Pakistan are Karakoram, Himalayas, Hindu Kush, Hindu Raj, Safed Koh, Suleiman range, Kirthar range, and Salt range. Besides Agriculture, mining is an important industry in Pakistan with deposits of several minerals including coal, copper, gold, chromate, mineral salt, and bauxite. A variety of precious stones and minerals are also present in Pakistan.

3.11.1 Human Resources

The population of Pakistan as per the 2017 census is nearly 212 million (Table 3.13). About 75.58 people live in urban areas in Pakistan which is 36.4% of the total population of the country (PBS 2017).

3.11.2 Land Resources and Uses

Pakistan has a total geographical area of 79.61 million hectares, out of which 72% is the reported area. Out of the reported area, around 8.12 mha can be used for agriculture by taking appropriate conservation measures and providing irrigation facilities. The area under forests, the area not available for cultivation, culturable waste, and the cultivated area i; 4.02, 22.88, 8.12, and 22.05 million ha, respectively, which forms about 5.1%, 28.7%, 10.2%, and 27.7%, respectively (Ahmed 2007).

The cultivated area in Pakistan has been further classified on the basis of canal command, tubewell command, Rod-Kohi, and Barani (Table 3.14). The country has a total cultivable area of around 33.74 mha, out of which around 8.12 mha are classified as culturable waste. So, the actual area for cultivation is only 25.62 mha (Government of Pakistan 2006a, b).

3.11.3 Water Resources and Use

The main water resources in Pakistan are rainfall, groundwater, and rivers. Water in Pakistan also comes from glaciers, ponds, lakes, streams, karez, wells, and also runoff from hills/mountains. On an average, the annual rainfall in Pakistan is 494 mm accounting for nearly 393 km^3 of water in the country. Surface and groundwater are estimated to be 47 km^3/annum and 55 km^3/annum, respectively. Surface water is mainly available from the Indus river draining 65% of the territory. One of the tributaries of the Indus river in Pakistan is Panjnad being formed by successive confluence or merger of the five rivers, namely Jhelum, Chenab, Ravi, Beas, and Sutlej. Two other rivers, Mashkel and Marjen in

Table 3.13 Population of Pakistan

Administrative unit	Population	Annual growth rate	% share
Khyber Pakhtunkhwa	30,523,371	2.89%	14.69
Tribal areas	5,001,676	2.41%	2.41
Sindh	47,886,051	2.41%	23.04
Punjab	110,012,442	2.13%	52.95
Balochistan	12,344,408	3.37%	5.94
Islamabad Capital area	2,006,572	4.91%	0.97
Pakistan	207,774,520	2.40%	100
Pakistan occupied Kashmir	4,045,366	–	–
Gilgit, Baltistan	922,745	–	–
Total	212,742,631	–	–

Source Pakistan Bureau of Statistics, Ministry of National Food Security & Research (2017)

Table 3.14 Land use classification of Pakistan

Land use	Area (million ha)	%
Area under irrigation	19.02	23.9
(i) Canal irrigation	7.06	8.9
(ii) Tubewell irrigation	7.98	10.0
(iii) Wells	3.76	4.7
(iv) Other types of irrigation	0.22	0.3
Rod-Kohi (hill torrents)	2.00	2.5
Sailaba Riverine	1.25	1.6
Barani (rainfed)	3.35	4.1
Total cultivated area	25.62	32.2
Total culturable waste	8.12	10.2
Total cultivable area	33.74	42.4
Area suitable for agriculture and forestry	4.00	5.0
Forests	37.76	47.4
Total geographical area	**79.61**	**100.0**

the basin of the Karan desert in Western Pakistan, provide water to 18% of the area. In the coastal zone, the main rivers are Hob, Porali, Hingol, and Dasht.

The total actual renewable water resources in Pakistan is 247 km^3/annum and per capita water availability has reduced to 935m^3 from 5260 m^3 in the last 70 years. According to International standards, the average availability of water per person is 1800 m^3 and if it is below 1000 m^3 per capita, the place is declared as suffering from water shortage (Ahmed 2007). Pakistan currently ranks 23rd among 167 countries facing water scarcity.

Total water withdrawal (Fig. 3.16) taking into account the surface (121.8 km^3) and groundwater withdrawal (61.6 km^3) is 184 km^3/year, and 74% of total actual renewable water resources is being utilized for drinking, irrigation, livestock, municipalities, and Industry. Pakistan has a no. of Dams created for hydroelectricity and water reservoir and Dam capacity is estimated to be around 24 km^3/year. Due to variability in seasonal rainfall, the river flows also undergo a variation during the Rabi and Kharif seasons.

Surface Water Resources

Western rivers, such as Indus, Jehlum, and Chenab, have an annual flow of 173 billion m^3 of surface water in an average year during the pre-storage period of 1937–67 (Table 3.15). Maximum flow has been recorded during *Kharif* (May_October), which was more than five times the flow in the *Rabi* season (November–April). Variability in flows of the eastern rivers, such as Ravi, Beas, and Sutlej, was even higher than the western rivers. Before the construction of Mangla and Tarbela storages, the eastern rivers have contributed about 26 billion m^3 of water to the Indus River system in an average year of which about 84% was in the *Kharif* season.

Table 3.15 Water and storage capacity in Pakistan

Parameter	Pakistan	Parameter	Pakistan
River flows (billion m^3)	181.4[1]	Storage Capacity (billion m^3)	22.95[5]
River flows (billion m^3)	195.29[2]	Storage Availability per Capita (m^3/person)	144
Geographical area (million ha)	79.61	Storage of water (m^3/ha)	288
Water availability (m^3/person/annum)	1223		

Source: Government of Pakistan (2008); World Bank (2005)

Surface Water Quality

Excellent quality water is available in the Indus river and its tributaries in Pakistan. It contains water having total dissolved solids (TDS) considered safe for irrigated agriculture, domestic, and industrial uses. There is a variation in TDS in the upper reaches during the period of high flow to low flow. The deterioration of water quality is observed downstream and TDS in the lower reaches of the river Indus becomes well above the permissible limit.

Groundwater Resources and quality

Groundwater contributed around 11.3 billion m^3 of water for irrigation in Pakistan during the pre-storage period, which was 10% of the total water available for agriculture (Ahmed et al. (2001). The estimated annual groundwater recharge in the Indus Basin is 67.9 billion m^3, out of which 44 billion m^3 occurs in areas of usable groundwater (Zuberi and Sufi 1992). The total annual groundwater potential in Pakistan is estimated at 67.9 billion m^3. About 79% of the area in Punjab and 28% of the area in Sindh have fresh groundwater suitable for agriculture (Afzal 1999; Bhutta 1999). It is a fact that by now, almost all easily exploitable surface and groundwater resources have already been tapped, and the future demand for water for agriculture or other uses will have to be met largely through rainwater conservation. The highest salinity > 3000 ppm is present in 4.28 mha water followed by 1000–3000 ppm salinity in 1.81 mha and less than 1000 ppm salinity in 5.75 mha. In addition to TDS, there are quality concerns in terms of sodium adsorption ratio (SAR) and residual sodium carbonate (RSC).

3.11.4 Livestock Resources

Livestock is a subsector of Pakistan agriculture that contributes approximately 56% of value addition in agriculture and nearly 11% to the gross domestic product (GDP). Livestock production makes a major contribution to agriculture value-added services. A large livestock population, well adapted to the local environmental conditions, is found in Pakistan. The current population (Table 3.16) of farm animals in Pakistan consists of 23.34 million buffaloes, 22.42 million cattle, 24.24 million sheep, 49.14 million goats, and 0.77 million camels. Pakistani buffaloes belong to two breeds namely, Nili-Ravi and Kundi. Nili-Ravi are considered to be the best dairy buffalo breeds in the world. Afzal and Naqvi (2004) reported that there are ten breeds of cattle found in Pakistan. Cattle breeds of Pakistan are Sahiwal, Red Sindhi, Cholistani, Dhanni, Tharparker, Bhagnari, Djal, Lohani, Rojhan, and Kankrej. There are 30 local breeds of sheep in the country. Like other countries of the HKH region, the rearing of livestock by every farming household in Pakistan is an important component of farming systems. It serves the purposes of plowing the land by draft animals, milk and dairy products, and organic manure to the fields to raise the fertility level of soils. The dung of the animals is also used for making dung cakes to be used as fuel. Livestock-production system is mostly mixed crop-livestock farming.

Table 3.16 Livestock population in Pakistan (Million)

Province	Cattle	Buffaloes	Sheep	Goat	Camel
NWFP	4.81	1.46	3.22	8.59	0.06
Punjab	9.67	14.20	5.89	18.25	0.14
Sindh	6.49	7.41	4.42	11.68	0.23
Balochistan	1.44	0.26	10.71	10.62	0.33
Pakistan	22.42	23.33	24.24	49.14	0.76

Livestock is a subsector of Pakistan Agriculture that contributes approximately 56% of value addition in agriculture and nearly 11% to the gross domestic product (GDP). Goat farming in Pakistan as the goat products have an excellent demand in the International market is being considered as one of the most lucrative and profitable businesses in Pakistan. In addition to goats, rearing of beef cattle is also a profitable activity. The most considerable constraints in Pakistan regarding livestock are nutrition, animal health, animal productivity/genetic makeup of the animal, finally finance provision to livestock farmers, and finally marketing.

Seasonal grazing in the mountain areas during the year follows a complicated structure, depending on the elevation (Schmidt 2000; Postigo et al 2008; Kreutzmann et al 2011; Kreutzmann 2012; Turner and Schlecht 2019). Ahmad et al. (2021) reported that the mountain pastoralists in Lotkuh Valley (Chitral, Pakistan) follow the seasonal use of their rangelands. During spring, sheep and goats graze on low-lying pastures and meadows; in summer, livestock are taken to a 3000 m elevational range to graze; and in autumn, the livestock are brought down from higher mountains to feed on field stubbles before the winter crops are planted.

References

ADB (Asian Development Bank) (2013) Myanmar: agriculture, natural resources, and environment initial sector assessment, strategy, and road map. 6 ADB Avenue, Mandaluyong City, 1550 Metro Manila, Philippines

Afzal M, Naqvi AN (2004) Livestock resources of Pakistan: present status and future trends. Sci Vis 9: 1–14. Pakistan Agricultural Research Council, Islamabad

Afzal M (1999) Water for agriculture. Paper for Water Vision, Pakistan

Ahmed S (2007) Land and water resources of Pakistan—a critical assessment. Pak Dev Rev 46:911–937

Ahmed S, Mulk S, Mohammad A (2001) Groundwater management in Pakistan. Country report. Prepared for the Global Water Partnership and Pakistan Water Partnership

Ahmad Z, Postigo JC, Rahman F, Dittmann A (2021) Mountain pastoralism in the Eastern Hindu Kush: the case of Lotkuh Valley, Pakistan. Mt Res Dev 41(4): R16-R28 (2021). https://doi.org/10.1659/MRD-JOURNAL-D-21-00007.1

Anonymous (2002) Basic statistics of North Eastern region, evaluation & monitoring cell. North Eastern Council Secretariat, Shillong, Meghalaya, India

Bahadur J (1999) The Himalayan environment (Dash SK, Bahadur J (ed)). New Age International, Pvt Ltd., New Delhi, pp 258–267

Bajracharya SR, Shrestha BR (2011). The status of glaciers in the Hindu Kush-Himalayan region. International Centre for Integrated Mountain Development (ICIMOD)

Barnett TP, Adam JC, Lettenmaier DP (2005) Potential impacts of a warming climate on water availability in snow-dominated regions (7066):303–309

Bhutta MN (1999) Vision on water for food and agriculture: Pakistan's Perspective. regional South Asia meeting on water for food and agriculture development. New Delhi

Biskaborn BK, Smith SL, Noetzli J, Matthes H, Vieira G, Streletskiy DA et al (2019) Permafrost is warming at a global scale. Nat Commun 10:264

Cao B, Zhang T, Peng X, Mu C, Wang Q, Zheng L, Wang K, Zhong X (2018) Thermal characteristics and recent changes of permafrost in the upper reaches of the Heihe River Basin, Western China. J Geophys Res Atmos 123:7935–7949

Chapagain PS, Ghimire M, Shrestha S (2017) Status of natural springs in the Melamchi region of the Nepal Himalayas in the context of climate change. Environ Dev Sustain 21:263–280

CBS (Central Bureau of Statistics) (2004) Stastical year book of Nepal. Kathmandu, Nepal

Chen Y, Fischer G (1998) A new digital georeferenced data base of grassland in China. IIASA Interim report. IIASA, Laxenburg, Austria: IR-98-062

Chhophel K, Chhophel GK (2006) The state of water resources in Bhutan. Hydromet services, National Environment Commission, Thimpu, Bhutan.

Dainese M, Martin EA, Aizen MA, Albrecht M, Bartomeus I, Bommarco R et al (2019) A global synthesis reveals biodiversity-mediated benefits for crop production. Sci Adv 5(10). https://doi.org/10.1126/sciadv.aax0121

Daniel D, Anandhi A, Sen S (2021) Conceptual model for the vulnerability assessment of springs in the Indian Himalayas. Climate 2021(9):121. https://doi.org/10.3390/cli9080121

Datta M, Laskar S (1992) Different reaction of local rice races to iron in Tripura. Oryza 29:388–391

Dennis P, Sheehy DP, Miller D, Johnson DA (2006) Transformation of traditional pastoral livestock systems on the Tibetan steppe. Sécheresse 17(1–2):142–151

FAO (1988) FAO publication year book, vol 42. Rome, Italy

FAO (2005) Global forest resources assessment 2005, progress towards Sustainable Forest Development,

FAO forestry paper 147, Country Report - Myanmar, WP 107
FAO (2008) Afghanistan-National livestock census, 2002–03, Food and Agriculture Organization of the United Nations, Rome, Italy
FAO (2010) FAOSTAT database. http://faostat.fao.org/
FAO (2012) Afghanistan: irrigation in Central Asia in figures - AQUASTAT survey–2012, FAO, Rome, Italy
FAO Aquastat (2011) Irrigation in Southern and Eastern Asia in figures – AQUASTAT Survey – 2011, FAO, Rome, Italy
FAO and Intergovernmental Technical Panel on Soils [ITPS] 2015 Status of the world's soil resources: main report FAO and ITPS Rome
Favre R, Kamal GM (2004) Watershed atlas of Afghanistan, FIRST edn. Working document for planners, Kabul, Afghanistan
Gerlitz JY, Banerjee S, Brooks N, Hunzai K (2014) An approach to measure vulnerability and adaptation to climate change in the Hindu Kush Himalayas, In: Handbook of climate change adaptation. Springer, Berlin, Heidelberg, Filho WL, pp. 1–22. https://doi.org/10.1007/978-3-642-40455-9_99-1
Giri N, Singh OP (2013) Urban growth and water quality in Thimpu. J Urban Environ Eng 7(1):82–95
Government of India (GoI) (2010) National mission for sustaining the Himalayan ecosystem under National action plan on climate change. Draft report, Department of Science and Technology, Government of India, 51p
GoP (Government of Pakistan) (2006a) Agriculture statistics of Pakistan. Economic Wing of the Ministry of Food, Agriculture and Livestock, Government of Pakistan, Islamabad, Pakistan.
GoP (Government of Pakistan) (2006b) Draft National water policy (Final Draft). Office of the Chief Engineering Advisor/Chairman Federal Flood Commission, Ministry of Water Resources, Government of Pakistan.
Government of Pakistan (GoP) (2008) Agriculture statistics of Pakistan. Economic Wing of the Ministry of Food, Agriculture and Livestock, Government of Pakistan
Guo D, Yang M, Wang H (2011) Characteristics of land surface heat and water exchange under different soil freeze/thaw conditions over the central Tibetan Plateau. Hydrol Process 25:2531–2541
Hrabovszky JP, Miyan K (1987) Population growth and land use in Nepal. "the great turn about". Mt Res Dev 7(3):264–270.
Jones JA, Liu C, Wu MK, Kung HT (2012) Regional hydrological response to climate change. Springer Science & Business Media, 360p
Karki M, Sharma S, Mahat TJ, Tuladhar A, Aksha S (2012) Sustainable mountain development in the Hindu Kush – Himalaya International Centre for Integrated Mountain Development (ICIMOD), Kathmandu, Nepal
Khan AH (1993) Farakka barrage: its impact on Bangladesh – an overview. Dhaka, Bangladesh, 13pp
Kreutzmann H (2012) Pastoral practices in transition: animal husbandry in high Asian contexts. In: Kreutzmann H (ed) Pastoral practices in High Asia agency of development effected by modernisation, resettlement and transformation. Springer, Berlin, pp 1–30
Kreutzmann H, Abdulalishoev K, Lu Z, Richter J (2011) Pastoralism and rangeland management in mountain areas in the context of climate and global change: regional workshop in Khorog and Kashgar, 14–21 July 2010. Deutsche Gesellschaft für Internationale Zusammenarbeit, Bonn, Germany
Kopittke PM, Menzies NW, Wang P, McKenna BA, Lombi E (2019) Soil and the intensification of agriculture for global food security. Environ Int 132. https://doi.org/10.1016/j.envint.2019.105078
Kumar P, Pradeep K, Sharma PK, Kumar P, Sharma M, Butail NP (2021) Agricultural sustainability in Indian Himalayan region: constraints and potentials. Indian J Ecol 48(3):649–662
Liu X, Chen B (2000) Climatic warning in the Tibetan plateau during recent decades. International J. of Climatology, 201: 729–742.
LRMP (1986) Land capability report and maps. Kathmandu: land resources mapping project/HMG (His Majesty's Government, Nepal). Kenting Earth Science, Ottawa, Canada
Lutz AF, Immerzeel WW (2013) Water availability analysis for the Upper Indus, Ganges, Brahmaputra, Salween and Mekong River Basins–final report. Future water report 127, Wageningen. http://www.futurewater.nl/uk/projects/water-availability-analysis-for-the-upper-indus-ganges-and-brahmaputra-riverbasins/
Lutz AF, Immerzeel WW, Shrestha AB, Bierkens MFP (2014) Consistent increase in High Asia's runoff due to increasing glacier melt and precipitation. Nat Clim Change 4:587–592. https://doi.org/10.1038/nclimate2237
Ma R (1996) Population and society in Tibet. Tongxin Press, Beijing (in Chinese)
Mahamuni K, Kulkarni H (2012) Groundwater resources and spring hydrogeology in South Sikkim with special reference to climate change. In: Arrawatia ML, Tambe S (ed) Climate change in Sikkim-patterns, impacts and initiatives. Government of Sikkim: Gangtok, India, pp 261–274. ISBN 978-81-920437-0-9
Mall RK, Bhatla R, Pandey SK (2007) Water resources in India and impact of climate change. Jalvigyan Sameeksha 22:127–147
Marahatta S, Dangoi BS, Gurung GB (2009) Spatial and temporal variability of climate change over Nepal (1976–2005). Practical Action Nepal Office, ISBN: 978-9937-8135-2-5
Ministry of Agriculture and Irrigation (2011) Government of Myanmar, Yagon, Myanmar

Ministry of Environment and Forest (2005) Economic survey–Fiscal year 2004–05. Ministry of Finance. Government of Nepal, Kathmandu, p 288

Ministry of Environment and population (MoPE) (2004) Initial National communication to the conference of parties of the United Nations framework on climate change. MoPE, Kathmandu, Nepal, p 153

Ministry of Livestock Development (MoLD) (2017) Livestock statistics of Nepal Government of Nepal, planning, monitoring and evaluation division, Singhadurbar, Kathmandu, Nepal

Mir JH, Blog ADB (2013). http//blog.adb.org/blog/Myanmar-must-face-environmentalcatch-challenges-stimulate-growth

MoA (1997) Agricultural census data. Ministry of Agriculture and Forests, Royal Government of Bhutan, Thimphu

MoAF, RGB (2016) RNR agricultural census data. Ministry of Agriculture and Forests, Royal Government of Bhutan, Thimphu

MoAF (2011) RNR agricultural census data. Ministry of Agriculture and Forests, Royal Government of Bhutan, Thimphu

MoEF, GoI (2010) Climate change and India: towards preparation of a comprehensive climate change assessment. Ministry of Environment & Forests, Government of India

MoWR (2010) The National Water Mission under National Action plan on climate change. Ministry of Water resources, Government of India

Mu C, Schuster PF, Abbott BW, Kang S, Guo J, Sun S, Wu Q, Zhang T (2020) Permafrost degradation enhances the risk of mercury release on Qinghai-Tibetan Plateau. Sci Total Environ 708:135127

Myint HT,WinYT, Maclean M, Thu A (2020) National livestock baseline survey of Myanmar. FAO, United Nations

Naing MM (2005) Paddy field irrigation systems in Myanmar. ftp://ftp.fao.org/docrep/fao/010/ai408e/ai408e01.pdf

National Livestock Census (2003) Interim report. Food and Agriculture Organization of the United Nations

Negi GCS, Joshi V (2004) Rainfall and spring discharge patterns in two small drainage catchments in the Western Himalayan Mountains, India. Environmentalist 24:19–28

Ni J (2002) Carbon storage in grasslands of China. J Arid Environ 50:205–218

NITI Aayog (2018) Report of working group I Inventory and revival of springs in the Himalayas for water security. NITI Aayog, New Delhi, India

Nitzbon J, Westermann S, Langer M, Martin LCP, Strauss J, Laboor S, Boike J (2020) Fast response of cold ice-rich permafrost in northeast Siberia to a warming climate. Nat Commun 11:2201

Noel S, Mikulcak F, Etter H, Stewart N (2015) Economics of land degradation initiative: report for policy and decision makers. ELD Initiative and Deutsche Gesellschaft für Internationale Zusammenarbeit (GIZ) GmbH, Bonn

Norbu T (2011) Climate change in Tibet and Asia. Vivekananda International Foundation (VIF), New Delhi, 06–07 September 2011

NSB (2014) Statistical year book. www.nsb.gov.bt/publication/files/pub9ot4338yv.pdf. Accessed 22 June 2018

Pakistan Bureau of Statistics (PBS) (2017) Pakistan year book 2017. Government of Pakistan, Islamabad

Pal RN, Parthasarathy VA (2001) Horticultural research in North-Eastern region-a reflection for future. In: Souvenir on approaches for increasing agricultural productivity in hills and mountain ecosystems. ICAR Research. Complex for NEH Region, 18–20 October 2001

Peng X, Zhang T, Frauenfeld OW, Wang S, Qiao L, Du R, Mu C (2020) Northern Hemisphere greening in association with warming permafrost. J Geophys Res Biogeosciences 125:125

Population and Housing Census of Bhutan (PHCB) (2005) Royal Government of Bhutan, Thimpu, Bhutan

Postigo JC, Young KR, Crews KA (2008) Change and continuity in a pastoralist community in the high Peruvian Andes. Hum Ecol 36:535–551. https://doi.org/10.1007/s10745-008-9186-1

Prakash A (2020) Retreating glaciers and water flows in the Himalayas: implications for governance. ORF Issue Brief No. 400, September 2020, Observer Research Foundation

Prăvălie R, Patriche C, Borrelli P, Panagos P, Roşca B, Dumitraşcu M et al (2021) Arable lands under the pressure of multiple land degradation processes. A global perspective. Environ Res 194:110697. https://doi.org/10.1016/j.envres.2020.110697

Qi Y, Li S, Ran Y, Wang H, Wu J, Lian X, Luo D (2021) Mapping frozen ground in the Qilian Mountains in 2004–2019 using Google earth engine cloud computing. Remote Sens 13(1):149. https://doi.org/10.3390/rs13010149

Ramakrishnan S (1998) Groundwater. Scitech publications Pvt. Ltd., G N Naidu Plaza, No.465/155, Velacherry Main Road, East Tambaram, Chennai-600059

Ran Y, Li X, Cheng G (2018) Climate warming over the past half century has led to thermal degradation of permafrost on the Qinghai-Tibet Plateau. Cryosphere 12:595–608

Rout B (2008) Water management, livestock and the opium economy. How the water flows: a typology of irrigation systems in Afghanistan. Afghanistan research and evaluation unit issue paper Series.

Sati VP (2006a) Forest resource management in mountain regions: a case for the Pindar Basin of Uttaranchal Himalaya. Lyonia: J Ecol Appl 11:75–84 www.lyonia.org

Sati VP (2006b) Role of off-season vegetables in the sustainable livelihood of hills' people—A case for the Pindar Basin (Uttaranchal Himalaya), published in the proceedings of international symposium towards sustainable livelihoods and ecosystems in mountainous regions, 7–9 March 2006b, Chiang Mai, Thailand

Web Link. https://www.uni-hohenheim.de/sfb564/uplands2006/contributions/PS-A-Vishwambhar-full.pdf

Sati VP, Singh RB (2010) Prospects of livestock farming in Uttrakhand Himalayas, India. Indian J Livest Sci 1 (1):9–16

Schmidt M (2000) Pastoral system in Shigar/Baltistan: communal herding management and pasturage rights. In: Ehlers E, Kreutzmann H (eds) High mountain pastoralism in Northern Pakistan. Franz Steiner Verlag, Stuttgart, pp 121–150

Schickhoff U, Bobrowski M, Böhner J, Burzle B, Chaudhary RP, Gerlitz L, Lange J, Muller M, Scholten T, Schwab N (2016) Climate change and treeline dynamics in the Himalaya. In: Climate change, glacier response, and vegetation dynamics in the Himalayas. Springer Cham, pp 271–306. https://doi.org/10.1007/978-3-319-28977-9_15

Scott CA, Zhang F, Mukherji A, Immerzeel W, Mustafa D, Bharati L (2019) Water in the Hindu Kush Himalaya. In: The Hindu Kush Himalaya assessment. Springer Nature Switzerland AG, Cham, Switzerland, pp 257–299

Sharma KP (1993) Role of rainwater in major river systems of Nepal. In: Young GJ (ed) International symposium snow and glacier hydrology. International Association Scientific Hydrology, Kathmandu, pp 113–122

Shastry SVS, Sharma SD, John VT, Krishnaiah K (1971) New sources of resistance to pests and diseases in Assam rice collections. Int Rice Commun Newsl 20:1–16

Sheehy D (2001) The rangelands, land degradation and black beach: A review of research reports and discussions. In: van Wageningen N, Sa Wenjun (eds) The living plateau: changing lives of herders in Qinghai. International Centre for Integrated Mountain Development (ICIMOD), Kathmandu

Shrestha AB, Agrawal NK, Alfthan B, Bajracharya SR, Maréchal J, van Oort B (eds) (2015) TheHimalayan climate and water atlas: impact of climate change on water resources in five of Asia's major river basins. ICIMOD, GRID-Arendal and CICERO

Stiegler C, Johansson M, Christensen TR, Mastepanov M, Lindroth A (2016) Tundra permafrost thaw causes significant shifts in energy partitioning. Tellus B Chem Phys Meteorol 68:1

Tambe S, Kharel G, Arrawatia ML, Kulkarni H, Mahamuni K, Ganeriwala AK (2012) Reviving dying springs: climate change adaptation experiments from the Sikkim Himalaya. Mt Res Dev 32:62–72

Tashi N, Xugang L, Shunxiang Y, Judson G (2005) Survey of the mineral status of livestock in the Tibet Autonomous Region of China. ACIAR Working Paper No. 59, Australian Centre for International Agricultural Research, GPO Box 1571, Canberra ACT 2601, Australia

Tiwari P (2008) Land use changes in Himalaya and their impacts on environment, society and economy: a study of the lake region in Kumaon Himalaya, India. Adv Atmos Sci 25:1029–1042

Tiwari PC, Joshi B (2011) Environmental changes and sustainable development of water resources in the Himalayan headwaters of India. Water Resour Manag 26:883–907

Turner MD, Schlecht E (2019) Livestock mobility in sub-Saharan Africa: a critical review. Pastor: Res Policy Pract 9:13. https://doi.org/10.1186/s13570-019-0150-z .

Uhl VW, Tahiri MQ (2003) Afghanistan: an overview of groundwater resources and challenges. Uhl, Baron, Rana Associates, Inc., Washington Crossing, PA, USA, and Basic Afghanistan Services, Kabul, Afghanistan

UNEP (2001) Nepal: State of the Environment 2001. United Nations Environment Programme and International Centre for Integrated Mountain Development, Kathmandu, Nepal, p 211

United Nations Environment Programme (2021) Becoming #GenerationRestoration: ecosystem restoration for people, nature and climate. Nairobi

Upadhayay DS (1995) Cold climate hydrometeorology. New age international pvt ltd., New Delhi

Valdiya KS, Bartarya SK (1991) Hydrogeological studies of springs in the catchment of the Gaula river, Kumaun Lesser Himalaya, India. Mt Res Dev 11:239–258

Vaidya R and Sharma E (eds) (2014) Research insights on climate and water in the Hindu-Kush Himalayas. ICIMOD, Kathmandu

Vohra CP (1981) The climate of the Himalayas, the Himalaya: aspects of change (Lall JS (ed)). Oxford University Press, New Delhi, India, pp 138–151

Wang S, Sheng Y, Wu J, Li J, Chen J (2015) The characteristics and changing tendency of permafrost in the source regions of the Datong River, Qilian Mountains. J Glaciol Geocryol 37:27–37

Wang T, Yang D, Yang Y, Piao S, Li X, Cheng G, Fu B (2020) Permafrost thawing puts the frozen carbon at risk over the Tibetan Plateau. Sci Adv 6:eaaz3513

Way RG, Lewkowicz AG, Zhang Y (2018) Characteristics and fate of isolated permafrost patches in coastal Labrador, Canada. Cryosphere 12:2667–2688

WECS, Government of Nepal (2011) Water resources of Nepal in the context of climate change (Aryal RS, Rajkarnikar G (ed)). Water and Energy Commission Secretariat, Government of Nepal, Singha Durbar, Kathmandu, Nepal

WECS (2003) Water resource strategy Nepal. Water and Energy Commission Secretariat Kathmandu, Nepal

World Bank (2005) Pakistan economy getting dry. Pakistan Water Strategy. World Bank

Wu Q, Hou Y, Yun H, Liu Y (2015) Changes in active-layer thickness and near-surface permafrost between 2002 and 2012 in alpine ecosystems, Qinghai-Xizang (Tibet) Plateau, China. Glob Planet Change 124:149–155

Yangchen U, Thinley U, Wallentin G (2015) Land use land cover change in Bhutan: 2000–2013. Occasional Publication No. 1, Centre for Rural Development Studies, College of Natural Resources, Lobesa, Royal University of Bhutan

Zhang Z, Wang M, Wu Z, Liu X (2019) Permafrost deformation monitoring along the Qinghai-Tibet Plateau engineering corridor using InSAR observations with multi-sensor SAR datasets from 1997–2018. Sensors 19:5306

Zuberi FA, Sufi AB (1992) State of art of groundwater exploration, exploitation, management and legislation. IWASRI, WAPDA, Lahore, 26p

4 Geology of Himalayan Soils

Abstract

The Himalayan mountain range is sub-divided into four principal tectonic zones, from south to north. These are the Sub-Himalaya, the Lesser Himalaya, the Higher Himalayan Crystallines, and the Tethyan Himalaya. The Sub-Himalaya, also known as the Shiwalik Range (250–800 m high), rises above the Indo-Gangetic Plains along the Main Frontal Thrust. The Higher Himalayas, also known as the Central Crystalline zone, are comprised of ductile deformed metamorphic rocks and mark the axis of orogenic uplift. Mica schist, quartzite, paragneiss, migmatite, and leucogranite bodies characterize this uppermost Himalayan zone. Corresponding mineral assemblages are dominated by biotite to sillimanite, representing greenschist to amphibolite facies deformation. The Higher Himalayan Crystalline rocks have imbricated thrust sheets, with the grade of metamorphism increasing from the Chail Group to the Vaikrita Group. The Chail Group consists of a metamorphosed sequence of greenschist facies including phyllites, phyllitic quartzite, psammitic schists, orthoquartzites, arkose, chlorite schist limestones, and metabasic rocks amphibolites. The Sub-Himalaya range consists predominantly of Tertiary and Quaternary sediments and is bounded to the north by northward dipping Main Boundary Thrust, separating it from the overlying Lesser Himalaya. The Lesser Himalayan Range, in general, is quite rugged and higher than 2500 m, however, in the Kashmir Valley (Northwestern Himalaya), the Pir Panjal Range rises to above 3500 m. The Lesser Himalayas consists of Precambrian and Cambrian sequences of the Damtha, Tejam, Jaunsar–Garhwal, and Mussoorie Group in Garhwal region of the western Himalaya. The Mussoorie Group is represented by a persistent horizon of conglomerate intercalated with graywackes and siltstones at the base, which pass into carbonaceous slates and varicolored limestone. This succession is followed stratigraphically by carbonate-limestone, marl, slate, dolomite horizon in the middle and shale, conglomerate interbedded with phosphatic carbonaceous–pyritous slates, black limestone, and pyritous and felspathic quartzite in the upper portion. The Tethyan Himalayas consist of thick, 10–17 km, marine sediments that were deposited on the continental shelf and slope of the Indian continent. The Shiwalik zone consists of clastic sediments that were produced by the uplift and subsequent erosion of the Himalayas and deposited by rivers. These rocks have been folded and faulted to produce the Shiwalik Hills that are at the foot of the great mountains. Sub-Himalayan rocks have been overthrust by the Lesser Himalayas along the Main Boundary Thrust Fault. It not only

U. C. Sharma et al., *Soils in the Hindu Kush Himalayas*, Geography of the Physical Environment,
https://doi.org/10.1007/978-3-031-11458-8_4

shows geological divisions within the mountain belt, but also structures and geologic relationships between rock types and structures.

Keywords

Duars (Piedmonts) · Geological features of HKH · Geological mapping · Quaternary deposition · Tectonic framework

4.1 General Description

The structures and lithologies in the HKH Region were formed by the closing and subduction of the Tethyan Ocean during the Paleozoic period, followed by the collision of continental landmasses. Multiplex nature distinguishes the mountains and surrounding regions, which exhibit a variety of deformed and collision-produced lithologies and encapsulates numerous phases of tectonic and deformational phenomena. The Himalayas are divided into six lithotectonic zones that run parallel to one other. The Trans-Himalayan batholith, Indus–Tsangpo suture zone, Tethyan (Tibetan) Himalaya, Higher or Greater Himalaya, Lesser (Lower) Himalaya, and Sub-Himalaya are the zones that make up these ranges (Fig. 4.1). Within these zones, several tectonic settings have been discovered. An island-arc margin in the western Himalayas and an Andean-type border in the central-to-eastern Himalayan areas resulted from a massive collision of plates around 45 million years ago (Windley 1995). The western trans-Himalayas are an island arc that originated on the Neo-Tethys' northern flank and became stuck between Asia and India.

A collision between the Indian plate and the Kohistan–Ladakh arc in the western Himalayas and the Tibetan Lhasa block in the east generated the Indus–Tsangpo Suture Zone (ITZS) (Windley 1995). Having spread for more than 2000 km, there are a variety of rock types, viz., Glaucophane schists, Olistoliths, Ophiolites, Limestones, red Sandstones, Cherts, Serpentinites, and Dunites with Mafic to felsic lavas thus indicating a complex collision of plate boundaries. The Tethyan (Tibetan) Himalayas, located south of the ITSZ, consist of marine sediments, 10–17 km thick. It is striking to note that the sediments are not metamorphosed having a clear deposition of fossils occurring in the east–west zone within strata. The presence of fauna could suggest a booming of life in this area of the Tethyan zone which had remained stationary between mid-Proterozoic and Eocene times. The growth of life with diversity could be due to the intervallic formation of land barriers (Sorkhabi 1999). The Trans-Himalaya zone is a plutonic complex that runs in a straight line. This is partly covered by forearc rocks and continental molasse

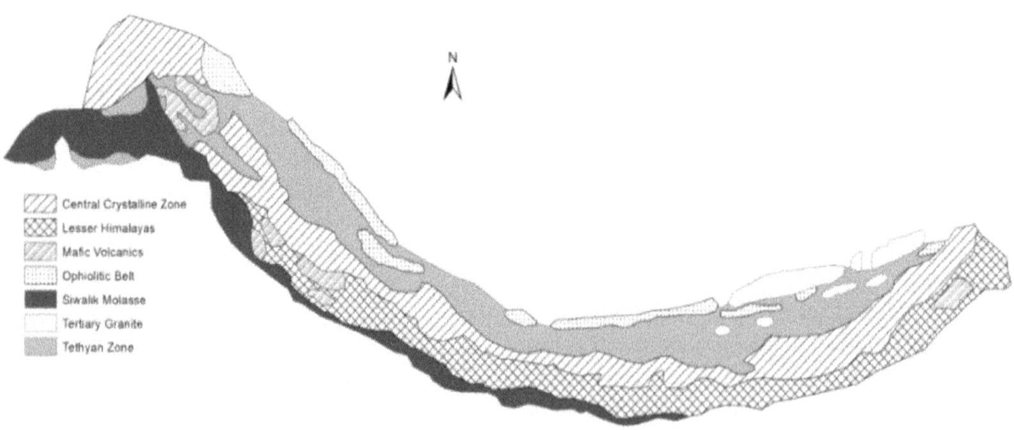

Fig. 4.1 Himalayan mountain ranges showing major tectonic sub-divisions

sedimentary rocks. These assemblages are the result of magmatic rocks being uplifted and then eroded. These magmas are assumed to have formed as a result of the partial melting of a subducting Neo-Tethyan slab beneath the Asian plate (Sorkhabi 1999). The western trans-Himalayas are an island arc that originated on the Neo-Tethys' northern flank and became stuck between Asia and India. Two stages of deformation were possible as a result of this. As the Tethys began to subduct beneath Asia, the first collision was with Eurasia, followed by a collision with India (Windley 1995). The cross-sectional view of the Himalayan zone was modeled by DeCelles et al. (1998).

4.2 Geology of Himalayas

Two important stages in the evolution of Himalayas are (i) India collided with Eurasia at some point of time between 65 million years ago (Ma) and 45 Ma, giving birth to the Indus Suture. There was an increase in pressure and temperature which resulted in the rocks recrystallizing, forming minerals, reducing friction, and helping further thrust. This happened at about 20 Ma. The present height of the Himalayas was, however, attained between 2.5 Ma and 1.8 Ma (Powell and Conaghan 1973). At this point in time, uplift of the Himalayas affected the climate circulation, and the emergence of the monsoon era in the region (Zhisheng et al. 2001). As a result of the uplift, the region began to experience high-altitude glaciation. The uplift of Himalayan ranges caused the ceasing of monsoonal moisture supply in Ladakh, resulting in glacial retreat (Owen et al. 2006) and low rains.

Bilham et al. (2001) were of the view that the earthquakes presented a major geo-hazard in the Himalayas, though the repeat time was more. The majority of India's and Eurasia's convergence is caused by earthquakes caused by fault slides in temporally localized episodes. The earthquakes of 7.9 and above on Richter scale can be of significant magnitude as was observed in the case of Wenchuan earthquake in the Longman Shan Region, which caused heavy losses of human lives and resources (Hubbard and Shaw 2009). Though there may be a possibility of economic benefits, a risk of landslides persists due to the mountains surrounding Ladakh. Glaciers and rivers undercutting rock, severe rains, melting glaciers, and flaws such as fractures and movement on faults are all factors that aid in the production of landslides. According to Dortch et al. (2008), 14 of 16 significant landslide deposits occurred during monsoons, implying that rainfall is a prevalent trigger mechanism. Earthquakes are common in the area, and they could be a key trigger (Dortch et al. 2008).

4.3 Classification of the Himalayan Ranges

4.3.1 Geographical

For geographical purposes, the Himalayan system is classified into three zones, each differing orographically from the other. These are (i) the innermost line of high ranges, rising above the line of permanent snow and, having a height of around 6500 m or more. Mount Everest, K2, Kangchenjunga, Nanga Parbat, Nanda Devi, etc. come in this zone; (ii) the middle ranges where the ranges are of lower elevation, generally up to 4500 m.; and (iii) The Outer Himalayas, or the Shiwalik ranges, lying between the Lesser Himalayas and the plains and having altitude varying from 800 to 1500 m amsl.

4.3.2 Geological

The Himalayas fall into three major stratigraphical zones as far as geological structure and age are concerned. These zones are

(1) The Tibetan Zone: this zone, which is behind the line of maximum elevation, is made up of a continuous series of highly fossiliferous marine sedimentary rocks that date from the early Paleozoic to the Eocene epochs.
(2) The Central or Himalayan Zone, which includes most of the Lesser or Middle Himalayas as well as the Great Himalayas, is primarily made up of crystalline and

metamorphic rocks such as granites, gneisses, and other metamorphic rocks, as well as ancient unfossiliferous sedimentary deposits.

(3) The Outer or Sub-Himalayan Zone: this zone corresponds to the Shiwalik mountains and is made up completely of Tertiary, primarily Upper Tertiary, sedimentary river deposits. The surrounding ranges that form the Indo-Afghan and Indo-Baluchistan borders are to the west, while Myanmar's mountain ranges are to the east. In the west are the Hindu Kush, Salt range, Suleiman range, Bugti range, and Kirthar ranges, and in the east, Assam ranges, Manipur ranges, Arakan range, and Tenasserim ranges (Wadia 1919). The Salt Range and Assam ranges are considerably distinct in that their rocks have seen more fracture and dislocation, and they are not as heavily composed of Tertiary rocks (Wadia 1919).

4.3.3 Almora–Dudhatoli Crystalline

The rocks of Srinagar area have been classified (Kumar and Agrawal 1975) into 3 groups and out of these, the Dudhatoli Group is represented by the phyllites and quartzites, forming the northing limb of the Dudhatoli syncline. The south dipping known as "North Almora Thrust" (NAT) has the rock composition of phyllite belonging to Dudhatoli group, plunged over the quartzite and argillocalcareous rocks of Garhwal group as reported by Kumar and Agrawal (1975). Along the NAT, neotectonic activities have been reported and phyllites in association with limestone are present in Chandipur group where phyllites of Krol group dominate over the subsequent gravels in the study area. Strikingly, this area is neotectonically active.

Crystalline Group.

a. Chrapataya formation

(i) Biotite Gneiss: The Panduakhal–Gaddikhal top of the basin is composed of biotite gneiss rocks. Some Quaternary deposits were reported in the upper valley. The medium to coarse-grained biotite-, quartz-, feldspar-bearing gniess is greenish, grayish black in color. This gneiss is thrust over the Pratapnagar quartzite around the hill top. The biotite gneiss is traversed by the veins of quartz measuring up to 5 cm in thickness.

Chandpur Group

The group which is exposed along the NAT consists of two formations. The trend of the phyllite having minor folds, mostly of recumbent to isoclinal types, is NW to SE with the location variation from SWE to NNW. It shows well-developed crenulation of strain–slip cleavage. The folded schistose phyllites are represented by ramifications of the quartz veins (5–10 cm) in thickness (Shandliaya 1984). This formation is exposed around Dang, Rampur, and Sili villages. The slate type of phyllite variety is well developed away from the NAT, especially near Kandi and Khark Villages. The Pratapnagar quartzite is comparatively more schistose, and shows the development of vertical folds. Generally, Pratapnagar quartzite does not contain metabasic rocks. This quartzite is well exposed around the Silkakhal in the western part of the study area. The current bedding and ripple marks are the main sedimentary structures present in the quartzites. Quartz veins and slickenside were observed in Pratapnagar quartzite (Saklani 1971) and are well exposed around Sarkeina. A thick pile of fine to coarse-grained quartzite characterized by the presence of a thin layer of the sericite schists and bands of dark grayish brown ferruginous quartzite occur in the major portion of the catchment.

Garhwal Group or Calc Group

Garhwal group consists of quartzite and metabasic rocks. The rocks have suffered low-grade metamorphism up to chloride grade. The rocks of the calc group in Garhwal group (Mehdi et al. 1972) occur as a window showing the tectonic contacts. Mainly quartzite, ferruginous in nature, is predominant in this zone. This quartzite can be differentiated from the ferruginous quartzite of the calc group as follows:

- The occurrence of metabasic rock is much less in the Pratapnagar quartzite than ferruginous quartzite of calc group.
- The Pratapnagar quartzite is comparatively more schistose in nature than the ferruginous quartzite.

In the area, a number of lensoid bodies of limestone/marble and slate occur in the basal part of the quartzite.

4.4 Metabasic Rocks

The basic rock in the area occur as intrusives and flows. The intrusives are dykes and sills. Based on mineralogical and textural variations, the various basic rocks have undergone metamorphism during the Himalayan orogeny possibly in its end phase. Generally, the metabasics in their core parts are compact while in their peripheral parts, these are schistose or crudely foliated. The metabasaltic flow occurs in the rocks of the Calc Group near Silkakhal and Phayalgaon. In Phayalgaon, thc mctabasic rocks occupy a 500 m. wide patch. These metavolcanics are comparatively more developed in the Calc group than those of the other rock groups. The metavolcanics with pronounced effects of metamorphism are amphibolite-schists and are frequently noticed in the ferruginous quartzite members of the Calc Group. The metavolcanics are also characterized by amygdaloidal structures. The main minerals of the basic rocks that occur in variable amounts are hornblende, chlorite, actinolite, plagioclase, and iron oxides. The rocks are well jointed and on weathering are characterized by the coating of iron oxides. The metavolcanics are pitted in appearance due to weathering. The metabasic terrains are highly dissected by the denudational processes and in those terrain gully erosion, landslide and slumping of high intensity were reported.

4.5 Quaternary Deposition

Colluvial and alluvial deposits of quaternary age were observed during the geological survey throughout the study area which spread over along the valley side from 100 to 500 m distance. The quaternary deposits are generally covered by dense forest cover. Big boulders are probably of glacial origin are found here and there. Colluvial fans and scree materials are forming the repose slopes. The Dhundsir basin presents a typical polycyclic landscape with a series of variegated erosions and depositional landforms. The erosional activities performed by the river are the accumulation of vertical corrosion as well as lateral plantation along the entire stretch of the basins. Landslides are common on phyllitic rocks.

4.6 Afghanistan

Afghanistan possesses the world's most complex and diverse geology. The Archean rocks discovered in Afghanistan are succeeded by rocks from the Proterozoic and Phanerozoic systems. Because of its location near the western edge of the Himalayas, the country has a complicated tectonic history. The Archean rocks are the oldest in Afghanistan, followed by rocks from the Proterozoic and every Phanerozoic period. Afghanistan's bedrock geology can be viewed as a jigsaw puzzle of crustal blocks separated by fault zones, each having its own geological history and mineral prospectivity.

4.6.1 Stratigraphy and Geologic History

Afghanistan is underlain by Precambrian metamorphic rocks, which form high regions in the country's center and in the Hindu Kush. The northernmost structural high, made mostly of metamorphic rocks, is located between Herat and Panjao. Precambrian rocks of Koh-i-Baba connect the higher area to the Hindu Kush. Phyllite, greenschist, garnet-mica schist, and partially melted gneiss that experienced anataxis are among the Precambrian rocks found in Afghanistan.

4.6.2 Paleozoic (541–251 Million years Ago)

In the Panjao region, 50–60 m thick limestone overlies about 70 m thick dolomite, including Late Cambrian trilobite fossils going back to the beginning of multicellular life. Non-fossil-bearing sandstone, shale, and limestone exist up to 680 m thick in the southwest of Kabul. The central Afghan bulge and the Hindu Kush have Early Devonian conglomerates with interlayered dolomite and marl. Afghanistan has a large number of Permian rocks. In the Logar Valley and close to Kabul, Devonian rocks unconformably cover earlier rocks, with calcareous sandstones up to 700 m thick. A 150-m dolomite and limestone series lie to the west of Heart. The western Hindu Kush, Kandahar, Gardez, Jalalabad, and Moqor are all covered in Permian rocks.

4.6.3 Mesozoic (251–66 Million years Ago)

In the east of Afghanistan, conformably overlying Triassic sediments, are the Jurassic limestone, marl, shale, and sandstone. To the west, Jurassic rocks are distinct, consisting primarily of three-kilometer-thick sandstone and shale with subordinate limestone and marl. Triassic rocks can be found to the north and south of the Hindu Kush central Afghan swell area. To the south, 1.8-km-thick limestone and dolomite rest right on top of Permian limestones. Thick volcanic rock sequences are typical in the southern half of the basin. Geologists discovered a 200-m sandstone, shale, and limestone strata from the Middle Jurassic atop coal near Karkar. Extensive Triassic fossils have been discovered in black shale near Badakshan. Cretaceous rocks, like Jurassic deposits, differ from north to south. Calcareous sediments interfinger with volcanic rocks between Asra and Khost. Paleozoic and Precambrian rocks are covered by 500 m of sandstone, shale, marl, and conglomerate in east-central Afghanistan.

4.6.4 Cenozoic (66 Million years Ago–Present)

Calcareous sedimentation around Khost lasted until the Eocene. Nummulitic limestone and volcanic rocks intercalate with sedimentary rocks toward the basin's margin. Paleocene and Eocene marls up to 300 m thick interbed with limestone and gypsum in northern Afghanistan, unconformably atop Cretaceous limestone. The Miocene and Pliocene are represented by a reddish conglomerate, sandstone, siltstone, marl, and claystone with minor amounts of travertine, gypsum, and limestone. In the Neogene, tectonic activity began to raise new mountains, causing bending and erosion in intermontane basins, rift valleys, and plains.

4.7 Bangladesh

Uplift of Himalayan Mountain and outbuilding of the large deltaic area by a major river system originating from the uplifted Himalayas affected the geological formations in Bangladesh. As a mega delta, Bangladesh covers the greater part of the current Bengal Basin with a huge thickness of fluvio-deltaic sediments, being formed by the merging deltas of the rivers Padma (Ganges), Jamuna (Brahmaputra), and Meghna. Major parts of the country, both onshore and offshore, are geologically built up of sediments deposited by these river systems, being active from Oligocene through to the present. All these rivers are coming from the rising Himalayan Mountains to the north and Burmese hills to the east. The delta building is still continuing into the present Bay of Bengal. Only the eastern part of this country has been uplifted into a hilly landform placing itself into the frontal zone of the Indo-Myanmar

Range. This was due to the Indian plate colliding with the Asian plate (Islam and Alam 2009).

4.7.1 Tectonic Framework

Bangladesh's tectonic framework is separated into two main tectonic units: the Stable Precambrian Platform in the northwest and the Geosynclinal Basin in the southeast. The "Hinge Zone," a third unit that runs practically the entire length of the country, connects the above two units.

4.7.2 Stable Precambrian Platform

The stable Precambrian platform, which is defined by a moderate thickness of sedimentary rocks atop a Precambrian igneous and metamorphic basement, is found in the Rajshahi, Bogra, Rangpur, and Dinajpur districts. In relative terms, this unit is geologically stable and has not been influenced by fold movement. The geosynclinal basin in the southeast is characterized by a massive thickness of clastic sedimentary rocks of roughly 20 km, predominantly Tertiary sandstone and shale. Greater Dhaka, Faridpur, Noakhali, Sylhet, Comilla, Chittagong, and the Bay of Bengal are all part of it. The fold belt in the east and the foredeep in the west split the geosynclinal basin into two halves. The sedimentary layers are folded into a succession of anticlines (upward folds) and synclines in the fold belt (downward fold). The hinge zone, which separates the Precambrian platform in the northwest from the geosynclinal basin in the southeast, is a 25-km broad northeast–southwest zone.

4.7.3 Structural Framework

The major structural elements of the Bengal Basin (Bangladesh) areas are identified by a combination of existing, observable surface features. To the north of Bengal Basin is the Shillong Plateau, a Precambrian block believed to have been moved upward during Pliocene–Pleistocene time. The Dauki Fault borders the southern margin of the plateau. The Indo-Burman Ranges define the eastern margin of the basin, comprising a folded, thrusted, and wrench faulted arc complex, developed along the edge of the Eurasian Plate due to subduction of oceanic crust and overlying sediments. In the Bengal Basin, the Eastern Fold Belt marks the outermost part of this compression zone. The Rangpur Saddle, a Precambrian basement high located between the Indian Craton and the Shillong Plateau, to the southwest and northeast respectively, separates the Bengal Basin from the Himalayan Foredeep. The Bogra Shelf area rests on a Precambrian surface, gradually dipping toward the southeast. The shelf itself is 60–130 km wide and its southeastern margin is defined by the continental slope.

4.7.4 Formations of Different Age Groups

The Barail group, which consists of sandstone, shale, and siltstone with occasional carbonaceous layers reflects Oligocene deposition under deltaic conditions. It's between 150 and 200 m thick. The Jamalganj Formation is made up of deltaic sandstone, shale, and siltstone and is around 400 m thick. The Surma Group is overlain by the Pliocene–Pleistocene Dupi Tila formation, which is made of medium to coarse-grained sandstone that is often pebbly and has a thickness of around 270 m. This unit depicts the river plain deposit that occurred after the basin's delta was filled. The Quaternary group is represented by Pleistocene reddish brown clay overlain by Bengal alluvial of the Holocene (Recent) fluvial plain deposits. Dihing Formation is composed of Pebbly sand to an occasional gravelly deposit of river origin below the alluvial deposits.

4.7.5 Geomorphology, Geology, and Tectonic Setting

Bangladesh is a riverine country, with the majority of the country's territory produced by the silt carried by various rivers and their tributaries. The main rivers are Brahmaputra, Burigangar, Dhaleswar, Feni, Gorai–Madhumati, Jamuna, Padma, Karnaphulli, Mahananda, and Meghna. Since the early Pleistocene, the river systems have delivered a huge amount of sediments constituting 80% of the landmass. The average altitude of the surface is not reaching more than 10 m above msl. Besides Chittagong Hills, the hydrological significance is mostly lowland geomorphology in Bangladesh and they are Braided river, Alluvial fans, Natural leeves, River meaders, Deltas, and Anastomosing channel.

Bangladesh's surface rocks and stratigraphy date from the Cenozoic epochs, with the exception of the Oligocene, and are poorly preserved (Khan et al. 1993; Hossain et al. 2018). Gneiss, granite, granodiorite, and minor sequences of schistose, amphibolite, and gabbro make up the Precambrian basement complex. Bangladesh was a part of the subcontinent Gondwana during the Permian period.

4.7.6 Geological Mapping

Geological mapping of the Northern region of the Chittagong district was done on a scale of 1:50,000 (Muminullah 1978), covering an area of about 1300 km^2. Sandstone, shale, and siltstone are the dominant rocks found in his zone. Rocks of the Bhuban (Middle and Upper), Boka bil, Tipan, Dupi tila, and Dihing are folded into an asymmetrical plunging antiline that extends from Chittagong city to beyond Zorarganj near the Feni river. The age of these rocks ranges from Miocene to Pliocene. Middle and Upper Bhuban formations are exposed in the steeper western flank, but due to a major fault some of the stratigraphic units of the sequence are missing from the western flank. The major fault is running straight from the NNW–SSE direction parallel to the general trend of the structure. It was reported that the calcareous sandstone contains 33–53% SiO_2, 3.6–4.8% Fe_2O_3, 27.6–43.4% $CaCO_3$, 4.31–6.06% $MgCO_3$, and 0.22–0.69% moisture.

4.8 Bhutan

Bhutan's landscape has been formed by the Himalayan uplift caused by the collision of the Indian and Asian continental plates. This uplift has altered the basic topographic structure, with the High Himalaya in the north and broad ranges of metamorphic rocks running southwards (Norbu et al. 2003). The valleys follow a roughly north–south axis, but their longitudinal profiles differ somewhat. The riverbeds in western and central Bhutan have tiered features, with the flatter areas forming Inner Valleys. The rivers in the east have been carved deeper, and their valleys are narrow and steep (Norbu et al. 2003).

The main source of information on topographic data is topographic maps produced by the Bhutanese and Indian Surveys at scales of 1:50,000 and 1:25,000, respectively. Many of the 1:50 000 sheets are old and were created before satellite images and global positioning systems were available. The previous data is now being replaced with more detailed mapping. Even the older sheets are of good quality, depicting landscape details clearly. Geomorphological elements of the forestry development area near Ura were described by Woehrer (1992). The longitudinal profiles of the country's principal rivers, interfluves, and relief were studied by Baillie et al. (2002).

The southern half of the country is divided into its own area, with the low hills at the southern end of the main mountain ranges forming the most expansive zone within it. The hills are drained here by east–west tributaries of the major rivers, as well as streams that go directly south to the duars (The Duars are the foothills of the eastern Himalayas in Northeast India around Bhutan). The region of Duar, which means "door," serves as an entryway to Bhutan from India. The Bhutanese people have 18 tunnels or gateways via which they can interact with the people of the plains. Lesser Himalayan rocks underpin this zone, which is located south of the Main Boundary Fault (MBF). Its topography is almost perpendicular to the east–west axis. Many of the metasediments and sedimentary rocks of the Lesser Himalayan range have low erosion resistance. These mountain ranges are also prone to high-intensity rainfall, resulting in significant runoff and surface erosion rates. Furthermore, the MBF, which runs along the northern boundary of this zone and is still active, goes through it. Although many of the low-order drainage patterns are dendritic, the big rivers show some structural control. As a result, the lower Amo and Puna Tsang have rectangular courses, with east–west and north–south reaches alternated. When saturated soils and severely worn rocks are subjected to significant rainfall, large landslides are fairly common. Some landslides are triggered by earthquakes. This region has one of the highest proportions of unstable slopes in the entire country. The lesser valleys meet deep and unstable landslip debris in numerous areas.

***Duars* (Piedmonts)**

The duars are the East Himalayan analogue of Nepal's terai region, located on the piedmont borders of the Brahmaputra alluvial plain (Chakrabarti et al. 1987). Around Geleyphu, the largest duar relic in Bhutan occupies over 60 km^2. The main topographical components around Geleyphu include gently sloping linear or slightly concave colluvial aprons and alluvial fans; broad braided riverbeds and floodplains with significant sand and gravel spills; and wide tilted river terraces at heights of 3–20 m above the floodplain. Clastic Quaternary alluvial deposits make up the duars. The upper borders of the smaller sections of duars in Bhutan are predominantly colluvial deposits near the base of the slopes, and the unstable alluvial fans of flashy streams like Patsakha. The braided floodplain of the Torsa can be seen from Sorchen, and there are beautiful views of these features from the Bhutan foothills. The floodplain and terrace elements that correspond to them are located south of the border.

4.9 Tibet (China)

The Himalayas in the Tibetan–Tethys region has originated from the top of the Higher Himalayan Zone. The metasediments that overlie the Higher Himalayan zone have a thickness presumed to be 7400 m (Fuchs et al 1988).

4.9.1 Geological Setting

The Tibetan Plateau, along with the Himalayas and neighboring mountains, constitutes the widest and highest orogenic system in the world. Rowley (1996) mentioned that the changes in its orography and the earth's climate are very closely linked to this collisional event. The south face of the Himalayas is covered by green vegetation of trees and grasses. The plateau, with an altitude variation from 4,000 to 5,000 m above mean sea level, is essentially a vast cold desert. It has an area of about 2.5 million km^2. The plateau attained its present altitude due to underthrusting of the Indian plate below Asia (Argand 1924). A multitude of microcontinents, flysch complexes, and island arcs make up the Tibetan Plateau. The ancient giant Indian plate shattered, resulting in the formation of tiny younger plates that broke away from the main plate and drifted toward Asia. This was a slow and steady process. During the Paleozoic Era, the microcontinents of the Tibetan plateau collided with the Eurasian plate and were subducted beneath it (Chang and Cheng 1973). The Qilian Shan, eastern Kunlun–Qaidam, Songpan–Ganzi, Qiangtang, and Lhasa terranes are the most important, and they are distributed from north to south, correspondingly. Stöcklin (1980) describes Tibet as a folded mountain area with a considerable number of east–west trending fold belts. Its crustal thickness frequently surpasses 70 km, nearly double the continental crust's usual thickness. The Indian bottom crust is thought to be underplated beneath Tibet. The lower crust was sliced and stacked numerous times throughout this process, resulting in Tibet's unusual crustal thickness, while the disconnected Indian upper crust was forced southward, generating the Himalayas.

The northernmost Qilian Shan terrane is made up of complexly deformed early Paleozoic arcs that developed near the south edge of the North China craton and were later offset by the Cenozoic Altyn Tagh fault (Yin and Harrison 2000). North of the Jinsha suture and south of the Anyimaqen–Kunlun–Muztagh suture is the triangular Songpan–Ganzi terrane. The Songpan–Ganzi flysch complex, which is really scattered over a much larger area than the Songpan–Ganzi terrane, is mostly made up of a thick, deep marine turbidity succession of Triassic age (Yin and Harrison 2000). They also proposed that the Jurassic succession, which is made up of volcanics and limestones, rests unconformably overlying the last phase. The Lhasa terrane is the world's southernmost microcontinent, with the Banggong–Nujiang suture on the north and the Indus–Tsangpo suture on the south. The Ordovician and Carboniferous to Triassic shallow marine clastic layers that make up the Lhasa terrane depend on a Mesoproterozoic to Early Cambrian basement (Yin and Harrison 2000). The Gangdese batholith, which intrudes into the rocks of the Lhasa terrane and forms its south margin from the Cretaceous to Tertiary period, is linked to the Indian plate's north-dipping subduction (Chang and Cheng 1973).

4.9.2 Himalayan–Tibetan Orogen

The Cretaceous to early Tertiary Xigaze forearc basin, lying to the south of the Gangdese batholith, is one of the best-exposed forearc sequences in the world (Durr 1996). Subduction dipping beneath India and Lhasa during the late Proterozoic and early Paleozoic produced Cambrian–early Ordovician granites in the Himalayas and the northern Lhasa terrane. As the Amdo gneiss complex, severe north–south shortening manifested itself in isoclinal folding and the formation of widespread foliation in the granites and their country rocks. This distortion could have occurred as a result of an early Paleozoic merger between India and Lhasa.

4.10 India

4.10.1 Geology of Eastern Indian Himalayas

Manipur valley

Due to its simple geology and structural features, the Imphal valley of Manipur in India provides an ideal place to test the significance of remote sensing and GIS techniques in geological studies (Khan et al. 2012). The Geographic Information System (GIS) is a set of tools for gathering, storing, retrieving, analyzing, integrating, and displaying real spatial data for a specific set of demands and goals. In contrast to traditional geological fieldwork methods, which are time-consuming, expensive, and need complicated logistics, Remote Sensing techniques are efficient, faster, and less expensive applications that can be used to support preliminary geologic–geomorphologic studies.

Precambrian

A basement of ancient rocks is seen over a large area in the Shillong plateau and in the Mikir hills and in smaller outcrops in the Brahmaputra valley north of Shillong plateau. These occurrences of crystalline rocks and associated ancient sediments have long been recognized as an outlying portion of Peninsular India. Deep drilling in the oilfields of the Upper Assam Valley shows that this basement complex extends for many kilometers northeast beneath the alluvium of the Brahmaputra valley.

Mesozoic

In the southern foothills of the Shillong plateau, the basement is locally overlain by a Jurassic basalt. Over part of the Shillong plateau, there are

Cretaceous sandstones and conglomerates. More generally Tertiary rocks rest directly on the basement.

Tertiary

The Assam-Arakan mountain range in Eastern India, as well as the Shillong–Mikir plateau, include a thick layer of Tertiary and Quarternary rocks. Individual stratigraphic units vary in thickness, with the Geosynclinal portions of the Arakan mountains having the thickest stratigraphic units. The region's Cenozoic rocks were deposited on top of Precambrian rocks and Lower Gondwana deposits. The Naga Metamorphics are a band of low- and medium-grade quartzites schists, limestones with sheared granite, and small serpentinites found in a linear overthrust belt along the Indo-Burma border. This sequence is unfossiliferous and it has been correlated with the Precambrian or Lower Paleozoic rocks of Burma. The periods of Upper Cretaceous, the lower and middle Eocene, are existing in Disang series in Upper Assam. On the other hand, the greater part of Eocene including Paleocene is in Jaintia series of southern Assam. Rocks of Upper Eocene and Oligocene ages are in Barail series containing coal seams and petroliferous beds. But considering Oligocene period, Barails differ from Surma series. Tipam series being Miocene also differs from Dupi Tila series. Dihings are mainly Pliocene.

4.10.2 Structure

The northeast Indian Himalayan region may be divided into the following six main divisions:

(i) the overthrust around eastern Himalayas,
(ii) the overthrust Patkai Kohima synclinorium, including the Naga hills,
(iii) the Assam valley, extending from Shillong plateau to Upper Assam,
(iv) Shillong plateau and Mikir hills,
(v) Surma valley, and
(vi) Arakam-Yoma folded belts.

Eastern Himalaya is composed of Paleozoic, Mesozoic, and Tertiary rocks. The general structure of eastern Himalayas is very complex. The tectonics of the eastern Himalayan mountains is dominated by extensive thrust sheets which have moved from the north toward south or southeast. This Eastern Himalayan zone is a part of the Eurasian seismic belt that extends from the Alpine to the East Indies.

The Naga Hills

The Naga hills consist of an entire Tertiary succession of Eocene to Paleocene rocks (Evans, 1964). It includes a few low ranges of hills extending into the plains of Upper Assam, Also included in this range is a part of the North Cachar hills. These hills consist of several overthrusts along which the rocks have moved northeastward relative to the "Foreland Spur" (Evans et al. 1997). Toward the southwest of Naga hills, the junction between alluvium and the hills is marked by the thrust called the Naga thrust. The extension of the Naga thrust to the southwest is called the Disang thrust which in turn extends as the Haflong fracture zone and joins east–west striking Dauki fault. Within the Naga hills, there is a feature named Patkai synclinorium.

Valley of Assam

The Assam valley was produced by sediments transported from the rising Himalayas to the north and the Naga hills to the south. The Precambrian basement complex, which is exposed in the Mikir hills, is overlain by a massive succession of Tertiary rocks (4–6 km thick) beneath the alluvium.

Mikir hills and Shillong plateau

The Shillong plateau and Mikir hills form a very prominent part of northeast India lying roughly in the east–west direction between the eastern Himalaya to the north and Surma valley to the south. The plateau is surrounded by the prominent Yamuna lineament to the west and Tertiary sediments of Upper Assam and also Kopili lineament to the east. The Mikir hills are detached from the Shillong plateau by the Kopili lineament. The Shillong plateau is mainly of sedimentary origin.

Surma valley

The Surma Valley is relatively simple in structure and seems to have been covered by a layer of alluvium. This zone of folded sediments is characterized by arcuate and sinuous ridges and valleys of low relief. NE–SW and NW–SE trending fractures are common which are closely spaced toward the eastern margin. It is limited by Dauki fault and to the east by the marginal faults of the Barail ranges (Nandy 1980).

Arakan – Yoma belt

The Arakan-Yoma belt, which runs roughly north–south along the Indo-Burma border, is made up of a sequence of thrusts and thrust slices (Brunnschweiler 1966). On the western side of the folded belt is a Paleogene flysch, while to the east is the Central Burma molasse basin (Irrawaddy) of the Quarternary age. Arakan-Yoma rocks, which are made up of a jumble of highly deformed sediments and fossiliferous (Cretaceous–Eocene) limestone, are profoundly exposed (Brunnschweiler 1966).

4.10.3 Geology of Western Indian Himalayas

The Ladakh Batholith is an intermediate rock created by Paleozoic to Cretaceous sediments, metasediments, granitic intrusions, and the rocks that make up the Ladakh region to the north of the Indus River (Dortch et al. 2008). Along the Karakoram fault, shear is occurring in Ladakh at a relatively slow rate. Because batholiths have been related to porphyry copper, tin–tungsten, and gold deposits, these minerals may be found in Ladakh's batholiths (Mitchel and Garson 1983). Jammu and Kashmir has an abundance of rubies as a result of high-grade metamorphism (Wilson and Wilson 2014).

The northern slopes are covered with a thick dense growth of forest vegetation, while the southern slopes are relatively bare, either containing snow deposits or supporting a sparse forest vegetation. The starting point of the Himalayan system is situated in the Pamir plateau ("the roof of the world"), which is a knot or nucleus from which also diverge all the principal ranges of Central Asia (Wadia 1919).

The Tethyan sequence, which ranges in age from Late Precambrian to Early Jurassic, lies unconformably overlying the Quaternary Karewa succession in the "Kashmir Basin." It is unconformably followed by Rishkobal Group deposits from the Early Ordovician to the Middle Silurian (Bhargava 2015). The Kashmir pre-Panjal Volcanic sediments have litho- and biofacies that are comparable to those found in the Spiti-Zanskar area. In comparison to the Spiti-Zanskar area, Kashmir's post-Panjal Volcanic sediments are shallow water deposits. The Himalayan Orogeny is responsible for around 35% of crustal shortening in the Kashmir region.

Having a length of 140 km and width of 40 km, the Kashmir valley is an oval-shaped basin, trending in the NNW–SSE direction and is surrounded by high and chiseled mountain chains of the Himalayas. The plateau-like terraces are called "Karewas" in the local language (Parray 2011). Karewa sediments represent a treasure of many human civilizations and habitations. The items it left behind are still buried in the plateau areas' deep earth horizons. The choppers, hand axes, and sharp stone blades are worth mentioning, forming a storehouse of the Stone Age factory. The Elephus remains found within the sediments provide evidence of his ability to hunt animals. In the past, correlations between exposures have been based primarily on lithological similarities and limited paleontological data. Magnetostratigraphy and fission track method dating allow a reliable chronology for the sediments deposited in the basin. The dated Karewa sediments span the interval from 3.0 to 0.4 Ma. The normal order of superposition of the rocks in the Lesser Himalayas of Himachal Pradesh state of India has been affected by later events of thrusting.

Srinagar–Garhwal region

Srinagar area located in the Lesser Himalayan Zone is characterized by superimposed thrust sheets. The area is demarcated by the North Almora Thrust (NAT) in the NE part which trends NW–SE, dipping at 40–500 toward SW. At places, much steeper dips are recorded ranging up to 800 toward SW. This thrust shows a variable strike and is undulating. The various rocks of the area belong to the Chandpur Group and Garhwal Group. The Chandpur Group consists of Srinagar phyllites associated with Kilkaleshwar metavolcanics. The Garhwal Group is represented by Marora Limestone, Garhwal Slate, and Koteshwar Quartzite associated with Chamdhar Metabasics. The phyllitic thrust sheet (Srinagar phyllite) is folded into major anticlines and synclines.

Lithostratigraphy

As mentioned earlier, various rocks of the area that belong to the Chandpur Group and Garhwal Group describe stratigraphy of the area. All the rocks are of Precambrian age. According to Kumar and Agarwal (1975), the Pauri phyllite forms a formation of the Dudatoli Group and is also included into the Kumaon Supergroup. The contact between the two groups of the area is thrusted and a general reversal in the stratigraphic succession of the area is observed in field. The succession as observed in field is Koteshwar quartzite (lowest unit) followed by Garhwal slate and Marora limestone. Tectonically, the topmost unit is Srinagar phyllite. Geological features of the rocks of the two Groups are described under respective subheadings in the following pages. The folds in phyllites are well developed in Srinagar and Srikot area. The axial planes of minor folds give rise to crenulation foliation. Pucker lineation in the phyllite is prominently seen along Srikot–Supana track and across the Alaknanda valley on Kilkaleshwar to Barkot track. The chevron folds are developed on Srinagar to Pauri, Srinagar to Kirtinagar, and Sumari road. These phyllites are associated with the greenish, grayish black metavolcanics which are medium to coarse-grained. The metavolcanics have been designated by the author as Kilkaleshwar metavolcanics.

Koteshwar Quartzite

It is medium grained grayish, pinkish buff to dark violet quartzite with intercalation of ferruginous shales/slates. It is exposed prominently to the south of Barkot, Supana, Koteshwar and toward the north and south of Chamdhar. There is an occurrence of epidioritic metabasics in it. The metabasics are medium- to fine-grained and commonly exhibit an abundance of greenish pyroxenes, amphibole, and white grains of plagioclase. The foliated type of metabasics shows hornblende, chlorite, and actinolite toward the marginal part. Joints in Koteshwar–Chamdhar area dip in the range of 30–800 toward ESE and WNW. These are occasionally coated by iron oxides.

The Garhwal Slate

It is fine-grained grayish green, brownish violet, and bluish black, exposed in the NE part of the area. It is associated with Koteshwar quartzite, Marora limestone, and the Chamdhar metabasics. The grayish green slate frequently shows stains of concentric reddish brown oxidation rings on the surface and exhibits an alternate arrangement of bluish black ferruginous bands. The slate strikes E/W and has steep dips (800) toward North. The secondary cleavage surfaces (S2) dip from 30 to 650 toward NE and SW directions. This slate overlies Koteshwar quartzite and occurs in the core of the syncline which pinches out near Koteshwar.

The Marora Limestone

It is bluish gray crystalline medium- to fine-grained and banded limestone showing ferruginous and siliceous impurities. It also exhibits characteristic elephant skin weathering. It occurs as a band along the NAT in NE part of the area. The general strike trend is E-W with a variation of NNW–ESE and has a dip in the range of 10–450 toward N and SSW. Intercalations of thin grayish shale impart a banded appearance to the rock. It is massive and shows a sugary texture. As one moves toward the east, the dimension of the limestone outcrop widens, and it gradually pinches toward the West. Moving uphill, it is found at higher elevation in the area showing tectonic contacts with the phyllites of Chamdhar Group along the North Almora Thrust (NAT). The contact is clear along the track from Chamdhar to Devalgadh Gad (at about 2 km from Chamdhar near Marora village). The limestone is overthrust by the Srinagar phyllite toward north of Marora characterized by many water seepages, which possibly caused solution cavities in the limestone. The geological cross section from Srinagar to Margaon shows the recumbently folded nature of limestone units.

4.11 Myanmar

4.11.1 Tectonic Setting of Myanmar

The Myanmar region was formed by two factors: first, the subduction of the northward moving Indian Plate beneath the Burma Plate along the Andaman Megathrust Zone at a rate of 4–6 cm yr-1; and second, the northward movement of the Burma Plate from the spreading center in the Andaman Sea at a rate of 2.5–3 cm yr-1. The majority of earthquakes in Myanmar have been caused by intermittent jerks along these major active faults. These seismotectonic processes continue to occur.

4.11.2 Regional Geological Features of Myanmar

Myanmar is a lovely country located east of the Andaman Sea. It is divided into seven administrative regions (Ayeyarwaddy, Bago, Magway, Mandalay, Sagaing, Tanintharyi (Tenasserim), and Yangon) as well as seven states (Chin, Kachin, Kayah, Kayin, Mon, Rakhine (Arakan), and Shan). It is located in the northwest corner of the Southeast Asian continent. Myanmar is around 2000 km long from north to south, with a temperate to subtropical north and a tropical south (Zaw et al. 2017). The Arakan or Rakhine Costal Strip, the Indo-Burman Ranges, the Western Inner-Burma Tertiary Basin as an inter-arc basin, the Central Volcanic Belt as an inner magmatic–volcanic arc, the Eastern Inner-Burma Tertiary Basin as a back-arc basin, and the Sino-Burma Ranges or Shan-Tenasserim Massif are the six trending tectonic domains. Between the Eastern-Burma Basin and the continental, ensialic Sino-Burma Ranges, the Sagaing Transform Fault serves as a tectonically significant border. Myanmar's topography consists of core lowlands surrounded by steep, impassable highlands. Mountain ranges mainly run north–south, with the Patkai Range, Naga Hills, Chin Hills, and Rakhine Yoma to the west, near the India–Bangladesh border (Zaw et al. 2017).

Myanmar's tectonic setting is located at the intersection of the Alpine–Himalayan Orogenic Belt and the Indonesian Island Arc System. The Himalayas and northern Myanmar's mountain ranges indicate the meeting point of the Indian subcontinent and the Eurasian continent's southern boundary. Debris from the Himalayas, transported by the rivers Ganges and Brahmaputra, has produced a massive delta into the Bay of Bengal on the Indian Ocean crust. The Indian Tectonic Plate, which carries India and the Indian Oceanic Crust, is currently migrating northeastwards at a rate of 5 cm per year, passing through Myanmar on transcurrent faults. At the Eastern Himalayan Syntaxis, the collision between the Indian continental crust and Eurasia continues, resulting in the Patkai Range and the Naga and Chin hills. Transcurrent faulting and transpression have influenced the overlaying strata of the Bengal Fan further south, forming a fold-and-thrust belt in the Rakhine Yoma. Along the Sagaing Fault, the Myanmar Microplate is currently advancing northwards at a rate of 18 mm per year relative to Southeast Asia. During the Mesozoic, this continental basement was thought to have been a separate crustal block known as Mount Victoria Land or the West Myanmar Block (Metcalfe 1996). During the Proterozoic and Early Paleozoic periods, the West Myanmar Block is thought to have been part of the northern edge of the megacontinent of Gondwana, which included all of the southern continents (Metcalfe 1996). This block, along with other crustal blocks, detached from the northern frontier of Gondwana in the north Australian region during the Jurassic, and collided with eastern Myanmar in the Cretaceous, according to Audley-Charles et al. (1988). Cai et al. (2016) identified West Papua as the source of the detrital zircons in these Triassic layers, however, Wang et al. (2016) recommended the Gondwanide Orogen in southeast Australia. Weathering and erosion of pre-existing rocks produce detrital zircons, which are part of the sediment. Many zircons are transported, deposited, and stored as detrital zircon grains in sedimentary rocks because zircons are heavy and resistant at Earth's surface.

4.12 Nepal

As a result of active lithospheric plate tectonics, Nepal's geology is very dynamic. The collision between the Indian and Eurasian plates, which occurred about 55 million years ago, is still active, and this process, together with accompanying phenomena, gave rise to Himalaya, of which Nepal is the most important part because it lies entirely within the Himalayas. The dynamics of this continent–continent collision have resulted in geological settings that are astonishingly diverse within a limited breadth of about 250 km. Nepal can be split into five geological zones, each of which is separated from the others by fault structures, the majority of which are thrust faults. From south to north, these five zones are the Indo-Gangetic Plain, Shiwaliks, Lesser Himalayas, Higher Himalaya, and Tibetan Tethys zone. Each of these zones has its own geological characteristics and has shaped the topography as well as the relief.

4.12.1 Geological Divisions

Like other Himalayas, Nepal Himalaya can be divided into five geological units that correspond to morphotectonic zones as well. Each of the zones is characterized by its morphological, geological, and tectonic features, and each of them is separated by distinct geological tectonic structures. From South to North, the five geological zones are

- The Indo-Gangetic Plain (Terai).
- The Sub-Himalaya (Churia/Shiwaliks).
- The Lesser Himalaya (Mahabharat/Midland Zone).
- The Higher Himalaya (Central Crystalline Zone).
- The Inner Himalaya (Tibetan Sedimentary Zone).

Middle Terai (Marshy) Zone

Situated in between the Northern Terai Zone and the Southern Terai Zone, the middle terai or marshy zone has a width of 10–12 km. This zone is composed of pebbly and sandy sediments with some clay. It has been found that the medium to coarse-grained sandy layers contain large amounts of groundwater.

Northern Terai (Bhabar Zone)

This Zone starts from the foothills of Shiwaliks and continues southward to the boundary with Middle Terai. Maximum width of this zone is 12 km. This zone acts as a recharge area for the groundwater of Terai. It has been found that most of the rivers lose their water while passing through this zone.

The Sub-Himalayan Zone (Shiwaliks/Churia)

The Sub-Himalayan or Shiwaliks region consists of young and fragile sandstones, mudstones, and conglomerates which are highly susceptible to weathering that converts rock into soil. The zone has noticeable relief change and the steepness of slopes is also considerable. According to Dhakal (2014), an increase in pore water pressure through joints and fractures raises pore water pressure, which raises the driving force in the rock creating slopes and causes slope failure. Because the Lower Shiwaliks are dominated by weak mudstones that are very sensitive to weathering, erosion is common in this zone. The areas in the boundary between Lower Shiwaliks and Middle Shiwaliks have witnessed big landslides. Middle Shiwaliks consist of alternating layers of relatively strong sandstone and weak mudstone, which provides a condition of differential weathering. Such weathering indicates that the support given by mudstone layers to the strong sandstone is lost after the former is washed away, and the sandstone beds got slided or overhanged. Plain failure, wedge failure, and toppling failure are possible in Middle Shiwaliks. Upper Shiwaliks consists of dominantly pebble conglomerates, which after the loss of cementing materials by weathering become vulnerable to slope failures and landslides. Bank scouring by rivers is also common mostly in the northern and middle parts of Shiwaliks, whereas flooding can occur in the lower and middle parts. The poor drainage management and improperly excavated roads further accelerate the landslide and other mass movement phenomena. The intensity of extreme precipitation has increased in this zone as a consequence of climate change indicating further land degradation in Shiwaliks and increased flood events in Terai. Fine-grained red, ash gray, gray, and reddish-brown sandstones are interbedded with purple, gray, and green shales and mudstones in the Lower Shiwaliks zone. Conglomerates are occasionally encountered in the middle and upper parts. Coarse conglomerates, sands, grits, and clays characterize the Upper Shiwaliks. This zone contains gneisses, schists, granites, and quartzites from the Higher Himalaya, as well as limestones, phyllites, slates, and sandstones from the Lower Himalaya.

The Higher Himalayan Zone

The Higher Himalayan Zone, also known as the Greater Himalayan Zone, is bordered on the south by MCT and on the north by the South Tibetan Detachment System (STDS). The Higher Himalaya of Nepal is exposed dominantly to rock falls, rockslides, and GLOF hazards as this zone is composed of hard crystalline metamorphic rocks with adverse orientation as well as many glacial lakes in threat of breaching. Further, the slopes are either free from soil cover or it is very thin. The Higher Himalayan rocks have also been subjected to significant weathering and distortion in numerous locations, posing the risk of large landslides. The river was dammed in 1998 by a landslide near Tatopani village along Kaligandaki Valley. This damming lasted only for 7 h that time and downstream people were evacuated on time; and therefore it didn't bring lots of devastation. Sunkoshi and Bhotekoshi stretch of Nepal is highly vulnerable to landslides and LDOF as it is in Kaligandaki River Valley.

Lesser Himalaya

The Lesser Himalaya is geologically and topographically highly heterogeneous indicating the varieties of slope processes and associated hazards in this zone. Landslides and slope failure are accelerated in the Lesser Himalaya by highly jointed metasedimentary rocks, as well as numerous folds and faults. Lesser Himalaya is sometimes divided into southern Mahabharat Range and northern Midland. The debris flow is accelerated in this zone because of heavy rainfall and a steep gradient of rivers. The presence of weak metasedimentary rocks like slates, phyllite, and schists as well as steep slopes and intense precipitation indicate the dominancy of large and deep-seated landslides along the highways, roads, and even in villages within natural slopes. Krishnabhir Landslide and Jogimara Landslide are two well-known landslides that have blocked the Prithvi Highway many times during the 1990s and 2000s. Landslides and slope instabilities in Mugling Naranghat Road section during the same period posed serious problems as well; this section is still unstable. Many other disasters arising from landslides have been recorded that have swept away whole villages or many cultivated lands. Debris flow and landslides from nearby high gradient rivers in this zone have been posing severe threats of river damming as well in the narrow and relatively low gradient rivers. Creeping is also one of the major modes of landslide in this zone.

4.13 Pakistan

Pakistan shares geological similarities with both the Indian and Eurasian tectonic plates. Sindh and Punjab provinces are located in the northwestern portion of the Indian plate, while

Baluchistan and the majority of Khyber Pakhtunkhwa are located in the Eurasian plate. The Pakistan Geological Survey and the United States Geological Survey performed geology and mineral investigations over a 5000 km^2 area from 1961 to 1965. The Hazara district was primarily targeted. The investigation indicated that the Hazara district's geologic strata range from Precambrian to Quaternary in age. It was discovered to contain sedimentary, igneous, and metamorphic rocks, as well as unconsolidated material in some areas. Precambrian and potentially early Paleozoic rock units, as well as early and middle Paleozoic rock units, are mostly clastic (Calkins et al. 1975). Carbonate deposition may be found in a fairly complete sequence of rock units with a thickness of roughly 1,650 m. In the middle and late Tertiary, clastic deposits became prevalent, indicating the start of the Himalayan orogeny, which may be responsible for deforming all rocks in this area. The rock units and faults continue around the circle and then curve in a broad arc toward the Indus River on the western side. The synorogenic intrusions of the Mansehra Granite were created by igneous activity that occurred during the Late Cretaceous and early Tertiary periods. The granite rocks represent the southern edges of the large granite intrusions in the Himalayas' axial zone. Mafic dikes and quartz veins are examples of postorogenic intrusive rocks (Calkins et al. 1975). The degree of regional metamorphism rises northward in general. Argillite, phyllite, slate, and quartzose schist are found in the southern section, together with basically unmodified shale, quartzose siltstone, and quartzite. Higher grade metamorphic rocks, such as garnet, kyanite, staurolite, sillimanite-bearing schists, and tremolite-bearing marble, are ubiquitous on the northern side of the Darband, Oghi, and Battal districts.

4.13.1 Stratigraphy

The southern two-thirds of the area appears to be made up of three structural blocks, each with its own stratigraphy. Within the bend of the Murree fault, the Garhi Habibullah syncline between the Murree and Panjal faults on the western limb of the syntaxis, and the western arc area west of the Panjal fault, the syntaxis' axial zone is situated. The main stratigraphic differences between the three structural areas are that the Tanawal Formation rocks in the western arc area are not present in the adjacent Garhi Habibullah syncline, and the entire section of Jurassic and Cretaceous rocks in the Garhi Habibullah syncline is not present in the adjacent axial zone of the syntaxis.

4.13.2 Precambrian to Ordovician Rocks: Salkhala Formation

The metamorphic rocks of the Salkhala Formation of Precambrian age are the oldest known rocks in the Hazara region of Pakistan. The Salkhala Formation is similar to the Jutogh Series of Shimla. The formation's main constituents include quartz schist, marble, graphite schist, and quartzo-feldspathic gneiss. Quartz schist is mostly fine-grained mica-chloritequartz schist, which makes up a substantial percentage of the formation. About 30% of the schist is chlorite and muscovite, with the balance being quartz. At several locations, modest amounts of clinozoisite, garnet, magnetite, and biotite were discovered. As the schist is close to layers of marble or graphite schist, calcite or graphite can be found in laminae and scattered grains. Calcsilicate rock has formed along the edges of granite layers as a result of interactions with granite. At Balakot, the quartzo-feldspathic unit contains strata of pure white fine-grained sugary quartzose sandstone.

4.13.3 Hazara Formation

Large parts of an arcuate belt are occupied by the Hazara formations. With a width of around 20 km, it stretches southward from the neighborhood of Garhi Habibullah through Abbottabad and then westward to Tarbela, combined with underlying younger strata. In the structural block west of the Panjal fault, the Hazara formation is overlain by the Tanawal formation. The Hazara formation is absent near the apex of the syntaxis, while the Salkhala formation lies exactly beneath the Tanawal formation. Locally, thickets of fine- to medium-grained graywacke sandstone can be found. In two different zones that stretch from Muzaffarabad to Kohala, limestone beds with a thickness of up to 150 m and a sequence of calcareous phyllite and gypsum with a thickness of 30–120 m are found. The Hazara formation also contains a few strata of limestone, as well as graphite, graphitic limestone, and calcareous shale in the Tarbela area.

4.14 Lithospheric Deformations in HKH

The Himalayas, including the Tibetan Plateau (TP), portray an active mountain range, with intercontinental collision. Jadoon et al. (2021) reviewed the geological and geophysical data for studying lithospheric deformation and active tectonics of this orogen. Shallow crustal deformation with the influence of thrusting along the margins of the Tibetan Plateau was elucidated with routine faulting in the center and strike–slip deformation with the lateral adaptation of blocks, over a wedge of ductile deformation. The seismicity is the linear concentration over the margins of the orogen to ~20 km depth with exception of the Hindukush and Pamir having seismicity to 300 km depth. The indention of the Indian into the Eurasian plate since about 55 Ma (Ding et al. 2016) has led to the uplift of the Tibetan Plateau to a height of ~5 km with north–south crustal stacking, east–west extension, and lateral extrusion of rigid segmented blocks along strike–slip faults (Feng et al. 2011; Bhatti et al. 2018; Molnar and Bendick 2019).

4.14.1 Active Deformation in HKH Region

Structural features in the NW Himalayas conform to the NW–SE Himalayan and NE–SW Hindukush leaning, having changed through the Nanga Parbat Haramosh Massif that displays an active north–south direction inducement of the lithosphere of the Indian plate at the western end of the Himalayas. The plutonic rocks with age varying from 01–10 Ma, along the length of the massif (Schneider et al. 1999; Zeitler and Chamberlain 1991), indicate crustal melting with the effect of deeper insight of crustal shears. Jadoon (2019) opined that the existence of hot springs aligned with faults and fault scarps in the Himalayas in north Pakistan shows zones of active deformation. The seismicity is experienced up to about 20 km depth in the Tibetan plateau, indicating shallow crustal deformation. Thrusting as a process accommodates about 90% of slip along the margin of the Tibetan Plateau in the Himalayas and the Altyn-Tagh (Kaneda et al. 2008), the central part being generally aseismic (Zhang et al. 2004). In the last 20 years, three major earthquakes were recorded in the Tibetan Plateau (Feng et al. 2011). The major earthquakes that occurred in the Himalayas are Kangra in1905 (moment magnitude or Mw 8), Dhubrin in 1930 (7.1 Mw), and 1934 Bihar in 1934 (8.3 Mw), all of which are located along the décollement, with an active narrow zone of steep (~30°) topography (Seeber et al. 1981) and, the 2015 earthquake in Nepal (7.8 Mw) with about ~9000 casualties (Avouac et al. 2015). The Himalayan chain with a narrow zone of steep topography may experience earthquakes of high intensity in future too (Bilham 2004; Bilham et al. 2001). Crustal seismicity in the Hindu Kush region appears to occur on multiple rather indistinct thrust faults that follow the Tajik Depression's boundary, according to Kufner et al. (2018), Perry et al. (2019), and Yeats and Madden (2003).

4.14.2 Slab Breakoff

Continental collision takes place due to firmness build-up between the deep and shallow slab, causing break-off of part of the down-going plate, for which resiliency is increased because of continental-crust subduction (Kufner et al. 2021).

Break-off occurs over a geologically short period of time and, therefore, real-time observations are rare, though it governs the subsequent orogenic evolution. Slab break-off is a continuous process in Hindu Kush, Afghanistan. Understanding the mechanics of the final detachment, as well as coupling these deep mantle processes to crustal deformation, is crucial to comprehending mountain-building processes (Braun 2010; Cloetingh and Willett 2013).

References

Argand E (1924) La tectonique de l'Asie. Congrès Géologique International, Belgique, Comptes Rendues de la 13ème Session, en Belgique 1922. H. Vaillant-Carmanne, Liège, pp 171–372

Audley-Charles MG, Ballantyne PD, Hall R (1988) Mesozoic-Cenozoic rift-drift sequence of Asian fragments from Gondwana land. Tectonophysics 155:317–330

Avouac J, Meng L, Wei S, Wang T, Ampuero J (2015) Lower edge of locked Main Himalayan Thrust unzipped by the 2015 Gorkha earthquake. Nat Geosci 8(9):708–711

Baillie IC, Tshering K, Dorji T, Tamang HB, Dorji T, Norbu C, Hutcheon AA, Baeumler R (2002) Regolith and soils in the Inner Valleys of Bhutan, Eastern Himalayas. Eur J Soil Sci 55(1):9–27

Bhargava ON (2015) Evolution of the Tethyan and Karewa successions in Kashmir: a synthesis. J Palaeontol Soc India 60(1):51–72

Bhatti Z, Zhao J, Khan NG, Hussain SST (2018) Structure of crust and upper mantle beneath NWHimalayas, Pamir, and Hindukush by multi-scale double-distance seismic tomography. Earth Planet Sci Lett 281:92–102

Bilham R (2004) Earthquakes in India and the Himalaya: tectonics, geodesy and history. Ann Geophys 47(839–858):2004

Bilham R, Gaur VK, Molnar P (2001) Himalayan seismic hazard. Science 293:1442–1444

Braun J (2010) The many surface expressions of mantle dynamics. Nat Geosci 3:825–833. https://doi.org/10.1038/ngeo1020; Brookfield ME (1993) The Himalayan passive margin from Precambrian to Cretaceous. Sedimentary Geol 84:1–35. Bull 3:25–27

Brunnschweiler RO (1966) On the geology of the Indoburman ranges. J Geol Soc Aust 13:137–194

Cai F, Ding L, Laskowski AK, Kapp K, Wang H, Xu Q, Zhang L (2016) Late Triassic palaeogeographic reconstruction along the Neo-Tethyan Ocean margins, southern Tibet. Earth Planet Sci Lett 435:105–114

Calkins JA, Offield TW, Abdullah SKM, Ali ST (1975) Geology of Southern Himalayas in Hazara, Pakistan. US Geological Survey Professional Paper 716-C.

Chakrbarti DK, Verma RP, Gautam AM, Narasimha S (1987) Geomorphological map of northeastern part of India – a landsat overview. Geol Surv India Rec 115

Chang CF, Cheng HL (1973) Some tectonic features of the Mt. Jolmo Lungma area, Southern Tibet, China. Sci Sin 16(2):257–265

Cloetingh S, Willett SD (2013) Topo-Europe: understanding of the coupling between the deep Earth and continental topography. Tectonophysics 602:1–14. https://doi.org/10.1002/2013eo050002

DeCelles PG, Gehrels GE, Quade J, Ojha TP, Kapp PA, Upreti BN (1998) Neogene foreland basin deposits, erosional unroofing, and the kinematic history of the Himalayan fold-thrust belt, western Nepal. Geol Soc Am Bull 110(1):2–21

Dhakal S (2014) Evolution of geomorphologic hazards in Hindu Kush Himalaya, In: Nibanupudi HK, Shaw R (ed) Mountain hazards and disaster risk reduction, Springer Japan

Ding L, Qasim M, Jadoon IAK et al (2016) The India-Asia collision in north Pakistan: insight from the U-Pb detrital zircon provenance of Cenozoic foreland basin. Earth Planet Sci Lett 455:49–61

Dortch JM, Owen LA, Haneburg WC, Caffee MW, Dietsch C, Kamp U (2008) Nature and timing of large landslides in the Himalaya and Transhimalaya in northen India. Quat Sci Rev 28:1037–1054

Durr SB (1996) Provenance of Xigaze fore-arc basin clastic rocks (Cretaceous, south Tibet). Geol Soc Am Bull 108:669–684

Evans JP (1964) The tectonic framework of Assam. J Geol Soc India 5:80–96

Evans JP, Forster CB, Goddard JV (1997) Permeability of fault-related rocks, and implications for hydraulic structure of fault zones. J Struct Geol 19:1393–1404

Feng M, An M, Zhao W, Xue G, Mechie J, Zhao Y (2011) Lithosphere structures of northeast Tibetan Plateau and their geodynamic implications. J Geodyn 52(5):432–442

Fuchs G, Widder RW, Tuladhar R (1988) Contributions to the geology of the Annapurana range (Manang area Nepal). Jahrb Geol Bundesanst 131:593–607

Hossain MS, Khan MSH, Chowdhury KR, Abdullah R (2018) Synthesis of the tectonic and structural elements of the Bengal basin and its surroundings. In: Mukherjee S (ed) Tectonics and structural geology: Indian context. Springer Geology. Springer, Cham, 455p

Hubbard J, Shaw JH (2009) Uplift of the Longmen Shan and Tibetan plateau, and the 2008 956 Wenchuan (M=7.9) earthquake. Nature 458(7235):194–197. https://doi.org/10.1038/Nature07837

Islam MS, Alam MJ (2009) Geological aspects of soil formation of Bangladesh. https://www.researchgate.net/publication/268391538

Jadoon IAK (2019) The behavior of Indian lithospheric mantle beneath the western Himalayan syntaxis: quaternary geology and landform in the western Himalayan mountains. Pakistan Science Foundation Project Report, Department of Earth Sciences,

COMSATS University Islamabad, Abbottabad Campus, Pakistan, North Pakistan
Jadoon IAK, Ding L, Jadoon SRK, Bhatti ZI, Shah STH, Qasim M (2021) Lithospheric deformation and active tectonics of the NW Himalayas, Hindukush, and Tibet. Lithosphere, 17pp. Article ID 7866954. https://doi.org/10.2113/2021/7866954
Kaneda H, Nakata T, Tsutsumi H et al (2008) Surface rupture of the 2005 Kashmir, Pakistan, earthquake and its active tectonic implications. Bull Seismol Soc Am 98(2):521–557
Khan MA, Jan Q, Weaver BL (1993) Evolution of the lower crust in Kohistan, N. Pakistan: temporal arc magmatism through early, mature and intra-arc rift stages. In: Treloar PJ, Searle MP (eds) Himalayan Tectonics. Geological Society, London, Special Publications, vol 74, pp 123–138
Khan A, Ahmad S, Kurashid S (2012) Geology and geomorphology of the Manipur Valley using digitally enhanced satellite image and SRTM DEM in the Eastern Himalaya, India. Int J Geosci 3:1010–1018
Kufner SK, Schurr B, Ratschbacher L, Murodkulov S, Abdulhameed S, Ischuk A, Metzger S, Kakar N (2018) Seismotectonics of the Tajik basin and surrounding mountain ranges. Tectonics 37:2404–2424. https://doi.org/10.1029/2017TC004812
Kufner SK, Kakar N, Bezada M et al (2021) The Hindu Kush slab break-off as revealed by deep structure and crustal deformation. Nat Commun 12:1685. https://doi.org/10.1038/s41467-021-21760-w
Kumar G, Agarwal NC (1975) Geology of the Srinagar – Nandprayag area (Alaknanda valley), Chamoli, Garhwal and Tehri Garhwal districts, Kumaun Himalaya, U.P. Himalyan Geol 5:29–59
Mehdi HS, Kumar G, Prakash G (1972) Tectonic evolution eastern Kumaun Himalaya: a new approach. Himalyan Geol 2:481–501
Metcalfe I (1996) Pre-cretaceous evolution of SE Asian terranes. In: Hall R, Blundell DJ (eds) Tectonic evolution of Southeast Asia. Geological Society, London, Special Publications, vol 106, pp 97–102. https://doi.org/10.1144/GSL.SP.1996.106.01.09
Molnar P, Bendick R (2019) Seismic moments of intermediate depth earthquakes beneath the Hindu Kush: active stretching of a blob of sinking thickened mantle lithosphere? Tectonics 38(5):1651–1665
Muminullah M (1978) Geology of the northern part of the Chittagong District, Bangladesh. Records of The Geological Survey of Bangladesh, vol II, Pt 3, pp 1–18
Nandy DR (1980) Tectonic patterns in Northeast India. Indian J Earth Sci 7:103–107
Norbu C, Baillie I, Tshering D, Tsheten D, Tamang HB, Kado T, Hutcheon A (2003) Provisional physiographic zonation of Bhutan. J Bhutan Stud 8(1):25–31. National Soil Services Centre, Semtokha, Ministry of Agriculture, PO Box 119, Thimphu, Bhutan
Owen LA, Caffee MW, Bovard KR, Finkel RC, Sharma MC (2006) Terrestrial cosmogenic nuclide surface exposure dating of the oldest glacial successions in the Himalaya orogen: Ladakh Range, northern India. Geol Soc Am Bull 118:383–392
Parray KA (2011) Karewas: a geological treasure and heritage of Kashmir. Curr Sci 100:805–806
Perry M, Kakar N, Ischuk A, Metzger S, Bendick R, Molnar P, Mohadjer S (2019) Little geodetic evidence for localized Indian subduction in the Pamir-Hindu Kush of central Asia. Geophys Res Lett 46:109–118. https://doi.org/10.1029/2018GL080065
Powell MCA, Conagan PJ (1973) Plate tectonics and the Himalayas. Earth Planet Sci Lett 20(1):1–12
Rowley DB (1996) Age of initiation of collision between India and Asia: a review of stratigraphic data. Earth Planet Sci Lett 145:1–13
Saklani PS (1971) Structure and tectonics of the Pratapnagar area, Garhwal Himalaya. Himalyan Geol 1:75–91
Schneider DA, Edwards MA, Kidd WSF, Asif Khan M, Seeber L, Zeitler PK (1999) Tectonics of Nanga Parbat, western Himalaya: synkinematic plutonism within the doubly vergent shear zones of a crustal-scale pop-up structure. Geology 27(11):999–1002
Seeber L, Armbruster JG, Quittmeyer RC (1981) Seismicity and continental subduction in the Himalayan arc. In: Gubta HK, Delany FM (eds) Zagros-Hindukush-Himalaya—geodynamic evolution. Geodynamic series, vol 3, pp 215–242
Shandilaya AK (1984) Studies on structure and morphotectonics between Srinagar Thrust and Alaknanda Fault, Garhwal Himalaya. PhD thesis of Garhwal University, Uttrakhand
Sorkhabi RB, Macfarlane A (1999) Himalaya and Tibet: mountain roots to mountain tops. In: Macfarlane A, Sorkhabi RB, Quade J (eds) Himalaya and Tibet: mountain roots to mountain tops, vol 328, pp 1–8. Geological Survey of America Special Paper
Stöcklin J (1980) Geology of Nepal and its regional frame. J Geol Soc Lond 137:1–34. https://doi.org/10.1144/gsjgs.137.1.0001
Wadia DN (1919) Geology of India – for students. Macmillan and Co., Limited ST. Martin's Street, London
Wang JG, Wu FY, Garzanti E, Hu X, Ji WQ, Liu ZC, Liu XC (2016) Upper Triassic turbidites of the northern Tethyan Himalaya (Langjiexue Group): the terminal of a sediment-routing system sourced in the Gondwanide Orogen. Gondwana Res 34:84–98
Wilson L, Wilson B (2014) Ore deposits. Welcome to the Himalayan Orogeny. http://www.geo.arizona.edu/geo5xx/geo527/Himalayas/ores.htm
Windley BF (1995) The evolving continents, 3rd edn. Wiley, Chichester, p 526p
Woehrer M (1992) Report on forest ecology/plant sociology and hydro-geomorphology. Integrated Forest Management Project, Ura
Yeats RS, Madden C (2003) Damage from the Nahrin, Afghanistan, earthquake of 25 March 2002. Seismol Res Lett 74:305–311

Yin A, Harrison TM (2000) Geologic evolution of the Himalaya Ttibetan orogen. Annu Rev Earth Planet Sci 28:211–280

Zaw K, Swe W, Barber JA, Crow MJ, New YY (2017) Introduction to the geology of Myanmar. Geol Soc Lond MemS 48:1–17. https://doi.org/10.1144/M48.1

Zeitler PK, Chamberlain CP (1991) Petrogenetic and tectonic significance of young leucogranites from the northwestern Himalaya, Pakistan. Tectonics 10 (4):729–741

Zhang PZ, Shen Z, Wang M et al (2004) Continuous deformation of the Tibetan plateau from global positioning system data. Geology 32(9):809–912

Zhisheng A, Kutzbach JE, Prell WL, Porter SC (2001) Evolution of Asian monsoons and phased uplift of the Himalaya-Tibetan Plateau since Late Miocene times. Nature 411:62–66

5 Soil Microbes and Biofertilizers

Abstract

Microorganisms are ubiquitous in nature and form vital components of all known ecosystems on earth. Their ubiquity is attributed mainly to their small size, easy dispersal, ability to survive and multiply in diverse habitats, including anaerobic and other extreme conditions, their metabolic versatility, and flexibility to utilize wide substrates as nutrient sources. One of the fascinating aspects of microorganisms is that some have evolved to thrive under conditions that are too harsh for animals as well as plants. The microorganisms that thrive under extreme environments, from polar deserts to geothermal springs, are referred to as extremophiles which are single-celled microorganisms belonging to two domains of life—bacteria and archaea. Soil microorganisms play an important role in nutrient cycling within soil–plant ecosystem. Microorganisms constitute < 0.5% (w/w) of the soil mass, yet they have a major impact on soil properties and processes. Microbes are also unique in the soil in that they can alter the solubility of soil mineral components, reduce organic compounds to essentially undetectable levels, modify soil structure, oxidize inorganic compounds, and use a variety of soil components as electron acceptors, viz., merely through growth and metabolism. Soil microbial biomass is a primary catalyst of biogeochemical processes as well as energy and nutrient reservoirs. The fertility of natural soil depends on the rate of turnover of organic matter brought about by the activities of soil microorganisms such as nitrogen fixation; mineralization of N, P, and S; and organic matter transformation. Large-scale use of microorganisms in agriculture may be considered as a useful proposition for they are capable of exerting multiple effects on fertilizers, pesticides, and plant growth regulators. Chemical fertilizers have played a significant role in the green revolution but excessive use of them has led to a reduction in soil fertility and to environmental degradation. Little is known of the mechanisms of Al toxicity to soil microorganisms despite the fact that this may be the major factor limiting microbial growth and activity in acid soils. Bacteria may be able to modify their local environment to produce favorable micro-sites. Fungi are generally more tolerant to soil acidity than bacteria, but the reason is not yet clear. Soil temperature, soil moisture, aeration, and the presence of heavy metals also have important bearings on the microbial population. The major groups of soil microorganisms observed in the Himalayan soils include representatives of bacteria, fungi, actinomycetes, protozoa, algae, viruses, nematodes, and mite populations. Despite this diversity, bacteria, fungi, and actinomycetes are the most prevalent in this region.

U. C. Sharma et al., *Soils in the Hindu Kush Himalayas*, Geography of the Physical Environment,
https://doi.org/10.1007/978-3-031-11458-8_5

Keywords

Biofertilizers · Enzyme activities · Phosphate solubilizing microorganisms · Soil fauna · Soil microbial activities

5.1 General Description

Forming vital components of known ecosystems on earth, the microorganisms are ubiquitous in nature. This is mainly due to their small size, easy dispersal, ability to survive and multiply in diverse habitats, including anaerobic and other extreme conditions, their metabolic versatility, and flexibility to utilize wide substrates as nutrient sources. The most interesting feature of microorganisms is that some of them can thrive under conditions that are too harsh for animals as well as plants. During the last three decades, it has been widely recognized that majority of such environments are inhabited by different and varied types of microbial communities. The microorganisms that survive under extreme environments, from polar deserts to geothermal springs, are referred to as extremophiles which are single-celled microorganisms belonging to two domains of life—bacteria and archaea. The microorganisms can tolerate extremes of temperature, pH, oxidation–reduction potentials, salinity and humidity, and various combinations of these, found in diverse terrestrial and aquatic habitats. The number of extremophiles known to exist in biotopes with environmental extremes has rapidly grown in recent years. This increase in the known microbial biodiversity is based both on laboratory cultures of isolated microorganisms and on the basis of characterization of 16S rDNA sequences recovered directly from the environment (Stolp 1988; Seckbach 2000; Budhiraja et al. 2002; Satyanarayana et al. 2005). Soil microorganisms play an important role in nutrient cycling within the soil–plant ecosystem. All macro and micronutrients that are essential for plant growth would have to be supplied from exogenous sources.

Agriculture in the Hindu Kush Himalayas (HKH) seems to be under stress in view of the pressure of the population. Nearly two-thirds of the population are rural and their survival depends mostly on agriculture in association with livestock, fishery, household industry, and other vocations. In the HKH region, the farmers cannot use the recommended doses of fertilizers, partly due to their non-availability at the right time and place and partly due to the capacity of the cultivators to purchase and apply full dose because of poverty. In this context, microbial inoculants are one of the viable options to bridge the gap between the demand and supply for various plant nutrients to sustain the crop productivity and soil health. Several studies conducted in India and elsewhere showed that *Rhizobium* inoculation of legumes helps in increasing yields by 15–30% with an additional yield and N uptake by 50–300 and 10–15 kg ha^{-1}, respectively (Rao 2003).

Because of the high increase in fertilizer prices and low purchasing power of the farmers of the region, it is very difficult to use chemical fertilizers in recommended quantities and balanced proportions. There is a need to exploit the beneficial impact of soil microbes on agricultural productivity to reduce dependence on fertilizer use.

5.2 Microbial Communities in Soils

For the cycling of carbonaceous, nitrogenous, and phosphorus compounds and for all plant-pathogenic reactions as well as biological transformations of xenobiotic compounds added to the soil, soil microorganisms are involved. Soils are natural storehouses of genetic information of soil organisms, many of which may be unculturable and hence not known to the scholars. Understanding the function of the soil ecosystem in relation to ever-changing soil conditions is the key to understanding the basic mechanisms of soil productivity. Soils of this region are diverse in their genesis, as well as chemical, physical, and biological characteristics. These variations in soils are reflected in many cases in their microbial composition. Factors like moisture, vegetation, pH, liming, use of organic manures, and management techniques show considerable effects on the distribution of different microbes in the soils. The importance of plant roots and soils in their immediate vicinity in governing the kind and number of microbes as well as their activities is widely recognized.

The soils of the Himalayas are rich in organic matter. These soils contain an appreciable amount of Al, Fe, and Mn in soil solution. The microbial component of the legume–*Rhizobium* symbiosis is more sensitive to acidity than the host plant, and Al appears to enter the bacterial cell and interfere with DNA replication (Wood 1989). Even then vigorous nodulation is found in the soybean, groundnut, and other legumes grown in the region. An acidophilus population may exist in a soil because of some protected micro-sites in the soils that have relatively higher pH values. The acidity of rhizospheric soils can be different from the acidity of non-rhizospheric soils and also localized increase of acidity occurs at the clay sites. Bacteria may be able to modify their local environment to produce favorable micro-sites. Fungi are generally more tolerant to soil acidity than bacteria, but the reason is not yet clear. Soil temperature, soil moisture, aeration, and presence of heavy metals also have important bearings on the microbial population.

Microbial biomass, that is the total sum of all microorganisms present in soils, serves as a temporary sink for nutrients including N and can be used as an index of soil fertility. Most of the attention has been given to bacteria and fungi as these groups of microbes are the most important with reference to energy flow and nutrient transformations.

Measurements of microbial biomass provide a sensitive indication of changes brought about by soil management practices, long before such changes can be detected in the organic matter content of the soil (Scholter et al. 2003). Measurement of soil microbial biomass is important because it acts as an agent of transformation of all substances entering into the soil and represents a relatively labile pool of C, N, P, S, and micronutrients. The availability of these nutrients is governed by the rate of turnover through microbial biomass. In most of the soils, the soil microbial biomass C comprises about 1–4% of the soil organic C (SOC) and there is a close relationship between microbial biomass C and amounts of soil organic C in arable, grassland, and forest soils (Banger et al. 1990). The soil microbial biomass increases or decreases in response to changes in soil management much more quickly than soil organic matter as a whole, where such changes can often take many years before being detected by chemical analyses. Thus, measurements of soil microbial biomass can be considered to act as an early warning of changing soil conditions. The relationship between microbial biomass C and total organic C can constitute to monitor heavy metal pollution. The major groups of soil microorganisms observed in the soils include representatives of bacteria, fungi, actinomycetes, protozoa, algae, viruses, nematodes, and mite populations. Despite this diversity, bacteria, fungi, and actinomycetes are the most prevalent in this region.

5.2.1 Bacteria

Bacteria are the smallest organisms in the soil. They are prokaryotes. The number of bacteria in 1 g of soil varies from one to three million. Bacteria are not uniformly distributed in the soil but are located in small colonies. One of the most important features of bacteria as a group is their biochemical versatility. They can metabolize a wide range of chemicals. Soil bacteria are considered saprophytic or parasitic in nature, with a few autotrophic (*Nitrosomonas* and *Nitrobacter*) and symbiotic (*Rhizobium*). Considerably less information is known about soil bacteria which are the most dominating community of soil microorganisms in the region. Talukdar and Barthakur (1981) studied the total number of bacteria as the number of *Azotobacter* sp. in the root suspension of certain rice varieties grown in soils of Assam. The highest count of bacteria was found in the root suspension of rice variety CH-63 under different treatments. The population of total bacteria in root suspension was 34.8, 44.0, 56.9, 49.8, and 62.2 million g^{-1} under control; application of N @ 120 kg ha^{-1}; application of N and organic matter; application of P as rock phosphate; and application of N, P, and organic matter, respectively. However, the highest bacterial count of 49.1 million g^{-1} was found in the root suspension of rice variety *Prasadbhog* when only organic matter was applied. The dose of organic matter was given through application of 20 g well-decomposed cow dung kg^{-1} of soil and that of P was 80 kg ha^{-1}. A similar trend was observed in the population of Azotobacter.

5.2.1.1 Nitrogen Fixation

A reliable source of fixed nitrogen is a requirement for sustaining all biomass components in the ecosystem in which the majority of this nutrient is provided by the mineralization process between organic and inorganic forms. Due to the inevitable losses of nitrogen from soils, some external supply of nitrogen is always required in terrestrial sites to maintain long-term productivity. Otherwise, fixed nitrogen reserves in the soil would slowly decline until biomass productivity is reduced or even precluded. Fortunately for the development of highly productive sustainable terrestrial systems, nitrogen inputs rectify nutrient losses. Total losses of fixed nitrogen from land ecosystems of approximately 160 to 225 kg yr^{-1} are replaced by inputs of 214 to 262 kg yr^{-1} (Knowles 1981). There are approximately 700 genera and about 13,000 species of legumes, only a portion of which have been examined for the capability to fix nitrogen. Rhizobial symbiosis with greater than 100 important legumes contributes half the annual quantity of biologically fixed nitrogen entering soil ecosystems.

5.2.1.2 Rhizobium

Seed inoculation with effective *Rhizobium* inoculant is recommended to ensure adequate nodulation and N_2 fixation for maximum crop productivity. Plant growth-promoting rhizobacteria (PGPR) were found to be synergistic with *Rhizobium* and their co-inoculation with *Rhizobium* showed an improvement in N_2 fixation, nutrient uptake, and yield of pulse crops. *Rhizobium* population in the soil is mostly governed by soil reaction, native soil fertility, and moisture content, and the nitrogen fixation by different legumes gets affected by their respective *Rhizobial* species. A sufficient number of species may not be present in the soil to cause effective nodulation of legumes; hence, seed inoculation with specific nodule bacteria before sowing is needed. The effectiveness of nitrogen fixation of the indigenous strains is generally higher than exotic strains. Performance of different locally isolated as well as exotic *Rhizobium* strains in an acidic sandy soil at Jorhat, Assam, showed that the strains from Pantnagar gave the highest yield of soybean, which was at par with the IARI-3 strain (Barthakur and Sarmah 1984). The strains IARI-3, JNKVV-2, and JNKVV-3 showed significantly higher performance in yield, respectively, and these strains are more adaptive to the acid soils of the Himalayan ecosystem.

The locally isolated strain of *Rhizobium,* i.e. RCI 04 recorded the highest nodule number (63.53), nodule dry weight (215.5 mg), and grain yield of soybean cv. PK-472 (1.693 t ha^{-1}) over the alien strains (TAL 377 and RCO 04) in an acidic soil at Iroisemba foot hills of Manipur (Chandra et al. (1995). Application of P increased the plant height and grain yield in both the inoculation treatments with RCI 04 and TAL 377.

The efficacy of different *Rhizobial* strains (TAL-1000, IGR-6, NC-92, and TCG-48) on groundnut in a Ultisol of Manipur showed that IGR-6 and NC-92 increased the pod yield and N and P uptake by pods and haulms significantly over control and TAL-1000 during 2002 (Raychaudhuri 2002), and similar results were reported during 2003 by Raychaudhuri and Raychaudhuri (2003).

Out of four strains of *Rhizobium* (T-169, JCG-48, T-1000, and NC-92) inoculated in groundnut to ascertain the infection/nodulation, TAL-1000 showed a good response in nodulation in maximum cultivars followed by JCG-48, T-169, and NC-92 (Raychaudhuri and Raychaudhuri 2003). However, inoculation of groundnut with T-169 resulted in more secondary roots per plant that may reflect in pod formation followed by T-1000 and NC-92.

Soil acidity, deficiency in macronutrients and Mo, and toxicity of Al and Mn adversely affect the N_2 fixation through symbiosis and liming regulates the growth of *Rhizobium* and its ability to nodulate legume crops. *Rhizobium* inoculation in soybean on an acidic clay loam soil (Typic Kanhaplahumult) of Manipur recorded an increase of 35–43% grain yield and nodule weight (Raychaudhuri et al. 1997b). Liming increased the Ca^{2+} levels in the soil which enhanced the root penetration of soybean into deeper soil layers and also induced the normal distribution of nodules on the tap root and lateral roots by *Rhizobium*. Lime pelleting provides a good environment for bacteria in the soil for their survival by neutralizing soil pH around the seeds. The strong acidic conditions affected *Rhizobium* growth and survival, which increased by 85% and improved the organic matter status of the soil and N and P uptake by soybean due to liming. Lime and *Rhizobium* interaction on yield and nutrient uptake was also found significant. It also showed an additive residual effect by increasing the mustard yield by 15%. Phosphorus use efficiency and benefit/cost ratio were maximum with the application of 13 kg P ha^{-1} along with *Rhizobium* inoculation, but the total grain yield and net returns were maximum (23.4 q ha^{-1} and Rs. 16,704, respectively) with 26 kg ha^{-1} along with *Rhizobium* inoculation (Raychaudhuri et al. 1997a).

Application of SSP in combination with *Rhizobium* inoculation in black gram (cv T-9) produced a 69.1% increase in nodulation over control in an acidic soil of Tripura (Datta and Laskar 1990). A maximum increase in nodule weight from 14.1 to 39.3 mg was recorded with the application of lime and inoculation of *Rhizobium*. Inoculation of both *Bacillus firmus* and *Rhizobium* increased the P solubilizing power of soil from 1.75 mg P to 2.58 mg P. Application of phosphate in combination with *Bacillus firmus* and *Rhizobium* could bring out a substantial increase in P solubilizing power of soil.

Combined application of lime + phosphate + *Rhizobium* (IGR 40) and PSB (*Bacillus polymyxa*) increased 31–45 and 18–22% in nodulation, while the nodule weight/plant increased by 50 and 72% over control at 30 and 60 DAS, respectively, in groundnut (Datta 2001). The pod and haulm yields of groundnut (cv JL-24), nutrient uptake, nodule number, and nodule weight were significantly increased with liming, phosphorus, and microbial inoculations in an Ultisol of Manipur (Raychaudhuri 2002). *Bradyrhizobium* (TAL-1000) and PSM (*Bacillus polymyxa*) combination showed an increase of 24% in pod yield over control. Maximum pod yield (19.9 q/ha) was recorded with 2.5 t ha^{-1} liming + 50 kg P_2O_5 ha^{-1} + *Rhizobium* + PSM inoculation. *Rhizobium* and PSM alone could not increase the nodule weight significantly but in combination with phosphorus and lime showed a positively significant response. All the strains of *Rhizobium* except TAL-1000 could increase the pod and haulm yield of groundnut (cv JL-24) during 2003 and the strain NC-92 recorded the highest mean pod yield of groundnut. *Rhizobium* inoculated plots added about 15 kg N ha^{-1} over uninoculated plots (5.2 kg N ha^{-1}) in the soil at harvest, however, no significant difference in respect of oil content due to *Rhizobium* inoculation. The residual effect of *Rhizobium* inoculation showed a significant effect on the yield of mustard over control.

Of all the demonstrations on *Rhizobial* inoculation, 65% of trials gave a 20% or more increase in yield of lentils over uninoculated control and 67% of demonstrations showed 15% more yield response in French bean. *Rhizobial* inoculation in soybean, groundnut, arhar, moong, broad bean, gram, and black gram showed a 10% higher yield in 72% field demonstrations over control (100% NPK).

The positive response of various legumes and non-legumes to microbial inoculants (*Rhizobium, Azotobacter, Azospirillum,* and phosphate solubilizing microorganisms) in different parts of the North Eastern Region indicates that neither soil pH nor strain source is the barrier to usage of these organisms in the region. However, there is always scope for better responses with more efficient strains which may be local or alien. Inoculation of legumes with *Rhizobium* consistently gave higher yield in most of the crops suggesting that at low pH also, the benefits of inoculation can be obtained. The nutrient dynamics in an acid soil have more influence on the legumes rather than inoculated *Rhizobium*, once they infect the roots and form nodules.

5.2.1.3 Bradyrhizobium

High numbers of rhizobia are important for successful nodulation, especially in soils where competitive native strains are present. Lime coating @ 25 g kg^{-1} seed for 30 s for *Bradyrhizobia* inoculated seeds in respect of soybean and ricebean and *Rhizobial* strains for French bean and broad bean in an acidic soil of Manipur and Sikkim (EIHR) showed a significant improvement in nodulation of these legume crops (Chandra et al. 1995, Sanjeev Kumar et al. 2001). Legume seed inoculation with *rhizobia* along with a sticking agent greatly increases the number of *rhizobia* adhering to the seed. Using adhesive materials like Gum Acacia recorded the highest number of nodules and nodule weight in soybean, French bean, broad bean, and ricebean over the other adhesives like sugar and molasses. Locally isolated *Bradyrhizobia* of *Glycine max* strain RCI 04 at Manipur showed nodule formation at 10 days after sowing and it was almost doubled at 15 days after sowing irrespective of the colony type. Symbiotic effectiveness of isolate in respect of total plant weight and nodule weight had not varied significantly within the colony type. Maximum root weight (0.31 g $plant^{-1}$) and N uptake (46.75 mg $plant^{-1}$) were registered with a small colony. The results showed that no relation can be drawn between the number of nodules occupied/plant and their N_2 fixation ability but nodule weight may be considered for the N_2 fixing ability.

The nodulation pattern in important legumes in Manipur indicated that 27% of the fields, out of 20 locations, showed excellent nodulation (>150) in French bean and other beans including cowpea, whereas 36% showed fair nodulation (25–50). In the case of soybean, 25% of the fields supported excellent nodulation, 50% showed fair to good (50–100) nodulation, and 25% poor (<25). Pea being grown abundantly, 80% of areas supported fair to good nodulation and 29% of the fields were excellent nodulated. The native *Rhizobial* population (cowpea group) varied between 3.17 and 1.27×10^3 g^{-1} of dried soil. The soils supporting higher nodulation were found to have greater *Rhizobial* count (Anonymous 2004).

5.2.1.4 Azospirillum

Inoculation of *Azospirillum* and phosphate solubilizing microorganisms (PSM) without any fertilizer input in paddy showed a 2–43% increase in yield over control at various locations in Manipur, Assam, and Sikkim. In all the demonstrations, inoculation of *Azospirillum* and PSM not only supplemented 15–30 kg N ha^{-1} but also gave higher yields over 100% NPK (Control). Low pH or anaerobic conditions due to submergence were not the limiting factors for building up of root–*Azospirillum* association in rice fields. These microbial inoculants not only provide additional returns but also minimize the expenditure on costly fertilizer inputs.

The response of *Azospirillum* + PSM in maize at different locations was recorded as an 8–15% increase in yield with 25–50% N reduction over 100% NPK control. Low or negative response in some locations might be due to the presence of high mineralizable N or native microflora in the soils.

Application of lime @ 1 t ha^{-1} + *Azospirillum* + 60 kg P_2O_5 ha^{-1} recorded the highest grain and straw yields (41.1 and 84.5 q ha^{-1}) over control (14.4 and 31.8 q ha^{-1}, respectively) in an acidic (pH 4.6) clay soil of Manipur (Raychaudhuri and Singh 2001). The lime + *Azospirillum* interaction was significant for grain yield. Inoculation of *Azospirillum/Azolla* in combination with a lower dose of N produced a significantly higher grain yield of rice (cv RCPL-1-87-8) over 100% NPK in an acidic soil of Meghalaya (Panwar and Hazarika 2002).

Isolation of non-symbiotic N fixers in rhizospheric and non-rhizospheric rice soils at 20 locations from the East Imphal district of Manipur indicated that *Azospirillum* populations were in the range of 10.3×10^2 to 6.27×10^4 g^{-1} of soil (Anonymous 2004). The total population of N fixers varied between 33.4×10^2 to 684×10^2 and 27.2 to 307×10^2 g^{-1} of dried soil. However, the rhizospheric soils contained 3 to 50 times more MPN of *Azospirillum*. The free-living aerobic N fixers were found to be higher in non-rhizospheric zones than in the rhizospheres (Anonymous 2004).

The N fixers as collected from the rhizosphere of rice soils of Manipur were found gram negative and catalase positive and 95% of the isolates utilize sucrose and 10% utilize glucose. Most of the strains of Azospirillum do not require biotin for their growth. Most of the isolates (71%) fixed less than 10 mg N g^{-1} of malate consumed. The strains I_8 and I_{11} fixed 12.8 and 18.6 mg N g^{-1} of malate consumed, respectively. The low N fixing capabilities of the majority of the strains indicate the scope for application of better strains having higher N fixing potentials in most of the soils for the higher benefit to the crops. No biotin requirement for the growth of most of the isolates suggests the possibility of prevalence of *Azospirillum brasilense* in the rice rhizospheres of Manipur soils over *Azospirillum lipoferum* (Anonymous 2004).

5.2.1.5 Azotobacter

The response to *Azotobacter*, a free-living N_2-fixer, is governed by the soil factors like pH, mineralizable N, organic carbon, temperature, etc. rather than plant factors, because it is non-specific to crops. Inoculation of *Azotobacter* in wheat showed an increase of 4–29% yield over control (100% NPK) at several locations in the NE region.

Azotobacter inoculation in combination with 50% NPK (40-30-20 kg N, P_2O_5, and K_2O ha^{-1}) and FYM @ 10 t ha^{-1} in a clay loam acidic (pH 4.8) soil at Langol Hill of Manipur resulted in highest yields in maize–mustard crop sequence and the yields were at par with 100% NPK (Raychaudhuri and Singh 2001). Intercropping with soybean has resulted in curtailing half of the recommended dose of NPK without affecting crop productivity. The residual effect on mustard yield was also found significant besides higher uptake of N and P by grains of maize and mustard. A significantly higher grain yield of paddy was recorded with the application of 60 kg N ha^{-1} + FYM @ 5 t ha^{-1} + *Azotobacter* (37.95 q ha^{-1}) followed by 60 kg N ha^{-1} + FYM @ 5 t ha^{-1} + *Azospirillum* (36.3 q ha^{-1}) in an acidic Alfisol of Meghalaya (Majumdar 2003). The available N content in the soil was improved with the inoculation of *Azotobacter* and *Azospirillum.* The field demonstrations with *Azotobacter* on mustard showed a wide variation in yield response ranging from 4 to 27% over uninoculated control.

The results of farmers' field demonstrations with *Azotobacter* + PSM on potato showed that the biofertilizer treatment supplemented 30 kg N with an increase in yield of 12.4% over 100% NPK (Control).

Inoculation of *Azotobacter* in cabbage and cauliflower at various locations in the NE region also showed a yield response of 2–28% over 100% NPK control besides supplementing 25–50% N fertilizers. Inoculation of *Azotobacter* strain (Mac-27) + *Azospirillum* (C_2) with maize in combination with FYM @ 5t ha^{-1} resulted in 20.4% higher grain yield of maize under Manipur conditions and a further increase of 13.8 and 29.1% yield when bacterial inoculation coupled with 50 and 100% of the recommended doses of NPK application (Raychaudhuri and Raychaudhuri 2003). Inoculation of *Azotobacter* + *Azospirillum* in combination with the application of FYM @ 10 t ha^{-1} recorded a significantly higher yield of maize (50.03 q ha^{-1}) in an acid soil of Manipur and become at par when coupled with recommended doses of NPK (52.21q ha^{-1}). The nutrient uptake also showed a similar trend as that of the grain yield. Among the local isolates, MK_4 gave maximum grain and stover yields of maize. The N fixing populations as enumerated in Nfb media at 45 DOS did not vary significantly, although a higher count was observed in inoculated plots than that in uninoculated. Of all the local isolates from rice rhizospheres, Azsp 3 and 5 were more efficient in biomass yield/plant (Anonymous 2004).

Presumably, *Azotobacter* and *Azospirillum* are more affected by the soil pH. The response of paddy, wheat, mustard, toria, potato, and vegetable crops has shown the economic viability of inoculation. However, there is still scope for better strains in terms of acid tolerance, survivability, compatibility, and other beneficial attributes local or outside the region. Thus, more vigorous and sincere efforts are needed till the most efficient strains are identified for the region and their cultural compatibility is fully justified for mass production. Nevertheless, the presently used available standard strains are competent enough to harvest significant benefits.

5.2.2 Fungi

Fungi perform an essential role in the functioning of the ecosystem (Doran and Parkin 1996; Hawksworth 2001). These are one of the most important functional groups of soil microbes. They are important for making the nutrients like C, N, P, and S accessible to the plants in the soil because of their capability to decompose complex macromolecules. Kennedy and Gewin (1997) reported on the significance of the fungal mycelium in playing an important role in the stabilization of soil, thereby helping in increasing the water-holding capacity of the soil. It is estimated that only 5% of fungal species have been described (Hawksworth 2001), even though the fungi are important in ecosystem functioning and actual species richness is thought to be much higher (Schmit & Mueller 2007). Studies on soil fungal diversity in relation to habitat, climate, and altitudinal gradient are scarce and limited in India (Pandey et al. 2006).

The future responses of biodiversity and functioning of ecosystems to environmental changes can be better interpreted if we have a clear understanding of the altitudinal distribution of microfungal assemblages. Bellis et al. (2007) reported that investigations on the analysis of vegetation changes along known environmental gradients would give good information on the interactions between community structure and the environment. The important plant species recorded from foothills of Himalayas were Altingia excels (*Hamamelidaceae*), *Dipterocarpus retusus (Dipterocarpaceae)*, *Duabanga grandiflora (Sommeratiaceae)*, *Mesua ferrea (Calophyllaceae), Terminalia indica (Combretaceae), Bombax ceiba (Malvaceae), Elaeocarpus sp. (Elaeocarpaceae), Quercus sp. (Fagaceae), and Gmelina arborea (Lamiaceae)*. In subtropical forests, *Actinodophne obovata (Lauraceae), Illicium griffithii (Schisandraceae), Quercus sp. (Fagaceae), Michelia oblonga (Magnoliaceae), Pinus roxburghii (Pinaceae),* and *Pinus wallichiana* (*Pinaceae*) were dominant. Plant species like *Abies spectabilis (Pinaceae), Acer pectinatum (Sapindaceae), Alnus nepalensis (Betulaceae), Castanopsis indica (Fagaceae), Pinus roxburghii (Pinaceae), and Rhododendron arboretum (Ericaceae)* were dominant in temperate forests. In subalpine and rhododendron forests, *Alnus nephalansis (Betulaceae), Rhododendron nivale (Ericaceae), Rhododendron anthopogon (Ericaceae), Sedum sp. (Crassulaceae), and Rhodiola sp. (Crassulaceae)* were common. Likewise, in degraded forest and abandoned jhum, plants like *Citrus sp. (Rutaceae), Callicarpa arborea (Lamiaceae), Macranga denticulate (Anacardiaceae), Clerodendrum sp. (Lamiaceae), and Chromolaena odorata (Asteraceae)* were common. Widden (1987) explained that higher fungal diversity in subtropical forests as compared to the tropical forests was due to the favorable physico-chemical properties of subtropical soil and stimulatory effects of vegetation on soil fungi (Widden 1987).

Shifting cultivation in the eastern Himalayan region drastically changes the microbial communities in the soils. During summer, in non-rhizosphere forested soil of East Garo Hill district of Meghalaya, *Fusarium* sp showed the highest occurrence followed by *Penicillium, Aspergillus, and Trichoderma* sp., while *Rhizopus*, *Mucor*, and *Alternaria* sp showed comparatively low occurrence. In jhum areas, *Aspergillus* sp showed the highest count followed by *Fusarium, Penicillium, Trichoderma, Mucor, and Rhizopus* sp, whereas *Alternaria* and *Geocladium* sp showed relatively low (Ahmed and Das 2002). In forested soil, rhizosphere fungi, viz., *Penicillium* sp showed the highest occurrence followed by *Trichoderma, Fusarium, Aspergillus, Rhizopus, Mucor,* and *Helminthosporium*, while *Penicillium, Trichoderma,* and *Aspergillus* sp were dominant in jhum areas.

During the winter season, *Fusarium* sp showed the highest count in non-rhizospheric soils under forest and jhum areas followed by *Penicillium, Trichoderma,* and *Aspergillus*. However, in the rhizospheric soils from forested and jhum areas, *Penicillium* showed the highest counts followed by *Fusarium* and *Trichoderma*. The pH values increased from 5.5 in the forest area to 6.04–6.5 in jhum area, which may influence the fungal microflora during the winter months. Since fungi are important degrading agents in the acid forest soils, the highest count of fungi in the forest soil does not interfere with the population growth of fungi, whereas the jhuming disrupts the whole setup and reduces the fungal population. In jhum areas, *Trichoderma, Penicillium,* and *Aspergillus* were considered early colonizers of burnt soil. These species have the greatest ecological tolerance to withstand extreme climatic variations. As a result of deforestation, the soil is exposed and is subjected to runoff loss. Fine clay particles are eroded more actively leaving the residual soil nutrient poor and resulting in low microbial growth (Ahmed and Das 2002).

Studies have shown that altitudinal gradient and bulk density have a negative effect, while soil temperature and soil pH had positive effects on the soil fungal communities (Sharma et al. 2015). They studied variations in fungal communities along 12 diverse vegetation zones and recorded 112 fungal types under 59 genera and 88 species from the selected soils. Subtropical evergreen forests showed maximum fungal diversity followed by tropical evergreen forests. Prominent fungal genera in most of the soils were *Oidiodendron* followed by *Acremonium, Cladosporium, Humicola, Aspergillus,* and *Penicillium.*

5.2.3 Azolla

The role of Azolla in curbing ammonia volatilization in flooded rice environments has been reported by Lejeune et al. (1999). *Azolla* stimulated rice root development and thereby CH_4 oxidation in the rhizosphere facilitates CH_4 transfer from water to air. Bharati et al. (2000) observed that *Azolla* intercropping decreased CH_4 emission due to an increased O_2 concentration in water. *Azolla* can also be used as a good digestible nitrogen source for pigs as well as the nutritive value and a positive effect on the yolk color of laying hens (Khatun et al. 1999). Besides fixing atmospheric nitrogen, *Azolla* controls weed population and reduces evaporation and *Azolla* flakes can be used as poultry feeds. Fresh *Azolla* has a good potential as a source of provitamin A. The growth and production parameters of fish were significantly improved by feeding *Azolla* in fish ponds. The results of the experiments also confirmed that rice–fish–*Azolla* system had positive effects on soil structure, chemistry, microflora, and crop productivity. Rice–fish–*Azolla* integrated farming system in China, rice–duck–*Azolla* system in Japan, and rice–fish–duck–*Azolla* system in the Philippines are being popular in terms of monetary returns and productivity of soils.

In the Northeastern India Himalayas, in spite of the abundance of perennial water sources, the potential of *Azolla* could not be exploited mainly because of the low biomass productivity of the local isolates and its poor survivability under cold winter months. Hazarika et al. (2003) reported that out of 13 *Azolla* species studied under agro-climatic conditions of Meghalaya, *Azolla caroliniana* was more promising with 95% survival and a biomass yield of 210 q ha^{-1}. Among the other varieties, *A. pinnata* (India) was also found better with 87% survival followed by *A. pinnata* (Thailand) with a biomass yield of 17.5 and 14.0 t ha^{-1}, respectively (Table 5.1).

A. caroliniana was found cold tolerant and can survive under acute moisture stress conditions (Hazarika et al. 2003). The addition of P through SSP and a water level of 15 cm are required to produce a higher biomass yield (17.2 q ha^{-1} $month^{-1}$). Application of 50% NPK + *Azolla* dual cropping produced the highest grain yield (28.15 q ha^{-1}) followed by 100% NPK (25.2 q ha^{-1}) and *Azolla* dual cropping alone (22.2 q ha^{-1}) in an acidic sandy loam soil of Meghalaya (Hazarika et al. 2003). Application of enriched *Azolla* compost (by adding rock phosphate @ 60 kg P_2O_5 ha^{-1}) @ 10 t ha^{-1} recorded a mean grain yield of 20.6 q ha^{-1}, with a yield response of 60% over control.

Table 5.1 Performance of different *Azolla* cultivars in different countries

Azolla cultivar	Survival (%)	Biomass yield (mean of 10 harvests)			Total yield (q* ha^{-1} yr^{-1})
		1993	1994	Mean	
A. pinnata (India)	87	16.0	19.0	17.5	175
A. pinnata (Bangladesh)	63	10.0	16.0	13.0	130
A. pinnata (Vietnam)	44	8.5	6.5	7.5	75
A. pinnata (Thailand)	79	13.0	15.0	14.0	140
A. pinnata (Baihata)	50	7.5	7.5	7.5	75
A. pinnata (Amingaon)	54	8.5	10.5	9.5	95
A. pinnata (Guwahati)	52	6.2	6.8	6.5	65
A. pinnata (Azara)	48	5.9	6.2	6.1	61
A. piliculoides	Nil	–	–	–	–
A. mexicana	Nil	–	–	–	–
A. ruba	Nil	–	–	–	–
A. microphylla	Traces	Traces	Traces	Traces	Traces
A. caroliniana	95	20.8	22.0	21	210
CD (P = 0.05)	–	2.8	3.1	–	–

* One q (quintal = 0.1 t ha^{-1})

Application of *Azolla* compost @ 2.5 t ha^{-1} and seed inoculation of PSM-27 individually recorded a pod yield of 19.1 and 20.3 q ha^{-1} of groundnut (cv GG-20) under agro-climatic conditions of Meghalaya (Hazarika 2003), whereas application of P_2O_5 @ 50 kg ha^{-1} + lime @ 2.5 t ha^{-1} + *Azolla* compost @ 2.5 t ha^{-1} produced maximum pod yield (26.9 q ha^{-1}). *Azolla* contains the following constituent on dry weight basis: N-4-5%, P-0.5-0.9%, K-2.0-4.5%, Ca-0.1-1.0%, Mg-0.5-0.65%, Fe-0.16-0.26%, Crude fat-3.0-3.36%, Crude protein-23-30%, soluble sugars-3.4-3.5%, starch-6.5%, Chlorophyll a 0.34–0.55%, and ash-10.5%. *Azolla* compost contains 1.51% N, 1.27–2.66% P, and 6.0–7.85% K (Hazarika 2003).

Azolla dual cropping with rice registered the highest grain yield of paddy (49.2 q ha^{-1}) with an increase of 28% over control in an acidic soil of Meghalaya during 2003, whereas *Azolla* green manuring and *Azolla* compost showed yield response of 11 and 16%, respectively, over control. *Azolla* dual cropping has contributed a net saving of 28 kg N ha^{-1} besides an improvement of 23% grain yield (Anonymous 2004).

5.2.4 Cyanobacteria and Algae

Cyanobacteria (blue-green algae) and *Chlorophyceae* (green algae) are mostly abundant close to the soil surface. Sufficient moisture and sunlight are essential requirements for their growth. Blue-green algae in association with *Azolla* fix atmospheric nitrogen in the rice fields. The North Eastern region has tremendous potential for exploration and utilization of blue-green algae as a biological nitrogen source.

Highest grain yields of 54.7 and 51.1 q ha^{-1} of rice (cv RCPL-1-87-4) were recorded in 2000 and 2001, respectively, in an Ultisol of Mizoram with the application of BGA (a mixture of *Anabaena variabilis, Nostoc muscorum, Aulosira fertilissima,* and *Tolypothrix tenuis*) + 80 kg N ha^{-1}, which was higher than 120 kg N ha^{-1} (Laxminarayana 2003). Application of straw-based culture of BGA alone showed non-significant increase in grain yield over control. The BGA inoculation also showed a significant response on nutrient uptake by grains and straw. Integrated application of BGA and fertilizer N improved the organic matter and available nutrient status of the soil and the BGA inoculation reduced the expenditure on fertilizer cost equivalent to 30–40 kg N ha^{-1}.

5.2.5 Phosphate Solubilizing Microorganisms

Microorganisms can affect P dynamics of the soils in many ways including the following: (1) an increase in biomass of phosphate solubilizing microorganisms (PSM) (Gyaneshwar et al., 2002); (2) Substantial release of P following the death of the microbial cells since microbes immobilize up to 10% of the total soil P (Richardson 2001); and (3) mineralization of organic P to inorganic P through phosphatase activity (Oehl et al. 2001). A large number of heterotrophic and autotrophic microorganisms, such as bacteria, fungi, actinomycetes, and cyanobacteria, are reported to solubilize insoluble phosphate. Among the soil bacterial communities, strains from the bacterial genera *Pseudomonas*, *Burkholderia*, *Bacillus*, *Rhizobium*, and *Enterobacter* and *Aspergillus* and *Penicillium* from fungal genera (Wakelin et al. 2004) have been reported as powerful phosphate solubilizers. Apart from these, the other bacterial genera reported as efficient phosphate solubilizers include *Agrobacterium*, *Achromobacter*, *Micrococcus*, *Aerobacter*, *Flavovobacterium*, *Erwinia* (Rodriguez and Fraga 1999), *Azotobacter*, and *Klebsiella* (Chung et al. 2005). The key processes involved in the solubilization of phosphate compounds by organic acids are attributed to acidification, chelation, and exchange reactions. However, many organisms that solubilize insoluble phosphates effectively in the laboratory are unable to do so in the natural field conditions (Richardson 2001).

Only about 25–30% of the applied phosphate is available to the crops in the year of its application and the remaining part is converted into insoluble forms. The efficiency of phosphatic fertilizers is very low due to chemical fixation within a short period of its application in the soil complex. In this situation, seed or soil inoculation of phosphate solubilizing microorganisms (PSMs) may benefit the crops by increasing the P availability from insoluble P sources. India's current reserves of rock phosphate are 140 Mt and conjunctive use of rock phosphate along with inoculation of P solubilizing microorganisms produces the highest crop yields and saves the P fertilizer cost (Dubey 1997). P solubilizing bacteria vary from soil to soil and constituted 1–50% and fungi 0.5–0.1% of the total respective population. Many microorganisms reduce the pH and solubilize P from rock phosphate.

Inoculation of P solubilizing bacteria (*Bacillus firmus*) in rock phosphate and poultry manure amended soils at Tripura caused a tangible increase in P to an extent of 4.27–8.70 and 8.45–16.73 kg/ha of Bray's P_1 and Bray's P_2, respectively (Datta and Laskar 1991). A significant increase is noted in P uptake by rice due to bacterial inoculation (Datta et al. 1992; Datta and Banik 1994) as well as inorganic P fertilization in poultry manure and rock phosphate amended soils. However, the P-fixation capacity of soil increased from 32% (127.2 mg P g^{-1}) in the normal soil to 62% (247.6 mg P g^{-1}) in SSP-treated soil and drastically reduced to 8% (31.6 mg P g^{-1}) under inoculation of *B. firmus* in combination with Purulia Rock Phosphate and poultry manure (Table 5.2) thereby suggesting the potency of *B. firmus* as an effective biofertilizer in P-deficient acid soils (Datta et al. 1982).

Datta and Banik (1997) showed that the bacterial inoculation in combination with rock phosphate is the best-suited phosphatic fertilizer for acid soils with the possible scope of increasing its efficacy by supplementing with organic manure.

Maximum yields of green gram were recorded with the application of rock phosphate and inoculation of P solubilizing bacteria (*Bacillus firmus*) under agro-climatic conditions of Tripura. It showed an increase of 49–64% in yield over control in various cultivars of moong. The efficiency of phosphobacterial strains (*Bacillus firmus*, *Pseudomonas striata,* and *Bacillus polymyxa*) on upland rice (cv TRC-87-251) under Tripura conditions showed that the seed inoculation of *Bacillus firmus* was more pronounced on grain and straw yields (24.8 and 58.0 qha^{-1}).

Table 5.2 Effect of phosphate in combination with phosphobactrin inoculation on various fractions (mg g^{-1}) of soil P

Treatments (50 kg $P_2O_5ha^{-1}$)	Saloid-P	Al-P	Fe-P	Ca-P	P-fixation capacity	P fixed/added
1. Control	Trace	14	32	28	12.72	0.32
2. SSP	0.80	16	56	34	247.6	0.62
3. Poultry manure	0.80	17	25	24	179.9	0.45
4. *Bacillus firmus*	1.07	16	23	23	78.4	0.19
5. PRP	1.07	14	32	35	55.8	0.14
6. MRP	0.54	16	28	31	78.4	0.19
7. PM + PRP	Traces	21	38	55	69.6	0.17
8. PB + MRP	Traces	28	45	39	163.3	0.41
9. PRP + BF	Traces	19	47	44	69.6	0.17
10. MRP + BF	1.07	14	30	30	60.4	0.15
11. PM + PRP + BF	1.33	13	22	34	31.6	0.08
12. PM + MRP + BF	1.61	15	28	45	64.9	0.16
CV (%)	–	23	29.9	25.3	59.9	61.8

Source Datta et al. (1982)

Spray inoculation of *Pseudomonas striata* and *Bacillus polymyxa* individually at both tillering and flowering stages of rice could produce a maximum grain yield of 28.3 and 27.9 qha^{-1}, respectively (Datta 2001). The residual effect of phosphobacterin through seed inoculation of rice showed a significant effect on grain yield of green gram (cv UPM-79-4-12). But the spray inoculation of phosphobacterin was of high efficacy in control compared to phosphate-treated plots. Bacterial inoculation of seed with PSM (*Bacillus polymyxa*) + *Rhizobium* (TAL-100) + PGPR-4 recorded 9.6 and 36.8 q/ha of pod and haulm yields of groundnut (cv ICGS-76) in an acidic soil of Tripura (Datta 2003). However, the groundnut productivity rose from 8.96 to 18.25 and 22.07 to 24.20 q ha^{-1} of pod and haulm yields, respectively, with the application of $N_{30}P_{50}K_{40}$ in combination with PSM, *Rhizobium,* and PGPR-4.

Seed inoculation with phosphate solubilizing microorganisms and application of gypsum @ 30 kg S ha^{-1} significantly increased the pod yield of groundnut and uptake of P and S by kernels in an Alfisol of Meghalaya (Venkatesh et al. 2001). However, the maximum pod yield (27.69 qha^{-1}) was recorded with the application of rock phosphate (50 kg P_2O_5 ha^{-1}). The highest pod yield (29.97 q ha^{-1}) of groundnut was recorded in an acidic soil of Meghalaya with the application of neem cake @ 15 qha^{-1} + *Rhizobium* + PSM, which was 8.2 and 9.3% higher over the yield recorded with the application of FYM @ 100 qha^{-1} and neem cake @ 30 qha^{-1} (Panwar et al. 2001).

Inoculation of PSMs along with P fertilizer significantly increases the grain yield of rice in an acid soil (pH 4.5) at Karimganj, Assam (Sarmah 2001). The grain yield of rice with 50% P + PSM inoculation was as good as 100% P as superphosphate. Increased levels of P addition up to 100 $kgha^{-1}$ and PSM inoculation enhanced significantly the P uptake by rice. Application of 40 kg P_2O_5 as rock phosphate + cow dung @ 5t ha^{-1} + PSB (*Bacillus polymyxa*) @ 20 g kg^{-1} seed inoculation resulted in a 47.9% increase in grain yield of rice in an acidic soil of Tripura. Application of *Gliricidia* leaf in combination with PSB and rock phosphate raised the available P from 16.28 to 20.82 kg/ha (Anonymous 2004). Application of poultry manure and PSB (*Bacillus polymyxa*) in an acidic soil of Tripura has recorded maximum productivity of cabbage with an increase from 3.09 t ha^{-1} under control conditions to 19.52 t ha^{-1} in treated plots. The application of Azotobacter and PSB (Bacillus polmyxa) alone increased the cabbage yield by 12.9% and 8.1%, respectively, over control; however, when poultry manure @ 5 tha^{-1} was applied along with PSB, the increase in yield was almost 6.3 times over control. The efficiency of different P solubilizers on crop productivity of bhindi (cv Najuka F-1) was found to be *Bacillus polymyxa* @ 1.00 g kg^{-1} seeds (20.63 q ha^{-1}), followed by *Bacillus firmus* (20.63 q ha^{-1}), *Pseudomonas straita* (20.44 q ha^{-1}), and *Bacillus circulans* (19.19 q ha^{-1}).

The efficiency of P solubilizing microorganisms on the yield performance of lowland rice (cv RCPL-1-87-75) in an acidic (pH 5.1) Ultisol of Mizoram showed that bacterial application to rice seedlings with *Pseudomonas straita* in combination with 60 kg $P_2O_5ha^{-1}$ recorded significantly higher grain yield of 5.53 and 5.39 t/ha during 2000 and 2001, respectively (Laxminarayana 2002) (Fig. 5.1). The efficiency of added P was improved with the inoculation of PSB along with externally added P and reduces the expenditure on fertilizer cost equivalent to 15–30 kg P_2O_5 ha^{-1}. Bacterial inoculation of three strains of *Pseudomonas* (HPR 6, RRLJ 008, and RRLJ 134) on French beans on a sandy loam soil at Jorhat (Assam) showed that all the strains induced a positive response on growth and physiological parameters resulting in a higher yield of *P. vulgaris* (Deka Boruah et al. 2003).

Thakuria et al. (2004) isolated three groups of rhizobacteria from the rhizosphere of rice grown at various locations in Assam. The isolates were identified as phosphate solubilizing bacteria (*Bacillus circulans, B. pantothenticus,* and *B. megaterium*) and Fluorescent bacteria (*Streptomyces anthocysnicus, Pseudomonas aeroginosa, P. pickettii,* and *Bacillus* sp). Three isolates of *Azospirilla* were identified as *Azospirillum brasilense* and the fourth isolate was identified as *A. amazonense*. Among the *Azospirilla* isolates, more quantity of IAA-like substances (6.0 ppm) was produced by *A. amazonense*. Among the PSB isolates, *B. megaterium* produced the maximum amount (6.4 ppm) and among the Fluorescent bacterial isolates, *P. pickettii* and *Bacillus* sp. produced the highest quantity of IAA-like substances (21.6 ppm). The bacterial strains showed wide variations in antibiotic resistance profile and nitrogenase activity (0.33–0.75 n mol of C_2H_4/h/100 cc root volume) in the inoculated roots with the highest in *A. brasilense*. Among the strains, *A. amazonense*, *B. pantothenticus,* and *Pseudomonas pickettii* recorded an increase of 55.5, 12.2, and 76.9% grain yield over uninoculated control and were found to be at par with or superior to the exotic strains of *A. brasilense* and *Pseudomonas fluorescence*.

5.2.6 Mycorrhiza

Mycorrhizae are beneficial partners of plants that help plant growth through increased nutrient and water absorption from the soil, protection from pathogens, increased tolerance to soil toxins, and root elongation. This is an unique fungus association with higher plants. Mycorrhizae are highly habitat-limited and found in the immediate vicinity of roots. Production of organic acids by fungi would certainly affect the availability of

Fig. 5.1 Effect of phosphate solubilizing organism on groundnut

acid-labile P from inositol phosphates and possess phosphatase activity to release P from soil organic matter. These can also increase the availability of Cu and Zn nutrient elements to the plant. Mycorrhizae are divided into two categories such as Endomycorrhiza and Ectomycorrhiza. A good ectomycorrhizal association between fungus and *Pinus kesia* is prevalent in the Upper Shillong soils of Meghalaya. In the case of an ectomycorrhizal association, the fungus forms a mantle around the exterior of the roots. The fungi of the endomycorrhizal association penetrate the host cell and from there they suck carbohydrates/other organic compounds thus supplying phosphorus to the plant. This association is very common in the plants of the family Orchidaceae as well as in fruit trees like citrus, coffee, and various legumes. The prominent fungi of endomycorrhiza which are found in the soils of this region are *Glomus tenuis, G. mosseae, G. etunicatum, and G. fasciculatum.* Inoculation of VAM (*Glomus mosseae*) with and without P showed a relatively lower response over PSB in an acid Ultisol of Mizoram (Laxminarayana 2002).

Vesicular Arbuscular Mycorrhizal fungi (VAM) remain in the symbiotic association with the roots of higher plants and help plants in acquiring nutrients, particularly P in nutrient-deficient soils. The external mycelium enters the soil up to a distance of 1.0–1.2 mm from the host root and it extended further and is capable of translocating labeled P at a distance of 70 mm from the host root (Rhodes and Gardemann 1975). Das et al. (1997) reported that the fungal genera *Glomus*, *Gigaspora,* and *Acaulospora* were dominant in the acidic soils of Assam (Eastern India Himalaya) (Das and Barthakur 1997).

Inoculation of VAM in black gram significantly improved the shoot dry weight and uptake of N and P by the plants. VAM inoculation in combination with externally added P @ 15.4 kg ha^{-1} as Mussoorie Rock Phosphate further aggravated the dry matter production and nutrient uptake by black gram (Barthakur et al. 1997a, b). The VAM spores collected from various locations and their inoculation on black gram showed that the Dohotia spore mixture resulted in highest shoot dry weight and nutrient uptake by plants. Results suggested that VAM inoculation in combination with P caused more infection in roots and thus enhanced the uptake of P and N in black gram. An increase in N uptake is due to higher N_2 fixation by the native *rhizobia* present in the soil as the adequate P supply to the plant is maintained by the inoculated VAM fungal spores. In all the spore mixtures under different P treatments, the number of spores belonging to *Glomus* species was the highest followed by *Gigaspora* and *Acaulospora.* The difference in mineral nutrition of test plants due to inoculated VAM spore mixture might be attributed to the differences in the nature of the infection.

A twenty-year-old fallow land at the Jhum cultivation site of Changki village in Nagaland was investigated (Nath and Datta 2013–14) to find the spore density and diversity of native arbuscular mycorrhiza (AM) under the project titled Impact assessment of jhumming on native plants and soil microbiota and restoration of sustainable jhum agro-ecosystem in Northeast India with the following observations. A total of 12,451 native AM spores of diverse morphotypes are extracted from Changki site. Native AM spore density ranged from 10 to 19 spores per gm of soil. Spore density was highest in the upper slope relatively followed by the middle and bottom slope. The overall gradient had the least influence on the spore population ($P \leq 0.05$). On preliminary investigation, the diverse AM morphotypes were categorized into three groups on the basis of spore diameters, viz., A-type with diameter > 250 μm, B-type with diameter 106–250 μm, and C-type with diameter 53–106 μm. The preponderance of the native AM spore was of the trend –B > C > A.

5.3 Actinomycetes

Actinomycetes are the transitional group between bacteria and fungi. These are numerous and widely distributed not only in soils but in a variety of other habitats such as compost, river muds, and lake bottom. They are present at the surface and sub-surface soils at considerable depth. The North Eastern region is rich in the actinomycetes population. A symbiotic relationship between *Frankia* sp. and non-leguminous plants like *Alnus nepelensis* and *Casuarina* sp. is found all over the region. *Frankia* sp in association with tree species fixes atmospheric nitrogen that is utilized by tree plants. Upon organic matter addition, the bacteria and fungal flora usually proliferate initially, particularly if nitrogen is plentiful and the response of actinomycetes frequently does not become pronounced until the later stage of decay (Alexander 1978).

5.4 Trichoderma

The application of antagonistic fungi to the rhizosphere of crop plants to be protected from any soil-borne disease requires mass production of the antagonists within a short time in a suitable medium. According to Barthakur et al. (1997b, a), *Trichoderma harzianum* recorded the highest spore conc. (319.7×10^5 spores g^{-1}) after 14 days of incubation followed by *T. viride* (310.0×10^5 spores g^{-1}) and *T. koningii* (288.0×10^5 spores g^{-1}). *T. viride* showed the highest propagules (34.8×10^7 c.f.u. g^{-1}) in wheat bran medium after 14 days of incubation followed by *T. harzianum* (33.7×10^7 c.f.u. g^{-1}). The results indicated that sporulation of *T. viride* and *T. harzianum* was significantly higher than T. koningii, whereas colony-forming units were significantly higher in *T. viride* than in *T. harzianum*. Wheat bran medium was found the best for better sporulation over rice bran, sawdust, maize meal sand medium, and potato dextrose agar.

5.5 Cellulolytic Microorganisms

There are several fungi that can enhance the decomposition of cellulogenous materials present in the soils. These are *Trichrus spiralis, Aspergillus* sp. *Trichoderma viride, and Paeciliomyces fusisporus* (Gaur and Mathur 1990). Substantial quantities of organic materials such as leaf litter, organic wastes, and crop residues are added to the soil of this region every year but the materials remain almost partially decomposed. By application of cellulose decomposing organisms, organic matter and fertility build-up of the cultivated soils can be increased.

Decomposition of forest litter is a very complex phenomenon and is achieved by different groups of microorganisms. The process of decomposition of plant residues plays an important role in the release and cycling of nutrients and the formation of soil organic matter. A major component of topsoil consists of different parts of plant materials and colonies of various groups of organisms thus initiating the decomposition of the debris. The fungal flora of the litter was maximum during July when the litter was moist and the rate of decomposition was rapid from May to September and then gradually slowed down in the winter. The leaching of soluble nutrients from the litter provides a suitable substrate for microbial growth and accelerates the rate of decomposition of litter. A total of 22 fungal species were identified from the decomposed litter under jhum fallows (Deka and Mishra 1982). The population of fungi was higher in 0- and 5-year-old fallows followed by bacteria and actinomycetes, which took a major role in the decomposition of litter. The dominant fungal genera were *Penicillium, Trichoderma, Aspergillus, Cladosporium,* and *Alternaria*. The bacterial population increased to 60×10^5 cells g^{-1} during June-July under a 10-year-aged fallow, reduced to $< 10 \times 10^5$ cells g^{-1} during winter months, and again rose to 30×10^5 cells g^{-1} during April–May. The loss of components was in the order: Sugar > amino acid > hemicellulose > cellulose > lignin.

Pine (*Pinus sp.*) forests are abundant in the North Eastern Region especially in Meghalaya. The annual pine litter fall ranges between 2.2 and 3.8 t ha^{-1} $year^{-1}$ depending on soil, plant type, and other edaphic factors. Pine needles are important source for their bio-conversion to humic substances by treating with microbes having lignolytic and cellulolytic activities. White rot *Basidiomycetes* are known to degrade lignin as well as cellulose. Pal and Singh (2002) reported that treating pine litter with white rot *Basidiomycetes* fungi lowered the C:N ratio and lignin content over the control. Composts prepared from pine litter helped in significant improvement of grain yield of mustard, finger millet, and French bean. Increase in soil microbial biomass carbon, available N, and P_2O_5 in sterilized soil amended with pretreated litter followed by inoculation of fungal isolates. Secretion of extracellular enzymes by the inoculated fungi to break the polymer of the lignocellulosic complex may also help in pine litter degradation (Michael et al. 1989; Pourique and Vande 1990). Higher levels of MBC, available N, and P_2O_5 in the soil through the application of compost may be due to the proliferation of a specific group of saprophytic microflora in addition to other microorganisms and absorption of nutrients on the organic complex added to the soil.

Inoculation of cellulolytic microorganisms hastened the composting of organic wastes and the usage of these organisms improves the quality of the compost. A higher rate of decomposition of residues was observed with earthworms *Eisenia foetida* over *Amynthas diffringens* at Jorhat Assam (Das and Talukdar 2001). Inoculation of cellulolytic fungal culture (*Aspergillus terreus*) @ 1 ml (10^5 cells ml^{-1}) with *Eisenia foetida* recorded the highest rate of decomposition (93.2%) followed by cellulolytic bacterial culture (*Pseudomonas* sp) @ 1 ml (10^7 cells ml^{-1}) with *Eisenia foetida* recorded 92.7% decomposition. Inoculation of cellulolytic fungal or bacterial culture along with earthworms (*A. diffringens* and *E. foetida*) increased the decomposition of wastes as well as improved the quality of vermicompost in respect of total N, P, and K; available N, P, K, and S; and exchangeable Ca and Mg. Cellulolytic microorganisms decomposed earthworms' excreta releasing more CO_2 which subsequently lowered the C:N ratio of the manure.

The microbial (*Aspergillus awamorii, Paecilomyces fusisporus, Azotobacter chrococcum,* and *Bacillus polymyxa*) enriched compost as prepared from paddy straw contain 38.4, 23.5, 45.0, 20.8, 162.1, and 21.3% higher values of total N, total P, NH_4-N, NO_3-N, water-soluble P, and citrate-soluble P, respectively, as compared to mineral enriched compost. Application of this microbial enriched compost @ 10 t ha^{-1} recorded 45% higher grain yield of paddy over ordinary paddy straw compost and 32% higher than 100% NPK (100-80-60 kg P_2O_5 and K_2O ha^{-1}) due to higher amount of available as well as total nutrient status of the compost (Anonymous 2004).

5.6 Enzyme Activities

Microorganisms existing in permanently cold habitats and at high altitudes are a great source of cold-active enzymes. Among these, cold-active proteases, lipases, amylases, alkaline phosphatases, etc. have a wide range of applications in food processing, dairy, and agricultural processes. Measuring dehydrogenase activity determines the metabolic activity of microbes in soil and their habitats. As microorganisms are an important source of soil enzymes, the activities of such soil enzymes should normally correlate with microbial activity, and therefore, measurements of the activities of these enzymes have been used as an index of the biological activity of soil. The organic matter status of the soil accounted for most of the variations in urease and dehydrogenase activities. Facultative and anaerobic bacterial population influences the dehydrogenase activity (Tiwari and Mishra 1989).

The dehydrogenase activity increased with the increase in altitude up to 1100 m and thereafter it declined sharply. The variations in urease activity, generally, coincided with the variations in dehydrogenase activity. The fungal and bacterial populations increases with an increase in altitude up to 1100 m and declined sharply thereafter. An increase in organic matter of the soil can be attributed to increased activities of dehydrogenase and urease enzymes. The decrease in organic matter at 1700 m altitude attributed to a slow rate of decomposition of organic matter resulting in a decrease in the activities of these enzymes. Significant positive correlations were observed between dehydrogenase activity, fungal and bacterial populations, and organic matter content of the soil.

Nannipieri et al. (2002) mentioned that soil enzyme activities are an index of microbial functional diversity and are one of the most important indicators of soil quality. Pascual et al. (2002) reported that enzymes are potential indicators of the extent to which soil disturbance by a given activity may affect the immediate environment. Caravaca et al. (2003) suggested that enzyme activity is an early and sensitive indicator for reflecting the degree of recovery of a soil system. Estimation of soil enzyme activities as a recovery indicator has an edge over other microbial parameters (Nannipieri et al. 2002).

5.7 Soil Fauna

The soil fauna contains protozoa and nematodes which are widely distributed in soils. The former generally occurs in the water film surrounding soil particles and the latter usually in the soil of the upper horizon among the plant roots. Cultivated soils of Tripura were found to have more nematode population (59 per 100 g soil), and the predominant parasitic nematodes observed were *Meloidogyne graminicola, Aphelenchoides besseyi, Meloidogyne incognita, and Meloidogyne hapla* (Dadhwal et al. 1987). Total nematode population negatively and significantly correlated with exchangeable Ca, H, and Al and exchangeable acidity of the soil; however, base saturation and exchangeable bases had a positively significant relation. Some animals such as earthworms are exceptional in that they frequently move from horizon to horizon and hence play an important part in the transport of materials in the soil system. Prior to the mixture of the organic residues with the soil, a large number of animals have taken part in the primary attack of the debris. On very wet or very dry soils in which few animals exist, surface litter tends to accumulate, and under these conditions, fungi will be the principal agents of decomposition. In soils where conditions are not favorable for the survival of soil fauna, the humus content of such soils is very low.

5.8 Microbiome of the Himalayan Ecosystem

Microbial diversity is the backbone of any ecosystem because it performs numerous functions which are essential for the biosphere. The management and proper exploitation of microbial diversity including mainly bacteria, fungi, and actinomycetes, which are mostly soil-inhabiting, have an important role in sustainable development toward large-scale industrial and commercial applications. The Himalayan mountain region is one of the richest bioresources of unique microflora particularly bacteria, fungi, and actinomycetes which are of immense biotechnological potential. The deep gorges which are covered with frozen snow throughout the year and alpine glaciers are a rich reservoir of extremophilic microorganisms particularly psychrophilic/psychrotolerant bacteria and actinomycetes.

The Himalayas rises from the Indian plains and stretches to the high plateau of Tibet, in the north, resulting in a complex series of parallel, converging, and diverging ranges (Fig. 5.2). Boetius et al. (2015) mentioned that microorganisms are the important life form of the Earth that occurs in all climatic areas, including even those which were once considered impossible to support life, i.e. the cold deserts like Arctic and Antarctic. The microorganisms have not only survived in alpine type of climate of the Himalayas but have also maintained their viability.

Mountain ecosystems are significant as biodiversity habitats (Myers et al. 2000) and as sources of raw materials and socio-cultural importance (Stepp et al. 2005) that create climatic, landscape, and ecological gradients that form the basis for species evolution (Spicer et al. 2017). Mountains in the world support about one-third of all terrestrial biodiversity

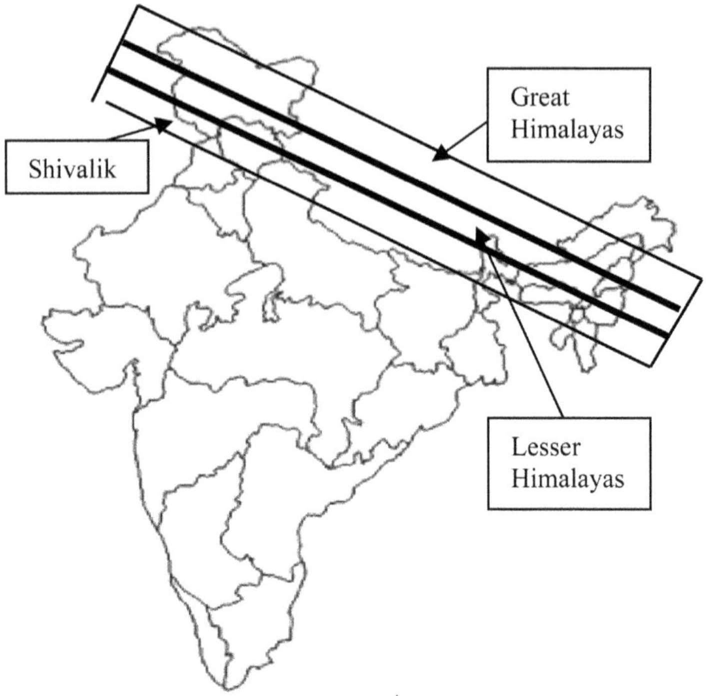

Fig. 5.2 Three parallel zones of the Himalayas

(Spehn et al. 2011). Major mountain ranges, including the Himalayas, Andes, and the Alps among others, are all known to be biodiversity hotspots (Mittermeier et al. 2004; Myers et al. 2000). Our understanding of the spatiotemporal maintenance of high biodiversity remains below par. The Himalayas have a complex mountain ecosystem with high topographic and climatic heterogeneity and inhabiting rich flora and fauna (Bambulwa et al. 2021) and there exists a high variance in species distribution, altitudinal gradients, and effect of climate change among plant species in the Himalayas.

The Himalayan mountain has its unique flora and fauna. Using 16S rRNA gene cloning and sequencing, the bacterial diversity of the Himalayan soils has been well investigated. Clone library analysis revealed the dominance of proteobacteria especially genus Pseudomonas and Rhizobium in the soil of high-altitude agroecosystems. Moreover, Bacteroidetes, Chloroflexi, Acidobacteria, Firmicutes, Nitrospira, Planctomycetes, Cyanobacteria, Gemmatimonadetes, Verrucomicrobia, Chlorobi, Actinobacteria, OD1, OP11, and BRC1 were also found in the Himalayan soils (Joshi et al. 2017). Microbes even thrive in extreme cold environments where higher plants cannot grow (Dhakar and Pandey 2020). This is due to the reason that microorganisms have many structural and functional adaptations to complete normal life processes under extreme environments, where these organisms are mainly inhabited (Margesin and Feller 2010), and rather, extremely low temperatures depend on these environment-specific microorganisms for their several ecological processes (Margesin and Collins 2019). Himalaya is a region having very low temperatures throughout the year. At higher elevations, near the sites of melting glaciers as a result of warming due to climate change, the soils represent a different type of ecosystem. The soils are changing with the colonization of various new microbial communities consisting of diverse microorganisms, heterotrophs to chemotrophs, that play a significant role in the biogeochemical cycles in the process of soil development. Schmidt et al. (2008) showed that the proliferation of cyanobacteria indicated their participation in increasing soil carbon and nitrogen content.

Similar conclusions were drawn by Franzetti et al. (2013) for heterotrophs as reported from the debris covering glaciers. The Tibetan Plateau contains glaciers with extremely cold and unique habitats for the establishment of microorganisms in the ice sheets (Xiang et al. 2004). Through the molecular methods, about 15 bacterial genera with the prevalence of Proteobacteria have been reported from the plateau (Liu et al. 2009; Zhang et al. 2008). The autotrophic microbial communities are mainly responsible for the biological processes and nutrient cycling in the semi-arid/arid ecological situations with scarce vegetation (Guo et al. 2015). Deng et al. (2014) mentioned that the wetlands of the Tibetan Plateau have been lowly studied for their microbial communities because of their heterogenous biogeography.

Glacial ice occupies about 10% of Earth's land surface (Choudhari et al. 2013). Gupta et al. (2020) found that the glacier soils contain taxonomically and metabolically diverse communities. Sub-zero temperature was earlier considered without any life but now glaciers and huge ice masses have been found to be home for microbes of diverse nature (Cowan and Tow 2004), forming the major component of the landscape of low to high temperatures and giving an appropriate environment for their survival (Rawat and Joshi 2018). As many as 9.61×10^{25} microbes have been reported by Priscu and Christner (2003) in a study from Greenland. Schuette et al. (2010) mentioned that microbes at low temperatures play a significant role in the development of soil and other biochemical processes. Next generation sequencing (NGS) has been developed as a novel method for determining the difficult samples for taxonomical biodiversity and metabolic pathways (Chan et al. 2015).

5.9 Microorganisms in Himalayas

The investigations with regard to biodiversity in the Himalayas started about 200 years (Rana et al. 2021). In 1808, Hooker published the first-known publication on Himalayan species on the mosses of Nepal (Hooker 1808). There were many publications in the nineteenth century when Hooker (1875) came out with the book "The Flora of British India" in seven volumes. Blanford (1832) studied in detail the animals and published the work as "The Fauna of British India" series. Microorganisms, evolving for nearly four billion years, can thrive in almost every habitat such as extremes of cold, heat, radiation, pressure, salt, acidity, and darkness (Joshi et al 2017). The Himalayan region is extensively explored for the study of microbial biodiversity, various plant growth-promoting rhizobacteria (PGPR), and economically important microorganisms (Suyal et al. 2014). Cold environments of Himalaya inhibit the diversity of microorganisms, which includes bacteria, archaebacteria, unicellular algae, and fungi. Suyal et al. (2015) studied soil bacterial diversity in the rhizosphere of *Phaseolus vulgaris* from Western Indian Himalayas and found that proteobacteria were the dominant group of bacteria with *Pseudomonas* as the predominant genus. Other bacterial groups present were *Chloroflexi*, *Actinobacteria*, *Acidobacteria*, *Gemmatimonadetes*, *Planctomycetes*, *Bacteroidetes*, *Firmicutes*, *Nitrospira*, *Cyanobacteria*, *Verrucomicrobia,* and *Chlorobi*. Yadav et al. (2015) reported soil bacterial diversity from different altitudes (3978–4069 m) of Rohtang Pass situated in the Northwestern Himalayas. *Proteobacteria* was the major group followed by *Firmicutes* and *Actinobacteria*, respectively. Other dominating groups of microorganisms inhabiting Himalayan soil other than bacteria are fungi. The fungal distribution in the soil ecosystem is affected by the soil organic content, soil texture, surface vegetation, and other physiochemical properties of soil.

Fungal diversity in the soil at higher altitudes of Sikkim and Uttarakhand Himalaya has shown that the *Penicillium* is the most abundant and diverse genus present over there. *P. raistrickii*, *P. janthinellum*, *P. pinophillum*, *P. javanicum*, *P. chrysogenum*, *P. oxalicum*, *P. purpurogenum,* and *P. aurantiogriseum* are some of the commonly occurring species of *Penicillium*. *Aspergillus*, *Epicoccum*, *Fusarium*, *Myrothecium*, *Cladosporium*, *Paecilomyces*, *Gangronella,* and

Trichoderma were other genera contributing to the fungal flora of this Himalayan region (Rai and Kumar 2015). Further, Devi et al. (2012) studied fungal diversity in the Eastern Himalayan range and found that the major diversity of fungi belonged to phyla *Zygomycota* and *Ascomycota*. *Eurotiales*, *Calosphaeriales*, *Hypocreales*, *Mortierellales*, *Pleosporales*, *Mucorales*, and *Capnodiales* were the seven dominant orders corresponding to these phyla.

Seven types of mycorrhizal associations are known to be present between fungi and plant roots, namely ectomycorrhizae, ecto–endo mycorrhiza, monotropoid, arbuscular mycorrhizae, ericoid, orchidoid mycorrhizae, and arbutoid mycorrhiza (Brundrett 2004). Singh (2014) described the diversity of ectomycorrhizae in the context of Himalayan trees and their plant growth-promoting potential. *Amanita*, *Russula*, *Boletus*, *Lactarius*, *Suillus*, and *Hygrophorus* are the common ectomycorrhizal fungal genera associated with oaks and conifers in the temperate forest of Western Himalaya (Wang et al. 2015). The existence of *Actinobacteria*, *Bacteroidetes*, *Firmicutes*, and *Proteobacteria* has been observed in the Northwestern Himalayas. Moreover, hot springs present in the Himalayan region represent an extreme environment for the growth of microorganisms because the outlet temperature of these hot springs ranges about 45 °C and above. Srinivas et al. (2011) studied bacterial diversity of Kafni Glacier in Kumaon Himalayas, at an altitude of about 3853 m using 16S rRNA gene clone libraries, and suggested that the majority of bacterial genera belonged to phylum *Proteobacteria*. In *Firmicutes*, *Bacilli*, *Clostridia*, *Erysipelotrichia*, and *Negativicutes* were dominant. Other important bacteria were *Cryobacterium roopkundense* (*Actinobacteria*), *Acidobacteria*, *Tenericutes*, *Chloroflexi*, *Chlamydiae*, and candidate phylum TM7s TM7a.

Roy et al. (2020) studied the geomicrobiology of boron-rich hot springs across the Himalayas, which have water temperatures around 70 °C, from a boratic microbialite called *Shivlinga*. The spring-water composition was dominated by borates, sodium, thiosulphate, sulphate, sulphite, sulphide, bicarbonate, and other macromolecule-stabilizing (kosmotropic) substances. It was found that the Bacteria, Eukarya, and Archaea formed >98%, ~1%, and <1% of *Shivlinga*'s microbiome, respectively.

Land use change is a major ecological problem as it changes natural ecosystems, degrades vegetation and soil, and results in loss of soil nutrients. Due to weak resilience, the degradation of high elevation vegetation is the beginning of the land degradation process in higher reaches of mountains. Therefore, degradation of vegetation, particularly in high-altitude ecosystems, has been the main factor in converting the land into a non-reversible degraded state. Vegetation cover is important for reducing runoff losses and minimizing nutrient loss and soil movements such as gully formation, topsoil loss, and mass wasting in the HKH region (Blaikie and Sadeque 2000).

Land use changes also affect biodiversity through the decline of natural habitats, elimination of species locally, and impacting ecosystem functioning. Araujo et al. (2010) investigated the effect of conversion of native savanna to *Eucalyptus grandis* forest on soil microbial biomass. They concluded that such changes in the soil changed the number and activity of soil microorganisms, which affected soil biochemical processes and influenced the plant growth. Zhang et al. (2006) observed a decrease in soil microbial biomass from undisturbed vegetation sites to severely degraded land and explained it as a consequence of the loss of vegetation cover and SOC from the soil surface. Soil organic matter content, soil nutrient status, and soil texture were found to influence plant–microbe interactions and thus affect the ecosystem processes. Several studies have demonstrated that microbial community composition and function are interrelated (Strickland et al. 2009) and shifts in the microbial community following a disturbance event can greatly affect microbial functions (Paterson et al. 2007).

The stability of a system depends on the resistance, i.e. the ability to withstand stress and resilience, i.e. the ability to recover to pre-disturbance level. A decrease in the diversity of microbes in the soil may cause a decline in resistance and resilience of the system. It is also important to consider the co-evolution between plants and their biotic environment, particularly microbial activities as plant cover degradation negatively affects soil microbial activity. Many changes in land use affect the amount of carbon held in vegetation and soil either releasing carbon dioxide into the atmosphere or removing it from the atmosphere. Upadhyay et al. (2005) reported that in fragile ecosystems, land use changes and soil degradation add to increased greenhouse gas emissions. Soil organic matter (SOM) is the largest terrestrial store of biologically active carbon. Land use change is known to be a major factor influencing the soil organic carbon (SOC) content.

Considering SOC as the most important source of C for soil heterotrophic microorganisms, cycling of C in the soil compartment is important during the ecosystem recovery period. Since soils contain about 75% of the C in the terrestrial ecosystem, knowing more about the microbial processes taking place, there will be a better understanding of long-term C storage in soils.

Many research organizations in India have explored the microbial diversity in the Himalayan ecosystem and the recorded species, among others, are *Pediococcus pentosaceus, Pseudomonas putida, P. synxantha, Serratia marcescens, Staphylococcus sp., Staphylococcus sciuri, and Weisella cibaria*. Thermophilic bacteria: *Bacillus licheniformis, B. megaterium, B. sporothermodurans, Brevibacillus thermoruber, Geobacillus pallidus, Geobacillus subterraneus, Hydrogenobacter sp., Paenibacillus sp., Pyrobaculum aerophilum, P. caldifontis, Thermus brockianus, and T. thermophilus. Halophilic bacteria: Halobacillus sp., Halomonas sp., Marinomonas sp., and Shewanella sp. Psychrophilic and psychrotrophic bacteria: Janthinobacterium lividum, Serratia quinivorans* A5-2, and S. quinivorans B8. *Actinomycetes: Frankia sp., Gordoniaterrae, Micromonospora endolithica, Nocardia globulera, Rhodococcus rhodochrous, Streptomyces griseus, and Streptomyces sp. Thermophilic cyanobacteria: Aphanocapsa, Calothrix brevissima, C. parietina, Chroococcus, Coelosphaerium, Entophysalis granulose, Leptochaete hansqirgi, Lynqbya, Mastigocladus, Microcytic stagnalis, Oscillatoria, Phormidium, Plectonema notatum, Scytonema leptobasis, Spirulina subsalasa, and Synechococcus elongates. Yeast: Candida bombicola, C. chiropterorum, C. glabrata, C. parapsilopsis, C. tropicalis, Debaromyces sp., Endomyces fibuligera, Geotrichum candidum, Kluyveromyces thermotolerans, Pichia anomala, P. burtonii, P. kudriavzevii, Saccharomyces cerevisiae, Saccharomycopsis fibuligera, Saccharomyces* spp., and *Zygosaccharomyces bisporus* (Thakur et al. 2015). Major genera and species of fungi: *Alternaria, Aspergillus, Cephalosporium, Cladosporium, Cunninghamella, Curvularia, Drechslera, Epicoccum, Fusarium, Gliocladium, Monilia, Mucor circinelloides, M. hiemalis, Rhizopus chinensis, Rhizopus oryzae,* and *R. stolonifer*. Mushrooms: *Agaricus compestris, Amanita sp., Boletus sp., Cantharellus cibarius, Gyromitra sp., Helvella compressa, Hericium erinaceus, Humaria hemisphaeria, Hygrophorus sp., Laccaria sp., Lactarius deliciosus, Lycoperdon sp., Macrolepiota procera, Morchella conica, M. deliciosa, M. esculenta, Pleurotus cornucopiae, Ramaria botrytoides,* and *Rhizopogon* sp.

Tang et al. (2020) found that the pH of the soil had an unrivaled direct impact on the variability of soil microorganisms. Vegetation also has a strong direct effect on fungal biomass and performs a significant role in habituating the rhizosphere microbiome (Berendsen et al. 2012). Walker et al. (2003) observed that areas of soil around plant roots are affected by root activity and root exudates and that each type of biomass helps a rhizosphere-specific microbial community (Philippot et al. 2013). An understanding of the kinship between vegetation types and soil microbial communities may help to protect the severe environment of the alpine ecosystem (Klimeš, 2006).

Shifting cultivation is practiced in the eastern Himalayan region. A lot of forest vegetation is felled, dried, and burnt in situ. A total of 35 actinomycetes in the *Streptomyces, Kitasatospora,* and *Nocardia* and several other genera were isolated from shifting cultivation plots of eastern Himalayas. Some of these isolates were capable of producing chitinases, glucanases, amylases, lipases, and proteases enzymes. Similarly, several bacterial species in the genera Bacillus and Pseudomonas were isolated from the fired plots. Another study showed that firing caused only a transient rise of temperature up to 100 °C within 0–5 cm with no significant effect on microbial diversity in the soil profile. The Himalayas being one of the mega-biodiversity hotspots of the world has natural ecosystems still prevalent in pristine conditions offering scope for bioprospection of novel organisms hitherto unknown to the scientific world. Further, the sacred groves in this region are rich patches of undisturbed virgin forests and serve as a natural habitat for many endemic, rare, and economically valuable species (Samati and Gogoi 2007).

Fungi in the soil samples collected from both Sikkim and Uttaranchal Himalaya (both Eastern and Western Indian Himalayas), species of *Penicillium*, namely *P. aurantiogriseum*, *P. chrysogenum*, *P. citrinum*, *P. janthinellum*, *P. javanicum*, *P. oxalicum*, *P. pinophillum*, *P. purpurogenum,* and *P. raistrickii* were recorded at higher altitudes. Species of *Aspergillus*, *Cladosporium*, *Epicoccum*, *Fusarium*, *Gangronella*, *Myrothecium*, *Paecilomyces,* and *Trichoderma* also contributed to the fungal flora. Baath (1981) reported the dominance of *Penicillium* spp. in pine forest soil in Central Sweeden, and Widden (1987) reported the dominance of *Penicillium* spp. along an elevational gradient ranging from 350 to 880 m in Northern England. Various actinomycetes, mainly those that produce diffusible pigmented metabolites, were also isolated from these soils.

Microorganisms and Himalayan grasslands

Zhang et al. (2021) mentioned that the 0–20 cm soil layer was the most active zone of microbial activity and numbers of bacteria, fungi, and actinomycetes significantly increased at the six to ten-year stage of restoration succession and after that a decrease in microorganism was found. Grassland gives many ecosystems services, such as forage production (Liu et al. 2011), erosion control (Sharafatmandrad and Khosravi Mashizi 2020), maintenance and improvement of soil and water quality (Oliveira et al. 2019), habitat for wildlife (Brambilla et al. 2020), and carbon accumulation (Ma et al. 2018); however, due to anthropogenic factors, such as overgrazing, complete degradation of grasslands has occurred (Wang et al. 2018). Nutrient cycling, soil health, and ecosystem productivity are strongly impacted by microorganisms (Moll et al. 2017; van Leeuwen et al. 2017).

5.10 Bioresource Utilization

Cyanobacteria and algae are photosynthetic microorganisms which by because of their remarkable features have emerged as promising bioresources for biotechnological applications. As compared to bacteria and fungi, these photoautotrophic microorganisms were not given their due importance and were ignored for their beneficial use in research and development. As far as the perspective of cyanobacterial diversity in Uttarakhand state of India in the western Himalayas is concerned, a total of 53 species belonging to 26 genera have been reported from aquatic habitats. In addition, 31 thermophilic cyanobacterial species belonging to 14 genera have recently been reported from geothermal springs (Bhardwaj et al. 2011). A total of 346 species belonging to 122 genera of algae have also been reported from Uttarakhand. There is a need to use the beneficial potential of microorganisms in agriculture, industry, and medical fields.

5.11 Success in Biofertilizer Promotion

The effectiveness of biofertilizers depends on the efficiency of the microbial strain, their shelf life in the carrier, and soil and weather parameters during crop growth. Although peat, charcoal, or lignite powders are suitable as carriers of biofertilizers, there is a scope for the development of better carriers to improve the effectiveness of biofertilizers. It is evident that even with best efforts, beneficial yield responses to inoculation of various microorganisms are not obtained in at least 1/3 of the field trials which could be due to.

- Spatial variability with respect to soil physical behavior and fertility gradients in the field as well as plant varieties.
- Survival and colonization of inoculated microorganisms in the rhizosphere.
- No knowledge of the previous history of plots, different fertilization of the previous crop, and not accounting for NO_3-N in the trial plots.
- Competition with native organisms.
- Insufficient nutrients in the rhizosphere to produce enough organic acids to solubilize soil phosphates.
- Inability of PSMs to solubilize soil phosphates.
- Adverse climatic factors.
- Improper site selection and shade effects of trees.
- Improper handling of microbial preparations and their application.

There are various constraints in biofertilizer technology that need to be addressed for success in the promotion of biofertilizers and their popularization among the farming community. The constraints are as follows:

- Physical and environmental constraints.
- Chemical constraints: acidity, salinity, etc.
- Biological constraints: protozoa, nematodes, pests, lepidoptera, dipterous insects, aphids, cyanobacteria, snails, mosquito larvae, etc.
- Technical constraints: Raw material, strains, suitable technology, and lack of quality assurance.
- Infrastructure constraints including electricity.
- Human constraints: technically qualified microbiologists and trained persons.
- Marketing constraints: weakness in the marketing network, retail outlet, storage facility, and standardization in packing, labeling, and prices of biofertilizers.
- Incentives/support/encouragement from Govt.

There are various researchable and policy issues in the area of biofertilizer promotion. These are as follows:

- Development of efficient strains of different biofertilizers through natural selection and genetic engineering.
- Enhancement of shelf life of the produce.
- Development of easy techniques for quality assurance.
- Cost-effective methods for mass production of arbuscular mycorrhizal fungi.
- Mass awareness about the use of biofertilizers.
- Provision of subsidy on biofertilizers.
- Strengthening financial support to public or private sectors for the production of biofertilizers.
- Training of farmers on the use of biofertilizers.
- Standardization of quality control aspects on biofertilizers.

5.12 Future Outlook

With the increasing cost of fertilizers and growing awareness of the ecological impact, the use of biofertilizers with the organic matter has become imperative. In order to harness the potential benefits of microorganisms in commercial agriculture in the Northeastern States, the improvement in the consistency of their performance is essentially needed. This requires research in many diverse areas such as the following:

- Identification of beneficial microbes in the entire region for their better exploitation.
- A detailed study should be carried out to know about the type of interaction between microbes and plant roots that will be helpful in the selection of efficient strains.

- To know about the survival and competitive ability of beneficial microbes and also the research should be carried out to enhance their performance under field conditions.
- Extensive research is needed on soil physical and chemical properties that govern the establishment of inoculated strain in the rhizosphere.
- Research to understand the mechanisms by which the microorganisms bring about the benefit to crops in acid soils.
- Although the microorganisms are known for their specific functions, e.g. N_2 fixation, phosphate solubilization, composting of the residues, etc., however, the evidence to prove these functions in the soil of this region is lacking.
- Mass multiplication of required microbial inoculants to meet the growing demands.
- Extensive studies on the enzymatic activities as influenced by various microbes for nutrient transformations in the soil–plant ecosystem.

References

Ahmed M, Das NK (2002) Certain effects of shifting cultivation in selected areas of East Khasi Hills district Meghalaya and effect on fungal microflora. Environ Ecol 20(4):839–843

Alexander M (1978) Introduction to soil microbiology. John Wiley and Sons. Inc, New York

Anonymous (2004) Annual report 2003–04. ICAR Research Complex for NEH Region, Umiam, Meghalaya, India, 14–154pp

Araujo A, Silva E, Cameiero RFV, Nunes LAPL (2010) Effect of converting native savanna to Eucalyptus grandis forest on soil microbial biomass. Land Degrad Dev 21(6):540–545

Baath E (1981) Microfungi in a clear-cut pine forest soil in central Sweden. Can J Bot. https://doi.org/10.1139/b81-179

Bambulwa MC, Milne R, Wu ZY, Spicer RA (2021) Spatiotemporal maintenance of flora in the Himalaya biodiversity hotspot: current knowledge and future perspectives. Ecol Evol 11(16):10794–10812

Banger KC, Kapoor KK, Mishra MM (1990) Soil microbial biomass: Its measurement and as a nutrient source. Indian J Microbiol 30:263–282

Barthakur HP, Rekha Das, Das R (1997a) Effect of vesicular-arbuscular mycorrhizal fungi on mineral nutrition of blackgram (*Vigna mungo*). J Agric Sci Soc North East India 10:1, 43–49; 21

Barthakur HP, Sarmah SC (1984) Indian J Agric Sci 54 (3):220–222

de Bellis T, Kernaghan G, Widen P (2007) Plant community influences on soil microfungal assemblages in boreal mixed-wood forests. Mycologia 99(3):356–367

Berendsen RL, Pieterse CMJ, Bakker PAHM (2012) The rhizosphere microbiome and plant health. Trends Plant Sci 17:478–486

Bharati K, Mohanty SR, Singh DP, Rao VR, Adhya TK (2000) Agric Ecosyst Environ 79:73–83

Bhardwaj KN, Tiwari SC, Pandey A (2011) Conservation of thermophilic cyanobacterial diversity of thermal springs of Tapoban geothermal field, Himalayan province (India). Natl Acad Sci Lett 34(3–4):77–82

Blaikie PM, Sadeque SZ (2000) Policy in high places: environment and development in the Himalayan Region. International Centre for Mountain Development, 209p

Blanford WT (1832) The fauna of British India, including Ceylon and Burma, vol I–VII. Taylor & Francis, London

Brambilla M, Gustin M, Cento M, Ilahiane L, Celada C (2020) Habitat, climate, topography and management differently affect occurrence in declining avian species: implications for conservation in changing environments. Sci Total Environ 742:140663. https://doi.org/10.1016/j.scitotenv.2020.140663

Boetius A, Anesio AM, Deming JW, Rapp JZ (2015) Microbial ecology of the cryosphere: sea ice and glacial habitats. Nat Rev Microbiol 13:677–690

Brundrett MC (2004) Diversity and classification of mycorrhizal associations. Biol Rev 79:473–495

Budhiraja R, Basu A, Jain RK (2002) Microbial Diversity: Significance, conservation and application. Sci Lett 25(7&8):189–201

Caravaca F, Alguacil MM, Díaz G, Roldán A (2003) Use of nitrate reductase activity for assessing effectiveness of mycorrhizal symbiosis in *Dorycnium pentaphyllum* under induced water deficit. Commun Soil Sci Plant Anal 34:2291–2302

Chan CS, Chan KG, Tay YL, Chua YH, Goh KM (2015) Diversity of thermophiles in a Malaysian hot spring determined using 16S rRNA and shotgun metagenome sequencing. Front Microbiol 5(6):177

Chandra K , Mukherjee PK, Karmakar JB (1995) Lime pellating of inoculated seed and their influence of the survival of rhizoibum in acidic soil of N.E.H. region. J Hill Res 8(2):199–202

Choudhari S, Smith S, Owens S, Gilbert JA, Shain DH, Dial RJ, Grigoriev A (2013) Metagenome sequencing of prokaryotic microbiota collected from Byron Glacier, Alaska. Genome Announc 2;1(2):e00099–13

Cowan DA, Tow LA (2004) Endangered antarctic environments. Annu Rev Microbiol 13(58):649–690

Dadhwal KS, Venkataraman S, Sen Gupt TK, Laskar S (1987) Nematode population in acid soil. J Indian Soc Soil Sci 35(1):141–142

Das BC, Roy SK, Bora LC (1997) Mass multiplication of Trichoderma species on different media. J Agric Sci Soc North East India 10(1):95–100

Barthakur HP, Rekha D, Das R (1997b) Effect of vesicular-arbuscular mycorrhizal fungi on mineral nutrition of blackgram (Vigna mungo). J Agric Sci Soc North-East-India 10:43–49

Datta M (2001) Efficacy of P solubilizers. In: Annual report 2000–01. ICAR Research Complex for NEH Region, Umiam, India, 122–123 pp

Datta M (2003) Organic farming in groundnut. In: Annual report 2002–03. ICAR Research Complex for NEH Region, Umiam, India, 43p

Datta M, Laskar S (1990) Response to phosphate and biofertilizers. In: Annual report 1989–90. ICAR Research Complex for NEH Region, Umiam, India, 46–47pp

Datta M, Laskar S (1991) Effect of phospho-bacterin inoculation on soil physico-chemical characteristics and P uptake in rice. In: Annual report 1990–91. ICAR Research Complex for NEH Region, Umiam, India, 21–22pp

Datta M, Banik S, Gupta RK (1982) Studies on the efficacy of phytohormone producing phosphate solubilizing *Bacillus firmus* in augmenting paddy yield in acid soils of Nagaland. Plant Soil 69:365–373

Datta M, Banik S, Laskar S (1992) Effect of inoculation of phosphate –dissolving bacteria on rice (*Oryza sativa)* in acid soil. Indian J Agric Sci 62(7):482–485

Datta M, Banik S (1994) Effect of poultry manure and phosphate dissolving bacteria on rice (*Oryza sativa)* in acid soil. Indian J Agric Sci 64:791–793

Datta M, Banik S (1997) Comparative efficacy of different phosphatic fertilizers and phospho-bacterium (Bacillus firmus) on rice (*Oryza sativa)* in an acid soil. Indian J Agric Sci 67(11):545–547

Deka Boruah HP, Rabha BK, Saikia N, Dileep Kumar BS (2003) Fluorescent *Pseudomonas* influences palisade mesophyll development and spatial root development in *Phaseolus vulgaris*. Plant Soil 256:291–301

Deka HK, Mishra RR (1982) Decomposition of bamboo (*Dendrocahmus hamiltonii* Nees.) leaf litter in relation to age of jhum fallows in Northeast India. Plant Soil 68:151–159

Deng Y, Cui X, Hernández M, Dumont MG (2014) Microbial diversity in hummock and hollow soils of three wetlands on the Qinghai-Tibetan plateau revealed by 16S rRNA pyrosequencing. PLoS ONE 9:e103115

Devi LS, Khaund P, Nongkhlaw FM, Joshi SR (2012) Diversity of culturable soil Micro-fungi along altitudinal gradients of Eastern Himalayas. Mycobiology 40:151–158

Dhakar K, Pandey A (2020) Microbial ecology from the Himalayan cryosphere perspective. Microorganisms 8:257. https://doi.org/10.3390/microorganisms8020257

Doran JW, Parkin TB (1996) Quantitative indicators of soil quality - a minimum data set. Methods for assessing soil quality (Doran JW, Jones AJ (eds)). SSSA Special Publication, vol 49. Soil Science Society of America, Madison, pp 25–37

Dubey SK (1997) Co-Inoculation of phosphorus bacteria with Bradyrhizobium japonicum to increase phosphate availability to rainfed soybean in vertisol. J Indian Soc Soil Sci 45(3):506–509

Franzetti A, Tatangelo V, Gandolf I, Bertolini V et al (2013) Bacterial community structure on two alpine debris-covered glaciers and biogeography of Polaromonas phylotypes. ISME J 7:1483–1492

Gaur AC, Mathur RS (1990) Organic manures. In: Kumar et al (eds) Soil fertility and fertilizer use IV. IFFCO, New Delhi, India

Guo G, Kong W, Liu J, Zhao J, Du H, Zhang X, Xia P (2015) Diversity and distribution of autotrophic microbial community along environmental gradients in grassland soils on the Tibetan Plateau. Appl Microbiol Biotechnol 99:8765–8776

Gupta V, Singh I, Rasool S et al (2020) Next generation sequencing and microbiome's taxonomical characterization of frozen soil of North Western Himalayas of Jammu and Kashmir, India. Electron J Biotechnol 45. https://doi.org/10.1016/j.ejbt.2020.03.003

Gyaneshwar P, Nareshkumar G, Parekh LJ, Poole PS (2002) Role of soil microorganisms in improving P nutrition of plants. Plant Soil 245:83–93

Hawksworth DL (2001) The magnitude of fungal diversity: the 1.5 million species estimate revisited. Mycol Res 105:1422–1433

Hazarika UK, Saxena DC, Panwar AS, Singh NP (2003) Effect of *Azolla caroliniana* and their nutrient sources on the growth and yield of wet land rice in mid hills of Meghalaya. In: Yadav AK, Chaudhuri SR, Talukdar NC (eds) Biotechnology in sustainable and organic farming. RBDC, Imphal, Manipur, India

Hooker JD (1875) The flora of British India, vols I–VII. L. Reeve, London. https://doi.org/10.5962/bhl.title.54393

Hooker WJ (1808) XXIV. Musci Nepalenses; or descriptions of several new mosses from Nepal. Trans Linnean Soc Lond 9:307–322. https://doi.org/10.1111/j.1096-3642.1818.tb00349.x

Joshi D, Kumar S, Sayal DC, Goel R (2017) The microbiome of the Himalayan ecosystem. In: Kalia VC, Shouche Y, Purohit HJ, Rahi P (eds) Mining of microbial wealth and metagenomics, pp101–116. https://doi.org/10.1007/978-981-10-5708-3_6

Kennedy AC, Gewin VL (1997) Soil microbial diversity: present and future considerations. Soil Sci 162:607–617

Khatun A, Ali MA, Dingle JG (1999) Comparison of the nutritive value for laying hens of diets containing azolla (*Azolla pinnata*) based on formulation using digestible protein and digestible amino acid versus total protein and total amino acid. Anim Feed Sci Technol 81:43–56

Klimeš L (2006) Alpine plant life. Functional plant ecology of high mountain ecosystems by C. Körner. Folia Geobot 41:454–455

Knowles R (1981) Denitrification. Soil Biochem 5:323–369

Laxminarayana K (2002) Effect of phosphorus solubilizing microorganisms on yield parameters of paddy. Indian J Hill Farming 15(1):22–25

Liu Y, Yao T, Jiao N, Kang S, Xu B, Zeng Y, Huang S, Liu X (2009) Bacterial diversity in the snow over Tibetan Plateau Glaciers. Extremophiles 13:411–423

Liu K, Sollenberger LE, Newman YC, Vendramini JMB, Interrante SM, White-Leech R (2011) Grazing management effects on productivity, nutritive value, and persistence of "Tifton 85" Bermudagrass. Crop Sci 51 (1):353–360. https://doi.org/10.2135/cropsci2010.02.0122

Ma T, Zu S, Wang Z, Chen D, Dai G, Feng B, Su X, Hu H, Li K, Han W, Liang C, Bai Y, Feng X (2018) Divergent accumulation of microbial necromass and plant lignin components in grassland soils. Nat Commun 9:1–9. https://doi.org/10.1038/s41467-018-05891-1

Majumdar B (2003) Integrated N management of upland rice. Annual report 2002–03. ICAR Research Complex for NEH Region, Umiam, India

Margesin R, Collins T (2019) Microbial ecology of the cryosphere (glacial and permafrost habitats): current knowledge. Appl Microbiol Biotechnol 103:2537–2549

Margesin R, Feller G (2010) Biotechnological applications of psychrophiles. Environ Technol 31:835–844

Michel HG, Wariish G, Khadar V (1989) Biocatalysts in agricultural biotechnology (Whitaker JR, Philips ES (eds)). American Chemical Society ,Washington, DC

Mittermeier RA, Gil PR, Hoffmann M, Pilgrim J, Brooks T, Mittermeier CG, da Fonseca GAB (2004) Hotspots revisited: earth's biologically richest and most endangered ecoregions. CEMEX

Moll J, Klingenfuss K, Widmer F, Gogos A, Bucheli TD, Hartmann M, vander Heijden MG (2017) Effects of titanium dioxide nanoparticles on soil microbial communities and wheat biomass. Soil Biol Biochem 111:85–93. https://doi.org/10.1016/j.soilbio.2017.03.019

Myers N, Mittermeier RA, Mittermeier CG, da Fonseca GAB, Kent J (2000) Biodiversity hotspots for conservation priorities. Nature 403:853–858

Nannipieri P, Kandeler E, Ruggiero P (2002) Enzyme activities and microbiological and biochemical processes in soil. In: Burns RG, Dick RP (eds) Enzymes in the environment. Marcel Dekker, New York, pp 1–33

Nath S, Datta M (2013–14) Annual report. ICAR Research Complex for NEH Region, Barapani, Meghalaya, India

Oehl F, Oberson A, Probst M, Fliessbach A, Roth H-R, Frossard E (2001) Kinetics of microbial phosphorus uptake in cultivated soils. Biol Fertil Soil 34:31–41

Oliveira CF, do Valle Junior RF, Valera CA, Rodrigues VS, Fernandes LFS, Fernandes FAL (2019) The modeling of pasture conservation and of its impact on stream water quality using partial least squares-path modeling. Sci Total Environ 697:134081. https://doi.org/10.1016/j.scitotenv.2019.134081

Pal SS, Singh U (2002) Bio-degradation of pine litter by a white rot fungus isolated from an inceptisol of Garhwal Himalayas. J Indian Soc Soil Sci 50 (4):467–471

Pandey A, Trivedi P, Chaurasia B, Palini LMS (2006) Soil microbial diversity from the Himalaya, need for documentation and conservation. NBA Sci Bull 5:28–60

Panwar AS, Hazarika UK (2002) Studies on efficient nutrient management. Annual report (2001–02). ICAR Research Complex for NEH Region, Umiam, Meghalaya, India, 12p

Paterson E, Gebbing T, Abel C, Sim A, Telfer G (2007) Rhizodeposition shapes rhizosphere microbial community structure in organic soil. New Phytol 173:600–610

Philippot L, Raaijmakers JM, Lemanceau P, Wh VDP (2013) Going back to the roots: the microbial ecology of the rhizosphere. Nat Rev Microbiol 11:789–799

Pourique J, Vande (1990) Biomass for energy and industry (Grassi G, Gosse G, Santos G (eds)), vol 2. Elsevier Applied Science, London, p 52

Priscu JC, Christner BC (2003) Earth's icy biosphere. Microb Divers Bioprospecting 7:130–145

Rana SK, Rawal RS, Dangwal B, Bhatt ID, Price TD (2021) 200 years of research on himalayan biodiversity: trends, gaps, and policy implications. Front Ecol Evol. https://doi.org/10.3389/fevo.2020.603422

Rai AK, Kumar R (2015) Potential of microbial bioresources of Sikkim Himalayan region. ENVIS Bull Himalayan Ecol 23:99–105

Rao DLN (2003) BNF and biofertilizers research in India: current status and new initiatives in ICAR network. Biofertiliser Newsl 11(2):3–10

Rawat N, Joshi GK (2018) Bacterial community structure analysis of a hot spring soil by next generation sequencing of ribosomal RNA. Genomics 111 (5):1053–1058

Raychaudhuri M, Raychaudhuri S (2003) Evaluation of different Rhizobial strains on groundnut. Annual report 2002–03. ICAR Research Complex for NEH Region, Umiam, Meghalaya, India, 37–51pp

Raychaudhuri M, Kumar K, Raychaudhuri S (1997a) Effect of Rhyzobium japonicum and phosphorus in the nutrient uptake in acid soils of Manipur. Indian J Agric Sci 67(10):459–462

Raychaudhuri M, Kumar K, Raychaudhuri S (1997b) Phosphate fractions in acid soils of Manipur. J Indian Soc Soil Sci 45(4):739–742

Raychaudhuri S (2002) Evaluation of different rhizobial strains on groundnut. Annual report 2001–02. ICAR Research Complex for NEH Region, Umiam, Meghalaya, India, 45–47pp

Rhodes LH, Gardemann JW (1975) Phosphate uptake zones of mycorrhizal and non-mycorrhizal onions. New Phytol 75:555–561

Richardson AE (2001) Prospects for using soil microorganisms to improve the acquisition of phosphorus by plants. Funct Plant Biol 28(9):897–990

Rodriguez H, Fraga R (1999) Phosphate solubilizing bacteria and their role in plant growth promotion. Biotechnol Adv 17:319–339

Roy C, Rameez MJ, Haldar PK et al (2020) Microbiome and ecology of a hot spring-microbialite system on the Trans-Himalayan Plateau. Sci Rep 10:5917. https://doi.org/10.1038/s41598-020-62797-z

Samati H, Gogoi R (2007) Sacred groves in Meghalaya. Curr Sci 93(10):1338–1339

Sanjeev Kumar S, Shivani, Patiram (2001) Effect of biofertilizers on productivity and quality of soybean. Annual Report 2000–01. ICAR Research Complex for NEH Region, Umiam, Meghalaya, India, 110–111pp

Sarmah AC (2001) J Agric Sci Soc North East India 14 (1):111–113

Satyanarayana T, Raghukumar C, Shivaji S (2005) Extremophilic microbes: diversity and perspectives. Curr Sci 89(1):78–90

Schmidt SK, Reed SC, Nemergut DR, Grandy AS et al (2008) The earliest stages of ecosystem succession in high-elevation (5000 metres above sea level), recently deglaciated soils. Proc Biol Sci 275:2793–2802

Schuette UM, Abdo Z, Foster J, Ravel J, Bunge J, Solheim B, Forney LJ (2010) Bacterial diversity in a glacier foreland of the high Arctic. Mol Ecol 19:54–66

Schmit JP, Mueller GM (2007) An estimate of the lower limit of global fungal diversity. Biodivers Conserv 16:99–111

Scholter M, Dilly O, Munch JC (2003) Indicators for evaluating soil quality. Agric Ecosyst Environ 98:255–262. www.Sciencedirect.com

Seckbach J (2000) Journey to diverse microbial worlds, vol 2. Kluwer Academic Publishers, Dordrecht

Sharafatmandrad M, Khosravi Mashizi A (2020) Investigating distribution of ecosystem services in rangeland landscapes: an approach based on weighted key functional traits. Ecol Indic 111:105971. https://doi.org/10.1016/j.ecolind.2019.105971

Sharma D, Gosai K, Dutta J, Arunachalam A, Shukla AK (2015) Fungal diversity of twelve major vegetational zones of Arunachal Himalaya, India. Curr Res Environ Appl Mycol 5(2):101–119

Singh SP (2014) Attributes of Himalayan forest ecosystems: they are not temperate forests. Proc Indian Natl Sci Acad 80:221–233

Spehn E, Rudmann-Maurer K, Körner C (2011) Mountain biodiversity. Plant Ecol Divers 4:301–302. https://doi.org/10.1080/17550874.2012.698660

Spicer R, Yang J, Herman A, Kodrul T, Aleksandrova G, Maslova N, Jin JH (2017) Paleogene monsoons across India and South China: drivers of biotic change. Gondwana Res 49:350–363

Srinivas TNR, Singh SM, Pradhan S, Pratibha MS, Kishore KH, Singh AK, Shivaji S (2011) Comparison of bacterial diversity in proglacial soil from Kafni Glacier, Himalayan Mountain ranges, India, with the bacterial diversity of other glaciers in the world. Extremophiles 15:673–690

Stepp JR, Castaneda H, Cervone S (2005) Mountains and biocultural diversity. Mt Res Dev 25:223–227

Stolp H (1988) Microbial ecology: organisms, habitats, activities. Cambridge University Press, Cambridge

Strickland MS, Lauber C, Fierer N, Bradford MA (2009) Testing the functional significance of microbial community composition. Ecology 90:441–451. https://doi.org/10.1890/08-0296.1

Suyal DC, Yadav A, Shouche Y, Goel R (2014) Differential proteomics in response to low temperature diazotrophy of Himalayan psychrophilic nitrogen fixing *Pseudomonas migulae* S10724 strain. Curr Microbiol 68:543–550. https://doi.org/10.1007/s00284-013-0508-1

Suyal DC, Yadav A, Shouche Y et al (2015) Diversified diazotrophs associated with the rhizosphere of Western Indian Himalayan native red kidney beans (*Phaseolus vulgaris* L.). Biotech 5:433–441. https://doi.org/10.1007/s13205-014-0238-5

Talukdar J, Barthakur HP (1981) Solubilization of rock phosphate by root suspension inoculum of certain rice varieties grown in Assam, India. J Res Assam Agric Univ 2(1):46–51

Tang M, Li L, Wang X et al (2020) Elevational is the main factor controlling the soil microbial community structure in alpine tundra of the Changbai Mountain. Sci Rep 10:12442. https://doi.org/10.1038/s41598-020-69441-w

Thakur N, Savitri Saris PEJ, Bhalla TC (2015) Microorganisms associated with amylolytic starters and traditional fermented alcoholic beverages of North Western Himalayas in India. Food Biosci 11:92–96

Thakuria D, Talukdar NC, Goswami C, Hazarika S, Boro RC, Khan R (2004) Characterization and screening of bacteria from rhizosphere of rice grown in acidic soils of Assam. Curr Sci 86:978–986

Tiwari SC, Mishra RR (1989) Seasonal variation in dehydrogenase and urease activity in hilly soils under grassland and forest. J. Ind. Soc. Soil Science 43:689–690

Upadhyay TP, Sankhayan PL, Solberg B (2005) A review of carbon sequestration dynamics in the Himalayan region as a function of land-use change and forest/soil degradation with special reference to Nepal. Agr Ecosyst Environ 105(3):449–465

van Leeuwen JP, Djukic I, Bloem J, Lehtinen T, Hemerik L, de Ruiter PC, Lair GJ (2017) Effects of land use on soil microbial biomass, activity and community structure at different soil depths in the Danube floodplain. Eur J Soil Biol 79:14–20. https://doi.org/10.1016/j.ejsobi.2017.02.001

Venkatesh MS, Majumdar B, Kumar K, Mishra VK (2001) Effect of rock phosphate, phosphate solubilizing microbes and gypsum on yield and nutrient uptake by groundnut. Annual report 2000–01. ICAR Research Complex for NEH Region, Umiam, Meghalaya, India 27p

Walker TS, Bais HP, Grotewold E, Jm V (2003) Root exudation and rhizosphere biology. Plant Physiol 132:44–51

Wang JT et al (2015) Altitudinal distribution patterns of soil bacterial and archaeal communities along Mt. Shegyla on the Tibetan Plateau. Microb Ecol 69:135–145

Wang L, Gan Y, Wiesmeier M, Zhao G, Zhang R, Han G, Siddique KHM, Hou F (2018) Grazing exclusion—an effective approach for naturally restoring degraded grasslands in Northern China. Land Degrad Dev 29 (12):4439–4456. https://doi.org/10.1002/ldr.v29.1210.1002/ldr.3191

Wakelin S, Warren RA, Harvey PR, Ryder MH (2004) Phosphate solubilization by Penicillium app. closely associated with wheat roots. Biol Fert Soils 40(1):36–43. https://doi.org/10.1007/s00374-004-0750-6

Widden P (1987) Fungal communities in soils along an elevational gradient in Northern England. Mycology 79:298–309

Wood M (1989) Soil biology. Blackie USA, Chapman and Hall, New York

Xiang SR, Yao TD, An IZ, Xu BQ, Li Z, Wu GJ, Wang YQ, Ma S, Chen XR (2004) Bacterial diversity in Malan ice core from the Tibetan Plateau. Folia Microbiol 49:269–275

Zhang S, Hou S, Wu Y, Qin D (2008) Bacteria in Himalayan glacial ice and its relationship to dust. Biogeosciences 5:1741–1750

Zhang Y, Li D, Wang H, Xiao Q, Liu X (2006) Molecular diversity of nitrogen-fixing bacteria from the Tibetan Plateau, China. FEMS Microbiol Lett 260:134–142. https://doi.org/10.1111/j.15746968.2006.00317

Zhang Q, Liu K, Shao X, Li H, He Y, Sirimuji, Wang B (2021) Microbes require a relatively long time to recover in natural succession restoration of degraded grassland ecosystems. Ecol Ind 129:107881. https://doi.org/10.1016/j.ecolind.2021.107881

Soils of HKH Region

6

Abstract

Soils are dynamic natural bodies comprising the upper layer of the earth. The parent material strongly influences soil composition both through its chemical characteristics, some of which are transferred to the derived soil, and through its physical constitution which influences leaching rate and soil aeration. The climate change, land use change, deforestation, and overgrazing are affecting mountains soils in HKH region. These soils are vulnerable to climatic changes, and the anthropogenic factors support and magnify these changes. Hazards such as floods, landslides, debris flows, and glacial lake outbursts are on the rise in HKH region, especially those areas with rapidly expanding populations and poor infrastructure. Climate change is amplifying the impact of hazards as it increases the frequency of extreme events, causing heavy rainfall, droughts, and glacier melt. Soils found in the Himalayas are diverse in character depending upon altitude, vegetation cover, slope, structure, and stage. The major soils groups in the Himalayas are brown hill soil (600–1700 m amsl), sub-mountain soils (formed under dense forest cover), mountain meadow soils (found in alpine and subalpine zones up to an altitude of 4400 m), and red loamy soils, apart from other less significant types. Most of the soils in the region are acidic in nature. The high-altitude meadow soils are found in high-altitude meadows near the snow line in all parts of the higher and trans-Himalayas. Since the texture of the soils is very coarse with high gravel content, they are prone to displacement due to slides and avalanches. These soils are dark in color having a high content of humus. Desert (Arid) soils are found in the cold desert area of Ladakh (India) and other mountain areas of the region with similar climate and altitude; red and black soils are found in isolated areas of Nepal, Bhutan, and Uttarakhand, and Himachal Pradesh state of India. Podzols are found in Western Himalayas. Apart from the above-mentioned soils, traces of alluvial and lateritic soils are found in some parts of the Himalayas. Mountain soils are highly dynamic and sensitive systems that react to environmental changes such as climate change and intense land use. Human-induced erosion rates are, in some mountain areas, much beyond soil production rates. Extensive erosion rates lead to rapid soil degradation and loss of areas for plant growth which, in turn, also negatively affects carbon sequestration. The Hindu Kush Himalayas is a densely populated area and human activities are the major cause of environmental and land degradation. Soil degradation and forest depletion are the most serious environmental issues in the region. The ecosystem of the great Himalaya Mountains is one of the most

U. C. Sharma et al., *Soils in the Hindu Kush Himalayas*, Geography of the Physical Environment,
https://doi.org/10.1007/978-3-031-11458-8_6

important life support systems on the earth. The rivers, which arise from the Himalayas, flow down to the plains and contribute to agriculture, industry, and energy sectors that sustain millions of people. The chapter mostly deals with soils of HKH region according to the new system of soil classification.

Keyword

Drivers of change · Future protocol of soil management · Mountain soil management · Networking of soil management · Regional collaboration in soil management · Soils resources

6.1 General Description

Comprising the upper layer of the earth, along with their mineral and organic components, water, and air, soils are dynamic natural bodies being formed because of atmospheric and biosphere forces acting on different parent materials under various relief and vegetation situations, over a period of time (Yaalon and Arnold 2000). The importance of soils can be gauged from the fact that they are the medium of plant growth and "no soils" means "no vegetation". Soils are of great significance to society as they provide various economic and cultural services to the dwellers and function as a support system for the growing plants. The human population depends on soil for growing its food, recycling its wastes, and building its structure. As early as in 1941, Jenny enumerated the important factors in the development of soil, such as parent material (geology), climate, topography, vegetation, and time or duration of action of various factors on the parent material. By allowing only one factor at a time, studies of chromosequences, biosequences, lithosequences, toposequences can be conducted, and it is possible to determine how these factors affect soil development (Jenny 1980).

The parent material strongly influences soil composition both through its chemical characteristics, some of which are transferred to the derived soil, and through its physical constitution which influences leaching rate and soil aeration. Generally, silicate rocks with high content of plant nutrient elements produce soil which is eutrophic or mesotrophic and highly favorable for plant growth (Patiram and Datta 2006). It is a known fact that climate indirectly influences soil temperature and soil water relationships and also has an indirect impact through vegetation. Temperature controls all biological, chemical, and physical processes in the soil and over ground. Weathering, neoformation of minerals, decomposition, humification, aggregate formation, ion exchange, and redox processes need water for all biological, chemical, and physical transformation processes. The water is of great importance as it affects profile differentiation through percolation, ascent (potential evaporation > rainfall), erosion, and waterlogging, and these all affect the translocation processes. Vegetation is not an independent factor like parent material and climate but results from the interaction rock, climate, and soil. Similarly, the time required for soil formation appears to be widely variable as it is strongly influenced by climate and parent material factors. However, the time factor, next to climate, is important in determining the final stage in the cycle of weathering.

Zanini and Romeo (2015) mentioned that the mountain peoples have learned to live on high altitudes and evolved resilient and sound systems to live under harsh climates and environments. This has helped in solving many problems and mitigating their impacts. These people are practising agriculture, but the cutting of forests and free-range grazing are affecting mountains in a big way (FAO 2015). The carbon stock potential is mainly influenced by vegetation density, aboveground biomass, species richness, and forest types (Choden et al. 2021) and subtropical forests have, generally, more carbon capacity and sequestration potential. The decline of forests in the Himalayas is complex, and the pattern of depletion is location-specific (Tucker 2021). The anthropogenic influence on forest and soil can alter sources and sinks for carbon (Candell and Raupach 2008), and an increase in demographic pressure has been identified as the major driver of deforestation (Oyetunji et al. 2020), besides infrastructure development and urban growth (Blanco-Canqui and Lal 2008; Barcena et al. 2014), and fossil fuel use (Andres et al. 2012), thereby triggering erosion of soil and its displacement. The estimates show that deforestation is responsible for a loss of forest cover at an annual rate of 9.4 million ha (Miettinen et al. 2011). The HKH countries, especially Bhutan, Nepal, India, and Pakistan are more prone to deforestation and, therefore, need a full knowledge of carbon dynamics in the soil.

Soil is an unexpandable resource that has a very important role for ecosystem services on a sustainable basis and human life, as well as ensuring enduring good environmental security and agricultural productivity. The soil is an important body for food security in the world because of its being a source of nutrients, water, and minerals required to nurture vegetation.

Most of the mountain soils have low depth and are generally poorly developed, and therefore are skeletal, shallow, acidic, and relatively with poor production capacity. The diverse climatic conditions, exposure, and slope variations cause more variability in these soils within limited areas. Many trees and crop species have adapted to the conditions of mountain soils. The soil surface vegetation greatly helps in controlling soil erosion by binding the soil particles.

The relationship between people and soil is fundamentally important in the mountain ecosystem. The method of management of crops and forests increase the pressure on the fragile ecosystem, resulting in soil erosion, soil degradation, and natural hazards. If the demographic pressure is high, the impact of natural calamities can be very destructive with consequences on people's livelihoods. The United Nations Environment Programme World Conservation Monitoring Centre (UNEP-WCMC 2002) developed a classification based on topographic criterion (elevation) that combines altitude, steepness of slope, and local elevation range. Environmental gradients that are key components of mountain environments were based on this classification. In all, six classes were standardized according to this. These classes are elevation $\geq$ 4500 m, elevation 3500–4500 m, elevation 2500–3500 m, elevation 1500–2500 m and slope 2°, elevation 1000–1500 m and slope $\geq$ 5° or local elevation range (LER) > 300 m, and elevation 300–1000 m and LER > 300 m.

Four major functions of the soils as highlighted by Romeo et al. (2015) are supporting, regulating, provisioning, and cultural. The "supporting function" provides the physical support for plants/trees' stand and natural regeneration of vegetation. The "regulating function" helps in water cycling, increased infiltration, and protection of water resources. It also regulates nutrient and organic matter cycling, climate regulation, improved carbon accumulation, and conservation of local biodiversity. The "provisioning" function means increased food production through improved soil health, and the "cultural function" of soil, however, ensures improved quality of life and enhanced potential for the attractiveness of ecosystem services (Bunning et al. 2015; Romeo et al. 2015). Almost similar views were expressed by Varga et al. (2015) explained that the soils are a fundamental medium of plant growth and perform various important functions such as (i) multiscale physical, chemical, biological, and genetic support, (ii) water, nutrient, biodiversity, and global climate regulation, (iii) basis for production of food, feed, fiber, bioenergy, and medicines, and (iv) basis for landscapes that provide recreation and ecotourism.

The diverse nature of cropping and production patterns is due to soil heterogeneity. The soil is said to be heterogeneous when it varies greatly in texture, fertility, topography, moisture content, drainage, etc. in a relatively small area. Soil is a three-phase dynamic system that performs many functions and ecosystem services (Fig. 6.1). There are many supporters of soil which help this body to perform its functions, soil microorganisms being one important component, supporting the soil. Generally, it has been found that we often consider soil health independently based on testing the soil for some parameters, without referring to interlinked soil functions. Soil tests are mostly performed in laboratories by simply testing organic carbon, available P and K, pH, and conductivity, for making fertilizer recommendations. Often, the physical condition of soil

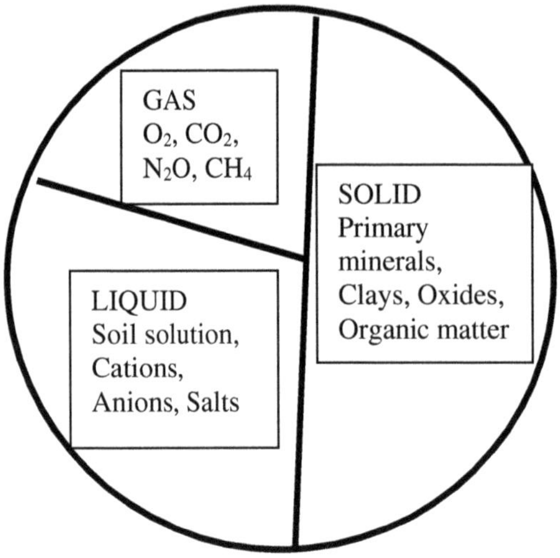

Fig. 6.1 Three phases of soil

and biological potentiality are ignored in soil health management. Taking cognizance of the importance of soil health in all dimensions and aspects, the 68th UN General Assembly declared the year 2015 as the International Year of Soils. India faces severe problems in agriculture, especially due to the poor health of soils, besides many other factors. It is estimated that about 52.8% of the total geographical area in India is degraded, producing less than 20% of its potential yield (Govt. of India 1990).

The Himalayan region, having an area of 430 million ha of land, encompasses forests, agriculture, and rangelands in 14%, 26%, and 60% of the area, respectively. The region comprises about 1% of water bodies and 5% of permanent snow and glaciers. The HKH boundary map is presented in Fig. 6.2.

Major issues facing Highland agriculture have been laid down in Fig. 6.3. Out of these issues, core issues are climate change and degradation induced by human interference. The interrelationship between various issues of mountain agriculture has been highlighted.

Fig. 6.2 HKH region boundary map

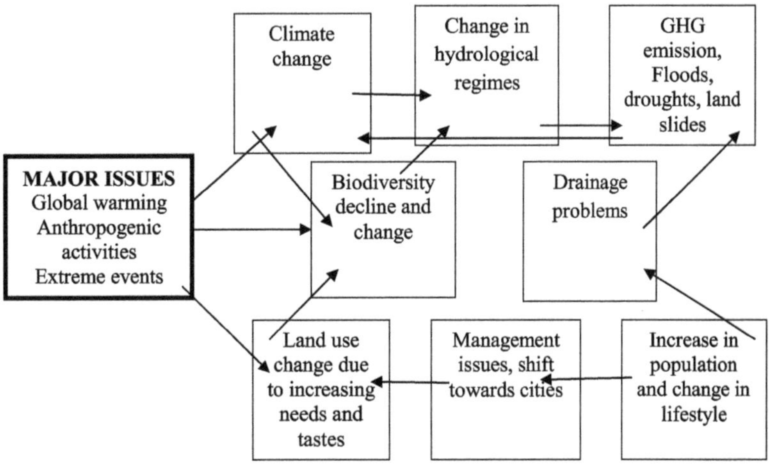

Fig. 6.3 Major issues facing high-altitude ecosystem. (Modified from Ning, 2013, ICIMOD, Kathmandu, Nepal)

6.2 Problem Soils

The problem soils are those soils which possess characteristics which limit vegetation productivity and make it uneconomical for the cultivation of crops without taking suitable reclamation measures. Soluble salts, soil reaction, ESP, waterlogging, aeration, etc. are all responsible for inducing some kind of problem to the soil. These soils are unfavorable for the cultivation of field crops because of one or more unfavorable soil properties. Although one cannot control many of the climate factors, many of the factors can and must be managed. More emphasis is required to be given to soil-related constraints because a majority of the problems can be changed to favorable conditions to increase soil productivity. Generally, four types of problems are encountered in soils, such as physical, chemical, biological, and nutritional.

6.2.1 Soil Physical Problems

Slow permeable/Impermeable soils

Occurrence and Causes

Soils are said to be slow permeable when the infiltration rate is <6 cm day^{-1}. This is mainly due to the very high clay content of the soil. The consequences are more runoff eventually leading to soil erosion and nutrient removal. Because of high capillary porosity, there is impeded drainage and poor aeration, resulting in reduced conditions. Remedial measures:

(i) Application of organics such as farmyard manure, compost, and press mud.
(ii) Formation of ridges and furrows.
(iii) Formation of broad beds in black clay soils to reduce the retention of water.
(iv) Providing open/sub-surface drainage.
(v) Contour bunding.
(vi) Application of soil conditioners like vermiculite.

The other major physical constraints are soil surface crusting, sub-soil hard pan, and shallow soils. Presence of colloidal oxides of iron and aluminum in soils cause surface crusting, which binds the soil particles under moist conditions. It not only is predominant in Alfisols but occurs in other soils too. Soil surface crusts prevent the germination of seeds, cause poor infiltration and aeration, and affect nodule formation.

Highly permeable soils

Highly permeable soils have inherent physical makeup of light texture, mostly sandy. These soils have very high hydraulic conductivity and infiltration rates. Due to a lack of finer particles, highly permeable soils have poor water-holding capacity, so much so that crops suffer due to restricted or no moisture supply to give potential yields. Due to a lack of sufficient amount of organic matter, there is weak formation of aggregates with very poor soil structure. So, the nutrients and water applied to these soils are not used fully by the crops and both are subjected to loss through leaching. Management technologies for such soils include uniform plowing, application of clay soil, crop rotation with green manure crops, split application of fertilizers, use of slow-release fertilizers, and irrigation with low quantity of water.

6.2.2 Chemical Problem Soils

Salt-affected soils

The salt-affected soils are saline soil and sodic soils. These soils are encountered in the arid and semi-arid regions where evapotranspiration greatly exceeds precipitation, due to the accumulation of ions of sodium, potassium, magnesium, calcium, chlorides, carbonates, and bicarbonates on the soil surface. The management options for these soils are the removal of salts through the processes of leaching with water and drainage. The construction of lateral and main drainage channels of suitable size can help in reclaiming these soils. For lowering the water table, removal of excess salts, and prevention of secondary salinization, sub-surface drainage will prove an effective method to mitigate the problem.

Physical Amelioration

Physical amelioration involves improving the soil's physical condition by improving infiltration and aeration and is not actually the removal of sodium from the exchange complex. This can be achieved through deep plowing in order to break the hard pan developed at sub-surface due to sodium and thereby improving free movement of water and air. Provision of drainage, applying sand to make the soil lighter, and profile inversion are other physical methods.

Chemical Amelioration

Reclamation of sodic soils is done to replace sodium ions with Ca^{2+} from the exchange complex of the soil. The chemical amendments can be gypsum, calcium carbonate, phosphogypsum, elemental sulfur, sulfuric acid, pyrites, $FeSO_4$, etc. Gypsum is mostly used as a chemical amendment. The choice of amendment material should be decided based on economy, availability, and efficiency.

Acid soils

On a global scale, the acidic soils are found in the northern belt, with a cold, humid temperate climate, and the southern tropical belt, with a warm and humid climate (Von Uexkull and Mutert 1995; Dalovic et al. 2012, Khan et al. 2001). Acid soils are formed due to parent material composition, biological weathering, leaching of salts like calcium, magnesium, potassium, etc. in the soil with rain or irrigation water, acid rain, use of some soil amendment, pollution, use of acid fertilizers, particularly ammonium-based fertilizers, and agricultural practices followed. Many ammonium fertilizers contribute to soil acidification even if the nitrogen is taken up by plants. The acid soils occupy nearly 40% of arable soils in the world, and this proportion is still increasing (Bian et al. 2013). Soil acidification can lead to nutrient deficiency, reduce water uptake, and heavy metal toxicity (Zhao et al. 2015), rendering soil unsuitable for production. Acid soils may also contain toxic amounts of copper and zinc. About 800 Mha area in the world has an acidic reaction. More than 90% of soils of Eastern Indian Himalayas are acidic in reaction, while about 30% of the geographical area of Jammu and Kashmir is acidic. In Himachal Pradesh, 0.33 Mha, and all hill soils of Uttarakhand come under acid soils. Aluminum and manganese in acid soils affect crop performance badly, resulting in crop failures.

Production constraints, management, and amelioration

Due to the fixation of P by sesquioxides in the soil and the increased toxicity of Al, Mn, and Fe, the productivity of crops is low on acid soils. Management of the acid soils is very important for the increase in crop productivity, either through the application of amendments to correct the soil constraints or through manipulation of the agronomic practices. This would depend upon the climatic and edaphic conditions. Lime has been recognized as an effective soil ameliorant as it reduces Al, Fe, and Mn toxicity and increases base saturation, and P and Mo availability of acid soils. Before making any recommendation, it is suggested that the economy of applying lime should be worked out.

Soil acidity is also a limiting factor affecting the growth and yield of many crops in the Himalayan region. The adverse effect of acid soils on plant growth is mainly related to nutrient imbalances and slow microbial activity. Low yield of crops grown on acid soil may be due to Ganeshamurthy et al. (2016a, b).

- Toxic concentration of aluminum,
- Toxic concentration of manganese,
- Iron toxicity in some soils,
- Deficiency of bases like calcium, magnesium, sodium, and potassium,
- Molybdenum deficiency,
- Nitrogen, phosphorus, and sulfur deficiency,
- Slow microbial activity and impairment of nitrogen fixation by legumes, and
- Slow organic matter decomposition.

6.2.3 Soil Fertility and Land Use in the World

Soil productivity depends on its available nutrient status, and its potential and ability to provide nutrients out of its own reserves and through external applications to crops for getting the optimum crop yields. Soil delivers ecosystem services that enable life on Earth. Percent of land used for farming in different continents of the world is given in Fig. 6.4. Maximum percentage of land is utilized for farming in Asia, while the minimum (6%) is in South America (Caon 2018).

.

The population of the world is increasing very fast and so, hunger is on the rise. To meet growing food demand, it is necessary to develop more productive and sustainable farming systems

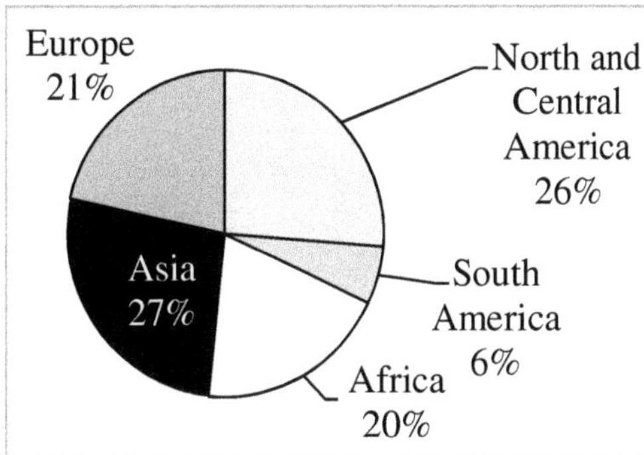

Fig. 6.4 Percent of land used for farming in different continents. *Source* Based on Caon (2018)

to feed about 815 million undernourished people. Caon (2018) reported that sustainable soil management could produce up to 58% more food. For this, the following suggestions were made:

- Land tenure rights;
- Targeted soil research;
- Effective education programs;
- International and regional cooperation on soils;
- Strengthening soil information systems;
- Strengthening of extension services;
- Soil testing before sowing/planting a crop;
- Use of amendments for problem soils.

6.2.4 Global Soils

Pretty et al. (2011) reported that the sustainable land and water management techniques, as also conservation agriculture, may help to restore and improve soil fertility through integrated soil fertility management. Resource-use efficiency can help in reducing input costs through decomposition and nutrient cycling, nitrogen fixation, and water storage and movement. Less fertilizer and pesticides may be needed if nutrient use efficiency is improved. The availability of water and nutrients to plants improves with the improvement in soil structure (Costanza et al. 1997). Improved soil management is a significant factor for enhanced crop productivity as it will improve soil physical condition, fertilizer use efficiency, proper use of external inputs, and economics of

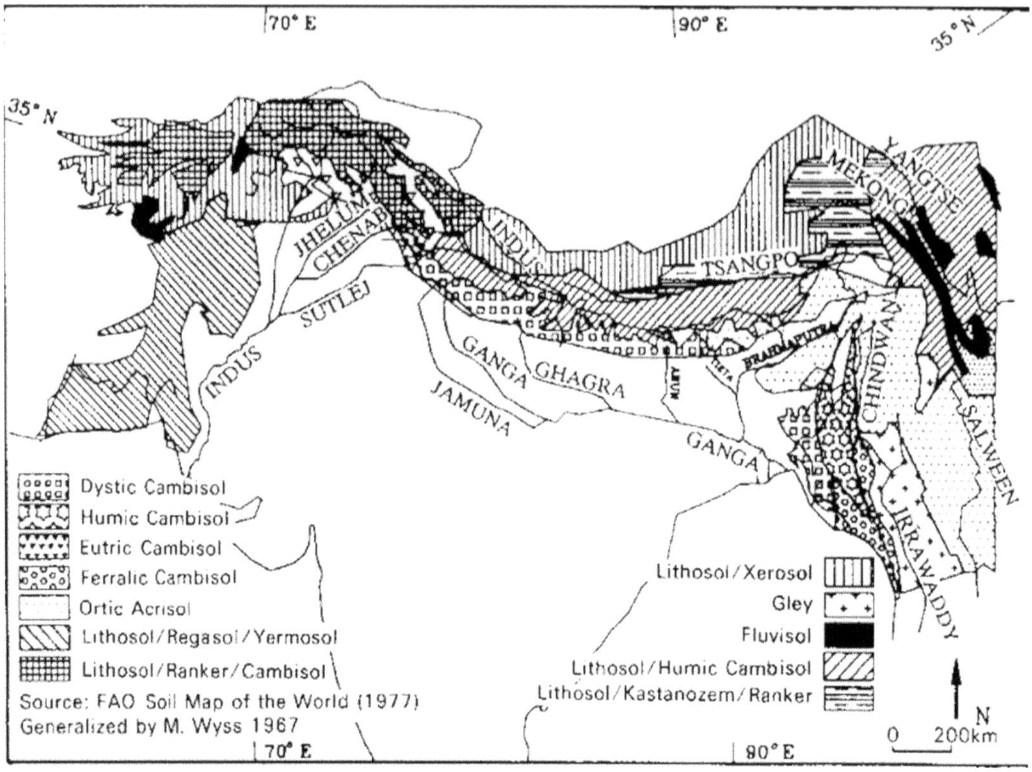

Fig. 6.5 FAO Soil map of the Himalayas (Adopted from FAO 2011)

crop production. Soil biodiversity determines the resource-use efficiency, as well as the sustainability and resilience of agro-ecological systems.

A soil map (FAO 2011) (Fig. 6.5) is given to understand different soils in the Himalayas.

6.3 Toxification of Acid Soils

Toxification of acid soils is caused by one or an amalgamated effect of many factors, which affect the crop growth, eventually ending with very low and uneconomic crop yields. The soil may be toxic when it contains high amounts of toxic heavy metal like iron, copper, manganese, zinc, and aluminum and lacks proper amounts of essential nutrients like phosphorus, magnesium, calcium, and potassium, and has low soil pH (Foy et al. 1978; Bian et al. 2013). High Al and Fe oxides and hydroxide in low soil pH are responsible for P fixation, thus making it unavailable to plants (Sanchez et al. 1997a, b; Oburo 2008; Kisinyo et al. 2013). Low soil pH may be due to excesses of aluminum, iron, and manganese, which hinder the availability of some essential crop nutrients, resulting in reduced crop productivity and production (Zeigler et al. 1995). Cognizance of all of these toxicities should form a basis during breeding programs for evolving crop varieties tolerant to low soil pH (Tandzi et al. 2018).

6.4 Mountain Soils

Relatively less information is available about the mountain soils compared to the plains. Nevertheless, the mountain soils are as important as those in plains, if not more. It is evident from the soil science literature that most of the soil survey programs have been undertaken on arable lands suitable for crop production. This was mainly due to the reason that many countries of the world suffered from food shortages, and land use planning was necessary to realize the full potential of the soils to achieve self-sufficiency in food grains. There is a need for an "integrated" mountain ecosystem management, which is found lacking in mountain landscapes. The varied role of soils requires an integrated approach to perform different agricultural operations as well as ecosystem management and development under mountain landscapes. In agriculture, management includes water regulation, production of food or non-food crops, sources of inputs, timely field operations, regulation or marketing of agricultural produce, activities to conserve soil resources from erosion, and land degradation. Mountain ecosystems have to be protected, particularly under the threat of climate change and its consequences.

Out of about 3500 km of the total length of the Himalayas, extending from Afghanistan to Myanmar, about 2400 km falls through India (from the state of Jammu and Kashmir to the state of Arunachal Pradesh). The altitudinal, physiographical, and locational differences in Indian Himalayas give an edge to India over other HKH countries in having large variations in soils and their suitability for growing number of crops and, plant and animal species.

6.5 Himalayan Soils

Himalayan soils are variable depending upon altitude, vegetation cover, slope, structure, and stage of evolution, major soil groups being brown hill soil, sub-mountain soils, mountain meadow soils, and red loamy soils. In general, the soils in the region are acidic in nature. The soils of the Himalayas have been broadly grouped as follows:

Brown hill soils

These soils are found at the altitudes of 600–1700 m, with a rainfall of 800–2000 mm. These are deep soils, with a pH of 6.0–7.6, and are loam to silty clay.

Sub-mountain soils

These are formed under dense forest cover. These are siliceous soils, having sandy loam texture, and are rich in humus.

Mountain meadow soils

These soils are found in alpine and subalpine zones up to an altitude of 4400 m. The texture ranges from sandy loam to silty loam. The entire Himalayan range has a wide variety of soil subtypes. It is variation in the soils of the Kashmir Himalayas, besides the different climatic conditions, that is responsible for the variation in the vegetation and the tree life in this Himalayan Range. The coniferous forest soils are spread along the hilly slopes and grow *Pinus wallichiana;* the hilly deciduous forest soil is covered under woods usually consisting of *Ulmus wallichiana, Corylus colurna, Padus cornuta,* and *Parrotiopsis jacquemontiana; the* moraine deciduous forest soil, formed mostly of washed down materials, grows trees, such as *Morus alba, Salix alba, Prunus cerasifera, Robinia pseudoacacia,* and *Rhus succedanea.* The upper layer of this soil is heavily impregnated with smooth boulders and stones. The fluvents, varying from sandy loam to loamy coarse sand, are found in the valleys of the Jhelum and the Indus.

The soils of the Kashmir Valley vary in origin from alluvial to lacustrine and glacial. The soil cover in the Valley is thick and rich in the low-lying areas along the Jhelum. However, the thickness and the fertility richness of the soil decrease with the altitude. The soils of low-lying areas and the Valley basin, being rich in nitrogen content, organic matter, and other plant nutrients are highly fertile. The soils of highlands are acidic in nature. In this type of soil flourish the deodar, silver fir, and blue pines. As against the above two soils, the Karewa soils, mostly composed of silts, are poorer as far as fertility is concerned. In Kashmir (Western Himalayas), Malik et al. (2021) found that the soils did not show much variation in texture and only two textural classes, such as sandy clay loam and sandy loam, were predominant. Soil supplies nutrition to vegetation and, in turn, the vegetation gives a cover to the soil to arrest soil erosion as well as additions of nutrients to the forest soil through litter fall and decomposition of vegetation (Singh et al. 1986; Eni et al. 2012). The vegetation affects the physical and chemical attributes of soil to a large extent (Shameem et al. 2011). Forest soils influence many forest attributes, like the composition of the forests, rate of tree growth, and efficacy of natural reproduction (Bhatnagar 1965). The vegetation in return enhances the soil structure, infiltration rate, water-holding capacity, hydraulic conductivity, and aeration (Champan and Reiss 1992). Physical properties of the soils show their suitability for different uses like agricultural, environmental, and engineering (Schoenholtz et al. 2000; Osman 2013; Phogat et al. 2015; Sarangzai et al. 2015). Himalayan forests contribute to preserving the soil, holding the hill slopes in position, and in buffering soil nutrients (Sharma et al. 2010; Saha et al. 2018). Soil serves as a physical base for trees and is a source of moisture and nutrients for vegetative growth (Zhou et al. 2015).

Desert (Arid) Soils

Desert areas of Ladakh, Lahaul Spiti, Kinnaur, Bharmour in Himachal Pradesh, pockets of northern Uttarakhand, and Sikkim have desert arid soils. These soils though at some places sandy-clay and at a few places loam or loamy-clay are generally classified as gray, light, and arid in nature. These soils are characterized by low fertility status and sparse plant life. The cold arid soils have very poor organic and nitrogen content, though they possess rich potassium and calcium contents. Because of the absence of organic matter, these soils are alkaline in nature.

Besides, there are Red and Black Soils and podzols. Podzols commonly exist in Western Himalayas of India, and in a few areas of Nepal, Sikkim, and Arunachal Pradesh. These soils develop under coniferous forests under humid temperate conditions, where the rate of precipitation exceeds the rate of evaporation.

6.6 Soil Degradation and Vegetative Cover

Land and soil degradation are the key environmental issues in mountain ecosystem, and their variation is dependent on space and time. Soil, as a part of the land, is actively engaged in this process. The intensity and frequency of soil degradation are controlled by land use and its management. The significance of vegetation cover on water infiltration, runoff, and soil erosion has been well documented by Wood and Blackburn (1981). Sharma et al. (1992a, b) reported that due to an increase in demographic pressure and subsequent high demand for food and forage, agricultural land is increasing significantly over the past decades at the cost of other land uses in the Himalayas.

6.7 Forest Cover in Himalayas

Due to demographic pressure, the Himalayan mountain ecosystems are under severe stress (Kapoor 1994; Myers 1986) which has been further amplified by climate change (Solomon 2007). This has resulted in a large scale encroachment of forest lands and land degradation (Shrestha et al. 2012; Bolch et al. 2012; Munsi et al. 2009). Total area under forest cover in western, central, and eastern Himalayas was 4.36, 8.95, and 7.67 million ha, respectively, during 1998 which changed to 4.29, 9.04, and 7.67 million ha in 2018, respectively (Gu et al. 2020). The total forest area in the Himalayas during 2018 was recorded as 20.99 and 21.01 million ha in 1998 and 2018, respectively.

6.8 Threats to Mountain Soils

The demographic pressure in HKH has accelerated agriculture expansion due to increase in food demand. Many forest areas on steep slopes have been put to agricultural use to meet the food requirements of local people and the remaining large forests are being over-exploited for fuel-wood, fodder, and litter collection (Schreier 2015). Human interferences by raising crops for food and cash have encouraged soil erosion triggered due to heavy rains. The production capacity of the soils has been reduced because of a decline in soil organic matter, nutrient deficiencies, and poor soil health. Less snow cover not only affects nutrient dynamics, but it also affects soil structure, which is a key factor for soil stability and erodibility (Freppaz et al. 2015). Mountains have been influenced by climate change caused as a result of GHG emissions. The soils of mountain areas are prone to degradation processes such as water erosion, loss of chemical and physical quality, and desertification. Indirectly, these processes may also affect surface and groundwater quality and quantity, soil as well as public health. When marginal areas are abandoned after over-exploitation for crop production, they become even more prone to natural hazards. Stanchi et al. (2015) explained that climate change will induce altitudinal shifts of plant species, threatening biodiversity, thereby compromising the interactions of soils, biosphere, and water cycle. The United Nations Convention to Combat Desertification (UNCCD) has suggested and promoted some robust measures for enhancing sustainable environmental and natural resource management to deal with desertification.

6.9 Afghanistan

Afghanistan is located in southwestern Asia and extends from about 965 km north to south and 1287 km east to west, with a land area of 652,237 km^2. It lies between 38° 26′ N and 29° 23′ N latitudes and 60° 34′ E to 74° 53′ E longitudes. It is a landlocked country with extensive mountain ranges, as well as desert and plains regions. The country comprises mostly the Hindu Kush mountains. The huge Hindu Kush mountains bisect the country into two prominent regions, the areas on the northern side and the rest of the country in the south. For administrative purposes, the mountain range has created three broad geographic regions, namely The Central Highlands, The Northern Plains, and the Southwestern Plateau.

6.9.1 The Central Highlands

Comprising about 70% of Afghanistan, the central highlands, have an area of about 414 736 km^2 (Saba 2001). This region of Afghanistan has deep, narrow valleys, as well as extremely high mountains, which have proved helpful in defending the country from outside invaders. Ranges of very high altitude are located in Afghanistan, and about half of the area of Afghanistan has an elevation greater than 2,000 m. The central highland region, a major part of the zone, has been classified as a sub-tropical desert and with grassland-steppe vegetation. One of the most famous routes from Afghanistan to India has been the Khyber Pass. The overall climate in this region is extremely dry. While the winters are very cold and harsh, the temperatures in the summer vary from 25 to 30 °C. The soil in this region is meadow-steppe and desert-steppe types.

6.9.2 The Southern Plateau

The southern plateau consists of plateaus of high altitudes and vast sandy deserts. This region covers about 129 500 km^2, and is transversed by many large rivers, including the Helmand. Within the Southern Deserts region, the average elevation of the southern plateau area in the southern desert region is about 900 m amsl (Saba 2001). This region receives very low precipitation. This region comes under the subtropical desert, and it is hot during summer and mild during winter (Microsoft 2001). Sparse vegetation and a very low human population are found in the southern desert of Afghanistan.

6.9.3 The Northern Plains

The northern plains of Afghanistan cover about 108 000 km^2 area of the country. These plain areas within the mountains of the HKH region are very fertile and productive. The Amu river runs through the edge of the foothills. The mean altitude of this area is about 600 m amsl. This region is famous for agriculture and Afghanistan is highly dependent on this part of the country for food security. This part of Afghanistan has about 15% of the country. Mid-latitude grassland-steppe is the major vegetation type of the northern plains of the country. The northern grasslands, having an area of about 103 684 km^2, include Afghanistan's most fertile soils and favorable climate (Saba 2001). Soils of the order Mollisols are present in this region. These soil characteristics are generally of prairies and grasslands. Aridisols are present where these conditions are not there. Short grasses and herbaceous forbs on moist sites (Mollisols) are generally the common vegetation in a mid-latitude steppe, whereas bunch grasses and dry shrubs are found in drier areas (Aridisols). This region is considered as the hub of Afghanistan's agriculture. Currently, however, only half of Afghanistan's agricultural lands are used as farmland (Saba 2001). Crops grown in drier areas need irrigation to survive.

6.9.4 Properties of Soils Under Rice Crop

Masunagai et al. (2014) investigated the properties of paddy soils in 20 villages in Nangarhar province, East Afghanistan. The sandy loam or loam soils were alkaline and contained carbonate ranging from 0.49 to 1.69% C. In the average total N content of 0.036% and organic C, 0.62% C was low compared to the average values in tropical Asian paddy soils. It was found that calcium was the predominant exchangeable cation followed by Mg, K, and Na in that order. The average available P (Olsen-P) content was 20.7 ppm P_2O_5 and did not vary significantly throughout the study area, while the available Si content (261 ppm) was comparable with the level in tropical Asia. The available P and Zn contents were deficient for rice growth. The properties of rice soils varied in Kabul and Kunar watersheds and were influenced by the spot distribution of different types of parent materials such as limestone, dolomite, and lava (Masunagai et al. 2014). The available Fe, Mn, Cu, and Zn (DTPA-TEA extraction) were 35.9, 9.6, 4.75, and 0.33 ppm, on average.

Decades of instability have destroyed basic infrastructure in Afghanistan and decimated its development. About 80% of the total population is engaged in agriculture (CIA 2012). The information available on the properties of Afghan soils is almost negligible. However, recently the situation is improving with the coming up of local experts and the efforts of the Government.

The results of the recent studies revealed that rice soils in the Nangarhar province were alkaline due to the accumulation of carbonate and were generally low in fertility, particularly available P and Zn. Local farmers and government officials in Afghanistan relied solely on the application of chemical fertilizers as a rice soil management strategy to improve fertility. However, the importance of micronutrients has not received due attention. The Zn content of soils is quite low. Wastewater treatment sludge such as sewage sludge high in Zn content (Beckett et al. 1983) is a viable alternative for Zn fertilizer, although its availability and the feasibility of using it in relation to culture and custom must be examined. Interrelationships among soil and environmental properties are affected by environmental conditions such as air and soil temperature, wind speed, and solar radiation (Kelley et al. 2010).

6.9.5 Soil Fertility in Afghanistan

Proper nutrient application and management are important components of sustainable farming. Poor soil fertility and low soil nutrient contents are often a major constraints on increased crop yields in agriculture. Soils in Afghanistan have low levels of nitrogen and micronutrient deficiencies of iron, zinc, copper, and boron (USDA & UC Davis 2013). Proper nutrient management is therefore crucial to improve nutrient inputs, reduce nutrient losses, and recycle nutrients on-farm (Helvetas 2014).

Soils

The calcium carbonate ($CaCO_3$) contents vary from 5 to 50% in cropped soils of Afghanistan (FAO 1973). The calcareous soils are abundant in Afghanistan. Due to the high content of Ca, soil pH (1:1 soil:water) varies from 7.5 to 8.8 and is uniformly high. Carbonate content was lowest in the Ghazni province and highest in the Balkh province. Sand content of soils ranges from 26 to 85%. The textural class of soils is loam to sandy loam. The soils in Afghanistan mainly belong to three orders, namely the Entisols, Inceptisols, and Aridisols. It has been found that most of the agricultural soils that are formed in alluvium or loess do not have a calcic horizon. So, there is no strong restriction on the root system of the crops (ICARDA 2002). Electrical conductivity was above 4 ds/m in only 0.77% of the soil samples analyzed, and only 7.75% of the samples had values of >1 ds/m. About 91.48% of the analyzed soil samples had an electric conductivity of <1 ds/m. Mean EC values ranged 0.4–0.7 ds/m for the five provinces of Afghanistan. The USDA of soil classification names these soils as Torriorthents, Xerorthents, and Camborthids and, are classified by FAO systems as Yermosols, Xerosols, and Fluvisols (FAO, 1973). Analysis of soil samples revealed that about 88.2% and 94.1% of samples fell under low status in the case of N and P, respectively. The rest of the soil samples were in the medium category. None of the samples had high fertility status. Diethylene triaminepent acetic acid (DTPA) extractable Zn was below the critical level of 0.5 kg^{-1} soil in about 67% of the irrigated soils in Ghazni and about 50% of the soils in Helmand, whereas the status of Zn status was, by and large, adequate in Herat, Baghlan, and Balkh provinces. Except for Balkh province, the zinc status in soils from rainfed farms was generally low.

Royal Government of Afghanistan made a request to FAO of the UN for assistance in fertilizer use, under the United Nations Development Programme, and appointed Dr. P. M. Tamboli as Technical Officer (Soil Fertility). Tamboli (1971) studied the following aspects of soils in Afghanistan:

- Fertilizer requirements of the main crops on important soil types under irrigated and rainfed conditions;
- Assemble, analyze, and report the results of soil fertility experiments, to interpret these results and develop fertilizer use recommendations;
- Correlate experimental yield data with soil tests and soil survey data, as an aid to develop practical recommendations.

The central and northeastern mountains consist of Pre-Ordovician phyllite, Paleozoic schist, slate, gneiss, granite, granodiorite, and diorite, and Cretaceous, Jurassic and Cenozoic limestones, sandy marls, and shales. Terrestrial (Muree and Shiwalik) formations are also represented in geologic columns. Outcrops of volcanic materials and igneous extrusions are scattered through much of the mountainous area. The lowlands of the north, south, and east are occupied primarily by Quarternary deposits including alluvial fans, playas, and sand dunes, with local loess coverings. Tamboli (1971) reported that the cultivable soils in Afghanistan occur scattered throughout some 40% of the total area, mostly at elevations of 1 500–2 700 m amsl. 60% of the country is in high mountain land, and 40% is an arid wasteland.

Mountain soils of Sierozem (Calciorthid) chestnut (Ustalf, Ustoll) and Podzolio (Udalf), Brown Forest (Eutrochrept), Alpine Forest (Cryoohrept), and Meadow (Cryumbrept) zones have formed from a complex of residual, alluvial and eolian material. Sierozem (Caloiorthid Haplargid) and associated soils on rolling hill land of lower altitudes near the cities of Baghlan, Kunduz, Mazar, Maimana, Herat, and Panjshir are formed from loess derived from glacial and alluvial materials. Glacio-fluvial, alluvial, and eolian (loesSial) sediments form the initial materials of the soil of (Camborthids, Haplargids, Ustifluvents, and Xerofluventa) the major river valleys. Desert (Camborthids, Haplargids) soils have developed from both alluvium and sedimentary rocks (Tamboli 1971). Tamboli (1971) further reported that the soils are mostly structureless, alkaline in reaction, and the soil fertility level, in general, is very low. The soils are high in calcium, low in organic matter, low in available nitrogen and phosphorus, and medium to high in available potassium. Nutrient-depleting crops such as wheat, corn, rice, and barley have been grown for many years without replenishing the nutrient removed. Lack of organic matter has resulted in the unfavorable physical condition of the soil for high crop production.

6.9.6 Fertilizer Imports and Use at Initial Stages

The position with respect to imports of chemical fertilizer in Afghanistan is given in Table 6.1 (Tamboli 1971).

6.9.7 Review of Fertilizer Trials on Wheat

In 1951, Dr. Hauser initiated a soil fertility investigation program in Afghanistan. On the basis of his studies, he concluded that (a) nitrogen is a limiting factor in production in most Afghan soils; (b) soils, if low in nitrogen, application of phosphorus alone will not help; (c) application of phosphorus on soils rich in nitrogen is beneficial; (d) potassium is not a limiting factor in Afghanistan; (e) soils having more than 5 ppm of water-soluble phosphorus will not respond to the application of phosphorus; and (f) one or two heavy applications of local manure did not correct deficiency of phosphorus.

After 1951, for a period of about 9 years, no work has been reported on fertilizer responses. Frutchey (1962) reported that on Darul Aman Farm, Kabul, a good response to the application of high doses of phosphorus was observed. He also found that the use of nitrogen alone decreased the yield. There was wide variability in the data reported by him and the results were not conclusive. Shuman (1960) worked on wheat in Kabul Province, and found that in Paghman, Chilstoon, and Siab areas, application of 15 pounds N and 15 pounds P_2O_5 per jerib (per jerib = 1953.6 m^2) resulted in a 149% increase in yield over control. In Lagman, Kabul, Kapisa, and Logar trials, the wheat yield under control (No NPK application) was 1.167, 1.655, 1.296, and 1.378 t ha^{-1}, respectively, which increased by 46.0%, 28.9%, 56.5%, and 22.2% over control when N–P_2O_5–K_2O were applied at 150, 150, and 100 kg ha^{-1}.

Nelson (1962) found that the application of 3 seers (1 seer = 0.928 kg) of urea, 6 seers of superphosphate, and 2 seers of potassium chloride per jerib (15–30–10 kg ha^{-1}) gave a 76% increase in wheat yield. Carter (1967) worked in Helmand Valley Authority and obtained 150 pounds increase over control (22%), by addition of 18 pounds N and 9 pounds P_2O_5 per jerib (4 020 kg/ha). Samin and Nielsen (1965–1966)

Table 6.1 Afghanistan fertilizer imports—metric tons 1961–1970

Fertilizer	1961–62	1962–63	1963–64	1964–65	1965–66	1966–67	1967–68	1968–69	1969–70
Ammon. Nitrate	5000	7500	7500	–	2000	1700	–	–	–
Urea	–	–	–	–	–	4200	19,000	6000	10,000
SSP	–	7500	7500	–	–	2600	4000	9900	–
DAP	–	–	–	–	–	–	6650	183	–

The two principal sources of fertilizer imports for Afghanistan were the U.S.S.R. and the U.S.A

conducted about 140 individual trials in the Jalalabad, Baghlan, and Kabul areas. They found that urea and triple superphosphate applied at the rate of 85 kg each of nitrogen and P_2O_5 per hectare increased wheat yield by 32 and 40% at the Ministry of Agriculture Stations of Taialabad and Baghlan, respectively. In the Kabul area, 85 kg of nitrogen and 165 kg of P_2O_5 increased yield by 10%. They also reported evidence of P-fixation problem in the Baghlan area, where soils with only 4.0 ppm available phosphorus did not respond to Phosphorus application. Mortensen (1967) reported the highest yield of 5966 kg ha^{-1} by application of 90 N + 45 P_2O_5 ha^{-1}, at the Jaïalabad Research Station.

6.9.8 Environmental Issues

Afghanistan has a harsh, rugged, and inhospitable terrain. Forests have been cut and burned. Agricultural lands are being degraded. The forests have been degraded, and erosion has increased. Pasture and agricultural lands have been subjected to increased overgrazing, salinization, and erosion. The soil degradation in Afghanistan will increase if destruction of forests, pastures, and agricultural lands would not be stopped immediately. There has been a decline in wildlife populations over the years, while water and air pollution are increasing. The prevailing environmental trends will eventually downgrade the quality of life of the people of Afghanistan.

6.10 Bangladesh

Located within 20°25′ to 26°38′ N latitude and 88°01′ to 92°41′ E longitude, Bangladesh is having a total land area of about 144,000 km^2, including 8,300 km^2 water bodies. Crops occupy more than 60% of the area of the country with a net cropped area of 91,300 km^2 (Ahsan and Karim 1988). Rice is the most dominant crop as it occupies about 80% of the total area under crops. Moslehuddin et al. (1997) mentioned that the major crops of Bangladesh are boro rice, wheat, maize, pulses, oilseeds, groundnut, spices, and a wide variety of winter vegetables.

6.10.1 Soils

There are three main physiographic zones of Bangladesh, such as northern and eastern hills of Tertiary formations, Pleistocene terraces of the Madhupur and Barind tracts, and recent floodplains mainly comprise alluvial sediments of the Ganges, Brahmaputra, and Meghna river systems. These zones cover 12%, 8%, and 80% of the area of Bangladesh, respectively. The floodplains can be further divided into Piedmont alluvial plains, meander floodplains, basin areas, estuarine floodplains, tidal floodplains, and sandy beaches (Saheed 1984). The soils were formed from various kinds of parent materials and then spread all over three major physiographic zones of the country. Soils are classified into 21 general soil types based on their formation process and physical appearance according to the Bangladesh soil classification system (Hussain 1992). Bangladesh, though a small country, has varied types of soil resources where a variety of crops can be grown throughout the year.

For judicious crop planning, based on physiography, soil, depth and duration of seasonal flooding, and climate, the country has been divided into 30 agro-ecological regions (FAO-UNDP 1988). Agriculture contributes more than 50% of the gross domestic product and provides employment for 74% of the rural people (Ahsan and Karim 1988). It may be mentioned that there was no use of chemical fertilizers in Bangladesh until 1950. During those days, the population was less per unit of land, and farmers did not need to use any chemical fertilizers to grow crops as enough organic manures like cowdung and crop residues were available for application to the land. Since sufficient arable land was available, the land was fallowed to conserve fertility, as the population pressure on the soil was not high. Application of organic manures like dung and urine of livestock was enough to grow local varieties of different crops. Though the yield was low, it was sufficient for a low population. With an increase in population, it has become necessary to apply inorganic fertilizers to enhance crop production to feed the burgeoning population.

6.10.2 Fertility Status of Soil

Organic matter

Organic matter contains almost all the nutrients required for optimum plant growth and enhancing crop productivity. The soil organic matter improves soil structure, water, and nutrient holding capacity, and enhances the activity of soil microbes, thereby helping in the successful growth of various crops. The soils of Bangladesh have been found to contain low organic matter, in general. Bhuiyan (1987) selected 17 soil series, each from 17 general soil types of Bangladesh, and compared the organic matter content of different soil samples. The peat had the highest organic matter content of 35.3%, followed by Acid basin clays and Acid sulfate soils which contained about 5.20% and 3.46% of organic matter, respectively. Out of the rest 14 soil series in Bangladesh, only 4 soil series had more than 2% organic matter content, while in the other 10 soil series, the organic matter content was below 2%. It has been concluded that the overall organic matter content has been found to be low in those soils in Bangladesh, where agriculturally important crops can be grown. As per the findings of Islam et al. (1992), organic matter ranged from 0.6 to 1.7% in different regions of the country. Earlier, Islam (1990) had reported that at least 2% organic matter should be present in the soil for optimum crop production. Karim et al. (1995) found that the addition of organic residues and an annual application of fresh cowdung at the rate of 25 tha^{-2} under the rice–wheat cropping system increased the organic matter content to the level of 1.1–1.3% from the initial level of 0.7% in the Shallow red brown terrace soil of Bangladesh. Mian (1993) reported a 9–46% depletion of soil organic matter in different regions of Bangladesh between 1970 and 1990, mainly due to intensive crop production. There could be several reasons behind the low organic matter content of Bangladesh soils, the major reason being the lack of organic recycling through the addition of crop residues and the negative balance between crop uptake of nutrients from the soil and their compensation through the application from organic sources. Moreover, a large number of aquatic weeds grow which add organic matter to these soils (FAO-UNDP 1988).

Nitrogen

Due to low N supplying capacity, Bangladesh soils have long been rated as poor in soil fertility (Islam 1983). Thompson and Troeh (1978) reported that the pale green color of N-deficient plants is the most common deficiency symptom exhibited by growing plants. Portch and Islam (1984) worked on 63 soil samples from different regions of Bangladesh and found that 100% of them were deficient in N. Most of the soils of Bangladesh have been rated as low in nitrogen (Islam et al. 1992; Moslehuddin 1993; Ali et al. 1981).

Bangladesh soils being low in N and organic carbon, the crop response to N was the highest among all nutrients in Ali (1991). The increase in yield with the application of nitrogen has been observed in almost all the crops, but it was more in crops such as rice, jute, sugarcane, and tea. The reasons for more response in some crops may be due to their large N requirement, deficiency of N in almost all soils of the country, and low N use efficiency. The effects of N application on the yields of various crops is presented in Table 6.7 (Islam et al. 1992; BRRI 1990, 1992). It has also been reported that soils of low-lying areas of most floodplains have a good amount of N due to the reason that these soils contain a good amount of organic matter as also nitrate–N accumulation in these soils through water movement from adjoining higher areas, where N-fertilizers are applied to grow crops. The floodwater accumulates in temporary water storages and leaves behind a lot of nutrients and organic matter after drying of water. This excess N is a cause of an ecological hazard for the soil and environment also (FAO-UNDP 1988).

Phosphorus

Bangladesh soils are, by and large, poor in available P. The soils of Bangladesh varied drastically in available P and, as reported by Bhuiyan (1988), it ranged from 2 to 14 ppm with a mean value of 12 ppm. The results obtained by Islam et al. (1992) found an almost similar range of available P. They analyzed 29 soil series from all over the country and found that the available P varied from 2 to 18 ppm. Most of these samples were below the critical level of 12 ppm. The findings of Portch and Islam (1984), who reported that 41% of the soils of Bangladesh contained P below the critical level and 35% of the soils contained P above the critical level, were not much different from Islam et al. (1992). Though 35% of the soil samples were above a critical level, these were below the optimum level, required for realizing the full potential of crops. The soils of the terrace area had the P content just above the critical level (Egashira and Yasmin (1990). They observed that the total P in these soils was enough to sustain the normal growth of rice crops. Looking into the status of available P in Bangladesh soils, the application of P from outside sources is necessary to compensate the deficiency of P in the soil for optimum and sustained crop production in the country. Ali (1991) has reported an increase of yield up to 25%, 16–34%, and 35–38% in rice, jute, and sugarcane, respectively with the application of recommended dose of P.

Potassium

It was earlier thought that Bangladesh soils do not need the application of potassium for higher crop yields. However, due to the intensification of agriculture, response to added potash fertilizer has been observed in many soils of the country (Islam et al. 1985). Response to potassium application in the soils of Bangladesh is less compared to the response to nitrogen and phosphorus. The terraced soils of Chittagong Hill Territory had a low content of available K than soils of floodplain areas (Ahsan and Karim 1988). About 50% of the area of Bangladesh was low in available K, based on the critical limit (0.17 cmol (+) kg^{-1}. Islam et al. (1992) analyzed 29 soil series from different areas of Bangladesh and found that the exchangeable K content varied from 0.07 to 4.7cmol (+) kg^{-1}. The K status of most terrace and Piedmont soils was found to be low to medium (Islam 1992).

Calcium, magnesium, and sulfur

Deficiencies of secondary nutrients like Ca, Mg, and S can slow down plant growth as much as deficiencies of major nutrients, such as N, P, and K (Potash and Phosphate Institute 1983). As early as in the 1980s, S deficiency was reported in Bangladesh soils when it was observed that S deficiency is getting widespread and acute (Islam 1983). Portch and Islam (1984) found that 68% of the soils were below the critical level for S. Bhuiyan (1991), on the other hand, reported that about 4 million ha of land is potentially deficient in S in Bangladesh. A detailed study of Bangladesh soils, Egashira and Yasmin (1991), showed that total S was not enough to sustain normal plant growth. In Bangladesh, studies have revealed that hill and terrace soils are severely depleted of Ca and Mg (FAO-UNDP 1971). The tuber yield of potato was reported to be increased significantly by the application of Mg. Islam et al. (1992) and BRRI (1990, 1992) reported that with the application of recommended dose of macro- and micronutrients, the yield of rice (boro), rice (Aman), wheat, tomato, chillies, turmeric, millet, maize, chickpea, and cauliflower increased by 24.3%, 5.2%, 16.1%, 129.8%, 72.5%, 192.1%, 107.1%, 44.0%, 78.9%, and 139.8% over control, respectively.

Micronutrients

Elements essential for plant growth but required in a small amount such as Zn, Fe, Mn, Cu, B, MO, and Cl are very important for the optimum growth of crop plants. The absence of any one of them in the soil can be a constraint in crop development, even though all the other necessary nutrients are available in sufficient quantity (Potash and Phosphate Institute 1983). The deficiency of Zn and B is most widely spread in Bangladesh while Cu and Mn are deficient in limited areas in the country (Islam et al. 1992).

Zinc deficiency was reported by many workers of Bangladesh (BRRI 1980; Mukhopadhayay et al. 1986). An increase of 27% in rice was found with the application of Zn. Though the Zn requirement of crops is very less, its deficiency can cause significant reduction in the yield of almost all the crops (Ali 1991). There could be many causes behind the deficiency of Zn in the soils of Bangladesh. The major factors are the decline of its availability to the crop plants because of high soil pH, high concentrations of calcareous materials, continuance of soil submergence with water during wet rice cultivation, and fixation of Zn with phosphate ion, due to interactional effects.

Actually, the soils of Bangladesh have exhausted native soil nutrients due to intensive crop cultivation using high yield varieties of the crop with imbalanced and low fertilizer doses, to feed the ever-increasing human population. Ali et al. (1997) supported this and mentioned that fertility of most of the soil of Bangladesh has declined over the years. This has resulted in showing a plateau in crop yields, and even a decline in crop productivity has been found in some cases (Cassman et al. 1999). On the other hand, the organic matter content of most of the Bangladesh soils is very low where the majority fall below the critical level of 1.5%. The main reason for poor soil fertility of the soils in Bangladesh is that a major portion of the organic residues is used as fuel by the rural people. According to Quamruzzaman (2005), who nicely analyzed the scenario of depletion of micronutrients from the soils of Bangladesh, the major reasons for the depletion of soil fertility are (i) for feeding the burgeoning population of the country, the farmers increased the cropping intensity on their small and marginal holdings, which caused a deterioration soil fertility; (ii) use of more fertilizers on lands with poorer soils under uncertain irrigation supply; (iii) Imbalanced use of N, P, and K resulting in low availability nutrient at the cost of the other; (iv) sudden emergence of deficiencies of secondary and micronutrients; (v) though slow, there has been a gradual decrease in soil organic matter; and (vi) an increase in soil degradation, which has restricted the cultivation of some crops due to unfavorable soil base, low availability of nutrients, and little response of crops to nutrients on such soils.

6.10.3 Chittagong Hill Tracts

Out of its total geographical area of 147,570 km^2, Bangladesh has about 11.75% (17,342 km^2) of its area under hills. This area is known as Chittagong Hill Tract (CHT), and comes under Hindu Kush Himalayas. This area was almost fully covered by forest before 1960, except some land for shifting cultivation and hutments. Due to increased demographic pressure and to cope with the rising demands for feed, fodder, and fuel, this area has been deforested largely. The main reasons for forest disturbance and deforestation in Bangladesh are encroachment (ADB 2002; Ali 2003; Rasul et al. 2004; PCB 2005), illegal cutting of trees and fuel wood collection (Rasheed 1995; ADB 2002), grazing, fire, uncontrolled and wasteful commercial logging (Khan 2001), internal armed conflict (FAO 2005), and wasteful development activities (ADB 1998; Ahamad 1998). The soils of CHT are mainly yellowish brown to reddish brown loams which contain broken shale or sandstones. The region soils have mottled sand at a variable depth.

The Chittagong Hill Tracts (CHT) is home to 11 indigenous groups, who have been practicing shifting cultivation (known as *jhum* locally), for generations. These shifting cultivators are generally poor as the food production from land does not meet their requirements. Food security has been further impacted by demographic, environmental, and policy pressures (Khisa et al. 2015). We are of the view that shifting cultivation has improved soil conditions and has remained a custodian of natural resources, including soils, for a long time, that is, from the Neolithic period to the middle of the twentieth century. The increase in population pressure, however, has made this practice unsustainable and the whole system started crumbling down. Unless suitable interventions are made, the shifting cultivation cannot ensure livelihood security in future. Moreover, the magnitude and pace of soil deterioration and degradation would increase if suitable policies are not enacted by authorities taking stakeholders into confidence. Agriculture is the main source of livelihood for most of the people in CHT. Tista floodplain soil of Bangladesh is comparatively fertile and well managed due to over-conscious and proper management (Ahmed et al. 2017). Poverty, caused due to traditional agriculture practices like shifting cultivation and environment degradation, can be reduced by introducing environmentally compatible and economically viable agricultural policies and programs in CHT (Thapa and Rasul 2005).

The soils of hilly areas are yellowish brown to reddish brown loams, having broken shale and mottled sand. At present, the most common phenomenon in the hill tract and undulated areas of Bangladesh is land degradation. Deforestation, soil erosion, and soil fertility depletion are common in CHT, which result in soil degradation. Of late, agroforestry is getting its due share and becoming popular with some tribes and other communities. Horticulture has also good potential in hilly areas of CHT. It is also considered to be environmentally sound and economically suitable. Some communities are also practising their agriculture by integrating trees and livestock with annual field crops. Horticulture is helpful in improving the economic benefits as well as ensuring enduring food scarcity and increasing the income of the cultivators (Khan and Khisha 1970, Roy 1995). Thapa and Rasul (2005) have categorized the agricultural systems into three major groups, such as extensive, semi-extensive, and intensive, in the mountain regions of the Chittagong Hill Tracts of Bangladesh.

6.10.4 Fertility Status of Forested Soils

A detailed study was conducted by Hossain et al. (2014) in forested soils of three hilly districts of Chittagong Hill Tracts, Bangladesh, to analyze the content of different nutrient elements in soils to understand their fertility status. Variation in land use, slope percent of the area, and altitude affected the contents of organic matter, total nitrogen, available phosphorus, and available potassium in the soil. Soil pH varied within a narrow range from 5.1 to 5.7. It revealed that almost all forest soils in CHT has an acidic reaction. This was mainly due to the continuous leaching of bases due to heavy rainfall in the area. But a wide variation in organic matter content of these forested hilly soils was found, varying between 1.0 and 2.44%. The litter fall was more in the area where OM was high. The soil organic carbon, cation exchange capacity, exchangeable magnesium, and total nitrogen differed among various sites indicating that slope, elevation, texture, moisture content, and temperature regimes greatly influence the soil properties of a place.

Physical Properties

Among the physical properties, mechanical composition and water-holding capacity have been studied. Sand, silt, and clay varied considerably and ranged from 37 to 67%, 7 to 36%, and 20 to 33%, respectively. Soil texture is less affected by management practices (Tilahun 2007). The texture of the soils was strongly impacted by the nature of parent materials. Huq and Shoaib (2013) categorized these soils as yellow brown to strong brown, permeable, friable, loamy, very strongly acid, and low in moisture holding capacity. The surface runoff is low to severe during the monsoon, depending on slope characteristics. Maximum water-holding capacity (MWHC) values varied between 41.17 and 51% (w/w). The mechanical composition of soils affected the water-holding capacity of the soils. Banglapedia (2012) reported that the Eastern Hills of Bangladesh are underlain by sandstone, siltstone, and shale of Tertiary and Quaternary ages. It was found that the soils developed on these parent materials are generally brown in color, usually loamy in texture, and exhibited very strongly acidic reactions.

Chemical Properties

Soil pH varied between 4.6 and 6.5 among these sites. According to the USDA classification (Soil

Survey Division Staff 1993), the mean pH of the presently studied soils can be placed broadly in two classes, such as strongly acidic (pH varying from 5.1 to 5.5) and moderately acidic (pH varying from 5.6 to 6.0). About 77% of soils showed a strongly acidic reaction, whereas only 23% soil had moderate soil pH. Studying soils of Chittagong hills at various sites, Hossain et al. (2014) found that soil pH, OM, and CEC [Cmol (+) kg^{-1}] ranged from 4.6 to 6.5, 0.15 to 3.27%, and 2.12 to 12.16, respectively. Organic matter content varied greatly between 0.15 and 3.32%. The organic matter was more on depression sites where runoff water got collected and remained there till its drying. The runoff might have brought forest litter along with. According to BARC (1989), CEC of agricultural soils in Bangladesh are classified as very high (>30 cmolc kg^{-1}), high (15–30 cmolc kg^{-1}), medium (7.5–15 cmolc kg^{-1}), low (3–7.5 cmolc kg^{-1}), and very low (<3 cmolc kg^{-1}). Out of 17 sites studied, 5 sites fall into low (<7.32 cmolc kg^{-1}), 6 sites fall into medium (between 7.32 cmolc kg^{-1} and 8.79 cmolc kg^{-1}), and the remaining 5 sites fall into the high category (>8.79 cmolc kg^{-1}). There was a close association between organic matter and CEC.

Nutritional properties

Total nitrogen contents in the soils of Bangladesh varied significantly between 0.03 and 0.24%. Total nitrogen content of soil was found significantly related to the content of organic matter. As nitrogen content of the soils is related to the organic matter content, generally, a site low in organic matter was also found to be low in total nitrogen. Interestingly, the C/N ratio varied within a very narrow range of 8–12. For getting good crop productivity, usually, deficiency of N is corrected in agricultural systems through the application of N fertilizers from outside sources. It is a known fact that the forest soils in most underdeveloped and developing countries are not fertilized, and nutrient requirements of forest vegetation are mainly met by nutrient recycling (Vitousek and Sanford 1986). Available P varied between 0.59 and 8.78 mg kg^{-1} at different sites of study. Hence, a great variation, among the different slopes, sites, and locations, was found in available P content. Available potassium varied between 21 and 375 mg kg^{-1}. Available calcium content in the studied soils ranged from 58 to 1674 mg kg^{-1} and available Mg varied between 31 and 756 mg kg^{-1}. The variation in nutrient status of the soils was due to differences in the parent material and losses due to erosion. Mesfin (1998) and Eyelachew (2001) have also reported that exchangeable Ca and Mg cations constitute the major proportion of total exchangeable cations, and contribute to the base saturation of most soils. Tilahun (2007) analyzed several hundred forests and agricultural soil samples and observed the highest average mean values of exchangeable Ca and Mg, as 10.75 cmol(+) kg^{-1} and 5.02 cmolc kg^{-1}, respectively. The CEC was 28.17 cmol(+) kg^{-1}). Soils of the forest plantations of Chittagong Hill Tracts appear to be relatively low.

Hassan et al. (2017a, b) conducted a study in three hill districts of Chittagong Hill Tracts of Bangladesh, to determine the chemical and nutritional properties of high, medium, and low hill soils. Chemical and nutritional properties varied among different hills as revealed through the results obtained. Mean values of soil pH, organic matter, total N, total P, Ca, Mg, and K of three different hill soils ranged from 5.00 to 5.35, 1.82 to 2.19%, 0.09 to 0.17%, 3.44 to 5.24 mg kg^{-1}, 2.19 to 2.82 mg 100 g^{-1}, 1.39 to 1.62 mg 100 g^{-1}, and 0.29 to 0.49 mg 100 g^{-1}. From the result, it was assessed that the soils are generally poor in organic matter and nutrients as well as poor in fertility status (Hassan et al. 2017a, b).

Generally, the farmers do not apply fertilizer to the hill soils in most developing countries, and nutrient demands are mainly met by nutrient recycling (Vitousek and Sanford 1986). In CHT, Bangladesh, total nitrogen contents ranged from 0.08% to 0.70% in high hill soils, 0.03–0.15% in medium hill soils, and 0.05–0.16% in low hill soils. Total P in the soils varied between 1.24 and 6.68 mg kg^{-1} among high hill soils, 1.50 and 10.38 mg kg^{-1} among medium hill soils, and 1.95 and 10.31 mg kg^{-1} among low hill soils. Total calcium content ranged from 1.0 to 5.50 mg 100 g^{-1} among high hill soils, 0.43 to

3.50 mg 100 g^{-1} among medium hill soils, and 0.10 to 6.01 mg 100 g^{-1} among low hill soils. Total magnesium content varied between 0.80 and 2.50 mg 100 g^{-1} among high hill soils, 50 and 2.50 mg 100 g^{-1} among medium hill soils, and 0.52 and 3.50 mg 100 g^{-1} among low hill soils. Total potassium ranged from 0.11 to 0.70 mg 100 g^{-1} among high hill soils, 0.15–0.47 mg 100 g^{-1} among medium hill soils, and 0.11–2.10 mg 100 g^{-1} among low hill soils.

Total S content varied between 1.00 and 12.47 mg kg^{-1} in high hill soils, 1.00 and 15.15 mg kg^{-1} in medium hill soils, and 1.00 and 20.36 mg kg^{-1} in low hill soils. Total Boron content ranged from 0.17 to 0.38 mg kg^{-1} in high hill soils, 0.12–0.36 mg kg^{-1} in medium hill soils, and 0.05–0.41 mg kg^{-1} in low hill soils. Total Copper content ranged from 0.04 to 1.47 mg kg^{-1} in high hill soils, 0.04–0.75 mg kg^{-1} in medium hill soils, and 0.08–1.36 mg kg^{-1} in low hill soils. Total Iron content varied between 10.60 and 73.95 mg kg^{-1} in high hill soils, 5.00 and 117.40 mg kg^{-1} in medium hill soils, and 8.00 and 336.80 mg kg^{-1} in low hill soils. Total Manganese content ranged from 5.00 to 46.20 mg kg^{-1} in high hill soils, 6.60–37.80 mg kg^{-1} in medium hill soils, and 4.20–26.80 mg kg^{-1} in low hill soils, and total zinc content varied between 0.08 and 1.72 mg kg^{-1} in high hill soils, 0.08 and 1.92 mg kg^{-1} in medium hill soils, and 0.38 and 1.40 mg kg^{-1} in low hill soils.

6.10.5 Development of Nutrient Deficiencies

The development of nutrient deficiencies in Bangladesh on a time scale has been beautifully documented by M. Fokhrul Islam, Adviser, The Netherland Development Organization (Rijpma and Islam 2003). In 1951, N deficiency was reported in Bangladesh and in 2000, the deficiency of N, P, K, S, Zn, B, Mg, and Mo has been reported. The deficiency of different nutrients appeared in Bangladesh soils with the passage of time from 1951 to 2018. This was due to the intensification of agriculture to produce more for feeding the increasing population. Starting with N deficiency reported in 1951, the deficiency of N and P was there in 1957, adding K to the deficiency list in 1960, S in 1980, Zn in 1982, B in 1995, and Mg and Mo in 2000.

6.11 Bhutan

Lying on the southern slopes of the Eastern Himalayas, Bhutan has a total geographical area of about 47,000 km^2. Though small in size, it is a beautiful country endowed with rich natural resources. In the south, its borders touch the floodplains of the Indian state of Assam and the Eastern state of West Bengal. From here, the tropical foothills rise with steep slopes until it reaches an area, which is less steep, having a basin-like terrain in most parts of the country. The country is located at altitudes between 2,500 and 4,000 m amsl. These inner valleys are the cultural representatives of Bhutan. The altitude increases again toward north, where there is a further increase in elevation, forming peaks of more than 7,500 m amsl and touching the border with neighboring China (Tibet) (Caspari 2005). The country is east–west aligned within the main chain of the High Himalayas.

6.11.1 Physiographic Zonation

The high steepness of the terrain from south to north gives a look of the big staircase to Bhutan consisting of consecutive E-W oriented altitudinal belts. This is the main basis and criteria for delineating different zones in Bhutan (Karan 1967; Eguchi 1987, 1991; Takada 1991). Bhutan is characterized by a predominantly N–S oriented landscape as compared to Central Himalayas which are NW–SE oriented. The most prominent north–south mountains in Bhutan include the Black Mountains, the Dagala, Pelela, Yotongla, Thrumsingla, and Korila ranges, which are separated from each other by the main rivers draining the country. Navara (1997) is among those who first studied this aspect. Norbu et al. (2003), integrating all of the main components of the natural environment, have suggested a different concept. Physiographically, they have classified Bhutan into three major zones, such as High Himalayas (with sub-zones as trans-Himalayan plateau, high peaks, and dissected plateau), North–South valleys and ranges (with sub-zones as Northern valleys and ranges, Inner valleys and passes, Eastern valleys and ranges and, Southern mountain and ranges), and Southern foothills (with sub-zones of front hills, S–E Bhutan, and Piedmont). The altitude of high mountains varies from 4000 to 7000 m amsl and soils are stony and not fully developed. North–South valleys vary in altitude from 500 to 4500 m amsl and have temperate forest soils, alpine meadow soils, and subtropical soils. The third zone, that is, the Southern part of Bhutan has an altitude varying from 100 to 2000 m amsl with deep, stony, and unstable; highly leached, and weathered soils (Norbu et al. 2003).

6.11.2 Redistribution of Local Sediments and Their Influence on Soil Formation

When various parent materials of soil formation are considered, loess like aeolian sediments appear to have played a dominating role on the southern slopes of the Himalayas, including Bhutan. The origin of loess-like sediments is thought to be, generally, due to cooler and drier climates, under limited vegetation cover conditions. The deposits of loess-like material are frequently dotted by paleosols. The paleosols are considered to have developed during stable phases under humid climates and luxuriant vegetation cover (Gerasimov 1973). Local signals of past environmental changes are represented by Loess-paleosol sequences (Kukla 1987; Kemp 1999). Its correlations with other proxies have been well demonstrated (Kemp 2001). It has been observed that most pedons found in Bhutan are deep and well-developed at many sites, polygenetic reflecting the repeated interruptions during pedogenesis (Baillie et al. 2004). Many scholars have mentioned this for different areas of the Himalayas from west to east (Gardner and Rendell 1994; Guggenberger et al. 1998; Bäumler 2001; Saijo and Tanaka 2002; Caspari et al. 2004).

6.11.3 Soil Survey and Classification

Karan (1967) was a pioneer in reporting about the general color, texture, and depth of soils in different parts of the Bhutan. He suggested the zonal distribution of soils from north to south of the country, based on the climatic and vegetation zones. Okazaki (1987) further detailed out this concept and suggested five major soil groups that are vertically distributed according to the altitude, that is, (i) 0–1200 m, (ii) 1200–2425 m, (iii) 2425–3100 m, (iv) 3175–4090 m, and (v) > 4090 m for dry areas, and (i) 0-1000 m, (ii) 1000–2025 m, (iii) 2025–2950 m, (iv) 2950–3950, and (v) > 3950 m for wet areas. The soils of the first group (both dry and wet) have thin A horizons; yellow-colored B horizons; are weakly acidic; and of low humus content, low CEC, and low base saturation. Yellow-colored B horizons compared to brown forest soils, slightly acidic, medium humus content, high CEC is observed in the second group. The soils of the third group are thin, porous A horizons; brown-colored B horizons; strongly acidic; and of high humus content, high P retention, and with low base saturation. Well-developed Podzol characteristics: acidic soil; low CEC; high humus content and P retention are found in the soils of the fourth group. Finally, the soils of the fifth group show shallow, weakly structured A horizons; sandy, brown-colored B horizons; medium humus content; low P retention and CEC (Okazaki 1987).

In order to fulfill the need for systematic information about the characteristics and spread of the soils of Bhutan, the Royal Government of Bhutan started the Soil Survey Project (BSSP) of Bhutan in 1996 with technical and financial help from Denmark, the Netherlands, and the European Union. Moderate to extreme weathering and leaching has been found in soil development in the temperate and inner valleys between 2,200 and 3,500 m. According to the WRB classification, the soils of this zone are generally Cambisols. However, orange-colored non-volcanic Andosols are also commonly found (Bäumler et al. 2005). Astonishingly, the Bhutanese soils are deep and well-developed in many places, which would not have been necessarily expected in the face of the steep and rugged terrain in combination with monsoonal rains and frequent earthquakes, as mentioned by Baillie and Norbu (2004). Podzol soils are mostly found in the subalpine zone from 3,500 m to the tree line at about 4,000 m. It has been observed that the leaching, acidification, and podzolization increase with altitude. Above that, one may find alpine turf soils and unweathered glacial deposits (Baillie and Norbu 2004), which could be considered as Regosols. The occurrence of permafrost has been observed from about 5,000 m amsl or even at higher altitudes (Takada 1991). The salient feature of Bhutanese soils is their short-range variability and regolith heterogeneity.

6.11.4 Soils as Indicators for Landscape History

In diversity of natural archives in the highlands of Central Asia, there is a paucity of terrestrial-based records, as found by the findings of the Himalayan Interdisciplinary Paleoclimate Project (HIPP) (Wake and Mayewski 1995). Following the "Glacier and glacial lake inventory within the Glacier Lake Outburst Floods Monitoring Programme (GLOF)", there is good knowledge of the present state of glaciers and glacial lakes. This program was undertaken as a joint exercise by the International Centre for Integrated Mountain Development (ICIMOD) and the United Nations Environment Programme (UNEP) in 2002. There are 677 glaciers and 2674 glacial lakes altogether, making up approximately 127 km^3 of ice reserves and covering an area of 1,317 km^2 (ICIMOD 2002). The glaciers in NW Bhutan extended as far down the valley as 3,550 m amsl, as indicated by findings of Meyer et al. (2001).

6.11.5 Status of Soils

As per FAO/UNESCO (1977) reports, about 27% of soils of Bhutan fall under either Cambisols (mid-altitude) or Fluvisols (southern belts), 45% of soils are acrisols, ferralsols, and podzols, and 21% of soils lithosols on steep slopes. Non-volcanic andosols have been reported to occur in a few pockets across the Bhutan, and soil types vary along different altitudes. Despite steep gradients, there are areas of deep and well-developed soils suitable for cultivation. Dorji and Dorji (2015) reported overall fertility status of Bhutan soils as follows:

- Soils are generally acidic to neutral in reaction.
- Exchangeable Al saturation is low to very low.
- No aluminum toxicity.
- Organic matter content of the soils varies from 0.8 to 1.5%.
- C:N ratio varies between 11.0 and 14.1.
- P and K content is low in most soils.
- Base saturation percent is low or very low in most soils.
- Most of the soils are low in CEC.

Despite steep gradients, Bhutan has deep and well-drained moderately or deeply weathered soils. Poor soils of Bhutan are not limiting the cultivation of crops as much as it is constrained by steep slopes, erosion risks, and climate (Baillie and Norbu 2004). Low pH, poor nitrogen and phosphate status, and imbalanced base nutrition are major concerns in the soils of the country. A nationwide soil fertility (SFM) survey was conducted in 1999–2000 to evaluate the basic assumption of "low productivity of land and unsustainable soil fertility management practices". About 376 soil samples were collected and analyzed. The main features of the soils as indicated by the analytical results are (Dorji 2008) as follows:

- Already having low pH, the application of acid-forming fertilizers has exacerbated the problems by enhancing soil acidity in the soils. Reduction in pH causes unavailability of many essential native and applied nutrients, thereby reducing crop productivity.

- In general, the nitrogen levels are very low to low, though carbon content is adequate.
- Available P and K are low in Bhutan soils. Decreasing use of FYM and inadequate use of P fertilizer application will result in declining crop yields.
- Base saturation percent and total exchangeable bases are low or very low in most soils of Bhutan.

The dryland soils have been found to contain higher soil nutrient levels than wetland soils (Dorji 2008). Geology is a significant factor affecting most variables. The relationship between management factors and soil nutrient status is weak but positive. Litter from six sources such as blue pine needles, blue pine needles + fern, fern, chir pine needles, oak, and oak + pine were tested for N, P, and K. The highest content of N (0.99%) and K (0.70%) was found in oak + pine litter, while the lowest N content (0.51%) was observed in blue pine needles and fern and that of K (0.22%) in chir pine needles. The highest P content of 0.08% was recorded in chir pine needles (Dorji 2008).

6.12 Qinghai–Tibet Plateau (China)

According to Chinese Soil Taxonomy, the soil of the permafrost region in the Qinghai–Tibet Plateau (QTP) can be divided into 6 orders. These six soil orders are Histosols, Aridosols, Gleyosols, Isohumosols, Cambosols, and Primosols. The soils of the zone come under 11 Suborders, 19 Groups, and 24 Subgroups (Table 6.9). The dominant soil types in the permafrost region are Cambosols, followed by Aridosols (Fang et al. 2015). In this zone, the soil type gradually changes from Cambosols to Aridosols, when moving from east to west of the plateau. Though the permafrost in the east of the plateau has shown some effect on the vertical soil zonality, the effect is little pronounced in the west. Huang (2000) had explained that the soil is a fundamental component of both the terrestrial ecosystem and the earth's surface system. The soil plays an important role in material circulation and energy transfer between the earth and the atmosphere (Tarnocai et al. 2009). It has been observed that the eastern region is dominated by Cambosols, with no obvious latitudinal zonality. From the south to the northwest of the western region, the dominance of Aridosols and Cambosols is found which changes gradually to Aridosols, presenting a latitudinal zonality. Precipitation and vegetation are the main factors that affect the zonal distribution of soils in QTP, as shown by the results. Fang et al. (2015) identified six orders to which the soils of QTP belonged. These are Histisols, Aridisols, Gleyosols, Isohumosols, Cambosols, and Primosols. The soils come under the suborders of Permagelic Histodols, Cryic Aridosols, Permagelic Gleyosols, Stagnic Gleyosols, Ustic Isohumosols, Aquic Cambosols, Ustic Cambosols, Gelic Cambosols, Alluvic Primosols, Sandic Primosols, and Orthic Primosols. The soil groups identified in the QTP region are Hemi-Permagelic Histodols, Calci-Cryic Aridosols, Hapli-Cryic Aridosols, Gypsi-Cryic Aridosols, Hapli-Permagelic Gleyosols, Hapli-Stagnic Gleyosols, Cryi-Ustic Isohumosols, Ochri-Aquic Cambosols, Molli-Ustic Cambosols, Hapli-Ustic Cambosols, Matti-Gelic Cambosols, Hapli-Gelic Cambosols, Permi-Gelic Cambosols, Aridi-Alluvic Primosols, Geli-Alluvic Primosols, Geli-Sandic Primosols, Aridi-Orthic Primosols, Usti-Orthic Primosols, and Geli-Orthic Primosols (Fang et al. 2015). They further sub-divided the soils of QTP into 25 subgroups.

Due to its unique location and climate, the Qinghai–Tibet Plateau (QTP) has a significant impact on the global climate system. As a result of its impact on global climate, the QTP is also one of the hot spots of research in the field of geographical science, earth science, atmospheric science, and cryospheric science (Wei et al. 2012). The basic data of soil taxonomy, distribution, and physico-chemical characteristics of this area are generally needed for ecological environment research (Yang et al. 2005), carbon cycle research (Wu et al. 2012; Baumann et al. 2009; Sommer 2006; Wang et al. 2002a, b), land surface modeling (Wei et al. 2014, 2013, 2012; Duan et al. 2012), as well as soil mapping, land resource development, and utilization. Permafrost degradation will result in the degradation of the ecological conditions (Wang et al. 2008; Yang et al. 2005) which eventually leads to soil degradation (Wang and Wu 2011; Yang et al. 2004). The permafrost region in China covers an area of about 1.5 million km^2 (Zhou et al. 2000a, b). Cheng and Liu (2008) reported that under the influence of climate change and global warming, the permafrost in the QTP is showing regional degradation.

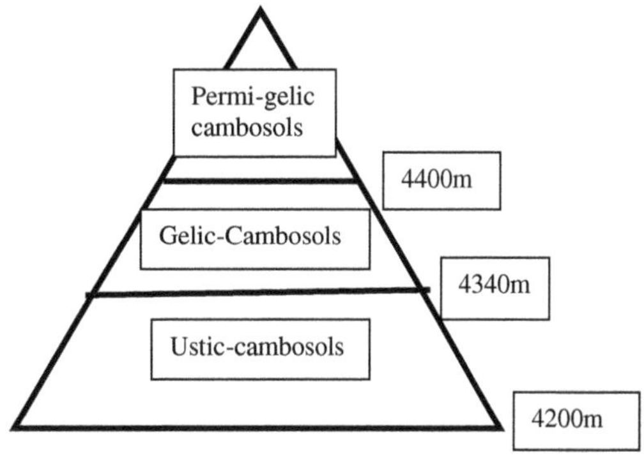

Fig. 6.6 Vertical distribution of Cambosols in Erla Mountain, Wenquan

6.12.1 Horizontal Distribution of Soil

Cambisols is the main order in the eastern region in the permafrost region on the Qinghai–Tibet, whereas Aridosols is the most dominant order in the western region, as is shown in the latest version of the soil map of China (Zhao and Zhang 2011). Highest precipitation has been recorded in the eastern part of the permafrost region (Zhao et al. 2004a, b). This is because this region has a profuse vegetation cover regime throughout the year. Alpine meadow is invariably the main vegetation type of the region, and the vegetation cover is, more or less, uniformly spread throughout this region and is in healthy condition. It has been concluded that the precipitation and vegetation in the permafrost region of the QTP are the main factors that influence the horizontal distribution of soil in this zone.

6.12.2 Vertical Distribution of Soil

The mountainous areas in QTP have a unique peculiarity; with an increase in altitude, the precipitation shows an increase and the temperature a decrease. It has been observed that vegetation is dependent on the changes in temperature and precipitation in the HKH region. Temperature and precipitation in HKH change with altitude and so does the vegetation. As the temperature and precipitation change, the vegetation types change from low to higher altitudes. Therefore, it is generally regarded that the soil types also alternate with altitude and show a vertical zonality (Huang 2000). Soil type distribution shows a certain vertical zonality in the eastern region. The Cambosols in the Erla Mountain in the Wenquan region are distributed at the altitude of 4200–4500 m, while the Isohumosols are distributed at the altitude of 3950–4400 m, and Gleyosol is distributed at the altitude of 4330–4600 m. On the level of the Suborders of Cambosols, Ustic Cambosols are mostly distributed below the altitude of 4340 m, while Gelic Cambosols are distributed above the altitude of 4340 m. Above 4400 m, there are the Permi-Gelic Cambosols (Fig. 6.6). Therefore, as the altitude rises, the distribution of the Suborders of Cambosols shows a certain vertical zonality (Singh et al. 2006).

6.12.3 Permafrost Distribution

Zou et al. (2017) provided more detailed information on the permafrost distribution and basic data for use in future research on the Tibetan Plateau permafrost. Due to its unique and extremely high altitude (>4000 m mean altitude) and low mean annual air temperatures (generally <2 °C), the Tibetan Plateau (TP) possesses the largest areas of permafrost in the mid- and low-latitude regions of the world (Zhou et al. 2000a, b; Zhao et al. 2004a, b, 2010; Yang et al. 2010). Permafrost, a major component of the cryosphere, is strongly affected and influenced by climate changes (Wu et al. 2000; Gruber 2012).

6.12.4 Soil Elemental Content

It has been established that the concentration of arsenic in cultivated soil of the Tibetan plateau was about double the average content of soils in China. However, it was far lower than the threshold value set by the Chinese environmental quality standards for soils. The molybdenum content was only one-third of the average concentration of soils in China (Zhang et al. 2010). They reported that the content of arsenic, cobalt, copper, molybdenum, selenium, manganese, and zinc in surface and sub-surface soils was 23.2, 14.6, 35.0, 0.71, 0.15, 803, and 98 mg kg^{-1} and 23.8, 15.4, 34.6, 3.72, 0.70, 0.17, 828, and 97 mg kg^{-1} of soil, respectively. The iron content was found to be 3.52 and 3.72%, respectively, and that of mercury was 59.6 µg kg^{-1} in both, surface and sub-surface soils.

6.12.5 Peatlands

When the soils remain at or below the freezing point of water (0 °C, 32 °F) for two or more years, then they are termed as permafrost. Total peat area in HKH region is 17,106 km^2 and degrading peat area is 8,236 km^2 (Fig. 6.7). The total area of high-altitude peatlands in TP was estimated at 14.4 × 10^3 km^2. It is about 43% of the total peatland area in China. Peatlands within the TP region are mainly scattered in the south, southeast, and eastern part of the plateau. The peatlands on the TP are one of the most important stores of carbon in the mountain regions storing 1,500–4,000 tones per ha or up to 8 to 20 times more than mountain forests and 50–100 times more than mountain grasslands. Total carbon stock is 1,594 Mt and total emission from degrading peatlands is 30.77 Mt CO_2 per year.

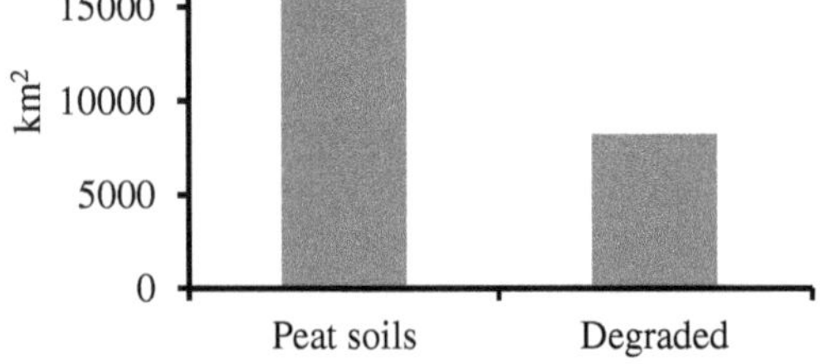

Fig. 6.7 Total peat and degraded lands in HKH region

The area under peats and degraded peat soils in HKH countries is presented in Fig. 6.8.

6.12.6 Permafrost Thawing

Simulations models evolved to understand and predict the fate of permafrost thawing reveal that the Tibetan permafrost will continue to retreat and thaw, and that the area of permafrost may decrease by 8.8–19% by 2049 and by 13.4–58% by 2099. There is an apprehension that warming-induced permafrost thawing would result in rapid carbon loss through the efflux of CH_4 and CO_2, and leaching of DOC (Chen et al. 2011). Wang et al. (2008) estimated the loss of SOC caused by permafrost degradation at 1.8 Tg C between 1986 and 2000.

In the Northern Hemisphere, the soil freeze–thaw cycle (FTC) is a conventional routine phenomenon and since the Qinghai–Tibet Plateau is located in the temperate zone with an alpine climate, the FTC is common here too (Ran et al. 2012). Predominant effects of freeze–thaw cycles on soil properties have been reported by many workers (Dagesse 2011; Dang et al. 2012; Liu et al. 2019), and it has been observed that FTCs can significantly change soil aggregate stability and density (Ma et al. 2021, Li et al. 2014), soil community structures, microbial activities, and extractable N concentration (Han et al. 2018; Sorensen et al. 2018, Watanebe et al. 2019) and organic mineralization (Liu et al. 2013). Most lead and chromium adsorption kinetics systems of freeze–thaw-treated soils were swift than in unfrozen conditions (Li and Wu 2021).

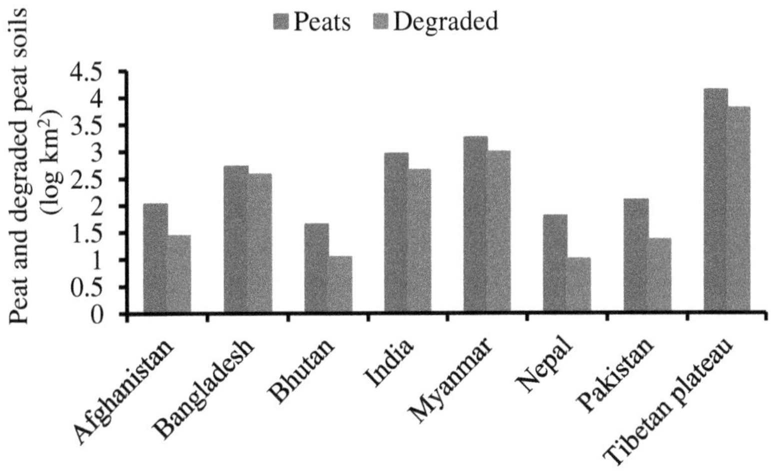

Fig. 6.8 Peat and degraded peat soils in HKH countries

6.12.7 Soil N Distribution in TP

Due to spatial heterogeneity and a lack of proper field observations, the estimated values of N may not be reliable, although it has been reported that the soils in the Tibetan Plateau shrublands store large amounts of total N. Spatially, it was observed that the soil N densities decreased from south to north and from east to west, and, vertically, the soil N in the upper 30 and 50 cm accounted for 42 and 64% of the total soil N stocks in the Tibetan Plateau. However, there were huge soil N stocks in alpine shrublands compared with desert shrublands, as reported by Nie et al. (2017).

Nitrogen is a significant stimulant of plant growth (Liu et al. 2017) and an important limiting element in terrestrial ecosystems (Vitousek and Howarth 1991). The levels of anthropogenic reactive nitrogen in the terrestrial ecosystem have increased in recent decades (Penuelas et al. 2013; Tian et al. 2016) and the accelerated industrialization and use of N fertilizer have significantly affected N cycles (Galloway et al. 2008; Chen et al. 2016). The availability of inorganic N has increased (Vitousek et al. 1997), and the rate of N flux to the biosphere has doubled (Lebauer and Treseder 2008). Currently, the increase in N deposition is expected to continue (Chen et al. 2016), and reports suggest that soil may function as an N sink in terrestrial ecosystems (Nadelhoffer et al. 1999, Chapuis-Lardy et al. 2007). Spatial heterogeneity in climatic conditions can cause dramatic variations in soil properties such as the N density (Liu et al. 2017; Xu et al. 2013), which varies widely by vegetation category (Tian et al. 2006). Therefore, an estimation of regional N stocks (Tian et al. 2006; Burke et al. 2002) in the Tibetan Plateau shrublands is necessary and would help to reduce uncertainties in the estimations of national and global soil N stocks.

The Tibetan Plateau is the largest high-altitude ecosystem on Earth (Ding et al. 2016). The mean elevation is about 4000 m, and it covers an area of about 26.8% of the total area of China (Zhang et al. 2002). Nie et al. (2017) gave the mean and standard deviation of soil N density and storage at various depths in shrublands on the Tibetan Plateau. A higher density of N was observed in 0–100 cm depth, and it increased with an increase in soil depth. A similar trend was found in N storage also. The N density was 2.40 kg m^2 in alpine shrubland and 0.34 kg m^2 in desert shrubland.

6.12.8 Soil Carbon

Patterns of soil C help to understand and estimate potential responses of soil C to global change in the future (Shi et al. 2012). The mean values of inorganic and organic carbon in the topsoil (0–20 cm) across the region were 0.38% and 3.63%, ranging from 0.00 to 2.92%, and 0.32 to 26.17%, respectively. Both, soil inorganic carbon (SIC) and soil organic carbon (SOC) in the Tibetan grasslands (0.51% and 5.24%, respectively) were higher than those in the Inner Mongolian grasslands (0.21% and 1.61%). The SIC stock in the top 1 m in the Tibetan alpine grasslands was approximately 15.2 billion tones C, with an average density of 13.46 kg C m^{-2}. In total, SIC stock in the Tibetan alpine grasslands was about 2.1 times the corresponding SOC stock, accounting for 28.5% of the total SIC stock in China (Yang et al. 2010). The SIC density at soil depths of 0–30, 0–50, and 0–100 cm in the Tibetan Alpine Grasslands was 5.70, 9.10, and 13.46 kg C m^{-2}, respectively and SIC stock was 6.43, 10.27, and 15.19 Pg C, respectively (Yang et al. 2010).

6.12.9 Soil Acidification

Zhou et al. (2010) were apprehensive that high nitrogen use in vegetable production around Lhasa, Tibet, may result in acidification, thereby affecting agricultural production on this soil. Bowman and Halvorson (1998) and Xu et al. (2002) found that the acidification may be more on naturally alkaline soils under arid conditions. Similar results were obtained by Wei et al. (2006) in China's Loess Plateau. Nitrogen fertilization significantly reduces soil pH by 0.3 and 0.5 units (Rodriguez et al. 2008). Biological fertility and its dynamics in degraded soil have been studied by Qian et al. (2006).

6.12.10 The Causes of Grassland Degradation

The causes of grassland degradation in the Qinghai–Tibet plateau are two major factors, such as ecosystem deteriorating human activities and natural disasters occurrence from time to time. Overgrazing is a fundamental cause of the increasing degradation of grassland in the region. A large number of livestock and their uncontrolled grazing lead eventually to degradation. The development of animal husbandry was associated primarily with increased numbers of animals, as was observed in this region. The grasslands are being over-exploited and overused, while little work has been done to improve the condition of grasslands by application of required amounts of nutrients through fertilizers. Rodent and insect infestations are big problems, which have been an important factor in causing grassland degradation in the region.

Anthropogenic interventions have spoiled the alpine meadows in the Qinghai–Tibet Plateau (QTP) resulting in their degradation. Soil degradation from overgrazing is the most direct cause of grassland degradation (Dong et al. 2021). They found that the bacterial Shannon diversity and richness in highly degraded grasslands were significantly lower than that in the lightly degraded ones, while no significant difference in soil fungal diversity among the different degradation levels was noticed. Soil fungal affluence decreased significantly from non-degraded to highly degraded grasslands.

Adverse environmental effects like desertification, dust storms, and soil erosion have been reported due to Grassland degradation (Bryan et al. 2018). About 38% of the grassland area in the Qinghai–Tibet Plateau (QTP) comes under alpine meadows, and thus forms a major part of the natural ecosystem (Zhang et al. 2019). The studies have shown that more than half of alpine meadows are degraded now, mainly due to human activities (Pan et al. 2017), such as overgrazing (Shen et al. 2013; Yuan et al. 2019). Overgrazing is a major cause of a decrease in above- and below-ground biomass (Li et al. 2016), decreased biodiversity (Lin et al. 2015; Li et al. 2019), as well as loss of carbon (Wang et al. 2002a, b). The changes caused by alpine meadow degradation can change the microbial community and its diversity (Zhou et al.2019). Soil water availability (Yang et al. 2018), pH (Yang et al. 2019a, b), bulk density (Yang et al. 2019a, b), and plant contents and vegetation (Wardle et al. 2004; Wu et al. 2014; Li et al.

2016) are major factors responsible for changes in microbial communities and their population in the soil. The attributes like OC, pH, water storage, BD, EC, nutrient availability, etc. are considered as indicators of soil quality (Lima et al. 2013; Askari and Holden 2015). Masto et al. (2008) mentioned that more than one indicator is, generally, needed for measuring the impacts of tillage management.

6.12.11 Prevention of Grassland Degradation

The following suggestions for the prevention of grassland degradation in the region are made:

(a) Undertaking awareness programs.
(b) Restricting the number of livestock according to the carrying capacity of the grassland.
(c) Judicious utilization of the grassland resource, such as establishment of rotation grazing plan.
(d) The inputs should be increased to improve the management of the grassland.
(e) Improvement of the degraded grassland through soil aeration, reseeding, irrigation, and fertilizer application. However, the measures used should be as per different characteristics and underlying causes of the grassland degradation.
(f) Suitable solution for prevention and control of rodents.
(g) The rich resource of Chinese medicinal herbs in the region is of great value and should be preserved scientifically.

6.12.12 Vegetation Carbon Pool

Among different study sites, the carbon pool in vegetation (both above- and below-ground biomass) ranged from 3.63 to 13.22 t Cha^{-1}, accounting for 7.54%, 7.14%, 6.67%, and 4.72% of the total carbon pool in the non-degraded (ND), moderately degraded (MD), heavily degraded (HD), and severely degraded (SD) grassland sites, respectively. Total carbon concentration in the above-ground biomass of grasslands of different degradation levels is shown in Fig. 6.9.

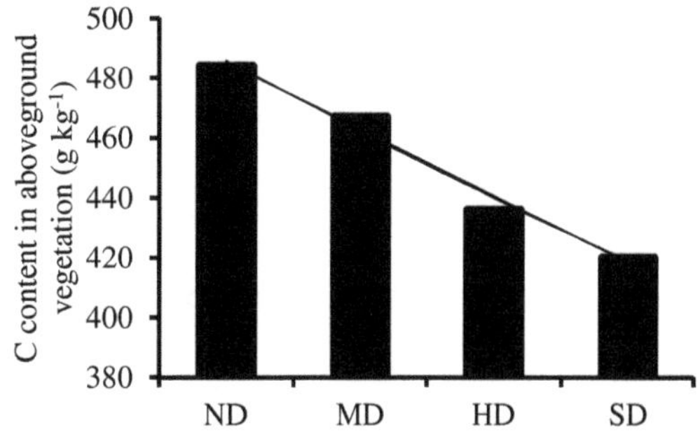

Fig. 6.9 Total carbon concentration in the above-ground biomass of grasslands at different degradation levels. *Source* Based on Wen et al. (2013)

6.13 India

Although India has about 18% of the world's human population and 15% of the world's livestock population, it has only 2.4% land area of the world. Despite its low proportional land area, India ranks second worldwide in farm output. The country has diverse climatic conditions as also other soil-forming processes. This has given rise to different types of soils, suitable for growing various types of crops. However, over-exploitation of this resource to feed ever-growing population, faulty management practices, deforestation, over-grazing, and gross human interferences have caused the degradation of soils, which is irreversible at many locations. Estimates show that 147 million hectare (M ha) soils of India suffer from one or other type of degradation. About 94 Mha of soils are degraded due to water erosion, 16 Mha from acidification, 14 Mha from flooding, 9 Mha from wind erosion, 6 Mha from salinity, and 7 Mha from a combination of two or more factors (Bhattacharyya et al. 2015). Natural causes of land degradation include earthquakes, droughts, avalanches, landslides, floods, and wildfires, whereas human-induced soil degradation causes are land clearing and deforestation, inappropriate agricultural practices, overgrazing, and mismanagement of forests.

6.13.1 Geographic Setting of Indian Himalayan Region

The Indian Himalayan Region (IHR) extends to eleven Indian states, namely Himachal Pradesh, Uttarakhand, Arunachal Pradesh, Assam, Manipur, Meghalaya, Mizoram, Nagaland, Sikkim, Tripura, and West Bengal and, two Union Territories, Jammu and Kashmir and Ladakh (Fig. 6.10). Starting from foothills in the south (Shiwaliks) region extends to the Tibetan plateau in the north (trans-Himalayas). The IHR (both, western and eastern Indian Himalayas) occupies the strategic position of the entire northern boundary (North–West to North–East) of the nation and touches almost all the international borders (Pakistan, China, Nepal, Bhutan, Myanmar, Bangladesh) with India. It contributes about 16.2% of India's total geographical area, mostly covered by snow-clad peaks, glaciers of higher Himalayas, and dense forest cover of mid-Himalayas. Compared to the rest of the country, the density of population in IHR is low and scattered. Physiographic conditions, poor infrastructure, low agricultural productivity, rugged terrain, inaccessibility, etc. are the likely causes for sparse population. However, the growth rate of population in IHR has an edge over that of the rest of the country.

Himalayas in India, from the state of Jammu and Kashmir to Arunachal Pradesh state, cover about 2400 km area. So, out of 3500 km of length of HKH region, India covers about 69% length-wise. The point in question is that a study

Fig. 6.10 Himalayan Region of India

of the Indian Himalayas is more or less a representative of the entire HKH region. Nevertheless, many variations exist.

6.13.2 Extent of Land Degradation in Indian Himalayan States

The extent of land degradation in Indian Himalayas has been estimated by National Bureau of Soil Survey and Land Use Planning (NBSS&LU) of Indian Council of Agricultural Research (ICAR). It indicates that total degraded area in Indian Himalayas is about 28.5 million ha, which is about 48.8% of the total geographical area of this region. The agents responsible for land degradation, that is, water erosion, wind erosion, waterlogging, soil acidity, and other complex problems, accounted for 15.4, 0.1, 1.7, 7.2, and 4.1 million ha or, 54.05, 0.35, 5.96, 25.26, and 14.38% of the degraded area, respectively. It reveals that more than half of the degraded area is through soil and nutrient erosion through water. Next to water erosion is the problem of acidity, which accounts for 25.26% of the total land degradation. Nearly 6.8 million ha of agricultural land is affected by varying degrees of salt problems in India, and the affected area is likely to increase in the near future (Abrol 1990).

6.13.3 Soil Resources

6.13.3.1 Extent and Distribution of Acid Soils in Indian Himalayas

Maximum area under acid soils is found in Eastern Indian Himalayan states. The Indian Himalayan Region has about 21 million ha of acid soils, the intensity of which is more in the Eastern Indian Himalayas (EIH). This is partly due to the parent material on which soils have formed and partly due to heavy rainfall received in this zone. In EIH, the maximum state-wise area under acid soils has been recorded in Arunachal Pradesh state, which is about 6.8 Mha. The states of Assam, Manipur, Meghalaya, and Mizoram have an area of 4.71, 2.19, 2.24, and 2.00 million ha under acid soils, respectively. The area on acid soils in Eastern Indian Himalayas as given above covers both arable and non-arable lands. A major chunk of the area in Arunachal Pradesh (4.8 Mha or 70.6% of total acid soils of the state) falls under the category of strongly acidic soils ($pH < 4.50$).

6.13.3.2 Problems Associated with Acid Soils in Indian Himalayas

Acid soils in India may be broadly grouped into two types, viz., (i) Truly tropical acid soils found in peninsular India, Islands, Orissa, West Bengal parts of Northeastern India and other parts, and (ii) those acid soils influenced by cold temperature during winters found in the Himalayan region. Among the adsorbed cations, Al and Fe take a larger portion. Even though the most important aim of liming in these soils is mitigation of Al toxicity, at $pH > 5.5$ the concentration of Al is so low that the crop growth is seldom hindered by Al. With high levels of Al and Fe oxides, when the pH of soils is increased through liming to >6.0, it leads to degradation of soil structure. There will be a progressive increase in soil erodibility, and sparingly soluble $Al(OH)_3$ is formed. It is wiser in these soils to focus lime application to correct Al and Fe toxicity rather than liming to increase pH to near neutral. Unlike tropical weathered soils which have variable surface charge, soils of temperate areas are less weathered and have a fixed negative charge. When the pH of the soil is increased through liming to levels >6.5, the soil structure improves in these acid soils. It is hence relevant in these soils to aim at liming to increase pH to near neutral to improve soil structure to prevent soil erosion, to enhance nutrient availability and crop growth (Ganeshamurthy et al. 2016b).

6.13.4 Soils of Eastern Indian Himalayas

The Eastern Indian Himalayas cover an area of 26.2 million ha, which forms about 7.98% of the total geographical area of India. On average, this region receives an annual rainfall of 2450 mm. Due to heavy rains, the EIH region is prone to the leaching of salts and nutrients, giving rise to soil acidity. Maji et al. (2010) reported that, in Arunachal Pradesh state, the total degraded and wastelands cover the largest area in the Along district, followed by Anini, Tezu, Bomdila, and Ziro districts. Waterlogging is a severe problem in Tezu and Khonsa districts of Arunachal Pradesh state. About 2.6 thousand ha area of Tezu district and 1.4 thousand ha in Khonsa are under waterlogged condition. In Assam state, the soils show variable behavior and belong to 4 orders, 9 suborders, 15 great groups, 26 subgroups, and 83 soil family associations. The Inceptisols are the dominant soils, followed by Entisols, Alfisols, and Ultisols, which occupy 41.4%, 33.6%, 11.3%, and 5.6%, respectively, of the total geographical area of the state. There are also miscellaneous land types belonging to marshy and riverine ecosystems, which have a total area of 8.1% (Sen et al. 1999). Soil classification of a few selected states is given in Table 6.2.

Covering about 38.4% of the area of Manipur state of India (Eastern Himalayas), Inceptisols dominate the landscape, followed by Ultisols, Entisols, and Alfisols which cover about 36.4, 23.1, and 0.2% of the area. About 2% of the area is under marshy land and lakes (Sen et al. 1997). Typic Hapludalfs, Typic Udorthents, Typic Haplaquepts, Typic Dystrochrepts, and Typic Haplohumults are the major soil subgroups. In Meghalaya state, Inceptisols occupy 46%, Ultisols 40%, Entisols 11%, and Alfisols 4% of the total geographical area. Singh et al. (1999a, b) classified the soils at the subgroup level. These are Ustic Hapludalfs, Typic Udifluvents, Aquic Udifluvent, Typic Humaquepts, Cumilic Humaquepts, Typic Haplaquepts, Aeric Haplaquepts, and Humic Haplaquepts. The Mizoram state covers an area of 21,081 km^2. Due to high precipitation and favorable condition of temperature and vegetation, weathering in the state is intense. Maji et al. (2001) explained that the Inceptisol soils are dominant in Mizoram state, followed by Ultisols, Entisols, and Alfisols. Agriculture is the main occupation of the people, and about 4% of the total area is under cultivation. About five decades earlier, shifting cultivation was practiced in more than 90% of the cultivated area of the state, which has now come down to about 50% of the cultivated area. Figures 6.11 show the parts of cultivated land in EIHR (Figs 6.11).

The Nagaland state has a geographical area of 16,579 km^2. Topographically, the state can be grouped into three major groups. These are (i) foothills with undulating to rolling topography of less than 100 m altitude, (ii) lower ranges and mid-ranges, with varying degree of slopes with an elevation of 1,000 m and above, and (iii) high hills and mountainous regions. In Nagaland state of India (Eastern Himalayas), Alfisols cover 2.9%, Entisols 7.3%, Inceptisols 66%, and Ultisols cover 24% of the total geographical area (TGA) of the state (Maji et al. 2000). With an area of 7096 km^2, the Sikkim state has three major soil orders. Soils belong to Inceptisols (43% of TGA), Entisols (42%), and Mollisols (15%). Haplumbrepts with 31% of TGA is the most predominant great group, followed by

Table 6.2 Classification of soils of Eastern Indian Himalayas

State	Orders	Suborders	Great groups	Subgroups
Assam	4	9	15	26
Manipur	4	8	13	22
Meghalaya	4	8	14	25
Mizoram	4	7	11	13
Nagaland	4	7	10	14
Sikkim	3	7	12	26
Tripura	5	7	9	19

Fig. 6.11 Images of Eastern Indian Himalayas

Fig. 6.11 (continued)

Fig. 6.11 (continued)

Udorthents (30%), Cryorthents (12%), and Hapludolls (12%). Das et al. (1996) reported that among subgroups, the most prominent ones are Typic Haplumbrepts (20%) and Cumulic Haplumbrepts (5%); Lithic Udorthents (19%) and Typic Udorthents (10%); Typic Hapludolls (5%) and Cumulic Hapludolls (2%) (Das et al. 1996). The 5 soil orders in Tripura state are Inceptisols, Entisols, Ultisols, Alfisols, and Histosols, covering 80.6, 8.1, 6.6, 4.5, and 0.2% of TGA. Generally, Inceptisols, Entisols, Ultisols, and Alfisols are observed in patches in hills.

Soil type

The National Bureau of Soil Survey and Land Use Planning (NBSS&LUP) of India has classified the soils of Eastern Indian Himalayas (EIH) into 5 orders, 29 great groups and 71 subgroups (Table 6.3). Inceptisols are the most dominant soil order which cover 11,431.5 thousand ha representing 43.60% of the region followed by Entisols by 7040.54 thousand ha (26.85%), Ultisols by 4346.3 thousand ha (16.58%), Alfisols by 1171.5 thousand ha (4.47%), and Mollisols by 81.66 thousand ha (0.31%). The distribution of soil orders indicates that Inceptisols are the dominant order in all the states. As far as different states of EIH are concerned, Ulfisols are the second dominant soils in Manipur (36.4%), Meghalaya (40.04%), and Nagaland (23.64%). The Sikkim state has 81.66 thousand ha (14.64%) area under Mollisols in the middle and lower hills. The depth of soils is quite variable in different places because of differences in physiographic characteristics and slope. The mass movement of soil due to water erosion is common, generally observed in the form of slips, mudflows, and solutions in the region. The situation arising out of the prevalence of shifting cultivation, unabated deforestation, and developmental works, in combination with high slopes, has further aggravated the problem of soil erosion to a larger extent.

Soil characteristics

The soils of the valleys in EIH are fairly rich in organic matter as most of the litter, humus, and organic matter from the hills flows down the slopes as rainwater and gets deposited. The practice of shifting cultivation has resulted in the loss of organic matter, and hill land surfaces suffer from erosion problems. Clay, silt, pH, and organic carbon content of soils are highly responsible for cation exchange capacity (CEC) (Kumar 1995). They reported that the CEC of acidic soils of this region ranged 21–54% and 6.2–37% of the total CEC in Sikkim and Manipur soils, respectively. The soils of Nagaland state and Cherrapunji have developed on shale, and siltstone with bands of weathered, rather soft, and micaceous and siliceous limestone are dominant in illite followed by kaolinite (Prasad et al. 1981).

The major soil orders in EIHR are Inceptisols (44%), Entisols (29%), Ultisols (14%), Alfisols

Table 6.3 State-wise distribution of soil orders in Eastern Indian Himalayas (Area in 000' ha)

Soil order	Arunachal Pradesh	Assam	Manipur	Meghalaya	Mizoram	Nagaland	Sikkim	Tripura
Entisols	2985.5* (35.64)**	2670.2 34.0	514.6 (23.1)	10.9 (0.49)	444.6 (21.06)	121.3 (3.65)	237.1 (42.52)	85.4 (8.13)
Inceptisols	3126.4 (37.36)	3216.8 41.0	858.2 (37.95)	992.2 (45.65)	1029.9 (48.80)	1094.7 (66.15)	238.9 (42.83)	845.9 (80.55)
Alfisols	22.8 (0.27)	862.0 11.0	3.8 (0.20)	80.7 (3.6)	82.4 (3.42)	47.8 (2.89)	–	47.1 (4.49)
Ultisols	1187.2 (14.16)	312.8 4.0	811.1 (36.4)	898.3 (40.04)	553.2 (18.16)	391.0 (23.64)	–	69.0 (6.56)
Mollisols	–		–	–	–	–	81.66 (14.64)	–

* Area in thousand hectare ('000 ha) ** Percentage of the state
Source NBSSLUP (2004)

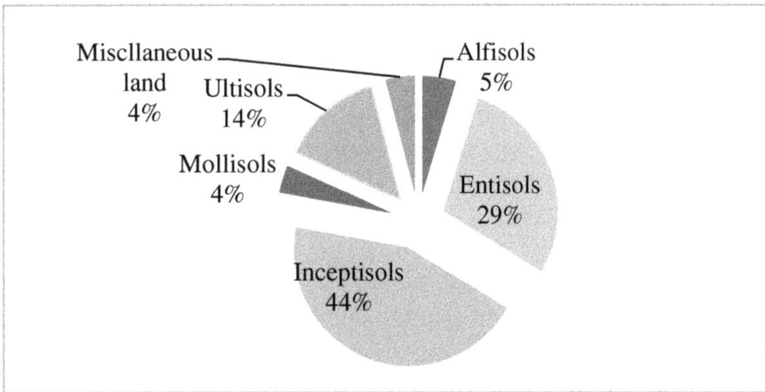

Fig. 6.12 Major soil types in Eastern Himalayas of India

(5%), Mollisols (4%), and others (4%) (Fig. 6.12). In the Himalayas, there is a large gap in crop production potential and its realization, possibly due to the following reasons (Venkatesh et al. 2006):

1. Non-availability of seed material of high-yielding varieties due to poor or no seed production program.
2. Very low application of a recommended dose of fertilizer nutrients so as to exploit the full potential of crop varieties.
3. Poor extension mechanism with respect to developed agricultural technologies.
4. Low irrigation facility.
5. Marginality, inaccessibility, degraded resource base, extreme poverty, and attitudinal constraints.
6. Lack of identification of niche areas for high crop productivity.
7. Mechanization in mountain agriculture has little scope. Due to the small size of holdings and uneven topography, use of a tractor for plowing and cultivation has a remote possibility.
8. Due to low returns from agriculture, the younger generation is migrating out to other areas in search of lucrative jobs. They have little interest in agriculture.
9. In some tribal areas, ownership/tenure rights of land are not conducive for investment in agriculture.
10. Fragmented land holdings.

Management acid soils in EIHR

The highly leached soils are generally known to be poor in fertility and their water-holding capacity is also low. It is a known fact that the soils with low pH (<5.5) are more problematic due to severe deficiencies of phosphorus, calcium, magnesium, and molybdenum and toxicities of aluminum and iron. The poor soil resource is one of the main reasons for low quality of life in the acid soil regions because of their low productivity and uneconomic returns. The addition of soil amendments like lime to these soils corrects soil acidity and creates favorable conditions for the biotic activity, release of nutrients, and their easy availability to plants (Shoemaker et al. 1961). The occurrence of acid soils in high rainfall mountainous regions is due to, primarily, the leaching of salts with rainwater.

6.13.5 Soils of Western Indian Himalayas

Three Indian states come under the Western Indian Himalayas. These states are the northernmost state of Jammu and Kashmir, Himachal Pradesh in the middle, and Uttarakhand state in the east. The soils of Jammu and Kashmir state have been classified into 4 orders, 9 suborders, 15 great groups, 27 subgroups, and 66 soil families. In this state, Entisols are observed to be the dominant soil, covering around 34% of TGA, followed by Inceptisols (6%), Alfisols (<1%), and Hollisols (<1%). Different soil subgroups in the state are Typic Hapludalfs, Aquic Hapludalfs, and Lithic Cryorthents (Rana et al. 2000).

The Himachal Pradesh state has an area of 55,673 km^2 and lies between 31°22′ to 33°12′ N latitudes and 75°45′ to 79°04’ E longitudes· The state of Himachal Pradesh has very high mountain peaks and very deep gorges. There are four major soil orders such as Entisols occupying the largest area (51%), followed by Inceptisols (20%), Mollisols (<1%), and Alfisols (<1%). The soils of the state belong to 6 suborders, 12 great groups, and 17 subgroups. The rock outcrops and glaciers also occupy a major chunk (28%) of the TGA of the state. Sidhu et al. (1997) mentioned that among the Entisols, the Udorthents constitute the major part (33%), followed by Cryorthents (15%), Ustifluvents (2%), Udifluvents (1%), Ustorthents (<1%), and Ustipsamments (<1%). Among Inceptisols, Eutrochrepts occupy the highest area (17%), followed by Ustochrepts (1%), Dystrochrepts (<1%), and Cryochrepts (<1%).

Figure 6.11 show the images of the Central and Western Indian Himalayas.

The state of Uttarakhand, lying between 28° 43′ to 31°27′ N latitudes and 79°33′ to 83°02′ E longitudes. And covering an area of 55,845 km^2, has a very rugged topography with deep gorges, narrow valleys, and high slopes. Singh et al. (2003) reported that the soils of summits and ridge tops in the state are shallow to moderately shallow, excessively drained, sandy/loamy-skeletal/loamy, and are exposed to moderate to severe erosion and moderate to strong stoniness. These soils are classified as Lithic/Typic Udorthents. They further mentioned that other soil formations are excessively drained, moderately shallow, fine-loamy/coarse-loamy, Dystric Eutrudepts, and Typic Hapludolls with slight to moderate erosion in association with sandy/loamy-skeletal Typic Udorthents. Slightly to moderately acidic soils are found in this area. The state has a rich forest cover in 64% of the area and 93% of its TGA is under mountains and hills.

6.13.6 Soils and Their Management in Western Indian Himalayan Region (WIHR)

The soils of WIHR suffer from a serious problem of soil erosion due to its fragile geology, steep topography, adverse climatic condition, overgrazing, deforestation, cultivation of slopy marginal lands, and increased human interferences. Soil erosion through water is a major cause of the decline in soil quality that threatens the sustainability of hill agriculture (Mina et al. 2006). Traditional hill agriculture system which integrates crop production, livestock, horticulture, and agroforestry needs to be adapted under climatic and edaphic conditions.

Soil characteristics in WIHR

The northwestern Himalayas have varied types of climate, soils, and management practices. The altitude of these mountains varies from just 305 m to more than 8000 m amsl. The rainfall in this Himalayan region varies from as low as 85 mm yr^{-1} in cold arid areas to as high as 2800 mm yr^{-1} in mid-hills sub-humid zone (Singh and Katoch 2000). The area under different soil orders is presented in Table 6.4 and soil degradation in Table 6.5 in WIHR. The major soil orders of the western Himalayan states are Entisols and Incetisols. The maximum percentage of degraded area is in Himachal Pradesh (75%), followed by Uttarakhand (71.6%) and Jammu and Kashmir (31.5%).

6.13.7 Soil Fertility

Soil fertility depends on the geological formation of the bedrocks from which the soil is derived, as well as nutrient additions and removal by the crops. Most of the soils of the Eastern Indian Himalayan (EIH) region are low to very low in P. However, the organic matter content of soils is high except for some parts. The status of available potassium varies from low to medium in all the states except Sikkim and some parts of Arunachal Pradesh. Due to heavy rainfall

Table 6.4 Area (km^2) and description of soil orders in NW Himalayas

State	Geographical area	Alfisols	Entisols	Inceptisols	Others
Jammu and Kashmir	222,240	1000 (0.5)	75,756 (34.1)	14,272 (6.4)	477 (0.2)
Himachal Pradesh	55,670	2380 (0.4)	25,843 (51.3)	10,999 (19.9)	4000 (0.8)
Uttarakhand	53,480	88 (0.17)	30,263 (56.5)	16,232 (30.3)	111 (0.21)

Source Sharma (2004a, b, c, d, e)
Values in parentheses are percent of GA

Table 6.5 Soil degradation in NW Himalayas

State	GA (km^2)	Degraded area (km^2)	Water erosion	Wind erosion	Physical deterioration
Jammu and Kashmir	222,240	70,200 (31.5%)	54,600 (24.5)	13,600 (6.1)	2000 (0.9)
Himachal Pradesh	55,670	41,780 (75.0%)	28,750 (51.6)	–	13,030 (23.4)
Uttarakhand	53,480	38,340 (71.6%)	15,540 (29.0)	–	22,800 (42.6)

received in the region, there is strong leaching of bases from the exchange complex, resulting in soil acidity. The production of crops in the Northeastern Indian Himalayan region is mainly based on the recycling of nutrients through added organic residues from animal feed, feces and bedding materials, and leaf litters from forests either for mulching or to produce organic manure. The consumption of fertilizer nutrients is very low in Eastern Indian Himalayan states. In fact, Sikkim, India's first "fully organic" state, won FAO'S Future Policy Gold Award. India's "100% organic state" the state of Sikkim has won the "Oscar for best policies", conferred by the Food and Agriculture Organization for the world's best policies promoting agro-ecological and sustainable food systems. Other states of EIH like Nagaland, Mizoram, and Arunachal Pradesh are organic by default or by tradition. Prasad et al. (1981) found that the critical limit of organic carbon for higher elevation (>1000 m) soils was 1.5%. The soils are not only deficient in native available P but, a major part of the applied P gets fixed in acidic soils, forming iron and aluminum compounds. Total and organic P content of soils increased with an increase in elevations (Singh and Datta 1987; Patiram et al. 1990). The critical limit of P availability was found between 0.34 and 1.2 µg P/ml, respectively.

The amount of available potassium (K) is usually low in soils derived from alluvium and sandstone parent materials. This is further aggravated by the high amount of rainfall in Eastern Himalayas. Therefore, most of the soils of this region are low in available and fixed K compared to the similar type of soils of other regions. Loss of top fertile soils from terraces and sloppy lands by runoff water also causes its deficiency in upland crops. By and large, the micronutrients were self-sufficient in soils, excepting Meghalaya state, where the deficiency of Zn and Cu has been reported. The soils of Sikkim are deficient in B and Mo due to light texture, acidic soil reaction, and leaching caused by excessive rainfall.

6.13.8 Soil Productivity Constraints

Soil acidity and associated problems

Soil acidity is a big constraint in crop production in EIH states. The soils of the entire region are acidic in reaction caused by heavy rainfall and leaching of bases to the lower horizon. As such, almost 81.1, 59.4, 31.9, 7.1, 98.0, 99.8, 97.2, 99.5, 75.9, 100.0, and 65.3% of soils of Arunachal Pradesh, Assam, Himachal Pradesh, Jammu and Kashmir. Manipur, Meghalaya, Mizoram, Nagaland, Sikkim, Tripura, and Uttarakhand, respectively, have soil pH below 6.5 and are considered acidic (Table 6.6). Interestingly, in EIH the acidic soils are 80.7% of the GA as compared to 20.6% in WIH. This can be mostly attributed to the parent material, amount of rainfall, and vegetation.

It may be imperative to take an account of soil acidity in the entire World. The world has 39.50 million km^2 of acidic soils which form 30% of the total geographical area. In India, 0.93 million km^2 of the geographical area has an acidic reaction and about 0.49 million km^2 of cultivated area has pH in the acidic range. Out of this, 0.26 million km^2 has a pH < 5.6 and 0.23 million km^2 has a pH value ranging from 5.6 to 6.5 (Motiramani 1971; Sarkar 2002). It is observed that acid soils of the world can be classified into eight orders, (i) Entisols, (ii) Inceptisols, (iii) Andisols, (iv) Spodosols, (v) Alfisols, (vi) Ultisols, (vii) Oxisols, and (viii) Histosols, out of which, Ultisols cover the maximum area of 21.8% (Table 6.7).

In accordance with pH variation as noted in soils of the world, soils having a moderate pH (4.5–5.5) cover the maximum (11.7%) area (Table 6.8).

Of the acid soils prevailing in Indian Himalayas, 21.4% fall under strongly acidic (pH < 4.5) category, 44.7% under moderately acidic (pH 4.5–5.5) category, and 33.9% under slightly acidic (pH 5.5–6.5) category.

Aluminum and iron toxicity

Aluminum toxicity is the other constraint in crop production in the EIH region. The most important associated problem of acid soils is when pH drops below 5.0. In the acid soils of the NE Himalayan region, the amount of exchangeable aluminum ranged from traces to as high as 7.1 meq/100 g soil of Nagaland (Prasad et al. 1985; Kumar 1995; Patiram 1990; Das et al. 1996; Sen et al. 1997) and percent aluminum saturation up to 85. Patiram et al. (1991) reported

Table 6.6 Acid soil in Indian Himalayan region (million hectares)

State	pH (<5.5)	pH (5.5–6.5)	Total acid soils	% of GA*
Arunachal Pradesh	6.52	0.27	6.79	81.1
Assam	2.33	2.33	4.66	59.4
Himachal Pradesh	0.16	1.62	1.78	31.9
Jammu and Kashmir	0.09	1.48	1.57	7.1
Manipur	1.87	0.32	2.19	98.0
Meghalaya	1.19	1.05	2.24	99.8
Mizoram	1.27	0.78	2.05	97.2
Nagaland	1.60	0.05	1.65	99.5
Sikkim	0.60	–	0.60	75.9
Tripura	0.81	0.24	1.05	100.0
Uttarakhand	1.18	2.30	3.48	65.3
Himalayan states	17.62	10.44	28.06	47.2
All India	31.0	58.94	89.9	27.3

* Geographical area

Table 6.7 Soil order-wise and continent-wise distribution (%) of acid soils in the world

Order	Area		Continent	Area	
	Million km^2	(%)		Million km^2	(%)
Entisols	8.24	20.9	Africa	6.59	16.7
Inceptisols	5.61	14.2	Australia and New Zealand	2.39	61
Spodosols	4.15	10.5	Europe	3.91	9.9
Alfisols	2.55	6.5	Asia	10.44	26.4
Ultisols	8.64	21.8	America	16.16	40.9
Oxisols	7.27	18.4	Total	39.50	100
Histosols	2.70	6.8			
Total	39.50	100			

Table 6.8 Distribution ('000 km^2) of acid soils in accordance with pH

Soil pH	Global		Southeast Area		India	
	Area	%	Area	%	Area	%
Slight (5.5–6.5)	14,233	10.8	1,704	4.9	462	14.1
Moderate (4.5–5.5)	15,400	11.7	1,916	5.5	434	13.3
High (<4.5)	9,871	7.5	3,253	9.3	41	1.2
Total	39,504	30.0	6,873	19.7	9.37	28.6

that the aluminum sensitive crops such as wheat, barley, and soybean may get damaged when aluminum saturation of cation exchange capacity exceeds 10%. Iron toxicity has been a problem in waterlogged soils. The waterlogging is caused due to underlying clay substratum restricting vertical drainage. Long-duration rice in such lands suffers from iron toxicity and bronzing. Microbial activities are reduced in strongly acid soils and aluminum toxicity and deficiency of phosphorus, calcium, and molybdenum in legumes also affect the efficiency of Rhizobium biofertilizer.

6.13.9 Physiographic and Soil Resources

The main problems associated with mountains are their peculiar geographical setting and the resultant physical constraints and human interferences. The Himalayas being young represent one of the most fragile mountain ecosystems on the globe. It has been observed that during the past few decades, a number of changes have emerged in the use of traditional resources due to population pressure and increasing demand for food, fodder, fuel wood, grazing areas, etc. in the Himalayas. Though their study has been very difficult as most of the areas are inaccessible to human and livestock population, management of natural resources in the Himalayas, especially soils, has become essential and urgent at present because people need goods and ecosystem services from this resource. Since soil organic carbon (SOC) influences various soil properties, it is an important component of agro-ecosystem (van Keulen 2001). A New period of remote sensing satellites has ushered in possibilities for this type of study (Ahuja et al. 1992; Saxena et al. 2000).

It has been observed that the soils of river valleys and piedmont plains are mainly medium deep to deep, loam to clay loam with structural B horizon and belong to Inceptisols followed by Entisols. Soils of peak tops contain more SOC stock in comparison to other physiographic conditions. The same trend was seen in higher elevations as in lower ones. In a study, undertaken to assess the soil organic carbon (SOC) stock in Kumaon Himalayan Mountains of India, Mahapatra et al. (2010) reported that the soils of summit tops and side slopes are mainly shallow to medium deep, coarse-textured, underdeveloped or partially developed, having A–C horizons (Entisols) followed by deep, fine textured, and comparatively developed soils (Inceptisols).

Soils of river valleys, occurring on different slopes, are deep, somewhat excessively drained, with low available water content (AWC) and nutrient retention capacity being developed on colluvium/alluvium material. Wheat, mustard, millets, and paddy are the major crops cultivated on these soils. It has been revealed through the data on organic carbon stock in the soils of major landforms that in general, all the soils are rich in organic matter in surface layers which decreased with depth (Sehgal 1973; Mahapatra et al. 2000).

6.13.10 Fertility Status of Soils of Eastern Himalayas

The hill soils of Eastern Indian Himalayan Region (EIHR) vary from sandy loam to clay loam in texture. Old alluvial soils of the Brahmaputra valley are silty loam to clayey, while new alluvial soils are mostly sandy, silt loam, or clay loam in texture (Chakravarty et al. 1978). Soil fertility indices show spatial variation and are presented in Table and fig. Soil fertility status of major nutrients shows that, by and large, the soils are medium to high in available N and K and, low to medium in available P. Due to soil acidity, P fixation is a major constraint in crop production. Organic carbon content of the soil in the entire region is high except in parts of Assam and Tripura states (Sharma 1989a, b; Sharma et al. 2006a, b). Deficiency of zinc, boron, and molybdenum has been experienced in the region. Soil fertility status of EIH states is given in Table 6.9.

The state-wise soil characteristics are briefly described below:

Arunachal Pradesh

More than 96% of the cultivated soils of the state are acidic in reaction and contain a high amount of organic matter and medium to high in available N (Table 6.10). Almost all soils of the state are low in available P, except East and West Kameng districts. Some important soil properties are mentioned in the table below.

Assam

More than 75% of the soils in non-riverine tracts are acidic in Assam. The pH of red loam soils varies from 4.9 to 6.6 with an average value of 5.4. N content varying from 0.013 to 0.217% in Brahmaputra valley soils. Different forms of K and their range have been given in Table 6.11. The Alfisols contain more exchangeable and non-exchangeable K compared to Entisols and Inceptisols, while water-soluble K content was higher in Entisols.

Manipur

Soils of Manipur are acidic in reaction with pH varying between 4.5 and 6.8. The OC content ranged from 1.4 to 5.1% in hills and 0.1–5.5% in the valley areas (Table 6.12). The difference in OM content in hills and valley soils was due to the nature of vegetation and intensity of cultivation. The soils are low to medium in available P, medium to high in available K, and high in OM content. Different forms of Al, such as exchangeable, amorphous, organic bound, crystalline and total ranged from 0 to 2.82, 0.03 to 7.56, 20.8 to 108.5, 2.2 to 33.2, 1.4 to 16.0 and 175 to 753 c mol $(p^+)\ kg^{-1}$, respectively.

Meghalaya

The OM content of Meghalaya soils is high in most of the soils varying from 0.9 to 7.4% in East Khasi Hills and 0.9–5.4% in West Khasi Hills district. Considering the critical limit of available P as 12 ppm, 60% of the soils in the

Table 6.9 Soil fertility status (N, P, K) in Eastern Indian Himalayas

Soil fertility class	Area
LLL	North Sibsagar district and parts of Jorhat and parts of Dibrugarh districts of Assam
LLM	Lunglei district of Mizoram and Saiha area of Assam
LML	Nowgong district of Assam
LMM	Parts of Dibrugarh district, southern parts of Kamrup and Goalpara districts of Assam
MLL	Northeastern parts of Dibrugarh district, western parts of Tripura state
MLM	East and West Khasi hill districts of Meghalaya state, Western Lakhimpur, Eastern Sonitpur district, northern parts of Kamrup and Karbi Anglong districts of Assam
MML	Sonitpur district, Eastern part of Sibsagar, and lower parts of Dibrugarh districts of Assam, Khowai Udaipur, Bagafa, Panisagar, Kanchanpur, and Kumarghat blocks of Tripura
MMM	Darrang and Golaghat districts of Assam, Amarpur and Dumbemagar blocks of Tripura
MMH	Aizawl district of Mizoram, Kokrajhar and parts of Lakhimpur district of Assam
MHM	Dhubri district of Assam
MHH	Mid-parts of Mizoram
HLL	Lower and upper Subansiri district of Arunachal Pradesh, parts of Karbi Anglong, Kohima and Phek districts of Nagaland
HLM	Ast and West Siang district of Arunachal Pradesh; Wokha, Mokokchung (Nagaland), Central Manipur, West Garo hills, parts of Sibsagar district of Assam
HLH	Dibang valley, Tezu area, Khonsa area, far eastern parts of East Siang and eastern part of Arunachal Pradesh, Mon strict of Nagaland
HML	Part of Lakhimpur, Barpe district (Assam), Southern parts of Kohima (Nagaland), Mao district (Manipur)
HMM	Kolasib and adjoining areas of Mizoram, parts of Silchar district of Assam, East and West Kameng district (Arunachal Pradesh), Northmost parts of Darrang, North of Dibrugarh and Kamrup districts of Assam
HMH	East Garo hills district of Meghalaya
HHL	Western parts of Aizawl (Mizoram), Zunhoboto, Tuensang districts of Nagaland, parts of Ukrul district of Manipur
HHM	Southwestern parts of Aizawl (Mizoram)

Source Prasad et al. (1981)

East Khasi Hills district and 50% of soils in West Khasi Hills are deficient in available P (Fig. 6.13). An appraisal of fertility status of Meghalaya showed that 13% of soils are deficient in OC, 83% of soils deficient in available P, and 5.5% of soils deficient in available K, below 700 m amsl while at high altitude (>1700 m above msl), the corresponding values were 5.5, 48 and 19%, respectively (Singh et al. 1988). The amount of total Zn, Cu, and Mn ranged from 10.0 to 47.2, 17.0 to 71.0, and 110 to 1230 ppm, respectively, in upland soils while their available contents varied 0.7–3.2, 0.6–3.0, and 2.3–162.0 ppm, respectively. About 40% of the soils of Meghalaya are rated low in available Zn and Mn (Patiram 1984). Iron toxicity was prevalent in soils of narrow valleys with a shallow water table (<40 cm and high in OM (>2.98%) status (Singh et al. 1992a, b).

Mizoram

The soils of entire Mizoram are acidic in reaction with high OM and total N. Due to the low mineralization rate, the availability of nitrogen varies widely in different parts of the state. About 40% of the soils are deficient in available P

Table 6.10 Characteristics of soils of Arunachal Pradesh

Site and elevation	pH in water (1:1)	pH in KCl (1:1)	Organic Carbon (g kg^{-1})	Effective CEC [e mol (p +) kg^{-1}]	Clay (%)	Texture
Bomdila (2150–2300 m)	4.8–5.5 (5.2)	4.4–5.3 (4.9)	50.6–56.7 (52.8)	9.1–14.1 (12.2)	8.9–11.1 (10.2)	Sl-l
Dirang (1475–1775 m)	4.5–6.3 (5.1)	4.0–5.5 (4.5)	3.4–18.3 (14.0)	2.2–6.3 (4.8)	10.9–23.2 (16.0)	l
Jerigaon (1725–1800 m)	4.8–5.4 (5.3)	4.7–4.9 (4.8)	29.6–46.7 (38.9)	8.5–13.2 (11.6)	11.0–17.0 (13.5)	Sl-l
Munna camp (1400 m)	5.7–5.8 (5.7)	5.4–5.5 (5.4)	16.0–21.6 (18.0)	5.6–13.5 (8.7)	10.9–23.2 (16.0)	Sl
Nafra (1000–1100 m)	5.9–6.3 (6.1)	5.5–6.0 (5.8)	11.7–50.3 (26.7)	5.9–16.4 (10.2)	5.4–11.8 (8.9)	Sl
Salari (1085–1225 m)	4.6–6.0 (5.4)	4.3–5.5 (4.9)	8.5–45.9 (17.7)	5.1–15.0 (7.4)	8.1–14.2 (11.4)	Sl-l
Bomdir (2225–2300 m)	4.5–4.6 (4.6)	3.8–4.0 (3.9)	28.0–41.7 (34.5)	–	9.7–12.4 (11.4)	l
Jung (2100–2200 m)	5.0–5.6 (5.3)	4.2–5.0 (4.6)	21.8–39.2 (30.6)	–	8.9–11.4 (10.1)	Sl-l
Tawang (2700–2775 m)	4.0–4.4 (4.2)	3.5–3.9 (3.7)	45.5–81.4 (59.4)	–	7.8–13.4 (11.1)	Sl-l

Table 6.11 Different forms of K in soils of Assam

Soil order	Forms of K		
	Water soluble	Exchangeable	Non-exchangeable
Entisols	7.9–15.0	29.2–54.1	220.5–860.3
Inceptisols	7.9–11.0	30.7–50.0	185.4–505.4
Alfisols	4.5–8.1	35.6–72.0	401.7–790.0

whereas, the majority of soils are medium in available K. Total N content varied from 0.052 to 0.162% with a mean value of 0.12%. Ammonical N was the dominant inorganic N fraction ranging 18.5–50.1 kg ha^{-1}, while nitrate N ranged from 2.5 to 15.0 kg ha^{-1} (Laxminarayana and Patiram 2004).

Nagaland and Sikkim

Soil nutrients in different land uses

A study was undertaken by Chase and Singh (2014) in Eastern Indian Himalayas to know the nutrient status in three land use systems such as natural forests (NF), shifting cultivation fallow

Table 6.12 Properties of Manipur soils

Area	pH	Org. C (%)	Av. P (ppm)	Exch. K [e mol (p+) kg^{-1}]	Exch. Ca+ Mg [e mol (p+) kg^{-1}]
Valley					
Imphal	4.9–6.1	0.51–3.66	1.31–17.06	0.127–1.202	3.50–19.10
Bishenpur	5.0–6.3	0.12–3.42	3.65–10.80	0.153–1.330	4.30–16.70
Thoubal	5.5–6.7	0.24–2.16	1.32–23.10	0.127–0.578	3.50–21.30
Hills					
Chandel	4.2–6.3	1.44–4.91	3.62–15.88	0.313–0.641	1.28–21.60
Senapati	4.3–5.5	1.83–3.77	2.97–26.72	0.235–1.517	1.79–13.61
Ukrul	3.8–5.7	1.44–5.10	0.48–9.42	0.241–2.820	2.54–18.48
Churachandpur	3.8–5.7	1.52–2.68	4.54–15.81	0.461–1.179	2.46–10.27
Tamenglong	4.0–4.8	1.84–3.12	2.49–8.19	0.256–1.256	3.86–8.41

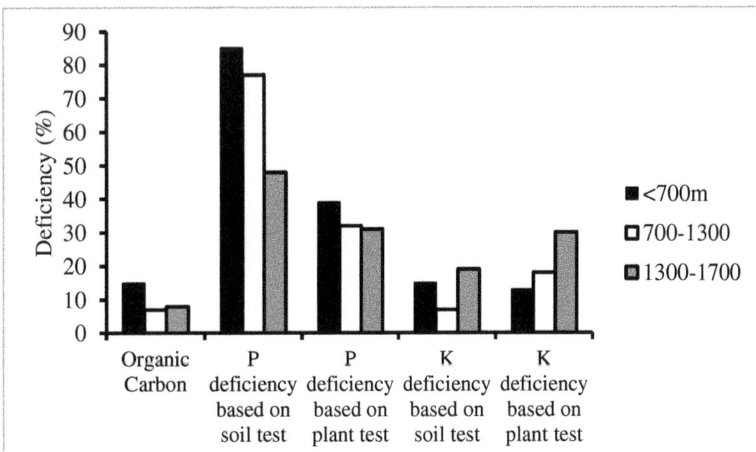

Fig. 6.13 Deficiency of OC, P, and K in soils of Meghalaya under varying altitudes

(SCF), and rice fields (RF) (Table 6.13). The study revealed that pH was higher (5.5) in RF as compared to NF (5.1) and SCF (4.8). More than 80% of the soils of this area exhibit acidic reaction. As a result of high rainfall received in the area, there is leaching of bases down the profile. The pH of rice fields is found to be favorable, and this has resulted in continued production of good crops with the use of farmyard manure and nutrient wash-out from the hill slopes without the use of inorganic fertilizers. As expected, the N, organic C, and OM were significantly higher in natural forests, and available P and exchangeable K were significantly high in SCF land uses compared to other land uses. More available P and exchangeable K in SCF were due to the rapid recycling of soil nutrients from Alder (*Alnus nepalensis*) trees.

The soils of Nagaland are low in available P due to low pH and high amount of exchangeable Al in the soil. About 50% of the soils are low in available P. However, the OM content of soil is generally high. The soils of Tuensang and Kohima districts are deficient in available K.

The soils of Sikkim are also low in available P. The critical limits of available P by Bray P1 and Mehlich extractants were 18 and 15 ppm, respectively, for maize in Sikkim (Table 6.14) (Patiram et al. 1990). The mean values of different forms of P and K are given in Table.

Table 6.13 Status of nutrients in different land uses in Khonoma (Eastern Himalayas)

Parameters	Natural forests	Shifting cultivation fallow	Paddy fields
pH	5.1	4.8	5.5
N (kg ha^{-1})	202	159	47
Available P (kg ha^{-1})	14	17	12
Exch. K (kg ha^{-1})	109	131	49
Organic carbon (%)	2.8	2.4	1.0
Organic matter (%)	6.7	4.1	1.8

Among the different inorganic P fractions, Fe–P was most dominant and their relative abundance followed the trend Fe–P > occluded Fe–P > Al–P > Ca–P > occluded Al–P. The availability of K also varied widely throughout the state. The OM content ranged from 1.44 to 7.59% in the state, whereas DTPA extractable available Zn, Cu, Fe, and Mn content varied from 4.0 to 6.5, 0.2 to 12.5, 52.1 to 110.5, and 2.0 to 25.5 ppm, respectively. The effect of altitudinal differences on the nutrient content of soils in Eastern Indian Himalayas is presented in Table 6.15.

Tripura

The soils of Tripura are low to medium in OM. Practically, entire soils of the state are deficient in available P. The soils of the South and North districts of Tripura are rated low (<125 kg ha^{-1}) in available K, whereas in Western Tripura, available K is medium and varies from 125 to 250 kg ha^{-1}. Distribution of micronutrients in the state varied with altitude (Datta and Munnaram 1993). The Zn content increased with depth in upland soils, while a decreasing trend was observed in lowland soils whereas Cu and Mn contents are found to decrease with depth in both the soils. The mean values of available and total micronutrient content in up and lowland soils of the state are presented in Table 6.16.

6.13.11 Oak and Pine Forest Soil in the Western Indian Himalayas

Soils of oak and pine forests are the common forest types in the Western Indian Himalayan (WIH) region. Negi et al. (2015) mentioned about the reason for oak forests being superior to pine forests and that these are important for sustaining agriculture and other associated ecosystem services and also, maintenance of biodiversity and enhancement of carbon stocks in vegetation and soil pools. Oak forests also serve as a large sink for atmospheric carbon dioxide (CO_2). Negi et al. (2015) emphasized that SOC stored in the surface layer is more vulnerable and

Table 6.14 Forms of phosphorus and potassium in soils of Sikkim

District	Forms of P (ppm)	Forms of K (ppm)					
	Total	Available	Organic	Inorg	Water soluble	Exch	Fixed
East	1196.1	19.0	441.4	729.6	7.4	121.7	522
West	1054.4	9.6	396.7	641.5	23.5	187.5	1247
North	–	27.5	1175.2	740.0	4.3	81.3	3780
South	989.8	11.7	341.3	624.7	14.5	141.3	874
Average	1118.7	17.0	413.9	684.0	12.4	133.0	755

Source Patiram (1990), ICAR

Table 6.15 Effect of altitudinal differences on nutrient content of soils in Eastern Indian Himalayas

Location	Altitude (m)	OM (%)	Total N (%)	Available nutrients (ppm)					
				P	K	Zn	Cu	Fe	Mn
Gangtok	1600	7.68	0101	20.3	12.3	2.5	0.2	98	5.0
Nagi	1500	6.96	0075	15.0	9.2	0.4	1.2	110	2.0
Namthang	1392	5.64	0054	14.1	17.6	1.0	2.7	89	12.6
Tarung	1290	6.22	0067	27.7	18.7	0.8	0.6	94	7.5
C. Pendum	510	5.43	0051	12.1	16.3	0.7	0.8	105	10.2
Narakjhora	503	4.35	0043	27.5	13.8	1.9	2.1	72	5.8
Singtum	391	4.05	0035	10.5	19.2	6.5	1.2	60	18.5
Majhitar	390	4.55	0030	7.1	4.3	2.0	2.5	76	3.9
Mining	360	5.03	0062	7.9	14.8	5.2	12.5	84	25.5
Rangpo	350	4.25	0025	15.7	11.6	3.1	3.8	52	15.7

Source Patiram and Prasad (1990b)

Table 6.16 Mean content of micronutrients in the soils of Tripura (mg kg^{-1})

Micronutrient	Available			Total
	Upland	Lowland	Upland	Lowland
Zinc	2.9	3.0	197	245
Copper	0.6	1.4	32.2	25.0
Iron	7.6	21.9	25,900	15,700
Manganese	21.2	2.4	380	407
Boron	1.1	0.8	89	138
Molybdenum	0.16	0.13	2.5	2.5

less stable than that in a deeper layer. Oak forests have higher water-holding capacity than pine forests and, therefore considered more fertile. These forests support a higher density of trees and hence, biomass productivity, diversity, species richness, and understory vegetation year-round than the pine forests.

6.13.12 Deficiency of Secondary and Micronutrient in Soils

Intensive cropping systems need a lot of nutrients from the soil to sustain crop yields. The soils get exhausted due to these systems if the loss/utilization of nutrients from the soil are not compensated by applying from outside through inorganic or organic sources. Therefore, nutrient deficiencies are bound to occur unless steps are taken to restore soil fertility. Crop yields are affected by the micronutrient deficiencies in soils. Analysis of 0.15 million soil samples across India has shown that about 47, 20, 18, 12, and 5% of soils were deficient in available Zn, available B, Mo, available Fe, and available Cu, respectively. It was revealed that Zn deficiency was most widespread followed by B, Mo, and Fe, in that order. Field-scale deficiency of Zn in crops is being increasingly reported from different areas, including mountains. The productivity of wheat could be restored by soil and foliar applications of micronutrients (Swarup and Ganeshamurthy 1998).

6.13.13 Types of Wastelands in Himalayas

According to National Remote Sensing Agency (NRSA), soils of India could be divided into 13 types:

1. Gullies and/or ravines.
2. With or without scrub uplands.
3. Marshy and waterlogged land.
4. Saline/alkaline land.
5. Lands where shifting cultivation is being practised.
6. Degraded forest land.
7. Degraded pastureland.
8. Degraded land under plantation crops.
9. Shifting sands—inland/coastal.
10. Industrial wastelands.
11. Wasteland.
12. Steep slopes.
13. Snow-covered area.

6.13.14 Hills and Mountain Ecosystem in India

Sidhu and Surya (2014) reported that hill and mountain ecosystem covers about 54 million ha area of the country, which mainly includes northern and northeastern states as well as some parts of central and southern states. The Himalayas have a great influence on the environmental conditions of northern India and the people living in adjoining Plains. Due to geological, topographical, climatic, and demographic reasons, inaccessible rugged terrain, and fragmented and small farm holdings, the mountain ecosystem is highly fragile and unsustainable. Soils of this ecosystem are less fertile with problems of soil acidity, low exchange capacity, and aluminum toxicity (Sen et al. 1997). However, hill and mountain ecosystems are also endowed with many opportunities to win over these challenges. These ecosystems have rich biodiversity, diverse nature of soils, and highly potential productivity as well as capacity in regulating the hydrology of river basins. The variation in soils of the Himalayas is mainly associated with altitude, land use, and management practices (Walia et al. 1999).

6.13.15 Western Indian Himalayas (WIH)

The WIH is located between 28° 43′ N to 37° 05′ N latitude and 72° 02′ E to 81° 02′ E longitude. This tract covers an area of approximately 33.12 million ha. The climate ranges from cold arid zone to temperate, subtropical temperate transitional, and low altitude subtropical zone. The mean maximum temperature ranges from 12.4 °C in Leh to 30.0 °C in Jammu and 34.5 °C in Shiwaliks, and the mean minimum temperature from −1.4 °C in Leh to 18.7 °C in Jammu and 5 ° C in Shiwaliks. Rainfall is highly variable from < 100 mm in the cold desert area of Ladakh to as high as >1600 mm in the Uttarakhand region. The temperature regime ranges from Cryic to Hyperthermic and the moisture regime from aridic cold in Greater Himalayas to humid in Lesser Himalayas and sub-humid in Shiwaliks. This tract is broadly divided into three distinct landform units, i.e. Greater Himalayas, Lesser Himalayas, and Shiwaliks. These units are sub-divided into 9 sub-units.

6.13.16 Soils of Greater Himalayas

The Greater Himalayan zone is highly rugged and dissected with snow-clad slopes, horned peaks, hanging valleys, and gigantic escarpments. Granites, phyllites, and mica schists are the major rock formations. The base of the sedimentary column in this region is formed by argillaceous metamorphic, where mica schists, rich in kyanite, staurolite, and garnets are found predominantly. The soils in this zone, cover eastern parts of the Ladakh plateau, including Leh and surroundings; ranges of Karakoram, Ladakh, Zanskar, western half of Ladakh plateau (Leh), Kargil, and Gilgit areas (Rana et al. 2000; Sachdev et al. 1995; Walia et al. 1999), the north–northeastern part of Lahaul–Spiti valley of Himachal Pradesh represent most of the cold desert (Sidhu et al. 1995a, b, 1997; Sehgal et al. 1993); northern parts of Uttarakhand bordering Tibet, commonly known as Himadri (Singh et al. 2004) show variable features as far as their fertility aspects are concerned. The winter snowfall accompanied by a cold climate (as low as −15 ° C) plays a crucial negative role in developing soil profile. However, during the following summer months, the snow melts, resulting in limited weathering and liberation of some bases toward the development of B-horizon in a favorable topographic position. Velayutham et al. (1999) reported that this region represents a cold arid to cool humid ecosystem. Due to high variability in relief, the mean annual temperature ranges from 2 to 8 °C. The Ladakh plateau is rugged mountainous terrain interspersed with narrow to nearly broad valleys. The soils of the Ladakh region are generally skeletal with A–C profile due to the effect of topography and frigid-like aridic climate that restrict the chemical weathering; the soils are seldom subjected to leaching due to limited moisture supply (Sehgal et al. 1993). The soils of Leh were characterized as Typic Eutrocryepts and that of Nubra valley as Typic Cryorthents (Yadav and Jagdish Prasad 2005). The soils in the area are in general shallow, mostly skeletal, coarse-textured, calcareous, and strongly alkaline (Sehgal et al. 1973).

Due to severe climatic limitations, higher elevation, and steep slopes, the majority of the area either remains barren or with sparse vegetation cover of dry alpines scrubs. Except for a few patches of Himalayan meadow, Ladakh and Spiti are almost devoid of vegetation cover. The dry alpine forest is mainly observed in the Lahaul and upper part of the Kinnaur areas. In valleys, deodar (Cendrus libani), Safedi (Poplar spp.), and willows (Salix spp.) are dominant trees. Nearly 5–10% of the geographical area is utilized as agricultural land for rainfed farming. The main crops grown are barley, wheat, buck wheat, pulses, peas, beans, kuth, potato, and cabbage. The temperature regime prevailing in the Himalayas limits the choice of crops due to short growing periods, steep slopes, excessive runoff and coarser texture of soils, low available water capacity (AWC), and rock outcrops. In the Western Himalayas, Sidhu and Surya (2014) delineated the soils of greater Himalayas into four landforms, such as (i) Soils of summits, ridge tops, cliffs, valley glaciers (soils on very steeply sloping summits and ridge are shallow to deep, excessively drained, sandy-skeletal to loamy-skeletal), (ii) Soils of side/reposed slopes (soils on steeply to very steeply sloping denuded side/reposed slopes are shallow to medium, excessively drained, loamy-skeletal and coarse-loamy with low available water capacity. They are slightly alkaline, calcareous, severely eroded soils with moderate to strong stoniness), (iii) Soils of Glacio-Fluvial valley (soils of gently to moderately steeply sloping glacio-fluvial valleys are occupied by medium deep to deep, well drained to excessively drained, loamy-skeletal and coarse-loamy to fine-loamy soils), and (iv) Soils of Fluvial Valley (soils are mostly slightly acidic to neutral soils, with low AWC. They are classified as Typic Udorthents/ Udifluvents and Dystric/Typic Eutrochrepts). Sidhu and Surya (2014) further mentioned that the area under these landforms in Jammu and Kashmir state is 5.12, 10.4, 0.01, and 0.63 million ha, in Himachal Pradesh it is 0.46, 2.01, 0.12, and 0.03 million ha, and in Uttarakhand state it is 0.74, 0.83, 0.03, and zero million ha in landforms (i)–(iv), respectively. Maximum area in Western Himalayas falls under the landform "soils of side/reposed slopes".

6.13.17 Soils of Indian Mid-Himalayas

Mesic temperature regime is mostly found in Kashmir valley. The important and extensive soils of Karewa landscapes are represented by Benchmark soils, viz., Gogji Pather and Wahthora series (Murthy et al. 1982) occurring in parts of Anantnag district of Jammu and Kashmir state. These soils are nearly neutral to mildly alkaline, deep, rich in organic matter, and base saturated. Floodplain of the Indus river system is mostly calcareous, neutral to slightly alkaline, and represented by Udifluvents and Haplaquepts. The plants' available water capacity is high (180–200 mm m^{-1}) (Sidhu and Surya 2014). All climatically adaptable crops can be grown under rainfed conditions.

6.13.18 Soils of Indian Lesser Himalayas

The soils of Indian lesser Himalayas, covering Pir Panjal range, Poonch, Rajouri, Udhampur, Doda, and Kishtawar of Jammu region and all the districts of Kashmir region, cover an area of 1.4 million (M) ha. In Himachal Pradesh, these soils cover Kinnaur, Shimla, Solan, Kullu, Mandi, Chamba, and northern parts of Kangra and Sirmour districts (Sidhu et al. 2007a, b; HPKKV 1991). The total area under this unit is 1.75 M ha (31.5%). In Uttarakhand, these soils cover an area of 3.02 M ha which is 65.3% of the TGA of the state. This region experiences subtropical and temperate climates. The temperature and moisture regimes are thermic and udic, respectively. The length of the growing period (LGP) is more than 150 days which is highly suitable for growing a wide range of fruit and other crops. The area has deciduous and coniferous thick forests. The soils are shallow to deep, well to excessively drained, sandy, loamy-skeletal, coarse-loamy and fine-loamy, and calcareous/non-calcareous. Typic Hapludalfs are also reported in the lesser Himalayas of the Kashmir Region (Kirmani et al. 2013). The soils occurring on different landforms of lesser Himalayas, across the Union territory of Jammu and Kashmir (J&K), and states of Himachal Pradesh and Uttarakhand are described hereunder.

Soils of summits and ridge tops

The soils of summits and ridges of mountains in the Indian Himalayas are neutral to slightly alkaline in reaction. These are severely eroded with low to medium available water capacity (AWC). The soils of summits and ridges are classified as Typic Udorthents and Dystric Eutrochrepts and are shallow to medium, well to excessively drained, coarse-loamy and fine-loamy.

Soils of side/reposed slopes

The soils are highly variable depending on the vegetation of the area and the geology of the parent material. The soils are shallow to deep, well to excessively drained, sandy, loamy-skeletal, coarse-loamy, and fine-loamy calcareous/non-calcareous. The soils are slightly acidic to neutral on high reaches and neutral to slightly alkaline on lower hills. The organic carbon content of the soils is high to very high. The soils have been classified as Lithic/Typic Udorthents, Typic/Dystric Eutrochrepts and Typic Hapludolls.

Soils of fluvial valleys

The soils in this zone have been classified as Typic Udifluvents, Typic Udorthents, and Dystric Eutrochrepts. These soils are medium deep to deep in depth, excessively drained, stratified, coarse-loamy, and fine-loamy, mostly non-calcareous. These soils are generally neutral to slightly alkaline in reaction. The organic carbon status of these soils is moderate to high and AWC is moderate.

6.13.19 Soils of Shiwaliks (Outer Himalayas)

Soils of Shiwalik region cover areas of south-western part of the Himalayas, comprising the Jammu region of J&K, districts of Kangra, Hamirpur, Bilaspur, Una, Sirmour, and southern parts of Solan and Chamba districts of H. P. state (1.1 M ha), and Uttarkashi, Dehradun, Tehri Garhwal, Pauri Garhwal, Chamoli, Pithoragarh, and Almora districts of Uttarakhand. The Shiwaliks predominantly consist of sedimentary rocks of Tertiary formations which comprise sandstone and shales, clays, and conglomerates (Gupta 2004; Gupta and Tripathi 1996). The climate is warm and humid to per humid. Length of the growing period ranges 210–300 days. The soils of Shiwaliks have been extensively studied by various workers (Gupta and Verma 1992;

Sidhu et al. 1998, 2000, 2008; Sidhu and Jain 1993). The denuded hills have sparse forest cover pertaining to tropical dry deciduous and tropical thorny forests and the soils of this region comprise brown forest and podzolic soils. The soils of the nearly level floodplain of the Indus river system are, by and large, calcareous, neutral to slightly alkaline, and represent Udifluvents and Haplaquepts (Sidhu et al. 1987). The dominant soil great group associations found in the area are Ustochrept—Udorthents, Ustochrept—Udorthents Haplustalfs, Hapludalfs—Entrochrept, and Hapludolls in association with glaciers. Sehgal et al. (1993) reported that the other great group associations found in pockets of Shiwalik hills are Ustochrepts—Ustorthents and Hapludalfs—Ustochrepts—Ustochrepts—Ustorthents, Ustorthents—Ustorthents with gullied land. The soils of summits and ridge tops in Shiwaliks have medium deep to deep, well to excessively drained, loamy-skeletal, severe to very severely eroded soils. The soils are slightly acidic to slightly alkaline with moderate to high organic carbon and low AWC (Lithic/Typic Udorthents). The soils on the side/reposed slopes are medium deep to deep, well to excessively drained, loamy-skeletal, coarse-loamy, and fine-loamy with moderate to severe erosion. They are slightly acidic to slightly alkaline (in some cases moderately alkaline) soils with low to moderate AWC and have been classified as Lithic/Typic Udorthents and Typic/Dystric Eutrochrepts. Soils occurring on very gentle to moderately sloping fluvial valleys are mostly medium deep to deep, well-drained, sandy, sandy-skeletal, coarse-loamy, and fine-loamy with moderate erosion. The soils are neutral to moderately alkaline with low to medium organic carbon and low to medium AWC. Sidhu and Surya (2014) found that the soils occurring on very gently to gently sloping piedmont plains are medium deep to deep, well-drained, coarse-loamy with moderate to severe erosion. The soils have been classified as Udic Ustochrepts and Typic Ustorthents.

These soils are developed in the alluvium of intermountain valleys. The clay content increases with depth whereas the surface horizon is rich in organic carbon. These are highly base saturated and agriculturally important soils under rainfed conditions. Sidhu et al. (1998) and Gupta and Chera (1996) reported that very deep, fine-loamy, slightly acidic to neutral and moderately eroded soils occur on Shiwaliks, and they are classified as Dystric Eutrochrepts and Typic Udorthents/Typic Dystrochrepts.

These soils are prone to severe to very severe erosion on steep slopes. Shallow depth in some soils, and deforestation on high slopes encourage soil erosion. Soil degradation is a problem in this area, and imperfect drainage conditions in valleys limit the choice of crops. This region is also characterized by frequent droughts, low available water capacity, and moderate to severe stoniness.

Soil degradation in northwestern Himalayas and its management Himalayan system are very fragile system due to steep landscape. Soil erosion is the main cause for soil and land degradation. The exposed land gets eroded by rains causing floods in the rivers and sedimentation in reservoirs. This has created the problem of siltation in the dams, thereby reducing their capacity and life span. For example, the annual rate of silting in the Bhakra dam, Bias unit II, and Panchet dams was assumed to be 4.29, 4.29, and 2.41 ha m km^{-2}, respectively, at the time of their construction. However, it has been found now that the observed rate is actually 6.10, 14.30, and 10.00 ha m km^{-2}, respectively (Dewan and Sharma 1985). The data show wide variation in sedimentation depending upon the topographic position, soils, and vegetation.

6.13.20 Soil Health Management in Himalayan Region

Limited availability of land for crop production has jeopardized the agriculture's ability to feed the increasing population in the mountains. Soil degradation has further aggravated the problem and raised doubts about the soil fertility and sustainability of agricultural production. Future strategies for increasing agricultural productivity will have to focus on efficient use of inputs and management of natural resources, including soil resource. A balance between the nutrient requirements of plants and the nutrient content in soils has to be achieved for getting and maintaining high yields and optimum levels of soil fertility, mitigating environmental contamination and degradation, and sustaining agricultural productivity and production on a sustainable basis. It has been observed that the soils of the Himalayan region are, generally, deficient in nutrients like nitrogen, phosphorus, calcium, magnesium, sulfur, zinc, boron, molybdenum, and iodine. To obtain optimum productivity, additional nutrient supplements have to be made through organic and inorganic sources of crop nutrients. The fertilizer used for supplying essential nutrients like N, P, and K is still low in Himalayan states of India. Besides inadequate use, fertilizer use is highly imbalanced with more emphasis on nitrogenous fertilizers.

6.13.21 Soil Erosion in Indian Himalayas

State-wise area under different soil loss classes due to water erosion in Indian Himalayan states is given in Table 6.17.

6.13.22 Micronutrient Status of Indian Himalayan Soils

The hill soils of the Indian Himalayan region vary widely in their kind and properties and are spread across the states of Jammu and Kashmir, Himachal Pradesh, Punjab, Uttarakhand, and Eastern Indian Himalayan states. The fertility of these soils suffers from various micronutrient deficiencies (Takkar and Jalali 2005).

Jammu and Kashmir

Analysis of DTPA extractable micronutrients in soils (Entisols) in the cold arid zone of Ladakh showed levels of Zn, Fe, Mn, and Cu ranging from 0.02 to 3.86, 1.56 to 14.6, 2.12 to 5.9, and 0.20 to 3.7 mg kg^{-1} soil, respectively, with 30% of soils deficient in Zn and nearly 26% deficient in Fe (Jalali et al. 2000). Subtropical, intermediate, temperate, and cold arid soils showed Zn deficiency in 22, 24, 18, and 30% of samples, respectively. In the Jammu region, 31, 26, 23, 19, 13, and 10% of samples in Jammu, Rajouri, Doda, Kathua, Udhampur, and Poonch districts, respectively, showed Zn deficiency. The

Table 6.17 State-wise area under different soil loss classes due to water erosion in Indian Himalayan states (>10 tonnes ha^{-1} yr^{-1})

State	TGA (km^2)	Moderate (10–15 t ha^{-1})		Severe (15–40 t ha^{-2})		Very severe > 40 t ha^{-2}		Total area	% of TGA
		Area km^2	Area (%)	Area km^2	Area (%)	Area km^2	Area (%)		
J&K	222,236	1400	0.6	4867	2.2	28,757	12.9	35,024	15.7
HP	55,673	3023	5.4	6208	11.2	8808	15.8	18,039	32.4
UK	55,845	4114	7.4	8688	15.5	18,267	32.7	31,069	55.6
Arunachal	83,743	4271	5.1	24,344	29.1	32,224	38.4	60,839	72.6
Assam	78,438	3592	4.6	25,814	32.9	22,198	28.3	51,604	65.8
Manipur	22,327	3,405	15.3	8493	38.0	–	–	11,898	53.3
Meghalaya	22,429	3315	14.8	8178	36.5	5980	26.6	17,472	77.9
Nagaland	16,579	678	4.1	3276	19.7	10,515	63.4	14,468	87.2
Sikkim	7096	64	0.9	639	9.0	1913	26.9	2616	36.8
Tripura	10,486	745	7.1	1416	13.5	1867	17.8	4027	38.4

Sources NBSS&LUP (2004)

available Zn content in these soils ranged from 0.1 to 6.1 mg kg^{-1} soil. Further investigations revealed that about 51% of soils in the mid-hills were also deficient in Fe as well as 24% in Zn (Jalali and Sharma 2002; Jalali and Pareek 2003). In the soils of Kashmir, available Zn was higher in the high altitude (kandi) soils (0.37–0.60 mg kg^{-1} soil) than in the valley basin (0.15–1.0 mg kg^{-1}) and karewas (Hapludalfs) soils (0.27–0.80 mg kg^{-1}). The Zn status in almost all the soils was in the deficient range and that of Cu, Fe, and Mn in the adequate range (Jalali et al. 1989). Thus, Zn deficiency is one of the major constraints in these soils for crop production. The content of available micronutrients decreased with pH and increased with soil organic matter (SOM) content. The magnitude of Zn deficiency decreased with an increase in SOM content and increased with an increase in soil $CaCO_3$ and pH (Jalali and Sharma 2002; Jalali et al. 1989).

The productivity of rice, wheat, maize, rapeseed, and mustard improved markedly with Zn application to the soils of the intermediate and subtropical zones in the Jammu region. Application of 4.4 kg ha^{-1} Zn as a top dressing gave maximum maize and rice grain yield responses of 48.5 and 17.9%, respectively; the basal Zn application lead to wheat and rapeseed yield increases of 11.8 and 7.6%. The residual effects in the paddy–wheat cropping system lasted for 2 years (Jalali et al. 1999; Sharma et al. 2003; Jalali and Pareek 2003). The soils of citrus, mango, and guava orchards contained low available (DTPA) Zn (0.26–2.48 mg kg^{-1}). Some 53% of soil samples were deficient in Zn, whereas all the soils contained adequate levels of available Cu (0.22–2.54 mg kg^{-1}), Fe (4.5–63.5 mg kg^{-1}), and Mn (1.92–36.2 mg kg^{-1}). Leaf analysis also showed Zn deficiency in 40% of mandarin samples, 56% of mangoes, and 30% of guavas. Deficiency of N, P, K, and S was also severe in orchard soils, up to 75, 45, 22, and 44% of samples, respectively; these appear to have masked the beneficial effects of Zn application to obtain optimum productivity. A strategy of correction of macronutrient deficiencies needs to be adopted while alleviating micronutrient deficiencies (Sharma et al. 2001a, b; Jalali and Pareek 2003).

Himachal Pradesh

In the wheat growing areas of Suwan and Nurpur, Chhota Bhangal, Bath and Kulu, Jahu, Janjethi, Shimla, and Ponta, 28% of soil samples and 12% of wheat plant samples were deficient in Zn. In the Palam Valley, 66% of soils and 31% of wheat samples suffered from Zn deficiency; in the remaining area, 20–25% of the soils were deficient in Zn. Seventeen percent of soil samples were deficient in Fe in the Kangra Valley; 1.3 and 20.0% were deficient in Cu in the Kangra and Bath valleys; and 20.0, 14.2, and 11.8% were deficient in Zn in the Bath, Kulu, and Kangra valleys, respectively (Tripathi et al. 1994).

In the major off-season vegetable-growing area in the Spiti valley and Kinaur district, 16–42% of soil samples were deficient in available Fe, 18–34% in Mn, and 11–65% in Zn (Parmar et al. 1999). The availability of Fe and Mn decreased significantly with an increase in soil pH and $CaCO_3$ content and that of Zn with an increase in pH. The availability of all three micronutrients increased with an increase in SOM content. The soils in the apple orchards of the Shimla Hills had 35 and 70% Zn deficiency, 30 and 94% B deficiency, and 50 and 35% Mo deficiency in surface and sub-surface soils, respectively. Thus, the nutrient status of sub-soil needs to be taken into account when ascertaining the fertility status of orchard soils. In another study, Sharma and Bhandhari (1995) observed that only available B was low in 57% of soil samples from a group of Delicious apple orchards. Leaf analysis indicated that 11, 5, and 40% of orchards were low in Zn, Cu, and B, respectively. The high deficiency of B is attributed to an acidic soil reaction coupled with the high humidity of the region. Nearly 76% of the brown hill soils of the potato-growing areas were deficient in Zn and 31% in Cu. In these soils, the response of potatoes to the addition of Zn, Cu, Fe, Mn, B, and Mo ranged from 1.0 to 7.6, 2.2 to 6.1, 1.1 to 4.0, 1.5 to 3.9, 3.2 to 4.4, and 1.0 to 3.5 t ha^{-1}, respectively. Zn efficient potato varieties have also been identified (Grewal and Trehan, 1990). In Palampur soils, wheat showed a significant response to Cu application of 107 kg ha^{-1}, and paddy to Zn and Fe of 456–764 kg ha^{-1}. In Kangra, Palampur, and Kulu districts, a significant response (242–276 kg ha^{-1}) of maize to Zn application was observed.

Uttarakhand

Zinc deficiency is a serious soil fertility constraint in the foothill Terai soils of Uttarakhand; field-scale deficiency of Zn in rice was recorded for the first time in these soils. A long-term fertilizer experiment (LTFE) with an intensive rice–wheat–cowpea cropping system on these Mollisol soils depleted the available Zn from adequate (2.77 mg kg^{-1}) to deficient level (<1.0 mg kg^{-1}) after 12 years. However, the application of 11 kg Zn ha^{-1} restored the soil Zn fertility status and improved the productivity of the system by 0.48 t ha^{-1}. Application of 15 t FYM ha^{-1}, annually, also maintained the Zn level in the adequate range and improved the productivity by 0.44 t ha^{-1} (Nand Ram 1998). Thus, the strategy to reduce the Zn deficiency problem on such Mollisols is to apply either Zn fertilizer or FYM. Clear Zn deficiency symptoms have been observed in maize, pulses, and citrus fruits in these areas. The available Zn content and distribution in six soil series of Nainital Terai soils representing poor, imperfect, and moderately well-drained soils were deficient, ranging from 0.08 to 0.60 mg kg^{-1} soil. Available Zn decreased with soil depth and was mainly regulated by the SOM content (Vittal and Gangwar 1974; Bhardhawaj and Prasad 1981). The micronutrients Zn, Cu, Fe, Mn, Mo, and B were deficient in 74, 20, 17.3, 20, 18.6, and 24% of soil samples, respectively, from the three hill districts of Almora, Pithoragarh, and Chamoli (Rawat and Mathpal 1981). The availability of these micronutrients decreased significantly with an increase in soil pH and $CaCO_3$, and increased with an increase in organic C content. The soils of the Bageshwar and Garur valleys of the Kumaon Division and Ranichauri area of Garhwal Division of UP contained 2.6 to 30.8 mg kg^{-1} of available Fe (NH_4OAc extractable) and were thus considered well supplied with Fe, taking 2.0 mg Fe kg^{-1} as the critical limit (Kumar et al. 1981).

A significant response of rice to the application of 5 kg Zn ha^{-1} was found in a rice–wheat system on Zn deficient deep alluvial soil in Doon Valley of UP, and of wheat grain to a residual level of 10 kg Zn ha^{-1}. Considering the net return and Zn use efficiency, Bhardhawaj and Prasad (1981) recommended 7 kg Zn ha^{-1} for rice–wheat rotation to achieve optimum yield and return. In acid soil at Ranichauri, Tehri Garwal, potatoes responded to the application of B (Dwivedi and Dwivedi 1992).

Meghalaya

Boron deficiency is a commonly observed micronutrient disorder in the Alfisols or acid soils in humid regions. With acute B stress in these soils, the application of B decided the success or failure of crops (Dwivedi et al. 1990b). Dwivedi et al. (1993) determined available B using six extractants. Of these, only four, mannitol-CaC12, hot-water soluble (HWS), HWS boiling reflex, and DTPA-NH_4HCO_3 were found promising with respective critical limits of 1.0, 0.7, 0.7, and 0.6 mg kg^{-1}. Soil testing below these critical limits responded significantly to B fertilization. About 35% of rice-growing acid soils were deficient in B and 17% in Zn. However, the content of available Mn, Fe, Cu, Co, and Mo was in the adequate range (Nongkynrih et al. 1996). Burning under bun cultivation led to a decrease in available Fe, Zn, and Cu, and increased Mn content four-fold (Venkatesh et al. 2003). Total micronutrients also increased on burning. However, the level of available and total micronutrients after 3 years of bun cultivation was almost the same as the initial status. The highest contents of available Fe, Zn, and Cu and lowest of Mn were observed in valley land soils. Submergence caused a gradual decrease in all the forms of Zn. Thus, the bulk of the applied Zn transformed into strongly bound forms and only a small portion remained available to plants. Plants got Zn largely from the Mn oxide fraction. The transformation of applied Cu in these soils was similar (Sharma 1987; Singh et al. 1999a, b, 1999a).

The direct and residual effects of liming (1.5, 3.0, and 6t ha^{-1}) and of B (1.5, 3.0, and 6.0 kg ha^{-1}) on the yield of pea and corn grown in a sequence were studied in a field experiment on a sandy loam acid Alfisol at Barapani, Meghalaya. The concentration of B in plant parts increased appreciably with increasing rates of B application but decreased with increasing rates of lime. Application of B significantly increased the hot-water soluble B from 0.24 to 0.89 mg kg^{-1} soil, but increasing rates of lime application significantly decreased B from 0.66 to 0.41 mg kg^{-1} (Dwivedi et al. 1990b, 1992, 1993). Dwivedi et al. (1992) reported that direct application of 1.5 kg B ha^{-1} significantly increased the pod and stover yield of peas. Liming accentuated B deficiency in the absence of applied B because of fixation of soluble B on Al and Fe oxides as a result of an increase in pH, but lime cured the toxic effect arising from a high rate of B application.

Rice responded to Zn and Cu significantly; application of 5 mg Zn kg^{-1} and 1.25 mg Cu kg^{-1} soil gave the optimum yield. The dry-matter rice yield, and Zn and Cu concentration in the plants were significantly correlated with initial DTPA extractable Zn and Cu. Soil testing below the critical level of 1.2 mg Zn kg^{-1} and 0.7 mg Cu kg^{-1}, and plant testing below 35.9 mg Zn kg^{-1} and 7.0 mg Cu kg^{-1} needs application of Zn and/or Cu to obtain optimum rice yield (Dwivedi et al. 1990a; Singh et al. 1999a, b; Singh and Nongkynrih 2000).

Tripura

Datta and Gupta (1984), Datta and Ram (1993) reported that the fertility status with respect to available Zn, Cu, Fe, and Mn in Tripura was higher in lowland than in upland soils, whereas B and Mo contents were higher in upland than in lowland soils. However, all the micronutrient cations were in the adequate range. Nearly 29% of soils were deficient in B and 50% in Mo. As soil pH increased, available Zn, Cu, and B content decreased significantly in upland soils. The available Cu and Fe content increased with an increase in organic C, while that of Zn and Mo decreased.

Assam

Micronutrient status was evaluated in seven soil profiles from old alluvium in Nowgong, new alluvium in Tejpur, forest soil in Guwahati, hill soils in Diphu and Nayabunglow, tea garden soil in Jorhat, and bheel soil in Cachar. Available B content in these profiles varied from 0.02 to 1.43 mg kg^{-1}, and all the surface soils contained sufficient B. The available Mo varied from 0.015 to 0.053 mg kg^{-1} soil and was in the range of deficiency to marginal fertility level. Available Cu ranged from 0.87 to 2.50 mg kg^{-1}. Cu content was the lowest and organic C the highest in bheel soils. The forest, hill, and bheel soils were rated low in available Cu, the result of the high intensity of weathering and leaching in these acid soils (Chakraborty et al. 1979, 1980, 1981). Available Cu increased significantly with an increase in pH ($r = 0.524^*$). Available B was less in high than in low rainfall areas of lateritic soils (Borkakati and Takkar 2000). Toria (rapeseed, *Brassica campestris*) did not respond to the application of B and Mo at the Regional Agricultural Research Station, Shillongani, and Nagaon, Assam. But their application significantly increased the oil content (Patgiri 1995).

Manipur

The fertility status of Zn, Cu, Fe, and Mn in seven profiles from Manipur indicated inadequate or marginally adequate levels of available Zn only (Sen et al. 1997).

Sikkim

The available micronutrient status in 44 soil samples (pH 4.2–5.9, organic matter 2.0–8.4%) from the eastern, western, and southern districts revealed adequate Zn, Fe, and Mn levels; their content ranged from 0.47 to 3.0, 5 to 210, and 0.05 to 6.2 mg kg^{-1}, respectively. The available Cu content ranged from 0.05 to 0.5 mg kg^{-1}; about 33% of the soils were Cu deficient (Khera and Pradhan, 1980). Total and available Cu and Mn and available B content decreased, and that of available Zn and Fe in soil increased markedly with increase in altitude. About 85 and 94% of soils were deficient in B and Mo, respectively. These soils were sufficient in Zn, Cu, Fe, and Mn (Yashoda and Avasthe 1995).

6.13.23 Iron Toxicity

Acid Alfisols are widely distributed in the states of Himachal Pradesh, Uttarakhand, Jammu and Kashmir, Meghalaya, and in the narrow valleys of Tripura. These soils generally exhibit widespread occurrence of bronzing disease in rice grown under wetland conditions. The visual symptoms of the disease are brown spots spread over almost the entire leaf surface, which later curls and dries. Under greenhouse conditions, the disease has been shown to result from an excessive concentration of Fe in diseased plants, 805 mg kg^{-1} compared to 480 mg kg^{-1} in the normal plants (Verma and Tripathi 1984). Submergence is mainly responsible for this disorder. The incorporation of 3t lime ha^{-1} delayed the bronzing symptoms and sustained the yield in HP. Singh et al. (1992a, b) reported that in Tripura and the east Khasi hill soils of the Meghalayas, rice bronzing disease is much more prevalent in soils with shallow water tables (<40 cm), high in organic C (>2.99%) and low in pH (<5.5). Severe Fe toxicity symptoms in plants were closely associated with a 3.1-fold level of Fe in diseased compared to normal plants and high active Fe (>1.2%) and organic C (2.99%) content, continuous submergence, and impeded soil drainage conditions (Singh et al. 1992a, b). Iron toxicity was responsible for the low yields of rice in these soils (Prasad and Ram 1985). In Assam, application of As-benomyl, a systemic fungicide, alone and with FYM, significantly retarded the decrease in redox potential which appreciably decreased available Fe and Mn and their total uptake in rice. As a result, Fe toxicity decreased and grain and straw yield of rice markedly increased (Bhattacharya et al. 1996).

6.13.24 Fertility Status of Mid-Himalayan Region

The soils of mid-Himalayan region are predominantly neutral in soil reaction with low soluble salts and high in organic carbon content (Annepu et al. 2017). Soil fertility assessment of crop fields showed the soils of the region are high in available K and P contents and relatively medium to low in available N content. Among the micronutrients tested, widespread deficiencies were recorded for available Cu, B, Zn, and Fe. The correlation studies between the soil properties and available nutrients showed that OC has a significant and positive correlation with the availability of N, P, S, B, and Mn indicating the significant role of OC in the maintenance of balanced soil health (Annepu et al. 2017).

The mid-hills zone of Himachal Pradesh extends from 651 to 1800 m amsl with a mild temperate climate, and it occupies 32% of the total geographical area and about 37% of the cultivated area of the state (Parmar 2014). Limited area under the fertile soils in the mountain landscapes is assuming extra significance because of the decreasing land availability for agriculture caused by the demand for alternate uses. Under these conditions, the high cropping intensity and unscrupulous use of chemical fertilizers adopted by the farmers to reach higher productivity levels have ultimately rendered the soils in this region with depleted nutritional status (Sharma 1988; Sharma et al. 2001a, b). Lack of understanding about soil health is leading to indiscriminate or imbalanced use of chemical fertilizers and exposing farm-based livelihoods to soil health-related risks. The soils of Himachal Pradesh in the western Himalayas have undergone depleting N from the system at an alarming rate. The situation of P and K seems to be slightly satisfactory with positive balance. However, information available on those aspects in the mid-hill conditions of Himachal Pradesh is highly sporadic and inconclusive. In such situations, assessment of site-specific soil fertility status and study of their relation with other physico-chemical properties would be of high significance. Muhr et al. (2017) delineated the soils having a pH < 5.5, 5.5–6.0, 6.0–6.5, 6.5–7.7, 7.5–8.5, and >8.5 as strongly acidic, moderately acidic, slightly acidic, neutral, moderately alkaline, and strongly alkaline, respectively; EC (dSm-1) < 1.0, 1.0–2.0, 2.0–3.0, and >3.0 as no effect on crop, critical for germination, critical for crops, and injurious to most crops, respectively. They classified the soil carbon content from 0 to 0.5%, 0.5 to 1.0%, and >1.0% into low, medium, and high categories. Arora (2002) classified available N ((kg ha^{-1}) as < 280, 280–560, and > 560; available P <10, 10–25, and >25 (kg ha^{-1}); and available K, < 118, 118–280, and > 280 (kg ha^{-1}) into low medium and high categories. This has been cited in Annepu et al. (2017).

6.13.25 Physico-Chemical Properties

The soil varies from dark gray to dark grayish brown in color, loamy to sandy loamy in texture, and neutral to slightly acidic in reaction. The pH values range from 6.59 to 7.81 with a mean pH value of 7.36 (Annepu et al. 2017). Higher rainfall and lower temperatures associated with increasing altitude result in lower pH (Katherine et al. 1985). EC of soil varies from 0.049 dS m-1 to 0.793 dS m^{-1} with a mean value of 0.426 dS m^{-1}. The EC values under the normal range (<1.0 dS m^{-1}) may be ascribed to the leaching of salts to lower horizons of soil due to its light texture. The soils of the study area are predominantly high in organic carbon (OC) which ranges from 0.15 to 1.98% (Annepu et al. 2017). It has been found that continuous low temperatures cause suppression of microbial and enzymatic activities, which results in low soil organic matter decomposition and its accumulation in surface soils (Bhattacharyya et al. 2008). Higher levels of OC were found in Central Himalayan Region, which ranged 0.80–2.30% (Khera et al. 2001). Maintenance of high cattle population per unit of cultivable land in mid-hills zone also upkeeps much organic matter in the surface soils (Bala et al. 2004).

6.13.26 Status of Macronutrients

The soils of the area are rich in organic carbon and N content. The available P content of mid-Himalayan soils varied from 7 to 9 kg ha^{-1} in soils with low P content and 10.06 to 25 kg ha^{-1} in soils with medium P content, and 26 to 45 kg ha^{-1} in soils with high available P content. Only about 2.8% of soils showed a deficiency of P. In the total study area, 86% of soils were in a medium category with respect to the availability of K and only 6.8% of soils were under the low range. Prevalence of clay minerals like illite and kaolinite in hills soils attributed to the sufficient levels of K. Available “S” content varied from 14.0 to 32.0 kg ha^{-1} with an average of 23.77 kg ha^{-1} (Table 6.18).

Table 6.18 Salient soil properties of study area

Soil characteristics	Range	Mean	Soil characteristics	Range	Mean
pH	6.59–7.81	7.36	Av. S (Kg ha^{-1})	14.0–32.0	23.7
EC (dS m-1)	0.049–0.793	0.426	Av. Zn (ppm)	0.29–5.07	1.25
Organic carbon (%)	0.15–1.98	1.09	Av. B (ppm)	0.21–0.98	0.54
Av. N (kg ha^{-1})	201.2–603.4	312.9	Av. Fe (ppm)	2.14–15.69	7.29
Av. P (kg ha^{-1})	7.0–45.0	21.4	Av. Mn (ppm)	0.45–8.87	2.36
Av. K (kg ha^{-1})	70.5–448.0	193.8	Av. Cu (ppm)	0.10–5.91	0.9

6.13.27 Sediment Yield from Indian Himalayas

Sharma and Sharma (2009) conducted a field investigation, involving seven land use systems such as livestock-based, forestry, agroforestry, agriculture, horticultural crops (fruit trees and vegetables), agri–horti–silvi-pastoral system and shifting cultivation to study the interactional effects of plant–livestock–soil–hydrology on soil and nutrient erosion, nutrient build-up, and crop productivity in Eastern Himalayas. The results of the study revealed that shifting cultivation on steep slopes encouraged soil erosion as, on an average, 36.2 tha^{-2} of sediment and 66.3% of in situ rainwater retention, in comparison to 0.16–0.82 t ha^{-2} of sediment yield and more than 90% of rainwater in situ rainwater retention in new land use systems. Poor land cover was mainly responsible for more soil loss in shifting cultivation. An appropriate soil surface vegetation cover can reduce the speed of the gushing water down the slope, thereby reducing the runoff and the runoff load, resulting in a significant reduction in the soil and nutrient erosion. It was observed from the study that the average sediment yield from the livestock-based, forestry, agroforestry, agriculture, agri–horti–silvi-pastoral and horticulture farming systems was only 0.44%, 2.68%, 1.47%, 0.31%, 0.73%, and 2.27%, respectively, of that of sediment load from shifting cultivation. The sediment yield varied according to the rainfall received and its intensity during a particular year, slope of the cultivated land, erodibility class, and the nature of vegetation in particular land use. The farming system with livestock was economically more sound than other farming systems. There was sufficient fertility build-up in the watersheds where animals were kept and after ten years of start of the study, there was no need for inorganic sources of nutrients as the dung and urine of animals kept in the system were sufficient to supply the necessary nutrients to the crop. The organic carbon, humus, exchangeable calcium, magnesium, potassium, and available phosphorus increased significantly in the new land use systems as compared to shifting cultivation, however, the Diethylene Triamine Penta Acetic acid (DTPA) extractable zinc, copper, iron, and exchangeable aluminum decreased over the initial soil status due to their more uptake by the crops because of higher biomass production in new land use systems. It showed that the vegetation in the new land uses had a smothering effect on the soil, and an increase in the nutrient elements with the application of organic livestock manure had a favorable impact on crop productivity.

Nitrogen, potassium, and phosphorus leaching losses in hill Alfisols, beyond 60 cm soil depth, varied depending on the amount of rainfall, cropland use, nutrients (native + applied), and soil texture (Sharma 1989a, b, 1990a, b, c, d, 1992a, b, c, 1999, 2018). No leaching of P was found when clay content exceeded 9% in the soil even at annual rainfall of over 2700 mm. Nitrogen was the maximum leached nutrient followed by K. The leaching of P in soils beyond 60 cm depth is negligible unless these are very light in texture.

Balanced use of nutrients is very important to get an optimum yield of crops. For this, the nutrient status of the soil before seeding/planting and of specific plant parts during crop growth is necessary. For judicious application and utilization of applied nutrients in Eastern Himalayas, methodologies were standardized, namely release pattern of nitrate–N from nitrogenous fertilizers under field conditions (Sharma 1991a), simple mathematical model to determine critical nutrient levels in soils and plants (Sharma 1991b), Diagnostic and Integrated Recommendation System for nitrogen, phosphorus and potassium foliar diagnostic norms (Sharma 1991c, d), use of boundary lines in establishing diagnostic norms (Sharma 1991d), and use of half-life technique to diagnose nitrate–N status in the plant petioles (Sharma 1996). Sharma et al. (2014) studied soil fertility, erosion, runoff, and crop productivity as affected by different farming systems and found that land use has a significant effect on the build-up of soil fertility, amount of runoff, and resultant soil erosion. Suitable land cover is important for soil health and preventing land degradation in mountain areas.

6.13.28 Soil Development in Indian Himalayas

Diasang, Barail, Surma, Tipam and Duptila series represent the rock formations in Nagaland, Manipur, Mizoram, and Tripura. In the Naga hills, the Eocene beds are the Disang shales—a great thickness of very well-bedded darl gray shales with thin well-cemented sandstones (Patiram and Datta 2006). Near Myanmar, the shales have become hardened and slaty and are associated with quartz veins and serpentine. Tipam series is widespread in Tripura, Nagaland, and Arunachal Pradesh. The rock type includes grits, sandstones, conglomerates, and clay with plant remains. The Tipam formation of Tripura consists of medium to medium to coarse-grained straw yellow color sandstone. The hill ranges of Tripura are formed from the rocks belonging to Surma and Tipam series. In the northeastern part of Assam, the Tipam series is entirely non-marine, consisting of ferruginous sandstones, mottled clay, and mottled sand.

The rocks formed under shelf faces occur in North Cachar, Jaintia hills, and along the southern slope of the Shillong plateau. They are known as Jaintia series and comprise sandstone, calcareous sandstones, and limestone. In Mizoram, the sedimentary rocks of Barail series consist of shales, locally enclosing bands of weathered micaceous, feldspathic sandstone, relatively soft medium-grained sandstone interbedded with siltstone. The Shillong series, including large parts of Garo, Khasi, and Jaintia hills, is a widely developed formation of an Archean gneissic complex comprising of a thick series of quartzites, slates, and schist masses of granitic intrusion and basic interbedded traps (Patiram and Datta 2006). The Shillong series is, for the greater part of its extent, overlain by horizontally bedded cretaceous sandstones. The Surma series has a wide extent in Naga hills, North Cachar Hills, and Surma valley of Assam. It is composed of sandstones, sandy shales, mudstones, and thin conglomerates, generally free from carbonaceous content. In Mizoram, the Surma series consists of hard, compact, medium-grained sandstones, shale, siltstones, and ferruginous current bedded sandstones. The oldest rocks found in the eastern part of Manipur belong to the cretaceous group and generally include variegated shale, slates, and siltstone with some amount of sandstone and quartzite. Limestone of cretaceous age occurs in various places around Ukhrul. The Barail series occurs over large parts of Senapati, Churachandpur, and Tamenglong districts of Manipur.

The metasedimentary sequences, developed extensively in Arunachal Pradesh, have been assigned various nomenclature, e.g. Potin, Bomdila, and Tenga which as a group could be correlated with the Daling–Buxa sequence and its equivalent described from Sikkim–Darjeeling–Bhutan. The dominant members are low-grade schist of the Daling status and the ortho quartzite–carbonate association equivalent to the Buxa. The rocks around Bomdilla consist of gneisses, mica schist, thin calesillicated, marbles, graphitic schist, metaphases, and quartzite. Most of the rocks of Subansiri comprise Precambrian schist, magmatites, and younger granites intensive of Se la group of Kameng at some places. Rocks of Siang district are mostly either carbonized or semi-carbonized. Lignite to peat materials is recognized in various sections across tertiary formations of the districts of West Siang and East Siang. The Lohit plain of the northeastern part has been chacterized as schistone metasediments, chloritic quartz, mica schists, carbonate rocks, lime rocks, etc.

The Daling series consisting of slates, chloritic and serisitic phyllites with hornblende-schists, and quartzites extend along the Teesta valley into Sikkim and then to Bhutan. Some loads of copper are associated with rocks, which are commercially workable in Sikkim. The Daling series of this region includes overlaying group schists copiously injected as with gneiss of Darjeeling series. The Daling and Darjeeling series of Sikkim are highly metamorphosed slates, magmatics, granites, and schists, containing such metamorphic minerals as garnet, sillimanite, and staurolite. The lower Gondwana in West Sikkim is associated with thick coal seams, underlying the dolomites and phyllites of Daling series.

6.13.29 Distribution of Different Soils in Eastern Indian Himalayas

The systematic soil survey for proper land use planning is imperative to provide the basic information on soil characterization, classification, and extent of distribution, and to highlight the problems and potential of soils. Knowledge of land resources and their potential is an essential prerequisite for the planning of optimum land use (Kassem et al. 1991). The steep slopes of the Himalayas encourage soil erosion and, to a large extent, result in a decrease in soil fertility. The soils of the region are quite variable (Table 6.19). It has been reported that there is limited downward migration of Fe and Al along with humus and absence of spodic horizon indicating the juvenile stage of podzolization.

Soil properties

The darkness of the soil color and coarseness in texture have the tendency to increase with the rise in altitude. The soils are acidic as a result of the humid climate and heavy rainfall. The soil organic matter content, CEC, and exchange acidity have the tendency to increase with altitude while pH has the reverse tendency (Gangopadhyay et al. 1990). Excepting at foothills and near the rivers and streams, which are colluvial and alluvial in origin, the soils are formed in situ. On hill terrain, the soils are light colored, highly leached, poor in bases, and low in CEC with low active clays. Though the soils are quite rich in organic carbon, there is a lot of difference in C content between surface and sub-surface soils. The surface soil contains more OC but it decreased with increasing depth. The soils in the valley and thick forest are rich in OM and well drained. Most of these acidic soils possess a high amount of exchangeable Al3 + throughout the soil column with an almost increasing trend, thus indicating that they still have preserved some weatherable minerals which liberate Al ions from the edges of clay minerals through weathering processes. The occurrence of a high amount of potential acidity would have resulted from a combined effect of OM, free sesquixide, and terminal hydroxyle groups associated with Si and/or Al, and/or the ruptured Si–O–Al bonds of

Table 6.19 Distribution of different soils in Eastern Himalayas

State	Order, Area (000' ha), % of TGA	Suborder area (000' ha), % of TGA	Local name	State	Order, Area (000' ha), % of TGA	Suborder Area (000' ha), % of TGA	Local name
Arunachal Pradesh				**Assam**			
	Alfisols 22.8 (0.27)	Udalfs 22.8 (0.27	Red and yellow soils		Alfisols 886.9 (12.3)	Udalfs 886.9 (12.3)	Red soils
	Entisols 2985.5 (33.64)	Aquents 69.9 (0.83)	Alluvial soil		Inceptisols 3245.3 (45.0)	Aquepts 898.4 (12.5)	Brahmaputra alluvial soil
		Psamments 84.1 (1.00)	Sandy soil			Ochrepts 2346.9 (32.5)	Old alluvial soil
		Fluvents 31.4 (0.37)	Recent alluvial soil		Entisols 2640.1 (36.6)	Aquepts 2107.0 (29.2)	Recent alluvial soil
		Orthents 2800 (33.44)	Gravelly hill soil			Psamments 59.1 (0.8)	Sandy soil

(continued)

Table 6.19 (continued)

State	Order, Area (000' ha), % of TGA	Suborder area (000' ha), % of TGA	Local name	State	Order, Area (000' ha), % of TGA	Suborder Area (000' ha), % of TGA	Local name
	Inceptisols 3126.4 (37.33)	Aquepts 62.4 (0.74)	Lohit alluvial soil			Fluvents 418.0 (5.8)	Recent alluvial soil
		Ochrepts 1717.9 (20.5)				Orthents 56.0 (0.8)	
		Umbrepts 1346.1 (16.1)	Brown forest soil		Ultisols 436.5 (6.1)	Udults 178.0 (2.5)	Red soil
	Ultisols 1187.6 (14.18)	Humults 989.3 (11.8)	Red soil/brown soil			Humults 258.5 (3.6)	
		Udults 198.3 (2.27)	Red soil	**Meghalaya**			
	Snow-covered mountains				Alfisols 80.7 (3.60)	Udalfs 80.7 (3.60)	
Manipur					Entisols 240.1 (10.71)	Fluvents 240.1 (10.71)	
	Alfisols 3.8 (0.2)	Udalfs 3.8 (0.2)			Inceptisols 1028.8 (45.65)	Aquepts 241.9 (9.58)	
	Entisols 515.6 (23.1)	Orthents 515.6 (23.1)				Ochrepts 654.1 (29.17)	
	Inceptisols 858.3 (38.4)	Aquepts 203.7 (9.1)				Umbrepts 154.8 (6.90)	
		Ochrepts 654.3 (29.3)			Ultisols 898.3 (40.04)	Humults 564.5 (25.17)	
	Ultisols 811.0 (36.4)	Udults 437.0 (19.6)				Udults 333.8 (14.87)	
		Humults 374.0 (16.8)		**Mizoram**			
Sikkim					Alfisols 82.4 (3.89)	Udalfs 82.4 (3.89)	
	Inceptisols 238.9 (42.84)	Ochrepts 56.09 (10.06)			Entisol 444.6 (21.07)	Orthents 370.1 (17.55)	

(continued)

Table 6.19 (continued)

State	Order, Area (000' ha), % of TGA	Suborder area (000' ha), % of TGA	Local name	State	Order, Area (000' ha), % of TGA	Suborder Area (000' ha), % of TGA	Local name
		Umbrepts 182.8 (32.78)				Aquents 63.6 (3.01)	
	Entisols 237.1 (42.52)	Arents 3.53 (0.63)				Fluvents 10.9 (0.51)	
	Mollisols 81.7 (14.64)	Udolls 81.7 (14.64)			Inceptisols 907.3 (43.06)	Ochrepts 907.3 (43.06)	
Tripura					Ultisols 553.6 (26.25)	Udults 399.9 (18.96)	
	Entisols 85.3 (8.1)	Orthents 84.2 (7.99)				Humults 153.7 (7.29)	
		Aquents 0.9 (0.08)		**Nagaland**			
		Psamments 0.2 (0.03)			Alfisols 47.8 (2.88)	Udalfs 47.8 (2.88)	
	Alfisols 47.1 (4.5)	Udalfs 47.1 (4.5)			Entisols 121.3 (7.32)	Fluvents 60.4 (3.65)	
	Ultisols 69.1 (6.6)	Udults 69.1 (6.6)				Orthents 60.9 (3.67)	
	Inceptisols 846 (80.6)	Ochrepts 649.2 (61.8)			Inceptisols 1094.7 (66.03)	Ochrepts 850.9 (51.32)	
		Umbrepts 63.8 (6.1)				Umbrepts 243.8 (14.71)	
		Aquepts 132.9 (12.7)			Ultisols 394.1 (23.77)	Humults 3.1 (0.18)	
	Histosols 2.4 (0.2)	Fibrists 2.4 (0.2)				Udults 391.0 (23.59)	

low active clays (Sen et al. 1997). The soils developed in Bomdila group (granite and gneisses) of rocks are rich in clay content as compared to others (Nayak and Srivastva 1995).

6.13.30 Soil Physical Behavior Under Different Land Uses in Eastern Himalayas

The northeastern region of India is characterized by heavy soil erosion, loss of soil fertility, and deforestation, causing acute environmental degradation and severe ecological imbalance (Sachchidananda 1989). Soil physical properties are a matter of great concern in the NE region, particularly at the moment of severe environmental degradation as soil texture, structure, air, water, etc. are becoming constraints to achieving the desired level of productivity. With changing requirements of high population pressure on land, *jhum* cultivation became very devastating in nature causing a drastic decline in crop yield, loss of forest wealth, biodiversity, and soil fertility degradation (Mishra and Saha 2006).

In Eastern Indian Himalayas, the average annual loss of topsoil, organic carbon, P_2O_5, and K_2O with runoff from shifting cultivation land was to the extent of 40,900, 703, 0.15, and 7.5 kg ha^{-1}, respectively. Land degradation is a very serious problem in the Himalayas and it can take many forms and shapes. The consequences or impact of land degradation depends on the nature of land degradation. It is usually associated with the deterioration of soil productivity. If productivity is to be maintained in any agricultural system, a satisfactory physical condition of the soil must be maintained. Fertilizer alone, even in conjunction with improved crop varieties and measures to control the pest and diseases, will not preserve soil productivity if significant deterioration of physical conditions occurs. The soil physical degradation takes into account the loss of finer particles, organic carbon, disintegration of soil aggregates, compaction, crusting, distortion in permeability, pore geometry/pore continuity, etc. This leads to a decrease in water storage capacity, reduces plant availability and movements of nutrients and water within the system, and as a whole decrease the soil productivity.

6.13.31 Soil Physical Behavior Under Traditional Farming Systems

Traditional Farming Systems

Traditional farming systems of this region have not changed much since their inception. Shifting cultivation, Bun system, Panikheti, tree-based farming, Apatani, and Zabo are location-wise popular systems among the farmers. These systems, based on low inputs, are of subsistence types. The degree of severity of soil erosion, water runoff, and soil productivity degradation under different systems are very high. In all the systems, farmers abandon the lands when crop yields are too low either because of the prevalence of pests, disease, or because of deterioration in soil quality.

In general, shifting cultivation practices deteriorate soil fertility due to huge soil loss of about 2–200 t ha^{-1} yr^{-1} (Singh and Singh 1978). This leads to deterioration in the soil physical condition. A minimum period of 10–15 years is very much essential to maintain soil fertility for sustainable crop production (Singh et al. 2003; Jha et al. 1979). Mishra and Saha (2004a) reported that only fallow periods are not enough to consider the sustainability of shifting cultivation. A proper ratio of cropping and fallow should be considered for its sustainability. They reported that the bulk density was maximum 1.20 Mg m^{-3} in phase IV (1 year cultivation:1 year fallow). On the other hand, the corresponding values were restored gradually in the soil of phase III (1 year cultivation:3 years fallow), II (1 year cultivation:5 year fallow), and I (1 year cultivation:10 years fallow) due to the recycling of more biomass under the wider ratio of cropping and fallow periods. Mean weight diameter (MWD) also decreased from 3.75 mm in reserve forest to 1.11 mm in phase IV soil and the micro- and macroaggregates ranged from 3.62 to 14.67

Table 6.20 Effect of land use factors on soil physico-chemical properties

Soil properties	Reserve forest	Land use		Factor (L)*	
		Phase I (>10)	Phase II (5–10)	Phase III (2–5)	Phase IV (<2)
Sand (%)	42.50	44.75	46.90	51.65	61.15
Silt (%)	33.45	34.10	33.50	31.15	23.50
Clay (%)	24.05	21.15	19.60	16.70	15.35
Bulk density (Mg m^{-3})	0.98	1.15	1.11	1.13	1.20
pH	4.55	4.85	4.80	5.10	5.00
Organic C (%)	3.32	2.33	2.09	1.42	1.18
Available N (kg ha^{-1})	478.50	447.00	425.00	342.50	300.00
Available P (kg ha^{-1})	11.85	9.80	8.30	6.00	5.50
Available K (kg ha^{-1})	301.50	283.50	259.00	184.5	171.50
MWD (mm)	3.75	3.24	2.85	1.52	1.11
Dispersion ratio	0.112	0.172	0.182	0.244	0.282
Macroaggregates (>0.25 mm)	74.24	65.65	61.44	48.38	36.26
Microaggregates (<0.25 mm)	3.62	5.25	7.62	10.43	14.67

* L = (C + F)/C, where L = Land use factor, C = Cropping period, and F = Fallow period, in years
Source Mishra and Saha (2006)

and 36.26 to 74.24%, respectively (Table 6.20). The dispersion ratio of soil in phases IV and III (land use factor of < 5) was relatively high, 0.282 and 0.244, respectively, against reserve forest (0.112).

Therefore, a minimum of 1:4 years of cropping and fallow periods could be a suitable cycle under *jhum* cultivation for the resilience of soil in this agro-climatic region. Jha et al. (1979) observed that the burning under *jhum* land use system of Tripura depleted the organic matter status, which led to low field and water-holding capacity. The effect of shifting cultivation on changes in the physico-chemical properties of the soils was evaluated in Karbi Anglong district of Assam (Singh et al. 2003). Results indicated that C and N in the soil may be among the most limiting factors for plant growth after a forest is cut and burnt. The highest percentage of organic carbon (1.92%) and total nitrogen (0.28%) were found in pre-burning stages of vegetation, and it significantly decreased in early fallow periods due to the rejuvenation of vegetation. Lowest and highest soil bulk densities were recorded in the land after burning and first year fallow, respectively. The negative effect of shifting cultivation on bulk density and water-holding capacity was also recorded in Manipur mid-hill soil by Kailash Kumar (1995). Mishra and Saha (2006) studied some physical properties of soil and reported that silt and clay content 1995of soil increased after the burning of forest vegetation, during cropping fallow period, and that of sand decreased significantly. The bulk density was highest (2.10 Mg m^{-3}) during the first year of fallow after cropping period and decreased subsequently with an increase in the fallow period.

Bun system

During the rainy season, the surface soil moisture status exceeds the level of field capacity (Fig. 6.14) in most parts of Meghalaya.

The vegetable crops suffer from excess soil moisture and are affected by disease and pests. It was observed that the bun system maintained the aerobic condition in the root zone area, which helps in plant growth and development particularly during the rainy season, when the moisture

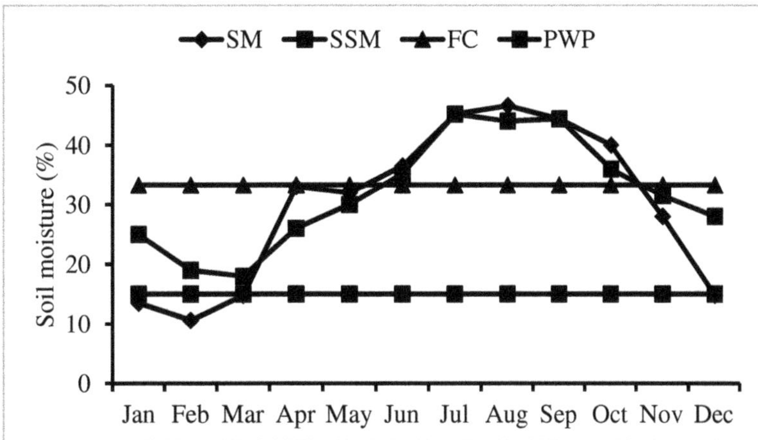

Fig. 6.14 Profile moisture distribution pattern in soils of Meghalaya (Barapani). Modified from Mishra and Saha (2006) (SM, Surface soil moisture; SSM, Sub-surface soil moisture; C, Field capacity; PWP, Permanent wilting point)

content exceeded the soil field capacity due to improvement in hydro soil physical behavior. The bulk density of soil decreased (Mishra 2005, Mishra and Sharma 2001) by 13.9% and MWD as soil structural index increased by 67.8% under bun over the flat method of cultivation.

6.13.32 Modified Farming Systems

Adoption of modified land use systems like Agri–horti–silvi-pastoral significantly increased the MWD (29.44%) and decreased in dispersion ratio (52.89%) over the shifting cultivation. Transmission pore (>50 μm) volumes increased under different land use systems in the order of forestry > natural fallow > livestock > agri–horti–silvi-pastoral 1 > agriculture > shifting cultivation. The available water content increased by 24.02–36.51% under the modified system over the shifting cultivation and decreased by 1.07–10.12% over the natural fallow system. The periodical soil moisture regime indicated that the moisture content below the permanent wilting point was started from October to March in shifting cultivation, whereas in modified systems like agri–horti–silvi-pastoral, agriculture, livestock-based land use system, it started from the month of November to February. The moisture contents under forestry and the natural fallow system never crossed the permanent wilting point as they might have high organic matter and better soil physical properties (Anonymous 2005).

The measurements of a few selected physical parameters may not serve the purpose of physical characterization of soils with respect to crop performance. The basic problem is that soil physical properties vary widely, and they are independent. It is therefore difficult to establish critical limits of physical properties in isolation and in relation to crop growth under different systems. Letey (1985) proposed a single value soil physical index, called the non-limiting water range (NLWR). The volumetric moisture content corresponding to 10% air-filled porosity (upper limit of NLWR) and 2 MPa SPR (lower limit of NLWR) and NLWR for different land use systems indicated that the soil moisture content corresponding to 10% air-filled porosity increased, while the moisture content corresponding to 2 MPa SPR decreased in the order of natural forest, agroforestry, fodder, and food agriculture and shifting cultivation. The soil water content associated with 2 MPa SPR revealed that the probable limit to root penetration in soil under shifting cultivation would occur much before the soil water content reached the permanent wilting point, whereas the agroforestry/fodder/food agriculture land use systems with adequate soil conservation measures decreased the upper limit from 26.2 to 23.6%. Consequently, the NLWR increased in modified land use systems in comparison to shifting cultivation. The NLWR was 117.4, 84, and 53.6% higher in agroforestry, fodder, and food agriculture, respectively than the shifting cultivation. The values of NLWR also appeared to be more sensitive to soil loss and other soil properties than PAWC. The NLWR (percent moisture) was exponentially, negatively, and significantly correlated with soil loss (Mishra and Saha 2004b).

6.13.33 Soil Properties Under Agroforestry Systems

Adoption of different models of agroforestry systems (Saha et al. 2003) in the NEH region could properly maintain the soil's physical status (Table 6.21). In agroforestry systems, the bulk density varied from 0.94 to 1.19 Mg m^{-3}, and it increased with soil depth in all the agroforestry systems. The decrease in bulk density in natural forest, multistoried AFS, and silvi–horti-pastoral systems was due to the effect of relatively high organic carbon content (1.23 to 2.70%). Higher carbon content in these agroforestry systems was because of heavy litter fall and their subsequent decomposition in the soil profile. Mean weight diameter (MWD) was higher in the surface soils in all the agroforestry systems. This, most probably, was due to the presence of a very high amount of sesquioxides, clay, and organic carbon content in the soil. The volume of residual pores (<0.5 μm) was maximum in natural forest (57.93%) followed by multistoried AFS (52.87%), indicating its higher capacity to retain soil moisture. About two times more erosion ratios were noted under khasi mandarin (4.46) and Assam lemon (4.26) systems in comparison to natural forest (2.10). This could be due to enhanced disintegration of the surface soils by different agricultural operations as compared to that of other AFS, having dense canopy cover throughout the year. The same trend was reflected in the case of erosion index and erodibility factor. Irrespective of soil characteristics, the rate of unsaturated hydraulic conductivity, and water diffusivity), invariably decreased with a decrease in their water content (Saha et al. 2003).

Tillage System

Tillage plays a major role in agricultural sustainability through its effect on soil properties (House and Pamelee 1985). Farmers in this region employ several traditional methods for seedbed preparation. Traditionally weed–bush re-growth is slashed manually and left on the soil as mulch or burnt in situ. The land is then hand hoed, often superficially. Farmers also make mound ridges. Mound ridges and other forms of raised beds are widely used in lowland rice fields for vegetable production during winter and rainy seasons. A slightly modified version of the raised bed (*bun* system—crop residue and weeds are buried and burnt in the raised bed) is under practice in Meghalaya. Zero tillage practice is also popular among the *jhumias* families for rice, maize, and millet cultivation. Wet and dry plowing through animal and power-drawn plow is being popularized among the farmers in valley land and in mild slopy condition.

Table 6.21 Effect of various Agroforestry systems on soil physical properties (Mishra and Saha (2006)

System	Bulk Density (Mg m^{-3})	MWD (mm)	Aggregate size		Pore size distribution (%)			Dispersion ratio
			Macroaggregate (>0.25 cm)	Microaggregate (<0.25 cm)	>50 μm	0.5–50 μm	<0.5 μm	
Arboretum	1.13	2.42	53.91	6.16	18.66	40.00	41.34	0.041
Khasi Mandarin	1.19	2.42	51.48	5.81	16.12	43.64	40.24	0.048
Assam Lemon	1.19	1.97	50.94	5.22	19.51	38.50	41.99	0.045
Alder + Pineapple + Fodder	0.98	2.43	55.06	6.25	21.94	34.95	43.11	0.033
Alder + Tea + Black pepper	0.97	2.65	56.32	6.37	20.08	27.95	52.87	0.039
Natural forest	0.94	3.13	60.78	6.87	30.12	11.95	57.93	0.025

Adopted from Mishra and Saha (2006)

Tillage is an integral component in cropping systems. It is referred to as the physical manipulation of soil to modify the edaphic environment to suit the specific needs of crop growth processes. Tillage influences the edaphic environment by affecting soil physical behavior like soil structure and water transmission properties. A well-structured soil provides ideal environments and proper bulk density for healthy plant growth. The bulk density of well-structured soil should be maintained below 1.25 Mg m^{-3}. One soil property that is always altered by tillage operations is the pore geometry of soil. The magnitude and direction of change in this property depend upon the change in bulk density, porosity, and stability of soil aggregates. It was observed that the bulk density of this region varied from 0.85 to 1.45 Mg m^{-3}. The continuous puddling in rice fields increased the bulk density from 0.98 to 1.29 Mg m^{-3}, whereas the zero tillage practices (Mishra et al. 2005) reduced its value up to 0.89 Mg m^{-3}. Tillage may also influence the stability and size of aggregates through changes in organic carbon content in the soil and the disintegration of soil aggregates. Tillage practices influence soil water transmission behavior through its effect on surface and sub-surface conditions, changes in number, continuity, and connectivity of pore caused by tillage as reflected in the values of infiltration rate and hydraulic conductivity of soil due to formation of macro pores as the result of biotic activity in zero tillage in comparison to conventional tillage.

Mishra et al. (2003) reported that the highest rate of water loss (2.18 mm ha^{-1}) in zero tillage and lowest (1.52 mm ha^{-1}) in power tiller puddled plots in foothill soil of Meghalaya. Mean water loss was highest in zero tillage (2.18 mm ha^{-1}) followed by Desi plow tillage (2.07 ha^{-1}), manual (1.93 mm ha^{-1}), and power tiller (1.52 mm ha^{-1}) land uses (Mishra and Saha 2006).

A great concern needs to be bestowed for the maintenance of a suitable physical environment. Integrated the farming system approach for soil fertility management is based on land use systems on a watershed basis. High physically degraded land due soil erosion could be maintained with the adoption of need-based different models of agroforestry. Soil conservation and agricultural production no longer should be regarded as separate activities. It must be an integral part of agricultural development and should start with improved farming systems and suitable tillage practice. A minimum of 1:4 years of cropping and fallow periods could be a suitable cycle under the jhum cultivation for the resilience of soil in this agro-climatic region.

6.13.34 Land Degradation in Eastern Himalayas

"Land" includes the soil, water, vegetation, landscape, and microclimatic components of an ecosystem, and "land degradation" refers to a temporary or permanent decline on the productive capacity of the land or its potential for environmental management. Scherr and Yadav (1996) explained that the net productivity change is both natural and human-induced processes of degradation and improvement. From a policy perspective, it is difficult to distinguish those lands that are undergoing degradation, in order to assess the need for action, and those which have reached a degraded state relative to their natural condition, long ago, but are currently in a stable or improved condition (Sharma et al. 2006a, b). If a higher level of inputs are required to maintain crop yields, then some type of degradation has probably occurred. The decline in productivity may be directly due to soil degradation such as depletion of soil nutrients, soil erosion, soil toxicity, and soil water-holding capacity or, indirectly, through infestation of degraded soils by persistent weeds that reduce yields. Environmental management of degraded lands refers to the recycling of nutrients, amelioration, and filtering of pollutants, transmission and purification of water, and encouraging negative growth to act as sinks for greenhouse gases (Scherr and Yadav 1996). Some people perceive that land degradation poses a potential threat to global food supplies in future (Brown and Kane 1994), while others consider that land degradation has been overestimated and, in any event, is relatively unimportant to global food supplies (Crosson 1994). Many others worry about the effects of land degradation on the livelihood of rural dwellers, particularly in the more marginal agricultural areas (IFAD 1992). Further expansion and intensification of food production could also have a potentially degrading effect on the environment (Pinstrup–Anderson and Pandya–Lorch 1994).

Global population, as per the United Nations forecast, will increase from the present 7.2 billion (2018) to 8.1 billion in 2025 and 9.6 billion in 2050, with the most growth in developing countries and more than half in Africa. This will increase the demand for food and other commodities for living and, therefore, resultant pressure on the natural resources for sustenance. There is a growing concern that inter-temporal degradation of agricultural soil resources, viz., decline in long-term productive potential is seriously limiting production in the developing world and the problem is getting worse (Lal 1990; UNCED 1992). Degradation is also associated with problems of sedimentation, carbon emission, and change in natural habitats leading to loss of genetic stock and biodiversity.

Degradation of land resources is a serious problem in Eastern Indian Himalayan Region. Hilly topography, high rainfall, primitive type of agriculture, excessive exploitation of forest wealth including deforestation, urbanization, construction of roads, population growth at a faster rate, its pressure on land, etc. are mainly responsible for land degradation in the region. Prevalence of shifting cultivation with primitive methods is a big threat to land resources and the environment and the problem is serious as far as normal living habitat in the region is concerned. There is an urgent need, therefore, to address various related land degradation issues to combat the problem.

Extent of Degradation

There is historical evidence of large-scale soil degradation in many parts of the world in the past 5000 years (Hillel 1991). About 2 billion ha of biologically productive land has been irreversibly degraded in the past 1000 years (UNEP 1991). Analysis of the global change in humus has revealed that there has been a loss of humus at a rate of 25.3 million tonnes per year, on average ever since agriculture began 10,000 years ago (Rozanov et al. 1990).

In the East Indian Himalayan region, the land degradation of agricultural lands began with the start of shifting cultivation in 7000 BC, and with an increase in population, the degradation has increased manifold at present (Fig. 6.15). It may be difficult to assess the actual extent of land degradation in the NE Region. Not until the past two decades have scientists or other agencies made a systematic attempt to assess the nature and extent of land degradation at a regional or global scale. However, of late, various methods including remote sensing have been used for assessing land degradation.

The transfer of soil and nutrient load along with runoff due to rainfall has many implications for the resource base and environment in the region (Sharma and Sharma 2004a). The prevalence of shifting cultivation over an annual area of 3869 km^2 has further aggravated the problem. The total land degradation area in Eastern Indian Himalayan states is 97190 km^2, out of which 76,410 km^2 (78.6% of degraded and 29.1% of geographical area) is degraded due to erosion, while 20,780 km^2 (21.4% of degraded and 7.9% of geographical area) is degraded by other means like soil acidity and biological degradation.

6.13.35 Soil and Nutrient Transfer

The transfer of soil and nutrient load with erosion from the East Indian Himalayan states is enormous causing environmental degradation (Table 6.22). The extent of soil and nutrients transfer from the region has been estimated to be about 601 million tonnes of soil and 685.8, 99.8, 511.1, 22.6, 14.0, 57.1, and 43.0 thousand tonnes of N, P, K, Mn, Zn, Ca, and Mg, respectively (Sharma and Sharma 2004a). Continuous degradation of land has affected the water resources,

Fig. 6.15 A view of degradation in hills

Table 6.22 Sinks for soil and nutrients transferred from the NE region

Sink	Soil (10^6 tonnes)	Nutrients (10^3 tonnes)						
		N	P	K	Mn	Zn	Ca	Mg
Rivers	86	77.5	11.6	57.0	2.5	1.9	7.7	5.6
Floodplains	46	48.7	7.5	36.5	1.6	0.9	3.6	2.6
Streams	66	72.0	10.3	53.1	2.3	1.8	7.0	5.5
Valleys	58	57.6	8.9	42.9	1.8	1.4	5.5	4.3
Temporary storage	30	27.4	4.9	20.0	0.8	0.6	2.3	1.9
Lakes and reservoirs	6	7.5	1.2	5.4	0.2	0.2	0.7	0.6
Sea	309	345.1	55.4	296.2	13.4	7.2	30.3	22.5
Total	601	685.8	99.8	511.1	22.6	14.0	57.1	43.0

which may lead to the crisis of enormous magnitude in the future. Planning for land resources, therefore, assumes great significance. Eutrophication of water bodies with agricultural runoff has caused drastic changes in biomass. Water pollution has been due to an increase in the concentration of chlorides, sulfates, nitrate, and aluminum. More than 95% of the soils are acidic due to the leaching of salts due to heavy rainfall.

6.13.36 Wastelands in Indian Himalayan States

The area under wasteland in the country is very high and works out to be 20.1% of the total geographical area. Five hundred and eighty-four districts in the country have been covered to estimate the area under wastelands, rendered non-productive due to various reasons. In India,

Table 6.23 Area under wastelands in Himalayan states of India (Area in sq. kms)

State	Number of districts covered	Area of districts (km^2)	Wasteland area (km^2)	% of total GA
Arunachal Pradesh	13	83,743	18,326	21.88
Assam	23	78,438	20,019	25.52
Himachal Pradesh	12	55,673	31,659	56.87
Jammu and Kashmir*	14	1,01,387	65,444	64.55
Manipur	9	22,327	12,949	58.00
Meghalaya	7	22,429	9,904	44.16
Mizoram	3	21,081	4,072	19.31
Nagaland	7	16,579	8,404	50.69
Sikkim	4	7096	3,570	50.30
Tripura	4	10,486	1,276	12.17
Uttarakhand	13	53,483	12,790	23.90
Total	109	472,722	188,413	39.85

(Anonymous 2000)

the maximum wasteland is found in Himalayan states with Jammu and Kashmir having the highest wasteland in the country (64.55%), followed by Manipur (58.3%), Himachal Pradesh (56.87%), Nagaland (50.69%), and Sikkim (50.3%); while minimum area (3.73%) under wasteland category is in Kerala state (Table 6.23). The Himalayan states are shaded in Table 6.41.

Wastelands in India

Maximum percentage of geographical area (6.13%) under wasteland is in the category of with or without scrubs, 4.44% in underutilized/degraded notified forest land, and 2.04% land is barren, stony, or rocky (Table 6.24). About 1.11% land of the total geographical area has been rendered waste due to shifting cultivation.

The wasteland in NE region is 78510 km^2 or 29.95% of the geographical area of the region. Wasteland under different categories in the Eastern Indian Himalayan region is presented in Table 6.25. Maximum area of 44.5% of total wasteland has been rendered waste due to the prevalence of shifting cultivation, followed by degraded notified forest area (15.6%), and land with/without scrubs (14.4%). Highest percentage of wastelands among NE states is in Manipur (58.0%), followed by Nagaland (50.7%) and Sikkim (50.3%) while the minimum is in Tripura (12.1%).

6.13.37 Land Use Systems in Eastern Indian Himalayas

The entire resource degradation process in the region is closely linked up with the land use systems. Forestry is the most dominant land use system in the region. Scientific studies and evaluations of some of the land use systems as practised in the region indicate that most of them are hazardous to resources and are not conducive to the aims of permanent agricultural systems with sustained production (Sharma et al. 2006a, b). Shifting cultivation in the past was a practical system of cultivation with a long cycle of operation for a unit hill terrain in a region having a sparse population. In recent years due to the increase in population and limited availability of land for agriculture, the jhum cycle has come down to 3–5 years from the old cycle of 20–25 years or more with consequent diminution in

Table 6.24 Category-wise wastelands of India (Area in Sq. kms)

Sr. No	Category	Total wastelands	% of TGA covered
1	Land of gullies and ravines	20,553	0.65
2	Land with or without scrub	1,94,014	6.13
3	Waterlogged and marshy land	16,568	0.52
4	Land affected by salinity–alkalinity in coastal/Inland area	20,477	0.65
5	Shifting cultivation area	35,142	1.11
6	Underutilized/degraded notified forest land	1,40,652	4.44
7	Degraded pastures/grazing land	25,979	0.82
8	Degraded land under plantation crop	5,828	0.18
9	Sands in Inlands/Coastal area	50.022	1.58
10	Mining/industrial wastelands	1,252	0.04
11	Barren rocky/stony waste/sheet rock area	64,585	2.04
12	Steep sloping area	7,656	0.24
13	Snow-covered and/or glacial area	55,788	1.76
	Total wasteland area	6,38,518	20.17

the supporting capacity of a unit of land. Cultivation under such conditions on hill slopes has led to accelerating soil erosion and land degradation. Change in landscape due to deforestation for shifting cultivation is presented in Fig. 6.16.

Assessment of Degradation

Estimates of the changes over time of the type, scale, and rate of physical soil quality at a sub-regional or higher scale are required for an assessment of the soil degradation with regard to productivity-related economic effects which are relevant to policy making and planning. These changes must then be linked to consequent changes in agricultural yield or production costs (Scherr 1999). Accelerated degradation is mostly human-induced, and it occurs in virtually all eco-regions, irrespective of social, economic, and political conditions. The most extensive effort in assessing soil degradation is the "Global assessment of human-induced degradation" conducted by the ISRIC under the auspices of UNEP (Anonymous 1999). A subsequent Assessment of the status of human-induced Soil Degradation in South and Southeast Asia (ASSOD) used a refined GLASOD methodology. Soil degradation was estimated in NE states by National Bureau of Soil Survey and Land Use Planning (NBSS&LUP), Nagpur, to assess the overall status of degradation.

6.13.38 Extent of Soil Degradation in Eastern Indian Himalayan States

The Eastern Indian Himalayan region of India is characterized by high rainfall and humidity, wide variations in altitude ranging from near sea level to 5000 m above msl, undulating topography with spread out hills interspersed by fertile plains and inhabited by a large number of ethnic groups with diverse socio-economic features. The region is characterized by slow growth with its deficiency in food grain production, lower agricultural productivity, heavy soil erosion, loss of soil fertility, and deforestation, thus causing acute environmental degradation and serious ecological imbalances (Anonymous 1985). The realistic estimates of various types of land degradation in the region are, however, rare, and most of the figures are based on secondary information. According to a survey report among the 10 leading states as far as wastelands are concerned,

Table 6.25 Various categories of wastelands in Eastern Himalayan states (Km^2)

State Category	AP	Assam	Manipur	Meghalaya	Mizoram	Nagaland	Sikkim	Tripura	Total	% of TGA
Land with/without scrub	3,327	844	1	4,191	0.0	1,596	1,073	287	11,319	14.4
Waterlogged/marshy	41	1,634	325	15	0.0	0.0	0.0	0.0	2,015	2.7
Shifting cultivation	3,088	8,391	12,014	2,086	3,761	5,224	0.0	401	34,965	44.5
Degraded notified forest	1,416	3,113	609	3,612	310	1,584	1,061	588	12,293	15.6
Degraded pasture	2,135	2,218	0.0	0.0	0.0	0.0	0.0	0.0	4,353	5.5
Sands	309	3,764	0.0	0.0	0.0	0.0	0.0	0.0	4,073	5.2
Others (Mining, snow, barren, steep slopes, etc.)	8,010 (85% under snow)	55	0.0	0.0	0.0	0.0	1,436 (99% under snow	0.0	9,501	12.1
Total	18,326	20,019	12,948	9,904	4,071	4,804	3,570	1,276	78,519	100.0
% of TGA	21.8	26.0	58.0	44.2	19.3	50.7	50.30	12.1	29.9	–

Sources Ministry of Rural Development, Govt. of India, New Delhi and National Remote sensing Agency, Govt. of India, Hyderabad (Anonymous 2000)

Fig. 6.16 Images of Central Himalayas

6 are from NE Region (Satapathy 2006). The region which had a forest cover of 67% lost 783 km^2 and 300 km^2 during 93–95 and 95–97, respectively, according to the forest survey of India. Drying up of wells, springs, and seasonal streams and lengthening of the dry season is noticeable in many areas. Reservoir siltation is another serious problem. High rainfall and steep hilly topography are always associated with the problem of severe soil erosion, particularly when the land use systems have biotic interference. Quantitative facts on soil erosion hazards from soil erosion studies in the region indicate that except for forest land use, none of the land uses are safe and lead to the land degradation process (Satpathy 2006). Excessive silt load in the river due to soil erosion and large-scale landslides in the hilly catchment areas due to intense rainfall on the fragile steep slopes result in instability of the river and its banks. In Assam, 10–12% of the total cultivated areas is almost regularly affected by floods (Satapathy 2005, 2006).

The region is globally recognized as a mega-biodiversity area from the point of view of its endowments in flora and fauna. It has already lost many valuable genes. Loss of genetic materials can be attributed to various factors such as environmental destruction caused by large-scale deforestation, shifting cultivation, soil erosion, floods, and earthquakes. There is also no scientific or systematic planning for mining operations. The natural resource of the ecosystem is seriously altered and disrupted when disturbed by surface mining. Of late, the government has laid special stress on the development of the region. One of the major constraints faced by planners and decision-makers is the lack of timely, reliable, and adequate information about the present resources and the dynamic changes over time and space. Given the remoteness, lack of infrastructure, and difficult topography, questions have been raised as to whether intensive agriculture is sustainable given the naturally high rate of erosion and relatively low availability of inputs. Very little long-term data on land use, soil fertility maintenance, erosion, and sedimentation processes are available to conceptualize the problem in a holistic sense. An integrated

Table 6.26 Status of soil degradation in Arunachal Pradesh

Soil degradation type	Degree of degradation (Area in '000 ha)				Total area
	Slight	Moderate	Strong	Extreme	
Water erosion					
Wt.—Loss of topsoil	60	689	3,699	55	4,503
Wd—Terrain deformation	–	112	1,630	–	1,742
Chemical deterioration					
Cn—Loss of organic matter, Loss of nutrients, Acidification, and Salinization	15	689	3,676	–	4,380
Physical deterioration					
Pc—Compaction, sealing, and crusting	–	–	–	–	–
Pw—Waterlogging	–	–	–	–	–
Pf flooding	60	8	23	55	146
Total area	60	689	3,699	55	4,503

approach to conservation of natural resources and their sustainability, therefore, should emphasize a land use policy with a multipurpose approach. This strategy would accurately identify problems of land degradation, increase the productivity of land resources, restore degraded lands, prevent further degradation, and make rational need-based and sustainable land allocations for different use.

Arunachal Pradesh

The data pertaining to the status of soil degradation in Arunachal Pradesh are presented in Table 6.26. Deforestation and the removal of natural vegetation are the main causative factors for soil degradation.

Soil degradation is strong in most of the cases, and it covers all 4 degrees, viz., slight, moderate, strong, and extreme depending on the slope of the effective terrain. Loss of topsoil (Wt) is present in all the soil units covering 54% of the total area of the state.

Assam

The causative factors of soil degradation in Assam are deforestation and unscientific agricultural practices and heavy rainfall under humid climates. Soil physiographic unit area on which soil degradation was found to occur accounted for 76.8%, but the actual extent of the area under degradation may be estimated to be 37.8% of TGA (Anil Kumar et al. 1994) (Table 6.27).

Manipur

The data pertaining to soil degradation status in Manipur are presented in Table 6.28. The actual extent of the degraded area is estimated to be around 15.1% of TGA in Manipur, though nearly 78% of the soil physiographic unit area is considered to be in a state of degradation. Here, the major share of soil degradation was constituted by chemical deterioration in combination with water erosion. The wasteland (mainly marshy) not fit for agriculture is around 1.8% of TGA (Anil Kumar et al. 1994).

Meghalaya

The state has been broadly divided into two physiographic zones, i.e. central plateau and lower plateau regions. In the upper reaches of central highlands, the loss of topsoil largely occurs through the surface wash and sheet erosion. Degradation is severe in the southern part of high land near Cherrapunji. It is moderate to severe in the western part around Nongstain and in the eastern part around Jowai. In the valleys, stagnation of water and flash floods are also experienced. The data on soil degradation status in Meghalaya is presented in Table 6.29 (Mukhopadhyay et al. 1994).

Table 6.27 Status of soil degradation in Assam

Soil degradation type	Degree of degradation ('000 ha)				Total area'000 ha under degradation	Percent of TGA
	Slight	Moderate	Strong	Severe		
Single problem						
1. Water erosion (Wt) loss of topsoil	127.6	829.3	67.6	367.4	13,391.9	17.73
2. Chemical Degradation Cn loss of nutrients and acidification	10.9	–	95.4	279.0	358.3	4.91
3. Physical Degradation	3.7	196.8	–	–	200.5	2.55
Pw—Waterlogging	–	–	–	–	–	–
Pf flooding	1.5	–	568.2	259.3	829.0	10.57
Complex problems						
Wt + Cn	326.9	298.5	402.1	115.5	1143.0	14.55
Wt + Pf	542.9	478.9	138.1	–	1159.9	14.77
Wt + Pw	36.9	36.9	–	–	73.8	0.94
Pw + Cn	150.6	111.5	50.6	165.8	478.5	6.10
Pf + Cn	–	131.5	125.0	111.4	367.7	4.68
Total	1201.0	2083.4	1447.0	1298.2	6024.6	76.80

Table 6.28 Status of soil degradation in Manipur State

Soil degradation type	Degree of degradation				Total area ('000 ha)	Percent of TGA
	Slight	Moderate	Strong	Severe		
Single problem						
Water erosion loss of topsoil (Wt)	–	128.1 (9.6)	5.3 (0.4)	–	133.4 (10.0)	6.0 (0.44)
Chemical deterioration (Cn)	–	–	-	–	–	–
Loss of nutrients and acidification	9.3 (0.02)	–	152.3 (11.5)	546.8 (114.0)	708.4 (125.7)	31.7 (5.62)
Physical deterioration waterlogging (Pw)	–	–	112.1 (6.3)	–	112.1 (6.3)	5.0 (0.28)
Complex problems						
1. Wt + Cn	4.4 (3.4)	333.9 (58.4)	39.7 (8.9)	303.5 (109.3)	681.5 (162.0)	30.5 (8.03)
2. Pw + Cn	23.2 (11.0)	–	68.8 (5.3)	–	92.0 (16.3)	4.1 (0.73)
Total	36.9 (14.6)	462.0 (68.0)	378.2 (32.4)	850.3 (223.3)	1727.4 (338.3)	77.3 (15.10)

Stable terrain under natural conditions (Sn) 467.2, 20.9
Wasteland not fit for agriculture (Marshy land) 41.4, 1.8
Values in parenthesis show the actual extent of degraded area

Table 6.29 Soil degradation status in Meghalaya ('000 ha)

Soil degradation type	Degree of degradation				Total area
	Slight	Moderate	Strong	Extreme	
Water erosion					
Loss of topsoil (Wt)	59.6	684.3	424.0	–	1167.9
Chemical deterioration					
Acidification (Ca)	39.8	1030.2	-	–	1070.0
Loss of nutrients (Cn)	262.5	885.8	30.1	–	1178.3
Physical deterioration					
Waterlogging (Pw)	6.0	–	–	–	–
Flooding (Pf)	–	34.1	–		34.1
Total area	367.9	2634.4	454.1	–	3456.3
Stable terrains under natural condition					44.2

Mizoram

Mizoram is a hilly state with narrow valleys in between the hill range of the northern and southern stretching. Loss of topsoil is encountered in all the soil units. Soil degradation is strong in most of the cases. Terrain deformation is moderate to strong (Table 6.30).

Nagaland

Nearly 60% of TGA is in a state of degradation mainly through water erosion and chemical deterioration (Table 6.31).

Sikkim

Out of 7,09,600 ha of TGA, about 2,34,401 ha (33.03%) is affected by water erosion causing various degrees of degradation. Water erosion also results in terrain deformation in 6,070 ha land (0.85% of TGA) (Table 6.32).

Tripura

Two categories of soil degradation found commonly in Tripura are (i) degradation by displacement of soil by water, and (ii) degradation due to loss of nutrients (Chemical deterioration)

Table 6.30 Status of soil degradation in Mizoram

Soil degradation type	Degree of degradation (Area in'000 ha)				Total area
	Slight	Moderate	Strong	Extreme	
Extreme water erosion					
Wt loss of topsoil	162.4 (7.7)	517.7 (24.54)	507.3 (25.89)	–	1187.4
Wd Terrain deformation	10.0 (0.47)	125.6 (5.98)	164.3 (7.79)	395.3 (18.76)	695.2 (33.0)
Chemical deterioration					
Cn loss of organic matter, loss of nutrients, acidification	871.1 (41.33)	1049.7 (49.8)	–	–	1920.8 (91.13)

Figures in parentheses indicate percent coverage of total geographical area

Table 6.31 Status of soil degradation in Nagaland

Soil degradation type	Degree of degradation (Area in '000 ha)				Total area
	Slight	Moderate	Strong	Extreme	
Water erosion					
Wt. Loss of topsoil	49 (2.9)	179 (10.8)	767 (46.3)	–	995 (60.00)
Wd Terrain deformation	–	66	667 (40.2)	–	733 (40.2)
Chemical deterioration					
Cn loss of organic matter, loss of nutrients, acidification	49 (2.9)	179 (10.8)	767 (46.3)	–	995 (60.0)
Total area	49 (2.9)	179 (10.8)	767 (46.3)	–	995 (60.0)

Figures in parenthesis indicate percent coverage of total geographical area

Table 6.32 Status of soil degradation in Sikkim

Type of soil degradation	Degree of degradation				Total area (000 ha)
	Slight	Moderate	Strong	Extreme	
Water erosion					
Loss of topsoil (Wt)	7.627	123.504	91.34	5.86	228.331 (32.18)
Terrain deformation (Wd)	–	–	6.07	–	6.07 (0.85)
Total area	7.627 (1.07)	123.504 (17.41)	97.41 (13.72)	5.86 (0.83)	234.401 (33.03)
Stable Terrain under natural conditions 214.979 (30.30) Rock outcrops and Cliff 81.67 (11.5) Misc. area (Ice cap) 90.36 (12.74)					
Total area 709.6 (100.0)					

Figures in parentheses indicate percent of TGA

or waterlogging and flooding (Physical deterioration). Nearly 60% of TGA are under various types of degradation processes (Table 6.33).

6.13.39 Soil Erosion and Water Resources

Due to its topography, rainfall, and climate in general, the Eastern Himalayan region is highly susceptible to soil erosion. Erosion index values of rainfall indicate wide variability in the distribution of rainfall erosivity. Erosion potential of rainfall is quite high toward northwest and southern part of the region, whereas the same is the minimum in Manipur Valley (Imphal). Average erosion index (EI) of five stations was found to be highest in June with a metric unit value of 169.2, followed by July (146.1) and August (130.4). Average minimum EI was 0.1 metric units in January.

Soils having a dispersion ratio of more than 10 are erosive in nature, thereby indicating that most of the soils in Tripura are highly erodible. Upland soil appeared to be more erodible than soils present in lowland topography. Surface soils of the river basin were also observed to be highly erodible. In general, the land and water resources in the region are under active process of degradation which may assume a serious threat to the environment and lead to shrinkage in cultivable land, increase in surface runoff during

Table 6.33 Status of soil degradation

Soil degradation	Degree of degradation			Total area (area in 000 ha)
	Slight	Moderate	Strong	
Water erosion				
Loss of topsoil (Wt)	113	312	203	628
Chemical deterioration				
Loss of organic matter, nutrients, and acidification (Cn)	113	32	203	628
Physical deterioration				
Waterlogging (Pw)	90	–	–	90
Flooding (Pt)	24	78	102	113
Total area	113	312	203	628

Since the degradation is a complex type in all the cases, the maximum area covered under a particular degree of degradation indicates the total area

the monsoon period, and drying up of streams and acute shortage of water with a result of desertic condition (Singh and Singh 1981).

6.13.40 Problems of Flood

The flood-prone susceptibility of NE region is due to its geophysical situation. The entire region is drained through the biggest river systems of India, namely the Brahmaputra, Barak, and Teesta. The sediment load in the river is of high magnitude (Table 6.34).

The abnormal flat of the rivers in the plain makes carrying away of silt impossible and the silt starts setting in bed right from foothill down below. There is a reduction in the water-carrying capacity of channels resulting in the extension of floodplains. Tables 6.35 and 6.36 give an indication of the enormity of the problem of displacement of soil and nutrients from the region, which ultimately settles down in various sinks (Sharma 2002).

Pre-monsoon rain, which is also mostly a regular phenomenon in this region, keeps the soil in a saturated condition before the onset of monsoon and thus the flood starts occurring during the initial monsoon rain itself. This causes a loss of agricultural land. As a result, there has been rapid encroachment of land in the reserved and protected forests, village grazing reserves, and professional grazing reserves. The vicious cycle of deforestation, consequent soil erosion, and silting effects have acted immensely to upset the ecological balance of the valley and its neighboring areas (Dutta 1984).

Table 6.34 Sediment and nutrient load in rivers

Rivers	Sediment (Million tonnes)	Nutrient ('000 tonnes)			
		N	P	K	Micronutrients
Brahmaputra	467	303	39	218	18
Barak	74	50	9	37	3
Other small rivers/streams	60	42	7	31	2
Total	601	395	55	286	23

Table 6.35 Sediment load (mg l^{-1}) in floodwater

Soil/nutrients	Range
Soil	1500–3000
NO_3–N	6.4–5.8
PO_4–P	2.3–8.5
K_2O	15.4–33.8
Zn	0.3–1.6
Mn	0.8–2.4
Cu	0.1–0.3
Fe	6.3–18.4
Ca	2.5–6.4
Mg	6.5–14.0
SO_4	5.0–8.5

Table 6.36 Nutrient status of temporary water storage

Nutrient	Before floods	After floods
N (kg ha^{-1})	170	253
P_2O_5 (kg ha^{-1})	6	14
K_2O (kg ha^{-1})	135	204
Zn (mg kg^{-1})	2.7	3.8
Mn (mg kg^{-1})	6.8	9.1
Cu (mg kg^{-1})	0.5	0.5
Fe (mg kg^{-1})	4.6	3.9
CEC [cmole (p^+)kg^{-1}]	6.7	11.5
Organic matter (%)	0.8	2.2

Effect on Climate

The climate of the earth fluctuates with time, but on a longer time scale. The name Meghalaya, as "the abode of clouds", is appropriate because the occurrence of clouds is a common phenomenon. But in recent years, it is noted that the density and frequency of cloud cover (overcast) have declined. Consequently, in Shillong annual rainfall decreased from 2415 to 2039 mm (i.e. 15.5%). Out of total rainfall, 96.42% occurs during summer and 3.58% in winter. In the last 12 years, average temperature of Shillong was 16.83 °C, which was 0.13 °C higher than the average annual temperature for 30 years from 1931 to 1960 (Hussain 1984). From the above facts and figures, the weather condition of Shillong may be inferred as follows:

- Very small increase in average annual temperature.
- Mild decrease in annual rainfall.
- Mild increase in humidity.
- Winter months have become less cool and summer months have become more warm.
- The number of frosty, foggy, rainy, and cloudy days have decreased.
- In ecology, there is a chain of changes, since constituents of an ecosystem function in a harmonious order.

- There is an increase in the growth of mosquitoes, insects, flies, etc. due to an increase in temperature and moisture and a decrease in rainfall (Hussian 1984).

6.13.41 Efforts to Check Degradation

Traditional Methods

Most of the agricultural activities in the region contribute considerably toward land and water degradation. From a prospective viewpoint, there are certain pockets where excellent local resource-based techniques have evolved which are potential but remain confined to limited areas only. Bench terrace cultivation in certain areas of Nagaland, Sikkim, and Manipur (Prasad and Sharma 1994), bamboo drip irrigation in Jowai District of Meghalaya (Singh 1989; Sharma and Prasad 1995), and water management technique evolved and are used at Apatani plateau of the Subansiri District and Lohit District of Arunachal Pradesh (Sharma and Sharma 2005a, b). The "Zabo" system of Nagaland (Sharma and Sharma 2004b) is based on local skill and resources. However, some traditional methods of cultivation like shifting cultivation and "bun" methods have resulted in land degradation and leaching of nutrients (Sharma 2000).

In technology development, the ICAR Research Complex for the NEH region since its inception in 1975 initiated research work with special reference to alternatives to shifting cultivation. The basic approach adopted for the management of land and water resources was the conservation of rainwater where it falls. An approach to watershed-based farming system was tested and a model of agri–horti–silvi pastoral land use was recommended as an alternative to shifting cultivation. The model based on local resource inputs for land and water management was found successful in retaining over 90% of the annual rainfall in the hill slope with soil losses below permissible level (12 t ha^{-1} yr^{-1}). This also provides a stable base for agricultural land use. The soil erosion is minimum in this land and most of the rain is retained in situ (Singh 1987; Singh and Prasad 1987).

Under the NEC's programs (Ministry of Home Affairs, Govt. of India), eight pilot schemes of soil conservation and jhum control were taken up in the NE region located in major catchment areas (Anonymous 1984). All these schemes were designed with a common objective of control of shifting cultivation, with a certain qualitative and quantitative difference in various components incorporated into them. The schemes of Manipur and Nagaland aimed at settling each jhumia family on 1–2 hectare of wetland terrace. In Tripura, the scheme was implemented both in the forest and agriculture sectors by engaging the tribals as wage earners in rubber plantation, settling them on forest land in small colonies. Dry as well as wetland terraced fields to the families in Mizoram and in Meghalaya had been allotted.

The jhum control schemes implemented by North Eastern Council and other State Governments could provide a means of deciding on the strategy to be followed in tackling the problem in NE region. The performance of these schemes would have to be judged from their effects on the qualitative characters of the problem and not on the quantitative achievements. The success of any jhum control scheme will have to be judged on the following:

- Whether the jhumias have taken to settle agriculture.
- Whether area under jhum in a particular location has been reduced.
- Whether there is an improvement in the socio-economic condition of the families settled under the scheme.

From the surveys conducted by different agencies and the visits by experts to the sites, a certain definite indication of the trends prevailing in such project areas has become available. Shifting cultivation has become a resource depletion practice with an increase in human and livestock population and the resultant increase in demands from the same piece of land. The mountain communities of HKH region have, however, now become aware of the bad effects of shifting cultivation and decline in crop productivity due to exhaustion of crop nutrients. Moreover, the labor requirement has also increased in shifting cultivation. The government has started investing more in new schemes either to eradicate shifting cultivation or improve it to make the practice economically viable. It is a fact that no single pattern can be evolved as a common solution for the entire region for dealing with the problems. The pattern of settlement of shifting cultivators will have to be devised for each location, based on climate, physiography, land use, and traditional occupation which, however, will follow one or the other pattern demonstrated through the pilot projects. Allotment of wetland terraces, particularly among paddy growing tribes, would ensure to go for settled cultivation. The shifting cultivators have not been able to derive maximum benefits from the government schemes and settled agriculture, mainly due to the reason that they remained ignorant of improved agricultural technology in absence of effective extension services.

6.13.42 Policy to Improve Agricultural Lands

The result-oriented response to land degradation will be to propose profitable incentives for farmers to care for their land and improving their inputs availability required for proper care of the land. The following recommendations are based on lessons learnt from the past successes and failures in dealing with land degradation (Scherr and Yadav 1996).

- Improve information systems.
- Development of research and technology.
- Investment in land improvement.
- Land tenure rights to cultivators so as to encourage investments on land.
- Participatory approach in planning and implementation.
- Local participation to manage their own resources.
- Market access.
- Credit facilities.
- Minimum support price.
- Rural income diversification techniques.
- Remove discrimination against marginal regains with regard to investment.

With the help of these, farmers can make short- and long-term decisions about land management.

Future Strategies

Given low-income levels and projected increase in the population of hills and mountains, the question left with the farmers is to choose the less degrading and more protective options. But in reality, there is no consistent relationship between population density and the frequency with which land is used for productive purposes and degradation of land. All the related factors must be taken into consideration in order to evolve a policy to get an ecofriendly system leading to less degrading and more productive options. Innovation in soil resource management under population or market pressure has been modeled by Scherr and Hazell (1994). There may be uncertainty about reaping the longer benefits of soil improvements due to tenure insecurity or price or climate risks. Under these conditions, targeted policy action is needed to slow or reverse soil degradation. Policy intervention may also be desirable to accelerate farmers' response in a situation where social benefits are greater than farmers' private benefits.

To implement productive soil and land use with a long-term sustainable perspective, the potential for resource degradation can be recognized, predicted, and avoided. Marginal lands of NE region are more vulnerable to soil degradation. Sustainable management of productive land should receive more attention than risky exploitation of new marginal lands, or continuing acceleration of land degradation will add to the overall scarcity of land resources. Decline in water resources has led to a decline in irrigation development, meaning that greater demands will be made on the productivity of land in rainfed areas, which are often vulnerable to degradation. This necessitates that more emphasis be placed on sustainable land use in these areas, which are more extensive than irrigated areas.

It is pertinent with respect to NE region that secure access to the land, preferably through tenure, would play an important role in how well the land is managed and conserved. Farmers with no long-term rights over land are unlikely to invest in improvement and conservation. To address the problem of land degradation, there is a need to enact suitable laws in the region for securing the permanent rights of farmers over the land for its judicious management. A strong relationship among land users, extension workers, and researchers is to be built. Farm family is an integral part of the farm in NE region, the women do most of the farm work. Generally, the extension workers are men, who tend to discuss and work with the male members of the family. Most of the schemes fail because the capabilities of women folk are not taken into consideration. More female extension workers are required to have meaningful interaction with the female family members. To reclaim degraded lands and control the useful lands from degradation, creation of necessary legislation, policy, institutional backup, and decision support system in place is necessary.

6.13.43 Soils and Soil Use Planning

The soils of the hilly region have not been surveyed thoroughly due to hilly and poor communication and accessibilities and dense forests. According to meager information, the area is covered dominantly by inceptisols (3%) followed by entisols (27.3%), ultisols (18.1%), and altisols (4.5%). Miscellaneous soils account for 6.8% that cover snow-capped areas, marshy lands, and rocky areas (Velayutham 1999). According to a classification, about 50% of the land area is under moderate slopes followed by strong and severe steep slopes (17.5 and 16%, respectively). Out of the total area, only about 1.67% is occupied by the valleys and the rest by mountainous and hilly tracts. Majority of the soils of the North Eastern Region are deep; however, patches of moderately shallow soils are also available in some portions of Meghalaya, Nagaland, Assam, and Arunachal Pradesh. Intense weathering due to high rainfall conditions and thick vegetative cover, under the influence of favorable tropical per humid climatic condition, has resulted in the formation of deep soils. Very poor and poor drainage accounts for 7.5% of the total area mostly confined to various river valley zones of the region resulting in huge losses due to floods. Soils are usually rich in organic matter and are acidic to strongly acidic in reaction. The low pH of the soil is attributable to leaching of bases under influence of high rainfall in hills. The soils of the NEH region are rich in exchangeable aluminum and iron and are affected by their toxicity under waterlogged conditions in some soils.

The degradation of the landforms is the main problem in the region. The causes of land degradation are the result of large-scale deforestation, high rainfall amount and intensity, undulating topography, developmental activities, increased population pressure, mining and querying, agricultural activities on lands not suitable for cultivation, shifting cultivation, etc. Cultivation of marginal land on steep slopes, excessive deforestation, and shifting cultivation are the main reasons for the degradation of steep hill slopes. According to the recent assessment, an area of approximately 7.7 million ha is degraded in the region, i.e. about 34.1% of the total geographical area of the region is suffering from the problem of land degradation of one kind or the other. Soil erosion due to water is the main form of erosion affecting almost the entire Northeastern region especially the hilly region (Table 6.37).

The states of Manipur, Mizoram, and Tripura have shown an increase in degradation as against a general trend of decrease. According to another survey report, among the 10 leading states as far as wastelands are concerned, six states are from North Eastern Region (Sethi et al. 1994). What is important for hills is the proportion of net area sown to the geographical area on the one hand and fallow land (both current and others) and culturable wasteland on the other that have got

Table 6.37 Effect of different land uses on soil erosion

Land use	Plot size	Soil loss ($t\ ha^{-1}\ yr^{-1}$)	Average annual rainfall (mm)
Shifting cultivation	Small	30.2–170.2	1600
Shifting cultivation	Field	5.1–83.3	1600
Tuber crops on raised bed (bun)	Medium	40.0–50.0	1800
Pineapple along the slope	Small	24.0–62.6	1600
Homestead areas	Field	16.8	1600
Mixed crop of maize and rice	Small	19.7–21.0	1600
Rice crop on slope	Small	32.9–45.0	1600
Bare fallow	Small	83.8	1600
Cropping systems	Medium	51.0–83.8	1600
Grass cover (planted)	Medium	10.83	1600
Natural bamboo forest	Field	0.04–0.52	1600

Areas of small, medium, and field size plots were in ranges of 2–5, 16.40, and 69,000 m^2, respectively. (*Source* Prasad et al. 1986)

Fig. 6.17 Land degradation in steep hills (*Source* Satpathy 2006)

bearing on shifting cultivation. Figure 6.17 shows the land degradation in Eastern Indian Himalayan Region.

Proportion of net area sown to the geographical area is negligible in Arunachal Pradesh (2.05% in 1988–89) and Mizoram (3.08 in 1988–89). It was 10.86% and 8.74%, respectively, in Nagaland and Meghalaya in the said years. Proportion of fallow and cultivable wastelands is extremely low in Arunachal Pradesh (0.99%) and Tripura (0.28%) but high in Nagaland (21%), Mizoram (24.5%), and Meghalaya (32.77%) in 1988–89 (Neog 1997). The soil erosion status of North Eastern Region of India has been presented in Table 6.38.

Soil Erodibility

High rainfall and steep hill topography are always associated with the problems of severe soil erosion problem. Most of the places of the region lie within the subtropical belt having a south Asiatic monsoon climate with a good amount of rainfall. The area can be divided into three distinct climatic regions (Borthakur et al.

Table 6.38 Soil erosion status of North Eastern Region

Class	Attribute	Area (%)	Area (km^2)
1	Slight	13.60	34,555
2	Moderate	46.94	1,19,141
3	Strong	17.41	44,186
4	Severe	15.74	39,946
5	Rocky mountains	4.16	10,552
6	Marshy lands	0.33	832
7	Rivers	1.82	4,624
Total		100	2,53,836

(*Source* Velayutham 1999).

1983) such as (i) the cold humid monsoon climate of the frontier hilly regions (about 2000 m), (ii) wet subtropical monsoon climate (covering southern Arunachal Pradesh, western Nagaland, Manipur, and Mizoram states of India in Eastern Himalayas), and (iii) humid mesothermal monsoon climate with heavy monsoon showers (covering the Brahmaputra valley, Meghalaya, the Barak valley including Tripura). So, magnitude of soil erosion problems are different in different zones depending on the amount and intensity of rainfall, slope percent, soil texture, vegetation cover, and management practices. The monsoon withdraws from the Eastern Himalayas almost abruptly in the last week of September or the second week of October and post-monsoonal rainfall is very scanty limiting the scope for agricultural activities during the rabi season. During winters, the soil erosion is much less as compared to the rainy season. As per estimates of soil eroded from shifting cultivation land on the hills of eastern Himalayas (1000 m amsl, 50% slope), mean soil erosion was found to be 55.2 t ha^{-2} from May to October (6 months) and, 4.1 t ha^{-2} from November to April (6 months), that is, in the ratio of 93.1: 6.9 (Sharma and Sharma 2003).

Although no systematic evaluation of the danger of rainfall erosivity has been made in the Eastern Himalayas, (Rambabu et al. 1978–1979) developed rainfall intensity duration return period equations and monographs as well as isoerodent (E I index) map for N.E. region. Sharma (1987) by superimposing the map of p2/p (rainfall aggressiveness: p = highest mean monthly precipitation, p = mean annual precipitation both expressed in mm) and annual rainfall kinetic energy categories the North East into areas of severe, high and moderate erosion risks. The severe erosion risk, according to this map, falls on the Monpuins in the south in Meghalaya, greater Himalayas in the south and northeast, the Nagaland hills in Central Nagaland, the eastern part of the Mizo hills, and also on the lower land in northeast Assam. The high-risk areas are found along the Brahmaputra river in Central Assam, Barak valley in southern Assam, parts of Nagaland and Manipur, western Mizoram, and the entire Tripura states.

Monthly and seasonal EI values are important to identify the critical periods as well as conservation practices to minimize the effects of highly erosive storms. The monthly erosion potential has a large variation in the area. Average monthly erosion index values for the monsoon months revealed that August is the most erosive month with an average EI of 168.31 tonne –cm ha -hr^{-1}. However, this month also showed the largest variation from year to year with a standard deviation of 110.19 mm. The observed maximum single storm erosion index was also the largest in the year.

No comprehensive and qualitative measurement of soil erodibility (erosion per annum per unit of the erosivity factor R) is done in the region. Awasthi (1984), however, determined soil erodibility factor at Byrnihat (Meghalaya) from an uncultivated bare fallow plot at a 65% slope. The average soil erodibility factor based on 3 years' data was worked out to be 5 kg ha^{-1}

unit^{-1} of rainfall. For mixed, sole cropping of rice and toposequential cropping, the "C" value ranged from 0.568 to 0.644. For the widely spaced crop of maize, the "C" value ranged from 0.857 to 1.0 under different management practices.

Quantitative facts on soil erosion hazards from soil erosion studies of the region indicate that except for forest land use, none of the land use practices are safe thus leading to land degradation. Micro-plot studies on bare fallow slope of 15, 30, 45, 60, 75, and 100% revealed that the soil loss and runoff varied from 13.15 to 21.54 t ha^{-1} yr^{-1} and 18.49 to 26.82% of rainfall, respectively (Singh 1985). The study also indicated that erosion and runoff increased up to 45% slope, thereafter they decreased linearly with further increase in the slope. Awasthi (1984) reported 17.80 to 205.6 t ha^{-1} yr^{-1} soil loss from 65% slope (bare fallow). In hills, development of new sites for building construction and development of roads are major sources of soil erosion. In a case study, (Singh and Singh 1984; Singh and Varma 1988) reported 16.8 t ha^{-1} yr^{-1} soil loss from a 6.9 ha hilly watershed under homestead land use. The hydrographic surveys of the Umiam reservoir (Meghalaya) conducted in January 1990 revealed a sediment deposition of 11,703 acre-ft during the last 25 years, giving an annual rate of sedimentation of 26.05 ha- m per 100 km^2 of catchment area (Anonymous, 1990). In shifting cultivation, biotic interference in jungle cutting, burning, clearing, and dibbling of seeds account for a considerable amount of loss of soil material. It is estimated that 88.3 million tonnes of soil are lost every year in this region due to this defective system of farming (Sharma and Prasad 1995).

6.13.44 Changes in Soil Properties Under Shifting Cultivation

Shifting cultivation (Jhuming) is an age-old system of cultivation, which is still widely practiced in the hilly areas of North Eastern Region. Shifting cultivation may be distinguished into jhuming (open burning) and bun (closed burning) cultivation (Venkatesh et al. 2001). The earlier includes cutting and burning of forest vegetation on sloping lands. On the other hand in bun cultivation which is being prevalent only in Meghalaya state, twigs and branches of forest tree species mainly *khasi* pine (*Pinus* sp.) along with existing weed biomass are kept in heaps at regular intervals along the slopes covered by a thin layer of soil. Bun is essentially a typical broad-based ridge and furrow system of cultivation. Each bun is usually 4–5 m long, 1–2 m wide, and 0.15–0.35 m in height spaced at 1–2 m intervals. The biomass is burnt underneath over a period of one month during January–March and ultimately converted to ash, which facilitates the spontaneous release of nutrients to the soil (Borthakur et al. 1983). Nutrition depletion takes place quite quickly, and productivity declines sharply in the first few years of cultivation. In hills and mountains, primary and other forest types are cut and burnt to add nutrients to soil for improving soil fertility and eventually crop productivity. The fastest technique to return the nutrients to the soil is by fire because more nutrients are stored in the biomass and not in the soil. Frequent burning can also prolong soil exposure and increase the potential for soil erosion. Therefore, fallowing is an integral part of the cropping system as it allows the forest to recover. Longer fallows give greater improvement in humus and nutrients in order to sustain "slash and burn" agriculture.

6.13.45 Acidity and Plant Disease Relationship

Plant disease caused by soil pathogens has been reported to be very complex (Hoper et al. 1995) and highly influenced by acidity and liming. Different soil pathogens need a particular soil pH environment for survival and proper growth and hence, the diseases caused by soil pathogens have a strong relationship with soil reaction. Siqueira and Moreira (1997) showed the effect of acidity on the host, symbionts, antagonists, and their interactions.

Characteristics of Soil Acidity

Reviews by McLean (1976) and Thomas (1977) could throw a deep insight into the developments in the concepts and theories of soil acidity. The major contributing factors for producing different kinds of acidities are exchangeable H^+, exchangeable and extractable Al, and organic matter. It is pertinent to mention that H^+ and Al^{3+} ions are not the only ions that contribute to soil acidity but Fe and Mn are perhaps the major hydrolyzable ions on the exchange sites of the soil complex, producing acidity after hydrolysis. The Al^{3+} ion, the role of which in soil acidity is important, is predominant below pH 4.7. Gradually, Al^{3+} undergoes hydrolysis (Chernov 1947) to monomeric and polymeric hydroxy-–aluminum complexes thus liberating hydrogen ions.

$$Al(H_2O)_6{}^{3+} + H_2O = Al(OH)(H_2O)_5{}^{2+} + H_3O^+$$

$$Al(OH)(H_2O)_5{}^{2+} + H_2O = Al(OH)_2(H_2O)_4{}^{+} + H_3O^+ \text{ and so on}$$

Acidity in the soil is of three types. (i) Active acidity referring only to H^+ ions in the soil solution; (ii) Exchange acidity including both H^+ and Al^{3+} which can be extracted by the salt solution of KCl/NaCl/$CaCl_2$; (iii) Non-exchange or residual/reserve acidity which takes into account the acidity caused by organic matter and bound Al. The total acidity estimated by $BaCl_2$-TEA comprises of pH-dependent or residual acidity and exchangeable acidity.

However, Coleman and Thomas (1967) have defined soil acidity in terms of KCl extractable and pH-dependent acidity. The first type of acidity is ascribed to isomorphous substitution from exchangeable positions while the second type owes to Fe and Al oxides and hydroxides and soil organic matter. The surface reaction of the variable charge components is regulated by the adsorption of potential determining protons or hydroxyls.

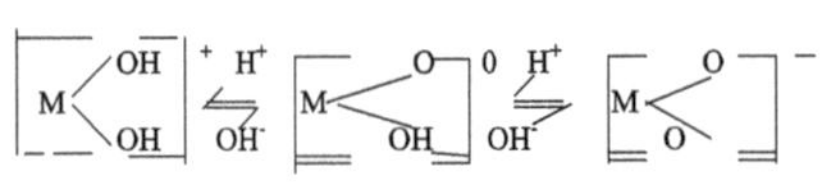

So, the predominance of +ve charge in the soil matrix leads to a raise in pH at which, point of zero charge (Parks and de Bruyn 1962) arrives and thus contributes pH-dependent acidity. Similarly, soil organic matter also possesses a number of functional groups containing H^+ ions that can contribute to the different kinds of pH-dependent acidities (Keeney and Corey 1963).

While studying different forms of acidity in major groups of soils in India, Sharma et al. (1990) reported that the electrostatically bonded H^+ and Al^+ comprised 21 and 79% of exchange complex, respectively, whereas pH-dependent exchange acidity accounted for only 71% of the total acidity. The rest of the unaccounted acidity was due to hydrolysis of Fe and Mn on the exchange sites of the soil complex. An attempt to evaluate characteristics of soil acidity needs to study acidity contributing factors in soils of NE states. Old alluvial soils are, in contrast to soils close to Brahmaputra ranges, more acidic with pH varying from 4.5 to 5.0. In soils of Toklai (Assam) and Aizawl (Mizoram), contents of exchangeable H^+, Al^{3+}, and extractable Al^{3+} were (Sharma et al. 1990) 0.15, 0.08 and 0.70, 0.57, and 1.8, 2.4 c mole (p^+) kg^{-1}, respectively. pH-dependent acidity comprising 60.1 and 50.0% of total acidity was 7.1 and 4.7 c mole (p^+) kg^{-1} in Tocklai and Aizawl, respectively. The relatively low contribution of exchangeable acidity toward total acidity is also reported by other workers (Pratt and Blair 1963; Mclean et al. 1964).

In soils of Meghalaya, pH, and exchangeable and extractable Al^{3+} underwent a variation (Ram et al. 1987) from 4.7 to 5.4, 0.04 to 3.53, and 0.06 to 5.44 c mole (p^+) kg^{-1}, respectively. Similarly, Singh et al. (1989), Nair and Chamuah (1993), and Dwivedi et al. (1993) also studied in detail the soil acidity in Meghalaya. High Al^3 saturation in excessively drained and severely eroded soils is indicative of the accelerated removal of bases and residual enrichment of colloidal complex with Al^{3+}. When the Al^{3+} saturation of the complex is 60%, the concentration of aluminum in only 1 mg kg^{-1} in soil solution creates toxicity although there is great variation in tolerance from one species to another and also within particular crop species. Soluble aluminum is reported to accumulate in the root tissue, preventing cell division/elongation and possibly inhibiting enzymes concerned with the synthesis of cell wall material (Rorison 1973). Furthermore, van Breemen (1973, 1976) reported Al^{3+} activity inversely related to pH and increased roughly tenfold per unit of pH decrease. Contents of exchangeable Al^{3+} in soils of Meghalaya were negatively correlated (r = 0.76) with soil pH. Besides Al^{3+} concentration, Fe-toxicity originated under the condition of high active Fe content (>1.25%), high organic matter (>2.98%), shallow water depth, and impeded drainage condition was also reported (Singh et al. 1989) in soils of Meghalaya.

Soils of Tripura (Laskar et al. 1983) were acidic in reaction. Upland soils were having more saturation of H^+ and Al^{3+} as compared to lowlands. Both available Fe and Mn contents were present in high amounts in lowland soils compared to uplands (Datta and Ram 1993). So iron toxicity (Datta and Laskar 1992) is reported to have a widespread occurrence in lowlands comprising about 40.05% of the gross cultivable area in Tripura.

Point of zero charge (pH^0) of soil samples of two soil profiles, Pedon 1 (Typic Dystrochrults) and Pedon 2 (Typic Haplaquent) collected from the soil series in upland and lowland of Tripura, respectively (Datta et al. 1991). The amounts of Fe and Al as extracted by the various extractants and pH^0 are also presented in Table 6.39.

Water soluble and NH_4OAC exchangeable Fe are appreciably high in pedon 2 as compared with upland soils (pedon 1). Aluminum, however, showed reverse trend possibly due to low pH in pedon 1. Oxalate extractable Fe, as expected, is markedly less than that of dithionite extractable one and the amounts were higher for pedon 2 as compared with that of pedon 1.

However, no such differences are observed for Al in the two pedons except that the oxalate extractable Al is somewhat higher than the amount extracted by the CBD method. The pH0 determined for soils of the two series varies from 1.90 to 3.42, thereby indicating the predominance of net negative charge. Surface electric potentials vary from –185.9 to –258.4 mV at 25 °C. Extractable Al^{3+} ranged from 0.23 to 0.95 with an average of 0.67 c mole (p^+)/kg. Available Fe and Mn contents varied (Datta and Gupta 1984) from 10.05 to 78.71 and 10.25 to 34.73 mg/kg in soils of Nagaland. Organic carbon varied from medium to high (0.55–3.09%) status. In soils of Sikkim, Gupta and Prasad (1983) reported that soil pH varied from 4.0 to 6.9 with a mean value of 5.2. Soils of the north district were highly acidic. Exchangeable aluminum varied from traces to 4.1 c mole (p^+)/kg. Percent of aluminum saturation underwent a variation from traces to 75% with an average value of 21%. Similar acidic nature of soils in Sikkim was also reported (Patiram and Prasad 1990a; Patiram 1991a, b). Lahiri and Chakravarti (1989) also reported that soils of Sikkim at various altitudes (400–1600 m MSL) are acidic in reaction and contain a considerably high percentage of organic matter in high-altitude regions.

In different hilly regions of Manipur, soil pH (Rao and Kumar 1991) varied from 4.6 to 6.8. Organic carbon varied from 0.8 to 5.1%. Exchangeable and extractable Al^{3+} ranged from traces to 8.18 and 0.64 to 10.80 c mole $(p^+)\ kg^{-1}$, respectively. Percent of Al saturation to CEC varied from trace to 74.12, the mean being 8.68. The pH-dependant acidity ranging from 3.22 to 23.22 c mole $(p^+)\ kg^{-1}$ contributed 60.9–99.3% of total acidity in the soils of Manipur.

Table 6.39 Amounts of iron and aluminum (mg kg^{-1}) in different extracts and pH of soils

Depth (m)	Fe				Al				pH^0	Surface electric potential (mv) at 25 °C
	1	2	3	4	1	2	3	4		
Pedon 1										
0–0.25	0.40	3.33	252.0	3669.2	2.50	5.17	170.3	160.4	1.90	– 206.5
0.25–0.46	0.35	1.66	228.3	3335.6	1.25	4.67	202.5	181.2	2.09	– 204.1
0.46–1.17	0.42	1.66	403.2	3039.1	0.50	3.67	204.2	195.2	2.60	– 215.4
Pedon 2										
0–0.10	2.90	16.60	572.2	2594.4	0.75	2.0	185.3	161.5	2.39	– 227.7
0.10–0.28	4.15	16.80	359.7	2446.1	0.50	2.83	180.7	140.9	2.27	–258.4
0.28–0.66	5.80	18.25	462.5	5596.4	Trace	2.20	205.3	167.2	3.30	– 185.9
0.66–0.91	3.70	15.50	421.0	5151.7	Trace	1.50	187.8	147.5	2.90	–247.8
0.91–1.37	3.50	16.24	151.2	5448.2	Trace	0.75	215.4	172.3	3.42	– 217.1

1. Water Soluble, 2. NH_4OAC Exchangeable, 3. Oxalate Extractable, 4. CBD Extractable, 5. KCl Extractable

6.14 Myanmar

6.14.1 Topography

Bender (1983) categorized Myanmar into four conspicuous Physiographic zones, having distinct characters and landforms. These four zones are the Sino-Myanmar Ranges, the Indo-Myanmar Ranges, the Rakhine coast Area, and the Inner Myanmar Tertiary Basin. The information about two mountainous physiographic zones, included in HKH region, is described as below.

Sino-Myanmar range

This zone includes the Shan or Eastern highlands, having high hills and mountains. The northernmost parts are ruggedly mountainous, undulating, and harsh. The highest peaks in this zone are higher than the regional snow-line altitude of about 4572 m. The highest peaks are Hkakaborazi and Gamlangrazi, which have an altitude of 5,881 m and 5837 m amsl, respectively. Between Myanmar and China, there are many passes, but the Tapeng and the Shweli are the most important as far as travel and communication are concerned. The central portion is the continuation of Yunan Highland and has an average height of about 915–1220 m. This part of the zone contains both Shan and Kayah plateaus. Many lofty peaks higher than 2135 m are found in this area. There is quite a good number of rift valleys and fault lines where many rivers and lakes are found in this zone. The arable land is limited in this area and so, the agricultural activities are limited to some flat land or/and comfortable slopes.

The Indo-Myanmar Ranges

The Indo-Myanmar Ranges encompass a series of low and medium hills. This zone is the Southeastern extension of the Himalayas. Patkai range is the important range of hills extending from Myanmar to India state of Nagaland. It is the northernmost range, which continues southward as Naga Hills with the Sarametik peak (3825 m). Toward the South are Lushai hills and Chin Hills with mount Victoria (3053 m) near Kanpetlet. Lushai hills are also part of Mizoram state of India. The Rakhine Yoma lies to the South of Mount Victoria with the highest peak of Hsung-Taung (1828 m). The mean altitude of Rakhine Yoma is over 1220 m in the North and about 915 m in the South. The Irrawaddy, the main river of Myanmar, flows from north to south through the Central Burma Basin. The river Mekong runs from Tibet to northeastern Myanmar into Laos.

The information about land use and water resources of Myanmar is presented in Table 6.40.

6.14.2 Soils

Myanmar is a vast country having different types of soil order and groups (Fig. 6.18). These soils

Table 6.40 Land use and water resources of Myanmar

Particulars	Value
Cultivable land	16.6%
Perennial vegetation	2.3%
Mountainous and forest lands	81.1%
Land with irrigation facility	21,100 km^2
Renewal water resources	1,168 km^3
Freshwater withdrawal	33.23 km^3
Use for domestic needs	10%
Use for industrial needs	1%
Use for agriculture	89%
Water withdrawal/capita	615.3 m^3/a (2017)

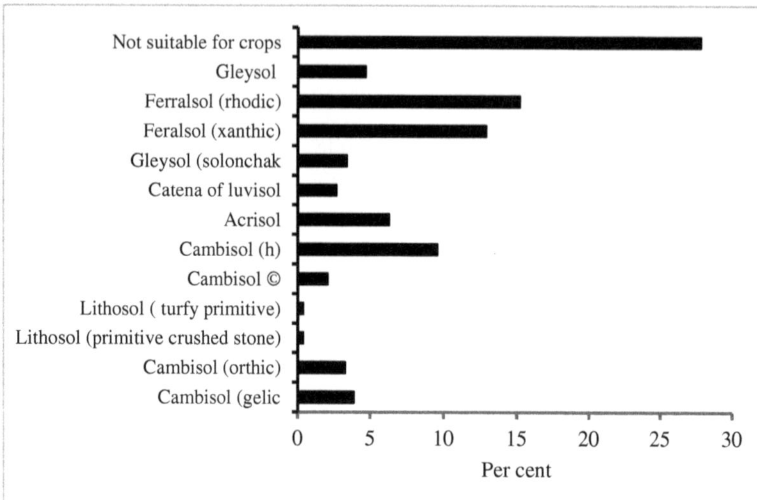

Fig. 6.18 Soil types of Myanmar

are formed as a result of variable parent material acted upon by different types of vegetation, climates, reliefs, and human interferences over a long period of time. Though there are many different soil groups, Ferralsols, Cambisols, and Acrisols are found widely. Besides these soil groups, other groups found are Fluvisols, Gleysols, Arenosols, Solonchaks, Nitosols, Vertisols, and Lithosols. Typical soils of the steep slopes of the Chin Hills at the elevation from 1220 to 1830 m are the Cambisols (orthic), or chin hill complex soils. Northern hill complex soils or Cambisols (orthic) are found in Myintkyina and in the northern mountainous region. Alpine complex soils or Cambisols (gelic) are observed on the highest snowcapped mountain peaks in Putao District. Ferralsols (Plinthic), or lateritic soils, occur mostly on the lower slope of Rakhine Yoma, Bago Yoma, and Dawna range. Ferralsols (Rhodic) soils are also known as red brown forest soil of Myanmar. These soils are found on the well-drained hill slopes at the elevation of about 305 m to 1220 m amsl. Cambisols (humic) and cambisols (Chromic) are the brown forest soils of mountains and mountainous red brown forest soil, respectively. These soils are especially found in the Shan plateau at the elevation from 1220 to 1830 m above sea level. On the lower slope of the hill in the Dry Zone of the Shan plateau, Nitosols are generally found. Catena of Luvisols on slopes and Vertisols in depressions mostly occur on the undulating slopes of the hills and gently sloping plains in the Dry Zone. It needs to be placed on record that suitable soil type is not the only criteria for good cultivation of crops for getting optimum productivity, but the amount of precipitation received within the area is very important for growing crops most successfully. Hence, the rainfall magnitude and the soil formation determine the level of agricultural production, as they are interrelated with each other.

According to the modern classification, Myanmar has 24 main types of soils. The characteristics of these soils are determined by.

- Composition of the parent material, both physical and chemical;
- Relief;
- Climate; and
- Vegetation.

In Myanmar, soil classification has generally been based on the distribution of the important land resources for agriculture. The different types of soils are described below (UNDP, nd).

(i) Alluvial Soils (Fluvisol);
(ii) Meadow Soils (Gleysol);
(iii) Meadow Alluvial Soil (Gleysol Fluvic);
(iv) Meadow Swampy (Humic Gleysol);

(v) Meadow Carbonate (Calcaric Gleysol);
(vi) Lateritic Soils (Plinthic Ferralsol);
(vii) Red Brown Forest Soils (Rhodic Ferralsol);
(viii) Yellow Brown Forest Soils (Xanthic Ferralsol);
(ix) Dune Forest and Beach Sand (Arenosol);
(x) Mangrove Forest (Orthic Solonchak);
(xi) Saline Swampy Meadow Gley (Gley–Gleysol);
(xii) Yellow Brown Dry Forest and Indaing (Orithic Cambisol);
(xiii) Light Forest Soils (Cinnamon) Nitisol;
(xiv) Red Brown Savannah Soils (Luvisol);
(xv) Dark Compact Soils (Vertisol);
(xvi) Red Earth and Yellow Earth (Acrisol);
(xvii) Mountainous Yellow Brown and Red Brown Soils (Histic Cambisol and Chromic Cambisol);
(xviii) Chin Hills Complex (Orthic Cambisol);
(xix) Northern Hills Complex (Orthic Cambisol);
(xx) Alpic Complex (Gelic Cambisol);
(xxi) Popa Complex (Andosol);
(xxii) Turfy Primitive and Primitive Crushed Stones (Lithosol);

6.14.3 Problem Soils

In Myanmar, there are many problem soils where the cultivation is associated with many constraints. An area of about 0.9716 million ha, accounting for about 5.3% of the total cultivable land area of 18.016 million ha, is occupied by soils having one or the other problem, impeding the smooth functioning of appropriate agriculture systems. The associated constraints are acid sulfate soils, degraded soils, peats, and swampy soils, constituting about 0.299 million ha, while saline and alkaline soils accounted for the remaining 0.672 million ha. Severely eroded soils are found in Shan State, Sagaing Division, and Chin State. Soil erosion in the upland regions of Myanmar is mainly due to cultivation on steep slopes.

6.14.4 Soil Characteristics

The main characteristics of the soils of Myanmar (Hadden 2008) are given hereunder:

- There exists a narrow strip of alluvial soil along the Bay of Bengal. In the wide Irrawaddy and Sittang river valleys, the mountain streams irrigate the land. This area has the most rich soils of Myanmar. Mainly silt and clay brought from the mountains and hills through runoff water are deposited in the lowland regions. These highly fertile deep soils are especially suitable for the cultivation of rice due to assured abundant soil moisture.
- The Dry Zone areas have clayey, sandy loam, and sandy soils that also contain gravel. Topography shows a great impact on these soils, and they vary with change in topography. Very low soil fertility and decreasing OM levels are found in all soil series in the Dry Zone, as indicated by soil survey data. The soils of this zone are also poor in N and K and crops and highly responsive to their application.
- The water-holding capacity of the soils of the Dry Zone is low and, therefore, crops suffer due to lack of moisture.
- Rich in calcium and magnesium, the red brown soils are found in dry areas of the central region of Myanmar.
- In some places in the Dry Zone, the soil has been almost completely removed by water and wind erosion because erosion is a big problem in this zone due to high slopes.
- In the upland areas of Kyaukpadaung and Chaung U, soil erosion is very severe due to high-intensity rainfall and high and fast surface runoff.
- The soils of the highland region of Myanmar are highly leached, iron-rich, dark red, and reddish brown.

Population growth in the Chin state is responsible for shortening field rotation (5–7 years instead of 8–10 years). Reduction in the period of field rotation has prevented the recovery of soil fertility and this has resulted in a shortfall in agricultural productivity (FAO/WFP 2009). Myanmar's Ministry of Agriculture and Irrigation has developed a table describing the soil types, locations, cultivation constraints, and suitable crops for each of eleven agro-ecological zones in the country. The fluvisols, gleysols, vertisols, acrisols, and gleysol-calcaric soils are most suitable for growing crops like pulses, chilies, onion, vegetable, groundnut, paddy, jute, maize, sesame, cotton, tobacco, and sorghum. Ferralsols are generally put under mango, rubber, tea, coffee, and forest plantations. The Cambisols, Andisols, and Lithosols soils are suitable for forests, and forestry is generally practiced in these soils in Myanmar.

6.14.5 Soil Classification

Htwe (2016) mentioned the soil classification of Myanmar soils made as follows::

- Classification of soils in Myanmar has been primarily based on data of aerial photo interpretations in 1955–1957 made by the Agricultural Planning Commission.
- The scientific classification was made by the Land Use Bureau with the help of Soviet Soil Expert in 1957. It was based on the Russian system of soil classification.
- There is a need to revise and redefine a new classification system which is particularly needed for adoption.
- The Land Use Division of Myanmar Agriculture service revised the old soil classification system in 1970 to correlate it with the FAO/UNESCO soil classification (Htwe 2016).

The classification of Myanmar soils according to FAO and USDA and the area falling under each class is given in Table 6.41.

Table 6.41 Classification of Myanmar soils by different agencies

S. no	FAO classification	USDA classification	Area (000' ha	%	S. No	FAO classification	USDA classification	Area (000' ha	%
1	Fluvisol	Fluvents	736	1.1	14	Gelic cambisol	Inceptisols	2596	3.8
2	Gleysol	Inceptisols and fluvents	3051	4.5	15	Histic cambisol	Inceptisols	6287	9.3
3	Gley–Gley	Inceptisols	555	0.8	16	Chromic cambisol	Inceptisols	1370	2.0
4	Calcaric Gley soil	Inceptisol	55	0.1	17	Plimthic ferralsol	Oxisols	588	0.9
5	Calcaric Gley soil	Inceptisol	2241	3.3	18	Lithosol	Lithic subgroup	241	0.4
6	Vertisol	Vertisols	482	0.7	19	Andosol	Inceptisols	46	0.1
7	Catena of Luvisol	Alfisols, Vertisols	1781	2.6	20	Humic gleysol	Inceptisols	203	0.3
8	Acrisol	Ultisols	4130	6.1	21	Solonchak	Inceptisols/Aridisols	42	0.1
9	Cambisol	Inceptisols	1085	1.6	22	Cambisol	Inceptisols	530	0.8
10	Radic Ferralsol	Oxisols	9971	14.7	23	Lithosol	Lithic Subgroup	290	0.4
11	Xanthic ferralsol	Oxisols	8363	12.4	24	Orthic cambisol	Inceptisols	2188	3.2
12	Arenosol	Entisols	244	0.4	25	Not suitable for crop		18,123	26.8
13	Othic cambisol	Inceptisols	2461	3.6	Total	Total		67,659	100

6.15 Nepal

Total land area of Nepal is 147, 181 km^2, measuring 885 kms from East to West and, 145–248 kms from North to South. The country has five physiographic regions, such as *Terai* (<300 m), Shiwaliks Mountain (300–700 m), Middle Mountain (700–2,000 m), High Himalayas > 2,500 m), and Mount Everest (8,848 m). Nepal is a mountainous country, with diverse and extreme relief features. The most common environment problems are land degradation, waste disposal, inland water pollution, air pollution, population growth, land acidification, and low agricultural production. Water is the most important natural resource in Nepal—the second richest country in the world, possessing about 2.27% of the world's water resources. Major sources of water are glaciers, snowmelt from the Himalayas, rainfall, and groundwater. Nepal is one of the world's least developed countries. More than 90% of the total population still lives in rural areas, under minimum physical facilities necessary to fulfill their basic needs, underdeveloped areas.

Many civilizations have disappeared from the world because of the degradation of soil resource and thc survival of civilizations depends on the performance of soils to provide food and essential goods for humans (Hillel 2007). Some important tools for proper soil management like use of GPS for collecting soil samples is very important for preparing thematic soil fertility maps (Mishra et al. 2013) and Geographical Information System (GIS) is a potential tool used for retrieval and manipulation of data.

6.15.1 Soil Classification of Nepal

LRMP (1986) listed 14 major soil groups covering 4 soil orders that are generally found in Nepal. According to USDA taxonomy, major soil orders of Nepal are Entisols, Inceptisols, Mollisols, and Alfisols. Other soil orders like Spodosols, Histosols, Ultisols, and Aridisols have been observed at some locations. It is observed that the soils produced as a result of weathering of soft rocks such as Phyllite, Quartzite, Sandstone, Granites, Gneiss, and Schists have a high degree of porosity and poor slope stability, shallow soil depth, coarse texture, and acidic reaction.

6.15.2 Specific Soil Problems

Like in any other region, the soils of Nepal also have some specific issues and problems which generally hinder crop productivity. Of late, the crop yields are showing a decline or stagnation, which is a matter of concern, challenging the food security of the country. Due to rugged topography and steep slopes, soil erosion is a serious problem. It has given rise to land degradation as soil conservation measures cannot be extended to all parts of the country due to economic reasons. The cultivable soil is being continuously mined for crop production in an effort to feed the increasing population. There has been a continuous decline in the organic matter and other essential nutrient content of the soils, leading to their exhaustion. Prevalence of unique climatic conditions in Nepal cause landslides and flooding during the monsoon rainy season. Because of the enhanced use of pesticides and fertilizers, the problem of soil pollution and contamination has also been found at some locations. Moreover, soil sealing and crusting have also been reported.

6.15.3 Soils

It has been found that the mountains in Nepal have steep to very steep slopes (>60%) and have very high ridges and narrow valleys all over the country except the *terai* zone. By and large, the soils are predominantly shallow, having Lithic and Typic subgroups of Ustorthents, Ustochrepts, Dystrochrepts, and Haplumbrepts. The mountains with middle ridges and narrow valleys have slopes varying from 15 to 70%. In some places, the soils are moderately deep to deep which are Typic Ustochrepts and Typic Kanhaplustalfs. About 5–60% varying slope is observed in the mountains with low ridges and hills and fans. In the main valley of Likhu Khola and in the tributary valleys, the soils are generally deep and belong to Anthraquic subgroup of Ustifluvents, Ustochrepts, and Fluventic Ustochrepts. Terraces have been found important in conserving soil moisture and protection of the land from erosion. Generally, the level terraces are used for rice cultivation, though some other crops like millets and maize are also grown in some areas. The forest type is pine (*Pinus roxburghii*) in association with broad leaf trees (*Schima wallichii*) at higher elevations and sal forest (*Shorea robusta*) at lower elevations, in the mountain areas with high peaks and narrow valleys. Soil Science Division, Nepal Agriculture Research Council (Soil Science Division 1992) conducted a detailed soil survey, and the results are based on soil data generated through this survey. Aerial photo interpretation of the sub-watersheds were based on geopedologic analysis (Zinck 1988).

Soil organic carbon

Carbon is a major constituent of organic matter in the soil and so, it adds to soil fertility as well as its sequestration in the soil. Crop productivity is highly dependent on soil fertility. Optimum organic carbon content in soil is important as it is needed to improve soil quality, increased productivity, and form stable soil aggregates. Organic carbon contributes not only to the reduction of atmospheric CO2, but also helps to increase food production, thereby enhancing food security (Shrestha et al. 2017). They suggested ways to enhance carbon in the soil complex. For mountain areas of the Hindu Kush Himalayan (HKH) region, the beneficial relationship between climate change mitigation and enhancing food security is even more appropriate. In the Himalayas, there is tremendous potential to increase CO2 sequestration and at the same time to increase food security in the long-standing acute food deficit areas. To compare the topsoil and sub-soil dynamics of the soil nutrients in Nepal, soil samples were taken from two depths: 0–15 cm and 15–30 cm below the soil surface (Shrestha et al. 2017).

6.15.4 The Nepal Himalayas and Land Use

The Nepal Himalayas form the central part of the HKH region. Agriculture is the main profession as about 65% of the country's population is directly engaged with this activity. The population of Nepal was 26.494 million according to the 2011 census. Since the population growth rate is 1.35%, annually, the expected population

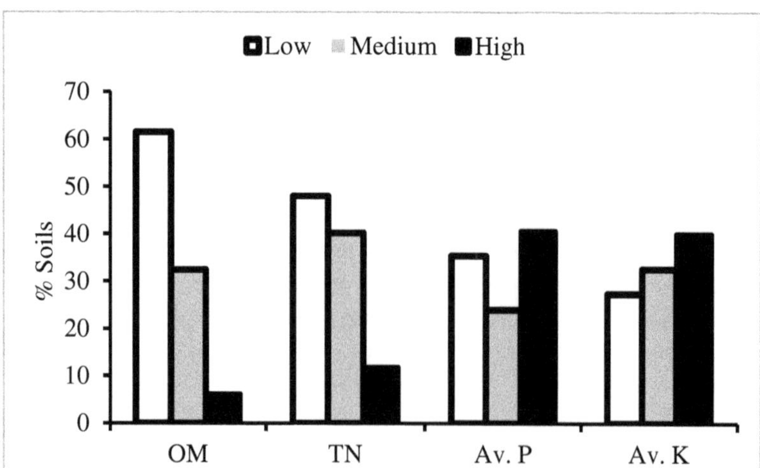

Fig. 6.19 General status of nutrients in Nepal soils. (Based on data from Regmi et al. 2005)

of the country should be about 30.295 million in 2021. The agricultural sector contributes to about 39% of the national GDP. The Middle Mountains of Nepal are densely populated and this region is particularly vulnerable to food insecurity. This region has only about 16% of the total area as arable land and is highly fragile ecologically and active geologically (Bajracharya and Sherchan 2009). The major cause of low yields is the low organic matter content of the soils and the very low use of fertilizers (Sherchan and Karki 2005).

6.15.5 Fertility Status of Soils in Nepal

For overall development, Nepal has been divided into five zones, namely Eastern Development Region (EDR), Central Development Region (CDR), Western Development Region (WDR), Mid-Western Development Region (MWDR), and Far-Western Development Region (FWDR). The fertility status of soils of Nepal is presented in Fig. 6.19. On average, in EDR, CDR, and MWDR, more acidic soils were reported from hills than plain areas. In plains, 46, 34, and 20% soil samples were found in pH categories of < 6.0, 6.0 to 7.5, and >7.5, respectively. The corresponding pH values in hills and mountain areas were 68, 20, and 12%, respectively. In plains, 63, 33, and 4% soil samples were having organic matter (OM) content of < 2, 2 to 4, and more than 4%, respectively; 53, 42, and 5% samples having total N (TN) content of 0.1, 0.1 to 0.2, and > 0.2%, respectively; 38, 27, and 35% samples had available P (AP) content of <50, 50 to 200, and 200 ppm, respectively; and 46. 37, and 17% samples had exchangeable K (EK) content of < 50, 50 to 200, and >200 ppm, respectively. The corresponding values of OM, TN, AP, and EK content in the case of soil samples from mountain areas were 57, 38, and 5%; 47, 42, and 11%; 28, 22, and 50 ppm; and 33, 27, and 40 ppm, respectively. It was observed that the OM, TN, AP, and EK were slightly higher in mountain soils than plains.

The status of micronutrients, such as boron, copper, iron, manganese, molybdenum, and Zn in soil, has been studied to some extent by Andersen (2000). Only Fe and Mn were adequate in Nepal soils while all other nutrients were low (Karki et al. 2005). Boron, copper, zinc, and molybdenum were observed to be low. Sapkota and Anderson (2005) reported that the mean concentration of B, Cu, Fe, Mn, and Zn in Nepal soils was 0.49, 2.52, 224, 45, and 1.62 mg kg^{-1} of soil. The highest concentration of B was found in soil samples of Palpa district, that of Cu, Fe, and Zn in Kathmaandu district soils, and Mn in Myagdi district soils.

6.15.6 Soil Fertility Variations

Tripathi and Shrestha (2000)and Tuladhar (1995) evaluated the soil fertility status of soils in relation to the altitudinal and physiographical regions in Nepal and found that the soils of the hills region of Nepal were more acidic in nature compared to those of the plains (Terai region). The mountain soils contained more organic matter than the soils of plains. Available P and exchangeable K levels were found to be high in mountain soils. In the Middle Mountains, elevation had no clear effect on extractable boron or zinc (Fig. 6.20a, b and c). In western Nepal, soil OM, TN, and available P all showed an increasing trend with elevation (Fig. 6.21a and b).

6.15.7 Influence of Cropping Systems on Soil Fertility

The cropping patterns on arable land affect soil fertility as well as agro-ecology as influenced by landscape positions, soil types, and slope aspect. Rice and wheat are typically grown mainly in the valleys and lowlands, while maize, millet, and barley are grown mostly in the sloping upland areas. Wheat, however, is grown both in irrigated lowland areas, as well as on upland terraces to considerably high elevations in eastern, central,

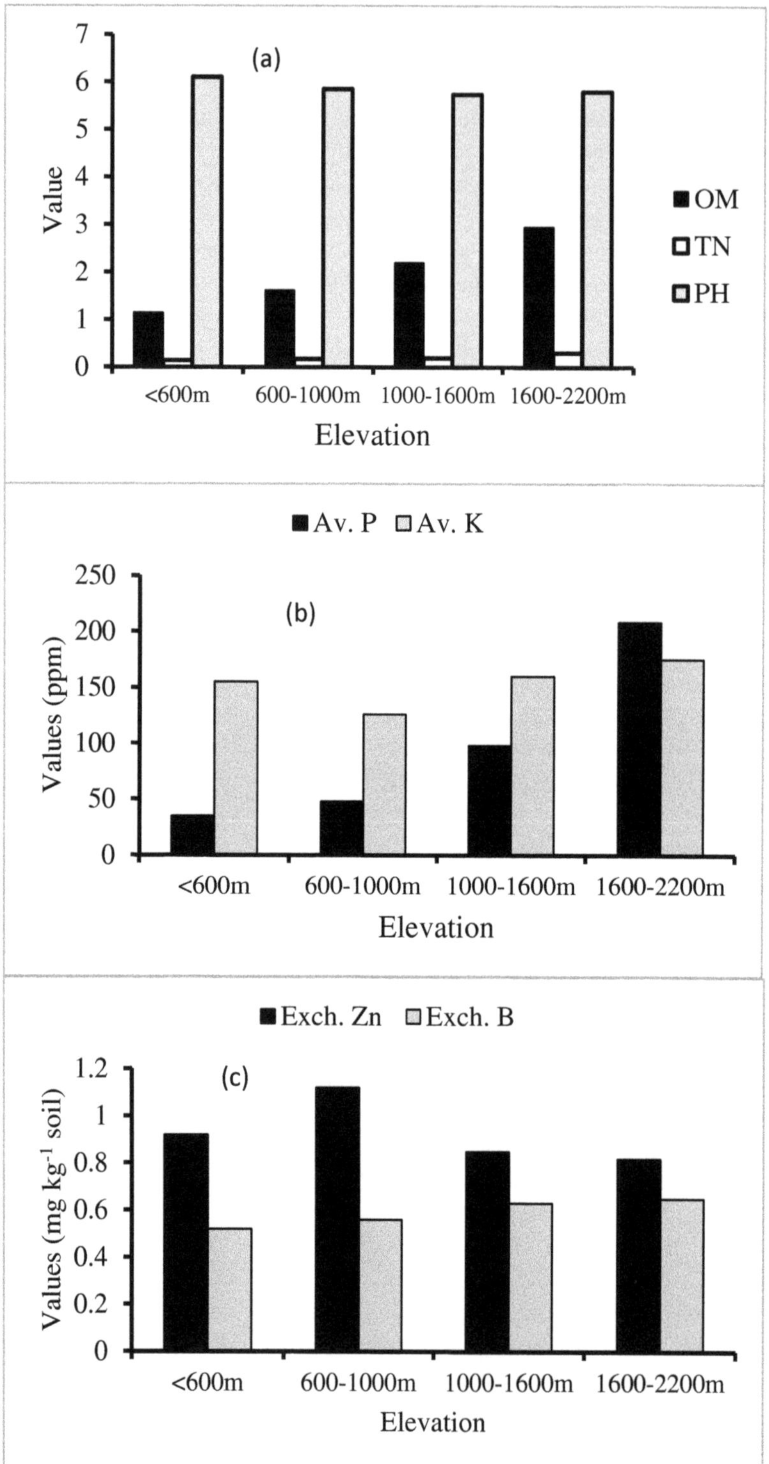

Fig. 6.20 Variation of soil fertility parameters with elevation in western Nepal. (Figures drawn based on data from, Tripathi 2001)

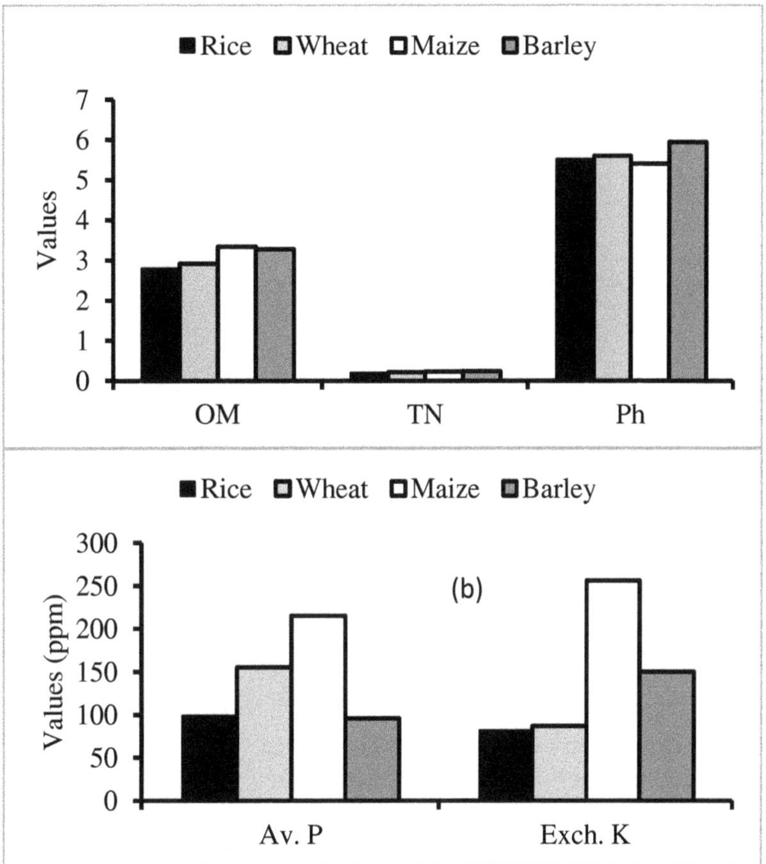

Fig. 6.21 Fertility status of soils under different crop types in western Nepal. (*Source* Tuladhar 1995)

and western Nepal (Sherchan et al. 1997a, b; Sherchan and Gurung 1998; Tripathi 2000; Tripathi and Shrestha 2000). Analysis of the soil fertility status by land use types in the mid-hills indicates forest and shrub land soils to have lower pH and higher OM contents than cultivated soils. A high variability observed in OM is due to forest condition (managed vs. degraded), stand density, and regular collection of leaf litter and fodder by local farmers particularly from unprotected forests (Shah and Schreier 1995; Schreier et al. 2000). Based on the studies (Bajracharya et al. 2006, 2007; Dahal et al. 2006; Sherchan et al. 2003; Baral et al. 1999; Tripathi 1999; Howell 1986), it was observed that forest soils were rich in organic matter (3.9%) and total N (0.17%) compared to upland and lowland soils. However, lowland soils contained highest available P (171 mg kg^{-1}) and upland soils highest exchangeable K (417 mg kg^{-1}) as compared to other soils.

6.15.8 Changes in Fertility Status

A FAO study revealed that the Nepal soils were, on the whole, neutral in reaction; low in OM, available P, boron, molybdenum, and zinc; medium in total N, exchangeable K, copper, and manganese; and high in calcium, magnesium, and iron. At various Regional Agriculture Research Centers, long-term fertility experiments showed an improvement in soil OM and total N, possibly due to better management practices. Some workers have also studied short-term changes in soil fertility status and crop productivity (Tripathi and Shrestha 2000, Sherchan and Gurung 1998). In western Nepal, Tripathi and

Shrestha (2000) assessed the change in soil fertility due to the application of organic manures and inorganic fertilizers. In a benchmark study, Sillanpaa (1982) reported that the mean values of pH, soil OM, CEC, total N, available P, exchangeable K, Ca, Mg, B, Cu, Fe, Mn, Mo, and Zn were 6.1, 1.72%, 19.7 cmole kg^{-1}, 0.119%, 21.6 mg l^{-1}, 80 mg l^{-1}, 1766 mg l^{-1}, 205 mg l^{-1}, 0.21 mg l^{-1}, 6.8 mg l^{-1}, 363 mg l^{-1}, 21.2 mg l^{-1}, 006 mg l^{-1}, and 1.6 mg l^{-1}, respectively. In comparison, however, an increasing trend in soil pH, OM, available P and K, and exchangeable Ca and Mg over a four-year period under both maize–millet and rice–wheat cropping systems was found in eastern Nepal (Sherchan et al. 1997a, b; Sherchan and Gurung 1998).

6.15.9 Alternatives for Fertility Enhancement

Long-term fertilizer trials in Nepal have indicated that continuous cropping of rice and wheat with chemical fertilizers decreased available P, exchangeable K, and some micronutrients like boron, molybdenum, and zinc. Tripathi and Shrestha (2000) and Tripathi (2001) reported that in farmer's fields of western Nepal, soil fertility also declined during the short-term trial between 1997 and 2000. During the past 2–3 decades, crop yields have been found to be declining (Maskey and Joshi 1991; FAO 1994; Brown et al. 1999) while food demands continue to increase, leading to a food deficit. However, integrated plant nutrient systems (IPNS) have been reported to be beneficial as shown by an experiment conducted at various locations. Sharma and Sharma (2004a, b, c, d, e) concluded that a combination of chemical fertilizers and FYM along with the inclusion of a legume in the crop rotation (rice–potato–mungbean) was highly effective in building up soil organic carbon, nitrogen, phosphorus, and potassium levels.

Soil pH

Soil pH affects many chemical processes in the soil (Havlin et al. 2010). High acidity is a major constraint in crop production. There were variations in soil pH and other nutrients. Panda (1987) explained that the cause of high acidity may be because of acidic parent materials like quartz, granite, rhyolite, etc. For optimum crop production, there is need to reduce soil acidity by application of amendments like lime in proper amounts. This would also help in improving soil fertility for sustainable soil fertility management and good soil health. The soil pH showed A low variability (8.05%) in soil pH was exhibited among the soil samples taken for study purpose. Khadka et al. (2018) analyzed the soil of Chungbang Farm, Pakhribas, Dhankuta, Nepal, and found a mean content of organic matter (1.32%), N (0.08%), P_2O_5 (54.88 mg kg^{-1}), and K_2O (104.1 mg kg^{-1} of soil).

Organic matter

Any material of biological origin that decomposes and forms a part of the soil is known as organic matter (Bot and Benites 2005). It is added to the soil through the use of farmyard manure, farm wastes, compost, green manuring, crop residues, and lopped succulent branches of trees. Many improper agriculture practices might be the cause of the low content of organic matter in the soil (Brady and Weil 2002). With the passage of time, the soils become poor in organic matter if only inorganic sources of nutrients are applied for a longer period of time. It is essential to add different materials which may contain high amounts of organic matter to the soil for build-up of OM in the soil and to obtain higher crop productivity. Adding organic materials like farmyard manure, compost green manure, oil cakes, farm and household wastes, and crop residues, and adoption of suitable crop rotation, mulching, necessary tillage, and good management practices are important for soil organic matter improvement in the soil. The soils of

Nepal are low to medium in organic matter content, though high contents have also been found in forest soils. This might be due to addition of forest litter in large quantities falling from the forest vegetation.

Total N, available P, and K

Like elsewhere, nitrogen is the most important element required for optimum crop growth in Nepal too. The total nitrogen content varied from 0.04 to 0.11% with a mean of 0.08% in the soils of Nepal. Brady and Weil (2002) explained that the low content of total nitrogen might be due to the low content of organic matter in the soils of Nepal as well as high leaching loss of inorganic nitrogen from the soil. It was due to the reason that nitrogen is a major component of organic matter. Low P availability of Nepal soils is the major crop productivity limiting factor. Low productivity of crops in Nepal has been attributed more to the deficiency of phosphorus, except nitrogen, than the deficiency of any other elements (Singh et al. 2016). Phosphorus is known to be a very immobile element in the soil. A full recommended dose of P is required to be applied to the soil in order to satisfy the needs of soil fixation with sesquioxides in acid soils as well as crop needs. Since potassium is involved in many physiological processes, it is an essential macronutrient for plants to realize the full potential of a crop (Amtmann and Francisco 2012). The available potassium had medium status in soils. About 60% of soils showed medium status whereas 40% 0f soil have high status of potassium. High variability (59.82%) in available potassium was found among the soil samples.

Secondary and micronutrients

Calcium is important secondary nutrient required for cell growth, division, elongation, and various essential biological functions (Hirschi 2004). Calcium, as a constituent of various lime materials, also works as an amendment on acidic soils. Low calcium status in the soils could be due to very high soil acidity. The availability of basic cations is reduced at very low or very high soil pH. Therefore, the amelioration of soil acidity is a very necessary solution for increasing calcium availability in the field. Moderate variability (36.77%) in available calcium was observed in the soil of Nepal. Due to its active role in metabolic processes such as DNA synthesis, respiration, and photosynthesis, iron is an essential micronutrient for crop growth (Rout and Sahoo 2015). However, it limits crop productivity also if it is present at toxic levels in the soil.

6.15.10 Status of Natural Resource Management in Nepal

- Forest resources are being seriously depleted every year.
- Steep topography—sensitive to erosion, acidification, and causing soil fertility depletion.
- Hill Agriculture system is complex.
- Scarcity of arable land due to high population density.
- Few employment opportunities.
- Market problems and weak institutional support.
- Annual food shortage in hilly districts.
- Encroachment upon forest lands, pasture, steep hill slops, and river fans (marginal lands).
- Mountainous regions—potential for livestock and horticulture production.

6.15.11 Soil Acidification and Its Impact on Nutrient Deficiency

Optimum soil pH for the majority of agricultural crops is between 5.5 and 6.5. Recent soil survey and experiments reveal soil pH of middle mountain ranges from 3.3 and 7.5 with a mean value of 5.8.

Reasons for the acidification of soils:

- Dominant bedrock—sandy stone, siltstone, and quartzite, all of which produce acidic soil materials.

- Cultivation practices—introduction of double and triple annual crop production, improper use of chemical fertilizer mainly ammonium sulfate and urea fertilizers.
- Less use of organic manure and no lime application.

6.15.12 Rating of pH in the Western Hills

Tripathi (1999) analyzed 261 soil samples from the Western Hills of Nepal and found that about 87.3% of the soils were acidic in reaction, 12.2% neutral, and 0.4% slightly alkaline.

6.15.13 Land Degradation

Though cultivation is carried out on high mountain slopes in Nepal, moch of Nepal's land is environmentally fragile and prone to erosion and degradation. So, a lot of surface soil is lost through runoff and settles in valleys and in the hills and plain areas too. Once shifting cultivation used to be common land use but over the years, this practice has been abandoned in most of the areas. This system of land use was compatible with the environment. This traditional land use system has not been able to cope with the ever-increasing population of both humans and livestock. The shifting cultivation encouraged deforestation, resulting in soil loss and land degradation. Recent decades have witnessed land degradation and degradation of mountain ecosystems in Nepal and, it is spreading fast. The traditional farming systems and cultivation on steep hill slopes have enhanced the rate of erosion and degradation.

Mountain agriculture and sustainability in a country like Nepal where the majority of the population have to depend upon mountain agriculture for their livelihood in Nepal. With increase in population, there has been an intensification of agriculture to satisfy their needs and demand. The mountain ecosystem, however, has not been able to bear the burden of the increasing population. The declining trend of farm productivity in the hills and mountains is much the result of the decline in soil fertility and land degradation due to soil erosion of fertile surface soils every year. There are no sound technologies also to avert the soil degradation processes. Maskey et al. (2003) mentioned the following issues facing mountain agriculture in Nepal:

- Low productivity due to land degradation.
- Lack of access to suitable and low-cost technology and lack of marketing facilities.
- Inadequate credit facilities.
- Lack of proper market facilities.
- Poor community organizations.
- Small and fragmented land holding.
- Natural factors like cyclones, landslides, floods, droughts, etc.
- Over-exploitation of natural resources.

Joshi et al. (1997) reported the following five causative factors for land degradation, relevant in the context of Nepal:

- Cutting of forests,
- Free-range grazing,
- Farm activities,
- Mismanagement of natural resources, and
- Bio-industrial and industrial activities.

6.15.14 Soil Quality in Nepal

Like other components of our physical environment, such as water or air, the quality of soil has, in recent years, become a topic of increasing concern and research (Bajracharya et al. 2007). The soils interact with all components of terrestrial ecosystems (Bajracharya et al. 2007). The agriculture sector faces the daunting task of meeting ever-increasing demands for food and fiber, while simultaneously conserving and developing the soil resource for future generations (Sturtz and Christie 2003). Hence, there is an undisputed need for regular, rapid, and cost-effective means of assessing the quality or "health" of soils. Two distinct approaches have been followed in the evolution of the soil quality concept, namely education and assessment (Karlen et al. 2003). The education aspect emphasizes increasing public and farmer awareness and understanding of the complex nature and multiple functions of soils and the importance of monitoring, maintaining, and enhancing its biological, chemical, and physical quality. The assessment of soil quality in different parts of the world, as influenced by land management practices and human intervention, also appears to take two approaches. The first is a research-based approach aimed at determining the usefulness, accuracy, and sensitivity of various soil parameters to serve as indicators of soil quality or sustainability of practices, while the other is a practical approach involving the development of assessment tools for farmer-based evaluation and education regarding the effects of crop and soil management practices on the soil resource, such as test kits and score-cards (Karlen et al. 2003; Karlen 2004).

The productivity and fertility of soils may be better reflected by SOC, forms of labile carbon, such as particulate organic carbon (POC), oxidizable C, and microbial biomass C, as well as by microbial biomass nitrogen, N mineralization, and soil respiration (Chen 2001). The two primary hazards facing soils in the steep, geologically active, and fragile Himalayan agroecosystems are soil erosion and fertility/productivity decline. It, therefore, follows that a soil quality index useful for farmers, conservation workers, and planners should encompass both these aspects of soil quality. The following is a review and analysis leading up to the suggestion of such an index through examination of physical, chemical, and biological aspects of soil quality in Nepal. Categorization and classification of soil types have been done only on a few occasions as a part of a broader land resource mapping (LRMP 1987) and for agro-ecological zoning for sustainable agricultural planning and development (Sharma et al. 1994). Sharma et al. (1994) described in some detail the soil units of Bagmati Zone, Central Development Region, Nepal, on the basis of inherent soil fertility, textural class, drainage, rooting depth, workability, flood risk, and erosion hazard. Studies on available micronutrients in western and eastern Nepal generally indicate the status of Zn and B to be low, while Fe, Mn, and Cu tend to be adequate (Tripathi 1999; Sherchan et al. 1998; Andersen 2000). Soil runoff and erosion are highly variable in the hills and mountain regions of Nepal (Chalise and Khanal 1997). In an extensive review of erosion processes in the Nepal Himalayas, Chalise and Khanal (1997). It has been observed that the runoff varied from as little as 0.5% to as much as 35% of incident rainfall in various parts of Nepal. High runoff proportions (20–25% of rainfall) were noted for shifting cultivation practices.

6.15.15 Biological Soil Quality

A variety of factors happening within the soil profile, at the surface or above the ground, that are associated with the living component of the soil ecosystem, or that derive from it, constitute the biological quality of soil. This includes the diversity and species composition of soil organisms, namely meso- and macrofauna, microorganisms, and flora (the types of plants, their root systems, and the vegetative litter produced at the soil surface). The amount and quality of soil organic matter (SOM) or organic carbon (SOC) is a reflection of the nature and abundance of soil flora and fauna and can be considered a biological soil quality indicator. Bajracharya et al. (2004) concluded that expectantly forest and shrub land soils had higher SOC contents in the top 30 cm, however, due to shallow soils and low density, forest soils had overall lower total OC stocks than other land uses. While cultivation tends to reduce OC contents of soils, practices such as mulching, minimum tillage retention/addition of organic residues, and cover crops could all enhance productivity and reduce erosion losses. Sitaula et al. (2004) gathered the available data on SOC status under various land uses in different watersheds of Nepal and reported that mean SOC contents in topsoil ranged from a low of 0.1% for severely degraded forests and grazing lands to a modest 4% under well-managed forest. For agricultural soils, the SOC contents fell mostly between 1 and 3% in topsoil. Median SOC pools in the upper 1 m soil profile was estimated to range from 2.6 kg C m^{-2} for Arenosols to a maximum of 17.8 kg C m^{-2} for Humic Acrisols (Sitaula et al. 2004). Shrestha et al. (2007) investigated SOC stocks and C sequestration in a mid-hill watershed in western Nepal and reported a significant influence of land use on SOC contents and distribution in various aggregate size classes. Among the cultivated land use types, upland soil had higher SOC, while natural forests had the highest overall OC stocks. A net loss of 29% of SOC stock in the upper 40 cm soil layer was calculated due to changes in land use over the period from 1978 to 1996 (Shrestha et al. 2007).

6.15.16 Soil Erosion in Nepalese Himalayas

Soil erosion is an acute problem under Nepal conditions where more than 80% of the land area is mountainous having very steep slopes. The magnitude of soil erosion has further been tremendously accelerated due to population increase, deforestation, intensive agriculture, and overgrazing, due to population pressure. The annual soil loss rates are the highest in the areas with rainfed cultivation on higher slopes (Morgan et al. 1984). The annual soil loss per ha was up to 1, 1 to 9, and 8 t ha^{-1} from dense forest, degraded forest, and grazing lands, respectively (Shrestha 1997). Landslides, collapse of man-made terraces, soil loss from steep slopes, and decline of forest areas are mainly responsible for resource degradation in the Himalayan region (ICIMOD 2006). Deforestation in the middle mountains is an issue of concern. Deforestation continues mainly for subsistence agriculture. However, despite population growth, the condition of the forest seems to have improved in some areas and, in some cases, the cutting of forests has been compensated by the plantation of more trees on agricultural land (Carter 1992; Gilmour 1991).

6.16 Pakistan

The soils of Pakistan have been identified and classified into six orders, namely Alfisol, Aridisol, Entisol, Inceptisol, Mollisol, and Vertisol. There are 21 suborders, 38 great groups, 112 subgroups, 353 families, and 894 soil series. Due to its geographical location, Pakistan has variable climates from tropical deserts to temperate deserts, perpetual snow cover on high mountains, variable landforms, and highly diversified landscape.

6.16.1 The Soil Survey of Pakistan

Since its inception in 1962, the Soil Survey of Pakistan (SSP) is working under the Ministry of Food and Agriculture, Govt. of Pakistan. The soil survey project was given the status of a regular department in July 1973. The main objective of the project were.

"To prepare an inventory of Pakistan's soil resources through standard reconnaissance soil surveys, in the form of reports and maps, which would assist the government in planning of projects aiming at new land development, irrigation extension, drainage improvement, land reclamation, soil conservation, forestry development, watershed management and range improvement, etc., and an overall economic development of the country". Area under different land uses in Pakistan is given in Tables 6.42 and 6.43.

6.16.2 Soils of Pakistan

Soil is the single-most essential natural resource on which the survival and welfare of humans depends. Pakistan has the distinction of having a large variety of soils, which vary significantly in nature and distribution. This allows adopting different pathways to optimally and sustainably use this resource. Optimal and justified use of soil may not only ensure permanent availability of the required human needs, but also improve

Table 6.42 Land use categories of Pakistan (000' ha)

Land use	Area	%	Land use	Area	%
1. Agriculture	21,733	27.3	6. Riverain forests	239	0.3
2. Rangelands	25,475	32.0	7. Wastelands (including area under snow and ice)	28,501	35.8
3. Coniferous forests	1353	1.7	8. Water bodies (rivers only)	1274	1.6
4. Irrigation plantations	80	0.1	9. Others	159	0.2
5. Scrub forests	796	1.0	Total	79,610	100

Table 6.43 Land use categories of Pakistan (Area in million ha)

Land use type	Area	%
Agriculture	21.73	27.3
Rangelands	25.48	32.0
Coniferous forest	1.35	1.7
Irrigated plantations	0.08	0.1
Scrub forests	0.79	1.0
Riverain forests	0.24	0.3
Wasteland including area under snow	28.50	35.8
Water bodies (rivers only)	1.28	1.6
Others	0.16	0.2
Total	79.61	100.0

the overall environment, quality of life, and livelihood security of the people (Mian and Mirza 1993). Productive soils that could be considered as prime agricultural land are scarce in Pakistan, despite having vast soil resources. In Pakistan, there is a need to protect prime and productive farmland from over-exploitation and mismanagement that may cause its degradation or loss through erosion. Pakistan has to rely on existing soil resources to get their full potential in order to satisfy human needs.

6.16.3 Soil Inventory

Soil Survey Project was undertaken by the Government of Pakistan and initiated the inventory of soils of the country in the year 1962. With the assistance of the United Nations Development Programme (UNDP), and the UN Food and Agriculture Organization (UNFAO), this project was actively and objectively started. The main objective of the government of Pakistan in starting this project, initially, was to help the government in strengthening other projects such as new land settlements, soil reclamation, irrigation extension and drainage schemes, soil and conservation, rangeland improvement, and afforestation programs. About 81% of the total area of the country has been surveyed through Soil Survey Project, including the Federally Administered Tribal Areas and the Northern Areas. The total area surveyed was 91,000, 347,000, and 73,000 km2 in NWFP and FATA provinces, Balochistan, and Galkit/Baltistan provinces, respectively. The total area surveyed stands at 511,000 km^2 (Soil Survey of Pakistan 2010).

There were two major purposes of reconnaissance reports; firstly, it provided basic data on soils for agricultural development planning and rational land use, and secondly to ensure that this resource is conserved while agricultural production is enhanced to bridge the food deficit gap. In total, there were 55 reconnaissance soil surveys. The inventory also included five integrated soil and vegetation reports, as well as maps of "Landforms", "Soils", "Land Capability" and "Present Land Use/Natural Vegetation" at 1:125,000 or 1:250,000 scale, and of "Generalized (Agricultural) Development Potential" at 1:750,000 scale. The agricultural operations based on soil survey data greatly helped in managing the soils on a sustainable basis and also increased food production for the benefit of the population.

6.16.4 Soil Potential

Reconnaissance soil surveys: province-wise coverage

Soil Survey of Pakistan was able to evaluate the soils of the surveyed area for sustained production of major crops as well as some special agricultural crops. It also helped in understanding the potential of soils for forestry and rangeland development. An exercise was also undertaken to integrate the data on soils with climatic, hydrological, plant, and other relevant information. About 750 different kinds of identified soils were then grouped into eight Land Capability classes (Class I–Class VIII) (Table 6.44), according to their agricultural potential or relative suitability for sustained agricultural use. Classes I–IV are meant for arable use where any crop can be grown according to the suitability of land, however, Classes V–VIII are appropriate for non-agricultural crops like forestry, grasslands, and rangelands. Broadly speaking, the land capability classification is broadly based on the relative degree of crop growing constraints, severity of various problems, and limitations of a particular soil group for agricultural use. Standard methods were followed to place various areas under different capability classes.

6.16.5 Soil Types

Pakistan has varied types of soils, which vary significantly from place to place according to the basic parent material, altitude, vegetation, etc. There is a need for an appropriate approach to dealing with these soils in order to use and manage them optimally and sustainably (MoE, GoP 2009). Although the country's soil resources are vast, the prime agricultural land for cultivation is limited in HKH part of Pakistan, though the country has vast soil resources. The extent of such soils can be increased only a little by improving a part of relatively poor-quality soil, though this would require heavy investment by the farmers or by the government through some big scheme. The suitable preposition would be to protect the arable agricultural land from misuse by the cultivators and other agencies through inappropriate management practices that may result in its degradation or loss.

6.16.6 Fertility Status

Mian and Mirza (1993) reported that the total area under cultivation in Pakistan is 21.59 Mha (27.1% of the geographical area of the country). Out of this cultivated area, only 5.34 m.ha. (i.e. 24.7% of cultivated area) has no specific soil limitations which may hamper agricultural productivity, and is suitable for intensive agriculture. The remaining agricultural land has a lot of various types of problems which affect crop performance at different stages of plant growth. This may include water erosion, waterlogging, formation of slow permeability, salinity and sodicity, and wind erosion. A permanent decline in agricultural productivity in Pakistan's HKH region has been observed due to nonmanagement or mismanagement of various soil limitations and suitably address them.

Khattak and Hussain (2007) conducted a study in NWFP and Punjab to evaluate the fertility status of the soils from orchards of apple,

Table 6.44 Land capability classification of Pakistan (000' ha)

Class	Province					Pakistan
	Punjab	Sindh	KPK + FATA	Balochistan	Galgit/Baltistan	
I	3486.4	1105.3	187.3	598.9	2.4	5380.3
II	3679.2	2336.2	524.4	481.0	145.3	7166.1
III	2395.1	1498.8	665.8	315.4	77.2	4952.3
IV	1439.9	838.5	581.6	929.2	105.5	3894.7
V	–	–	70.1	–	101.1	171.2
VI	261.8	8.3	827.0	84.6	114.6	1296.3
VII	4,610.6	2454.2	2603.8	9294.7	869.4	19,832.7
VIII	4159.7	3372.3	2974.0	22,699.5	4364.1	37,569.6
Total	20,032.7	11,613.6	8434.0	34,403.3	5779.6	80,263.2

Soil Survey of Pakistan, Lahore, 1988, 2010

peach, pear, citrus, and plum. The physico-chemical properties of soils showed that the texture of all soil samples collected from different areas were in the loam and silt loam categories. This was because of the removal of clay content from the upper soils due to soil erosion leaving the higher contents of sand and silt. The soils were found to be moderately calcareous in their properties, and OM was medium to high in most of the soil samples. Nitrogen content was poor and more than 50% of soil samples were found deficient in N. About 35.4, 46.2, and 53.8% samples of Birote, Boi, and Thai areas contained low to medium concentrations of P (<10 mg kg^{-1}). The low content of N and P might be due to their loss through leaching or runoff, as the texture of the soils was mostly loam or silt loam. Having K-bearing minerals in the parent material, Potassium concentrations were sufficient in all the soils. Almost all the soils were sufficient in Cu, Fe, and Mn concentrations with 4.2 and 23.1% of soils in Birote and Boi area having deficient Zn levels (<1.0 mg kg^{-1}). Major threats to Pakistan agriculture are soil health, farm inputs, climate change, agriculture credit, farm mechanization, agricultural marketing, and extension services.

Out of all the factors, resulting in the decrease of agriculture crop production in Pakistan, the status of nutrients in the soil or soil fertility is the most important factor (Khattak and Hussain 2007). Perveen et al. (2010) reported that the yields as well as quality of almost all crops in Pakistan are well below their potentials as compared to the other developed and developing countries of the world. The soils of Pakistan are poor in essential nutrients content and, thus, unable to support optimum levels of crop productivity due to the poor status of available nutrients in the soils (Ahmed and Rashid 2003). It has been observed through a series of studies that widespread micronutrient deficiencies have emerged in Pakistan because of the rise in intensity of cultivation, adoption of high-yielding varieties, and the increased use of high analysis fertilizer materials as well as the basic nature of Pakistan soils. Numerous researchers reported the deficiencies of macro- and micronutrients in KPK soils (Khattak and Hussain 2007; Rashid and Bhatti 2005). The nutrients like P, N, and K are required in larger and on regular basis to the crops, while micronutrients like Fe, Cu, and Zn are required in smaller amounts as compared with macronutrients but these perform an important role in various physiological functions of crops (Imran and Gurmani 2011).

Jamal and Jamal (2018) conducted a study on fertility status of Swabi district, Khyber Pakhtunkhwa, Pakistan. The results of the study indicated that the soil of Swabi was moderately to strongly calcareous in nature and alkaline in reaction. The OM content was found to be deficient in 75% and marginal in 25% of soil samples. Soi samples taken from 11 locations showed that pH varied from 7.7 to 8.6 with a mean value of 8.2. The EC varied from 0.13 to 0.56 dSm^{-1} with a mean value of 0.27 dSm^{-1}. The mean values of CaCO3, OM, P, K, Fe, Zn, and Cu were 7.16%, 0.82%, 9.22 mg kg^{-1}, 137 mg kg^{-1}, 7.40 mg kg^{-1}, 1.88 mg kg^{-1}, and 5.8 mg kg^{-1} of soil (Jamal and Jamal 2018).

Other than abiotic factors like total ion content, acidity, carbon, total nitrogen, and total phosphorous, a few other factors like climate, landform, topography, soil texture, soil moisture, and plant community composition also affected soil properties (Maria et al. 2004; Takata et al. 2008). Strongly correlating with the composition of plant communities are the concentrations of soil carbon, nitrogen, total ion content, and acidity. The soil geochemistry and the distribution pattern of vegetation in an area are independently affected by the total ionic contents and acidity influence.

To assess the seasonal and spatial changes in the soil physico-chemical characteristics and nutrient availability, Ahmad et al. (2011) conducted a study at six ecologically diverse sites in the sub-mountainous Himalayan tract of Pakistan. A significant variation in soil physico-chemical characteristics was observed due to both, sites and seasons. Nitrogen and phosphorus contents were significantly higher during the summer season. Mean soil composition (mean of four seasons, that is, Autumn, Winter, Spring, and Summer) for field capacity, EC, pH, N, P, and K varied from 27 to 51%, 1.7 to 5.0 dSm^{-1}, 7.8 to 8.0, 0.08 to 0.14 mg g^{-1}, 8.2 to 15.2 mg g^{-1}, and 0.02 to 0.08 mg g^{-1}, respectively (Ahmed et al. 2011).

6.16.7 Soil Erosion

Soil Erosion is defined as the displacement of the land surface soil by detachment and transport of soil/rock materials through the action of any agent such as moving water, wind, ice, or other geological agents. Soil erosion is detrimental to soil health. This will reduce the productivity potential of soil. The process of soil erosion is classified into two categories, (i) "Geologic" and (ii) "Accelerated". "Geologic" is due to natural conditions and is mainly responsible for the evolution of the natural landscapes and landforms. "Accelerated" is due to anthropogenic factors mainly as a result of disturbance of the natural landscape through human activities. Apart from the land degradation and many other direct adverse economic consequences, it affects the quality of life through environmental degradation, especially in terms of water and air pollution. Water erosion is responsible for large-scale loss of soil and nutrients from the slopes of hills and mountains, while wind erosion is a characteristic feature of areas of dry, loose, bare sands, or finer materials subject to strong air currents. Mian and Mirza (1993) reported that the total area involved in soil erosion (water and wind erosion) in Pakistan during 2007 was 6.455, 29.459, 19.959, 2.282 million ha in slight to moderate, moderate to severe, severe to very severe, and bank erosion categories, respectively. Total area involved in all categories of erosion was 58.155 million ha. Though Balochistan has a semi-arid to arid type of climate, water erosion is the prime cause of soil loss involving 3.17 times more area than the wind erosion. The percent of land involved in slight to moderate, moderate to severe, severe to very severe, and bank erosion categories was 11.1, 50.7, 34.3, and 3.9%, respectively (Mian and Mirza 1993).

Basically, all these types of erosion are geologic in nature, occurring in undisturbed natural environments. Soil erosion type has been further split into a number of classes based upon the severity of the problem. These classes are of practical value in the assessment as well as amelioration/management of the problem (Mian and Mirza 1993).

6.16.8 Annual and Seasonal Distributions of Runoff

The runoff pattern was observed to vary among different land uses in Hilkot watershed in Pakistan (Fig. 6.22). The results of the study

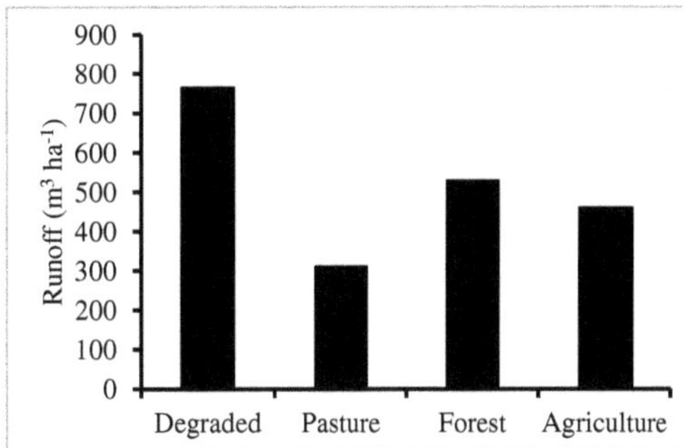

Fig. 6.22 Annual runoff from different land uses in Hilkot watershed, Pakistan

revealed that the mean annual maximum runoff was 674 m^3ha^{-1} from the degraded plots. This was mainly due to less plant cover on the soil surface in the degraded plots. The mean annual minimum runoff of 310 m^3ha^{-1} was observed from the pasture plot due to its good vegetative cover because all the soil surfaces remained well protected. Data also reveals that an annual runoff from the degraded land ranged from 429 to 814 m^3ha^{-1}, while the range for the pasture, forest, and agriculture plots was 181 to 372 m^3ha^{-1}, 377 to 685 m^3ha^{-1}, and 235 to 732 m^3ha^{-1}, respectively.

6.16.9 Soil Loss Through Erosion

The experiments conducted in Hilkot watershed in Pakistan revealed that annual soil losses measured on different land uses varied from 0.3 to 16.3 t ha^{-1} (Fig. 6.23). On average, annual maximum soil loss of 6.5 t ha^{-1} was found from the degraded plot. This was mainly due to the reason that this land use had scarce vegetation cover, giving rise to more runoff. Good vegetation cover in pasture plots helped in arresting soil erosion. The average soil loss from the pasture watershed was 1.8 t ha^{-1} only. The average annual soil loss from agriculture and forest plots was about 3.0 t ha^{-1}. These plots had a medium type of vegetation cover. Carver and Schreier (1995) mentioned that the pre-monsoon storms were more harmful, which led to main soil loss from agricultural fields due to sowing and weeding practices. During sowing and weeding operations, the soil becomes loose and, therefore, more prone to erosion. Seasonal vegetation cover is essential for controlling soil erosion from various land uses on hill and mountain slopes.

6.16.10 Surface Runoff

A study was conducted in Hilkot watershed, Pakistan, Zokaib (2011), for measurements of runoff and resultant soil erosion. There were four major land uses in the watershed, such as degraded land, forest land, agricultural land, and pasturelands. On average, annual rainfall received in the study area was 1160 mm. In all the erosion plots, almost 50% of the runoff and soil loss occurred during the monsoon period. The mean maximum runoff was from the degraded plot (674 m^3 ha^{-1} yr^{-1}), while the minimum was observed at the pasture plot (310 m^3 ha^{-1} yr^{-1}). The average runoffs on other land uses were 529 and 460 m^3 ha^{-1} yr^{-1} from the forest and agriculture plots, respectively.

6.16.11 Impacts of Land Uses on Runoff and Soil Degradation

Surface runoff and soil erosion are regulated up to a great extent by surface vegetation cover by increasing infiltration and reducing velocity of runoff water (Karvonen et al. 1999). Hofer (1998a) reported that soil conservation is important in reducing soil loss rates, which is quite visible in micro-scale catchments. Many researchers highlighted the importance of

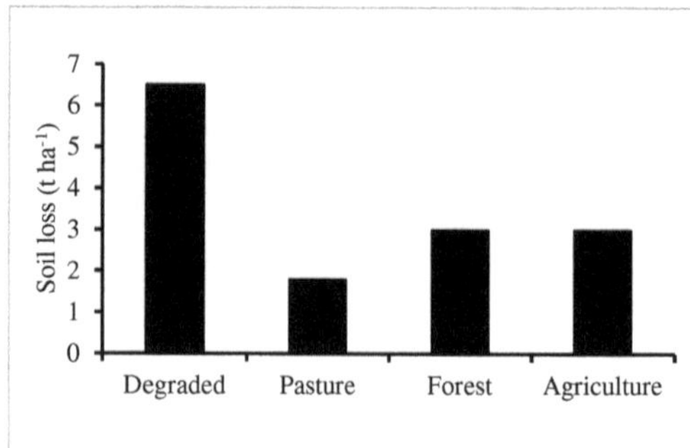

Fig. 6.23 Annual soil loss (t ha^{-1}) from different land uses in Hilkot watershed, Pakistan

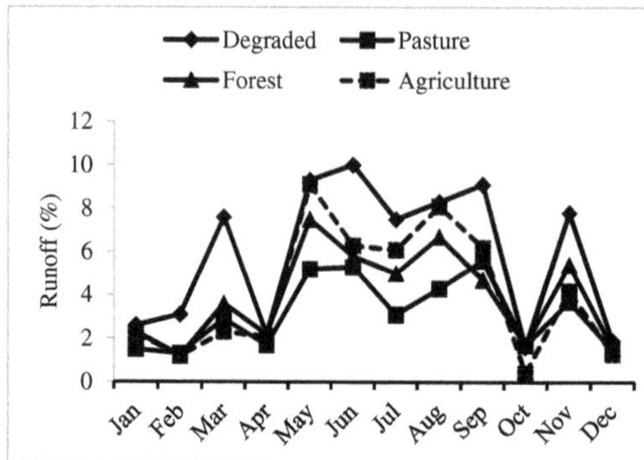

Fig. 6.24 Runoff percentage from all land uses

vegetative cover and land use management practices in runoff and soil loss generation (Bryan and Campbel 1986; Kosmas and Danalatos 1997). Sharma et al. (1992a, b) reported that the land use effect is critical in the prediction of sediment and nutrient budget. The month-wise runoff from four land uses, such as degraded, pasture, forest, and agriculture, is shown in Fig. 6.24. Maximum average runoff was recorded from May to September, that is, during the monsoon season, thereby showing a positive relationship between the amount of water received through rainfall and the runoff generated.

Figure 6.25 presents average monthly rainfall for four annual quarters and soil loss km^{-2} for four land uses (Degraded land, Pastureland, Forest land, and Agricultural land) at plot level at Hilkot watershed in Pakistan. The percentage of runoff water from the degraded land was comparatively higher than that of all other land uses.

6.16.12 Forest Cover Distribution and Change Patterns in Pakistan

There are high variations of forest cover distribution across the HKH region of Pakistan (Qamer et al. 2016). The Khyber Pakhtunkhwa province of Pakistan has the largest area of forest cover. The forest vegetation contains a broad range of forest types including dry temperate, subtropical pine forests, and moist temperate as the most dominant forest types. Galgit-Baltistan has the smallest forest cover area among all the administrative units. Between 2010 and 1990, a decrease of 15.1% and 2.3% has been found in Khyber Pakhtunkhwa and Galgit-Baltistan provinces, respectively (Qamer et al. 2016). Though a decrease of 112.1 and 35.1 thousand ha in forest cover due to deforestation and degradation was recorded between 1990 and 2010 in Khyber Pakhtunkhwa, there was a regeneration of forest

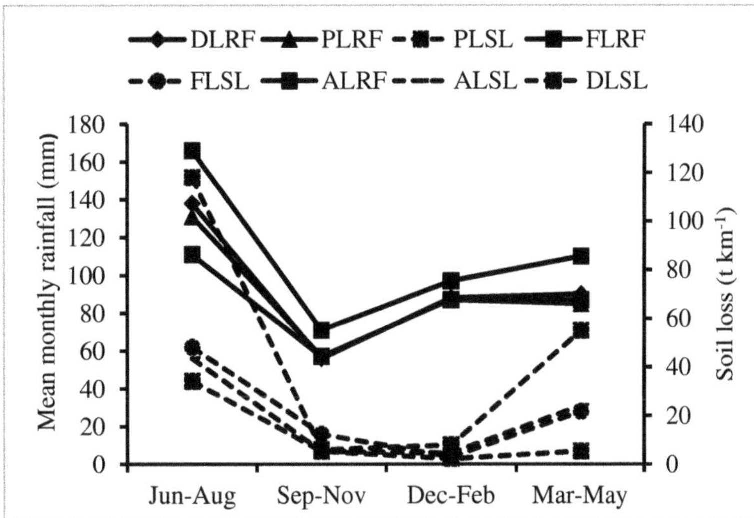

Fig. 6.25 Average monthly runoff and soil loss from all land uses in Hilkot watershed, Pakistan (DL, Degraded land; PL, Pastureland; FL, Forest land; AL, Agricultural land; RF, Rainfall; SL, Soil loss)

cover also to the tune of 9.9 thousand ha. Similarly, in Galgit-Baltistan provinces, a decrease of 7.68 and 2.70 thousand ha in forest cover was found during this period; there was a regeneration of forests in about 2 thousand ha. Overall, an annual change of about −0.42% and −0.31% has been found in Khyber Pakhtunkhwa and Galgit-Baltistan provinces.

6.17 Practices for Sustainable Soil Management and Utilization in HKH

6.17.1 Drivers of Sustainability

There are many facets of sustainability which need to operate in HKH to ensure resource sustainability, particularly the soils, major being human, social, economic, and environmental sustainability. The logic behind the four pillars of sustainability indicates that there is a need to solve sustainability problems completely and then be maintained in future. Sustainability has three dimensions or it is based on three pillars. For ensuring the availability of food on a sustainable basis and that the needs of the present generation are met without compromising the ability of future generations to meet their own needs, it would be essential that the development occurs in a socially and culturally appropriate and equitable manner. Sustainable development can be explained through "8 P's" such as people, planet, prosperity, peace, participation, partnerships, principles, and persuasion. Musson (2013) considers sustainability as the organizing principle for sustaining limited resources in order to provide for the needs for resources and desirable living conditions of the coming generations. Soil is subject to severe and fast degradation processes, particularly in the HKH region, because of the pressure of human activities and their interferences. The significance of land use in soil organic matter and nitrogen dynamics under variable soil and climatic conditions has been well recognized.

6.17.2 Sustainable Soil Management

We need to ensure (i) not to go for land use change (urban and industrial development) at the expense of productive lands, and inappropriate land use intensification, (ii) the farmers should not go for such management practices which are detrimental to the environment under specific agro-climatic conditions, (iii) unsustainable land management practices like no vegetation as soil cover, application of inappropriate doses of fertilizer nutrients and their poor management, poor water management, and forestry practices that excessively disturb the soil, leading to its degradation, (iv) not to engage ourselves in deteriorative soil management practices which trigger GHG emissions leading to climate change, degrade soil organic matter and loss of natural resources, and biodiversity, (v) public awareness about soils, their role in providing ecosystem services, and sustainable soil management strategies, (vi) introduction of state-of-the-art technologies soil management, (vii) to advance the research infrastructure for soil-data storage, analysis, and sharing, storing large volumes of data and developing more sophisticated predictive models, (viii) conducive hydrological regime in soils for optimum crop production, (ix) land use planning according to soil suitability classification for production of different crops, (x) to prioritize programs and technical assistance designed to promote sustainable land management practices and to minimize unsustainable land management practices, and (xi) management of problematic soils like acid soils, alkaline/saline soils, eroded soils, etc.

6.17.3 Maintaining and Improving Soil Productivity

Managing soil nutrients

Judicious nutrition of the crop plants and increasing the use of fertilizers depends on the application of nutrients at the time of their actual requirement by the crop. Correct method of nutrient application is as important as the amount and time of nutrient application. Correct plant nutrition management depends on many factors and their interactional effects. It should be borne in mind that while deciding about the method of tillage or no-till agricultural systems, one must consider the soil quality, soil organic matter, and nutrient availability. The nature and amount of weed is also an important factor in deciding about the tillage methodology to be followed. Soil quality indicators in minimum datasets include total organic carbon, volume weight, aggregate resistance, usable moisture content, pH, and EC. Proper soil management is extremely important for ensuring fertilizer use efficiency, development of crop root system, availability of nutrients and optimum moisture, and proper crop stand in the field.

Managing soil physical conditions

- Avoid unnecessary tillage operations.
- It is essential to follow the agricultural practices that help in enhancing soil organic matter and maintaining proper soil structure.
- Grazing should be regularized.
- For healthy and optimum growth of the plants, ensure a sufficient amount of moisture in the root zone of the soil during the plant's growing season. If the rainwater received is not sufficient, the water needs should be satisfied through an assured and controlled irrigation system. A proper moisture regime in the soil is also necessary for the availability of nutrients to the plants in proper amounts.

Jahangir et al. (2018) mentioned that soils provide essential ecosystem services such as regulation of above-ground biodiversity, primary production function, regulation of biogeochemical cycles, water filtration, and resistance to diseases and pests. Adoption of improved management practices is key to maintaining sustainable agricultural production and adaptation to global climate change through improved soil quality. We can evaluate and monitor an agricultural management system for sustainability, by ensuring that the system will not exhaust the resource base on which the future of the next generations depends, optimizing soil conditions and reducing food production vulnerability, while at the same time maintaining or enhancing crop productivity. One has to consider that advanced soil management practices for economic, social, and environmental sustainability need better management of the soil–crop–nutrient system. Sustainable soil management includes (i) soil management and conservation for combating desertification, soil erosion assessment, soil fertility enhancement, fertilizer recommendation, and soil conservation measures, and (ii) technologies and practices for sustainable use and management of water for optimizing irrigation systems and enhancing crop water productivity. Increase in the use efficiency of inputs is one of the strategies for greenhouse gas emission reduction.

6.17.4 Policy Tools for Soil Protection

Soil is an essential ingredient for life but its unsustainable use threatens the quantity and quality of soils in HKH region. Environmental policies can help to induce sustainable use of this limited resource. Global soil policies got a shot in the arm after the 2012 Rio + 20 United Nations conference on sustainable development. The conference highlighted the urgency for international soil protection commitments, and reaffirmed the long-term socio-economic relevance of such actions. Soil has a number of important ecological functions that should be considered for decision-making. The application of fertilizer nutrients to the soil in larger than required amounts by the crop will affect the buffering and filtering capacity of the soil and its ability to providc nutrients for plant growth. Application of excessive nutrients even through animal manure will be harmful as it will increase the emission of GHG from the soil, particularly the nitrous oxide. This will have a significant impact on the environment in general and on soil in particular.

Achieving food security with minimal risks to the environment appears to be unimaginable under the present situation when large amounts of fertilizers, pesticides, and other agrochemicals are being used to enhance agricultural production. This will be a big challenge, particularly for developing countries (Lal 2000). High risks of soil degradation by a wide range of degradative processes have to be addressed suitably with available technologies. The resource-poor farmers, having weak institutional support, require proper and timely guidance for solving the problems faced by them. Soil-specific technologies for agricultural intensification will have to be developed and adopted to take care of ecology and the environment. Under the circumstances, some of the priority issues for future will be to generate technologies to arrest soil degradation and find practices that would restore degraded soils, ensure increase in carbon sequestration to mitigate the greenhouse effect, and find solutions for decreasing the risks of eutrophication of surface water and contamination of groundwater. FAO (2019) was very much concerned about deteriorating soil health conditions and its degradation in many regions of the globe, especially in mountain areas, due to soil erosion by water, wind, and some irrelevant tillage practices. It continues to be the greatest threat to soil health and ecosystem services even after almost a century of research and extension efforts. Poor management of soils through unsustainable agricultural practices will help in releasing soil carbon into the atmosphere in the form of carbon dioxide. This will substantially contribute to climate change and global warming with negative impact on agriculture at many places.

Soils were and still are the primary medium for agricultural production. Agricultural intensification is a change in the cropping or livestock-rearing practices that make use of a fixed area of land more frequently or intensely than previous traditional practices. Though intensification of agriculture results in higher returns, it is often achieved through the use of chemical fertilizers and synthetic pesticides, which have far-reaching and long-term consequences for the health of humans and the ecosystem. The pressures on natural resources in the HKH region will be extreme due not only to the increasing population but also because of the fragile ecosystem of the region. The uncertainty of future changes in the productive capacity of the land will be the unpredictable impacts of climate change. The possible measures and technologies to be in place for sustainable soil management and agricultural intensification are presented in Table 6.45.

Appropriate and essential steps are required to sustain mountain environments and improve livelihood in the Hindu Kush Himalayan (HKH) region which is one of the world's greatest mountain systems, home to the world's largest peaks, unique culture, diverse flora and

Table 6.45 Technologies for sustainable intensification of agriculture in the Himalayan region

Resource category	Measures
Soil management	Improvement in soil health through efficient use of fertilizers, both organic manures and inorganic sources of nutrients. There is a need to adopt measures for soil carbon sequestration, neutralize soil acidity by use of amendments, use biofertilizers, and optimize rhizosphere microbial activity. Nitrogen fertilizer use efficiency has to be improved to prevent losses due to leaching and erosion
Farm energy management	Effective energy efficiency measures can help agricultural producers save energy without harming productivity, and prevent compaction of soils through farm traffic
Water management	Increased water use efficiency through adequate but controlled irrigation. micro-irrigation, and timing of application; water recycling, wastewater reclamation, and reuse and desalinization are required
Maintenance of carbon stocks	Sustainable soil management techniques could enhance the provision of ecosystem services by storing carbon and decreasing greenhouse gas emissions
Crop/agro-biodiversity resource management	More efficient and sustainable farming systems should be in place. Agroforestry with high-value crops should be encouraged
Policy and institutional support	Institutional support and capacity building policies are required to encourage conservation and sustainable soil management. Activate extension services
Tillage	There is a strong need to decide about proper tillage operation such as conventional or conservation or minimum tillage operation as per situation or go for "no till"

fauna, and a vast reserve of natural resources. Being the source of 10 major Asian rivers encompassing 4.2 million Km^2 area, HKH provides essential resources, especially water and biodiversity to 1.9 billion people in the river basins having an area of 8.9 million Km^2. From the Hindu Kush Valleys in Afghanistan to the variable Hill and mountain systems of Mynamar, one may experience a lot of ethnic diversity and cultural identities which have played an important role in shaping the ecosystems of the HKH region. There are the arid and semi-arid regions of the Pamir and Karakoram mountains, the high Himalayas of India, Nepal, Pakistan, and Bhutan, having perpetual snow cover and Chittagong Hills of Bangladesh getting high rainfall during the monsoon season, the Tibetan plateau of China and the three parallel rivers of the Salween, Mekong, and Yangtze in the far eastern region located in Yunnan Province of China. In CHT of Bangladesh, there are landslides, land subsidence, and cyclones of high intensity during the rainy season. The HKH region features great heterogeneity from North to South and East to West in relation to precipitation, vegetation, and human livelihoods. There is a need to mitigate the impacts of climate change and global warming by following appropriate practices and means. Carbon sequestration has to be encouraged through suitable practices which may increase the C content of soils of the region. Though per capita emission of fossil fuel CO_2 from the HKH countries is one-sixth of the global average, the region heavily suffers from the impact of climate change.

The soil management activities should focus on various methods which can reduce soil erosion and degradation, enhance soil fertility, and improve soil health on a sustainable basis. Affordable technologies for marginal and poor farmers have to be generated and adopted. The major practices for potential soil management are conservation farming, agroforestry, SALT, green manure, shelter/protection belts, and composting.

6.17.5 Conservation Farming

Conservation farming is a system that helps in preventing the loss of arable land and encourages regeneration of degraded lands. It enhances biodiversity and natural biological process above and below the ground surface, increases water and nutrient use efficiency, and promotes sustained crop production. Conservation farming in the true sense implies minimum mechanical soil disturbance, and the application of agrochemicals and organic or inorganic plant nutrients in optimal amounts and in ways and quantities that do not interfere with or disrupt the biological processes. There are two types of strategies to reduce the detrimental environmental effects of the agricultural activity. The first is to maintain a proper vegetational ground cover to avoid the impact of rainfall and the second is to construct barriers, terraces, ditches, etc. which may reduce soil erosion and runoff. Protection of native vegetation can help in avoiding environmental damage caused by agricultural activities. This strategy can also provide other important advantages such as perpetuating a reliable supply of water for domestic use, and reserves for native fauna and flora, which also become increasingly pivotal as thc human population grows in number. Using local tree species or invasive and exotic species adapted to the local conditions, reforestation can be undertaken. These tree species must have some desirable characteristics, like a fast growth rate, the ability to fix nitrogen, forage or wood value, etc.

Strategies in Cultivation Systems

The erosion from hills can be reduced or controlled by contour planting of a crop and contour strip cropping techniques which act as live barriers. The windbreaks are also an important live barrier to avoid erosion or crop damage due to wind rather than water. It has been found that dead barriers function like live barriers, the difference being that they are composed of rocks, plant residues, or other non-living materials (Fig. 6.26). It is easy to construct dead barriers if stones are present in a field. These barriers are very useful in controlling soil erosion and more in situ retention of rainwater. The live barriers have an edge over the dead barriers in the sense that these can be lopped to provide green forage to livestock. It is useful to construct these and in the process make the soil easier to work with. The dead barriers can be completed during the dry season, indicating that they are constructed to make them functional at the start of the rainy season.

Construction of raised beds for planting and contour furrows may also prove useful contour barriers (Fig. 6.27). The raised beds are important in the cultivation of vegetables or any raw crops in which the soil is cultivated intensively.

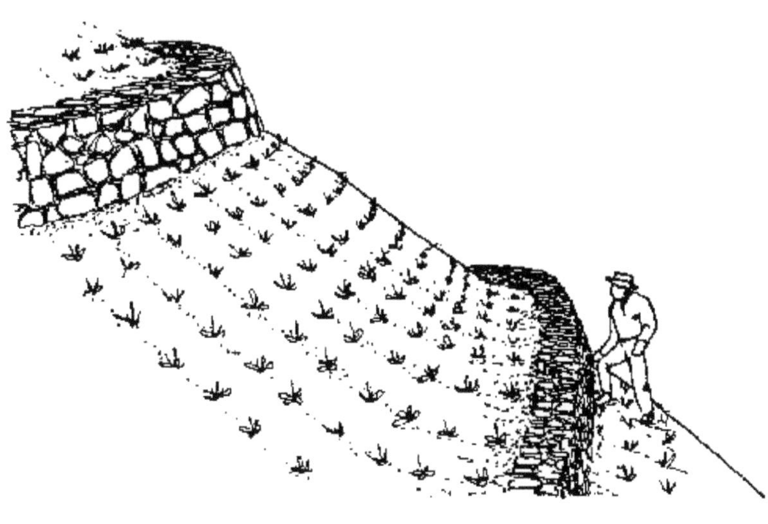

Fig. 6.26 Contour dead barriers using rocks

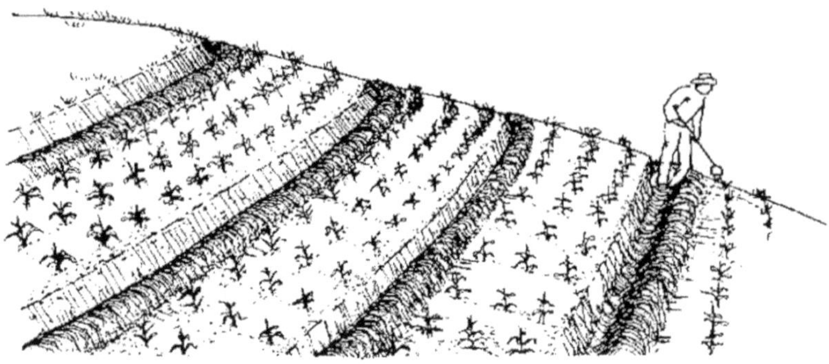

Fig. 6.27 Contour planting beds

The contour trenching helps in reducing the speed of gushing rainwater down the slope and, therefore, results in reduced soil erosion, permitting drainage of excess soil moisture from the planting bed and avoiding compaction of the planting bed surface by providing grassed waterways. The width between furrows is variable and depends on the crop and any irrigation requirements.

Contour ditches

Contour ditches have similar function as that of contour barriers. Contour ditches completely stop downslope water movement as the water falls into the ditch. These structures need less construction work compared to terraces and are very useful for small and marginal farmers. The size of the ditch may be 30 cm wide and 30 cm deep or could be any convenient size. The removed earth is placed in the lower lip of the ditch and a suitable live barrier may be planted to stabilize the ditch. It is very much useful to plant the live barrier first, well in advance or one season in advance, especially on high degree slopes. This way the ditch will be adequately protected after it is built. Water can be allowed to come to a drainage point to increase the water infiltration and in this way, soil erosion can be minimized.

6.17.6 Terraces

The terraces are constructed and supported with strong risers depending on the slope of the land and size of the slope. Contour lines are drawn by separating the topsoil upward so that sub-soil is exposed for a firm foundation. After removing the topsoil, the sub-soil below the proposed riser is dug and heaped up for compaction. The leveling of the terrace is done by the cut and fill method. When the terrace is leveled, the removed topsoil is spread uniformly over the surface of the terrace. Perfect leveling of the terraces and well-managed irrigation-cum-drainage system prevent soil erosion and loss of fertility (Sonowal et al. 1989; Sharma and Prasad 1994). Terraces serve the functions of stopping downslope soil and water movement and also provide a flat surface for the planting of crops, thereby further reducing soil erosion. The terraces generally have a diameter of 1–1.5 m and should be thoroughly compacted to prevent collapsing. Annual crops can be grown in the spaces between the individual terraces. Bench terraces are stair-like structures which help in in situ retention of rainwater and supply moisture to plants for a longer time. They reduce erosion because the terraces can provide a good land cover.

The alder-based farming system is a self-sustainable farming practiced in Nagaland in North East India. Alder (*Alnus nepalensis*) provides not only fuel wood to the farmers, but also fixes atmospheric N, which provides nutrients to the agricultural crops. It can rejuvenate the vegetation and soil fertility at a faster rate in shifting cultivation (SC) lands. Alder trees can grow in degraded land that is the reason it is one of the best suitable tree species for rejuvenating SC areas. The shaded tree leaves from the alder tree help in mulching the soil and retaining moisture naturally. Alder trees about 10 years old are pollarded at the average height of 1.7–2.0 m from the ground level. The cutting process is done before the start of the cultivation. After the slash and burnt process is over, paddy and other vegetable crops are planted.

6.17.7 Gully Prevention and Control

Gullies are formed when a large amount of water flows, taking away the vegetation, topsoil, and sub-soil from an area. The process continues and deep waterways are formed eventually. Gullies form naturally, but sometimes, human activities are also responsible for this when water courses are redirected to change drainage patterns to protect roads, buildings, or fields. The gullies tend to enlarge both in length and width, if proper control measures are not adopted. The most effective control technique is to capture and redirect water suitably in order to control soil erosion. Once water is prevented to enter the gully, the soil is often stabilized, the area is revegetated naturally, and filling in occurs naturally without any further care. It is essential to protect the gully from further detriment due to cattle trampling and overgrazing, or due to continuous cultivation. Check dams are also useful in controlling water flow, which can be made of rocks, wood, wire mesh, etc. and serve to reduce the velocity of any water still flowing. The gully can also be more quickly revegetated by planting different types of trees, shrubs, or grasses, rather than waiting for the regeneration of vegetation to take place naturally. The soil conservation plan in the HKH region can be designed by selecting a plot, its slope characteristics, problems relating to erosion, drainage and drought, selecting conservation strategy, marking contour lines, adopting plant protective barriers followed by crop planning with a periodical review of structures, and maintenance to have a proper long sustaining approach.

6.17.8 Green Manure Crops

Green manure crops are nitrogen-fixing plants, which are added to the soil so as to add organic matter in the soil along with macro- and micronutrients to enhance soil fertility. The added organic material is eventually decomposed and nutrients are made available to the crops, subsequently grown on the land. In the process of growing green manuring crops, there are methods which allow for the production of these crops without affecting the normal crop production cycle. Some species, such as Subabul (Leucaena leucocephala), Gliricidia (Gliricidia maculata), Indian Coral tree (Erythrina variegate), Flemingia macrophylla, Golden shower (Cassia fistula), Desmodium rensonii, Jackbean (Canavalia ensiformis), Velvet bean (Mucuna pruriens), Dolichos lablab (Lablab purpureus), Butterfly pea (Clitoria ternatea), Tropical kudzu (Pueraria phaseoloides), Sunn hemp (Crotalaria Juncea), Dhaincha (Sesbania bispinosa), Indigofera (Indigofera tinctoria), and Cowpea (Vigna unguiculata), can be grown with other crops. Green manure crops serve as useful vegetative soil surface cover or living mulch which reduces rainfall impact and runoff, eventually reducing soil erosion.

6.17.9 Ecosystem Services

The ecosystem services from the Himalayas can be divided into four areas, such as provisioning (food, fodder, fuel, freshwater, minerals, raw materials, pollination, and genetic resources); regulating (climate regulation, water regulation, and biological control); supporting (recharging, groundwater, soil formation, erosion control, water retention, and soil nutrient cycling); and cultural (traditional knowledge, soil, conservation practices, erosion control measures, and agro-biodiversity). The Himalayan ecosystem services are interdependent in the HKH region. For instance, flood generated in Nepal will cause floods in India; glacial-like outburst in China

(Tibet) can affect hydropower stations in Nepal; erosion in one country will produce sediment in another country in HKH region. In addition to providing surface and groundwater, Himalayan rivers carry soil and nutrients to downstream areas making floodplains fertile leaving behind the hills under severe land degradation thus attracting attention for ameliorative measures as a part of effective soil management.

6.17.10 Water, Energy, and Climate

Rice and wheat are the staple food in the HKH region and 50% of dietary energy comes from the two crops. But these crops require a huge amount of water. Rice consumes about 4000–5000 L of water per kg of grain produced. Crop production depends on water availability in the dry season and on the irrigation facility which depends on rivers of the HKH region. The world's largest irrigation concentration is in the Indo-Gangetic plains. Ganges river system is the main source of freshwater in India, Bangladesh, and also in Nepal. Freshwater stress is growing in the HKH region particularly in India with per capita annual water availability declining from 5,177 m^3 in 1951 to 1,508 m^3 in 2014 and is likely to reduce further to 1,235 m^3 by 2050. From an energy perspective, the major challenges relate to supplying enough energy for producing food and other economic activities as well as domestic use without using carbon intensity. Hydropower from HKH river systems can enhance energy security in the region and also can provide quality energy for agriculture and food production. The hydropower potential in the HKH region is more than 500 GW with the contribution of hydroelectricity to commercial energy being 50% in Bhutan, 17% in Nepal, 13% in Pakistan, 6% in India, and 4% in Afghanistan. On the other hand, the Himalayan mountains also carry out climate regulation by regulating microclimate as well as wind and monsoon circulation in the region. The Himalayas block moisture-laded monsoon winds from traveling further northward thus facilitating timely and heavy precipitation in the region. During winter, the Himalayas pose a barrier to storms coming from the West thus receiving snow at higher elevations and rainfall at lower elevations.

6.18 Change of Soil Under Human Activity and Climate Change

Usually vulnerable to erosion and landslides even without human interference, the HKH region with young and rising mountains is geologically fragile. It has been observed that the region is undergoing rapid change driven by stressors such as climate change and human conflicts, and factors like globalization, infrastructure development, migration, tourism, and urbanization. The countries of the HKH region are vulnerable to the impacts of global climate change. The region, the biggest mountain area in the world, faces enormous challenges, and there is a strong need to identify the challenges, develop plans, and implement them judiciously in an effort to mitigate the impacts of climate change. The mountain ecosystem of the Hindu Kush Himalayas is recognized as one of the most vulnerable ecosystems due to climate change. Urgent steps are required to sustain mountain environments and improve livelihoods in the HKH region. Spreading length-wise for about 3,500 km over eight countries, such as Afghanistan, Pakistan, India, China, Nepal, Bhutan, Bangladesh, and Myanmar, the HKH is one of the world's unique mountain systems. The HKH directly supports the livelihoods and ecosystem services for over 240 million people living in its hills and mountains, and it provides many benefits to the 1.65 billion people living downstream. Though HKH is important for livelihood security, the anthropogenic factors and climate change are immediate threats to the region's livelihoods, biodiversity, and sustainability. Changes in the highest mountains of the world are having and will continue to have major consequences, not only for the region but also globally. To save the unique ecosystem and sustain the mountain environment, local, national, regional, and global actions are urgently needed, focusing on substantially, increased investments, and more robust regional cooperation. It is to be ensured that the climate investment reaches the marginalized and vulnerable populations such as mountain communities to help them prevent ecological, economical, and social damages and adapt to rapid climate changes and help in mitigating the change. The impacts of land use and climate change would increase in the future due to the increased deforestation and clearing of lands for agriculture. More attention is required to looking into the way in which the future climate will interact with the land use changes and affect the water balance in the watersheds. The resultant effects of biodiversity loss due to climate change are thought to be extremely dangerous for the people who exclusively depend on natural resources.

6.18.1 Models Used to Project Climate Change

Magnitude of future emissions of greenhouse gases, aerosols, and other natural and human activities will determine the climatic conditions in times to come. The pressures from human activities are external to the climate system and affect how it behaves. Earth's response to those pressures is of paramount importance as it shapes the future climate, as well as the internal variability inherent in the system. Generally, simple models can be adequately applied to process large-scale average data, like global average temperature, but minute details, like regional precipitation, can be simulated and processed only through complex models. In some unique cases where very long period, say pertaining to one or more centuries, simulations are needed or where numerous understandings and interpretations are required, a faster model with lower resolution, or a simplified description of some climate processes, may be used.

The recent advancement in Precision agriculture is a much-talked-about topic these days, and its development and furtherance as well as the need for the spatial assessment of soil properties have increased enthusiasm in soil Visible and Near-InfraRed (V-NIR) reflectance spectroscopy. Evaluation of soil nutrient content has raised the interest of the farmers in order to adjust practices such as tillage, fertilization, and irrigation in precision agriculture (Ahmadi et al. 2021). A good understanding of the soil properties can help farmers with regard to their choice of agricultural operations and treatments in soil management (Khan et al. 2020; Roberton et al. 2020). However, standard methods of analysis like wet chemistry need special expensive instruments and may require more time, especially when dealing with a high spatial sampling density (McCarty et al. 2002; Cozzolino and Morón 2006). As an alternative to the standard wet chemistry, soil Visible and Near-InfraRed (V-NIR) reflectance spectroscopy has proven to be a fast, cost-effective, nondestructive, environmental-friendly, repeatable, and reproducible analytical technique (Viscarra et al. 2006, Soriano-Disla et al. 2014; Nocita et al. 2015).

Bayer et al. (2021) found that although a general leaning for agricultural expansion exists under different conditions, where and how particular land use and land use changes (LULUC) are realized differ widely in the use of various models. However, these models are useful in determining climate change mitigation and adaptation plan in terms of total land demand, investment costs, and direct and indirect socio-economic and ecological effects (Humpenöder et al. 2014; Popp et al. 2014; Reilly et al. 2012). The variability in ecosystem service indicators was generally high for vegetation carbon stocks and crop production, especially in tropical forest regions because of extreme diverging LULUC change projections along with high biomass yield, and declining or enhancing the effects of climatic change. The IPCC Special Report on Climate Change (IPCC 2019) signified unprecedented rates of land and water use, and ecosystem diversity loss due to socio-economic, ecological, and climatic challenges such as increasing food consumption, land degradation, floods, and droughts. The Intergovernmental Science-Policy Platform on Biodiversity and Ecosystem Services (IPBES) global assessment report (IPBES 2019) also indicated the qualitative and quantitative decline in most ecosystem services (ESs) because of human activities.

The Hindu Kush Himalayas is a place of different overlaid natural hazards, but the risks involved and the people exposed to them are remotely understood (Rusk et al. 2021). They evolved high-resolution models showing the magnitude of exposure to natural hazards of the HKH communities. The 6th Assessment of the IPCC Report indicates that many regions are undergoing multiple changes due to climate change (IPCC 2021). Arias et al. (2021) found that mountain areas are likely to face the fallout of floods, landslides, and other hazards due to increased warming and high precipitation. Exacerbating the risks because of these hazards will be increased urbanization (Ehrlich et al. 2021).

6.18.2 Soils and Carbon Stocks

Soils serve as the largest sink of terrestrial carbon. Soils are able to play an important and major role in climate change adaptation as well as mitigation by carbon sequestration and decreasing greenhouse gas emissions in the atmosphere. For this, the soils should be managed using Sustainable Soil Management (SSM) techniques. This would also help in increasing the provision of ecosystem services. Promising Sustainable Land Management (SLM) methods can be used to improve crop productivity and contribute to reducing greenhouse gas emissions through increased carbon sequestration. It has been established that the soils contain more organic carbon than there is in the atmosphere and vegetation together combined (Fig. 6.28). Mostly in the form of CO_2, with much smaller amounts of methane, the atmosphere contains approximately 750 PgC. Plants, animals, soils, and microorganisms constitute the terrestrial ecosystem carbon. The Earth's plants, collectively, store approximately 560 Pg C. So, the total C stored in the atmosphere and vegetation is approximately 1,310 Pg C. The total amount of carbon in the world's soils is estimated to be 1500 Pg C, which is 190 Pg C more than vegetation and the atmosphere put together.

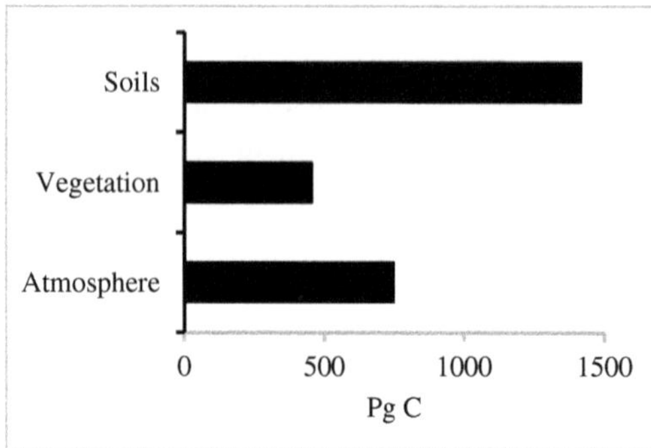

Fig. 6.28 Carbon in soils, atmosphere, and vegetation

6.18.3 Influence of Climate Change on Soil Processes

Brevik (2013) mentioned that the soils are very much linked to the atmospheric and climate system through the cycles of carbon, nitrogen, and hydrology. Due to this reason, a changed climate will have a great impact on soil processes, properties, and their interactional behavior. The way soil organic matter levels react to changes in the C and N cycles will affect the ability of soils to help crop growth, which has significant implications for food and livelihood security. Organic matter is an important constituent of soils as it influences structure formation and maintenance, water-holding capacity, cation exchange capacity, and as a storehouse of nutrients. Soils which have a sufficient quantity of organic matter are more supportive and high-yielding than the ones that are poor in OM. Climate change will have a great impact on soil processes and properties. This would show how eventual changes in the C and N cycles will impact soils in a particular area. Climate change and anthropogenic activities have induced soil erosion and, soil and environmental degradation in HKH region.

In the HKH region, having a vast area including forests, grass, farm, and wetlands, anthropogenic activities have led to a large rise in

GHG emissions (Usman et al. 2021). Over the last century, the atmospheric CO_2 has gone up from 277 to 409 ppm in 2019, rising at the mean rate of 2.3 ppm per year from 2010 to 2019 (Friedlingstein et al. 2020). Atmosphere warming due to the rise in CO_2 concentration can catalyze plant growth by increasing net primary productivity leading to enhanced carbon uptake (Delpierre et al. 2009; Sullivan et al. 2008; Wu et al. 2011). Droughts may increase in commonness, span, and asperity due to increase in atmospheric temperature (Trenberth et al. 2013). Therefore, understanding the effect of the global rise in temperature on forests is becoming more important as vegetation response to climate change will result in changes in net carbon uptake, water use efficiency, carbon biomass allocation, and interaction with disturbances (Urban et al. 2017).

6.18.4 Crop Production

In HKH, the climate change will negatively impact the four major factors of food security, that is, its availability, access to healthy food, its proper utilization and stability, and the possible interactions of these factors. Earlier, it was expected that increased atmospheric CO_2 would lead to increased plant productivity coupled with increased C sequestration by soil. Rapid increase in population in the region will outpace food production, thus increasing the dependence of mountains for food on the plain areas. Around 240 million people live in the Hindu Kush Himalayan (HKH) mountain areas (Rasul et al. 2018), within an area of 3.441 million km^2. Mountain communities, almost, completely depend on local resources at various altitudes and locations for livelihoods and food security, although specific agro-ecological and livelihood potentials vary considerably according to site specificity and local agro-climatic conditions. Hussain (2016a, b; 2018) reported that the main sources of livelihood in mountains are agriculture, livestock, and horticulture. Now, livestock rearing is becoming more important than arable agriculture at higher elevations in the HKH region. However, in recent years, climatic and socio-economic factors have contributed to the depletion of the natural resource base across the region (Rasul and Hussain 2015).

The mountainous areas of the HKH region have very limited area suitable for agriculture, and so, it has to be used very carefully to derive maximum benefit in the shape of agriculture production to feed the Himalayan population (Kumar et al. 2021). Complex biophysical and socio-economic conditions have resulted in the vulnerability of agricultural communities in this region (Shukla et al. 2016). Due to significant changes in climatic parameters, the region has been termed as climate change hotspot (Szabo et al. 2016). There is a need to exploit area-specific opportunities to get the full potential of crop productivity and production, particularly cash crops (Szabo et al. 2018). It is also very important to evaluate the current potential of land in terms of agriculture production to reduce food insecurity in the region and arrest soil degradation (Li et al. 2015). Climate change has affected the availability of irrigation water, so necessary for crop production (IMI and FAO 2019; Pramanik et al. 2020b). Himalayan states have a high niche potential for organic agriculture (Mishra et al. 2015), horticulture, and production of livestock, which, if judiciously harnessed, can improve the quality of life of the farmers in the Himalayas (IMI and FAO 2019).

More multidimensional and holistic approaches are required to meet the challenges faced by environmental, socio-cultural, and economic changes in the HKH region. Drivers of change and their complex interactions need individual identification and more comprehensive and integrated strategies to be adopted. It will be better if the respective Governments of the region take strong, suitable, and timely actions to strengthen the socio-cultural and environmental dimensions of sustainability. For the success of various programs, economic growth in the HKH is very necessary and interventions should be made with more inclusive, participatory approaches in natural resource management.

6.19 Future Protocol of Soil Management in HKH

6.19.1 Ecosystem Management in the HKH

The Hindu Kush Himalayan (HKH) region has many types of ecosystems, which help the mountain communities with multiple goods and services to a large population living in the region as well as in adjacent river basins. Production of food and forest products, regulating services such as the regulation of climate and water flows, nutrient cycling, and cultural services such as spiritual and recreational benefits are the major ecosystem services being provided by the HKH mountains. It is important for the inhabitants of the region to manage these ecosystems in a judicious and suitable manner and sustain the flow of goods and services to ensure human welfare in the region. Due to the fast growth of the population and other socio-economic processes such as accelerating globalization, and the need for ecosystem goods and services is growing at a high rate. The soil is an important basic resource providing a wide range of ecosystem services. Some of the pressing issues pertaining to sustainable management of ecosystems in the region include (i) the forests, agro-ecosystems, grasslands, and wetlands, the most important ecosystems, are changing rapidly in their structure and function; (ii) there has been great change in biological diversity of the region and at some places it has disappeared completely or partly; (iii) loss of traditional knowledge due to its non-transferring to next generations, mostly due to lack of interest of the new generations; (iv) there is increased food shortage and insecurity due to poor land use practices and inappropriate farming systems; (v) local communities are mainly depending on forest biomass resources for their daily needs which is a cause of concern; (vi) lack of alternative resources of income and livelihoods; (vii) deterioration and fragmentation of natural habitats, natural resources; (viii) spread of alien invasive species, (ix) forest fires, and (x) fast socio-economic, socio-cultural, and demographic changes (Yi et al. 2017).

Soil is essential to human life, the environment, and the development of a sustainable ecosystem. Its importance is not only due to the fact that it is providing most of the food for human and livestock consumption, but also due to its critical role in ensuring water quality and availability. It helps in the mitigation of climate change, and affects biodiversity important for ecological resilience. There is a need for every human to care for the soil. Due to their importance, the soil must be protected from degradation, otherwise we may lose important ecosystem services. Soil is a medium of plant growth, sustaining plant stands; a storehouse of nutrients responsible for the nourishment of plants for optimum production, and a perpetual resource for the sustenance of ever-increasing human and animal population. Given the widespread concern about the deterioration of soils with regard to quality and health in the HKH region, the desirable policy response is to revive the productive potential of soils, using both local and scientific knowledge. The Ten-step Protocol for sustainable soil management in the Hindu Kush Himalayas is depicted in Fig. 6.29.

6.19.2 Protocols for Assessing Sustainable Soil Management in HKH

Bazza (2019) developed some protocols for sustainable soil management (SSM). In addition to some of these protocols, a few protocols relevant to the HKH region are listed as follows:

- What is expected of protocols for sustainable soil management in the region and identifying the areas of management desired under the prevailing socio-economic and climatic conditions.
- Identifying the biophysical and socio-economic context of the HKH ecosystem and designing ecosystem management strategies and plans.

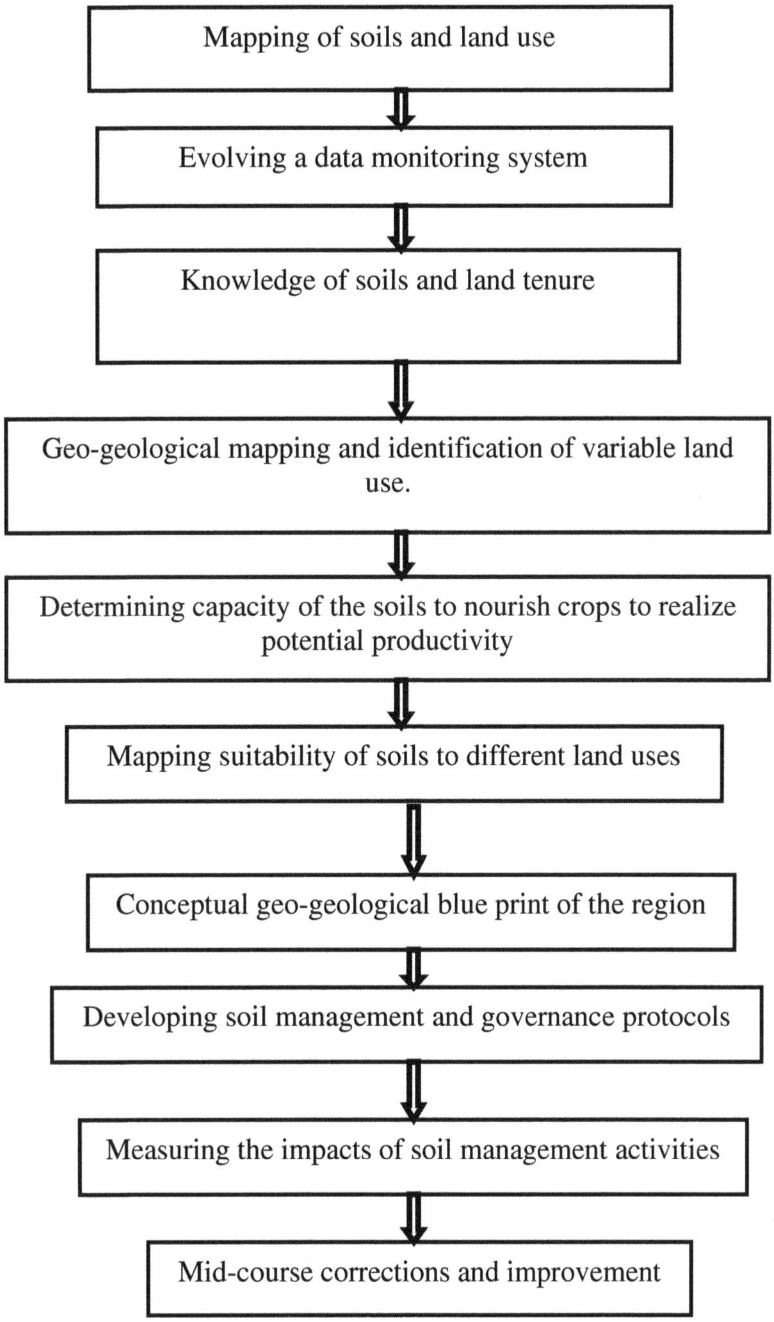

Fig. 6.29 The ten-step protocol for sustainable soil management in the HKH

- To make people aware of the main soil threats and what can be done to ward off the probable danger.
- To indicate a beginning point for the regional and local assessments of sustainable soil management. There is need to understand whether the present soil management practices in HKH region are sustainable or not, and where practices are not sustainable, to identify potential actions to improve the condition of soils.
- To provide guidance to select and apply soil indicators to develop ongoing assessment of the sustainability of soil management practices for any particular site or area.
- General description of the location, soil, and land use, identification of natural and off-site threats, description of current soil management practices, selection, measurement, and recording of sustainable soil management indicators, interpretation of the results and recommendations of improved practices to promote SSM and design and implementation of ongoing management and monitoring plans.
- Identify case studies at regional and national levels where SSM is successfully implemented based on the SSM assessment protocol, mapping of successful projects, and monitor where areas under SSM are increasing or decreasing.
- Formulation of policies on management of rangelands and rangeland soils.
- Study on carbon stocks in the soils of HKH region.
- Managing agro-ecosystems for conservation of soil, water, and nutrients.
- Managing critical watersheds and upper catchment areas of rivers for hydrological functions.
- Developing/strengthening community-based natural resources management institutions.
- Women are a formidable force for farm work in the HKH region and so, strengthening social institutions should get priority for sustainable soil management.

6.20 Change Velocity of Soil Under Different Land Use Scenarios

When compared to more densely populated areas, direct human and climate influence on the HKH mountain soils is relatively low. However, the HKH mountain region is strongly affected by various complex local drivers of change. Interventions made by different types of stakeholders working at different levels impact the overall sustainability in this vulnerable landscape. Decisions and actions made outside the region have significantly impacted the soils and land use in the HKH region. Soil quality and health have always been an important feature to enhance crop productivity. Only higher production can help in meeting the food requirements for the current and future population in HKH countries as their livelihood mainly depends on agriculture. Serious development challenges in the region are visualized due to soil degradation. The problem of food insecurity in the region has been aggravated due to soil degradation induced by mismanagement of soil resource as well as high dependency on erratic precipitation. Though land use and climate change are the greatest threats to sustainable soil management, it is difficult to understand their combined effects. It has been hampered by modeling and data limitations. This has resulted in part from the different scales at which land use and climate processes operate. Climate change has affected the soils and has proved as an increasingly important driver of soil change. There are very limited studies undertaken in this remote HKH region to document the impacts of long-term cropping and land use systems on soil properties and nutrient stocks in the region.

The dynamics of terrestrial ecosystems depend on interactions between a number of biogeochemical factors, such as carbon, nutrient, and hydrological cycles. Human activities affect these interactions. Human activities have been found to impact carbon stocks in these pools and the atmosphere through land use types, land use changes, and forestry, among other activities.

With the introduction of new technologies, there has been heavy use of external inputs, and these changes in land use have affected the amount of carbon held by vegetation and soil, thereby, either it releasing carbon dioxide to, or removing it from, the atmosphere. The fluxes of carbon occur when forests are cut and converted to open lands. An important role is played by soil organic matter and organic carbon in increasing crop productivity and mitigating greenhouse gas emissions. It is necessary to maintain adequate soil organic matter content and proper soil structure for the sustainability of the land use systems.

Due to the combined effect of natural and anthropogenic factors, the soil system in the world is now changing at an unprecedented pace. Significant changes have been found in many soils in providing essential ecosystem services, such as food for inhabitants, clean water, fiber, and wood, climate regulation, and suitable habitat for a wide variety of plants and soil organisms. Policymakers, people, and the society should understand the changes for making informative, science-based decisions that impact the conservation and sustainability of the world's natural resources. However, changing demographic pressures, socio-economic and socio-political policies, cultural and religious beliefs, technological advances, and educational awareness affect indirectly societal drivers (Arnold et al. 1990).

Landscapes developed long before the appearance of man. Arnold et al. (1990) mentioned that mankind has been altering the earth's environment for at least two million years. During most of this period, man's influence has been nominal, extending to local situations and having positive and negative effects. Recent studies show that, of late, man's impact on changing soils and environment has increased many folds. Many activities of man have led to deterioration, degradation, and destruction of the system. Man's ability to change the natural environment is creating uncertainty in our understanding of the earth as a system. Soil is an important natural resource for growing food, feed, timber, and fiber. In addition to the physical support of life, the soil has many other important functions. Soils function not only as a plant-growing media but also as redistributors and regulators of most of the important fluxes of matter and energy. Soils change because of their intensifying use. The changes may be gradual, rapid, or even catastrophic. Such changes in the pedosphere affect the immediate carrying capacity of the land, through their influence on the vegetation and land use types, runoff, evaporation, and groundwater quality. Directly or indirectly the soil changes have a substantial effect on global climatic conditions, which in turn influence soils. Changing climate will play an important role in the future availability of food, fiber, and forest products on the earth. The soil aspect is particularly important in estimating the future agricultural potential of crops because soils are unsuitable for the growth of crops my dominate regions which are otherwise climatically suitable.

Soil resources in HKH region are affected in two ways, viz., reduction of arable land by conversion to other purposes and reduction of arable land by adverse processes. The adverse processes like acidification of soils, contamination of soils with heavy metals, pesticides, biomass transformations, and soil changes have induced a slow but steady change in soils of HKH region. The future changes of soil depends very much on the policies established and implemented by the governments of HKH nations. Soil properties are not constants and may undergo reversible or irreversible changes due to abiotic and biotic transport and transformation processes under the influence of pedogenic factors. Soil regimes express these temporal changes, which can be systematic or random, fluctuations or trend changes, reversible or irreversible, natural or man-induced, at one time or other, if proper sustainable management practices are not adopted.

Soil changes can be observed (i) in the total quantity of various soil constituents, such as organic matter, salts, and nutrient moisture content; (ii) their vertical profile distribution: leaching or accumulation; formation or destruction of diagnostic soil horizons; (iii) horizontal distribution due to surface runoff and erosion; (iv) CEC; (v) relative quantities and ratios; (vi) quality and functions, like stability, organic matter, solubility, humus, and other organic compounds. However, time changeability of soil properties and regimes exhibit a great variation in HKH region. Soil changes can be irreversible or nearly irreversible such as physical and chemical weathering, water and wind erosion; clay illuviation; slow reversible changes, such as development of diagnostic horizons and their characteristic sequence, and the soil profile; decomposition of organic matter; destruction of clay minerals; salt accumulation in heavy-textured, highly alkaline, and Na -saturated swelling clays; moderately reversible changes, such as structure destruction, aggregate failure; leaching; P and K fixation in heavy-textured soils; reversible changes, such as wetting and drying; flocculation and dispersion soil compaction; dissolution and precipitation of carbonates, soil temperature and, changes in the composition of soil air.

Soil changes impact four major soil components, viz.; soil forming processes; soil physics through erosion, drainage, and water balance; soil chemistry through nutrient contents, salinity, acidity, nitrification and denitrification and soil biology through organic matter, biological activity, and methane formation. The time dimension over which these changes will occur differs widely and may be within a short time or within a few years such as due to temperature, precipitation, and/or biological activities. The changes may occur within decades, hundreds, or thousand of years. In order to assess the magnitude, timing, and impact of soil changes at the HKH scale, addititional research is required in this data-starved region to obtain the necessary basic data, to ensure comparability of information and sound base to reach a justified judgment.

However, it would be unwise to recognize all soils when subject to minor changes. Cultivation of the original natural soils has gradually decreased the organic matter and biomass carbon in the soil environment, and fertilization has frequently changed drastically the pH of the clay. Thus, we may infer that most soils covering the present landscape in HKH region have been subject to minor fluctuations in the present environment. At the beginning, when mankind was not yet widespread, environmental changes did not differ from those preceding human interference. Later, however, when the population increased in HKH and, the effects of their activities became detrimental, the natural environment began to alter at a much more accelerated rate. One of the methods for identifying and evaluating the role of soils in HKH region is to have asoil and terrain digital database which is authentic, approachable, and easily accessible for analysis in combination with other resource databases, like climate, geology, topography, land use, land cover, etc. The possibilities given by this amalgamation will open new horizons for our knowledge of soil resources, terrestrial ecosystems, and their roles in soil change and environmental degradation. The significant loss of forest cover in the Himalayas is responsible for changes in soil regimes. Out of the 4.3 million km^2 geographical areas, 873,650 km^2 (20.20%) was under the forest in the year 2000 and 11,250 km^2 was lost during 2000–2010 (Behera et al. 2018).

6.20.1 Drivers of Change in HKH

Most of the area of HKH region remains covered with snow throughout the year. The landscape is dominated by treeless scenario and estimates show that about 79.8% of the total area falls in this category, followed by forests, scrub, and grasslands covering 17.7%, 2.2%, and 0.3% of the total area, respectively. In the HKH region, forests, scrub, and grasslands have an area of 767,200 km^2, 92,850 km^2, and 13,600 km^2, respectively. Myanmar tops the HKH countries in forest loss over the years and alone accounted for 51% forest loss, followed by India where 24% forest loss occurred.

The population and economic growth have remained the primary drivers of changes in soils in the HKH region. There has been a tremendous jump in both, population and economic growth in the region, and it is possible that economic growth will continue to be a major driver of soil change for the next three or four decades at least. Some of the best and productive soils in the region have been lost due to urbanization, industrial growth, and other anthropogenic activities. If we scan the past data related to the twentieth century with regard to population and economic growth, and an associated revolution in agriculture, it is revealed that phenomenal changes have taken place. Between 1961 and 2000, the growth percent in population and food production in the world was 98 and 146, respectively, during the four decades spanning 1960–2000. Per capita food production increased by 24% during the same period. The other factors of soil change in HKH were soil erosion, nutrient imbalance, soil carbon and biodiversity losses, soil sealing. soil acidification, contamination, salinization, and soil compaction. However, due to the prevalence of low temperature in HKH, many compositional changes have been slow. Direct and indirect human disturbance is a major soil change driver in HKH region.

6.20.2 Issus and Future Line of Action

With contexts to the situations in which poverty and ecosystem degradation exist and persist in the HKH region, the major drivers of change or factors responsible are fast changes in land use, population explosion, globalization, urbanization, climate change, and migration. It is a fact that all these drivers of change have impacted and posed challenges to traditional livelihood strategies and ecological stability, but nevertheless they have also provided opportunities in which mountain people can adapt to changes, build resilience, and make progress equitably. A need has been felt for approaches that integrate multi-stakeholder priorities with scientific data. This has to be done in such a way that trade-offs between conservation and development can be properly evaluated, promoting cross-country and regional acceptance of common frameworks, information sharing, and collaborative management, and contributing scientific input to regional policy processes.

Threats to soil and emerging changes can be taken care of by following specific and suitable management practices such as (i) increased crop nutrition through balanced measures which should include crop rotations with N-fixing crops, judicious use of organic and inorganic fertilizers in a balanced amount of N, P, and K, and suitable amendments like lime and gypsum to deal with specific soil chemical conditions such as soil acidity or sodicity/alkalinity, (ii) minimize soil disturbance by avoiding mechanical tillage, (iii) use of conservation tillage or no-till system, and (iv) maintain a proper protective organic cover on the soil surface using cover crops and crop residues. It is of great significance that these management practices are mostly interlinked and, in the long term, will minimize or reduce specific soil threats such as soil erosion by wind, water, and tillage, soil organic carbon loss, and soil compaction and physical deterioration. There is a strong need to minimize soil degradation in the region and for this, the drivers of soil degradation, such as climate, soil, topography, vegetation cover, land use, and hydrology, require special attention and management. It is important to apply nutrients, from both organic and inorganic sources, to sustain soil fertility for good crop production. Nutrient application can increase plant growth and maintain appropriate nutrient status in the soils as well as addition of organic matter to improve soil physical condition, which in turn can improve soil quality. It is important to apply a balanced dose of nutrients.

6.21 Potential Network of Soil Management in HKH

6.21.1 Recognition of Traditional Knowledge

Many ethnic tribes inhabit the HKH region. Due to their indigenous technical knowledge (ITK), these tribes have been able to survive in difficult mountain environments by taking appropriate mitigating measures. Of late, the HKH countries have begun tapping in the ITK to address some of the recent concerns related to climate-supported changes in cropping patterns, and surviving the extreme weather conditions. For inter-generational enrichment and transfer of resources, institutionalizing ITK can be a critical element.

6.21.2 Regional and Global Networking

On the basis of Agenda 21, Chapter 13, mountains have now received global attention. In 1995, ICIMOD started the Asia–Pacific Mountain Network (APMN) with the strong support of the Swiss Agency for Development and Cooperation. As a platform for knowledge sharing, the APMN connected through dialogue and networking the mountain regions and individual members whosoever wanted to join. It has proved quite useful. During the recent decades, Asia–Pacific Mountain Network has made purposeful and significant endowments in raising state-of-the-art repositories and expertise through participation and networking with its very large members and stakeholders. The network enshrines, enriches, and spreads information from a number of sources on mountain research and development issues in and for the Asia–Pacific region. The National Spatial Data Infrastructure (NSDI) supports investments in technology to generate and make available the geospatial knowledge needed to know, protect, and encourage our national and global interests. Its goals are (i) Create potentiality for National Shared Services, (ii) Confirm responsibility and purposeful development and management of geospatial resources, and (iii) Arrange get-together of leadership of the national geospatial community.

6.21.3 Institutional Networking

Georgidou and Groot (2002) mentioned that in the developing countries, the organizational, institutional, and human resources are considered as the "poor link", and capacity building for technically skilled human resources is a basic requirement for enhancing the benefits of spatial data infrastructure (SDI). A good number of countries are involved in SDI development, which takes into account the development of geospatial services that support public service delivery. This development ranges from local to global level. The SDI components are grouped into four classes, such as Policy and Legal Issues, Technical, Funding, and People. Through inter-disciplinary and cross-sectoral partnerships, ICIMOD has been engaged in promoting regional cooperation, partnerships, and networking among the national and regional institutions. Their capacity building has been one of the major strategies in the process. The ISCN's mission is to improve the understanding of carbon dynamics in soils across the world.

The networking and picture archiving and communications systems (PACS) are two different areas that work in such a way which requires a very special set of skills and know-hows. The International Board for Soil Research and Management approach shows the significance of the multidisciplinary strategy, involving soil scientists, agronomists, and socio-economists. To implement these networks on real and important issues on a realistic base, three regional network programs have since been highlighted, which include (i) Land development and management of acid tropical soils in Africa, (ii) Land development and soil management in monsoon Asia, and (iii) Management of Vertisols under semi-arid conditions in Africa and South-west Asia. To implement this goal, the International Board for Soil Research and Management approach is to help cooperators, through soil management networks, to conduct the investigations necessary for the practical adaptation and validation of improved soil management and land use practices. Soil management, in the International Board for Soil Research and Management view, should be based on a multidisciplinary approach, which must combine inputs from soil science, agronomy, and socio-economics.

Soil organic matter (SOM) governs many physical and chemical characteristics of soils, and it is one determinant of a soil's capacity for fertility, ecosystem productivity, and CO_2 sequestration. On a global scale, SOM is one of the largest actively cycling carbon (C) reservoirs, and direct human activities impact over 75% of C stocks in the upper meter of soil. But there are even more challenges. Soils affect everything on earth, which makes it a great challenge not only to conserve and preserve it for future generations but also to collectively manage judiciously to get optimum ecosystem services at all times.

6.21.4 Networking in HKH Region

Major characteristics on which networking is possible in HKH region are listed in Table 6.46.

6.21.5 Problems and Prospects for Regional Collaboration in HKH Region

Experts are of the view that the future may remain uncertain in the HKH region, because collaborative actions of the countries encompassing the region are lacking for developing region-wide policies. The growth of extreme geopolitical positioning has, as a consequence, made it more difficult for nation-states to develop region-wide policies. Several problems exist between different nations of the region for which the solutions are a distant dream. Some countries have a conflictual situation which prevents them to come to the table for a just decision. Water sharing is one of the major problems. Climate change and the critical need for water sharing among various countries of the region is a basic problem.

Table 6.46 Major characteristics on which networking is required in HKH region. (modified over Harden et al. 2017)

Characteristics	Functional problems	Restorative practices
Physical		
Microaggregate stability	Erosion, compaction	Conservation tillage (CT), no till
Water infiltration rate	Low infiltration, erosion	Residue inputs, cover crops, CT
Low soil temperature	Delayed germination,	Suitable crop cultivars
Poor moisture during winter	Poor crop stand	Soil moisture conservation techniques
Water-holding capacity	Arid region water management	Organic matter additions
Chemical		
Potentially mineralizable N	Poor fertility	Fertility management
Available P	Poor fertility	Fertility and pH management
Soil acidity	Low availability of P, Zn	Use of amendments
Micronutrient deficiency	Low crop yields, diseaces	Fertility management
Biological		
Microbial biomass C	Limited soil life	Residue inputs, cover crops, CT, "no-till"
Fungal and viral diseases	Poor yields,	Control measures, IPM

In order to meet challenges of climate change, disaster risks, and sustainable development in the Hindu Kush Himalayas, the International Centre for Integrated Mountain Development (ICIMOD) supports regional and transboundary cooperation (Molden et al. 2017). Sustainability in HKH has the potential and capacity to directly improve the lives of more than 1.5 billion people who are directly or indirectly benefitted by Hindu Kush Himalayas. However, difficulties arise in facilitating cooperation and policy coherence among the countries of the HKH region using and sharing natural resources, as there is a persistent challenge in a region that is prone to conflicts among the countries and is highly variable regarding the road to development. It is true that ICIMOD has been able to foster collaborative work and enhance the knowledge available on the HKH; it is recognized that official policy discussions and negotiations around these issues are needed to foster deeper, long-term understanding and collaboration.

6.21.6 Why Networking in Soil Management Could Be Uncertain in HKH Region

There are many factors which are responsible for non-starting a potential networking program for sustainable soil management in HKH region.

1. Almost all countries in HKH region are under some turmoil. Perpetual conflicts between India and Pakistan, India and China, and Pakistan and Afghanistan are being the most conspicuous, which can have unpredictable consequences. The conflict categories are internal in nature and most of the conflicts in the region are as a result of past spill-over effects.
2. Boundary disputes among different countries of HKH region are a source in blocking to share soil (geographical) information among different countries of the region. Perhaps these countries believe that sharing of information on sustainable soil management would lead to sharing some strategically important information, which is not in their interest.
3. Water sharing has always been a problem among different countries of the region. Though "Indus Water Treaty" between India and Pakistan has been successful up to a large extent, of late a shadow is looming large about its fate and continuity. India has also problems with Bangladesh and Nepal.
4. Many intergovernmental issues have become a stumbling block on having potential networking in the region.
5. Difficulties in using existing information and no access to basic geographic information have led to non-cooperation among member countries of HKH region. There is a need to be able to access, integrate, and use spatial data from different sources in supporting decision-making on sustainable soil management.

6.21.7 Way Forward

There is a need and an opportunity to (1) take cognizance of the datasets to identify and understand various processes, mechanisms, and properties of the ecosystem and the area that impact and guide soil organic carbon sequestration and its magnitude of change; (2) recognize and spread available datasets; (3) methodology and arrangements for sharing models, data, and management practices for soil organic carbon; and (4) enhance the relationship between soil management and global C cycle. It may be quite difficult to have a uniform approach as the levels of GIS used in the different agencies participating in "National Spatial Data Infrastructure" (NSDI) are at different stages. Involvement of many governmental and non-governmental institutions is required for NSDI initiatives. Its implementation with visible impacts takes a long time due to some basic requirements. The countries in HKH region have initiated NSDI activities in different forms such as short-term projects and the involvement of national and international stakeholders. There could be some challenges at different levels of any initiative. The major challenge is to mainstream the use of Geographic Information System and the related technologies in the regular decision-making processes, apart from the technical infrastructures and institutional coordination.

The broad recommendations needed for Himalayan soils, forest, water, farmlands, and rangelands on which the sustainability and stability of the flow of the esosystem services depend, are as follows:

- Synchronize programs and policies in the Himalayan ecosystem in the long run security of soil, water, energy, and food in the region.
- To increase resource-use efficiency and mitigate or reduce environmental impacts, there is a strong need for integrated planning of land, energy, water, forest, agriculture, and food security.
- Judicious management of demand for water and energy in order to make efficient use of these natural resources for food production.
- Strengthen coordination mechanisms between upstream and downstream countries to protect Himalayan ecosystems, catchments, watersheds, and headwaters.
- Control black carbon emissions by providing clean energy options to the rural people to organize, manage, and sustain ecosystem services of HKH region glaciers in providing freshwater to the communities in local as well as downstream areas.
- The ecological, environmental, and social implications of hydropower structure development should be given due importance while exploiting the possibility of hydropower potential.
- For multiple uses of water for irrigartion, energy, navigation, fisheries, and domestic uses, create a harmonized framework.
- For understanding the dynamics of food, water, and energy nexus, encourage integrated research and development programs to promote synergies among the concerned sectors.

References

Abrol IP (1990) Problem soils in India. Problem Soils of Asia and the Pacific. FAO/RAPA, Bangkok, pp 153–165

ADB (Asian Development Bank) (2002) Country Assistance Plans. Bangladesh: III. Sector Strategies. via http://www.adb.org/documents/caps/ban/0301.asp

ADB (Asian Development Bank) (1998) Forestry master plan, Volume 1. Government of Bangladesh. In: Ministry of Environment and Forests. Asian Development Bank, Manila, Philippines, 162 pp

ADB (2006) Environmental Assessment of Nepal: Emerging Issues and Challenges. Kathmandu, Nepal, 224p In: Advancing regional and transboundary cooperation in the conflict-prone Hindu Kush–Himalaya, Mountain Research and Development, 37 (4):502–508

Ahamad FU (1998) Environment and Development: Some lessons from a social forestry programme in Bangladesh. South Asian Anthropologist, 19: 11–18

Ahmad I, Ahmad MSA, Hussain M, Ashraf M, Ashraf MY (2011) Spatio-temporal variations in soil characteristics and nutrient availability of an open scrub type rangeland in the sub mountainous Himalayan tract of Pakistan. Pak J Bot 34:565–571

Ahmed AA, Hassan M, Ahmed SU, Rahman MK (2017) Fertility status on some Tista floodplain soils of Bangladesh. Int J Sci Eng Res 8(6):579–582

Ahmadi A, Emami M, Daccache A, He L (2021) Soil properties prediction for precision agriculture using visible and near-infrared spectroscopy: a systematic review and meta-analysis. Agronomy 11:433. https://doi.org/10.3390/agronomy11030433

Ahsan E, Karim Z (1988) Soil and management research on upland soils of Bangladesh. In: Proceedings of the international conference on the management and fertilization of upland soils in the tropics and subtropics, 247–251pp

Ahuja RL, Manchanda ML, Sangwan BS, Goyal VP, Agarwal RP (1992) Utilization of remotely sensed data for soil resource mapping and its interpretation for land use planning of Bhiwani district, Haryana. J Indian Soc Remote Sens 20(2–3):105–120

Ali M I (1991) Crop response to different fertilizer elements in Bangladesh (rice, jute, sugarcane and tea). Paper presented in the conference on Production and Use of Multinutrient Fertilizer in Bangladesh at Bangladesh Agricultural Research Council, Dhaka, Bangladesh, 25–26 November 1991

Ali MI, Bhuiya EH, Rahman MM and Habibullah AKM (1981) Nutrient status in Bangladesh soils evaluated through soil test and pot experiments using tracer technique. Research Report No. 40, INA, Bangladesh

Ali MH (2003) Agriculture produces. Banglapedia. Volume 1. Asiatic Socity of Bangladesh, Dhaka, Bangladesh, 70–82pp

Ali MM, Shaheed SM, Kubota D (1997) Soil degradation during the period 1967–1995 in Bangladesh. II. Selected chemical characters. Soil Sci Plant Nutr 43:879–890

Amtmann A, Francisco R (2012) Potassium in plants. In: Biogeography, vol 13. Wiley, England, pp 345–358

Andersen P (2000) Beyond local knowledge and institutional reach: micronutrient disorders in Hill agriculture. Paper presented at a regional conference on physical mobility and development in the mountains, Kathmandu, Nepal, March 15–17 2000

Andres RJ, Boden TA, Breon FM, Ciais P, Davis S, Erickson D, Gregg JS, Jacobson A, Marland G, Miller J, Oda T (2012) A synthesis of carbon dioxide emissions from fossil-fuel combustion. Biogeosciences 9(5):1–4

Anil Kumar KS, SenTK, Baruah U, Chamuah GS, Sehgal J (1994) In: Annual Report, NBSS & LUP, Nagpur, India

Annepu SK, Shirur M, Sharma VP (2017) Assessment of soil fertility status of mid Himalayan region, Himachal Pradesh, India. J Ecol 44(2):226–231

Anonymous (1984) Seventh five year plan (1985–90), North Eastern Council, Shillong, Meghalaya, India

Anonymous (1985) Report of the working group on development of north Eastern Region, North Eastern Council, Shillong, India 83–84 pp

Anonymous (1990) Report of the hydrographic survey and sedimentation studies: Umiam Reservoir Project. Meghalaya, India

Anonymous (1999) State Forest Report. In: Forest Survey of India, Ministry of Environment and Forest, Dehradun, India

Anonymous (2000) Wasteland Atlas of India. In: Ministry of Rural Development, Department of Land Resources, Govt. of India, New Delhi & National Remote Sensing Agency, Department of Space, Govt. of India, Hyderabad, India, 1–81 pp

Anonymous(2005) In: Annual Report, ICAR Research Complex for NEH Region. Meghalaya, India

Arias PA, Bellouin N, Coppola E, Jones RG et al (ed) (2021) Climate Change 2021: The physical science basis. Contribution of working Group I to the Sixth Assessment Report of the Intergovernmental Panel on climate change. Cambridge University Press (In Press)

Arnold RW, Szabolcs I and Targulian VO (1990) Global soil change Report of an International Institute for Applied System Analysis, International Society of Soil Science, United Nations Environmental Programme, CP-90–2 In: Task force on the role of soil in global change international Institute for applied systems Analysis, Laxenburg, Austria

Askari MS, Holden NM (2015) Quantitative soil quality indexing of temperate arable management systems. Soil Till Res 150:57–67

Awasthi RP, Prasad RN (1984) Efficient cropping system for rainfed terraces in North Eastern Region of India as an alternative to shifting cultivation. Status paper presented in National symposium on efficient cropping system zone of India held at the University of Agricultural Sciences, Bangalore, India 7–10 January 1984

Baillie I, Norbu C (2004) Climate and other factors in the development of river and interfluves profiles in Bhutan. J Asian Earth Sci 22:539–553

Bajracharya RM, Sherchan DP (2009) Fertility status and dynamics of soils in the Nepal Himalaya: a review and analysis. In: Lucero DP, Boggs JE (eds) Soil fertility, Nova Science Publishers, Inc.

Bajracharya RM, Sharma S, Sitaula BK, Dahal BM, Rokaya K and Jeng A (2006) Assessment of soil quality using physiochemical and biological indicators in a mid-hill watershed of Nepal. Paper presented at International. Seminar on Environmental and Social Impacts of Agricultural Intensification in Himalayan Watersheds, Kathmandu, Nepal, Oct 15–17, 2006 , 105-114pp

Bajracharya RM, Sitaula BK, Sharma S, Jeng A (2007) Soil quality in the Nepalese context—an analytical review. Int J Ecol Env Sci 33(2–3):143–158

Bajracharya RM, Sitaula BK, Shrestha BM, Awasthi KD, Balla MK, Singh BR (2004) Soil organic carbon status and dynamics in the central Nepal Middle Mountains. J Inst Forest 12:28–44

Bala B, Sharma RK, Sharma SD (2004) Size and composition of bovine population in Himachal Pradesh—a district wise analysis. Indian J Animal Res 38(1):8–13

Banglapeida (The National Encyclopedia of Bangladesh) (2012) In: Asiatic Society of Bangladesh. Dhaka, Bangladesh

Baral SR, Malla MB, Howell J (1999) Reclaiming rato mato: a study of the rehabilitation of red clay soils in the Nepal Middle Hills. Dept. of Roads, HMG/Min. of Works and Transport, and Ecological Society, Kathmandu, 44 p

BARC (1989) Fertilizer recommendation guide. In: Karim Z, Ali MI , Md Miah MU and Hussain SKG (eds.). Soils Publ. No: 32

Barcena TG, Kiær LP, Vesterdal L, Stefánsdóttir HM, Gundersen P, Sigurdsson BD (2014) Soil carbon stock change following afforestation in Northern Europe: a meta-analysis. Glob Change Biol 20(8):2393–2405

Baumann F, He JS, Schmidt K, et al (2009) Pedogenesis, permafrost, and soil moisture as controlling factors for soil nitrogen and carbon contents across the Tibetan Plateau. Global Change Biol 15: 3001–3017. https://doi.org/10.1111/j.1365-2486.2009.01953.x

Baumler R, Caspari T, Totsche KU, Dorji T, Norbu C, Baillile IC (2005) Andic properties in soils developed from nonvolcanic materials in Central Bhutan. J. Plant Nutr. Soil Sci. 168: 703–713. https://doi.org/10.1002/jpln.200521793703

Bäumler R (2001) Pedogenic studies in aeolian deposits in the high mountain area of eastern Nepal. Quatern Int 76(77):93–102

Bayer AD, Fuchs R, Mey R, Krause A, Verburg PH, Anthoni P, Arneth A (2021) Diverging land-use projections cause large variability in their impacts on ecosystems and related indicators for ecosystem services. Earth Syst. Dynam 12:327–351. https://doi.org/10.5194/esd-12-327-2021

Bazza Z (2019) Protocol for assessing sustainable soil management. Food and Agricultural Organization of the United Nations, Rome, Italy

Beckett PHT et al (1983) Cu and Zn in soils treated with sewage sludge: their 'extractability' to reagents compared with their 'availability' to plants. Plant Soil 70:3–14

Behera MD, Murthy MSR, Das P, Sharma E (2018) Modelling forest resilience in Hindu Kush Himalaya using geoinformation. J Earth Syst Sci 127:95. https://doi.org/10.1007/s12040-018-1000-x

Bender F (1983) In: Geology of Burma. Gebriider Borntraeger, Stuttgart

Bhardhawaj SP, Prasad SN (1981) Response of rice-wheat rotation to zinc under irrigation condition in Doon Valley. J Indian Soc Soil Sci 29:220–234

Bhatnagar HP (1965) Soils from different quality sal (*Shorea robusta*) forests of Uttar Pradesh. Trop Ecol 6:56–62

Bhattacharyya R, Ghosh BN, Mishra PK et al (2015) Soil degradation in India: challenges and potential solutions.sustainability 7:3528–3570. https://doi.org/10.3390/su7043528

Bhattacharyya T, Sehgal J, Sarkar D (1996) Soils of Tripura for optimizing land use. In: NBSS Publ. 65a, c, NBSS&LUP, Nagpur, India, 154p

Bhuiya ZH (1987) Organic matter status and organic recycling in Bangladesh soils. Resour Conseruation 13:117–124

Bhuiyan NI (1988) Co-ordinated project on potassium studies. In: Progress Report (1987–88). Bangladesh Rice Research Institute, Gazipur, Bangladesh

Bhuiyan NI (1991) Appropriate nutrient ratios for soils and crops of Bangladesh. Paper presented in the conference on Production and Use of multinutrient fertilizer in Bangladesh, at Bangladesh Agricultural Research Council, Dhaka, Bangladesh, 25–26 November 1991

Bian M, Zhou M, Sun D, Li C (2013) Molecular approaches to unravel the mechanism of acid soil tolerance in Plants. Crop J 91–104

Blaikie P (2016) The political economy of soil erosion in developing countries: Routledge. London, 200p. https://doi.org/10.4324/9781315637556, eBook ISBN 9781315637556

Blaikie PM, Muldavin JSS (2004) Upstream, downstream, China and India: the politics of environment in the Himalayan region. Ann Assoc Am Geogr 94 (3):520–548

Blanco-Canqui H, Lal R (2008) No-tillage and soil-profile carbon sequestration: an on-farm assessment. Soil Sci Soc Am J 72(3):693–701

Bolch T, Kulkarni A, Kääb A et al (2012) The state and fate of Himalayan glaciers. Science 336:310–314

Borkakati K and Takkar PN (2000) Forms of boron in acid alluvial and lateritic soils in relation to ecosystems and rainfall distribution. Paper presented at international conference on management of natural resources for sustainable agricultural production in the 21st Century, New Delhi, India, 12th February 2000

Borthakur DN, Singh A and Awasthi, RP et al (1983) Shifting cultivation in North Eastern Region, Shillong, Meghalaya, India

Bot A, Benites J (2005) The importance of soil organic matter: key to drought-resistant soil and sustained food production, FAO Soils Bulletins: 1–94

Bowman RA, Halvorson AD (1998) Soil chemical changes after nine years of differential N fertilization in a no-till dryland wheat–corn–fallow rotation. Soil Sci 163:241–247

Brady NC, Weil RR (2002) The nature and properties of soils, 13th edn. Pearson Education, New Jersey

Brevik EC (2013) The potential impact of climate change on soil properties and processes and corresponding influence on food security. Agriculture 3: 398–417. https://doi.org/10.3390/agriculture3030398

Brown S, Schreier H, Shah PB, Lavkulich LM (1999) Soil nutrient budget modeling: An assessment of agricultural sustainability in Nepal. Soil Use Manag 15:1–15

Brown LR, Kane H (1994) Full house: reassessing the earth's population carrying capacity. In: Norton WW (ed), The World Watch Environmental Alert Series, New York

Bnnning S, Vargas R, Vita A (2015) Mountain soils and agriculture. In: Romeo R, Vita A, Manuelli S, Zanini E, Freppaz M, Stanchi S (eds) Understanding mountain soils: a contribution from mountain areas to the international year of soils 2015, FAO 2015, Rome, Italy

BRRI (1980) Annual report. Soil chemistry division, Bangladesh Rice Research Institute, Gazipur, Bangladesh

BRRI (1990) Annual internal review. Bangladesh Rice Research Institute, Gazipur, Bangladesh

BRRI (1992) Annual internal review. Bangladesh Rice Research Institute, Gazipur, Bangladesh

Bryan RB, Campbell IA (1986) Runoff and sediment discharge in a semi-arid drainage basin. Zeitschrift Fur Geomorphologic 58:121–143

Bryan BA, Gao L, Ye Y, Sun X, Connor JD, Crossman ND, Stafford-Smith M, Wu J, He C, Yu D, Liu Z (2018) China's response to a national land-system sustainability emergency. Nature 559(7713):193–204

Burke IC, Lauenroth WK, Cunfer G, Barrett JE, Mosier A, Lowe P (2002) Nitrogen in the central grasslands region of the united states: current anthropogenic additions of nitrogen to ecosystems of the US central grasslands far outweigh loss of nitrogen through crop removal, resulting in increased nitrogen fluxes with the potential to alter regional-scale biogeochemical cycling. Bioscience 52:813–823

Canadell JG, Raupach MR (2008) Managing forests for climate change mitigation. Science 320(5882):1456–1457

Caon L (2018) Soil fertility in mountain soils. Global soil partnership, food and agricultural organisation of the United Nations. In: Expert consultation workshop on

Hindu Kush Himalayan Mountain Soils, Khatmandu, Nepal, 20–21 March 2018
Carter CP (1967) Helmand-Arghandab valley authority. In: Extension sponsored corn demonstration report. SALO, Bost, Afghanistan. ditto and mimeo, 8 p
Carter EJ (1992) Tree cultivation on private land in the middle hills of Nepal: lessons from some villagers of Dolakha district. Mt Res Dev 12(3):241–255
Carver M, Schreier H (1995) Sediment and nutrient budgets over four spatial scales in the Jhikhu Khola Catchments: implications for land use management. In: Schreier H, Shah PB, Brown S (eds) Challenges in mountain resource management in Nepal. Processes, trends, and dynamics in middle mountain catchments. Proceedings of a workshop held in Kathmandu, ICIMOD, 163–170pp, 10–12 April 1995
Caspari T (2005) The soils of Bhutan: parent materials, soil forming processes, and new insights into the palaeoclimate of the Eastern Himalayas. Technische Universität München, Germany, Lehrstuhl für Bodenkunde
Caspari T, Baumler R, Dorji T, Norbu C, Tshering T, Baillie IC (2004) Pedochemical characteristics and landscape history of the Thagbi river terrace system, central Bhutan. Zeitchrift Fur Geomorphologie 48 (2):145–165
Cassman KG (1999) Ecological intensification of cereal production systems: yield potential, soil quality, and precision agriculture. Proc Natl Acad Sci 96 (11):5952–5959
CBS (2011) National Population and Housing Census 2011 (National Report), Central Bureau of Statistics, Government of Nepal, Kathmandu, Nepal
Chakraborty SK, Sinha H, Prasad R (1979) Copper status of soils of Assam. J Indian Soc Soil Sci 27:83–84
Chakraborty SK, Sinha H, Prasad R (1980) Distribution of boron and molybdenum and their relationship with certain properties of soils from Assam. J Indian Soc Soil Sci 30:92–93
Chakraborty SK, Sinha H, Prasad R (1981) Zinc status of soils of Assam. J Indian Soc Soil Sci 29:377–378
Chakravarthy DN, Sehgal JL, Dev G (1978) Influence of climate and topography on the pedogenesis of alluvium derived soils of Assam. Indian J Agric Chem 11:77–97
Chalise SR, Khanal NR (1997) Erosion processes and their implications in sustainable management of watersheds in Nepal Himalayas. In: FRIEND'97—regional hydrology: concepts and models for sustainable water resource management. Proceedings of Postojna, Slovenia Conference, International Association of Hydrologic Sciences, Publ. No. 246, Sept–October 1997, pp 325–334
Champan JL, Reiss MJ (1992) Ecology principals and application. Cambridge University, UK, p 294
Chapuis Lardy L, Wrage N, Metay A, Chotte JL, Bernoux M (2007) Soils, a sink for N_2O? A review. Glob Chang Biol 13:1–17
Chase P, Singh OP (2014) Soil nutrients and fertility in three traditional land use systems of Khonoma, Nagaland, India. Resour Env 4: 181–189. https://doi.org/10.5923/j.re.20140404.01
Chen BS (2001) Cultivation of forage grasses and forage crops. China Agricultural Science and Technology Press, Beijing
Chen D, Li J, Lan Z, Hu S, Bai Y, Niu S (2016) Soil acidification exerts a greater control on soil respiration than soil nitrogen availability in grasslands subjected to long-term nitrogen enrichment. Funct Ecol 30:658–669
Chen XP, Cui ZL, Vitousek PM et al (2011) Integrated soil–crop system management for food security. Proc Natl Acad Sci USA 108(16):6399–6404. http://www.pnas.org/cgi/doi/10.1073/pnas.1101419108
Cheng ZG, Liu XD (2008) A preliminary study of future permafrost distribution on the Qinghai-Tibet Plateau under global warming conditions. Areal Res Dev 27: 80–85. https://doi.org/10.3969/j.issn.1003-2363.2008.06.020
Chernov VA (1947) On the nature of soil acidity. Academy of Sciences, Moscow, 170 pp
Choden K, Suberi B, Chettri P (2021) Terrestrial carbon stock potential in selected forest in Bhutan, India and Nepal. Asian J Environ Ecol 15:21–32. https://doi.org/10.9734/ajee/2021/v15i130219
CIA (2012) The World Factbook. https://www.cia.gov/library/publications/the-world-factbook/geos/af.html
Coleman NT, Thomas GW (1967) The basic chemistry of soil acidity. In: Pearson RW, Adams F (eds) Soil acidity and liming. The American Society of Agronomy. Madison, Wisconsin, pp 1–41
Costanza R, Arge de, Groot R, Farberk S et al (1997) The value of the world's ecosystem services and natural capital. Nature 387(15):253–260. https://doi.org/10.1016/S0921-8009(98)00020-2
Cozzolino D, Morón A (2006) Potential of near-infrared reflectance spectroscopy and chemometrics to predict soil organic carbon fractions. Soil Tillage Res 85:78–85
Crosson P (1994) Degradation of resources as a threat to sustainable agriculture. First World Congress of Professionals in Agronomy, Santiago, Chile
CSO (Central Statistics Organization) (2018) Afghanistan Living Conditions Survey 2016–17, Kabul, Afghanistan, CSO
Dagesse D (2011) Effect of freeze-drying on soil aggregate stability. Soil Sci Soc Am J 75(6):2111–2121
Dahal BM, Sitaula BK, Bajracharya RM, Atreya K, Jeng A (2006) Soil and nutrient losses from intensified agricultural land in Ansikhola watershed in Kavre Distric of Nepal. In: Proceedings of International seminar on environmental and social impacts of agricultural intensification in Himalayan Watersheds, Kathmandu, Nepal, 15–17 Oct 2006, 120–128pp
Dalovic IG, Dorde S, Jockovic DS, Goran J, Dugalic GJ, Bekavac GF, Purar B, Seremesic SL, Jockovic MD (2012) Soil acidity and mobile aluminum status in pseudogley soils in the Cacak-Kraljevo Basin. J Serb Chem Soc 77(6):833–843. https://doi.org/10.2298/JSC110629201D

Dang X, Zhang Y, Yu N, Zhang Y (2012) Cadmium adsorption-desorption of brown soil with freeze-thaw cycles in northeast China. Int J Environ Pollut 49(1–2):89–99
Das TH, Thampi CJ, Sehgal J, Velayutham M (1996) Soils of Sikkim for optimizing land use. NBSS Publ. 60b. NBSS&LUP, Nagpur, India, 44p
Datta D, Sarkar D, Banerje SK (2004) J Indian Soc Soil Sci 52:56–62
Datta MG, Gupta RK (1984) Studies on P, Ca, Zn and Fe content in soil and plant tissues as influenced by pressmud and lime. J Indian Soc Soil Sci 32:450–458
Datta M, Laskar S, Prasad RN (1991) Appraisal of charge characteristics in some soils of Tripura. J Indian Soc Soil Sci 39:767–769
Datta M, Laskar S (1992) Different reaction of local rice races to iron in Tripura. Oryza 29:388–391
Datta M, Ram M (1993) Status of micronutrients in some soil series of Tripura. J Indian Soc Soil Sci 41:776–777
Delpierre N, Soudani K, François C, Köstner B et al (2009) Exceptional carbon uptake in European forests during the warm spring of 2007: a data-model analysis. Glob Change Biol 15:1455–1474. https://doi.org/10.1111/j.1365-2486.2008.01835.x
Dewan ML, Sharma S (1985) People's participation as a key to—Himalayan eco-system development. Report of the NGOs Meeting at the Centre for Policy Research, Published by Centre for Policy Research Dharma Marg, Chanakyapuri, New Delhi, 56 pp
Ding J, Chen L, Zhang B, Liu L et al (2016) Linking temperature sensitivity of soil CO_2 release to substrate, environmental, and microbial properties across alpine ecosystems. lob. Biogeochem Cycles 30:1310–1323
Dorji KD, Dorji T (2015) The status of national soil resources of Bhutan. In: Asian soil partnership consultation workshop on sustainable management and protection of soil resources, Bangkok, Thailand
Dorji KD (2008) Agriculture and soil fertility management in Bhutan: An overview. Paper presented in the meeting of Asia-Pacific net on integrated plant nutrient management and international workshop on sustainable nutrient management: technology & policy, Shijiazhuang, Hebei, China, May 27–31 2008
Duan ZH, Zhang Y, Zhang W et al (2012) Application research of four cold regions land surface and hydrological model to Qinghai-Tibet Plateau frozen soil region. J Water Resour Water Eng 23:43–50
Dutta AK, Das AL, Pandey S, Hirekerur LR (1992) Bulletin, National Bureau Soil Survey and Land use Planning, Nagpur, India 2
Dutta P (1984) Geographical dimensions of flood in the Brahmaputra valley. In: Jaswal IJS (ed) Status of ecology. Punjab Publishing Hosue, Patiala, pp 93–102
Dwivedi GK, Dwivedi M (1992) Efficacy of different methods of application of copper, zinc and boron to potato. Annals of Agricultural Research, 13: 1–6
Dwivedi BS, Ram M, Singh BP, Das M, Prasad RN (1990a) Evaluation of soil and tissue tests for predicting response of rice to Zn in acid alfisols. Indian J Agric Sci 60:122–127
Dwivedi BS, Ram M, Singh BP, Das M, Prasad RN (1990b) Differential response of crops to boron fertilization in acid alfisols. Indian J Agric Sci 60:122–127
Dwivedi BS, Ram M, Singh BP, Das M, Prasad RN (1992) Effect of liming on boron nutrition of pea (Pisum sativum L.) and corn (Zea mays L.) grown in sequence in an acid alfisol. Fertilizer Res 31:257–262
Dwivedi BS, Ram M, Singh BP, Das M, Prasad RN (1993) Comparison of soil tests for predicting boron deficiency and response of pea to boron application on an acid alfisol. J Indian Soc Soil Sci 41:321–325
Egashira K, Yasmin M (1990) Total and available phosphorus of some floodplain soils of Bangladesh. Bull Inst Trap Agr Kyushu Univ 13:127–137
Egashira K, Yasmin M (1991) Sulfur status of Bangladesh floodplain soils. Bull Inst Trop Agr Kyushu Univ 14:15–23
Eguchi T (1987) Topographic features in the central part of the Bhutan Himalayas. In: Ohsawa M (ed) Life zone ecology of the Bhutan Himalaya. Laboratory of Ecology, Chiba University, Japan, pp 185–208
Eguchi T (1991) Regional and seasonal change in precipitation in Bhutan—analysis of daily precipitation. In: Ohsawa M (ed) Life zone ecology of the Bhutan Himalaya II. Laboratory of Ecology, Chiba University, Japan, pp 1–20
Ehrlich D, Melchiorri M, Capitani C (2021) Population trends and urbanisation inmountain ranges of the world. Land 10:255. https://doi.org/10.3390/land10030255
Eni DD, Iwara AI, Offiong RA (2012) Analysis of soilvegetation interrelationships in a south-southern secondary forest of Nigeria. Int J For Res 1–8. https://doi.org/10.1155/2012/469326
Eylachew Zewdie (2001) Study on physical, chemical and mineralogical characteristics of some Vertisols of Ethiopia. In: Wondimagne C, Engda M (eds) Proceeding of the 5th Conference of the Ethiopian Society of Soil Science (ESSS). The Ethiopian Society of Soil Science (ESSS). Addis Ababa, Ethiopia, 30–31 March 2000, 87–102 pp
Fang HB, Zhao L, Wu XD, Zhao YG, Zhao YH, Hu GJ (2015) Soil taxonomy and distribution characteristics of the permafrost region in the Qinghai-Tibet Plateau, China. J Mountain Sci.https://doi.org/10.1007/s11629-014-3133
FAO (1973) FAO soils Bulletin 21—calcareous soils. FAO, Rome. http://www.fao.org/docrep/x5868e/x5868e00
FAO (1977) The state of food and agriculture. Food and Agricultural Organization of the United Nations, Rome, Italy
FAO (1994) The collection and analysis of land degradation data in Nepal. In: Report of the Expert Consultation of the Asian Network on Problem Soils. Food and Agric. Organization/United Nations, Bangkok, Thailand, Publication 1994/4

FAO (Food and Agriculture Organization) (2005) State of the World's Forest 2005. Food and Agriculture Organization, Rome, Italy

FAO (2005) Current world fertilizer trends and outlook to 2009/2010, Italy, Rome

FAO (2011) The state of the world's land and water resources for food and agriculture (SOLAW) managing systems at risk. Food and Agriculture Organization of the United Nations, Rome and Earthscan, London

FAO (2015) Understanding mountain soils: a contribution from mountain areas to the International Year of Soils 2015. In: Romeo R, Vita A, Manuelli S, Zanini E, Freppaz M, Stanchi S (eds) Rome, Italy

FAO (2019) Soil erosion: the greatest challenge to sustainable soil management, Rome, 100 pp

FAO/WFP (2009) In: Special report—crop and food security assessment mission to Myanmar. http://documents.wfp.org/stellent/groups/public/documents/ena/wfp196368.pdfM10.1080/07900627.2015.1030494

FAO UNDP (1971) Bangladesh: soil resources. Technical Report No. 3, FAO, Rome

FAO UNDP (1988) Land resources appraisal of Bangladesh for agricultural development: agroecological regions of Bangladesh. Technical Report No. 2, FAO, Rome

Foy CD, Chaney RL, White MC (1978) The physiology of metal toxicity in plants. Ann Rev Plant Physiol 29:511–566

Freppaz M, Williams W (2015) Mountain soils and climate change. In: Romeo R, Vita A, Manuelli S, Zanini,E, Freppaz M, Stanchi S (eds) Understanding mountain soils: a contribution from mountain areas to the International Year of Soils 2015, FAO 2015, Rome, Italy

Friedlingstein P, O'Sullivan M, Jones MW, Andrew RM et al (2020) Global carbon budget 2020. Earth Syst Sci Data 12:3269–3340. https://doi.org/10.5194/essd-12-3269-2020

FRMD (2017) Land use and land cover of Bhutan 2016, maps and statistics, Forest Resources Management Division Department of Forests & Park Services Ministry of Agriculture and Forests Thimphu, Bhutan. ISBN: 978-99936-743-2-0

Frutchey GW (1962) Semi-annual Report USOM, Wyoming, Kabul, Afghanistan, 17 p

Galloway JN, Townsend AR, Erisman JW et al (2008) Transformation of the ntrogen cycle, recent trends, questions, and potential solutions. Science 320:889–892

Ganeshamurthy AN, Kalaivanan D, Satisha GC (2016a) Management of vegetable crops in acid soils of India. In: Peter KV (ed) Innovations in horticultural sciences. New India Publishing Agency, New Delhi, India, pp 559–584

Ganeshamurthy AN, Kalaivanan D, Satisha GC (2016b) Management of fruit crops in acid soils of India. In: Peter KV (ed) Innovations in horticultural sciences. New India Publishing Agency, New Delhi, India, pp 539–558

Gangopadhdhyay SK, Das PK, Mukhopadhyay SS, Bansal SK (1990) J Indian Soc Soil Sci 38:93–99

Gardner RAM, Rendell HM (1994) Loess, climate and orogenesis: implications of South Asian loesses. Zeitschrift Für Geomorphologie NF 38(2):169–184

Gerasimov IP (1973) Chernozems, buried soils and loesses of the Russian plain: their age and genesis. Soil Sci 116:202–210

Gilmour DA (1991) Trends in forest resources and management in the middle mountains of Nepal. In: Workshop proceedings: soil fertility and erosion issues in the middle mounts of Nepal, Topographical Survey Branch, HMG, Nepal

Government of Pakistan (1991) The Pakistan National Conservation Strategy, The World Conservation Union (IUCN), Karachi, Pakistan

Gruber S (2012) Derivation and analysis of a high-resolution estimate of global permafrost Zonation. Cryosphere 6: 221–233. https://doi.org/10.5194/tc-6-221-2012

Gu C, Zhao P, Chen Q, Li S, Li L, Liu L, Zhang Y (2020) Forest cover change and the effectiveness of protected areas in the Himalaya since, (1998) Sustainability 12: 6123–6146. https://doi.org/10.3390/su12156123

Guggenberger G, Bäumler R, Zech W (1998) Weathering of soils developed in eolian material overlying glacial deposits in eastern Nepal. Soil Sci 163(4):325–337

Gupta PK (2004) Soil plant, water and fertilizer analysis. Shyam Printing Press, Agrobios, India, p 38

Gupta RD, Tripathi BR (1996) Mineralogy, genesis and classification of soils of North-West Himalayas developed on different parent materials and variable topography. J Indian Soc Soil Sci 44(4):705–712

Gupta RK, Sanjay, Chera RS (1996) Soil characteristics as influenced by slope aspects in Middle Siwaliks. Agropedology 6:43–48

Gupta RD, Verma SD (1992) Characterization and classification of some soils of Kandi Belt in Jammu Siwalik Hills. J Indian Soc Soil Sci 42:809–815

Gupta RK, Prasad RN (1983) In: Annual Report, ICAR research Complex for NE Hill Region, Meghalaya, India

Hadden RL (2008) The geology of Burma (Myanmar): an annotated bibliography of Burma's geology, geography and earth science, Topographic Engineering Center (TEC), Engineer Research and Development Center (ERDC), 7701 Telegraph Road, Alexandria, VA 22315

Han Z, Deng M, Yuan A, Wang J, Li H, Ma J (2018) Vertical variation of a black soil's properties in response to freeze-thaw cycles and its links to shift of microbial community structure. Sci Total Environ 625:106–113

Harden JW, Hugelius G, Ahlstrcom A et al (2017) Networking our science to characterize the state, vulnerabilities, and management opportunities of soil organic matter. Global Change Biology. Wiley. https://doi.org/10.1111/gcb.13896

Hassan M, Ahmed AA, Hassan MA, Nasrin R, Rayhan AB, Salehin SM, Rahman MK (2017a) Changes of soil fertility status in some soil series of Tista Floodplain Soils of Bangladesh, during 1996–2016. Asian Res J Agric 5(3):1–9

Hassan M, Hassan R, Mahmud MA, Pia HI, Hassan MA, Uddin MJ (2017b) Sewage waste water characteristics and its management in urban areas—a case study at Pagla Sewage Treatment Plant. Dhaka. Urban Regional Plan 2(3):13–16

Havlin HL, Beaton JD, Tisdale SL et al (2010) Soil fertility and fertilizers—n introduction to nutrient management, 7th edn, India, PHI Learning Pvt. Ltd

Helvetas (2014) Conceptual note SLMP training of resource persons on sustainable agriculture and livestock management, 4 p

Hillel D (2007) Soil in the environment: crucible of terrestrial life. Academic Press, MA, p 320

Hillel D (1991) Out of the earth: civilization and the life of the soil. Free Press, New York

Hirschi KD (2004) The calcium conundrum Both versatile nutrient and specific signal. Plant Physiol 136:2438–2442

Hofer T (1998) Hydro-meteorological measurements and analysis in interdisciplinary watershed projects a strategy paper prepared for PARDYP Project. ICIMOD, Kathmandu

Hoper H, Steinberg C, Alabouvette C (1995) Involvement of clay type and pH in the mechanism of soil suppressiveness to fusarium wilt of flax. Soil Biol Biochem 27:955–967

Hossain MI, Kashem MA and Osman KT (2014) Fertility status of some forested soils of Chittagong hill tracts, Bangladesh. Int J Latest Res Sci Technol 3: 82–87. http://www.mnkjournals.com/ijlrst.htm

House GJ and Pamelee RW (1985) Comparison of soil arthropods and earthworms from conventional and no tillage agroecosystems.Soil and Tillage Research 5: 351–360.

Howell JH (1986) Soils of the Forestry Research Project Trial Sites: Chautara, Panchkhal, Thulo Sirubari and Kadambas. FRP trail site report No. 8. Soil Section, Forestry Research and Information Centre, Babar Mahal, Kathmandu, Nepal

HPKKV (1991) Status Report on Kinnaur and Spiti Catchment of Sutlej River in Himachal Pradesh, HPKVV, Palampur

Htwe TN (2016) Soil types and general soil information in Myanmar. The second meeting of the National Soil and Nutrients Management Experts Group, ASEAN guidelines on soil and nutrient management, Yangon, Myanmar

Huang CY (2000) Pedology. Beijing China Agriculture Press, 3–142pp. (In Chinese)

Humpenöder F, Popp A, Dietrich JP, Klein D, Lotze-Campen H, Bonsch M, Bodirsky BLBL, Weindl I, Stevanovic M, Mueller C, Müller C (2014) Investigating afforestation and bioenergy CCS as climate change mitigation strategies. Environ Res Lett 9:064029. https://doi.org/10.1088/1748-9326/9/6/064029

Huq I, Shoaib JUM (2013) The soils of Bangladesh. Springer, Dodrecht, The Netherlands

Hussain A, Rasul G, Mahapatra B, Tuladhar S (2016a) Household food security in the face of climate change in the Hindu-Kush Himalayan region. Food Sec 8:921–937. https://doi.org/10.1007/s12571-016-0607-5

Hussain A, Rasul G, Mahapatra B, Tuladhar S (2016b) Household food security in the face of climate change in the Hindu-Kush Himalayan region. Food Secur 8 (5):921–937

Hussain A, Rasul G, Mahapatra B, Wahid S, Tuladhar S (2018) Climate change-induced hazards and local adaptations in agriculture: a study from Koshi River basin, Nepal. Nat Haz 91(3):1365–1383

Hussain MS (1992) Soil classification with special reference to the soils of Bangladesh. University of Dhaka, Dhaka, Bangladesh

Hussain Z (1984) Some ecological observations on climatalogical data of Shillong, Calcutta Research India Publications, 11–19 pp

ICARDA (2002) Needs assessment on soil and water in Afghanistan. International Center for Agricultural Research in the Dry Areas, Sponsored by the United States Agency for International Development, Aleppo, Syria

ICIMOD (2006) In: Annual report 2005, ICIMOD, Kathmandu, Nepal, 92 p

ICIMOD (2007) MTAP-II, Medium Term Action Plan 2008- 2012. International Centre for Integrated Mountain Development, Kathmandu, Nepal

ICIMOD/UNEP/RGoB (2002) Inventory of glaciers, glacier lakes, glacier lake outburst floods monitoring and early warning system in the HKH Region, Bhutan

IFAD (1992) A report prepared for IFAD by the Centre for Development Cooperation Services. Free University, Amesterdam

IMI and FAO (2019) Integrated Mountain Initiative and FAO Report on State of the Himalayan farmers and farming: A study about social, economic and ecological factors changing the mountain agriculture scenario. https://www.mountaininitiative.in/images/IMI-FAO

Imran M, Gurmani ZA (2011) Role of macro and micro nutrients in the plant growth and development. Sci Technol Dev 1:129–131

IPBES (2019) Summary for policymakers of the global assessment report on biodiversity and ecosystem services of the Intergovernmental Science-Policy Platform on Biodiversity and Ecosystem Services, IPBES, Bonn, Germany

IPCC (2019) Climate change and land. IPCC special report on climate change, desertification, land degradation, sustainable land management, food security, and greenhouse gas fluxes in terrestrial ecosystems, Intergovernmental Panel on Climate Change (IPCC), Geneva, Switzerland

IPCC (2021) Summary for policymakers. In: Masson-Delmotte V, Zhai P, Pirani A, Connors SL et al (ed) Climate change 2021: the physical science basis.

Contribution of working Group I to the sixth assessment report of the Intergovernmental Panel on Climate Change. Cambridge University Press (In Press)
Islam A (1992) Review of soil fertility research in Bangladesh. In: Hussain MS, Huq SMI, Iqbal MA, Khan TH (eds) Proceedings of the inter-congress conference of Commission IV of ISSS on improving soil management for intensive cropping in the tropics and sub-tropics, pp l–18
Islam MS (1983) Soil management in agricultural research in Bangladesh, Bangladesh Agricultural Research Council, Dhaka, Bangladesh, 105–109 pp
Islam MS (1990) Soil characteristics and their effects on sulfur and zinc deficiency problems in Bangladesh. Paper presented at a workshop on Sulfur and Zinc Management of Soils and Crops and Balanced Fertilization in Bangladesh, Dhaka, April 1990, 10–11 pp
Islam, MS, Altamash S, Sarker NI, Hossain KM (1985) Potassium responses in greenhouse and field studies in Bangladesh. In: Proceedings of the international symposium on potassium in agricultural soils, soil science society of Bangladesh, Bangladesh Agricultural Research Council, 70–89 pp
Islam MS, Amin MS, Anwar MN (1992) Integrated soil fertility management in Bangladesh. In: Hussain MS, Huq SMI, Iqbal MA, Khan TH (eds) Proceedings of the inter-congress conference of Commission IV of ISSS on improving soil management for intensive cropping in the tropics and sub-tropics, 147–156 pp
Ives JD, Messerli B (1989) The Himalayan dilemma: reconciling development and conservation. Routledge, London and New York
Ives Jack D (2004) Himalayan perceptions: environmental change and the well-being of mountain peoples. Routledge, New York
Jahangir MMR, Jahan I, Mumu NJ (2018) Management of soil resources for sustainable development under a changing climate. J Environ Sci Nat Resour 11:159–170
Jalai VK, Sharma MP (2002) Status of available micronutrient cations in soils of mid-hill intermediate zone of Jammu region. Indian J Agric Sci 72:616–618
Jalal VK, Gupta JP, Gupta RD, Khajaria BB (1999) Chemical pools of zinc in intermediate zone of Jammu region and response of maize to zinc. Appl Biol Res 1:167–169
Jalali VK, Gupta JP, Razdan HK (2000) Distribution of micronutrient cations in soils of cold arid zone of Ladakh. Indian J Agric Sci 70:128–130
Jalali VK, Pareek N (2003) Need of micronutrient research in hill soils of western Himalayas with special reference to Jammu and Kashmir. Indian Life Sci 4(1&2):1–10
Jalali VK, Talib AR, Takkar PN (1989) Distribution of micronutrients in some benchmark soils of Kashmir at different altitude. J Indian Soc Soil Sci 37:465–469
Jamal A, Jamal H (2018) Assessment and distribution of macro and micro nutrients in different soil series of district Swabi, Khyber Pakhtunkhwa, Pakistan. J Hortic Plant Res 2:23–32
Jenny H (1941) Factors of soil formation. McGraw Hills, New York
Jenny H (1980) Soil resources: origin and behaviour. Springer-Verlag, New York
Jha MN, Pandey P, Pathak TC (1979) Studies on the changes in physio-chemical properties of Tripura soils as a result of *jhuming* (shifting cultivation). Indian Forester 105:436–443
Joshi D, Pandey SP, Maskey RB (1997) Status of land degradation in Nepal. In: Ghimire MP, Uprety BK (eds) Combating Desertification: Report of the Seminar on Desertification and Land Management. Ministry of Population and Environment, HMG/N in collaboration with Secretariat of the UNCCD, Kathmandu, Nepal
Kailash Kumar (1995) Nutrient status of soils of Manipur. Annual Report ICAR research Complex for Northeast Region of India, Umiam, Meghalaya, India
Kapoor AK (1994) Ecology and man in the Himalayas, MD Publications Pvt. Ltd., Delhi, India
Karan PP (1967) Bhutan—a physical and cultural geography. University of Kentucky Press, Lexington, KY, USA
Karan Pradyumna P (2009) The-new Tibet. FOCUS on Geogr 52 (2):7–13
Karim AJ, Egashira MSK, Yamada Y, Haider J, Nahar K (1995) Long-term application of organic residues to improve soil properties and to increase crop yield in terrace soil of Bangladesh. J Fat Agr Kyushu Uni 39:149–165
Karki KB, Tuladhar JK, Uprety R, Maskey SL (2005) Distribution of micronutrients available to plants in different ecological regions of Nepal. In: Andersen P, Tuladhar JK, Karki KB, Maskey SL (eds) Proceedings of international workshop on micronutrients in South and Southeast Asia, Kathmandu, Nepal, 8–11 Sep 2004, 17–29 pp
Karlen DL (2004) Soil quality as an indicator of sustainable tillage practices. Soil Tillage Res 78 (2):129–130
Karlen DL, Ditzler CA, Andrews SS (2003) Soil quality: why and how? Geoderma 114(3–4):145–156
Karvonen T, Koivusalo H, Jauhiainen M (1999) A hydrological model for predicting runoff from different land use areas. J Hydrol 217:253–265
Kassam AH, Van Velthuzan HT, Fischer GW and Shan MM (1991) Agroecological and resource assessment for agriculture development planning. A case study of Kenya world soil resources. Report 71, FAO, Rome
Katherine AB, Logan MJS, Floate, (1985) Acidity in upland and hills soils: cation exchange capacity, pH and lime requirement. J Sci Food Agric 36:1084–1092
Keeney DR, Corey RB (1963) Factors affecting the lime requirements of Wiscosin soils. Proc Soil Sci Soc Am 27:277280. https://doi.org/10.2136/sssaj1963.03615995002700030019x
Kelley JR, Wakeley LD, McKenna JR, Jackson SS (2010) US Army Corps of Engineers Engineer Research and Development Center Geotechnical and Structures

Laboratory 3909 Halls Ferry Road Vicksburg, MS 39180
Kemp RA (1999) Micromorphology of loess-palaeosol sequences: a record of paleoenvironmental change. CATENA 35:181–198
Kemp RA (2001) Pedogenic modification of loess: significance for palaeoclimatic reconstructions. Earth Sci Rev 54:145–156
Khadka D, Lamichhane S, Bhantana P, Ansari AR, Joshi S, Baruwal P (2018) Soil fertility assessment and mapping of chungbang farm, Pakhribas, Dhankuta, Nepal. Adv Plants Agric Res 8:210–227. https://doi.org/10.15406/apar.2018.08
Khan NA (2001) Final report of the Bangladesh case study-2011. Regional study on forest policy and institutional reform. Asian Development Bank, Manila, The Philippines
Khan H, Farooque AA, Acharya B, Abbas F, Esau TJ, Zaman QU (2020) Delineation of management zones for site-specific information about soil fertility characteristics through proximal sensing of potato Fields. Agronomy 10:1854
Khan FK, Khisha AL (1970) Shifting cultivation in East Pakistan. The Oriental Geographer, 14: 24–43
Khanal A, Fox GA, Al-Madhhachi AT (2016) Variability of erodibility parameters fromlaboratory mini Jet erosion tests. J Hydrol Eng 21.
Khattak RA, Hussain Z (2007) Evaluation of soil fertility status and nutrition of orchards. Soil and Environ 26 (1):22–32
Khera NA, Kumar JR, Tewari A (2001) Plant biodiversity assessment in relation to disturbances in mid elevational forest of Central Himalaya. India. Tropical Ecology 42(1):83–95
Khera MS, Pradhan HR (1980) Micronutrient status of soils of Sikkim and some inter-nutritional relationship. J Indian Soc Soil Sci 47:374–377
Khisa SK, Mohiuddin M, Cherrier J (2015) Shifting cultivation: soil fertility and food security issues in Chittagong Hill Tracts, Bangladesh. In: Romeo R, Vita A, Manuelli S, Zanini E, Freppaz M, Stanchi S (eds) Understanding mountain soils: a contribution from mountain areas to the international year of soils 2015, FAO, Rome, Italy
Kirmani NA, Wani SP, Mushtaq A, Sofi JA (2013) Characterization and classification of Alfisols under lesser Himalayan temperate region. Agropedology 23:118–121
Kisinyo PO, Othieno CO, Gudu SO et al (2013) Phosphorus sorption and lime requirements of maize growing acids soil of Kenya. Sustain Agric Res 2:116–123
Kosmas C, Danalatos N (1997) The effect of land use on runoff and soil erosion rates under Mediterranean conditions. CATENA 29:45–59
Kukla G (1987) Loess stratigraphy in Central China. Quatern Sci Rev 6:191–219
Kumar Kailash. (1995) Characterization of soil acidity and lime requirement of mid hill soils of Manipur with special reference to *Glycine max* and *Vigna radiata* crops. PhD thesis
Kumar A, Pramanik M, Chaudhary S, Negi MS (2021) Land evaluation for sustainable development of Himalayan agriculture using RS-GIS in conjunction with analytic hierarchy process and frequency ratio. J Saudi Soc Agric Sci 20(1):1–17. https://doi.org/10.1016/j.jssas.2020.10.001
Kumar S, Omanwar PK, Sachan RS, Sharma RB (1981) Iron status of some soils of Kumaon and Garhwal, Himalaya. J Indian Soc Soil Sci 29:266–267
Lahiri T, Chakravarti SK (1989) Characteristics of soils of Sikkim at various altitudes. J Indian Soc Soil Sci 37:451–454
Lal R (2000) Soil science soil management in the developing countries. Soil Sci 165(1):57–72
Lal R, Stewart BA (ed) (1990) Soil degradation, vol. 11, Advances in Soils Science. Springer–Verlag, New York
Laskar S, Dadhwal KS, Prasad RN (1983) Bulletin ICAR research complex for NEH region, Meghalaya, India, p 9
Laxminarayana K, Patiram (2004) Studies on carbon-nitrogen relationship to predict N availability in rice soils of Mizoram. J Hill Res 17:43–51
Lebauer DS, Treseder KK (2008) Nitrogen limitation of net primary productivity in terrestrial ecosystems is globally distributed. Ecology 89:371–379
Letey J (1985) Relationship between soil physical properties and crop production. In: Steward BA (ed) Advances in soil science, vol 1. Springer, New York, pp 277–279
Li SD (1996) Cheng GD (1996) Map of frozen ground on Qinghai-Xizang Plateau. Press, Lanzhou, China, Gansu Cult
Li GY, Fan HM (2014) Effect of freeze-thaw on water stability of aggregates in a black soil of northeast China. Pedosphere 24(2):285–290
Li P, Feng Z, Jiang L, Liao C, Zhang J (2014) A review of swidden agriculture in South-East Asia. Remote sensing, 6: 1654–1683
Li H, Gao J, Hu Q, Li Y, Tian J, Liao C, Ma W, Xu Y (2019) Assessing revegetation effectiveness on an extremely degraded grassland, southern Qinghai-Tibetan Plateau, using terrestrial LiDAR and field data. Agr Ecosyst Environ 282:13–22
Li Y, Li Y, Westlund H, Liu Y (2015) Urban–rural transformation in relation to cultivated land conversion in China: Implications for optimizing land use and balanced regional development. Land Use Policy 47:218–224. https://doi.org/10.1016/j.landusepol.2015.04.011
Li Y, Wang S, Jiang L, Zhang L, Cui S, Meng F, Wang Q, Li X, Zhou Y (2016) Changes of soil microbial community under different degraded gradients of alpine meadow. Agr Ecosyst Environ 222:213–222
Li L and Wu J (2021) Lead and chromium immobilization process subjected to different Freeze-Thaw treatments in soils of the Northeastern Qinghai-Tibet Plateau.

J Chem.Article ID 5286278, 11 p. https://doi.org/10.1155/2021/5286278
Lima ACR, Brussaard L, Totola MR, Hoogmoed WB, de Goede RGM (2013) A functional evaluation of three indicator sets for assessing soil quality. Appl Soil Ecol 64:194–200
Lin L, Li YK, Xu XL, Zhang FW, Du YG, Liu SL, Guo XW, Cao GM (2015) Predicting parameters of degradation succession processes of Tibetan Kobresia grasslands. Solid Earth 6(4):1237–1246
Liu J, Wu F, Yang W et al (2013) Effect of seasonal freeze-thaw cycle on net nitrogen mineralization of soil organic layer in the subalpine/alpine forests of western Sichuan, China. Acta Ecol Sin 33(1):32–37
Liu Y, Wang Q, Liu S et al (2019) Experimental investigation of the geotechnical properties and microstructure of lime-stabilized saline soils under freeze-thaw cycling. Cold Reg Sci Technol 161:32–42
Liu Y, Wang C, He N, Wen X et al (2017) Global synthesis of the rate and temperature sensitivity of soil nitrogen mineralization: latitudinal patterns and mechanisms. Glob Chang Biol 23:455–464
LRMP (1986) Land resource mapping project (Main Report). Land Resource Mapping Project, HMG/Nepal
LRMP (1987) Land systems report and economic report of land resources mapping project. Kenting Earth Sciences Ltd., Ottawa, Canada
Ma R, Jiang Y, Liu B, Fan H (2021) Effects of pore structure characterized by synchrotron-based micro-computed tomography on aggregate stability of black soil under freezethaw cycles. Soil Tillage Res 207. Article ID 104855
Mahapatra SK, Sharma JP, Martin D, Sharma RD (2010) Soil resource assessment of Kumaon Himalayan Mountains of India. In: 19th World Congress of Soil Science, Soil Solutions for a Changing World, 1–6 August 2010, Brisbane, Australia
Mahapatra SK, Rana KPC, Singh SP, Vellayutham M, Sehgal J (2000) Soils of Delhi for optimizing land use. NBSS Publ. 72. NBSS&LUP, Nagpur, India, 70 p
Maji AK, Obi Reddy GP, Sarkar D (2010) Degraded and wastelands of India—status and spatial distribution. Indian Council of Agricultural Research and National Academy of Agricultural Sciences, New Delhi, India
Maji AK, Dubey PN, Verma TP, Chamuah GS, Saigal JL, Velayutham M (2000) Soils of Nagaland for optamising land use. NBSS and LUP publication 67, p 28, , Nagpur, India.
Maji AK, Dubey PN, Sen TK, Verma TP, Marathe RA, Chamuah GS, Sehgal J, Velayutham M, Gajbhiye KS (2001) Soils of Mizoram for optimizing land use. NBSS Publ. 75b. NBSS&LUP, Nagpur, India, 28 p
Malik ZA, Haq MA, Bussmann RW, Bhat JA (2021) Altitudinal variation in soil properties with reference to forest structure and composition in Western Himalaya. Indian Forester 147(3):288–301. https://doi.org/10.36808/if/2021/v147i3/151484
Maria BJ, Garcia S, Adrian E (2004) Landscape- and field-scale control of spatial variation of soil properties in Mediterranean montane meadows. Biogeochemistry 69:207–225
McCarty GW, Reeves JB, Reeves VB, Follett RF, Kimble JM (2002) Mid-infrared and near-infrared diffuse reflec-tance spectroscopy for soil carbon measurement. Soil Sci Soc Am J 66:640–646
Maskey RB, Joshi D (1991) Soil and nutrient losses under different soil management practices in the middle mountains of central Nepal. In: Shah PB, Schreier H, Brown S and Riley KW (eds) Soil Fertility and Erosion Issues in the Middle Mountains of Nepal. Proceedings of the workshop on international center for integrated Mountain Dev., Kathmandu, Nepal, 22–25 April 1991, 105–120p
Maskey RB, Binod P, Sharma BP, Madhav Joshi M (2003) Human dimensions in sustainable land use management in degraded land areas of Nepal. Paper presented at the Open Meeting of the Global Environmental Change Research Community, Montreal, Canada
Masto RE, Chhonkar PK, Singh D, Patra AK (2008) Alternative soil quality indices for evaluating the effect of intensive cropping, fertilisation and manuring for 31 years in the semi-arid soils of India. Environ Monit Assess 136(419–435):2008. https://doi.org/10.1007/s10661-007-9697-z
Masunagai T, Kamidohzono A, Nezam AW, Azmar S (2014) Paddy soil properties in Nangarhar Province, East Afghanista. JARQ 48(3):299–306. http://www.jircas.affrc.go.jp
McLean EO (1976) Chemistry of soil aluminum. Commun Soil Sci Plant Anal 7(7):619–636. https://doi.org/10.1080/00103627609366672
McLean EO, Hourigan WR, Shoemaker HE, Bhumbla DR (1964) Aluminum in soils: V. Form of aluminum as a cause of soil acidity and a complication in its measurement. Soil Sci 97(2):119–26
Merz J (2000) Water and erosion studies in an integrated watershed management project in the Hindu Kush-Himalaya. International Centre for Integrated Mountain Development, Nepal
Abebe M (1998) Nature and management of Ethiopian soils. Alemaya University of Agriculture, Ethiopia, p 272
Meyer MH, Ha¨usler D, Skuk LS, Wangda D (2001) Glacial geological history and glacial lake outburst floods in Lunana, NW Bhutan, J. Asian Earth Sci 19 (3A), Special Abstract Issue, 45
Mian A, Mirza MYJ (1993) Pakistan's soil resources, Pakistan's National Conservation Strategy Paper No. 4, Jointly Published by IUCN Pakistan and Environment and Urban Affairs Division, Government of Pakistan
Microsoft (2001) Afghanistan. Microsoft Encarta Interactive World Atlas. http://www.microsoft.com/encarta
Miettinen J, Shi C, Liew SC (2011) Deforestation rates in insular Southeast Asia between 2000 and 2010. Glob Change Biol 17(7):2261–2270
Mina BL, Bhattacharrya R, Kumar N, Kundu S (2006) Hill soils and their management for sustainability. In:

Sustainable Production Agricultural Watersheds in Northwest India, (ed.) Gupta HS, Srivastva AK and Bhatt JC, VPKAS, Almora, India
Ministry of Environment Government of Pakistan (2009) Land use Atlas of Pakistan, Government of Pakistan
Mishra A, Das D, Saren S (2013) Preparation of GPS and GIS based soil fertility maps for Khurda district, Odisha. Indian Agric 57(1):11–20
Mishra AK, Sharma UC (2001) Traditional wisdom in range management for resource and environment conservation in northeastern region of India. ENVIS Bull Himalayan Ecol 9(1):263–269
Mishra AK, Deep S, Choudhary A (2015) Identification of suitable sites for organic farming using AHP and GIS Egypt. J Remote Sens Sp Sci 18:181–193. https://doi.org/10.1016/j.ejrs.2015.06.005
Mishra RV, Roy RN, Hiraoka H (2003) On-farm composting methods. Land and Water Discussion paper 2, FAO, Rome, Italy.
Mishra VK, Saha R (2006) Soil physical behaviour under different land use systems. In:Soils and their management in Northeast India (eds) Sharma UC, Datta M, Samra JS, Indian council of agricultural research complex for Northeastern Hills Region, Meghalaya, India, 324–335 pp
Mishra VK, Saha R, Bujarbaruah KM (2005) Zero tillage technique for transplanted rice in high rainfall ecosystem, Scientific leaflet, ICAR Research Complex for NEH Region, Umiam, Meghlaya, India
Mishra VK (2004) In: Annual Report (2003–04), AICRP Project on water Management
Misra UK, Saithantuaanga H (2000) J Indian Soc Soil Sci 37:451–454
MOCHTA (2011) Ministry of Chittagong Hill Tracts Affairs, Government of Bangladesh. Int J Plant Prod http://www.mochta.gov.bd/modification and crop diversification for enhancing productivity of a Vertisol
MoE, GoP (Ministry of Environment, Government of Pakistan) (2009) Land Use Atlas of Pakistan. Islamabad, Pakistan
Molden D, Sharma E, Shrestha AB, Chettri N, Shrestha Pradhan N, Kotru R (2017) Advancing regional and transboundary cooperation in the conflict-prone Hindu Kush-Himalaya. Mt Res Dev 37(2017):502–508
Morgan RPC, Morgan DDV, Finney HJ (1984) A predictive model for the assessment of soil erosion risk. J Agric Engng. Res 30:245–253
Mortensen E (1967) Review of rice research conducted at Shisam Bagh Experimental Station Jalalabad, MAI USAID, Kabul, Afghanistan
Moslehuddin AZM, Laizoo S, Egashira K (1997) Fertility status of Bangladesh soils—a review. J Fat Agr Kyushu Univ 41:257–267
Moslehuddin AZM (1993) A detailed summary of the fertility status of Bangladesh agricultural university Farm Soils. MSc thesis, Soil Science Department of Bangladesh Agricultural University, Mymensingh, Bangladesh
Motiramani DP (1971) In: Soil and water research in India, retrospect and Prospect, New Delhi, India, 98 p
Mukherjee JN, Chatterjee B (1942) Indian J Agric Sci 12:105
Mukherjee JN, Chatterjee B (1945) Nature. London
Mukhopadhayay DM, Eunus, Huq MM (1986) Response of major crops to balanced fertilizer application. Department of Agricultural Extension and Food and Agriculture Organization, Dhaka, Bangladesh
Mukhopadhyay S, AnilKumar KS, Baruah U, Chamuah GS, Sehgal J (1994) In: Annual report, NBSS & LUP, Nagpur, India
Munsi M, Malaviya S, Oinam G, Joshi PK (2009) A landscape approach for quantifying land-use and land-cover change (1976–2006) in middle Himalaya. Reg Environ Chang 10:145–155
Murthy RS, Hirekerur LR, Deshpande SB, Venkta Rao B (1982) Benchmark soils of India, NBSS&LUP, Nagpur, India, 379 p
Musson A (2013) Combining sustainable development and economic attractiveness: towards an indicator of sustainable attractiveness. Int J Sustain Dev 16 (1/2):127–162
Myers N (1986) Environmental repercussions of deforestation in the Himalayas. J World Forest Resour Manag 2:63–72
Nadelhoffer KJ, Emmett BA, Gundersen P, Kjønaas OJ, Koopmans CJ, Schleppi P, Tietema A, Wright RF (1999) Nitrogen deposition makes a minor contribution to carbon sequestration in temperate forests. Nature 398:145–148
Nair KM, Chamuah GS (1993) Exchangeable aluminium in soils of Meghalaya and management of Al^{3+} related productivity constraints. J Indian Soc Soil Sci 41:331–334
Nand Ram (1998) Effect of continuous fertilizers use on soil fertility and productivity of a mollisol. In: Swarup, A, Reddy, D, Prasad RN (eds) Proceeding of national workshop on long-term soil fertility management through integrated plant nutrient supply. Indian Institute of Soil Science, Bhopal, India, 335 p
Navara G (1997) The lay of the land. In: Schicklgruber C, Pommaret F (eds) Bhutan: mountain fortress of the gods. Bookwise, Delhi, India, pp 31–42
NBSSLUP (2004) Soil resource management reports. National Bureau of Soil Survey and Land Use Planning, Nagpur, India
Negi GCS, Joshi G, Sundryal RC, Dhyani PP (2015) Oak and pine forests soil in the Western Himalayan region of India. In: Romeo R, Vita A, Manuelli S, Zanini E, Freppaz M, Stanchi S (eds) FAO 2015, Understanding mountain soils: a contribution from mountain areas to the International Year of Soils, Rome, Italy
Nelson EM (1962) Wheat fertilizer trial results, Kabul Area. USAID Agriculture, Kabul, Afghanistan, 2 p
Neog AK (1997) Transforming Hill Agriculture in Tribal Areas of Northeast India - Trends in Agrarian structure in the hills of Northeast India (ed book). Common Wealth Publishers, New Delhi, pp 99–108
Ni J (2002) Carbon Storage in Grasslands of China J Arid Environ 50:205–218

Nie X, Xiong F, Yang L, Li C, Zhou G (2017) Soil nitrogen storage, distribution, and associated controlling factors in the Northeast Tibetan Plateau Shrublands. Forests 8: 416. https://doi.org/10.3390/f8110416

Ning Wu, Rawat GS, Joshi S, Ismail M, Sharma E (2013) High-altitude rangelands and their interfaces in the Hindu Kush Himalayas, ICIMOD. Kathmandu, Nepal

Nocita M, Stevens A, van Wesemael B, Aitkenhead M, Bachmann M et al (2015) Soil spectroscopy: an alternative to wet chemistry for soil monitoring. In: Advances in Agronomy; Academic Press, Cambridge. MA, USA 132:139–159

Nongkynrih P, Dkhar PS, Khathing DT (1996) Micronutrient elements in acid alfisols of Meghalaya under rice cultivation. J Indian Soc Soil Sci 44:455–457

Norbu C, Baillie IC, Tshering D, Tsheten D, Tamang HB, Tshering K, Hutcheon AA (2003) A provisional physiographic zonation of Bhutan. J Bhutan Stud 8:54–87

Oburo PA (2008) Effects of soil properties on bioavailability of aluminium and phosphorus in selected Kenyan and Brazilian Soils. PhD Thesis, Perdue University, West Lafayette, USA.

Okazaki M (1987) Soils of the Bhutan Himalaya. In: Ohsawa M (ed) Life zone ecology of the Bhutan Himalaya. Laboratory of Ecology, Chiba University, Tokyo, Japan, pp 145–184

Osman KT (2013) Forest soils: properties and management, Springer International Publishing, Switzerland, pp 19–28. https://doi.org/10.1007/978-3-319-02541-4

Oyetunji PO, Ibitoye OS, Akinyemi GO, Fadele OA, Oyediji OT (2020) The effects of population growth on deforestation in Nigeria: 1991–2016. J Appl Sci Environ Manag 24(8):1329–1334

Pan T, Zou XT, Liu YJ, Wu SH, He GM (2017) Contributions of climatic and non-climatic drivers to grassland variations on the Tibetan Plateau. Ecol Eng 108:307–317

Panda N (1987) Acidic soils of eastern India- their chemistry and management. J Indian Soc Soil Sci 35:568–581

Parks GA, de Bruyn PL (1962) The zero point of charge of oxides. Journal of Physical Chemistry, 66: 967–972

Parmar DK, Sharma V, Sharma KD, Sharma TR (1999) Micronutrient status of vegetable growing pockets in cold desert area of Himachal Pradesh. J Indian Soc Soil Sci 47:280–283

Parmar DK (2014) Yield produce quality and soil health under vegetable cropping systems as influenced by integrated nutrient management im mid-hill zone of Himachal Pradesh. Journal of the Indian Society of Soil Science, 62: 45–51

Patgiri DK (1995) Effect of boron, sulphur and molybdenum on Toria (Brassica campestris L) in a Typic Haplaquept. J Indian Soc Soil Sci 43:295–296

Patiram (1984) Annual Report, ICAR Research Complex for Northeastern Region, Umiam, Meghalaya, India

Patiram (1990) Forms of aluminium in some acid soils of Sikkim. J Indian Soc Soil Sci 38:652–655

Patiram (1991a) In: Annual Report, ICAR Research Complex for NEH Region, Meghalaya, India, 137 p

Patiram (1991b) J Indian Soc Soil Sci 39:178

Patiram, Datta M (2006) Geology, mineralogy and genesis of soils. In: Sharma UC, Datta M, Samra JS (eds) Soils and their management in Northeast India. ICAR Research Complex for NEH Region, Umiam, Meghalaya, India, pp 280–323

Patiram RN, Prasad RN (1990a) Phosphate absorption by acid soils from different altitude. J Indian Soc Soil Sci 38:602–608

Patiram RRN, Prasad RN (1990b) Forms of soil phosphorus and suitable extractants for available phosphorus in acid soils of Sikkim. Journal of Indian Society of Soil Science 38(237):242

Patiram RRN, Singh KP (1991) Application of domolitic limestone enhances crop yields in acid soils of Sikkim. Indian Farming 41:18–20

PCB, Planning Commission of Bangladesh (2005) Unlocking the Potential. National Strategy for Accelerated Poverty Reduction (Draft), Government of Bangladesh

Penuelas J, Poulter B, Sardans J et al (2013) Human-induced nitrogen–phosphorus imbalances in natural and managed ecosystems across the globe. Nat Commun 4:2934

Perveen S, Malik Z, Nazif W (2010) Fertility status of vegetable growing areas of Peshawar, Pakistan. Pak J Bot 42:1871–1880

Phogat VK, Tomar VS, Dahyia R (2015) Soil physical properties. In: Rattan RK, Katyal JC, Dwivedi BS, Sarkar AK, Bhattachatyya T, Tarafdar JC, Kukal SS (eds) Soil science: an introduction (Indian Society of Soil Science, New Delhi, pp 135–171. https://www.researchgate.net/publication/297737054

Pinstrup–Anderson P, Pandya–Lorch R (1994) Food, agriculture and environment, Discussion Paper No. 1, Int. Food Policy Res. Inst Washington, DC

Popp A, Humpenöder F, Weindl I, Bodirsky BL, Bonsch M, Lotze-Campen H, Müller C, Biewald A, Rolinski S, Stevanovic M, Dietrich JP (2014) Land-use protection for climate change mitigation. Nat. Clim. Change 4:1095–1098. https://doi.org/10.1038/nclimate2444

Portch S, Islam MS (1984) Nutrient status of some of the more important agricultural soils of Bangladesh. In: Proceedings of international symposium on Soil Test Crop Response Studies, Bangladesh Agricultural Research Council and Soil Science Society of Bangladesh, 97–106 pp

Potash and Phosphate Institute (1983) Soil Fertility Manual. Potash and Phosphate Institute, Atlanta, USA

Prasad RN, Ram M (1985) Soils of Northeastern Hill Region and their management. In: Soils of India and their management. Fertilizer Association of India, New Delhi, pp 266–283

Prasad RN, Sharma UC (1994) Potential indigenous farming systems of North Eastern Region, ICAR Research Complex foe NEH Region, Barapani, Meghalaya, India, 56p

Prasad RN, Singh A, Varma Arun (1986) Problems of hill lands and their management in North Eastern India
Prasad RN (1985) Degradation of soil and water resources and crop production in NEH region. India 1(1):1–8
Prasad RN, Patiram, Ram M (1985) Forms of aluminium in soils of east Khasi hills, Meghalaya. J Indian Soc Soil Sci 33:523–527
Prasad RN, Patiram, Barooah RC, Ram M (1981) Soil Fertility management in North Eastern Hills Region, Research Bulletin No. 9, ICAR Research Complex for NEH Region, Shillong, India
Pratt PF, Blair FL (1963) Buffer method for estimating lime and sulphur applications 444 for pH control of soils. Soil Sci 93:329–331
Pretty J, Camilla Toulmin C, Williams S (2011) Sustainable intensification in African agriculture. Int J Agric Sustain 9(1):5–24. https://doi.org/10.3763/ijas.2010.0583
Pretty J, Hine R (2011) Reducing food poverty with sustainable agriculture: a summary of new evidence. University of Essex, Essex, UK
Qamer FM, Shehzad K, Abbas S, Murthy MSR, Xi C,, Gilani H, Bajarcharya B (2016) Mapping deforestation and forest degradation patterns in Western Himalaya Pakistan: The case of Swat and Shangla districts. Journal Mountain Science, 9: 1–10
Qian C, Peng YL, Jia JY, Wei N (2006) Biological fertility and its dynamics of degraded soil in Tibet Plateau. Chin J Appl Ecol 17:1185–1190 (in Chinese with English abstract)
Quamruzzaman M (2005) Integrated nutrient management for sustaining crop productivity and improvement of soil fertility in Bangladesh agriculture. In: Improving plant nutrient management for better farmer livelihoods, food security and environmental sustainability. In: Proceedings of a regional workshop, Beijing, China, Regional Office for Asia and the Pacific, Bangkok, 12–16 December 2005
Ram M, Prasad RN, Patiram (1987) Nutrient management in soils of Northeastern Region. Journal Indian Society Soil Science, 35: 207–212
Rambabu, Tejwani KG, Agrawal MC, Bhusan LS (1978) Rainfall Intensity- Duration—return period and nomographs of India Bulletin No. 3, CSWCRTI, Dehradun, India
Ran Y, Li X, Cheng G et al (2012) Distribution of permafrost in China: an overview of existing permafrost maps. Permafrost Periglac Process 23(4):322–333
Rana KPC, Walia CS, Sidhu GS, Singh SP, Velayutham M, Sehgal J (2000) Soils of Jammu and Kashmir for optimizing land use. NBSS Publ. 62. NBSS&LUP, Nagpur, India 71 p
Rao KVP, Kumar K (1991) In: Annual report, ICAR Research Complex for NEH Region, Meghalaya, India
Rasheed KBS (1995) Participatory forestry as a strategy for reforestation in Bangladesh. GeoJournal 37(1):39–44
Rashid M, Bhatti AU (2005) Mapping of spatial variability of macro and micronutrients for site specific management. Soil and Environment 24:34–52
Rasul G, Thapa GB, Zoebisch A (2004) Determinants of land-use changes in the Chittagong Hill Tracts of Bangladesh. Appl Geogr 24(3):217–240
Rasul G, Sharma B (2016) The nexus approach to water–energy–food security: an option for adaptation to climate change. Clim Policy 16(6):682–702. https://doi.org/10.1080/14693062.2015.1029865
Rasul G, Hussain A (2015) Sustainable food security in the mountains of Pakistan: towards a policy framework. Ecol Food Nutr 54(6):625–643
Rasul G, Saboor A, Tiwari P, Hussain A, Ghosh N, Chettri G et al (2018) Food and nutrition security in the HKH. In: Chapter 9, Hindu Kush Himalayan monitoring and assessment programme (HIMAP). International Centre for Integrated Mountain Development (ICIMOD), Kathmandu, Nepal
Rawat PS, Mathpal KN (1981) Micronutrient status of some soils of UP hills. J Indian Soc Soil Sci 29:208–214
Regmi BD, Poudel C, Tripathi BP, Schulz S, Dhital BK (2005) Managing soil fertility problems of marginal agricultural lands through integrated plant nutrient systems: experiences from the hills of Nepal. In: Andersen P, Tuladhar JK, Karki KB, Maskey SL (eds) Micronutrients in South and Southeast Asia, Proceedings of International workshop held at Kathmandu, Nepal, 8–11 Sep 2004, 109–119 pp
Reilly J, Melillo J, Cai Y, Kicklighter D, Gurgel A, Paltsev S, Cronin T, Sokolov A, Schlosser A (2012) Using land to mitigate climate change: hitting the target, recognizing the trade-offs. Environ Sci Technol 46(5672–5679):2012. https://doi.org/10.1021/es2034729
Rijpma J, Fokhrul Islam M (2003) Nutrient mining and its effect on crop production and environment in Bangladesh. Paper presented at seminar on "Soil Health Management"; DAE—SFFP Experience, Bangladesh
Roberton JM, Lobsey BJM, Bishop TF (2020) Assessing the sensitivity of site-specific lime and gypsum recommendations to soil sampling techniques and spatial density of data collection in Australian agriculture: a pe-dometric approach. Agronomy 10:1676
Rodriguez MB, Godeas A, Lavado RS (2008) Soil acidity changes in bulk soil and maize rhizosphere in response to nitrogen fertilization. Commun Soil Sci Plant Anal 39:2597–2607
Romeo R, Vita A, Manuelli S, Zanini E, Freppaz M, Stanchi S (2015) Understanding mountain soils: a contribution from mountain areas to the International Year of Soils, FAO, Rome, Italy
Rorison JW (1973) The effect of soil acidity on the nutrient uptake and physiology of plants. In Dost 1:233

Rout GR, Sahoo S (2015) Role of iron in plant growth and metabolism. Rev Agric Sci 3:1–24

Roy RN (1995) FAO-Soil fertility and integrated plant nutrition programmes. In: Nuclear techniques in soil-plant studies for sustainable agriculture environmental preservation, International Atomic Energy Agency, Vienna, Austria, 107–129 pp

Rozanov BG, Targulian V, Orlov DS (1990) Soils in the earth as transformed by human action. In: Turner BL, Clark WC, Kates RW, Richards JF, Mathews JT, Meyer WB (eds) Cambridge University Press, Cambridge, UK

Rusk J, Maharajan A, Tiwari P, Chena THK, Shneiderman S, Turin M, Seto KC (2021) Multi-hazard susceptibility and exposure assessment of the Hindu Kush Himalaya. Sci Total Environ 804:150039

Saba D (2001) Geography: land and resources. http://www.afghan-web.com/geography/environment.html. Accessed January 2001

Sachchidananda (1989) Shifting cultivation in India. Concept Publishing Company, New Delhi, p 256

Sachdev CB, Tarsem Lal, Rana KPC, Sehgal J (1995) Soils of Haryana for Optimizing Land Use NBSS Publ. 44 (Soils of India Series). National Bureau of Soil Survey & Land Use Planning, Nagpur, India, 59 p + 2 sheets soil map (1:50,000 scale)

Saha R, Mishra VK (2007) Long term effect of various land use systems on physical properties of silty clay loam soil of NE hills. J Indian Soc Soil Sci 55(2):112–118

Saha R, Mishra VK, Majumdar B, Laxminarayana L, Ghosh PK (2010) Effect of integrated nutrient management on soil physical properties and crop productivity under a Maize (Zea mays)-mustard (Brassica campestris) cropping sequence in acidic soils of Northeast India. Commun Soil Sci Plant Anal 41 (18):2187–2200

Saha R, Mishra VK, Patiram (2003) Long-term effect of agroforestry systems on soil hydro-physical behaviour. Annual Report, ICAR Research Complex for Northeaster Region,. Umiam, Meghalaya, India

Saha S, Rajwar GS, Kumar M (2018) Soil properties along altitudinal gradient in Himalayan temperate forest of Garhwal region. Acta Ecol Sinica 38(1):1–8. https://doi.org/10.1016/j.chnaes.2017.02.003

Saheed SM (1984) Soils of Bangladesh. In: International Symposium on Soil Test Crop Response Studies, Bangladesh Agricultural Research Council and Soil Science Society of Bangladesh, Dhaka

Saijo K, Tanaka S (2002) Palaeosols of middle Holocene age in the Thakkhola Basin, Central Nepal, and their paleoclimatic significance. J Asian Earth Sci 21:323–329

Sanchez PA (1997) Changing tropical soil fertility paradigms: from Brazil to Africa and back. In: Moniz AC, Furlani AMC, Schaffert RE, Fageria NK, Rosolem CA, Cantarella H (eds) Plant-soil interactions at low pH: sustainable agriculture and forestry production, Brazilian Soil Science Society, Brazil

Sanchez PA, Shephard KD, Soule MJ, Place FM, Buresh RJ, Izac AN, Mokwunye AU, Kwesiga FR, Ndiritu CG, Woomer PL (1997a) Soil fertility replenishment in Africa: an investment in natural resource capital. In: Buresh RJ et al (eds) Replenishing soil fertility in Africa. SSSA, Madison, WI

Sanchez PA, Shepherd KD, Soule MJ, Place FM, Buresh RJ, Izac AMN, Mokwunye AU, Kwesiga F, Nderitu CG, Woomer PL (1997b) Soil fertility management in Africa: an investment in natural resource. In: Brush RJ et al (eds) Replenishing soil fertility in Africa, SSSA Special Publication, 51, Madison, WI, USA, 1–46 pp

Sapkota K, Andersen P (2005) Commercial horticulture farming and its effects on soil fertility: a case study from peri-urban agriculture in the Kathmandu Valley. In: Andersen P, Tuladhar JK, Karki KB, Maskey SL (eds) Micronutrients in South and Southeast Asia, Proceedings of international workshop held at Kathmandu, Nepal, 153–165 pp

Sarangzai AM, Siddiqui MF, Ahmed M, Hussain ML, Laghari SK, Ahmed A (2015) Relationship between soil properties and natural regeneration pattern of Juniperusexcelsa forest in Ziarat, Balochistan, Pakistan. J Botany 47(3):905–910

Sarkar AK (2002) In: 3rd Annual ICAR network project report for the year, 2001–2002, soil characteristics and resource management of acid soil regions for increasing productivity, Department of Soil Science and Agricultural Chemistry, Birsa Agricultural University, Ranchi

Satapathy KK (2005) Runoff production on hill slopes under different land use systems, in Agroforestry in North East India: Opportunities and Challenges. In: Bhatt BP, Bujarbaruah KM (eds) ICAR research complex for NEH region. Umiam, Meghalaya, India, pp 451–459

Satapathy KK (2006) Soil Conservation and Management Strategies. In: Sharma UC, Datta M, Samra JS (eds) Soils and their management in Northeast India Indian Council of Agricultural Research Complex for Northeastern Hills Region, 372–388 pp

Saxena RK, Verma KS, Chari GK, Srivastava R, Bhartawal AK (2000) RS IC data application in watershed characterization and management. Int J Remote Sens 21(7):3179–3208

Scherr SJ (1997a) People and environment: what is the relationship between exploitation of natural resources and population growth in the south? Forum for Dev Stud 1:33–58

Scherr SJ (1997b) Is soil degradation a threat to developing country's food security? Agriculture and Resource Economics Department, University of Maryland, College Park, Maryland

Scherr SJ (1999) Soil Degradation—a threat to developing country's food security. Int Food Policy Res Inst, Washington, DC

Scherr SJ, Hazell PA (1994) Sustainable agricultural development strategies in fragile lands, EPTD

discussion paper no.1, International Food Policy Research Institute, Washington, DC
Scherr SJ, Yadav S (1996) Land Degradation in the Developing World : Implication for Food, Agriculture and Environment to 2020. International Food Policy Research Institute, Washington, DC
Scherr SJ , Bergeson G, Pender J, Barbier B (1996) Policies for sustainable development in fragile lands. EPTD paper, International Food Policy Research Institute, Washington, DC
Schild A (2008) ICIMOD's future agenda: a way forward'. In: ICIMOD and the Himalayan region-responding to emerging challenges. Kathmandu, Nepal
Schoenholtz SH, Van Miegroet H, Burger JA (2000) A review of chemical and physical properties as indicators of forest soil quality: challenges and opportunities. For Ecol Manage 138(1–3):335–356. https://doi.org/10.1016/S0378-1127(00)00423-0
Schreier H (2015) Mountain soils and human activities. In: Romeo R, Vita A, Manuelli S, Zanini E, Freppaz M, Stanchi S (eds) Understanding mountain soils: a contribution from mountain areas to the international year of Soils, FAO 2015, Rome, Italy
Schreier H, Shrestha B, Brown S, Shah PB (2000) Forest dynamics in Nepal: quality, quantity, and community forestry issues in middle mountain watersheds. In: Allen R, Schreier H, Brown S, Shah PB (eds) The people and resource dynamics project: the first three years (1996–1999). International Center for Integrated Mountain Dev., Kathmandu, Nepal, 129–139 pp
Sehgal JL (1973) Pedogenic calcite accumulation in acid and semiarid region of the Indo-Gangetic alluvial plains of erstwhile Punjab (India), their morphology and origin. Geoderma 8:59
Sehgal J, Karan S, Ram J, Sidhu GS, Gajbhiye KS (1993) Cold desert soils of India—their taxonomic problems. Agropedology 3:13–28
Sen TK, Nayak DC, Singh RS, Dubey PN, Maji AK, Chamuah GS, Sehgal J (1997) Pedology and edaphology of Benchmark acid soils of north eastern India. J Indian Soc Soil Sci 45:782–790
Sen TK, Chamuah GS, Sehgal J, Velayutham (1999) Soils of Assam for optimizing land use. NBSS Publ. 66. NBSS&LUP, Nagpur, India, 51p.
Sethi RM, Nair GK, Rao S, Singh RR (1994) Status of land degradation in India. J Soil Water Conserv 38 (3&4):163–172
Shah PB, Schreier H (1995) Maintaining soil fertility in agriculture and forestry. In: Scheier H, Shah PB and Brown S (eds.) Challenges in mountain resource management in Nepal: Processes, trends and dynamics in middle mountain watersheds. In: Proceedings of workshop on International Center for Integrated Mountain Dev, Kathmandu, Nepal, 10–12 April 1995, 171–182 pp
Shameem SA, Kangroo NI, Bhat GA (2011) Comparative assessment of edaphic features and herbaceous diversity in lower Dachigam National Park, Kashmir, Himalaya. J Ecol Nat Environ 3(6):196–204
Sharma CM, Gairola S, Ghildiyal SK, Suyal S (2010) Physical properties of soils in relation to forest composition in moist temperate valley slopes of the Central Western Himalaya. J Forest and Environ Sci 26(2):117–129
Sharma E, Sundriyal RC, Rai SC, Bhatt YK, Rai LK, Sharma R, Rai YK (1992a) Integrated watershed management. Gyanondays Prakashan, Nainital, India
Sharma E, Sundriyal RC, Rai SC, Bhatt YK, Rai LK, Sharma R, Rai YK (1992b) Integrated watershed management: a case study in Sikkim. G.B. Pant Institute of Himalaya Environment and Development, Koshi, India
Sharma JC, Bhandari AR (1995) Mineral nutrient status of apple orchards in Himachal Pradesh. J Indian Soc Soil Sci 43:236–241
Sharma PD, Baruah TC, Maji AK, Patiram (2006a) Management of acid soils in NEH region. Indian Council of Agricultural Research, NRM Division, Krishi Anusandhan Bhavan, India
Sharma PK, Sharma SP, Jain PK (2001a) Nutrient mining in different agro- climatic zones of Himachal Pradesh. Fertilizer News 46(8):69–73
Sharma R, Sharma E, Purohit AN (1994) Dry matter production and nutrient cycling in agroforestry systems of cardamom grown under Alnus and natural forest. Agrof Syst 27:293–306
Sharma RK, Jalali VK, Gupta JP, Pareek N, Wali P (2003) Zinc status and its response to various crops in soils of different agro-climatic zones of Jammu and Kashmir. J Res Skuast (j) 2(1):75–80
Sharma RK, Jalali VK, Koul RK (2001b) Nutrient indexing of (Citrus reticulata Blanco) Life Sci Reporter 3: 34–36
Sharma UC (1987) Potato yield as affected by zinc, iron and their interactions with phosphorus. Journal Indian Potato Association, 14: 21–25
Sharma UC (1989a) Influence of soil texture on leaching of potassium and its recovery by potato. Indian J Agric Sci 59:713–717
Sharma UC (1989b) Potato yield and nutrient uptake as influenced by fertility levels at different locations in Meghalaya. J Indian Potato Assoc 16:91–95
Sharma UC (1990a) Relative efficacy of methods of Zn, Mn, Fe and Cu application to potato. Indian J Agric Sci 60:806–809
Sharma UC (1990b) Evaluation of two P sources on acidic hill soil of Meghalaya with potato as test crop. J Indian Soc Soil Sci 38:46–50
Sharma UC (1990c) Effect of sources and methods nitrogen application on the yield and N uptake of potato in Meghalaya. Indian J Agric Sci 60:119–122
Sharma UC (1990d) Relative efficacy of zinc, manganese, iron and copper application with potato as test crop. Indian J Agric Sci 60:806–809
Sharma UC (1991a) Release pattern of nitrate-N from nitrogenous fertilizers under field condition in Meghalaya and its effect on the yield of potato. Indian J Agric Sci 61:561–566

Sharma UC (1991b) Response of potato to phosphorus and farmyard manure on acidic soils of Khasi hills of Meghalaya. J Indian Soc Soil Sci 39:386–388
Sharma UC (1991c) A simple mathematical model to determine critical nutrient levels in soils and plants. J Indian Soc Soil Sci 39:509–513
Sharma UC (1991d) Diagnostic and Integrated Recommendation System for nitrogen, phosphorus and potassium foliar diagnostic norms for poatato. J Indian Soc Soil Sci 39:482–487
Sharma UC (1992a) Effect of soil texture and precipitation on phosphorus leaching in Alfisols of Meghalaya. J Indian Soc Soil Sci 40:413–416
Sharma UC (1992b) Effect of soil texture and precipitation on phosphorus leaching in Alfisols of Meghalaya. J Indian Soc Soil Sci 40:413–416
Sharma UC (1992c) Critical limit of P for potato in Alfisols of Meghalaya. J Indian Soc Soil Sci 40:211–212
Sharma UC (1992d) Potassium requirement of potato on acid soils. Indian J Agric Sci 62:149–151
Sharma UC (1993) Use of boundary lines in establishing diagnostic norms for potato. Indian J Agric Sci 63:19–22
Sharma UC (1996) Use of half-life technique to diagnose nitrate-N status in the petioles of potato during the growing season. Indian J Agric Sci 66:1–7
Sharma UC (1999) Loss of N through leaching and runoff from two-potato land use systems on different soils. In: Heathwaite L (ed) Impact of land use change on nutrient loads from diffuse sources , Birmingham, IAHS Publ . No. 257. IAHS Press, Wallingford, UK, 27–32pp
Sharma UC (2000) Loss of N through leaching and run off from two potato based land use systems. In: Healthwaite L (ed) Impact of land use change on nutient load from diffuse sources. IAHS Press, Wallingford, UK, pp 27–32
Sharma UC (2002) Land use options for the flood plains of NE Region of India. In: Dyer FJ, Thomas MC, Olley JM (eds) The structure, Function and Management Implications of fluvial Sedimentary system, IAHS Press, Wallingford, UK. IAHS Publ. No. 276, 239–244 pp
Sharma UC (2006) Water quality and sedimentation control. Invited lecture in a special meeting of Japan Aerospace Exploration Agency (JAXA), held in Bangkok, Thailand
Sharma UC, Arora BR (1987) The effect of nitrogen, phosphorus, potassium and FYM on the yield of potato. J Agric Sci (Camb) 108:321–329
Sharma UC, Arora BR (1988) Effect of nutrient application to potato on the starch, sugars and protein content of potato tubers. J Food Chem (England) 30:313–317
Sharma UC, Arora BR (1989) Critical nutrient ranges for potassium in potato leaves and petioles. J Hortic Sci (England) 64:47–51
Sharma UC, Sharma V (2003) Mathematical model for predicting soil erosion by flowing water in ungauged watersheds. In: Boer DH, Froehlich W, Mizuyama T, Pietroniro A (eds) Erosion prediction in ungauged basins: integrating methods and techniques. IAHS Press, Wallingford, UK , Publ. 279 , 79–83 pp (Paper presented at International Union of Geodesy and Geophysics, IUGG)
Sharma UC, Sharma V (2004a) Implications of soil and nutirents transfer with runoff in NE Region of India. In: Golosov V, Belyaev V, Walling DE (eds) Sediment transfer through the fluvial system, IAHS Publ. no.288, IAHS Press, Wallingford, UK, 488–493 pp
Sharma UC (2004b) Integrated catchment management in NE region of India to control fluvial sediment transport. In: Webb B (ed) Hydrology—Science and Practice for the 21st Century, London, UK, vol II, 474–479 pp
Sharma UC, Sharma V (2004c) The 'Zabo' water management and conservation system: tribal beliefs in the development of water resources. In: John C, Rodda, Lucio Ubertini (eds) Basis of civilization-water science? Rome, Italy, IAHS Publ. 286. IAHS Press, Wallingford, UK, 184–191 pp
Sharma UC, Sharma V (2004d) Implications of nutrient and soil transfer with runoff in the northeastern reghion of India. In: Golosov V, Belyaev V and Walling DE (eds) Sediment transfer through the fluvial system, IAHS. (Moscow, Russia), IAHS Press, Wallingford, UK, Publ. No. 288, 488–493 pp
Sharma UC, Sharma V (2004e) The 'Zabo' soil and water management and conservation system in northeastern India: Tribal beliefs in the development of water resources and their inpact on society– An historical account of a success story. In: Rodda JC, Lucio Ubertini (eds) The basis of civilization: water science, IAHS Pub. No. 286, pp 184–191, IAHS Press, Wallingford, UK
Sharma UC, Sharma V (2005a) Traditional wisdom of tribal people in Integrated water resources management and environmental quality. In: Takara K, Tachikawa Y, Nawarathna NMNSB (eds) Prediction and Mitigation of Water Related Disasters, Kyoto University, Kyoto, Japan, Monitoring, pp 704–712
Sharma UC, Sharma V (2005b) Socio-economic aspects of ecological aspects of headwater region of Chenab and Tawi rivers in Jammu Province of India. Paper presented in Bergen, Norway, in the conference on Problems in water resources management and development in Headwater region, Bergen, Norway, 21–25 June, 2005
Sharma UC, Sharma V (2009) Plant-livestock-soil-hydrology interactions in the north-eastern region of India. Biologia 64(3):460–464
Sharma UC, Datta M, Samra JS (2006b) Soils and their management in Northeast India Indian Council of Agricultural Research Complex for Northeastern Hills Region. Meghalaya, India, p 545
Sharma UC, Prasad RN (1994) Potential Indigenous Farming Systems of Northeastern Region of India. ICAR Reseach Complex for NE Region of India, Shillong, Meghalaya, India, 64p.

Sharma UC, Prasad RN (1995) Socio-economic aspects of acid soil management and alternate land use systems for northeastern states of India. In: Date RA, Grundon NJ, G Rayment GE and Probert ME (eds) Plant-soil Internations at Low pH, Principles and Management, Kluwer Academic Press, The Netherland,s 689–696 pp

Sharma UC, Datta M, Sharma V (2014) Soil fertility, erosion, runoff and crop productivity affected by different farming systems. Ecopersia 2:629–650

Sharma UC (2018) Assessment of nitrogen leaching in potato (Solonum Tuberosum l.) cultivated humid alfisols and its impact on crop productivity. Innov Techn Agric (Scientia recerca) 3(2): 597–611

Sharma UC, Sharma V (2016) Modeling sediment dynamics from catchment to coasts. J Water Resour Arid Environ 5:579–596

Sharma SN, Sharma SK (2004e) Role of crop diversification and integrated nutrient management in resilience of soil fertility under rice-wheat cropping system. Arch Agron Soil Sci 50:345–352

Shen HH, Wang SP, Tang YH (2013) Grazing alters warming effects on leaf photosynthesis and respiration in Gentiana straminea, an alpine forb species. J Plant Ecol 6:418–427

Sherchan DP, Karki KB (2005) Pland nutrient management for improving crop productivity in Nepal. In: Improving plant nutrient management for better farmer livelihood, food security and environmental sustainability, Proceedings Regional Workshop, Beijing, China, 12–16 December, pp 41–57

Sherchan DP, Bajracharya RM, Tiwari KR, Balla MK (2003) Soil survey report of Pokhare Khola sub-watershed, Dhading, Nepal. Himalayan Degradation Project, Institute of Forestry, Tribhuvan Univ./Agic. Univ. of Norway, Pokhara, Nepal, 52 p

Sherchan DP, Gurung GB, Resham KC (1997a) Effects of long term application of chemical fertilizer and manure on soil properties and crop yields under maize-millet cropping pattern: the results of 1996/97 season. PAC Working Paper No. 193, Pakhribas Agric. Center, Dhankuta, Nepal, 6 p

Sherchan DP, Gurung GB (1998) Effect of long term application of chemical fertilizer and manure on crop yields and soil chemical properties under Rice–Wheat cropping pattern (The Results of 8 Cropping Cycles). PAC Technical Paper No. 183. Pakhribas Agricultural Centre, Dhankuta, Nepal 20 p

Sherchan DP, Gurung GB, Resham KC (1997b) Effects of longterm application of chemical fertilizer and manure on soil properties and crop yields under maize-millet cropping pattern: the results of 1996/97 season. PAC Working Paper No. 193 Pakhribas, Dhankuta, Nepal 6 p

Shi Y, Baumann F, Ma Y, Song YC, Uhn PK, Scholten T, He JS (2012) Organic and inorganic carbon in the topsoil of the Mongolian and Tibetan grasslands: pattern, control and implications. Biogeosciences 9:2287–2299. https://doi.org/10.5194/bg-9-2287-2012

Shoemaker HE, McLean EO, Pratt PF (1961) Buffer methods for determining lime requirement of soils with appreciable amounts of extractable aluminum 1. Soil Sci Soc America J SSSAJ 25. https://doi.org/10.2136/sssaj1961.03615995002500040014x

Shrestha HL, Bhandari TS, Karky BS, Kotru R (2017) Linking soil properties to climate change: mitigation and food security in Nepal. Environments 4: 29–39. https://doi.org/10.3390/environments4020029

Shrestha UB, Gautam S, Bawa KS (2012) Widespread climate change in the Himalayas and associated changes in local ecosystems. PLoS ONE 7:e36741

Shrestha BM, Singh B, Sitaula BK, Lal R, Bajracharya RM (2007) Soil aggregate- and particle-associated organic carbon under different land uses in Nepal. Soil Sci Soc Am J 71(4):1194–1203

Shukla R, Sachdeva K, Joshi PK (2016) Inherent vulnerability of agricultural communities in Himalaya: a village-level hotspot analysis in the Uttarakhand state of India. Appl Geogr 74:182–198. https://doi.org/10.1016/j.apgeog.2016.07.013

Shuman FH (1960) Extension with Wheat. ICA Report. Kabul, Afghanistan

Sidhu GS, Jain SP (1993) Soil survey and evaluation of ORP villages Boothgarh, Jhandupur & Majhot, district Hoshiarpur, Punjab. Report No. RCD 91 NBSS&LUP, Nagpur, India

Sidhu GS, Sharma JP (2008) Soil resource data for perspective land use planning in Shiwaliks of Himachal Pradesh. In: YSPUH and F Nauni State Level Workshop on Catchments Treatment of River Valley and Flood Prone Areas in Himachal Pradesh," (Solan), NRM Div. of Min. of Agric. and H. P. State Department of Forests (ISBN 978-81-211-0676-4)

Sidhu GS, Rana, KPC, Tarsem Lal (1987) Soil survey and land evaluation of sub-watershed area "Meer-Noara" Udhampur district, J&K. National Bureau of soil survey and land use planning. Scientific Report No. 490 (ICAR), Nagpur, India

Sidhu GS, Rana KPC, Sehgal J, Velayutham M (1997) Soils of Himachal Pradesh for optimizing land use. NBSS publ. 57. (Soils of India Series). National Bureau of Soil Survey & Land Use Planning, Nagpur, India, 44 p + 2 sheets soil map (1:50,000 scale)

Sidhu GS, Rana KPC, Lal T, Mahapatra SK, Verma TP, Rao RVS, Ram J, Singh SK, Singh SP (2007a) Soils of Himachal Pradesh: inventory for land use planning. J Indian Soc Soil Sci 55:203–208

Sidhu GS, Rana KPC, Lal T, Mahapatra SK, Verma TP, Rao RVS (2007b) Soils of Himachal Pradesh—II: land capability classification and assessment of soil degradation status for suggested land use. J Indian Soc Soil Sci 55:335–339

Sidhu GS, Rana KPC, Walia CS, Mahapatra SK, Lal T (1998) Characteristics and classification of dominant soils of Jammu region for land use planning. Agropedology 9:22–29

Sidhu GS, Walia CS, Rana KPC, Sehgal J (1995a) Soils of Punjab for optimizing land use. NBSS publ. 45 (Soils of India Series). National Bureau of Soil

Survey & Land Use Planning, Nagpur, India, 75 p + 2 sheets soil map (1:50,000 scale)
Sidhu GS, Walia CS, Sachdev CB, Rana KPC, Dhankar RP, Singh SP, Velayutham M (2000) Soil resource of N-W Shiwaliks for perspective land use planning. In: Mittal SP (ed) Fifty years of research on sustainable resource management in Shivaliks, India
Sidhu GS, Walia CS, Tarsem Lal, Rana KPC, Sehgal J (1995b) Soils of Punjab for optimizing land use, p 67. NBSS Publ. 45b. NBSS&LUP, Nagpur, India
Sidhu GS, Surya JN (2014) Soils of North-Western Himalayan eco-system and their land use, constraints, productivity potentials and future strategies. Agropedology 24:1–19
Sillanpaa M (1982) Micronutrients and the nutrient status of soils: a global study. FAO Soils Bulletin No. 48. Food and Agriculture Organization of the UN, Rome, Italy 242–250pp
Singh A, Singh MD (1981) Effect of various stages of shifting cultivation on soil erosion from steep hill slopes. Indian Forester 106(2):115–121
Singh A (1989) Bamboo drip Irrigation. Res. Bull., ICAR Research Complex for NEH Region, Shillong, Meghalaya, India
Singh AK, Prasad A, Singh B (1986) Availability of Phosphorus and Potassium and its relationship with some important physico-chemical properties of some forest soils of Pali Range, (Shahdol, MP). Indian Forester 112(12):1094–1103
Singh G, Sharma M, Manan J et al (2016) Assessment of soil fertility status under different cropping sequences in district, Kapurthala. J Krish Vigyan, J Agri Sci 5:1–9
Singh S, Parkash B, Rao MS, Arora M, Bhosle B (2006) Geomorphology, pedology and sedimentology of the Deoha/Ganga–Ghaghara Interfluve, Upper Gangetic Plains (Himalayan Foreland Basin)—Extensional tectonic implications. Catena 67: 183–203. https://doi.org/10.1016/j.catena.2006.03.013
Singh A, Singh MD (1984) A case study on farm stead development in hill relation to soil and water conservation from steep hill slopes. Indian Forester 106 (20):115–121
Singh A, Varma A (1988) Land and water management in hills. Indian J Hill Farming 1A(2):1–9
Singh A (1987) Studies on some aspects of soil and water in relation to resource management in NE Region. Faculty of Agriculture, BCKVV, West Bengal
Singh A, Prasad RN (1987) Alternative to shifting cultivation, ICAR, New Delhi
Singh A, Singh MD (1978) Effect of various stages of shifting cultivation on soil erosion from steep hill slopes. Indian Forester 106:115–121
Singh AK, Khan SK, Nongkynrih P (1999a) DTPA-extractable Zn, Cu, Mn, and Fe in wetland rice soils and their availability to rice under submergence. J Indian Soc Soil Sci 47:381–383
Singh AK, Khan SK, Nongkynrih P (1999b) Transformation of zinc in wetland rice soils in relation to nutrition of rice crop. J Indian Soc Soil Sci 47:248–253
Singh AK, Nongkynrih P (1999) Distribution and transformation of copper in wetland soils and its availability to rice plant. J Indian Soc Soil Sci 47:452–457
Singh AK, Nongkynrih P (2000) Critical limit of copper for predicting response of rice to copper application on wetland rice soils of Meghalaya. J Indian Soc Soil Sci 48:406–408
Singh BP, Das M, Ram M, Dwivedi BS and Prasad RN (1989) Micronutrient status of Alfisols of Meghalaya.. Journal Indian Society of Soil Science, 37: 591–595
Singh BP, Madhumita Das, Ram M, Dwivedi BS, Prasad RN (1992a) Characterization of iron toxic soils and affected plants in Meghalaya. J Ind Soc Soil Sci 40: 326–328
Singh K, Sharma UC (1987) Response of potato to magnesium on acidic hill soil. J Hill Res 1:88–95
Singh K, Sharma UC (1989) Nutrients needs of maize-potato cropping sequence in acidic soils of Meghalaya. Indian J Agric Sci 59:159–161
Singh G, Ram B, Narain P, Bhushan LS, Abrol IP (1992b) Soil erosion rates in India. Ind J Soil Conserv 47:97–99
Singh MD (1985) Rainfall and soil relationships on varying degree of slope. M.E. dissertation, Dept. of Civil Engineering, Assam Engineering College, Assam, Guwahati
Singh OP, Datta B (1987) Phosphorus status of some hill soils of Mizoram in relation to pedegenic properties. J Indian Soc Soil Sci 34:600–705
Singh OP, Datta B (1988) Organic carbon and nitrogen status of some soils of Mizoram occuring at different altitudes. J Indian Soc Soil Sci 36:414–420
Singh SK (1998) Soils of Himachal Pradesh: their interpretation for thematic mapping and land use planning. Presented and abstracted in International Symposium on Sustainable Agriculture in Hilly Areas, held at HPKV, Palampur on October 29–31, HPKKV, Palampur, India
Singh SP, Jagat Ram, Walia CS, Sachdeva, CB,, Dhankar RP, Rana KPC, Saigal JL, Velayutham M, Gajbhiye KS (2004) Soils of Uttra Pradesh for Optimizing Land Use. NBSS & LUP, Nagpur, India Publication No. 68
Singh TV, Kaur J (1989) Studies in Himalayan ecology and development strategies. Himalayan Books, New Delhi, India
Siqueira JO, Moreira FMS (1997) Microbial population and activities in highly weathered acidic soils: highlights of the Brazillian research. In: Moniz AC, Furlani AMC, Schffert RE, Fageria NK, Rosolem CA, Cantarella H (eds) Plant-Soil Interactions at low pH: sustainable agriculture and forestry production. Brazilian Soil Science Society, Brazil, 139–156pp
Sitaula BK, Bajracharya RM, Singh BR, Solberg B (2004) Factors affecting organic carbon dynamics in soils of Ncpal/Himalayan region—a review and analysis. Nutr Cycl Agroecosyst 70:215–229

Soil Science Division (1992) Soils of Bore, Mahadev, Dee, Khahare and Jogi & Bhandare Khola sub-watersheds. Land use, soil conservation and water resources management Project, Likhu Khola watershed area, Nuwakot district, Nepal

Soil Survey of Pakistan (2010) Soil survey reports, Lahore, Pakistan

Soil Survey of Pakistan (1988) Reconnaissance soil survey reports from 1965–88, Lahore

Solomon S (2007) Climate change: the physical sciences basis. In: Processings of the contribution of working Group I to the fourth assessment report of the IPCC, San Francisco, CA, USA, 10–14 December.

Sommer M (2006) Influence of soil pattern on matter transport in and from terrestrial biogeosystems—a new concept for landscape pedology. Geoderma 133:107–123. https://doi.org/10.1016/j.geoderma.2006.03.040

Sonowal DK, Tripathi AK, Krirekha V (1989) Zabo, an indigenous system of Nagaland. Indian J Hill Res 2:1–8

Sorensen PO, Finzi AC, Giasson MA, Reinmann AB, Sanders-DeMott R, Templer PH (2018) Winter soil freezethaw cycles lead to reductions in soil microbial biomass and activity not compensated for by soil warming. Soil Biol Biochem 116:39–47

SRDI Staff (1965–1977) Reconnaissance Soil Surmey Reports of Bangladesh. Soil Resources Development Institute, Dhaka, Bangladesh.

Soriano-Disla JM, Janik LJ, Rossel RAV, Macdonald LM, McLaughlin MJ (2014) The performance of Visible, Near-, and Mid-Infrared reflectance spectroscopy for prediction of soil physical, chemical, and biological properties. Appl Spectrosc Rev 49:139–186

Stanchi S, Frepazz M, Walter S, Wanderwegen L (2015) Mountain soils and threats. In: Romeo R, Vita A, Manuelli S, Zanini E, Freppaz M, Stanchi S (eds) Understanding mountain soils: a contribution from mountain areas to the International Year of Soils 2015, Rome, Italy

Sturtz AV, Christie BR (2003) Rationale for an holistic approach to soil quality and crop health. Soil Tillage Res 72(2):105–106

Sullivan PF, Arens SJT, Chimner RA, Welker JM (2008) Temperature and microtopography interact to control carbon cycling in a high arctic fen. Ecosystems 11:61–76. https://doi.org/10.1007/s10021-007-9107-y

Swarup A, Ganeshamurthy AN (1998) Emerging nutrient deficiencies under intensive cropping systems and remedial measures for sustainable high productivity. Fertilizer News 43(7): 37–40 , 43–50.

Szabo S, Hossain MS, Renaud F, Traore D, Hussain A, Matczak P, Ahmad S, Singh DR, Neumann B, Mathews Z (2018) Accelerating progress toward the zero hunger goal in cross-boundary climate change hotspots. Environ Sci Policy Sustain Dev 60(3):18–27

Szabo S, Nicholls RJ, Neumann B, Renaud FG, Mathews Z, Sebesvari Z, AghaKouchak A, Bates R, Ruktanonchai CW, Kloos J, Foufoula-Georgiou E (2016) Making SDGs work for climate change hotspots. Environ Sci Policy Sustain Dev 58(6):24–33

Takada M (1991) Landform and Quaternary geohistory of the Bhutan Himalaya. In: Ohsawa M (ed) Life zone ecology of the Bhutan Himalaya. Chiba University, Japan, Laboratory of Ecology, pp 41–88

Takata Y, Funakawa S, Akshalov K, Ishida N, Kosaki T (2008) Regional evaluation of the spatio-temporal variation in soil organic carbon dynamics for rainfed cereal farming in northern Kazakhstan. Soil Sci Plant Nutrition 54:794–806

Takkar PN, Jalali VK (2005) Soil fertility problems and strategies to reduce them in the Himalayan region of India. In: Andersen P, Tuladhar JK, Karki KB and Maskey SL (eds) Micronutrients in South and South-east Asia, Procs. International Workshop, Kathmandu, Nepal, 133–142pp, 8–11 Sep 2004

Tamboli PM (1971) FAO Report to the Government of Afghanistan on Soil Fertility and Fertilizer Uses. United Nations Development Programme, Report No. TA 3016 , 77 p

Tandzi LN, Mutengwa CS, Ngonkeu ELM, Vernon Gracen V (2018) Breeding maize for tolerance to acidic soils: a review. Agronomy 8:84. https://doi.org/10.3390/agronomy8060084

Tarnocai C, Canadell J and Schuur E et al (2009) Soil organic carbon pools in the northern circumpolar permafrost region. Global Biogeochemical cycles 23: 1–11. Available via DOI: https://doi.org/10.1029/2008GB003327

Thapa GB, Rasul G (2005) Patterns and determinants of agricultural systems in the Chittagong Hill tracts of Bangladesh. Agric Syst 84:255–277

Thomas GW (1977) Historical developments in soil chemistry: ion exchange. Soil Sci Soc Am J 41:230–238

Thompson LM and Troch FR (1978) Soils and Soil Fertility. McGraw -Hill Pub., New York, USA

Tian D, Niu S, Pan Q, Ren T, Chen S, Bai Y, Han X, Whitehead D (2016) Nonlinear responses of ecosystem carbon fluxes and water-use efficiency to nitrogen addition in Inner Mongolia grassland. Funct Ecol 30:490–499

Tian H, Wang S, Liu J, Pan S, Chen H, Zhang C and Shi X (2006) Patterns of soil nitrogen storage in China. Glob Biogeochem Cycles 20

Tilahun G (2007) Soil fertility status as influenced by different land uses in Maybar areas of south Wello zone, North Ethiopia. M.Sc. Thesis, Department of Plant Sciences, School of Graduate Studies, Haramaya University, Ethiopia

Tisdale SL, Nelson WL, Beaton JD (1985) Soil Fertility and Fertilizers MacMillan Publishers Co. New York, USA, 754p

Trenberth KE, Dai A, van der Schrier G, Jones PD, Barichivich J, Briffa KR, Sheffield J (2013) Global warming and changes in drought. Nat Clim Change 4:17–22. https://doi.org/10.1038/nclimate2067

Tripathi BP, Shrestha SP (2000) Nitrogen content in farm yard manure and its effects on the productivity and soil

properties of rice-wheat, upland blackgram and maize-fingermillet systems. Lumle Working Paper No. 2000/14, Lumle Agricultural Research Center, Kaski, Nepal, 8p
Tripathi BP (1999) Soil fertility status in farmer's fields of Western Hills of Nepal. Lumle Seminar Paper No. 99/4, Lumle Agricultural Research Center, Kaski, Nepal, 8 p
Tripathi BP (2001) Enhancement of fertility and crop productivity for acidic soils in the hills of Nepal. Lumle Technical Paper No. 2001/2, Agriculture Research Station, Lumle PO Box 1, Pokhara, Nepal
Tripathi D, Singh K, Upadaya GP (1994) Distribution of micronutrients in some representative soil profiles of Himachal Pradesh. J Indian Soc Soil Sci 42:143–145
Tripathi RD, Tripathi P, Dwivedi S, Kumar A, Mishra A, Chauhan PS, Norton GJ, Nautiyal CS (2014) Roles for root iron plaque in sequestration and uptake of heavy metals and metalloids in aquatic and wetland plants. Metallomics 6:1798–1800
Tucker RP (2021) Dimensions of deforestation in the Himalaya: the historical setting. In: Choden et al (ed) Terrestrial Carbon Stock Potential in Selected Forest in Bhutan, India and Nepal. Article no.AJEE. 6886429, Mountain Research and Development, 1987, pp 328–331
Tuladhar JK (1995) Fertility status of farmer's field soil within the extension command area of Lumle Agricultural Research Centre. LARC Working Paper No. 95/47, Kaski, Nepal, 27p
UNCED (1992) Agenda 21, United Nation Commission on Environment. United Nations Conference on Environment & Development Rio de Janerio, Brazil, 3 to 14 June 1992
UNDP (United Nations Development Programme) Digital Agriculture Atlas of the Union of Myanmar, Ministry of Agriculture and Irrigation, Myanmar. UNDP, FAO, Rome, Italy, 15p
UNEP (2001) Nepal: State of the Environment. UN Environment Programme in collaboration with HMG/Min. of Population and Environment, Int'l. Center for Integrated Mountain Dev. and United Nations Envir. Prog., Thailand, 211p
UNEP/WCMC (United Nations Environment Programme-World Conservation Monitoring Center) (2002) Mountain Watch: environmental change and sustainable development in mountains, Cambridge, UK.
Urban J, Ingwers MW, McGuire MA, Teskey RO (2017) Increase in leaf temperature opens stomata and decouples net photosynthesis from stomatal conductance in Pinus taeda and Populus deltoides x nigra. J Exp Bot 68:1757–1767. https://doi.org/10.1093/jxb/erx052
USDA and UC Davis (2013) Afghan Agriculture Portal. http://afghanag.ucdavis.edu/
Usman H, Pugh TAM, Ahlstrom A, Baig S (2021) Climate change projections of terrestrial primary productivity over the Hindu Kush Himalayan forests. Earth Syst Dynam 12:857–870. https://doi.org/10.5194/esd-12-857-2021
van Breemen N (1973) Dissolved aluminum in acid sulphate soils and in mine waters. Proc Soil Sci Soc America 37:694–697
van Breemen N (1976) Genesis and solution chemistry of acid sulphate soils in Thailand. Pudoc, Wageningen
van Keulen H (2001) Tropical soil organic matter modelling: problems and prospects. Nutr Cycl Agroecosyst 61(1–2):33–39
Varga R, Castro A, Ziadat F (2015) Mountain soils and ecosystem services. In: Romeo R, Vita A, Manuelli S, Zanini E, Freppaz M and Stanchi S (eds) Understanding mountain soils: a contribution from mountain areas to the International Year of Soils, FAO 2015, Rome, Italy
Velayutham M (1999) Soil Resources of India-potential, problems and strategies for the future. National Seminar on Strategies for Agricultural Research in North East, Shillong, Nov 10–12, 33p
Velayutham M, Mandal DK, Mandal C, Sehgal J (1999) Agro-ecological subregions of India for development and planning. Publ. 35, NBSS & LUP, Nagpur, India, 452p
Venkatesh MS, Majumdar B, Kumar K, Patiram (2003) Status of micronutrient cations under various land use systems of Meghalaya. J Indian Soc Soil Sci 51:60–64
Venkatesh MS, Mishra AK, Satapathy KK, Patiram (2001) Effect of burning on soil properties under bun cultivation in Meghalaya. J Hill Res 14(1):21–25
Venkatesh MS, Kumar K, Raychaudhuri M, Datta M, Sharma UC (2006) In: Sharma UC, Datta M, Samra JS (eds) Soils and their management in Northeast India. ICAR Research Complex for NEH Region, Umiam, Shillong, Meghalaya, India, 545p
Verma TS, Tripathi BR (1984) Bronzing disease of rice, its causes and remedial measures in acid alfisols of Himachal Pradesh. J Indian Soc Soil Sci 32:504–506
Viscarra Rossel RA, Walvoort DJJ, McBratney AB, Janik LJ, Skjemstad JO (2006) Visible, near infrared, mid in-frared or combined diffuse reflectance spectroscopy for simultaneous assessment of various soil properties. Geoderma 131:59–75
Vitousek PM, Sanford RL Jr (1986) Nutrient cycling in moist tropical forest. Annu Rev Ecol Syst 17:137–167
Vitousek PM, Mooney HA, Lubchenco J, Melillo JM (1997) Human domination of Earth's ecosystems. Science 277:494–499
Vitousek PM, Howarth RW (1991) Nitrogen limitation on land and in the sea: how can it occur? Biogeochemistry 13:87–115
Vittal KPR, Gangwar MS (1974) Zinc in soil profiles of nainital Tarai. J Indian Soc Soil Sci 22:151–155
Von Uexkull, Mutert E (1995) Global extent, development and economic impact of acid soils. In: Date RA, Grundon NJ, G Rayment GE, Probert ME (eds), Plant-soil internations at low pH, principles and management. Kluwer Academic Publishers, The Netherlands

Wake CP, Mayewski PA (1995) Report from the International Himalayan/Tibetan Plateau Palaeoclimate Workshop in Kathmandu 2–7 April 1995. Himalayan Interdisciplinary Paleoclimate Project–Science and Implementation Plan. http://www.ccrc.sr.unh.edu/hipp/KduWkshp_contents/KduWkshp_rep.html

Walia CS, Rana KPC, Sidhu GS Mahapatra SK, Lal T (1999) Characterization and classification of some soils of Ladakh region for land use Agropedology 9:16–21

Wang GX, Qian J, Cheng GD et al (2002a) Soil organic carbon pool of grassland soils on the Qinghai-Tibetan Plateau and its global implication. Sci Total Environ 291:207–217. https://doi.org/10.1016/S0048-9697(01)01100-7

Wang JF, Wu QB (2011) Influences of the vegetation degradation on the shallow cryic soil environment in the wet meadow areas on the Qinghai-Tibetan Plateau. J Lanzhou Univ (Natural Sciences) 47:39–45

Wang Y, Zhou G, Jia B (2008) Modeling SOC and NPP responses of meadow steppe to different grazing intensities in Northeast China. Ecol Model 217:72–78

Wang GX, Qian J, Cheng GD, Lai YM (2002b) Soil organic carbon pool of grassland soils on the Qinghai-Tibetan Plateau and its global implication. Sci Total Environ 291:207–217. https://doi.org/10.1016/s0048-9697(01)01100-7

Wardle DA, Bardgett RD, Klironomos JN, Setala H, Van Der Putten WH, Wall DH (2004) Ecological linkages between aboveground and belowground biota. Science 304(5677):1629–1633

Watanabe T, Tateno R, Imada S et al The effect of a freezethaw cycle on dissolved nitrogen dynamics and its relation to dissolved organic matter and soil microbial biomass in the soil of a northern hardwood forest. Biogeochemistry 142 (3): 319–338

Wei SG, Dai YJ, Duan QY et al (2014) A global soil dataset for Earth System Modeling. Journal of advances in modeling earth systems 6: 249–263. https://doi.org/10.1002/2013MS000293

Wei SG, Dai YJ, Liu BY et al (2012) A soil particle-size distribution dataset for regional land and climate modelling in China. Geoderma 171:85–91. https://doi.org/10.1016/j.geoderma.2011.01.013

Wei SG, Dai YJ, Liu BY et al (2013) A China data set of soil properties for land surface modeling. J Adv Model Earth Syst 5: 212–224. https://doi.org/10.1002/jame.20026

Wei XR, Hao MD, Shao MG, Gale WJ (2006) Changes in soil properties and the availability of soil micronutrients after 18 years of cropping and fertilization. Soil Tillage Res 91:120–130

Wen L, Dong S, Li Y, Wang X, Li X, Shi J, Dong Q (2013) The impact of land degradation on the C pools in alpine grasslands of the Qinghai-Tibet Plateau. Plant and Soil 368:329–340. https://doi.org/10.1007/s11104-012-1500-4

Wood MK, Blackburn WH (1981) Grazing systems: Their influence on infiltration rates in the rolling plains o Texas. Journal of Range Management. 34: 331–335

Wu GL, Ren GH, Dong QM, Shi JJ, Wang Yl (2014) Above- and belowground response along degradation gradient in an alpine grassland of the Qinghai-Tibetan Plateau. Clean-Soil Air Water 42:319–323. https://doi.org/10.1002/clen.201200084

Wu XD, Zhao L, Chen MJ et al (2012) Soil organic carbon and its relationship to vegetation communities and soil properties in permafrost areas of the central western Qinghai-Tibet Plateau, China. Permafrost Periglacial Processes 23:162–169. https://doi.org/10.1002/ppp.1740.

Wu Q, Zhu Y, Liu Y (2000) Evaluation model of permafrost thermal stability and thawing sensibility under engineering activity. Cold Reg Sci Technol 34:19–30

Wu Z, Dijkstra P, Koch GW, Peñuelas J, Hungate BA (2011) Responses of terrestrial ecosystems to temperature and precipitation change: a meta-analysis of experimental manipulation. Glob Change Biol 17:927–942. https://doi.org/10.1111/j.1365-2486.2010.02302.x

Xu ZJ, Liu GS, Yu JD (2002) Soil acidification and nitrogen cycle disturbed by mall-made factors. Geology-Geochemistry 30:74–78 (in Chinese with English abstract)

Xu X, Thornton PE, Post WM (2013) A global analysis of soil microbial biomass carbon, nitrogen and phosphorus in terrestrial ecosystems. Glob Ecol Biogeogr 22:737–749

Yaalon DH, Arnold RW (2000) Attitudes toward soils and their societal relevance: then and now. Less Soil Sci 165: 5–12

Yadav Janardan and Jagdish Prasad (2005) Characteristics of some typical pedons from Leh. Ann Arid Zone 44:195–198

Yang JP, Ding YJ and Chen RS (2005) NDVI reflection of alpine vegetation changes in the source regions of the Yangtze and Yellow Rivers. Acta Geographica Sinica 60:467–478. https://doi.org/10.3321/j.issn:0375-5444.2005.03.013

Yang C, Li J, Liu N, Zhang YJ (2019a) Effects of fairy ring fungi on plants and soil in the alpine and temperate grasslands of China. Plant Soil 441:499–510

Yang C, Liu N, Zhang YJ (2019b) Soil aggregates regulate the impact of soil bacterial and fungal communities on soil respiration. Geoderma 337:444–452

Yang C, Zhang FG, Liu N, Hu J, Zhang YJ (2018) Changes in soil bacterial communities in response to the fairy ring fungus Agaricus gennadii in the temperate steppes of China. Pedobiologia 69:34–40

Yang MX, Wang SL, Yao TD et al (2004) Desertification and its relationship with permafrost degradation in Qinghai-Xizang (Tibet) Plateau. Cold Regions Sci Technol 39:47–53. https://doi.org/10.1016/j.coldregions.2004.01.002

Yang Y, Fang J , Ji C, Ma W, Su S, Tang Z (2010) Soil inorganic carbon stock in the Tibetan alpine grasslands. Global Biogeochem Cycles 24:GB4022. https://doi.org/10.1029/2010GB003804

Yang YH, JY, Fang DL, Guo C, Ji J and Ma WH, (2010) Vertical patterns of soil carbon, nitrogen and carbon: nitrogen stoichiometry in Tibetan grasslands. Biogeosci Discuss 7:1–24

Yashoda A, Avasthe RK (1995) Altitudinal distribution of micronutrients in the soils of Sikkim. J Indian Soc Soil Sci 47:374–377

Yi S, Rawat GS, Wu N, Bubb P, Chettri N (2017) Framework for integrated ecosystem management in the Hindu Kush Himalaya. ICIMOD Working Paper 2017/10. Kathmandu, Nepal

Yuan ZQ, Xiang XJ, Liu GJ, Jin HJ, Wu QB (2019) Response of soil organic carbon and nutrient stocks to human induced grassland degradation in a Tibetan alpine meadow. CATENA 178:40–48

Zanini E, Romeo R (2015a) Introduction In: Romeo R, Vita A, Manuelli S, Zanini E, Freppaz M and Stanchi S (eds) Understanding mountain soils: a contribution from mountain areas to the international year of soils 2015, Rome, Italy

Zanini E, Frepazz M, Stanchi S, Bonifacio E, Egli M (2015b) Soil variability in mountain areas. In: Romeo R, Vita A, Manuelli S, Zanini E, Freppaz M and Stanchi S (eds) Understanding mountain soils: a contribution from mountain areas to the international year of soils 2015, Rome, Italy

Zeigler RS, Pandey S, Miles J, Gourley LM and Sarkarung S. (1995) Advances in the selection and breeding of acid-tolerant plants: Rice, maize, sorghum and tropical forages. In: Plant soil interactions at low pH; Kluwer Academic Publishers, Dordrecht, The Netherlands, 391–406pp

Zhang Y, Li B, Zheng D (2002) A discussion on the boundary and area of the Tibetan Plateau in China. Geogr Res 21:1–8 (in Chinese)

Zhang WJ, Xue X, Peng F, You QG, Hao AH (2019) Meta-analysis of the effects of grassland degradation on plant and soil properties in the alpine meadows of the Qinghai-Tibetan Plateau. Glob Ecol Conserv 20: e00774

Zhang B, Tang L, WangW, Li Y and Li H (2010) Quantification and comparison of soil elements in the Tibetan plateau Kaschin-Beck disease area : A case study in Zamtang County, Sichuan Province, China. Biology of Trace Element Research

Zhao L, Ping CL and Yang DQ et al (2004a) Changes of climate and seasonally frozen ground over the past 30 years in Qinghai-Xizang (Tibetan) Plateau, China. Global Planetary Change 43:19–31. https://doi.org/10.1016/j.gloplacha.2004a.02.003

Zhao YG, Zhang GL (2011) Chinese soil map. Beijing Science Press, China

Zhao L, Ping CL, Yang D, Cheng G, Ding Y, Liu S (2004b) Changes of climate and seasonally frozen ground over the past 30 years in Qinghai-Xizang (Tibetan) Plateau. China Glob Planet Change 43:19–31

Zhao L, Wu Q, Marchenko SS, Sharkhuu N (2010) Thermal state of permafrost and active layer in Central Asia during the international polar year. Permafrost Periglac 21:198–207

Zhao YC, Wang MY, Hu SJ et al (2018) Economics- and policy-driven organic carbon input enhancement dominates soil organic carbon accumulation in Chinese aroplands. Proc Natl Acad Sci 115:4045–4050

Zhou YW, Guo DX, Qiu GQ et al (2000a) Geocryology in China. Beijing Science Press, China, p 44p

Zhou X, Zhou Y, Zhou C, Wu Z, Zheng L, Hu X, Gan J (2015) Effects of cutting intensity on soil physical and chemical properties in a mixed natural forest in Southeastern China. Forests 6(12):4495–4509. https://doi.org/10.3390/f612438

Zhou Z, Guojian Zhang G, Xiaomei Song X, Shuhan Du S and Gangcai Liu G (2010) Soil acidification with various planting patterns in Lhasa, Tibet, China. Paper presented at 19th World Congress of Soil Science, Soil Solutions for a Changing World, Brisbane, Australia. Published on DVD

Zhou Y, Gu D, Qiu G, Cheng G, Li S (2000b) China Permafrost. Science Press Beijing, China, pp 145–151

Zhou H, Zhang D, Jiang Z, Sun P, Xiao H, Yuxin W, Chen J (2019) Changes in the soil microbial communities of aline steppe at Qinghai-Tibetan plateau under different degradation levels. Sci Total Environ 651:2281–2291

Zhuang Q, He J, Lu Y, Ji L, Xiao J and Luo T (2010) Carbon dynamics of terrestrial ecosystems on the Tibetan Plateau during the 20th century: an analysis with a process-based biogeochemical model. Global Ecol Biogeogr 19: 649–662

Zinck JA (1988) Physiography and soils ITC lecture notes on soil survey course

Zokaib S (2011) Rainfall, runoff and soil degradation in the Hindu Kush-Himalayas—a case study in Hilkot Watershed Pakistan—a thesis submitted in partial fulfillment of the requirements for the degree of Master of Applied Science in Civil Engineering, The University of British Columbia (Okanagan Campus)

Zou D, Zhao L, Sheng Y et al (2017) A new map of permafrost distribution on the Tibetan Plateau The Cryosphere 11: 2527–2542. https://doi.org/10.5194/tc-11-2527-2017

7 Land Use and Management

Abstract

The Hindu Kush-Himalayan (HKH) region is one of the largest and most diversified mountain settings in the world, having several parallel mountain ranges. Mountains cover nearly 24% of the land surface in the World and directly support over 12% of the world's population that lives in these areas. About 20% of the world's population derives a wide variety of ecosystem services from the mountains, including food, water, energy, timber, bioresources, and fodders for the livestock. Because of their geographical isolation and singular biophysical setting, many HKH areas exhibit high levels of endemism and rapid evolutionary processes. Mountain agriculture is complex and interacts with biophysical, socio-cultural and agro-ecological factors for making food and social life security decisions. Fragmented and small holding size, limited access to inputs, credits, demands driven technologies and services further constrain agricultural development in the mountains of HKH. The farmers across the HKH region practise subsistence farming that depends on forest resources. This is mostly due to severe climatic conditions and lack of market facilities. Improper land use and climate change have increased stress on mountain ecosystems in the region. Overgrazing is widespread and plant biomass is releasing carbon from peat land and underground soils due to livestock pressure. The high alpine meadows are particularly under threat as topsoil cover is being eroded by water and strong winds, making restoration of vegetation more difficult. The degradation of the rangelands is also forcing nomads to change their profession. Some case studies conducted in Afghanistan, Nepal and Pakistan suggest that decreased rainfall in rangelands has often resulted in increase of degraded area and accelerated desertification, affecting the livestock development. The Tibetan Plateau is highly prone to heating, drying, and more extreme weather conditions, which is having pronounced impacts on rangelands besides influencing the Asian monsoon system. Range and pasture lands account for over 60% of the surface of the HKH mountains. The impact of climate change has caused sharp declines in fodder productivity and animal carrying capacity and has also endangered the socio-economic fabric and wellbeing of pastoral communities. On acid soils, highly responsive crops like cotton, soybean, pigeon pea, groundnut, French-bean, etc. should be grown in the first year of liming. This should be followed by medium responsive crops like maize and wheat in the subsequent seasons. Cultivation of lowest responsive crops like millets, rice, barley, linseed, etc. should be done in the last season/year prior to commencement of the next liming schedule. While choosing crops

U. C. Sharma et al., *Soils in the Hindu Kush Himalayas*, Geography of the Physical Environment,
https://doi.org/10.1007/978-3-031-11458-8_7

for mountains, one major point needs to be considered is the local needs and preferences. The horticultural crops are relatively more suited in HKH mountains. Food security in the Hindu Kush Himalayas depends mainly on local production and purchasing power of the people. Fragile environment, depleted natural resource systems, limited availability of suitable land for large-scale production, physical inaccessibility, and poor local infrastructure are responsible for food insecurity in the Himalayas.

Keywords

Conservation agriculture · Crop production · Fertilizer use · Forces of change · Integrated plant nutrient management · Land use and management · Mountain agriculture · Natural resources

7.1 General Description

It has been estimated that mountain landscapes have nearly 24% of the land surface of the world but supports only about 12% of the population that resides in these regions (Sharma et al. 2010a, b). About 20% of the world's population derives a wide variety of ecosystem services from the mountains, as also feeds and fodders for the livestock. Koerner (2003) was of the view that the mountains are storehouse of biological diversity. Mountain ecosystems, on the other hand, are among the most vulnerable in the world, and climate change poses a serious threat to them (Ning et al. 2013). The HKH is separated into two areas in terms of physical features: the mountainous area, which has a wide variety of forest types, and the Tibetan plateau, which is located above 4,000 meters and includes grasslands, wetlands, and desert steppe. Overuse of natural resources in these locations can have a significant impact on the flow of ecosystem products and services out of the mountains (Sharma and Yonzon 2005; Sharma et al. 2010a, b; Tse-ring et al. 2010). Total mountain area in HKH region has been estimated to be about 3.4049 million km^2, out of which sloping lands between 8–30% constitutes 20.7% of the total mountain area and >30% sloping lands constitute 43.0% of the area. Mean area under agricultural land in the HKH countries is 5.36% (Pratap 2001; Rasul and Kollmair 2008). The highest percent area under 8–30% sloping land category is in Bangladesh (60.1%), followed by Myanmar (37.4%), Afghanistan (35.1%), India (30.7%), Pakistan (29.3%), Bhutan and Nepal (12.7% each) and China (10.0%). In >30% category of sloping lands, highest percent of land is in Bhutan (88.4%), followed by 66.3% in Nepal, 50.7% in China, 41.9% in Afghanistan, 35.6% in Pakistan, while the lowest percent land in Bangladesh (12.2%). In HKH region, the area under agricultural land is low because of hilly terrain. The highest percent of agricultural land is in Nepal (18.0%) and the lowest in China (1.2%) (Pratap 2001; Rasul and Kollmair 2008). The highest percent of per capita land available for cultivation is 0.325 ha in Myanmar, followed by Bhutan (0.163 ha), Afghanistan (0.159 ha), Nepal (0.102 ha), Pakistan (0.096 ha), while lowest per capita land availability is in Bangladesh (0.069 ha). The technologies need to be designed in accordance with the slope of the land, climatic conditions in the area, and availability of inputs, for getting potential productivity from the land.

Due to steep slopes, rainfall, deforestation, and anthropogenic factors, the Himalayan mountain's ecosystem is prone to degradation and erosion. Jodha (1993) mentioned that the inaccessibility, fragility, marginality, diversity are important factors associated with mountain agriculture and which affect and limit agricultural productivity in HKH region. However, there are some niche conditions available which need to be exploited for better returns. Crop land security, soil erosion and falling soil fertility, rising poverty, water scarcity, and the negative consequences of climate change due to global warming are all predictable concerns that must be addressed effectively to improve the region's quality of life (Pratap 2003, Jodha and Shrestha 1994, Eriksson et al. 2009a, b).

Due to limited research on mountain agriculture, the understanding of the problems associated with the hill agriculture is scant. Mountain agriculture is complex and interacts with biophysical, socio-cultural and agro-ecological factors for making food and social life security decisions (Demaine 1998). The constraints in agricultural development in the mountains of HKH in general are small holding size, limited access to inputs, credits, demands-driven technologies, and services further (Pratap 2003, Jodha and Shrestha 1994). Rasul and Karki (2007) suggested that the best practices need to be introduced, and a more recently evolved interventions are necessary based on agro-ecological potentials and access to the market (Rasul and Karki 2007).

The farmers across the HKH region practise subsistence farming that depends on forest resources. This is mostly due to severe climatic conditions, lack of market facilities, and small and fragmented holdings. Because of low returns, the mountain people feel it difficult to make both ends meet and therefore move within or outside the region for livelihood (Karki et al. 2012a, b). This is further aggravated by shrinking population-land ratio, reduced economic viability, and decreasing interest in agriculture among the new generation youths. Crop land security, soil erosion and falling soil fertility, rising poverty, water scarcity, and the negative consequences of climate change due to global warming are all predictable concerns that must be addressed effectively to improve the region's quality of life (Pratap 2003, Jodha and Shrestha 1994, Eriksson et al. 2009a, b).

7.2 Land Management

There has been a renewed interest in soil resources of the world for food security after the 2008 food crisis. Available fertile soil resources for food production are limited and fixed (about 1600 million ha) and due to increase in demographic pressure, the availability of arable land per capita is decreasing fast. As per Stoddart et al. (1975), range and pasture lands account for over 60% of the surface of the HKH mountains. Climate-induced negative changes are thought to be the cause of multiple threats to rangeland's ecosystems services throughout the HKH region. According to Yi and Sharma (2009), more than 90% of rangelands in China have been damaged, with nearly a quarter of them facing severe deterioration or desertification. For their daily survival, more than 100 million people rely on rangeland resources, conducting various sorts of animal husbandry and collecting rangeland goods. Climate change has resulted in dramatic decreases in fodder output and animal carrying capacity, putting the socio-economic fabric and wellbeing of pastoral communities under jeopardy.

7.2.1 Management of Mountain Acid Soils

Acid soils can be controlled by growing crops that are suited to a specific soil pH or by amending the soil with amendments such as lime. Liming elevates the pH of the soil. The faster the reaction, the higher the soil moisture. Acidity is produced when H^+ ions react to form weakly dissociated water, and calcium ions undertake exchange processes to replace more exchangeable H^+. The amount of liming material that must be added to raise the soil pH to a predetermined value, usually in the range of 6.0–7.0, is known as the lime need of soil. The following is the reaction that occurs when lime is applied (Swarup and Purakayastha 2008):

$$CaCO_3 + H_2O + CO_2 \rightarrow Ca(HCO_3)_2$$

$$Ca(HCO_3)_2 \rightarrow Ca^{2+} + 2HCO_3$$

$$\text{H-Soil} + Ca^{2+} \rightarrow \text{Ca-Soil} + 2H^+$$

$$H^+ + HCO_3 \rightarrow H_2CO_3 \leftrightarrow H_2O + CO_2$$

Use of Liming Materials

Lime is the most important material which is used to correct soil acidity. Lime when added to soil neutralizes soil acidity, increases nutrient availability and improves the physical condition of soil. Calcite and dolomite are important among the naturally occurring lime sources.

Agricultural grade limestone powder and commercial lime are effective, but they may not be readily available in all areas, and in such cases, industrial byproducts, notably paper mill waste, could be used as a low-cost liming resource.

Use of Organic Manures

In acid soils, regular application and incorporation of well-decomposed organic manures such as farmyard manure, compost, or green manuring are critical to avoid abrupt pH changes. It boosts soil buffer capacity, increases P availability, and decreases Fe and Al toxicity in acid soils. On acid soils, mulching is also recommended. The application of weed biomass also significantly minimizes the negative effects of soil acidity. In the acid soils of India's Eastern Himalayan states, phosphorus deficit is more pronounced. The problem of P deficit in acid soil regions must mostly be addressed by applying an adequate amount of a suitable type of P fertilizer and implementing effective management methods. Iron toxicity is a problem in some parts of Eastern India Himalayan Region under low land rice conditions.

7.2.2 Crop Scheduling

It is always preferable to cultivate crops or crop varieties that are acid tolerant on acid soils. It is an effective weapon for combating this soil condition, and the development of such cultivars is particularly important for achieving higher production, especially in locations where liming is not a viable option (Table 7.1). Crops can be categorized based on how well they function in various pH ranges of soil.

While choosing crops for acid soils, one major point that needs to be considered is the local needs and preferences. Farmers generally grow rice on acid soils, irrespective of the type of land, as the crop has certain degree of tolerance to soil acidity. However, the crops to be grown on acid soils should be based on the degree of their tolerance to acidity. The following points need to be considered:

(i) Slightly tolerant to acid soils;
(ii) Moderately tolerant to acid soils;
(iii) Highly tolerant to acid soils.

The horticultural crops are relatively more suited on acid soils. The soil acidity tolerance of some horticultural crops in Eastern Indian Himalayan region is given in Table 7.2.

7.2.3 Phosphorus Deficiency

Applied P is converted into relatively unavailable forms in acidic soils. P deficiency is difficult to diagnose and severe yield decrease may often occur without showing leaf deficiency symptoms. Some of the major effects of P deficiency are as follows:

(i) Plant maturity and fruit set are delayed, stems are thin and woody, and growth is slower than normal,
(ii) Leaves are smaller and darker green, with a reddish-purple coloration on the undersides, and
(iii) Fibrous root formation is limited. The interference of Al with phosphorous causes phosphorus deficiency in vegetable crops grown on acid soils or in nutrient solutions (Ganeshamurthy et al. 2016a).

Table 7.1 Relative tolerance of crops to soil acidity

Crop	pH range	Crop	pH range
Maize, sorghum, wheat, barley	6.0–7.5	Groundnut	5.3–6.6
Millets	5.0–6.5	Sugarcane	6.0–7.5
Rice	4.0–6.0	Cotton	5.0–6.5
Oats	5.0–7.7	Potato	5.0–6.5
Field beans, soybean, pea, lentil, etc	5.5–7.0	Tea	4.0–6.0

Table 7.2 Tolerance of fruit and vegetable crops to soil acidity

Crop	Slightly tolerant pH 6.0–6.8	Moderately tolerant pH 5.5–6.8	Highly tolerant pH 5.0–6.8
Fruit crops	Mango, citrus, banana, Guava, papaya, cashew, apple, kiwi, peach	Pineapple, orange, litchi, passion fruit, Jack fruit	Strawberry, bel, gooseberry, plum
Vegetables	Beet, Broccoli, cabbage, cauliflower, celery, spinach, Chinese cabbage, leek	French bean, lima, bean, carrot, cucumber, brinjal, knol khol, pea, chilli, pumpkin, radish, summer squash, winter squash, tomato. colocasia	Potato, sweet potato, chow chow, yam, dioscoria

Source Ganeshamurthy et al. (2016a)

7.2.4 Correcting Soil Acidity

Liming

Agriculturally, liming is the addition to the soil of any calcium or calcium and magnesium-containing compound that is helpful in reducing soil acidity. The improvement of soil reaction with suitable amendments would assist in the conversion of certain unavailable forms of plant nutrients to available forms. Liming improves the soil's physical condition for proper plant growth. It also influences the activities of soil microorganisms which play an important role in the soil fertility. It is important, however, that only that much amount of amendment is applied which raises the pH to a suitable level when most of the nutrients are available in proper amounts for optimum crop growth. When the pH of such soils falls, the solubility of micronutrients and, most likely, aluminum (Al) rises to levels that are harmful to plants. Liming is not a cure for acidity in the soil, and it is not always cost-effective. As a result, the best technique for overcoming Al toxicity is to select or breed genotypes that are resistant to aluminum toxicity. Soil and water testing should be a part of any fertilizer management program to enhance overall production efficiency. Liming materials' effectiveness varies depending on their composition and the crops to which they are applied. Without integration, liming materials only move a small amount into the soil. By mixing all lime compounds into the rooting zone, tillage improves their effectiveness. Information about liming materials and their neutralizing values is given in Table 7.3.

Nutrient Management

Organic manures in combination with inorganic fertilizers and lime in appropriate ratios have proven to be a good solution for increasing the productivity and health of acid soils. By convention, a fertilizer's acidity is expressed as the amount of pure limestone necessary to neutralize the acidity produced by the fertilizer's reaction. The most frequent potassium compounds are neutral salts with no effect on the pH of the soil (Ganeshamurthy et al. 2016a, b).

Most of the acid soils in HKH region are still covered by primary or secondary forest. It is always necessary to consider short as well as long term costs and benefits and possible damage to the environment before any area under forest is to be put under plow. von Uexkiill and Bosshart (1989) opined that in future, we may require the acid soils more as a source of timber and a sink for carbon dioxide than for food production. Once the soils lose their fertile top layer, even with the best management, they cannot be returned to their former productivity level in less than ten years. There appear to be no easy and cost effectives alternatives to preserve and protect the present soils while still earning a good livelihood, now or in the future from these soils. It is a known fact that low-yield agriculture is neither profitable nor sustainable, because it results in soil degradation and many a time abandonment of the land. If the productivity, even with the use of recommended inputs has to be sustained, there is strong need that the present-day management of agricultural soils must protect them against erosion and leaching losses.

Table 7.3 Liming materials for applying to soils with their neutralizing values

Material	Formula	Amount of Material to be used to equal 1 ton of calcium carbonate (kg)	Neutralizing value %[a]
Calcium carbonate	$CaCO_3$	1000	100
Calcium-magnesium carbonate, dolomite	$CaCO_3{\cdot}MgCO_3$	840	109
Calcium oxide, burnt lime	CaO	500	179
Calcium hydroxide, hydrated lime	$Ca(OH)_2$	680	36
Calcium silicate, slag	$CaSiO_3$	1065	86
Magnesium carbonate	$MgCO_3$	760	119

[a]The higher the neutralizing value, the greater the amount of acidity that is neutralized per unit weight of material

Due to land clearing for agriculture use, land and environmental degradation have occurred in temperate regions. Most of the people do not think that major differences exist between the temperate and tropical situations (von Uexkiill and Bosshart 1989) due to the following reasons::

(a) Temperate and tropical regions have very different climate and soils;
(b) Tropics have more land and environmental degradation than the temperate regions;
(c) In a temperate forest ecosystem, organic carbon is roughly evenly distributed between soil and above-ground biomass, but in a tropical forest ecosystem, organic carbon is concentrated almost entirely in the biomass (over 85%) and very little in the soil; however, temperate forests contain more than 45% carbon (Fig. 7.1).

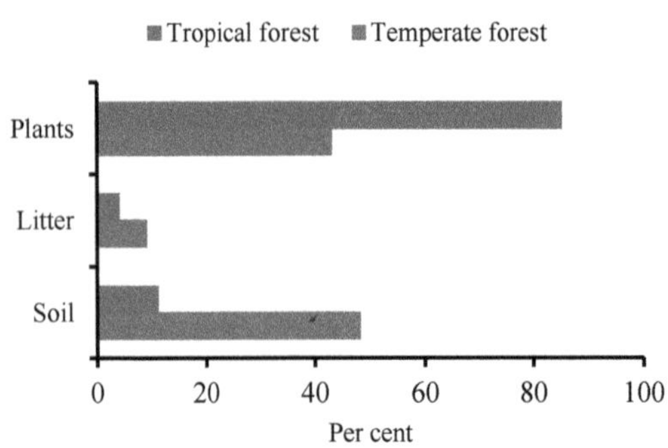

Fig. 7.1 Distribution of organic carbon in temperate and tropical forest ecosystem (Kira and Shidei 1967)

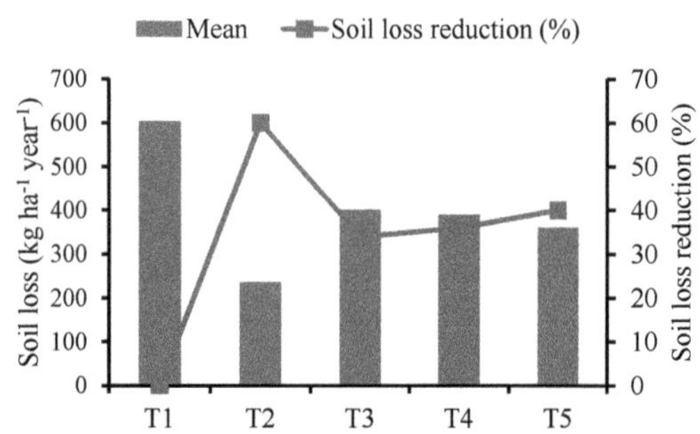

Fig. 7.2 Soil loss and its reduction under various treatments T1 = Farmers' practice (FP), T2, FP + Hedgerow (HR) (Dalbergia sissoo); T3, HR and without fertilizer; T4, FP + HR (Desmodium rensonii); T5, FP + HR + Fruit tree (Mango)

Combined application of lime and manure, applied at the rate of 2 t ha^{-1} each, increased the soil pH by 0.2 units (Schreier and Shah 2000; Schreier et al. 1995). Soil loss due to erosion and its reduction with various treatments is shown in Fig. 7.2.

7.2.5 Soil Fertility Declination in Middle Mountain

There has been soil productivity declination by 20–30% over the last 25 years in the middle mountains of Nepal due to the following reasons:

- Acidification of soil;
- Forest cutting;
- Decline in livestock numbers;

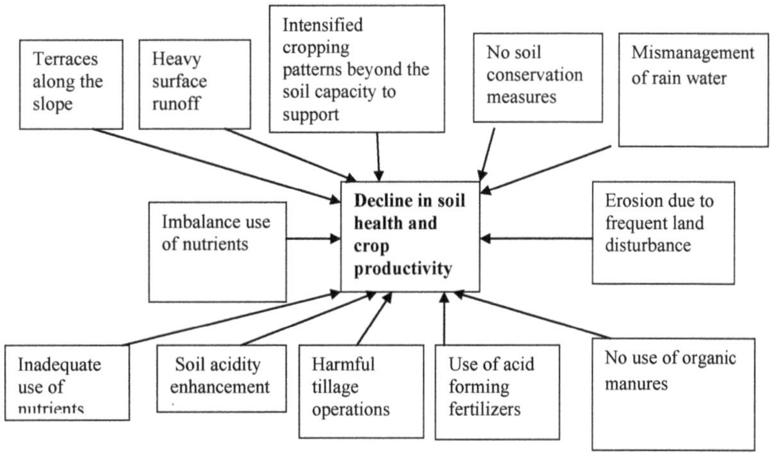

Fig. 7.3 Causes of poor soil health and crop productivity

- Intensifying crop production;
- Use of improved varieties;
- Lack of crop residue management;
- Shortage of organic manure;
- Lack of understanding of the need to address soil acidity.

Adoption of proper agronomic and soil management measures, as well as the application of lime. To reduce nutrient losses from the soil surface, soil erosion management methods should be implemented (Mg and Ca). Long-term soil fertility can be maintained by combining organic and inorganic fertilizers. To increase physico-chemical characteristics for sustainable agriculture productivity, integrated plant nutrient management (IPNM) is the best solution (Fig. 7.3). Threats to sustainable soil management in mountain areas are given in Fig. 7.4.

7.2.6 Threats to Sustainable Soil Management in Mountain Areas

Sustainable management and development of the mountains of HKH region is a challenging task because of a host of problems associated with this. Variability in altitudinal and climatic differences makes it all the more difficult in tackling various issues involved. Figure 7.4 highlights some major problems of soil management in the region.

7.2.7 Techniques for Managing Soil Fertility

Low-input agriculture depletes the soil of various soil nutrients, resulting in a reduction in

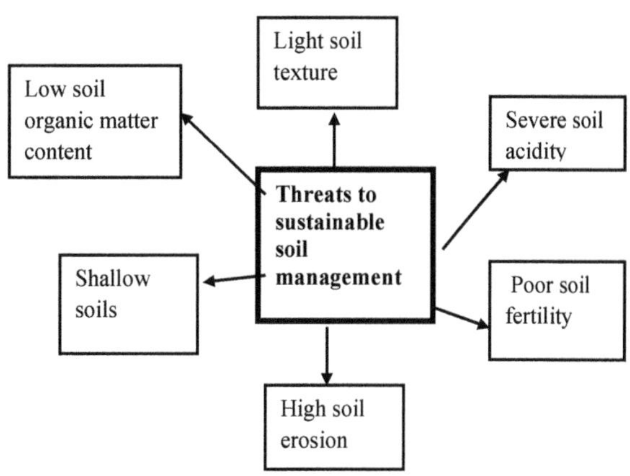

Fig. 7.4 Threats to sustainable soil management in mountain areas

agricultural production, resulting in non-sustainable farming systems. Mineral fertilizer, in combination with other treatments for improving soil health, has shown beneficial in restoring and strengthening soil fertility, as well as raising crop yields, when administered in appropriate amounts and with proper methods. Integrated soil fertility management is a strategy for increasing crop output, preventing soil degradation, and reducing nutrient loss by using both organic and inorganic plant fertilizers. Nutrient delivery via organic inputs such as compost, manure, and inorganic fertilizer, as well as the integration of nutrient-fixing crops, are the major factors. There are numerous benefits to combining organic and inorganic inputs in crop production. Tillage, on the other hand, should be avoided for long-term soil health benefits (FAO 2011a, b, c). Crop residues, animal manure, composting of organic wastes, or biological fixation through leguminous crops, green manures, or nitrogen-fixing trees can all be used to add organic sources of plant nutrition to soil. Crop residue recycling reduces losses, but it does not compensate for nutrients lost, nor does it increase the total quantity of nutrients available. Organic fertilizers must be used in conjunction with other nutrition sources in order to maximize crop yield (FAO 2011a, b, c, 2015a, b).

Legumes are important crops to be grown for enhancing biological nitrogen fixation. While the amounts of nitrogen fixed by legumes under experimental conditions and at farmer's level have been investigated, less data is available from the studies conducted on farmer fields. Furthermore, phosphorus deficit in the soil limits the effectiveness of nitrogen fixation. Farmers typically plant a variety of legumes for food, with the most prevalent crops being phaseolus beans, cowpeas, pigeon peas, groundnuts, and relay or mixed cropping (FAO 2011a, b, c, 2015a, b). Diversity in cropping systems lessens the detrimental impact of monocropping on soils and can aid in increasing soil health, soil quality, nutrient cycling, and biodiversity preservation. Agroforestry systems are known to have contributed to soil fertility. The partially degraded areas, not suitable for cereal and other crops, can be successfully used for agroforestry.

7.2.8 Major Soil Constraints to Crop Production

The number and severity of problems increases rapidly if soil has been put in different situations like left without cover, eroded, burnt, or has been compacted by machinery. The general problems in acid soils may include (von Uexkiill and Bosshart 1989): Major soil constraints in crop production are sensitivity to erosion, low pH, low microbial activity, low water-holding capacity, sensitivity to compaction by heavy machinery, low permeability to air, water, and roots (high soil strength), slow water infiltration rate, low base saturation percentage, high exchangeable Al concentration, high Al saturation percentage, low available P and high P fixation capacity, mineral complexes, high available Mn concentration, low and pH-dependent CEC, low concentrations of exchangeable Ca, Mg, and K, low organic matter, and absence of OM-clay.

7.2.9 Changing Tropical Soil Fertility Paradigms

Soils produce food, fiber, feed, and fuel as well as play a central role in determining the quality of environment (Sanchez 1997). Most of the technologies developed on the basis of understanding have focused on intensive agriculture systems where working soil fertility paradigm for years has been (Sanchez 1997): "*Overcome soil constraints through the application fertilizers and amendments to meet plant requirements*". Fertilizer research has primarily focused on nutrient intake by crops, with little attention paid to what happens to nutrients that accumulate in the soil or are lost to groundwater and/or the atmosphere. When pollution concerns were highlighted, the paradigm encompassed the filtering function. To tackle these issues, a second paradigm has eventually emerged (Sanchez 1994), which is "*Optimize biological efficiency; rely more on biological processes to optimize nutrient cycling, minimize external inputs and maximize the efficiency of their use*".

The application of second paradigm involves two dimensions, attaining sustainability in marginal lands prior to degradation and reclaiming already degraded lands. Efficient nutrient management is the underlying factor of both dimensions. Increase in soil fertility, stabilizing soil organic matter, provision of soil cover to reduce erosion, and improvement in soil physical condition are imperative for optimum crop growth. Therefore, the second paradigm needs special focus on nutrient management without neglecting soil and water conservation (Sanchez 1997).

Nutrient Deficiencies and Toxicities in Different Soil Groups

Acid soils cover approximately 30% of the world's ice-free land area, or 3.95 billion hectares. On acid soils, deficits of N, P, Ca, Mg, Mo, Zn, and toxicities of Al, Mn, Fe, and H elements are the most common limiting factors for plant growth (Baligar and Fageria 1997). Generally, Andosols are deficient in P, Ca, Mg, B and Mo and toxic in Al; Podsols are deficient in N, P, Ca and micronutrients and toxic in Al; Histosols are deficient in Si and Cu; Arenosols are deficient in K, Zn, Fe, Cu, and Mn; Gleysols are deficient in Mn and toxic in Fe and Mo; Vertisols are deficient in N, P, and Fe, but rich in S; Xerosols are deficient in Mg, K, P, Fe, and Zn but toxic in Na; and Solonchaks are toxic in B, Na, and Cl (Dudal 1976, Clark 1982).

Many of the mineral deficiencies and toxicities are related to soil parent material. The soil forming processes and climatic factors contribute to mineral stress problems. Acidifying impacts of acid-forming N fertilizers, crop removal of cations, increased cation leaching and runoff, and leguminous crops (N_2 fixation) have all contributed to a decrease in natural soil pH. (Baligar and Fageria 1997). The efficiency of nutrient acquisition, transport, and use by plants in soil is determined by two factors: (i) the soil's ability to deliver nutrients, and (ii) the plants' ability to absorb, use, and remobilize the nutrients. Supply, procurement, transportation, and use efficiency vary depending on soil, genotype of various crop varieties, and environmental conditions. Baglihar and Fageria (1997) reported that liming acid soils reduce Al, which helps to increase the ECEC, thus reducing the potential leaching losses of K. However, if liming materials are applied in forms other than carbonates or oxides, more leaching of K may occur because Ca and/or Mg may compete with K on the exchange sites. If the anion associated with cations is chloride or sulfates, K might be removed from soil (Baligar and Bennett 1986a). To minimize N losses, as NO_3-N, several slow-release forms of N fertilizers have been proposed. Hauck (1985) grouped slow-release N fertilizers as (i) limited solubility (metal ammonium phosphate), (ii) limited solubility but microbial degradable (urea forms, oxamides), (iii) water soluble that gradually decomposes (guanylurea), and (iv) materials with permeable or semi-permeable coating (polyolefin-coated fertilizers, POCF). Sulfur-coated urea is a promising form of N, especially for submerged rice fields (Baligar and Bennett 1986b).

Based on the external inputs applied (mineral fertilizers, machinery, pesticides), such agricultural systems are sub-divided into low external input agriculture (LEIA) and high external input agriculture (HEIA). When sustainability is included, these terms change to LEISA (low external input sustainable agriculture) and HEISA (high external input sustainable agriculture) (Keift 1993; Baligar and Fageria 1997). The LEIA is heterogenous with poor infrastructure, small scale, self-supporting, and labor intensive farming systems. In this case diversity is high, external inputs are minimum with less risk and the overall output is low to medium. Against this, the HEIA is heterogenous but with good infrastructure, large scale, market-oriented and capital intensive farming systems. The diversity is low but output is high fetching more income (Kieft 1993).

7.3 Integrated Plant Nutrient Management System (IPNMS)

The IPNMS was proposed by the United Nations Food and Agriculture Organization with the goal of restoring, maintaining, and improving soil fertility for long-term crop productivity by integrating all sources of plant nutrients from inorganic and organic sources, including biological N_2-fixation, under specific ecological, social, and economic conditions (Roy 1995; Zapata and Hera 1995). (Zapata and Hera 1995; Roy 1995). The IPNMS is defined as a comprehensive set of practices for manipulating the plant growth environment in order to provide critical nutrients to a crop in an appropriate amount and proportion for optimal production while preserving natural resources.

7.3.1 Integrated Plant Nutrient Management

There are several definitions for integrated plant nutrition management. It is the balanced application of organic and inorganic types of nutrients to give critical crop nutrients as needed for optimum productivity of a specific crop. In integrated nutrient management, all sources of plant nutrients are exploited, including crop wastes, green manure crops, manures, composts, and artificial fertilizers. Organic compounds, in addition to providing nutrients, have favorable effects on soil structure and water-holding capacity, soil chemical properties such as cation exchange capacity and soil microbial activity. Integrated nutrient management includes climate-smart agriculture, which comprises resilient, diversified, and adaptive multi-cropping systems.

7.3.2 Nutrient Use Efficiency in Acid Soils

Most important acid soil constraints can be high H^+ and Al^{3+} activities in soil solution, low soil cation exchange capacity, high capacity of soil solid phase to adsorb anions, especially phosphate ions, and low activities of the organic and biological soil fraction (Goedert et al. 1997).

The acid soils have some common properties or characteristics which are strongly related to the availability of soil nutrient to plants and to the condition of plant growth. The first property that affects this system is the level of H^+ in soil solution, which is high in acid soils. This situation is reflected by a low value of pH (usually between 4.5 and 5.5) and a high percentage of Al^{3+} saturation of the exchange complex. The nutrients applied through fertilizers go to soil solution and are then adsorbed on the soil solid phase. The nutrients from the soil solution are taken up by the growing crop and also lost through leaching (Goedert et al. 1997).

Sanchez (1976) mentioned that the magnitude of the soil CEC results from the mineral and organic colloids and pH of the soil. The clay fraction of Oxisols and Ultisols is usually determined by the sesquioxides, gibbsite, kaolinite, and intergrade minerals. These compounds have a low intrinsic quantity of negative charges and, therefore, most (85–95%) of the CEC of these soils depends on organic matter and is dependent of the soil solution pH value. At low pH, the number of negative charges developed at solid surface is low. As a consequence such soils exhibit a strong relationship between charge and pH. In some cases, these soils may exhibit net positive charge at low pH, which affect some nutrient availability (Sanchez 1976).

The nutrient use efficiency in acid soils is strongly dependent on soil management practices, especially those which aim to (i) decrease or neutralize the soil acidity, (ii) increase the CEC and decrease the capacity of irreversible adsorption of anions, (iii) minimize nutrient losses, mainly by leaching and volatilization, and (iv) monitor and manage the organic fraction and the biological activity. The main purpose of liming is to neutralize the exchangeable aluminum and increase the level of exchangeable calcium and magnesium, which is normally accomplished by raising the pH to 5.3 (Sousa et al. 1986).

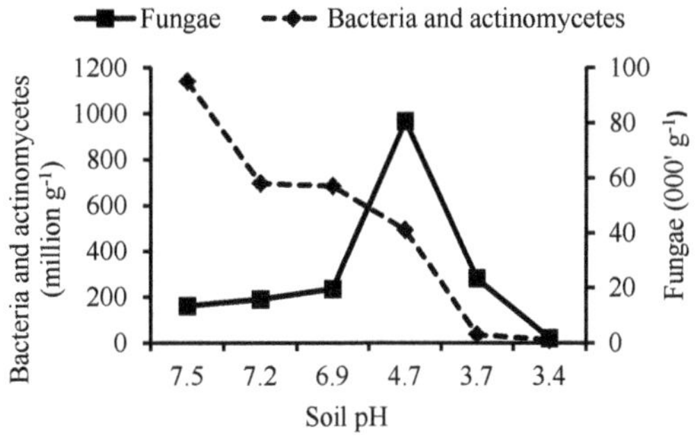

Fig. 7.5 Number of microorganisms as a function of soil pH (Goedert et al. 1997)

In acid tropical soils, the organic matter plays an important role in nutrient availability, because most of the soil exchange capacity is due to organic colloids. Furthermore, this exchange capacity is totally pH dependent, the CEC being very low at low pH. In addition, absorption of nutrient by plants is connected to the activity of soil organisms which show low activity at high soil acid conditions (Fig. 7.5).

7.3.3 Microbial Population and Activities in Soils

Soil is a key natural resource for the maintenance and sustainability of any ecosystem that needs to be managed effectively (Woomer and Swift 1994). Several strategies can help to achieve more sustainable cropping system (Siqueira and Moreira 1997):

- Selection of soil and suitable production system;
- Use of crop genotypes more tolerant to soil and climate limitations;
- Adoption of low-cost and environmentally sound soil management practices, such as minimum tillage and erosion control;
- Correct management of soil acidity, phosphate fertilization, and use of biocides;
- Maximize the use of biological processes such as nutrient mineralization and cycling, biological nitrogen fixation, mycorrhizal symbiosis, and biological control.

Microbial Populations

Soil acidity controls the soil microbial populations and regulates biomass and biological processes through either direct or indirect effects on microbial cells, microbial interactions or processes, therefore affecting soil quality and sustainability of the ecosystem (Siqueira and Moreira 1997). Detrimental effects caused either by low nutrient availability or metal toxicities are the major factors stressing soil microbial population in the soil (Foy 1984). Strain tolerance to low pH has been found for different groups of microorganisms, including rhizobia, actinomycetes, and arbuscules. In general, microorganisms exhibit tolerance to a wider range of pH in controlled environment than in nature. Critical pH values, in general, for growth of selected microorganisms and selected microbial processes as compiled by Siqueira et al. (1994), are shown in Table 7.4. While certain fungi and bacteria can grow at pH 3.0 or below, others, such as nitrifying bacteria, *Streptomyces scabies,* and *Glomus mossae*, are limited to soils above pH 5.5 (Siqueira and Moreira 1997).

Biological Nitrogen Fixation

Nitrogen can be taken up by the plants from inorganic, organic sources, fertilizers, manures, or biological N_2 fixation. Biological nitrogen fixation is done by a variety of N_2-fixing organisms, ranging from free living bacteria (including blue-green algae) to high evolved symbiotic systems such as rhizobia-legume associations (Siqueira and Moreira 1997). The soil acidity affects nodulation and BNF depends on acidity-related components of the soil, host, and bacterial tolerance to them. The relationships between microbial population and activities with soil sustainability are still a challenging question. It has been impossible to distinguish between functional and populational diversities in soil ecology. In part, this results from the complexity of the soil microbiota and its inter-relationships with the edaphic environment. Acidity is a major limiting ecological factor for the microorganisms in highly weathered soils, or that liming enhances their activity. Major liming induced biological

Table 7.4 Generalized critical pH values for selected microorganisms and processes

Processes	pH	Microorganism
Nitrification	>5.5	*Derxia, Azotobacter*
Nodulation and BNF		Bacteria in general
Potato scab		*Glomus mossae, Rizopogon*
		Stroptomyces scabies
Denitrification	4.0–5.5	*Rhizobium*
Arbuscular mycorrhizae		Majority of AM fungi
		Community structure
Decomposition	3.0–4.0	*Bradyrhizobium*
Ammonification		*Beijerinckia*
Soil respiration		*Ectomycorrhizal* fungi
Lactic fermentation	<3.0	*Pisolithus tinctorius*
Soil enzymes		*Thiobacillus spp*
Pyrite formation		*Acetobacter diazotrophicus* Oxidizing autotrophs
Sulfur oxidation		

Source Siqueira et al. (1994). (Given in Siqueira and Moreira 1997)

effects include enhancement of soil organic matter decomposition and root-microbial symbiosis.

About half of the mountain area in the World is not suitable for growing crops, rearing livestock or raising and managing forests. This condition has necessitated the mountain peoples to adopt diversified farming systems (Romeo et al. 2015). Many mountain people are practicing shifting cultivation or intensified farming to make the best use of soils, and many (nomads) are shifting their livestock to make the best use of pastures during different seasons, particularly summers and winters. Mountain people have developed strategies to live in harsh environment and imbibed the indigenous knowledge that take care of soil, vegetation, water, and crop husbandry. With change in land use in future, the degradation of mountain soil and vegetation cover is possible. The combined effect of steep slopes, undulating topography and alternating dry and wet spells requires suitable strategical measures to counter. Restoration of soil organic matter is needed on sloping lands to ensure good soil health for optimum productivity.

7.4 Soil Constraints Limiting Agricultural Productivity in Mountains

Unfavorable soil pH (<5.5 or >8.0) induces deficiencies of many macro and micronutrients like phosphorus, calcium, magnesium, copper, zinc, manganese, and iron. Moody and Cong (2008) mentioned that serpentine origin may show boron toxicity. Excess sodium in relation to other cations has been found to promote soil dispersion. Because the soil disperses when wet and hardens when dry, it has a negative impact on water infiltration. Sodicity and salinity are caused by rising saline water tables or seawater inundation. Crops incur significant water stress as a result of the high soluble salt concentration in the soil solution, which causes withering and possibly death. There are, however, a variety of crops that can survive soil salt to a certain extent. In acidic soils, the low ability of the soil to hold exchangeable Ca, Mg, and K is compounded by high exchangeable acidity (H^+ plus Al^{3+}), which reduces the fraction of basic cations on the soil's limited cation exchange sites even further.

In the HKH region, phosphorus deficit is a serious barrier to crop output, which is worsened in soils with high P fixing potential. Because it is strongly sorbed onto iron and aluminum oxyhydroxides in the soil, the availability of additional P is greatly decreased, resulting in a low soil solution P concentration. When soil is waterlogged, there is no air-filled pore space, limiting the amount of oxygen available to crop roots. Toxic ionic forms of iron and manganese may be reduced, limiting crop growth. Temporary water storages are formed on top of the compaction layer and lateral sub-surface drainage can occur. Silty soils are particularly susceptible to compaction.

The HKH Region is mostly mountainous and so highly subjected to soil erosion, landslides, landslips, and degradation. Erosion, beginning as sheet erosion, rills and then gullies form. Sheet erosion occurs even on low slopes if the soil is highly erodible. Susceptibility of all soils to erosion greatly increases with increase in slope, and there are large areas of soils in the Himalayan region on slopes greater than 50%.

7.5 Best Soil Management Practices (BSMP)

Climate-Smart Agriculture

Sound soil management practices include the long-term usage of soil in accordance with agricultural production methods. The soil is a net neutral or net positive sink for carbon sequestration and is not a significant source of nitrous oxide emissions, which is a critical performance indicator for climate-smart agriculture. Paustian (2016) enumerated some sound land management practices to satisfy this indicator, and are given as below:

- To achieve maximize soil nutrient use efficiency by applying nutrients according to soil potential and crop requirement. Applying nutrients in excess of crop demand results in (i) loss of nutrients through leaching or runoff, which may affect surface and groundwater quality and (ii) create soil nutrient imbalances that restrict crop growth and production;
- To prevent the loss of soil organic matter and nutrients, erosion menace needs to be controlled;
- Use of locally available organic materials as nutrient sources in combination with fertilizers and cultivation of leguminous crops to fix atmospheric N;
- Efficient management of soil water, including effective water harvesting and storage to mitigate drought;
- Combine conservation agriculture and organic agriculture principles into the farming system to maintain/improve soil health.

7.6 Indigenous Knowledge and Local Wisdom

Farmers in the HKH region have developed local indigenous knowledge for the maintenance of soil fertility due to their ingenuity and skill. For example, in Indian Himalayan Region farmers consider that black soils adjacent to hills are fertile, with a high water-holding capacity, and high organic matter content. These soils are good for crop production. Red and light brownish soils are most commonly described as poor soils because they are considered to be associated with low fertility, poor water retention, and lower yields. The presence of sand stones on the soil surface is also an indicator of poor soil. Sound soil management outcomes are highly dependent on implementing good land management practices because of complex soil–plant interaction. There is strong need to test various cropping systems under local agro-climatic conditions. Several cropping systems may show their sustainability and the productive capacity of the soil.

7.7 Mountain Agriculture

The hunger in the world has greatly reduced due to marked growth in food production in the last few decades (Godfray et al. 2010; FAO 2015a, b). This has drastically decreased the proportion of malnourished people at global level (Godfray et al. 2010). Hasler et al. (2017) explained that the population of the world, expected to stabilize at approximately 9 billion people by 2050, will face greater competition for land, water, and energy. To provide livelihood to such a huge population, the agricultural production needs to be significantly increased to a higher level.

Climate change, growing population, stagnation in crop productivity, and increasing food prices have highlighted the question of whether there will be enough food in the future to feed people, particularly in underdeveloped and developing countries. According to Kurvits et al. (2014), climate-related disasters are among the main drivers of food insecurity and markets have shown themselves to be very sensitive to recent extremes in climate.

One of the major causes of food insecurity in the HKH countries is widespread poverty which does not allow the poor farmers to take sound decision to improve agriculture by investing in quantity and quality of desired inputs. Many children are affected by malnutrition, and the rate of stunting is high across much of the region, with remote mountain areas having more impact. Women also suffer from higher levels of malnutrition as a result of the higher energy demands required from them but having poor food availability. Prediction indicates that the Himalayan glaciers will melt and release more water in near future, followed by a gradual decrease. It has been projected that more extreme weather conditions and increasing rainfall variability will lower agricultural productivity and production in the coming years. Achieving food security under climate change situation will require the following:

- Plugging knowledge gaps;
- Encouraging younger generations to choose farming as a profession;
- Small-scale farming needs to be diversified;
- More gender-sensitive farming approaches;
- Integrating food security and policies addressing climate change adaptation.

Food security in the Hindu Kush Himalayas is mostly determined by local production and household purchasing power. For food, the inhabitants are extremely reliant on local natural resources and subsistence farming. Poverty is prevalent in the HKH, which has a direct influence on their ability to produce and purchase food. While poverty may be higher in some areas of the plains, the difficulties in the mountains are more complicated. When poverty is present, people are left with few options for dealing with change, and their food security position can quickly deteriorate. Mountain societies have gradually moved away from subsistence farming and toward market-based agricultural output and cash crops, becoming more integrated into regional and global markets as a result of their desire to improve their lives (ICIMOD 2008). As a result of improved economic position and rapid population expansion, the world's food supply has been unable to keep up with growing demand. While global cereal demand has climbed gradually, global cereal output has leveled off, and global cereal consumption has been higher than production since 2000 (ICIMOD 2008).

In the Himalayan countries, undernourishment remains very high. According to China's Medium- to Long-Term Food Security Plan (2008–2020), (Government of the People's Republic of China 2008), 350 kg of cereal per capita are required annually to achieve minimum food requirements. In the production of rice, wheat, and corn, China considers itself to be mostly self-sufficient and food secure (Blas and Dyer 2009). Population increase, economic growth, rapid loss of agricultural areas due to urbanization, water and air pollution, and soil degradation have all been cited as important issues in terms of food security. Growth in the agricultural sector has been attributed to the extensive use of chemical fertilizers and pesticides, which are now the country's most significant contributors of water contamination. Rangeland deterioration and desertification are serious concerns in China's mountainous terrain. Grasslands cover 84 million hectares of the Tibetan plateau's 160 million hectares of rangeland. This area is rapidly deteriorating because of overgrazing and overcultivation, which is aggravated by climate change. According to Han et al. (2008), overgrazing in the Tibetan Plateau's arid temperate rangelands increased drastically from zero in 1990 to 30% in 1999 over a nine-year span. During the same time period, the amount of degraded land was projected to have increased slightly from 14 to 15%. The nomadic pastoralists who rely primarily on livestock husbandry on the Tibetan Plateau would lose their livelihoods as a result of these environmental impacts.

Twentith century witnessed the beginning of scientific commercial agriculture in the Himalayas. This was the only way to keep pace with increasing population in the Himalayan region. Fighting hunger required increase in the intensity of food production in mountain agroecosystem that leads to sustainable growth and equip the people with future challenges as a result of climate change (Shoemaker et al. 1961). If we fail in our efforts to increase agriculture production in fragile mountain ecosystems, the poverty will continue to grow. The possibility of growing more food required over next generations has to be achieved through increase in yield per unit of land. Shoemaker et al. (1961) stressed that the yield increase has to come via existing available technologies and those at the advance stage of breakthroughs. Adoption of the latest agricultural technologies would lead to economic development, improvement in living standard and social order. It will also solve the environmental problems that may appear as a result of cultivating land unsuited to crop production.

7.8 Pentagon of Livelihood Security in Mountain Ecosystem

To ensure livelihood security in the HKH region, five major issues need to be focused. The new challenges in the resolution of critical pentagon (Fig. 7.6) that fulfil the aspirations of the people are economic growth, the reduction of poverty, sustainable crop production, conservation of natural resources, and ecosystem health. Solutions may not be readily available in remote mountain areas to face the challenges. A comprehensive policy framework and attitude of stakeholders is necessary at various levels to achieve the objective of self-reliance. A focused and integrated approach is needed to accomplish the objectives.

Policy makers and the people working in the development departments in the HKH region are in a fix as how to increase agricultural productivity and reduce poverty without degradation and/or loss of the natural resources on which mountain farmers largely depend. The technological interventions by the extension people in the HKH for enhancing agriculture productivity have met with little success and were not very effective as were mostly focused on green revolution's technologies and services for plains. To attain this goal, it is vital to understand mountain agriculture's characteristics, specificities, socio-cultural and policy components, and niche potentials for leveraging comparative advantages to sustain livelihoods and safeguard the environment. For making long-term impacts, Rasul

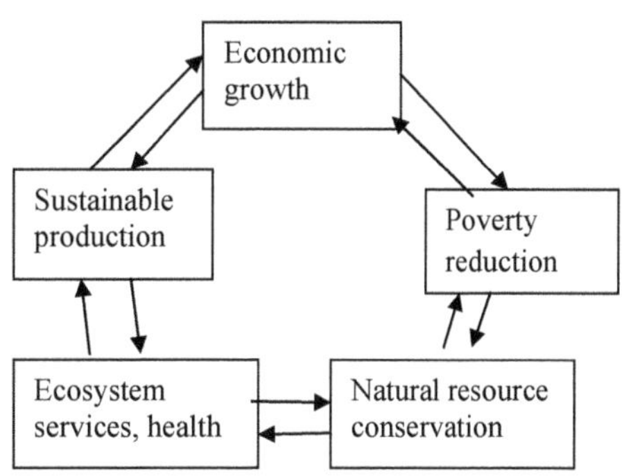

Fig. 7.6 Pentagon of livelihood security in mountain ecosystem

and Kollmair (2008) underlined the importance of analyzing essential issues, gaps in policy, programs, implementation strategy, demand-driven technology, and services.

7.9 Policy Issues in Mountain Agriculture

There is need to remove policy biases, provide security of tenure, remove policy and institutional constraints, provide inputs at doorsteps as the terrain is difficult, provision of requisite incentives and redefine support services to poor. In mountain areas, we are required to promote integrated programs like agricultural and rural income diversification-niche potentials, strengthen support services-participatory research, extension, credits, markets, develop sustainable physical and socio-economic infrastructure, understand mountain specificities, bio-physical and socio-economic conditions and mountain farmers' livelihood options.

Demand-driven, participatory efforts with adequate consideration of mountain specificities and niche potentials have been lacking in technological interventions in mountain research and development. While niche commodities such as fruits, vegetables, spices, medicines, NTFPs, honey, mushrooms, and other such enterprises have had some successful impacts through agricultural transformation in parts of India (Himachal), Nepal, Bhutan, China, and Pakistan, other successful impacts have been seen through adoption of niche commodities such as fruits, vegetables, spices, medicines, NTFPs, honey, mushrooms, and other such enterprises (Banskota 2006). Policy reforms are critical for guaranteeing food security in the HKH region since farmers are impoverished and have limited access to technologies, inputs, markets, and policy decisions.

Jodha (1995) reported a significant loss of natural resources of water, land, and forests that have affected people's livelihood in HKH region. People's livelihoods are heavily reliant on these resources. New agricultural technology must be introduced with care so that they do not harm the environment or product quality (Tulachan 2001) and can help to capture niche potential while preserving the ecosystem. Conservation agriculture, including soil, nutrient, and water management, is essential for long-term viability. Forest and agro-biodiversity conservation, as well as their wise use for the mountain community's economic and environmental benefit, offer a ray of hope for long-term sustainability. Aside from that, the indigenous knowledge of how to manage and use this important resource is another aspect of science and technology for research and development. The pullars and pushers of mountain agriculture and pathways to sustainable agriculture development (Rasul and Kollmair 2008) are presented in Fig. 7.7.

7.9.1 Food Security Challenges in HKH Countries

Rasul and Joshi (2011) reported about the wide range of challenges faced to achieve food security in HKH countries. They have appropriately identified the challenges in food security in different HKH countries (Table 7.5).

Nutrition is particularly poor in the HKH region (Rasul et al. 2016), and a considerable portion of the population suffers from food insecurity and hunger. In Pakistan, roughly 67% of the population lives in poverty in the mountains, compared to 40% in the plains; mountain areas have a food deficit of about 50%, compared to an 18% surplus in the plains. Food insecurity affects 60% of people in Nepal's mountains and 53% of people in the hills, compared to 48% in the plains. In India, rural mountain people consume 2,098 kcal/day per capita, less than the national average of 2,147 kcal/day; protein intake is also less than the national average. The statistics point towards high food insecurity in the mountain areas of the HKH region.

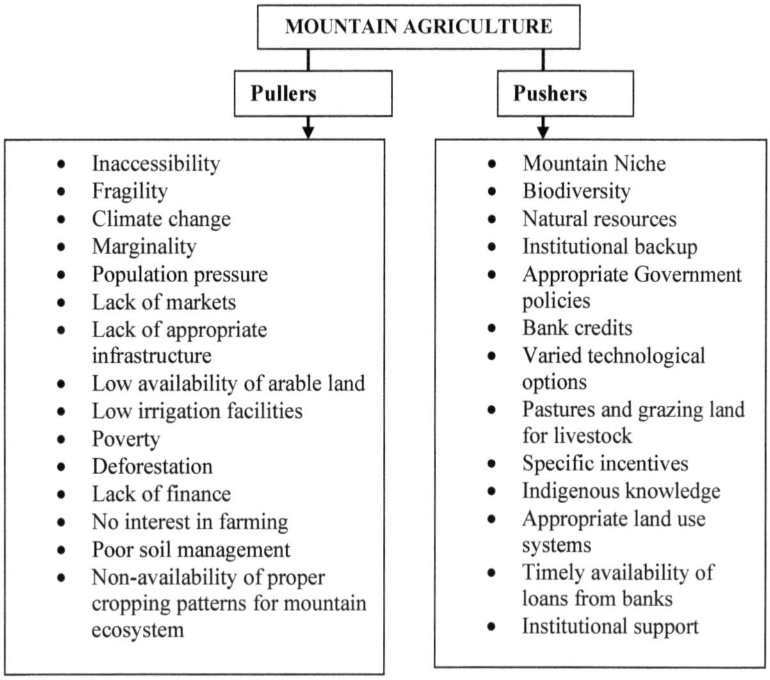

Fig. 7.7 Pullars and pushers of mountain agriculture (Modified over Rasul and Kollmair 2008)

Table 7.5 Food security in HKH countries (Modified over Rasul and Joshi 2011)

Country	Food security challenges
Afghanistan	Unstable environment, limited resources, financial constraints, high population growth, lack of proper food policy, lack of location-specific research on food crops relevant to the climatic and topographical conditions, lack of adequate and controlled irrigation facilities, etc.
Bangladesh	High rate of population growth, natural disasters (floods, cyclones), climate change impacts, limited per capita land resources, stagnation of rainwater in fields in many places, etc.
Bhutan	<5% of land is suitable for cultivation, whereas 80% of population depends on agriculture, rugged terrain, low crop productivity, major part of cultivable land remains fallow
India	Though India is self-sufficient in food grains, a decline in growth rate of production and stagnation in crop productivity is a matter of concern. Climate change impacts causing frequent floods, droughts, and natural disasters. For food security, growth rate in food production has to be higher than population growth. Heavy wastage of food and food products has to be avoided
Nepal	Food security is a concern in remote areas of Nepal, stagnation of crop productivity during about last three decades, marginal farmers and low-income people have been mostly affected
Pakistan	Foodgrains production is erratic and is not keeping pace with high population growth rate. About 10–15% population has less food than required and situation may further deteriorate. Non-availability of water for irrigation in certain traditional assured areas, and droughts are a common feature affecting foodgrains production

7.9.2 Change in Agricultural Land Use

The rapid socio-economic developments in the HKH are also impacting changes in agricultural methods, cropping systems, crop types, land use, and consumption patterns, according to Rasul et al (2016). Barley, sorghum, and millet were widely grown in Pakistan's Khyber Pakhtunkhwa province, but their area under cultivation has decreased by nearly half in recent years (Fig. 7.8). Between 2009 and 2013, India's sorghum production dropped by 43% (from 11.4 million tons to 6.4 million tons) (FAOSTAT 2013). The area under pulses in most of India's hilly regions, particularly Jammu and Kashmir, Himachal Pradesh, Sikkim, and Tripura, fell significantly in 2011/12 compared to 1990–91. (GoI 2013). The decrease in the production of these high-protein foods is likely to have an impact on food security. According to many nutritional studies, millet, sorghum, and buckwheat have a better nutritious value than white rice (Gopalan et al. 1989). The high calcium, dietary fiber, polyphenols, and protein content of millet sets it apart from other grains (Devi et al. 2011). It also includes more fat than maize, rice, or sorghum, as well as high amounts of essential amino acids, especially sulfur-containing amino acids (methionine and cysteine) (Obilana and Manyasa 2002). Millet is a good source of natural oil that is high in linoleic acid and tocopherols, according to several studies (Liang et al. 2010). The calcium and potassium content in finger millet is high compared to all other cereals (Gopalan et al. 2009) and total dietary fiber is higher than that of wheat, rice, maize, and sorghum (Shobana and Malleshi 2007). Sorghum is abundant in micronutrients including iron (more than 70 ppm) and zinc (more than 50 ppm), both of which are important for lowering anemia and stunting (ICRISAT 2011) and can aid with micronutrient deficiencies.

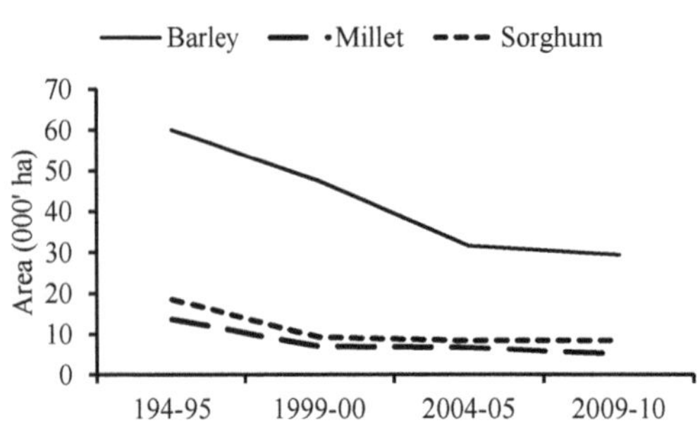

Fig. 7.8 Decline in area under cultivation of underutilized crops in KPK (Khyber Pakhtunkhwa province, Pakistan)

7.9.3 The Potential of Himalayan Traditional Crops

The rural poor in the Himalayan foothills face dwindling livelihood possibilities as the global food security base narrows. Underutilized crops have a particular place and a well-defined role in supplying food and nutritional security, according to both historical records of cropping patterns and evidence of dietary preferences of people in mountain and hill environments. Traditional crops, according to recent studies, play an essential role in people's lives, particularly for those living in marginalized areas and belonging to marginalized social groups (Giuliani 2007). Gruere et al. (2006) found that certain plant species can be locally plentiful and have a high local use, but their full potential has yet to be achieved. Lack of awareness of these species among producers and consumers, a lack of markets and/or marketing facilities, a lack of policies, and a lack of information on the uses and advantages are among the obstacles. Various experts, notably Maikhuri, have emphasized the importance of traditional crops and the necessity for concerted research efforts to help them develop (1996). Gupta et al (2013) mentioned about several traditional agricultural species in the Himalayas, such as Amaranth (*Amaranthus spp.*), quiona (*Chenopodium spp.*), buckwheat (*Fagopyrum esculentum*; *F. tataricum*) in pseudo cereals; sorghum (*Sorghum bicolor*), pearl millet (*Pennisetum glaucum*), barnyard millet (*Echinochloa utilis*), finger millet (*Eleusine coracana*), proso millet (*Panicum miliaceum*), little millet (*Panicum sumaternese*), kodo millet (*Paspalum setaceum*), foxtail millet (*Setaria italica*) in millets; sword bean (*Canavalia spp.*), hyacinth bean (*Lablab purpureus*), grass pea (*Lathyrus sativus*), horse gram (*Macrotyloma uniflorum*), velvet bean (*Mucuna spp.*), winged bean (*Psophocarpus tetragonolobus*), faba bean (*Vicia faba*), moth bean (*Vigna aconitifolia*), adzuki bean (*Vigna angularis*), rice bean (*Vigna umbellata*) in legumes and pulses; elephant foot yam (*Amorphophallus paeoniifolius*), taro (*Colocasia esculenta*), yams (*Dioscorea spp.*) in roots and tubers; Aibika (*Abelmoschus manihot*), leafy amaranth

(*Amaranthus spp.*), brassica spp., kangkong (*Ipomoea aquatica*) in vegetables.

Minor millet grown in the Himalayas is exceptionally rich in micronutrients, particularly iron and calcium, and is also high in dietary fiber. Pseudocereals have a well-balanced amino acid (protein) composition that is superior to that of conventional cereals, with a high amount of essential amino acids (Aubrecht and Biacs 2001). Amino acids lysine, which is found in insufficient proportions in maize, wheat, rice, and other cereals, as well as sulfur-containing amino acids, which are generally limited in legumes, are notably abundant in amaranth protein (Schoenlechner et al. 2010). Protein bioavailability in pseudocereals is also high, and it has been demonstrated in multiple studies to be superior to that of conventional cereals and comparable to animal protein quality (Gamel et al. 2004; Li and Zhang 2001). Millet and pseudo cereals are gluten-free cereal and thus suitable raw materials for the manufacturing of foods and beverages for people suffering from coeliac disease (Alvarez-Jubete et al. 2010).

7.10 Shifting Cultivation in HKH Region

Shifting cultivation, variously termed as "slash-and-burn agriculture", "swidden," and "rotational bush fallow agriculture," is called "jhum" in India. It is practiced in the wet tropics and covers ~30% of the world's uplands from Africa, Latin America, Oceania, and South and Southeast Asia (FAO 2010). Representing the dominant land use in the mountainous regions of South and Southeast Asia and being the oldest land use of the world, it involves clearing mountainous land by cutting the vegetation (tree or bushes), leaving the biomass in situ for drying, and finally burning it for production of charred material for soil fertility enrichment. Subsequently, multiple crops (annual and perennial) grown simultaneously on the same land until soil fertility are depleted, and farmers move to a new forest patch and repeat the cycle (Ramakrishnan 1992).

Because of the vast forests and limited people, these mountain villages were able to conduct a variety of shifting farming practises, allowing them to survive in relative harmony with their surroundings (Sharma and Bujarbaruah 2017). Shifting agriculture has thrived in the hills of north-eastern India for decades, owing to excellent energy efficiency (energy output/energy input) that ranged from 34 in a 30 year cycle to over 47 in 10 year and 5 year cycles (Toky and Ramakrishnan 1982). According to Sharma (1993a, b), the region has a great agricultural potential if appropriate, environmentally friendly, and socially acceptable land use techniques are adopted. Despite not being opposed to newer technology, farmers who practise shifting cultivation do not accept them for socio-economic reasons.

Soil degradation in the northeastern region is mostly caused by shifting agriculture (jhum), deforestation, incorrect methods of cultivation on hill slopes, road development, and other factors. Existing good techniques for preserving productivity without soil degradation in a few pockets of the region require further research in order to be expanded to other agro-climatic zones. The improved/enriched short fallows and combination of trees, shrubs with crop and livestock farming may be helpful in preserving the soil health as a promising improvements/alternatives to shifting cultivation. Soil degradation involves a number of interrelated processes, leading to deterioration of the chemical, physical and biological condition of soils and affecting the socio-economic pattern of the farming community (Theng 1991). Only those soil conservation practices as developed with the consent and cooperation of the local community, whose socio-economic aspirations receive due consideration, can only be accepted.

Shifting agriculture with a fallow period of more than 10 years is sustainable from ecological, economical, and energy efficiency points of views; it should continue. The germplasm of native crops should be preserved for a long time preferably by cryopreservation. It is necessary to document traditional knowledge of cultivation, resource management, and plant use. Home gardens, which contain a wide variety of native trees, shrubs, and herbs, should be bolstered. Species diversity of the region which is at present is threatened due to heavy pressure on the natural forests should be preserved for future generations. To take the pressure off the lands under short cycles of 5 years or so, the valley cultivation should be strengthened for cereal production. *Apatani* way of wet rice cultivation as prevalent in Arunachal Pradesh, effectively recycling the domestic wastes, should be strengthened in other areas also. Irrigation of crops during dry periods using native technology of bamboo irrigation should be used extensively. Short-cycle degraded lands should be repaired with cash-producing crops like fruit, spices, and short-rotation woods (Toky and Ramakrishnan 1983a).

7.10.1 Status and Extent

Shifting cultivation has been practiced in forests around the world for thousands of years, particularly in the tropics, where it provides a source of income for rural inhabitants. According to estimates (FAO/UNEP 1982), 500 million people out of a total agricultural population of 1200 million were dependent on shifting agriculture in 1980 (FAO/UNEP 1982). Approximate area under shifting cultivation in HKH Himalayas has been estimated to be 35,496 thousand ha with fallow periods varying from 3 to 30 years. The practice is found in almost every section of Bhutan, although it is most common in the dzongkhags (districts) of Zhemgang, Pema Gatshel, Samdrup Jongkhar, Mongar, Samtse, and Chukha (Wangchuck and Tashi 2004). In Bhutan, shifting cultivators are mostly indigenous communities (such as the Doyaps and Monpas), who have a thorough understanding of their surroundings and a deep regard for natural resources. Shifting agriculture is more common among indigenous populations in Nepal than among dominant populations, but it is still practiced by numerous cultural groups in the Chittagong Hills Tracts (CHT) of Bangladesh as a key source of subsistence. In Bangladesh, the study was limited to ten villages in the CHT's Rangamati and Bandarban districts. The CHT has an area of 13,184 km^2, with highlands accounting for 92%, medium highlands for 2%, medium lowlands for 1%, and homesteads and water bodies for 5%. The CHT has a total population of around 1.3 million people, 51% of whom are indigenous people from 11 ethnic groups. The Chakma, Marma, and Tripura ethnic groups are the most prominent. Mru, Tanchainghya, Bawan (or Bom), Khumi, Khyang, Lushai (or Mizo), Pankho, and Chak are other major ethnic groups (or Sak). The subtropical forest is found in the foothills, at elevations ranging from 200 to just over 1,000 masl, in places with 2,500–5,000 mm of annual rainfall. Warm broad-leaved forest is found at higher elevations (1,000–2,100 masl) and receives less rainfall (2,300–4,000 mm per year), with a mix of evergreen and deciduous broad-leaved plants.

Soil erosion of up to 130.2 t ha^{-1} has been documented in the north-eastern hill region due to shifting agriculture on slopes with a slope of 70%. Shifting cultivation entails the entire nexus of people's religious belief, attitude, self-image, and tribal identity, not just a set of agricultural activities. Because of these interconnections between many aspects and sectors of social life, growers are unable to abandon shifting cultivation until viable and acceptable alternatives are offered. The NE region loses 88.3 million tons of soil each year due to shifting cultivation, 90.7 million tons due to other agricultural lands, and 422.7 million tons due to non-agricultural land. Annual losses of N, P, K, Mn, Zn, Ca, and Mg are projected to be 686.0, 100.2, 511.0, 22.7, 14.0, 57.3, and 43.4 thousand tons, respectively (Sharma and Sharma 2003b).

A large variation in population density is observed in eight eastern Indian Himalayan states. The population census report of 2011 shows that the density of population was 340 persons km^{-2} in Assam as compared to 13 persons km^{-2} in Arunachal Pradesh. The lowest percentage of area affected by shifting cultivation is in Arunachal Pradesh (3.7% of GA) while the highest percentage of area is in Manipur (538% of GA) (Fig. 7.9). Families involved in shifting

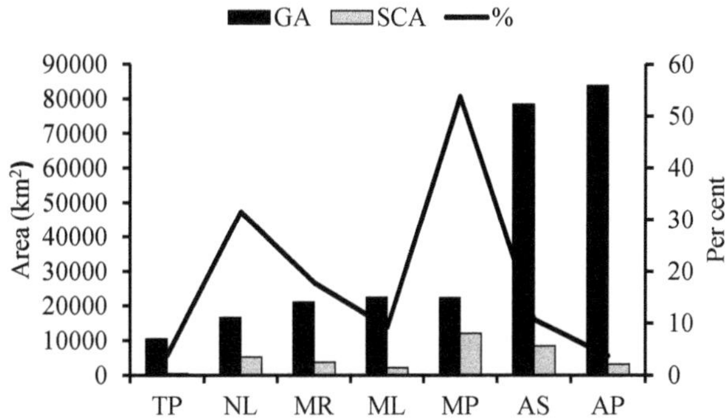

Fig. 7.9 Shifting cultivation area (SCA) as percent of geographical area (GA) in IEH states*. *AP, Arunachal Pradesh; AS, Assam; MP, Manipur; ML, Meghalaya; MR, Mizoram; NL Nagaland; SK, Sikkim; TP, Tripura

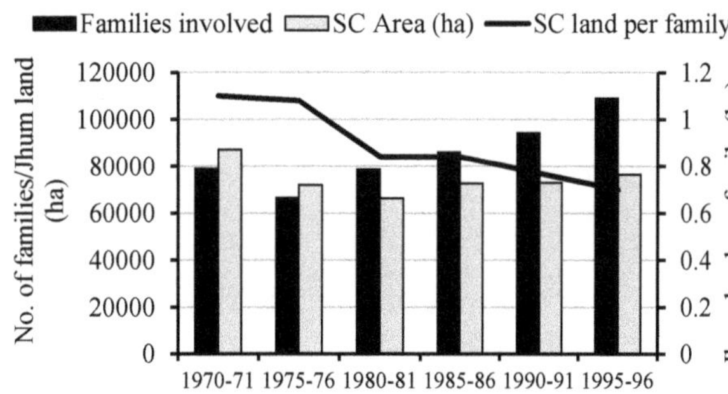

Fig. 7.10 Change in shifting cultivation area, families, and cultivated area per family in EIH

cultivation and the area under the practice is given in Fig. 7.10. The area per family is constantly decreasing, which has affected the food security in the region and the people involved are finding other avenues of income for sustenance. The area per family has decreased from about 1.1 ha to 0.74 between 1970–71 and 1995–96, and now (2018), it has decreased to 0.42 per family.

7.10.2 Social Security

There are as many as 65 or more ethnic groups in India practicing shifting cultivation. Various tribes practice shifting cultivation in mountain areas and their economy was mostly dependent on hunting, nomadism, pastoralism, shifting cultivation, and now, settled cultivation up to some extent. For a long time, the social sanctions and belief systems in these areas maintained a balance between resource potential and usage, but an imbalance has been established due to increased demographic pressure and indiscriminate use of natural resources. People have mismanaged and exploited natural resources to meet their requirements as the population has grown, putting pressure on the natural resources responsible for food production. There has been widespread land and environmental degradation as a result of manmade and natural factors such as shifting agriculture, land tenure systems, free range grazing, deforestation, and severe rains. To sustain the fast-growing population, there is strong need for integrated approach for natural resources management and livelihood security. The most important issue is to promote natural resource conservation and sustainable use while also being egalitarian and environmentally friendly (Sharma and Bujarbaruah 2017).

The shifting cultivation also provides full social security to the members of a clan. In earlier days of hunting and food gathering, the entire expedition used to share whatever was collected, though the brave and the leaders could get a major share, but those who were left behind such as old and infirm or women, had also a well-defined share in the collection (Sharma and Bujarbaruah 2017). The individual's welfare was the social responsibility of the society and it continues to be honored in shifting cultivation phase also. The clearing of forest is a community operation, where everybody contributes according to his ability. The old may only accompany the group and tender advice out of their past experience, while the youth may provide bulk of the physical labor. The old, infirm, and widows are given fields in the middle and the young on the periphery. Many operations in the shifting cultivation are community based, where everybody works according to his capacity but gets according to his needs. Thus, they are not dependent on their own children for their needs.

7.10.3 Transformations in Shifting Cultivation

Perhaps there is a misunderstanding that shifting cultivation is a subsistence, commercially unviable, and environmentally damaging practice. As a result, the HKH countries primarily pursued framed policies and promoted settled agricultural techniques such as wet rice production, horticulture, and cash crops as alternatives to shifting cultivation. Governments worked to foster the economic development of communities that practiced such agriculture, making it easier for them to integrate into the mainstream of economic development. This process of assimilation has also resulted in an accelerated penetration of market forces into areas under shifting cultivation, often at rates much faster than that of agricultural transformations, making cash generation essential for shifting cultivators in order to keep pace and avoid further marginalization and deprivation. Development changes and the effect of market assimilation have also simultaneously fuelled aspirations of such communities to enhance their purchasing power and participate in market economies, consequently making them desperately seek options for change (Leduc and Choudhury 2012).

7.10.4 Soil Properties

Effects of Shifting Cultivation on Soil Nutrient Status

Shifting cultivation causes loss of flora and fauna, including rare species of plants, shrubs, medicinal plants, and minor forest products. The practice results in deforestation, drying-up of springs, wild animals' shelter, flooding, soil erosion, siltation of riverbeds, and ecological imbalance. There are changes in physical, chemical, and biological properties of the soil after burning. Immediately after burning, the soil properties undergo a complete change. The pH rises initially, favoring the release of soil nutrients for the benefit of the succeeding crop (Table 7.6). There is substantial increase in organic carbon and, available P and K. However, this increase depends on the amount of vegetation burnt (Sharma and Bujarbaruah 2016). Therefore, sufficient time has to be given between two cropping phases to help in rejuvenation of sufficient vegetation for the success of the practice. Much of the microbial population may perish during burning due to intense heat and high temperature in the surface layers of soil. Some species of microbes may resist this heat. The perished microbes are likely to be there in deep layers of the soil and may re-establish in the surface soil with the passage of time. Detailed studies are required on this aspect.

Table 7.6 Changes in soil properties after burning of vegetation

Stage	pH	OC (%)	Av. P (kg ha^{-1})	Av. K (kg ha^{-1})	Exch. Ca (me/100 g)
Before burning	4.9	0.51	5.2	185	5.6
After burning	5.6	0.82	9.6	530	7.2

7.10.5 Soil Carbon

Shifting agriculture at lower elevations involves high-intensity burn depending upon quantity of slash under different length of fallow period from 5 years to sometimes 30 years. The severity of the burn has a major impact on soil qualities such as organic carbon and pH. In comparison to medium and longer cycles, the short jhum cycle of 5 years starts with a low carbon budget (Ramakrishnan and Toky 1981). Due to poor litter return (1.2 ton/ha/year) and accelerated oxidation at the surface soil, soil carbon depletion persists for up to 22% during the first year of cropping and continues for up to 5 years during the early stages of succession. One of the strongest arguments against a short cycle is this. The improvement of soil carbon is maximum at 10 year age fallow due to production of high litter fall (7.1 ton/ha/year), of which bamboos (*Dendrocalamus hamiltonii*) are the main contributor (Toky and Ramakrishnan 1983a). This is one of the reasons that soil fertility is replenished in a fallow of 10 year which is the optimum cycle for the yields of cereals and legume against a short cycle of 5 year where the yields of rice and legumes are much reduced. Higher level of organic carbon maintained after one year of cropping in medium and longer cycles may be able to support another year of cropping. However, there was a net loss of carbon throughout cropping phases and fallow phases up to 10 years irrespective of one or two years of cropping or the types of fallows (Ramakrishnan and Toky 1981). The keystone tree species such as bamboos (*Dendrocalamus hamiltonii*) in the lower elevations and alders (*Alnus nepalensis*) in the higher elevations are extremely important for conservation of potassium and fixation of atmospheric nitrogen in *jhum* fallows. In some parts of Eastern Himalayas, around Shillong, the farmers are not burning the slash completely. The slash of branches and needles of *Pinus kesiya* trees and other herbaceous weeds are pulled in rows, the soil is put on the slash and then it is slowly charred to retain maximum carbon in the soil. Instead of slash-and-burn agriculture, it is slash and char. They also add some inorganic fertilizers to raise the potato crops. These are more and less permanent systems (Mishra and Ramakrishnan 1983).

7.11 Rangelands in HKH Region

Nearly 60% of the Hindu Kush Himalayas is covered by rangelands (Miller 1996). They are vital to country economies, environmental protection, and cultural legacy. According to Wilkes (2008), roughly half of the Tibetan Plateau grasslands have been degraded. Rangeland resource management and suitable use are critical for long-term sustainability. Most pastoralists are impoverished and reliant on rangeland resources in general. Traditional management practices were sustainable, but increased pressure on land development policies is causing significant tracts of rangeland to deteriorate.

The distribution of different rangeland types in the region is presented in Table 7.7.

Pastoralism in the HKH region is under severe strain due to a fourfold increase in human

Table 7.7 Extent of rangelands within the HKH portion of the countries of the HKH Region

Country	Area (km^2)	Area %
Afghanistan	291,880	7.57
Bhutan	17,419	0.45
China	1,545,542	40.09
India	169,381	4.39
Nepal	77,826	2.02
Pakistan	188,118	4.88
Total area (%)	2,290,169	59.41

population and a doubling of animal population during the last four decades. As a result of rising human and animal populations. Heterogeneous rangeland ecosystems, as well as their integrity, are critical for providing services to far-flung settlements.

7.11.1 Ecosystem Services

The key biological functions of rangelands in HKH Region are fodder for domestic livestock, food for human, biofuels, fiber, fishing, hunting, biochemicals, and genetic material. The major hydrological functions are drinking water, water for irrigation, flood mitigation water bodies for recreation and tourism, C sequestration, clean energy, wind, and hydropower. The other functions of rangelands include cultural, spiritual, tourism, historical and scientific studies.

7.11.2 Carbon Stocks in Range Lands

Litter pools account for a modest portion of the total carbon store of rangelands. On the Tibetan Plateau, there is a large disparity in carbon stocks between degraded and non-degraded grasslands. Low vegetation cover and biomass are common in degraded grasslands. Increased vegetation cover will enhance organic matter inputs into grassland soils and limit soil respiration, allowing carbon to be sequestered in the soil. FRMD (2017) reported that soil carbon stocks in lightly degraded, medium degraded, and severely degraded rangelands of alpine grasslands in Tibetan plateau varied from112.7 to 130.4 t C ha^{-1}, 43.8 to 56.1 t C ha^{-1}, and 33.8 to 54.9 t C ha^{-1}, respectively, which includes above and below ground biomass carbon also. The mean soil C alone was 115.1 t C ha^{-1}, 44.6 t C ha^{-1}, and 40.3 t C ha^{-1} in lightly degraded, medium degraded, and severely degraded rangelands of Tibetan plateau.

Country-wise information in HKH region regarding Land use and Management is mentioned hereunder.

7.12 Afghanistan

The current population of Afghanistan is 37,090,897 as of 2019, based on the latest United Nations estimates, which is equivalent to 0.48% of the population of the World. The total land area of the country is 652237 km^2, thus having about 57 persons per km^2.

7.12.1 People and Farms

Farm Households and Rural Population

Estimates of approximate number of farm households (Table 7.8) and rural population have been made and documented by FAO (2003a, b). It refers to the total settled rural population in

Table 7.8 Rural farming households and population in Afghanistan

Agro-ecological zone	Farm households	% of farm households	Farm population	% of farm population
Badakhshan mountains	35,346	3.3	416,274	3.4
Central mountains	167,168	15.7	1940,369	16.1
Eastern mountains	177,322	16.6	2024,525	16.7
Southern mountains	79,426	7.5	1070,265	8.9
Northern mountains	281,308	26.4	3006,333	24.8
Total mountain areas	740,570	69.5	8457,766	69.7
Plains, lowlands, etc	324,954	30.5	3646,198	30.3
Total	1065,524	100.0	1210,3964	100.0

Source FAO (2003a, b)

Table 7.9 Total arable land (irrigated or rainfed) in Afghanistan (ha)

Agro-ecological zone	Arable land	Irrigated	Rainfed	% Irrigated
Badakhshan mountains	374,900	78,154	296,746	20.8
Central mountains	522,797	274,135	248,662	52.4
Eastern mountains	245,571	226,458	19,112	92.2
Southern mountains	607,309	532,078	75,231	87.6
Northern mountains	2,823,010	368,667	2,454,343	13.0
Turkistan plains	789,777	589,958	199,819	74.7
Herat-Farah lowlands	445,957	392,716	53,240	88.0
Helmand River valley	727,650	683,206	44,444	93.9
Total	6,536,971	3,145,373	3,391,597	48.1

Source FAO (2003a, b)

households engaged in agriculture (growing crops, rearing livestock, pisciculture, etc.), regardless of land ownership. The number of such households was estimated at about 1.06 million, with a total population of about 12.1 million people or more than 11 persons per household.

7.12.2 Land in Agriculture

A survey was done to determine the total area of irrigated and rainfed arable land in the farms. All farmers were asked about the overall area under their control, which included land they owned or leased from others under a tenancy agreement. The total agricultural land in different agro-ecological zones is given in Table 7.9 and, percent area under different sizes holding in Fig. 7.11.

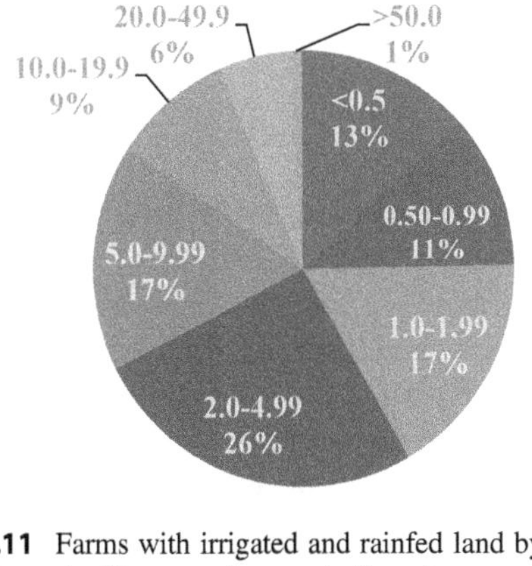

Fig. 7.11 Farms with irrigated and rainfed land by farm size (Percent of farms and percent of area)

7.12.3 Land Tenure and Ownership

Due to a long period of war and political instability, land ownership and access rights in Afghanistan are very unique, but also complex. According to a recent study of land rights in Afghanistan, "At any given time, a single farmer may be owner, tenant, sharecropper and mortgagor—and may be in transition from one status to another with respect to one or more of his plots" (Alden Wily 2003). In Afghanistan, private land ownership implies full ownership of the soil, but not the minerals found in, under or upon it. This is a form of tenure that which can be called as "freehold". At household, clan and community levels, part-ownership or shared ownership also exists with respect to both arable and non-arable pastureland. Land leases are allowed under the civil code, however formal leases appear to be uncommon. Tenancy, in the form of renting, crop-sharing, and/or mortgaging, is such an essential part of rural land access in Afghanistan that "tenancy" is often described as the prevalent "landholding" norm (Alden Wily 2003).

7.12.4 Importance and Constraints in Agriculture

In Afghanistan, after the service sector, agriculture is the key productive sector, contributing 31.4% of GDP in 2009/2010 and is one of the main pillars of the economy. According to 2009 estimates, about 76% of people in the country live in rural areas and 78% of the active population is employed in agriculture. Afghanistan has about 55% of its households engaged in farming, and 68% of household have some type of livestock (FAO 2012). Agriculture still plays a key role in the livelihood of the country's population, even though opportunities for the rural youth are inadequate. FAO (2012) remarked that farming and pastoral activities are often the only feasible alternative to unemployment in the country. During 2010–11, dried fruits and nuts made up to 27% of total exports, while fresh fruit accounted for 7.2%. Raisins are the number one agricultural export, followed by almond, pistachio, and walnuts. The agricultural production and livestock population decreased by 3.5% and 1.7% a year between 1978 and 2004, respectively. Agriculture growth is dependent on weather fluctuations in Afghanistan, coupled with political situation.

In Afghanistan, people rely on cereal eating, with wheat being the most common. The country has never been self-sufficient, with self-sufficiency ranging from 53% in 2004 to 96.6% in 2009, when the country benefited from extraordinarily good weather (FAO 2012). As a landlocked country with seasonal water flows and irrigation water diversion, fisheries do not make a significant contribution to the economy, yet they can be evaluated in the context of water use and withdrawals from a biodiversity standpoint. Major constraints experienced in agriculture are as follows:

- **Limited arable land**: Only around 12–15% of the country's overall geographical area is arable land, whereas 45% is "rangeland". Irrigated land accounts for 3.2 million ha of arable land (although one-third is left fallow due to water scarcity), whereas non-irrigated land accounts for 6.1 million ha (rainfed). Irrigated land is used to cultivate about 80% of agricultural crops (FAO 2012).
- **Limited access to water**: The average annual rainfall in Afghanistan is 300 mm. When there is no need for irrigation, almost 70% of the rainfall falls throughout the winter. Snowfall is the most common kind of precipitation at high altitudes, and it serves as a natural reserve to partially meet crop water requirements in lowlands throughout the growth stage of crops. In semi-arid locations, reliable and timely water delivery is a major barrier to agricultural output and food security. The situation has deteriorated as a result of recent climate change, deforestation, and other factors. Afghanistan uses only approximately 30% of the 57 billion m^3 annual surface water resources available for agriculture, and it does so inefficiently. Traditional irrigation techniques waste a lot of water.
- **Low productivity**: Even after several improvement initiatives, low agriculture, livestock, and forestry yield is widespread. Small per-capita holdings, insufficient regulating water resources infrastructures, poor irrigation and water management, inadequate land management, inadequate land preparation, low use of improved seed, fertilizer, and pesticides, limited crop diversification, and low input quality are all major constraints.
- **Poor post-harvest operations and lack of infrastructure**: Production issues are caused by improper handling, threshing, cleaning, and storage; a lack of skills for quality processing; poor quality control; and restricted transportation facilities.
- **Shortfalls in agricultural product commercialization:** They include limited value addition, a lack of business planning, management, and marketing skills, weak links between farmers/producers, wholesalers, and consumers or exporters, limited access to and use of market information, and insufficient access to credit and financial services, all of which contribute to farmers' low income.

7.12.5 Food Insecurity and Poverty

Afghanistan is a food insecure country due to insufficient production to meet the demand and lack of infrastructure to fully safeguard and judiciously distribute whatever is produced. According to FAO (2012), there is widespread unequal distribution of grain consumption and poverty; the wealthy 20% of the population consumes 39% of overall consumption, while the poor consume only 9% of total consumption. More than a third of the country's 34 million people live in poverty, unable to meet their basic needs and nutritional requirements. According to the findings of a fast nutrition assessment conducted in 2008, global acute malnutrition affects 16.7% of the population, with 4.7% of those suffering from severe acute malnutrition. This is higher than the WHO's emergency threshold of 15% for acute malnutrition, which is used to determine the severity of a country's nutritional difficulties.

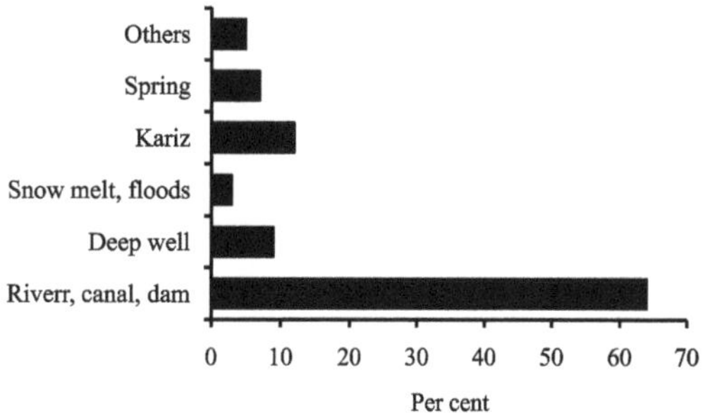

Fig. 7.12 Irrigation percent through different sources

7.12.6 Agriculture and Livestock

Agriculture, notably livestock, is the backbone of Afghanistan's economy, as it provides the primary source of income for about half of the country's families. Agriculture employs 40% of the workforce, making it the most important industry in the country. With a diverse geology and topography, only around 12% of the entire land area is arable, 3% is forested, 46% is under permanent pasture, and 39% is mountainous and unsuitable for cultivation (CSO 2012).

7.12.7 Irrigated Land

Irrigated Land Tenure

About 1.4 million households or approximately 38% of all households in Afghanistan own irrigated land. The majority of these landowners (58%) have a farm size of less than 0.8 ha, whereas the median and mean irrigated land size is, respectively, 1.2 and 0.6 ha. Compared to 2007–08, the share of landowners with small landholdings (0.8 ha) has increased.

7.12.8 Crop Production from Irrigated Land

Wheat is the most important crop produced on irrigated land. The percent of area under irrigation from different sources is given in Fig. 7.12. About 60% of the area under irrigation is served through canals/river water. According to the households reporting in the National Risk and Vulnerability Assessment (NRVA) 2011–12, 2.1 million metric tons of wheat was harvested in the spring cultivation season of both years (Table 7.10).

7.12.9 Livestock

The cattle industry is a vital part of Afghanistan's economy. Table 7.11 displays the findings of the 2002–03 census and (National Risk and

Table 7.10 Crop production from irrigated land in spring cultivation season prior to the survey (2012)

Crop	Production (MT)	Crop	Production (MT)	Crop	Production (MT)
Total cereals	2,393,223	Potatoes	156,711	Beans	13,400
Wheat	2,090,587	Fodder	328,719	Fruit & nuts	5882
Rice	187,623*	Cotton	145,831	Opium	3434
Maize/sorghum	153,204	Onions	84,421	Vegetables	51,222
Barley	61,295	Watermelon	55,813	Other crops	27,100
Millet	2935	Tomato	52,366	–	–

Table 7.11 Livestock numbers in Livestock Census 2002–03 and NRVA 2011–12 ('000)

Type of animal	Census (2002–03)	NRVA (2011–12)
Cattle	3715	2854
Oxen/yaks	NA	474
Horses	142	102
Donkeys	1588	15,191
Camels	175	481
Goats	7281	10,445
Sheep	8772	18,018
Chickens	12,156	13,176
Other poultry	1022	1367

Vulnerability Assessment (NRVA)-based estimates on various types of cattle in 2011–12. According to the NRVA data, the number of cattle has decreased in the years since the livestock census. The number of small ruminants, particularly sheep, has increased, though. In addition, according to the NRVA 2011–12 data, 39% of Afghan households possess one or more cattle. Households with goats and sheep account for 29 and 31% of all households, respectively.

7.12.10 Crops

Grown on 2.5 to 3.0 million ha, wheat is the most important crop that is sown almost throughout Afghanistan. Variation in the precipitation received and drought conditions impact the annual area under wheat. Irrigated wheat covers about 1.2 million hectares, while the rest is rainfed. The majority (almost 90%) of this wheat is planted in the fall, with barely 10% planted in the spring. Maize, rice (both cultivated on around 300,000 hectares), cotton, potatoes, sorghum, and vegetables such as cucumbers, grapes, melons, apples, pomegranates, apricots, almonds, and walnuts are the next most important crops (Roy 2003).

7.12.11 Fertilizers

Fertilizer use in Afghanistan is going on since 1962 with the import of 5,000 tons of ammonium nitrate for crops like cotton and sugar beets grown in Konduz, Takhar, and Baghlan (IFDC 2002). Starting with the Ministry of Agriculture with importing and distributing fertilizers, a company namely 'Afghan Fertilizer Company (AFC)' was later created to handle this business. This company was renamed 'Afghan Fertilizer and Agricultural Services Enterprise (AFASE)' after a few years. This firm imported, stored, and distributed fertilizers across Afghanistan, selling to farmers directly as well as through licensed dealers and commission agents. In 1974, a urea plant was developed in Mazar-e-Sharif with Soviet aid and using local gas. This factory had a 105 thousand ton capacity. Over 60% of fertilizer is used on wheat, 15% on cotton, and the remaining 25% on grapes, potatoes, rice, vegetables, sorghum, and other crops, according to estimates.

In 2001/02, Afghanistan consumed roughly 170,000 tons of fertilizer, with urea accounting for more than 80% of that. Diammonium phosphate (DAP), nitrophosphate (NP), single superphosphate (SSP), calcium ammonium nitrate (CAN), and ammonium nitrate (AN) were among

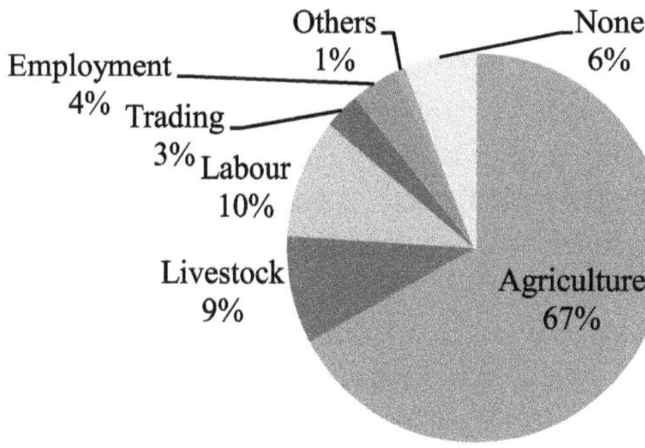

Fig. 7.13 Main sources of household income

the other fertilizers used (AN). The percent income from different sources in households is given in Fig. 7.13.

Optimum rainfall and water management, availability of high-quality seeds, high-yielding crop varieties, and best management practices are helpful in the increase of yield of different crops in Afghanistan (FAO 2003a, b). The trend in the yield of wheat, rice, and maize from 1990 to 2002 is given in Fig. 7.14 (FAO 2002).

7.12.12 Threats to the Environment and Natural Resource Management

Afghanistan is a semi-arid landlocked country with mountainous terrain dividing the north's fertile plains from the central plateau and southern lowlands, as well as deserts. Drought, floods, land erosion, top fertile soil erosion, and earthquakes; insufficient stabilization of national vegetation due to destruction of riparian vegetation and subsequent riverbank erosion; mismanagement of natural resources; massive destruction of forests leading to erosion and desertification; overgrazing in rangelands and conversion of rangeland to rainfed agriculture; biodiversity losses; mismanagement of natural resources. The environment has been harmed by the rapid depletion of natural resources and their long-term misuse. Natural resource environment in the Afghanistan are facing threat from extreme flood events, an additional danger.

7.12.13 Range Lands of Afghanistan

About 30 million hectares area (roughly 45% of the country's geographical area) is covered by rangelands in Afghanistan. The total area suitable for grazing (including marginal lands) is estimated at 70–85% of the total land area. It provides habitat and forage for nearly 35 million livestock as well as numerous wild animals (Joshi et al. 2013). They reported that over the last three decades, although livestock production remains the "backbone of Afghanistan agriculture" and "crucial for the country's socio-economic growth," the number of sheep and goats in Afghanistan has decreased from more than 30 million to around 16 million. Kuchi pastoralists, who account for 20% of the rural population, rely heavily on rangelands. Biomass is used for fuel and forage in the rangelands. Livestock and wildlife use the rangelands for fodder and habitat. The biophysical functions of

Fig. 7.14 Food crop productivity in Afghanistan. *Source* FAO (2002)

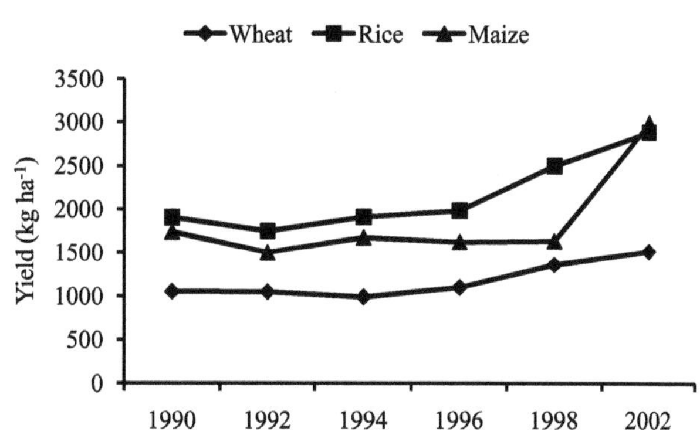

Afghanistan's rangelands are vital as a watershed, a soil erosion control mechanism, a carbon sink, and a wildlife home. Building materials, fruits and nuts, meat from wild animals, and medicinal plants are some of the other socioeconomic productive functions of Afghanistan's rangelands.

Human activities including grazing, agriculture, and deforestation have a significant impact on forest vegetation. Reduced pressure on already deteriorated natural pastures, according to Ali and Shaoling (2013), is critical for the rangeland ecosystem's recovery and vitality. In locations where there is a potential for fodder cultivation, drought-resistant plants should be cultivated. The need for fuelwood in the target area is enormous, and increasing and diversifying rural energy and improving energy efficiency in local communities have the potential to supply a significant portion of this demand. Multipurpose forestry, which supplies fuelwood, timber, soil and water conservation, bank stabilization, and wind barriers, has the potential to be very profitable.

7.12.14 Agriculture and Rural Development in Afghanistan

Agriculture and rural development have been designated as a key priority area by the Afghan government in its development agenda. USAID has asked the International Center for Research in the Dry Areas (ICARDA) to organize and supply high-quality seed and other technical support to help Afghanistan restore its agricultural productive capacity and food security (ICARDA 2002).

7.12.15 Irrigation in Afghanistan

Because of the poor and irregular rainfall, agricultural cultivation in most of the country is impossible without irrigation. The Afghan people strictly observe the customs and traditions of their villages when it comes to water and land allocation. Irrigation systems in Afghanistan can be categorized into two types, according to ICARDA (2002): (i) traditional irrigation systems and (ii) contemporary irrigation systems. Water is provided by stream flow, which is diverted with the use of temporary weirs, which are frequently positioned in remote valleys along a stream or river. Water rights are decided and acknowledged in a similar fashion in these systems, which are built and maintained in a traditional informal manner on a community basis. These water management systems are mainly located in the plains along the main river valleys. The villages located in the vicinity can share water from such a system. According to 1981 water regulations, the amount of water required for irrigation is determined by the area under cultivation, the type of crop, the irrigation regime, the water rights document, local practices, and the amount of water in the source. Water is distributed according to local tradition and agreements between farmers, mirab (water master or water controlling authority), and the government in practice. Each village has at least one water master (mirab), who delegates his authority to sub-water masters who are in charge of allocating water to the scheme's various fields (ICARDA 2002).

Shallow Wells (arhad) System

This system is followed while using groundwater. The water is lifted from shallow wells with the help of wheel (*arhad*) with animal power supplying irrigation water to the fields of an individual farmer. The total number of shallow wells in the year 2012 in Afghanistan was 6598 that irrigated around 12,060 ha of land (CSO 2012).

Springs

When the groundwater table rises above the surface, it begins to flow on the surface, forming springs. In Afghanistan, there are roughly 5558 springs that irrigate about 188,000 hectares of land. When the groundwater level drops, as it does during drought years, the output from springs decreases. Some of the country's most drought-stricken districts are in places where irrigation is mainly reliant on spring water.

During dry years, crops suffer and productivity is drastically decreased.

Karez (qanat) Systems

Karezes are underground passageways that collect groundwater from alluvial fan aquifers. Water is transported from the source to the areas under cultivation via underground tunnels with mild slopes. The karezes can be many kilometers long despite their modest size. Their discharge ranges from 10 to 200 l s^{-1} on average, but can exceed 500 l s^{-1} in extreme circumstances. Both irrigation and drinking water are used in the Karez system. It is one of the most cost-effective ways to use groundwater for irrigation. It is harmless for the environment, as water is pulled by gravity. In Afghanistan, there are 6741 karezes. Approximately 163,000 hectares of land are irrigated by these karezes. One of the karezes' drawbacks is that there is no method to stop water from flowing during the winter or when irrigation is not required. Around 25% of water is wasted with the karez irrigation process (ICARDA 2002).

Modern Irrigation Systems

The irrigation department of Afghanistan operates and maintains a fixed intake facility. The irrigation scheme's management follows the same regulations as large-scale classical surface water schemes. The important difference is that water flow regulation in the system is dependent on a joint decision between government authorities and village residents. The creation of large-scale organized irrigation systems is a relatively new project that began roughly five decades ago. Five large-scale modern irrigation systems had been completed and were in service by the late 1970s. Traditional systems of land tenure were also distinct. Some of the schemes were run under private land ownership agreements, while others were run as "government-owned" State farms. These schemes were extensively subsidized by the government, and farmers had relatively limited options for crop choices or farming practices.

Canal Irrigation

Canal irrigation is by far the most important and widely used technique of irrigation in Afghanistan, with various types of irrigation systems providing water to around 85% of the total farmed area. Canals irrigate roughly 75% of Afghanistan's land, or 1.9 million hectares. The North, West, and Southwest of the country have the most canal-irrigated land. The water in these canals comes from the region's melting rivers (ICARDA 2002). Small diversion facilities have been erected along the river to shift water from the river to the irrigation canals. Water is transferred from these canals into small irrigation channels.

7.12.16 Rainfed Agriculture

About 1.3 million ha of Afghanistan's cultivated land is rainfed, out of a total of 3.9 million ha. Despite the fact that irrigated areas produce more than 85% of all agricultural products, rainfed areas play an important role in meeting the food needs of large populations living in areas where irrigation is either not possible due to topographical conditions or irrigation facilities have not been built (ICARDA 2002). Over most of the country, where terrain and soils are suited for agriculture, rainfall is scarce and highly erratic. In Afghanistan, there are just a few regions where rainfed crops may provide a stable source of income for the inhabitants.

7.12.17 Recommendations for Food Security

For food security in Afghanistan, there is need to carry out more extensive survey of soil fertility status in order to develop province or watershed level nutrient recommendations as well as rates of water input. The soils are deficient in zinc and boron and their deficiency has to be met through fertilizers. Wheat is the most important crop of the country, and modern technologies have to be used to enhance the productivity of this crop. Improvement in N use efficiency has to be ensured to realize potential crop yields.

7.13 Bangladesh

Bangladesh is the largest deltaic floodplain in the world with a total area of 147 570 km^2 (Roy and Farid 2011). It has 8.3 million ha of arable land with an average cropping intensity of about 185%, supporting about 149 million people. Agriculture has played an important role in the economy of the country, despite significant contributions from the garment industry and remittances from abroad. The agricultural sector contributes about 22% to the country's Gross Domestic Product (GDP). About 69% of this comes from the main driver of sectoral growth, the crop, and horticulture sector. About 84% of the total population live in rural areas and are engaged in a wide range of agricultural activities in one or the other way. About 87% of rural households are landless, marginal, and small; 63% are employed in agriculture (about 53% in crop agriculture); and about 33% in non-farm activities with backward and forward linkages with agriculture (FAO 2011a, b, c).

The major challenges facing Bangladesh agriculture are as follows:

- Increase in productivity and profitability;
- Instability reduction;
- Increasing resource-use efficiency;
- Ensuring equity;
- Improving product quality; and
- Diversification and commercialization of agriculture.

The major weaknesses identified by the National Agriculture Plan are as follows:

- Comparative weakness of the agricultural marketing system;
- High post-harvest losses;
- Farmers' inadequate capital for agricultural activities;
- Limited access to institutional agricultural credit;
- Inadequate and ineffective farmers' organizations;
- Low input use (water, fertilizer, pesticides) efficiency;
- Inadequate technology to meet export market requirements;
- Insufficient technologies to cope with an unfavorable environment;
- Insignificant private sector investment in Research and Development;
- Inadequate numbers of trained scientists and infrastructural facilities for advanced agricultural science;
- Low level of agricultural diversification;
- Weak quality control mechanism for agricultural inputs;
- Minimal coordination between public, private, and university research;
- Almost non-existent use of information and communication technology (ICT) in the extension system;
- Inadequate opportunities for farmers' and entrepreneurs' training; and
- Persistently inadequate production and supply of quality inputs (e.g. fertilizer, seed).

In order to produce more food for an increasing population at the rate of about two million per year and raw materials for agro-industries, the only option left is to expand the use of yield-contributing inputs combined with efficient scientific management practices. Percent area under different cultivated crops in Bangladesh is shown in Fig. 7.15 (FAO 2011a, b, c). In order to increase rice production, support is needed in the form of timely adjustment in fertilizer and diesel prices, increased availability of fertilizers, agricultural input cards and enhanced finance from banks, including collateral free credit to sharecroppers. Such programs are at various phases of implementation (Roy and Farid 2011).

Fig. 7.15 Percent of cultivated area under different crops in Bangladesh

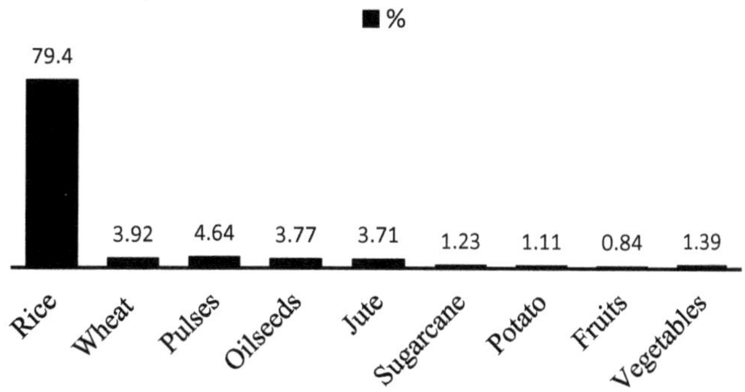

7.13.1 Food Security Situation

Despite the fact that Bangladesh has made significant gains in food production (Table 7.12), poverty and widespread malnutrition remain common. According to estimates from the Bangladesh Bureau of Statistics (BBS) in 2007, over two-fifths of the 149 million people are calorie-poor, consuming fewer than 2122 kcal per capita per day, and one-fifth are hard core poor, consuming less than 1805 kcal per capita per day. The foodgrain balance (production/demand) in Bangladesh, which used to be in negative up to 1997–98, became positive from 1998 to 1999. However, this balance is showing narrowing trend from 2004 to 2005 and is a cause of concern. However, foodgrains situation improved during 2016–17 when the production was 38.5 million tones (Mt) against a demand of 33.9 Mt, thus registering a surplus of about 4.6 Mt (Fig. 7.16).

7.13.2 Sustainable Soil Fertility and Fertilizer Management Policies

Agricultural production must attain continuous growth in order to maintain the level of food supplies or further improve it. The role of fertilizers and other agricultural inputs in enhancing agricultural production is crucial in view of the limitations on bringing more land under cultivation. About 53% of arable land is severely

Table 7.12 Foodgrain demand, supply, and balance

Year	Demand (000' t)	Supply	(000' t)	Availability (000' t)	Balance (000' t)
		Production[a]	Import		
1997–98	20,696	18,599	1933	20,532	−164
1998–99	21,027	19,631	5486	25,117	4090
1999–00	21,358	22,416	2104	24,520	3162
2000–01	21 689	24,083	1554	25,637	3948
2000–02	22,020	23,315	1799	25,114	3094
2002–03	22,351	24,025	3221	27,246	4895
2003–04	24,549	24,699	2798	27,497	2948
2004–05	24,945	23,520	3373	26,893	1948
2005–06	25,309	24,539	2562	27,101	1792
2006–07	25,686	25,250	2421	27,671	1985
2007–08	28,422	26,798	3212	30,010	1588

Source BBS (HIES) (1996, 2006, 2007, 2009)
[a]Gross production is less than 10% for seed, feed, and wastage; rice is calculated on the basis of milled rice

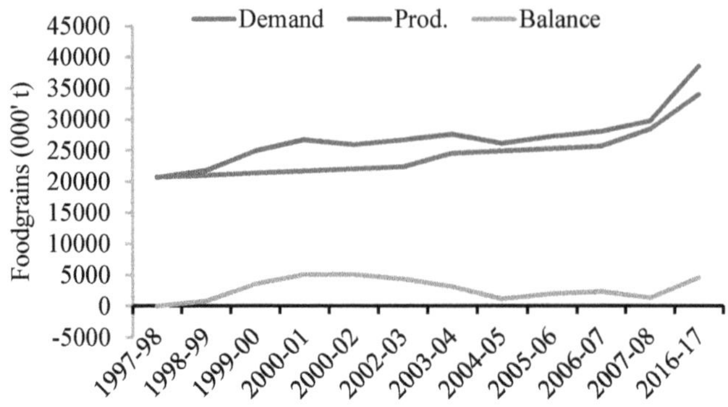

Fig. 7.16 Foodgrains demand, production, and balance in Bangladesh. *Source* Roy and Farid (2011)

deficient in organic matter. This is an alarming situation. The important task is to follow sound fertilizer use and soil fertility management practice to upgrade the biological potential of nutrient-depleted soils through increased application of organic residues, manure, and compost along with balanced use of mineral fertilizers.

7.13.3 Crop Production

Rice is the major crop of Bangladesh. The number of households planting the crop, the total area planted, total production, and grams of the crop available per household per day are given in Table 7.13. About 78% of the cultivated land is under rice crop and rice constituted 69% of total food crop production in metric tons (Fiedler 2014).

7.13.4 Land Use in Chittagong Hill Tract (CHT)

Out of 1303,808 ha area of CHT, about 800,000 ha is under forests (Fig. 7.17) and 90,000 ha is net cropped area. About two-thirds of the net sown area is single cropped and the rest is either double or triple cropped. The area under forests has declined in CHT but area under cereals has increased over time. Increase in population and subsequently food demand was

Table 7.13 Area under different crops and production in Bangladesh

S. No	Crop	Area	planted	Production	Availability	
		(000' ha)	%	(000't)	g/HH/day[a]	%
1	Rice	7819.3	78.45	29,252.6	2427	77.41
2	Wheat	226.4	2.27	537.9	45	1.42
3	Maize	66.2	0.66	391,1	20	0.65
4	**Total cereals**	**8111.9**	**81.38**	**30,181.6**	**2492**	**79.48**
5	Sugar	65.6	0.66	2411.7	20	0.64
6	Pulses	344.3	3.45	309.8	26	0.82
7	Oilseeds	279.5	2.81	298.6	25	0.79
8	Vegetables	470.9	4.73	5390.1	345	11.02
9	Fruits	527.2	5.28	2924.9	138	4.38
10	Others	168.0	1.69	1177.8	89	2.87
11		9967.4	100.00	42,694.5	3135	100.00

[a]HH, household; Grams per household per day, adjusted for refuse. Includes all Bangladesh households. *Source* Fiedler (2014).

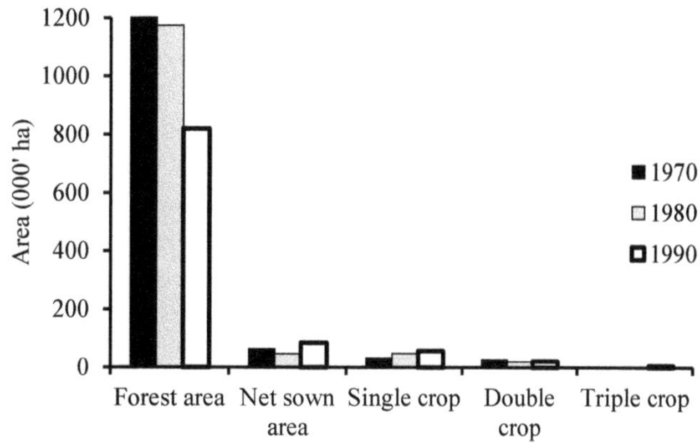

Fig. 7.17 Change in area under different land uses in CHT

responsible for decline in forest area and increase in area under cereal crops. More and more of forest land is being deforested every year for agricultural use.

7.13.5 Fertilizer Use in Agriculture in Bangladesh

Fertilizer is one of the most important inputs for raising crop yields, accounting for nearly half of all agricultural production. The delivery of this crucial input to farmers' doorsteps in order to maintain agricultural output has recently become a major challenge. At the start of each harvest season, farmers across the country make urgent requests for fertilizers. With the purchase of 2,698 tons of ammonium sulfate in 1951, mineral fertilizers were first used in food-deficient East Pakistan (now Bangladesh) agriculture. After the introduction of contemporary cultivars to feed millions of people, consumption climbed gradually over time, peaking at around 1.7 million tons NPK in 2007–08. In Bangladesh, the fertilizer is substantially subsidized so that farmers can afford it. Soil fertility and production, the country's most valuable natural resource, must be maintained at their highest levels. This is important in order to meet the challenges of increasing food production in a country where food security is critical for an ever-increasing population.

7.13.6 Soil Fertility and Soil Health

Demographic pressure in Bangladesh has demanded more production of food grains from a limited arable land area. Intensive agricultural practices have resulted in land degradation, and a wide range of environmental threats. There are 46 important economic crops grown in the country. High and medium high lands constitute about 65% of the country's area, where the cropping intensity is as high as 300–400%. According to the Soil Resource Development Institute (SRDI) in Bangladesh, an area of about 8.44 million ha is deficient in phosphorous and about 7.14 million ha deficient in sulfur. A soil fertility monitoring database, using data generated at 40 different sites throughout the country over the last 17 years by SRDI, shows that phosphorous availability has decreased by 35–65%. An FAO-financed project on Zn and Sulfur deficiency in Bangladesh indicated that most soils are deficient in sulfur. Sulfur may be considered the second nutrient element limiting rice yields. Because of intensive farming and the non-return of organic leftovers, many soils are severely depleted of certain nutrients and organic matter. Varied agro-ecological regions have different levels of depletion. However, extensive phosphorus, sulfur, and organic matter deficit are severely restricting. Phosphorus, potassium, and sulfur deficiencies affect around 1.8 million, 0.4 million, and 1 million hectares of land, respectively. This demonstrates the significant deterioration of the soil resource. If this problem is not solved, agricultural production will be unable to support the country's rising population.

Organic matter, an important component of soil, is critically low in about 53% of the Bangladesh's arable land, being <1% organic matter. The main causes of this depletion are Intensive land use, unbalanced use of mineral fertilizers, little or no use of organic manure/waste and removal of stubble/residues from the field after harvest, are the major causes of soil organic matter (SOM) depletion. The cultivation of two harvests of high-yielding variety (HYV) rice, each yielding more than 10 tons per year, removes 350 kg of nutrient from the soil each hectare. In Bangladesh, the input–output balance sheet for nutrients shows a high net annual removal of nutrients (mining) from the soil, which exceeds two million tons per year from arable land (Karim et al. 1995). Another arable land degrading problem is the salinity which is spreading in more and more areas.

7.13.7 Soil Fertility and Crop Response

Cropping intensity in Bangladesh is more than 180% (BBS 2006) with high-yielding varieties of crops. In intensive cropping areas, the arable soils are being mined which poses a serious threat to sustainable production. Yields of many crops are either stagnating or declining. Fertilizer is a key input and has contributed significantly (>50%) to the increase in agricultural production. Recent analysis of research findings indicates that 1 kg of fertilizer produces 26 kg of HYV Boro rice, 43 kg of potato, or 41 kg of maize.

7.13.8 Fertilizer Use

Bangladesh state authority launched a 'Grow More Food' campaign to feed the country's increasing population in 1960 (Rahman and Zhang 2018). Growth in fertilizer use and crop production has been quite remarkable in Bangladesh. About 60% of N is used on Boro rice. Urea is cheap and has an immediate greening effect on crops, so, it is the farmer's first choice. The comparatively lower use of P and K is leading to nutrient imbalance. The preferred ratio of use for N, P_2O_5, and K_2O is 4:2:1, whereas the actual ratio in 2007–08 was 7.8:1.3:1. In 2008–09, this deteriorated to 23.7:1.6:1. The deficiency of secondary and micronutrient is also fast increasing. Growth of fertilizer use from 1980–81 to 2009–10 is presented in Fig. 7.3. Consumption of nutrients (NPK) remained around 1.5 million metric tons from 2006 to 2008. As a result of the very high international price of fertilizers during 2008–09, consumption dropped to about 1.3 million tons.

7.13.9 Other Nutrient Resources

Rice straw is very cheap and a valuable source of carbon, potassium, silicon, and organic manure. It is also very cheap. The straw is primarily used as fodder and fuel. An estimate of national availability of straw, dung, and litter is presented in Fig. 7.18. The use of animal wastes/manure is also very limited, dung being used primarily as fuel. Despite the use of green manure in rice cultivation having been well demonstrated and validated by research organizations (Farid et al. 1994), it is not widely used. Recommendations are made for using poultry manure, cow dung, rice straw, and green manure based on field trials and validated through on-farm trials at BARI and BRRI.

It has been shown in a 12 year study how rice straw could be properly utilized for soil health improvement without interfering with cropping time, particularly in an intensively rice-cropped area (Farid and Hossain 1988). In other studies, the utilization of poultry manure and cow dung combined with mineral fertilizers showed sustainability of yield and fertility (Farid et al. 1998b). Compost and vermicompost have been commercially explored and accepted by entrepreneurs in Bangladesh. Some farmers are using organic fertilizer for their fruit and vegetable gardens. Although a lot of research has been

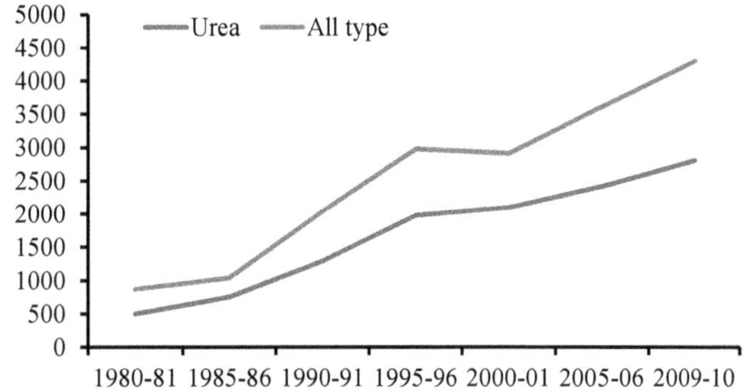

Fig. 7.18 Growth of fertilizer use in Bangladesh. *Source* Roy and Farid (2011)

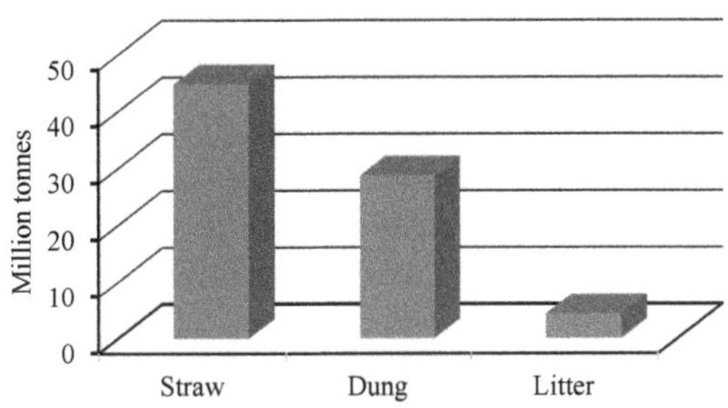

Fig. 7.19 Yearly generation of agricultural waste (straw, dung, and litter)

done by NARS and significant technologies have been developed, there has been no breakthrough in the commercialization of rhizobium, azospirillum, BGA, or Azolla. Yearly generation of agricultural waste (straw, dung, and litter) in Bangladesh is shown in Fig. 7.19.

7.13.10 Fertilizer Use in Bangladesh

The production and use of fertilizers in Bangladesh has been shown in Fig. 7.20. The figure depicted the use of nitrogen, phosphorous, and potassium in the country from 1981 to 2010 and trends in rice + wheat production.

7.13.11 Integrated Nutrient Management

The integrated nutrient management need to be based on cropping system as a whole rather than individual crops. The productivity of rice and wheat is affected by the application of nutrients and the yield of hybrid maize is influenced by lime and boron application (Table 7.14).

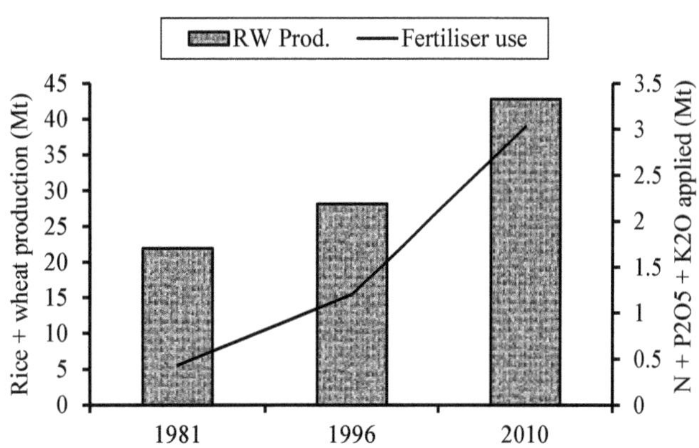

Fig. 7.20 Rice and wheat production and fertilizer use in Bangladesh

Table 7.14 Grain yield of wheat and rice obtained in integrated nutrient management at Dinajpur, 1994–97 (mean values)

Treatment	Wheat yield (t ha^{-1})	Rice yield (t ha^{-1})
Control	0.62	3.29
100% NPKSZn	3.60	4.57
75% NPKSZn	2.58	3.92
GM + 50% NPKSZn	2.95	3.94
GM + 75% NPKSZn	3.25	4.32
FYM + 50% NPKSZn	2.83	3.75
FYM + 75% NPKSZn	3.74	4.54

Source BARI annual report

7.13.12 Comparative Use of Fertilizer

Comparative information on average fertilizer consumption (kg ha^{-1}) in the world and Bangladesh between 2000 and 2015 is given in Fig. 7.21. Bangladesh consumes fertilizers at amounts higher than the global average. It consumed 188 kg ha^{-1} in 2002, increasing to 275 kg ha^{-1} in 2012, compared to global average of 103 and 135 kg ha^{-1}, respectively. In 2014, fertilizer nutrients use in China was 565.26 kg ha^{-1}, and in India 138.041 kg ha^{-1} (World Bank 2015) which was also more than the global average.

7.13.13 Soil Organic Matter Management

Organic matter in the soil is an essential source for soil fertility. In Bangladesh, the loss of soil organic matter is a key stumbling block to increased agricultural output. In around 3.63 million hectares of land, organic matter is below 1.7%. Depletion of organic matter is caused by increased cropping intensity and a faster rate of organic matter breakdown in the current hot and humid climate. Furthermore, organic manures such as farmyard manure, compost, and/or green manuring are rarely used in the country.

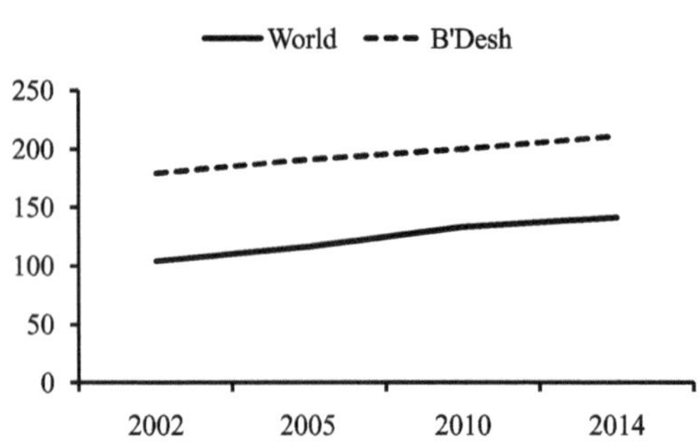

Fig. 7.21 Fertilizer consumption (kg ha^{-1}) in Bangladesh and the World (2000–2015)

7.13.14 National Priorities for Sustainable Soil Management

For sustainable management of soils in Bangladesh, the important activities that need to be undertaken are (i) soil OM management; (ii) soil fertility and fertilizer management; (iii) soil fertility evaluation and GIS mapping every five years; (iii) improve nitrogen usage efficiency; (iv) integrated nutrient management for crop and cropping patterns; (v) liming to raise soil pH to desired level; (vi) balanced fertilizer use and high-quality fertilizer; (vi) land resources inventory, including a semi-detailed soil survey to update an existing database and a detailed soil survey to develop a "Land and Soil Resources Utilization Guide" at the union level (lowest administrative unit), as well as soil salinity surveys every five years; (vii) soil management in ecologically challenging ecosystems; (viii) the influence of climate change; (ix) creation of GIS-based land, soil, and soil fertility information; (x) disseminating technologies; (xi) soil management and awareness programs; and (xii) development of human resources.

7.13.15 Land Rights of the Indigenous Peoples of Chittagong Hills Tract

Indigenous peoples from the Bawm, Sak, Chakma, Khumi, Khyang, Marma, Mru, Lushai, Uchay, Pankho, Tanchangya, and Tripura tribes live in the Chittagong Hill Tracts of Bangladesh (CHT), which are found in south-eastern Bangladesh. The Chittagong Hill Tracts are located in Bangladesh's south-eastern area (Kalindi Roy 2000). The Chittagong Hill Tracts (CHT) have a total surface area of 13,189 km^2, comprising river and reserve forest regions, with a total land area of 3,685 km^2. With an area of 6,089 km^2, 4,502 km^2, and 2,590 km^2, the CHT included the current three Hill Districts of Rangamati, Khagrachari, and Bandarban. The rivers Feni, Karnaphuli, Matamuri, and Sangu, as well as their tributaries, split the Chittagong Hill Tracts into four valleys (Kalindi Roy 2000).

7.13.16 Policies in CHT

The CHT's development strategies have taken a two-pronged approach:

(i) Financial gain for state administrative powers, and
(ii) Assimilation of indigenous peoples into mainstream national culture.

Although British policies were not focused for cultural integration throughout their administration, these two goals are not always mutually contradictory. Their operations were solely for financial gain (Kalindi Roy 2000).

Table 7.15 shows the area and % of agricultural land used in different soil classifications. The soil was evaluated by forest experts based on its condition.

7.13.17 Problem Soils of Bangladesh

Chemical Properties

- Acid Sulfate soils (Chittagong coastal plains, Mangrove forest areas);
- Saline and ii. Alkali soils;
- Peat Soils (Gopalganj-Khulna Beels, Sylhet Basin);
- Soils with nutrient toxicity; and
- Nutrient deficiency and Arsenic contaminated soil.

Physical Properties

- Steeply sloping soils (Hilly, Valley areas)—Land Slides;
- Coarse textured soils (Piedmont Plain, Active Floodplain);
- Shallow soils (Eroded, summits of the hills);
- Poorly drained (devoid of O_2, gases affect upland crops but suitable for rice);
- Heavy textured soils and f. Soils with plow pan.

7.13.18 Agriculture-Related Economic Activities

Over half of all Chittagong Hill Tract (CHT) households' annual net income originates

Table 7.15 Land use classification for best management practices in CHT, Bangladesh

Category	Land use	% of land mass	Area (ha)
Class A	Agriculture	3.2	30,586
Class B	For terraced agriculture/horticulture	2.9	27,148
Class C	Horticulture/afforestation	15.5	146,649
Class C-D	Afforestation/horticulture	1.4	12,810
Class D	Forestry only	77.0	726,797

Source Kalindi Roy (2000)

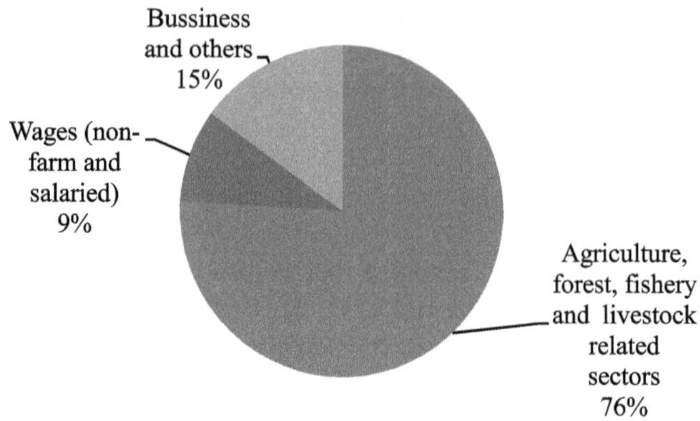

Fig. 7.22 Different agriculture-related economic activities in Chittagong Hill Tracts. *Source* UNDP (2009)

from various agriculture-related sources in Bangladesh (Fig. 7.22, UNDP 2009). A vast proportion of rural households in CHT rely on agricultural-related activities to produce food and earn money. Local ethnic households have a higher share of agriculture-related annual income than Bengali households (63% versus 49%) (UNDP 2009). The Khumi, Marma, Mro, and Bawm ethnic groups generate more than 67% of their net yearly household income through agricultural activities (UNDP 2009).

7.14 Bhutan

7.14.1 Land Use

In Bhutan, drylands cover 60.6% (68,255 ha) and wetlands have a share of 28.3% (31,911 ha) of the area cultivated for field crops. The percent area under apple, citrus, cardamom plantations, areca nut, and other horticultural crops is 1.8 (2,081 ha), 4.9 (5,488 ha), 3.2 (3,600 ha), 1.1 (1,199 ha) and 0.1 (about 50 ha), respectively (MoAF 2014). As per constitution of Bhutan, 60% area of the country need to be covered by forest. Currently, trees such as blue pine, broadleaf, conifer, Chir pine, fir, and permanent conifer forest cover roughly 71% of the country's land area. Agriculture covers 2.93% of the remaining land. Marshy, meadows, shrubs, barren ground, snow cover, water bodies, and constructed and unbuilt up areas make up the rest of the terrain.

7.14.2 Bhutan Agriculture

With highly undulating terrain, Bhutan has about 7.8% area under cultivation. Agricultural land is mostly confined to remote villages on steep slopes and valleys, average size of holding being 1.75 ha. Farming is generally subsistence, labor intensive, low farm inputs and mixed including crops, livestock and forest with limited scope for mechanization (Norbu 2011; Dorji 2015). The authorities have fixed a target to reduce poverty from 23 to 15% and 65% self-sufficiency in rice by 2020, promotion of high-yielding and disease/pest resistant varieties and value addition to agro-products through post-harvest Storage and processing. Now, traditional subsistence farming is slowly evolving into commercial ventures and orchard land use gaining more importance (Dorji 2015). The climatic zones range from subtropics in the south with considerable annual rainfall (5500 mm) to alpine in the north with low rainfall (400 mm) (Dorji 2008). Arable agricultural land according to land use type is given in Table 7.16.

The economy of Bhutan is still dominated by agriculture, though there has been sudden drop in share of agriculture in GDP, between 1980 and 2006, from roughly 56% to 22%. However, it is the single largest sector in Bhutan, providing employment to about 69% of the population. This could be owing to a variety of factors, including the expansion of other sectors such as energy, transportation, and communication. These industries have produced job opportunities in more lucrative industries such as construction

Table 7.16 Arable agricultural land by land use type

Classification	Area (ha)	%
Irrigated	27,649	17.7
Rainfed	106,806	68.3
Pangshing	11,470	7.3
Orchard	9191	5.8
Kitchen garden	1391	0.9
Total	156,507	100.0

Source Dorji (2008)

and manufacturing, resulting in rural–urban migration and a labor scarcity on farms (Dorji 2008). Within the previous few decades, hydropower's contribution to GDP has increased significantly. Because of the small land holding sizes on steep slopes and the lack of farm automation, farming in Bhutan is difficult and labor consuming. Farm households have an average land holding size of less than 1.0 hectare. About 14% of households own less than 0.4 ha of land, while 56% own between 0.4 and 2.0 ha. The remaining households do not own any agricultural property and rely on farm laborers or renters on other people's land to make a living (Fig. 7.23).

Bhutan has two types of shifting cultivation. The first is known as tsheri and is mostly practiced in subtropical and tropical broadleaf woods, while the second is known as pangshing tsheri and is mostly practiced in subtropical and tropical grasslands. The following are some of the common properties of these systems:

- There are no household relocations, and farmers cultivate tsheri land from their permanent houses near their permanently cultivated field.
- Depending on altitude, the fallow period might last anywhere from 5 to 12 years. The longer the fallow, the higher the altitude.
- Shifting cultivation intensity is proportional to population density; the higher the population, the greater the intensity of shifting cultivation.

Dorji and Dorji (2015) reported that forests occupy 70.46% of the total area of the country while cultivation is done only in 2.93% of the area (Fig. 7.24). Shrubs, meadows, water bodies are in 10.43%, 4.10%, and 0.72% of the area of Bhutan. About 7.44% area remains covered with snow and 3.20% remains bare and the rest is under other uses.

7.14.3 Land Categories

Bhutan has five categories of agricultural land such as wetland, dryland, tsheri land (shifting cultivation), orchard land, and kitchen garden land. Percent land under wetland, dryland, tsheri land (shifting cultivation), orchard land, and kitchen garden is 23.5, 51.8, 9.3, 14.4, and 1.0, respectively. In Bhutan, highest area is under dry land category (65,638 ha), followed by wetlands

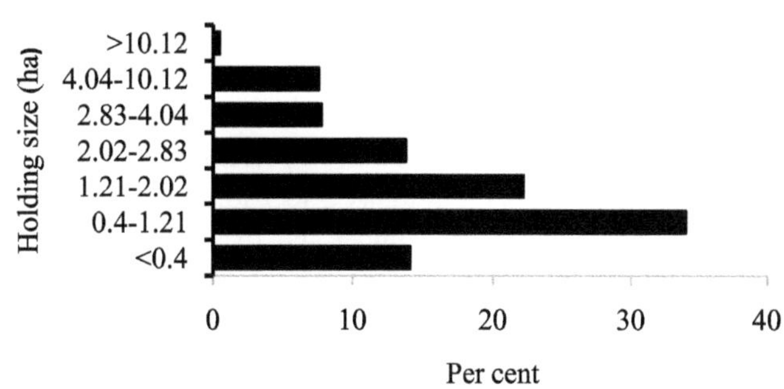

Fig. 7.23 Distribution of land-holding size in Bhutan

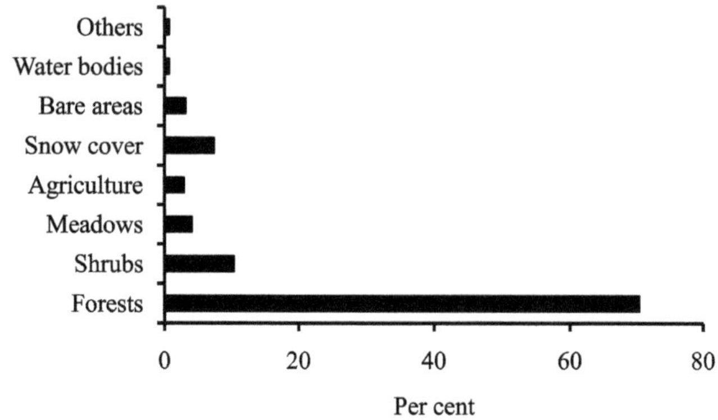

Fig. 7.24 Land use in Bhutan

(29,721 ha), Orchard land (18,216 ha), Tsheri land or shifting cultivation land (11,830 ha) and kitchen land (1280 ha) (Ghalley 1990).

7.14.4 Soil Fertility Management (SFM) in Bhutan

Many times, locally available organic supplies such as forest litter, crop residues, green legumes, fodder residues, compost, and animal wastes are used for crop nutrition, or it might be claimed that organic farming is practised by default. Farmers can use them alone or in combination with inorganic sources to replace nutrients lost by crops. SFM is based on the integrated use of the forest as a source of feed and leaf litter, livestock for dung, and crops for crop residue in the traditional approach. Farm labor shortages, decreases in animal size, and increased distances to forests for leaf litters are all complicating the SFM system and posing a threat to its long-term viability. Household labor availability and livestock number and management system are important for sustainability of the traditional SFM system. Interventions from the line departments have made the farmers aware of the benefits of using inorganic fertilizers, increased access to fertilizers and farmers' enhanced purchasing power have changed the SFM practice to a certain level. In general, roughly 5 to 7 tons of FYM are applied per ha^{-1} under the existing SFM method. A typical farmyard manure (FYM) contains 1.38% nitrogen, 0.29% phosphorus, 1.97% potassium, and 2.35% calcium, equating to around 41 kg nitrogen, 9 kg phosphorus, 59 kg potassium, and 71 kg calcium per hectare. Inorganic fertilizer supplementation is required since the amount of nutrients provided by FYM alone is insufficient to assure high yields. In comparison to yields in many other nations, national averages for several crops are still quite low (Dorji 2008). Table 7.17 shows the area and production of several crops in Bhutan in 2016. (Royal Government of Bhutan 2016).

7.14.5 Inorganic Fertilizer

In Bhutan, the fertilizers use has increased over the years mainly due to their increased availability with the improvement in road access and improvement in distribution systems; effective awareness programs among the farmers (Norbu and Floyd 2001). Subsidy incentives are being provided to the farmers through indirect government support to the national marketing and distribution system (Dorji 2008). The import of inorganic fertilizers increased from about 250 MT in 1960s to about 11,436 MT in 1997. Despite concerted efforts of the Royal Governent of Bhutan, the use of inorganic fertilizer in Bhutan is still very low as compared to many other countries.

As far as use of fertilizers application for different crops is concerned, the potato is the most preferred crop. More than half of fertilizers (about 53%) are used in potato, followed by

Table 7.17 Area and production of different crops in Bhutan

Crop type	Crop	Area (ha)	Production (Mt)	Yield (kg ha^{-1})
Cereals	Paddy	21,480	85,090	3,961
	Maize	22,919	82,035	3,579
	Wheat	1,505	2,521	1,675
	Barley	992	1,702	1,715
	Buckwheat	2,792	3,705	1,327
	Millets	1,314	1,714	1,304
	Total	51,002	176,766	
Oilseeds	Mustard	969	892	920
	Groundnut	84	149	1774
	Soybean	220	254	1154
	Sunflower	6	7	1160
	Pyrilla/naam	16	12	750
	Total	1,295	1,314	
Spices	Cardamom	4,488	2,736	609
	Ginger	1,932	10,871	5,627
	Total	6,420	13,607	
Legumes	Rajma bean	633	984	1554
	Mung bean	385	492	1280
	Total	1,019	1,476	
Roots/tubers	Sweet potato	13	29	2231
	Tapioca	112	415	3705
	Total	125	444	
	Vegetables	–	56,298	–
	Potato	5,926	58,820	9925
	Grand total		308,725	

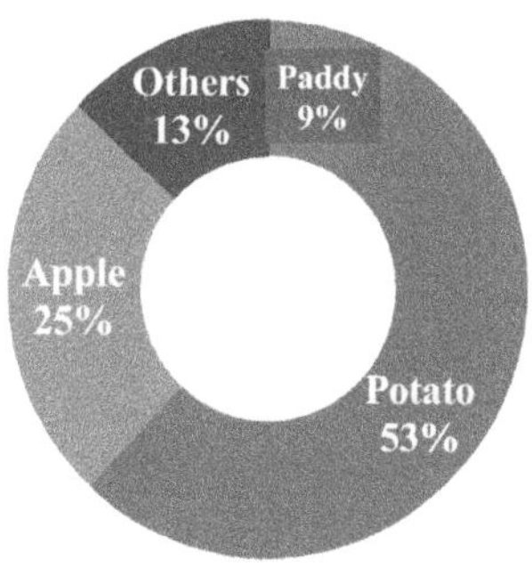

Fig. 7.25 Percent of fertilizer applied to different crops

apples (25%), paddy (9%), and others (13%) of the total consumption (Fig. 7.25). Eastern region of the country uses the highest amount of fertilizer nutrients which is about 39% of the total use, followed by the western region (24%) and then by west central (19%) and east central (18%) regions. The use of higher quantities of fertilizers in potatoes (eastern region) and apple orchards (western region) is the cause of high fertilizer usage in the eastern and western areas (western region). Overall use of fertilizer use in this Himalayan Kingdom is still low and the cultivators still prefer the organic sources of nutrients through the materials like decomposed droppings of livestock, compost, farm wastes, etc.

The difference between N and P, K is expanding as nitrogenous fertilizer distribution expands, from roughly 3:1:1 in early 1990 to

7:1:1 in 2005. Imbalance nutrient utilization is primarily because of urea's lower cost and availability compared to complex fertilizers. Plant nutrient imbalance is a significant topic that is being addressed at the extension level, particularly through training, demonstrations, and on-farm trials (Dorji 2008).

7.14.6 Organic Programs

Apart from offering diversity in food crops, diverse cropping patterns such as crop rotation and intercropping with leguminous crops are being encouraged to maintain and improve soil fertility. Composting and green manuring are also encouraged, particularly in the country's southern and eastern regions, where FYM output is lower. The Ministry of Agriculture is responsible for The Ministry of Agriculture established the National Organic Program (NOP) to preserve the agricultural system, which is still primarily natural. Bhutan's Organic Farming Program was created to promote organic farming. Organic farming, according to Dorji (2008), could be a viable alternative for Bhutan if the following factors are taken into account:

- Due to the steep terrain, the majority of the farmers are small and cannot afford to invest extensively in agricultural inputs.
- There is limited opportunity for agricultural land expansion.
- Bhutan is unable to compete on a commercial level with its neighbors China, India, and Bangladesh.
- Bhutan has a strong policy on environmental protection.
- Bhutan can exploit niche market for organic crops.
- Import of fertilizers will become expensive and so organic sources for crop nutrition will be a good option.

7.14.7 Nutrient Status of Soils of Bhutan

The most major factors determining variance in soil nutrient status in Bhutan are land use and agro-ecological zone (Dorji and Dorji 2015). Different soil fertility management strategies used by farmers have an impact on variations in soil nutrient status. Soil fertility management in Bhutan is mostly accomplished through an integrated method that incorporates both organic and inorganic plant nutrients. According to Dorji and Dorji (2015), the traditional SFM system's long-term viability is highly dependent on household labor availability and livestock numbers.

7.14.8 Threats in Land Management

With increasing population, the already limited arable land is dwindling both in size and productivity (Dorji and Dorji 2015). Shortening of fallow period of shifting cultivation and burning crop residues have caused land degradation on fragile hill slopes. Intensification of agricultural production systems for market-oriented farming has caused the problem of groundwater pollution and environmental degradation due to use of inorganic fertilizers. In regions with subsistence farming and poor soil fertility management, nutrient mining is a severe concern. In the absence of effective land suitability and capacity classification systems, land is not being used optimally and sustainably.

Table 7.18 shows that alpine scrub accounts for 3.39%, bushes for 9.74%, and cultivated agricultural land and meadows for 2.75% and 2.51%, respectively (FRMD 2017). Water bodies, built up areas, non-built up areas, landslides, and moraines each account for less than 1% of the total, while snow cover accounts for 5.35% and rocky outcrops for 4.15%.

Table 7.18 Land cover class and the area in Bhutan (2016)

Land cover	Area (km^2)	%
Alpine scrubs	1300.97	3.39
Built up	74.57	0.19
Cultivated agriculture	1056.82	2.75
Forests	27,171.61	70.77
Landslides	37.30	0.10
Meadows	962.73	2.51
Moraines	143.93	0.37
Non built up	5.95	0.02
Rocky outcrops	1594.55	4.15
Shrubs	3740.32	9.75
Snow and glacier	2053.43	5.35
Water bodies	251.75	0.65
Total	38,394.00	100.00

Source FRMD (2017)

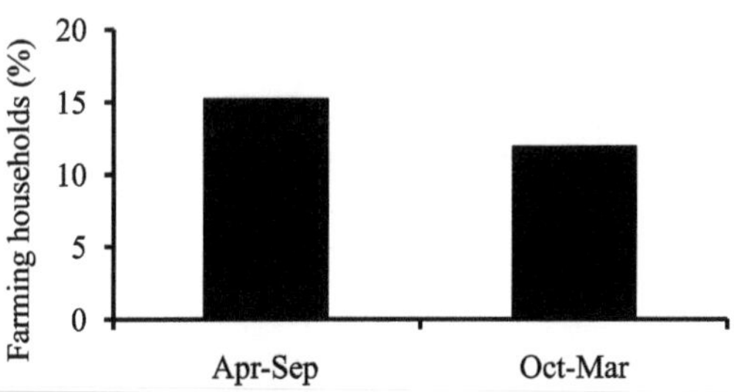

Fig. 7.26 Estimated proportion of farming households facing food (agriculture crops) shortage in the year 2016

7.14.9 Agriculture and Climate Change

The climate change will have a great impact on agricultural productivity and production in Bhutan. The agriculture in mountains is dependent on prevailing temperature and amount of precipitation received. The change in climate would affect the precipitation pattern and normal temperatures. In spite of concerted efforts of the Royal Government of Bhutan and the line departments, many farming households face food shortage during the year. Estimated proportion of farming households facing food (agriculture crops) shortage during summer (15.2%, April–September) and winter months (11.9%, October–March) in the year 2016 is given in Fig. 7.26 and proportion of HHs using various coping mechanisms to address the food shortage is given in Fig. 7.27.

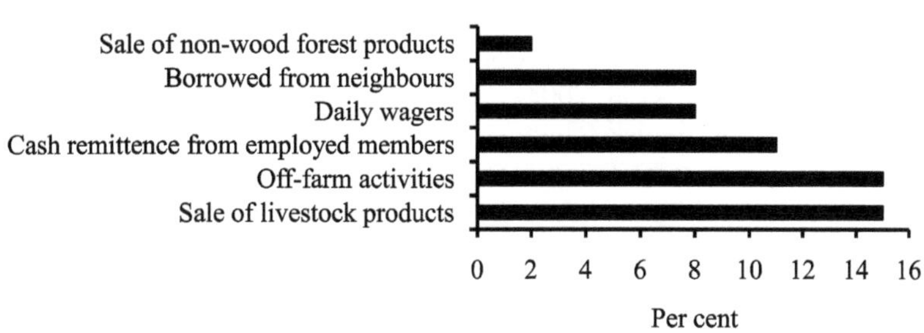

Fig. 7.27 Proportion of HHs using various coping mechanisms to address the food shortage in 2016

7.14.10 Nutrient Flow from the Forest

In spite of constraints, the farmers of Bhutan produce reasonably good crop yields with minimal inputs of chemical fertilizers and this has been possible due to continuous flow of plant nutrients from the forest ecosystem to agricultural fields (Roder et al. 2003). Phosphorus is the most limiting plant nutrient in the higher regions where soils are generally derived from granite gneiss. Methods used to facilitate and to hasten the nutrient flow include:

- Animals as agents for collection and acceleration of nutrient flow;
- Leaf, needle, and decomposing litter collecting from the forest; and
- Enhancement of nutrient availability through the burning of soil organic matter (Roder et al. 2003).

The P content ranged from 0.07 to 0.14% for herbaceous plant biomass of forest or permanent grassland and from 0.22 to 0.53% for dung produced by animals grazing these resources. Roder et al. (2003) reported that the cattle transfer about 300 to 700 tons P annually from the forest to agricultural land. Through the blue pine biomass used to ignite the soil organic matter, about 6 kg P is added to the system. Soils on mountain slopes inherently show low fertility due to high erosion potential, limited soil depth, and nutrients poor parent material. At elevations above 2000 m, soils are generally derived from granite gneiss and are extremely poor in available phosphate. In these soils, P is the most important limiting plant nutrient and without P inputs, agriculture is not possible (Roder et al. 2003).

The most important species are oak (*Quercus grifithii, Q. lanata, and Q. semecarpifolia*), blue pine (*Pinus wallichiana*), chir pine (*P. roxburghii*), *Castanopsis indica,* and *Schima wallichii* (Roder et al. 2003). Semi-decomposed leaf material from broadleaf forests is also widely used as a source of organic matter and plant nutrients in apple orchards and ornamental gardens (Roder 1990). Rates applied are 10–30 kg per tree at the time of tree planting and 10–20 kg $tree^{-1}$ $year^{-1}$ in established orchards (Roder 1990).

7.14.11 Grass Fallow Systems

Roder et al. (1993) explained that a grass fallow system is traditionally used at elevations above 2500 m to produce buckwheat (*Fagopyrum esculentum* and *F. tataricum*) on soils low in available P. The various steps in soil preparation, consisting of cutting the topsoil, making mounds with the dry sods of the topsoil and spreading the soil after burning demand high labor inputs. For the cultivation of one ha of buckwheat about 150–400 labor days are required (Roder et al. 1993). The burning process reduces organic C and total N; reduces the C/N ratio; and increases soil pH, available K, and available P.

7.14.12 Agricultural Transformation in Bhutan

Bhutan remained isolated for a long time, and it wasn't until the 1960s that the country was finally opened up to the rest of the world (Mathou 1999). Agricultural growth has gotten a boost thanks to the Royal Government of Bhutan's (RGoB) strong support and imaginative policies, among other sectors. Bhutan's agricultural modernization initiatives began with the country's first five-year plan, which ran from 1961 to 1966. (Kobayashi et al 2015). It has been discovered that the elements influencing agricultural transformation are multifaceted and vary in time and space. As shown in Fig. 7.28, Dendup (2018) constructed a framework encompassing significant components of agricultural transformation based on expert recommendations and his personal expertise and understanding of the field.

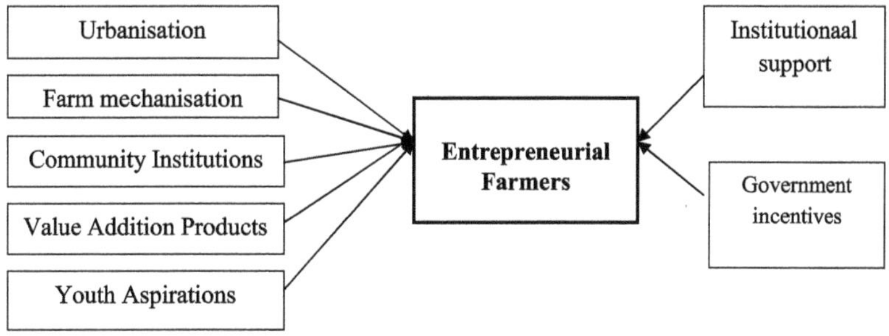

Fig. 7.28 Components of agricultural transformation in Bhutan (Modified over Dendup 2018)

The improved infrastructure in Bhutan with regard to agriculture has helped in enhancing agricultural productivity up to a large extent. As the population density is low, better employment opportunities in other sectors are on the rise in urban areas and the demand for skilled workers is rising to handle other sectors too, though more than half of the country's population is still employed in the agriculture sector (NSB 2017). More than 35% of Bhutan's population is now employed in the service sector and 8.7% in the industry sector, resulting in migration from rural to urban areas. This impacts both positively and negatively the agriculture development (Bajaj 2014).

Now the farmers have better access to modern agricultural technologies, due to improved infrastructure. They cultivated only enough crops to feed their families, and commercial farming is still a distant dream. To supplement their living, they keep animals such as cattle, horses, chickens and pigs for milk, meat, eggs, labor and manure. Due to low population of the country, the demands are also limited and so are the marketing facilities. The farming practices are based on indigenous knowledge passed down through generations (Dendup 2018). Moreover, Bhutan has mountainous terrain, and 49.5% of its total geographical area has a slope greater than 50%, making it vulnerable to soil erosion, landslides, extreme runoff, and unproductiveness. The rugged and difficult terrain left the rural communities inaccessible to one another and, so, the spread of new technologies in agriculture is taking much time than genuinely required (MoAF 2015).

7.15 Tibet (China)

The population of this region was 32.5 million in 2015 and is expected to reach 38.8 million in 2020. With an average elevation of around 4000 m amsl and a surface area of 2.5×106 km^2, the Qinghai-Tibetan (QT) Plateau is the world's highest and largest plateau (Zheng 1996). Its vegetation is thought to be extremely sensitive to climate change and susceptible (Zheng 1996). Because of its high elevation and unique climate, the plateau has an impact on terrestrial ecosystems in western China (Zhang 1983).

Supporting the food and livelihood for human and livestock and impacted by climate change, the agro-ecosystem is the most important artificial ecosystem (Tubiello et al. 2007). The effect of climate change on agricultural ecosystems has been an important subject of research due to its significance for food security (Fuhrer, 2003). Agro-ecosystems influence climate change through altering global carbon, water, and nutrient cycles (Bondeau et al., 2007); on the other hand, current climate change, particularly warming (IPCC 2007a, b), has significant impacts on agro-ecosystems by changing heat, water, and other environmental elements. Climate change's influence on cropping systems has not been adequately documented (Feng and Hu 2004). Climate change has changed cropping intensity as well as crop growth in different parts of the world. In China, the principal crops have moved to the northern regions, where cultivation was formerly impossible due to unfavorable weather conditions (Xu et al. 1999).

Table 7.19 Area (Million ha), yield (kg ha^{-1}) and production (Million tons) of major crops in China (2017–18)

Crop	Area	Yield	Production
Wheat	24.51	5.48	134.33
Corn	42.40	6.11	259.07
Rice	30.75	4.84	148.87
Coarse grains	44.57	5.96	265.72
Oilseeds	24.08	2.47	59.48
Soybean	8.25	1.85	15.28
Cotton seed	3.40	3.18	10.80
Peanuts	4.61	3.71	17.09
Oats	0.48	1.16	0.55

7.15.1 Land Use in China

Around 75% of China's agricultural land is dedicated to cultivating food crops. Rice is the most significant crop, accounting for nearly a quarter of all farmed land (Table 7.19). Wheat is the second most important grain crop in the United States, and it is grown in various parts of the country. Maize and millets are grown in north and northeast China, while oat is grown in the Himalayan region, which includes Tibet. Sweet potatoes, white potatoes, and a variety of other fruits and vegetables are among the other crops. Northern Liaoning and Shandong are known for their apple and pear production. Citrus is a key cash crop in southern China, with production strewn across the Yangtze River valley and to the south.

In its first fifty years, that is, from 1949 to 1999, China greatly increased agricultural production through technological improvements. The ACGR of food grains was 3.06% between 1949 and 1978 and 2.46% between 1978 and 1999 (Fig. 7.29). However, the ACGR remained only 0.42% from 1999 to 2017, indicating a drastic decline in growth rate over the earlier 50 years. The ACGR of oilseeds and cotton between 1949 and 1978, 1978 and 1999, and 1999 and 2017 was 2.58%, 7.94% and 4.71%, and 5.84%, 2.76% and 11.52%, respectively. The grain production increased from 113.2 million tones (Mt) in 1949 to 548.9 Mt in 2017. The corresponding figures for above period were 2.56 and 59.6 Mt for oilseeds and 0.44 and 27.3 for cotton.

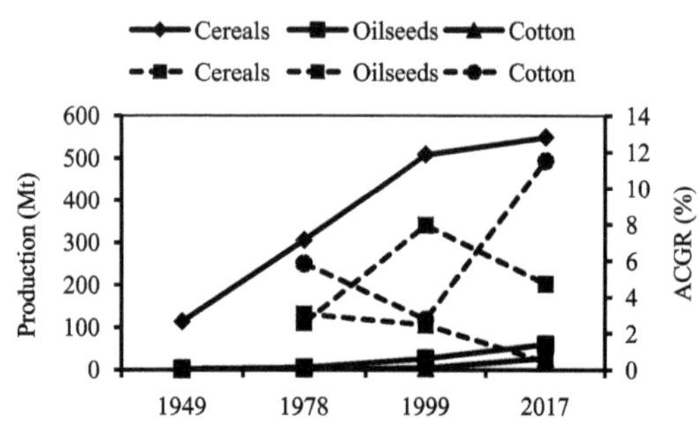

Fig. 7.29 Production of cereals, oilseeds, and cotton in China and annual compound growth rate (ACGR) of cereals

7.15.2 Land Use in Tibet

The Tibetan agricultural system is, more or less agropastoral, a mixed crop-livestock system that involves rearing livestock across a range of elevations combined with field agriculture (Yi et al. 2007). Field crops include several cereals, such as Tibetan barley, wheat, maize, and buckwheat; root crops such as turnips and potatoes; and garden fruit and nut trees such as apples and walnut (Salick et al. 2005). Livestock include yak, cattle, and pianniu (a crossbred of yak and cattle); horses, donkeys, and mules; as well as goats, sheep, pigs, and chickens (Yi et al. 2007). In high Tibetan villages (>3000 m amsl) and those distant from roads, there is greater overall plant diversity (Salick et al. 2005) including that of traditional Tibetan crops such as barley, buckwheat, and turnips.

About 65% of the area of China is covered by hills and mountains. These areas cover about 1/3 of China's population and cultivable land. The mountain region of south-west China covers the Tibetan and Yun-Gui Plateaus. Tibet lies on the main Tibet-Qinghai plateau at an average elevation of above 4000 m amsl. Tibet has 0.221 million ha of cultivated land and 55.3 million ha of grasslands. Tibet can conveniently be divided into five agro-ecological zones, namely, freezing cold, cold, cool, temperate, and warm-hot.

Agriculture and pastrolism are common in cool and temperate climate areas at elevation varying from 3000 and 4500 m amsl (Table 7.20). Yak, sheep, cows, and goats are common animals in the area. Agricultural crops like maize, wheat, rape seed, barley, and horticultural crops such as apples, peach, oranges, etc. are generally grown in these zones.

The area under three major crops such as maize, wheat, and rice in Tibet and Yunnan is presented in Table 7.21. Foodgrains production is the main farming activity in Tibet and wheat and barley are grown in more than 90% of the area under grain crops. However, in Yunnan and Sichuan Mountain provinces, rice, wheat, and maize are the major food crops. In Yunnan, maize is the predominant crop, followed by rice and wheat; while in Sichuan, area under rice is more compared to wheat and maize. Other major crops of the mountain areas in China are potatoes, cotton, tea, sugarcane, apples, pears, peaches, etc.

7.15.3 Livestock

Livestock are an integral part of mountain agriculture in Tibet, Yunnan and Sichuan provinces of China. Cattle, pigs, sheep, goats, horses, etc.

Table 7.20 Agro-ecological zones of Tibet and their characteristics

Agro-ecological zone	Altitude (m amsl)	Average temp. (°C)	Livestock	Crops	Farming system
Freezing cold	>5000 m	<6	Yak, sheep. goat	–	Grazing
Cold	4500–5000 m	6–10	Yak, sheep. goat	Barley	Predominantly agro-pastoral
Cool	4000–4500 m	10–18	Yak, sheep. cow	Spring barley, wheat, rape seed, peas	Agro-pastoral and one crop
Temperate	3000–4000 m	10–18	Yak, sheep. cattle	Spring barley, winter wheat, maize, apple, peach	Spring and winter crops
Warm hot	<3000 m	>18	Cow, pig, goats	Winter wheat, maize, rice, apples, oranges	Mixed agro-pastoral-forestry

Table 7.21 Area (000'ha) and production (000'tones) in mountain provinces of China (1997)

Crops	Tibet		Yunnan		Sichuan	
	Area	Production	Area	Production	Area	Production
Rice	1.0	5.0	939.2	5361	3020.1	21,823
Wheat	51.0	249.0	664.3	1467	2364.9	7203
Maize	3.0	13.0	993.8	3692	1762.1	7169

Source Statistical year books of Tibet, Yunnan and Sichuan (1997)

Table 7.22 Livestock population (million) in mountain provinces of China

Livestock	Tibet	Yunnan	Sichuan
Cattle	5.100	8.022	11.333
Pigs	0.226	23.344	70.816
Goats	5.832	6.538	9.717
Sheep	11.098	1.228	3.637
Horses/donkeys/mules	0.508	1.823	0.762

Source Statistical Year Books of Tibet, Yunnan and Sichuan

are the major livestock reared by the rural people. The highest number of cattle, pigs, and sheep is reared in Sichuan Province while Tibet leads in total number of sheep (Table 7.22).

7.15.4 The Tibetan Steppe

Compared to the plain areas of the world, where change in ecological zones might be seen after thousands of kilometers, because of altitudinal variances, the Tibetan Plateau and Himalayan foothills are very varied over short distances. The Tibetan Plateau spans about 3.1 million km^2, the majority of which is above 4000 m above sea level (Chang 1983). Temperature and rainfall gradients are important factors in establishing biological zones on the Plateau, and they have a substantial impact on the crops that can be produced there. The Tibetan Plateau is separated into four economic geographic zones: high Western Tibet, Central Tibet's River valleys, Northern Tibet's Changtang Plateau, and Eastern Tibet. The distribution of acceptable crops to be grown is influenced by the severe boundaries between ecological zones.

7.15.5 Low Nutrient Use Efficiency

China is one of the world's largest users of chemical fertilizers. However, in certain parts of the country, difficulties with imbalanced nutrient application and low utilization efficiency are frequent. It has been shown that the amount of fertilizer utilized varies greatly across the country. It is widely used in southeast China, but not so much in central and western China. Horticultural crops are the preferential crops for the fertilizer use, especially for greenhouse vegetable cultivation. Zhang and Shen (2013) reported that although nitrogen and phosphorus have been used in luxurial amounts, potassium use has not been used in desired amounts. Environmental problems have cropped-up due to overuse of fertilizers at many locations. Guo et al. (2010) mentioned that the water and atmospheric environment are also affected by overuse of pesticides and fertilizers and soil acidification rate is fast.

7.15.6 Increasing Cropping Intensity in Tibetan Plateau

Though the consequences of global warming on agriculture have been thoroughly documented all over the world, the influence of climate change on agricultural in alpine regions has received less attention (Zhang et al. 2013). The Tibetan Plateau (TP) has a low agricultural output, but it is increasing, and as a result, it is playing an increasingly vital role in regional food security as the population grows. Rapid warming has had a significant impact on agro-climate resources, altering cropland patterns and cropping intensity. From the downstream valleys of the Lhasa River and Nyang Qu River of the tributaries of the Brahmaputra to upstream valleys, the area suitable for single cropping expanded from 1.91 million hectares in the 1970s to 2.0 million ha in the 2000s. From 900 ha in the 1970s to 15,000 ha in the 2000s, the area suitable for double cropping has gradually increased (Zhang et al. 2013).

7.15.7 Soil Acidification and Planting Patterns in TP

Vegetable production in peripheral Lhasa (Tibet), China, by applying high doses of nitrogen fertilizer may cause accelerated acidification and thus affecting soil productivity (Zhou et al. 2010). Experiments with three land use patterns, namely grain, openfield vegetable, and greenhouse vegetable, were conducted in order to better understand the diverse patterns of soil acidity in Lhasa. The soil ecosystems in the semi-arid Tibet Plateau are delicate and sensitive. According to Zhou et al. (2009), since the Second National Soil Survey in the 1980s, changes in cropland soil pH have occurred in the semi-arid central Tibet Plateau, with average pH values decreasing from 7.70 in the 1980s to 7.22 in 2008. Despite the fact that industrial operations have had little impact on the Tibetan Plateau, land use is changing, with a growing trend toward vegetable land. This was mainly due to increase in demand for vegetables caused because of improved living standards, urban population increase, and change in their lifestyle.

7.15.8 Insufficient Water Resource and Low Use Efficiency

Water shortage has been developed in China as a result of increased water demands and low agricultural use efficiency, restricting agricultural productivity. China's average water availability per capita is only 2173 m^3, or one-fourth of the global average. Rainfall during the monsoon season is not only scarce, but also irregular. Water scarcity is compounded by its low use efficiency, with irrigation water efficiency ranging from 40 to 50%, compared to 70 to 80% in many wealthy countries (Xia et al. 2013).

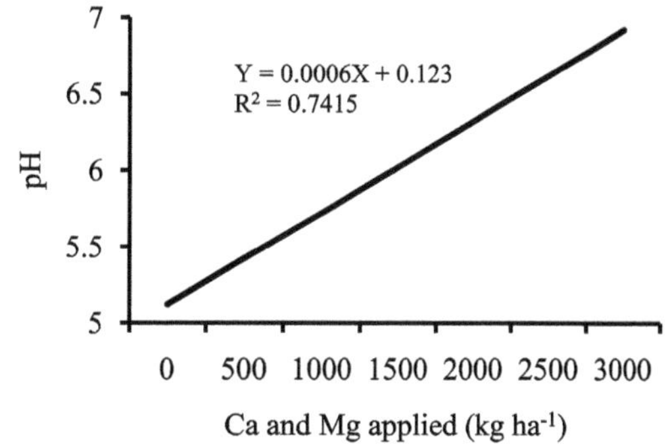

Fig. 7.30 Relationship between total Ca and Mg applied (kg ha^{-1}) and soil pH measured in NaCl. *Source* (Zhao et al. 2007)

7.15.9 Effect of Fertilizer and Lime Treatments on Soil Properties

When the total application of Ca and Mg surpassed 2000 kg ha^{-1}, soil pH changed considerably (P0.01) from 5.4 (measured in 2001 prior to planting) to >6.5. (Fig. 7.30). When administered at similar rates of Ca and Mg, the source of the base cations had a considerable impact on pH change, with Ca and Mg derived from quick lime having a far larger impact on soil pH than Ca, Mg, P. The relationship between quick lime base cations and soil pH explained 80% of the change in soil pH, indicating that a target pH of 6.5 could be achieved with 1300 kg Ca ha^{-1} applied in either a large single application (1500 kg CaO ha^{-1}) or two or more applications (1000 kg CaO ha^{-1}) over several years. For 1 or 2 years, either application approach kept soil pH over 6.5. (Zhao et al. 2007).

Overall, application of Ca, Mg, P fertilizer significantly increased ($P < 0.05$) total soil P from the pre-treatment level of ~380 mg kg^{-1} to 519 mg kg^{-1} when the highest total amount of Ca, Mg, P (354 kg P ha^{-1}) as applied (Fig. 7.31),

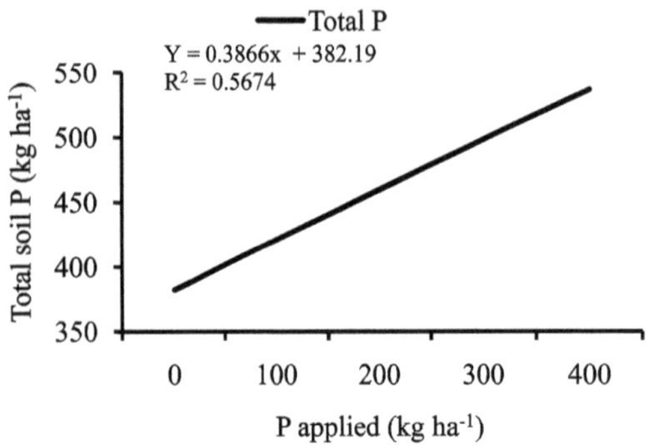

Fig. 7.31 Relationship between total P applied in fused Ca, Mg, P fertilizer, and total soil P

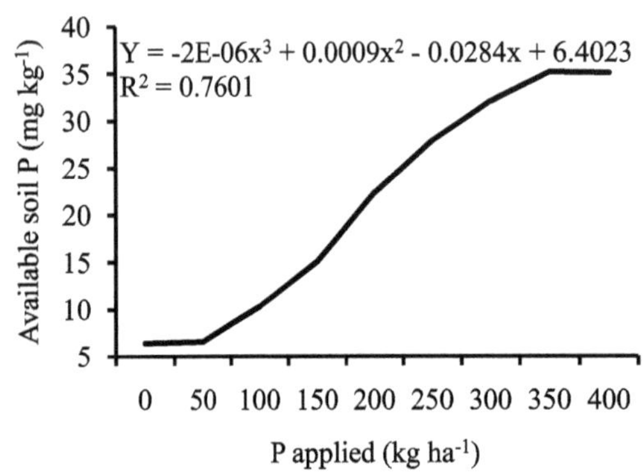

Fig. 7.32 Relationship between total P applied (kg ha^{-1}) and available soil P (mg kg^{-1} Bray P). *Source* (Zhao et al. 2007)

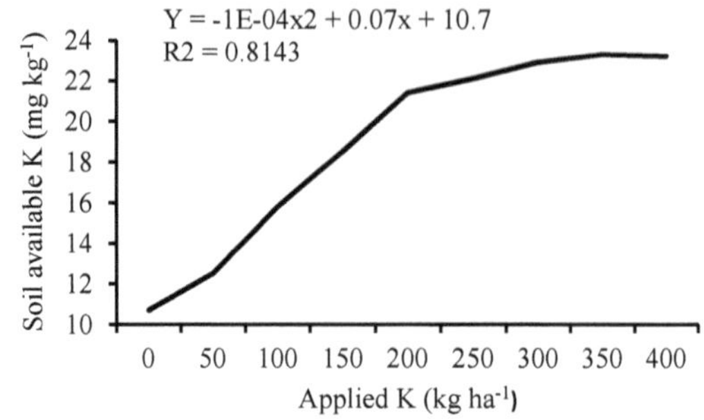

Fig. 7.33 Relationship between total K applied (kg ha^{-1}) and available soil K (mg kg^{-1})

but only about 15% of this increase in total soil P was available for plant growth. The fertilizer P (R^2 = 0.76) provided the majority of the available soil P, with data indicating that available soil P rose by roughly 0.1 mg kg^{-1} for each kilogram of P applied (Fig. 7.32). However, the cubic relationship between available soil P and fertilizer P applied revealed that any meaningful increase in available soil P required an initial application of >50 kg ha^{-1} of fertilizer P. The difference in reaction to the initial and later P applications could be related to the positioning of P, with the first being banded in the drill rows and the latter being surface spread. Due to the random soil sampling approach adopted, this could have resulted in an underestimating of available soil P in the first year response. In plots where no Ca, Mg, or P was administered, there was no substantial change in accessible soil P due to plant uptake or soil fixation and re-release mechanisms.

The addition of 200 kg K ha^{-1} quadrupled the amount of accessible K in the top 7.5 cm of soil. When no K fertilizer was provided, available soil K decreased dramatically (P0.01) from 12 mg kg^{-1} in 2001 to 5 mg kg^{-1} (Fig. 7.33).

Pasture species responded to P application in varied ways, with considerable increases in white clover and cocksfoot production with low P doses and no reaction in perennial ryegrass. Because legumes are such an important source of biologically fixed nitrogen (N) in pasture systems where no N fertilizer is used, P application rate recommendations should be based primarily on white clover response (Michalk and Huang 1992).

7.15.10 Agricultural Systems of Xizang

Three major factors govern the agricultural systems of Xizang such as elevation, remote geographic location, and sparse population. In Xizang, about 77.7% land is above 4500 m amsl, 8.4% between 4000 to 4500 m amsl, and the rest below 4000 m (Cheng Hong 1990). Only a small area of valley and basin land is cultivable, while uplands are mostly used for pasture. The rest of the land is covered by glaciers, snow, rocks, desert, etc. Altitude differences are responsible for wide changes in temperature and have an tremendous impact on livestock husbandry, nature of crops grown, variation in soils and water resources, and crop production. He categorized the agricultural systems of Xizang in two parts such as (i) animal husbandry or a combination of animal husbandry and crop production and (ii) crop production as the major activity. The land is generally not cultivated in higher mountain areas due to severe cold conditions. Livestock rearing is the only option. The climate is warmer below 4,200 m in eastern Xizang and below 4,500 m in western Xizang; these places represent the upper limits of farming. From May to September, the majority of crops are grown. The four principal crops of the highlands are barley, wheat, peas, and rape, which account for 95% of the entire area under cultivation.

7.16 INDIA

7.16.1 Land Resource in Indian Himalayas

India has a wide range of physical characteristics that are distinct, diverse, and complex. There are mountains, hills, plateaus, and plains, giving humans a wide range of options for how to exploit these geographical resources. Hills and mountains cover over 30% of India's land area, with the northern mountains and hills falling within the Himalayan environment. Despite the fact that plateaus cover 28% of the total land area, only around 25% of them is suitable for farming. The plains, which make about 43% of the overall land area, have around 95% of it suitable for farming. As a result, half of the country's surface area is under cultivation. This is one of the highest ratios in the world. The use of land in India is depicted in Fig. 7.34.

The area under the non-cultivable land category has decreased from 50.7 million hectares in 1960–61 to 40.8 million hectares in 1990–91. The percentage of fallow land has decreased slightly from 9.9% in 1950–51 to 7.5% in 1990–91. Between 1950–51 and 1990–91, there was a 34% decrease in cultivable wastelands. In 1950–51, India had 118.7 million hectares under cultivation, which climbed to 142.4 million hectares in 1990–91 and around 146 million hectares in 2020. The fact that only 5% of the land in a country with the world's highest bovine population is under permanent pastures and grazing is cause for concern. With the rapid rise of the economy, the amount of land used for non-agricultural purposes has expanded.

A comparison of area under various crops in Eastern and Western Himalayas of India shows that while rice is the predominant crop in the former, wheat and maize are the major crops in the latter (Fig. 7.35). Rice is grown in about 64% of the total area under food grain crops in the EIHR, followed by pulses (13%), maize (11%), oilseeds (8%), millets (3%), and wheat (1%). This order for WIHR is wheat (36%); rice and maize (23% each); millets (9%); oilseeds (5%); and pulses (4%) (Gupta 2006, Gupta et al. 2006) (Table 7.23).

In land utilization, the Indian Himalayan Region (IHR) has 59% of the total geographical area under forests compared to all India percentage of 22%, while net sown area is only 12%

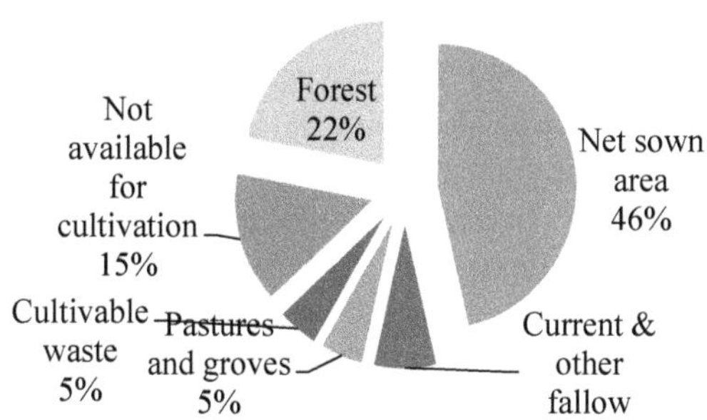

Fig. 7.34 Land utilization in India

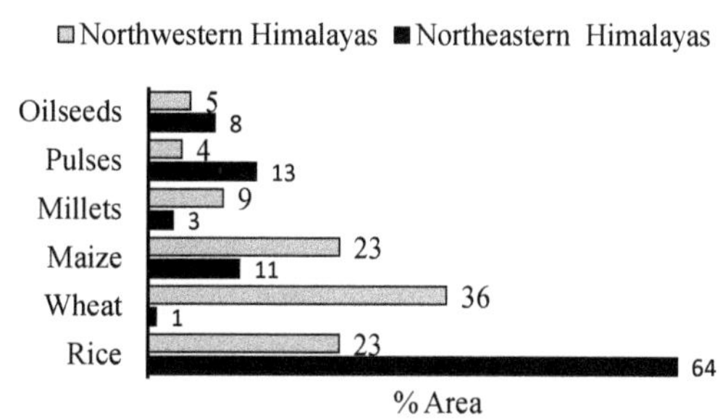

Fig. 7.35 Percent area under different crops in Northeastern and Northwestern Indian Himalayas

Table 7.23 Food grains area (000 ha), production (000 t), and yield (kg ha^{-1}) in Indian Himalayan states (2014–15)

State	Rice			Wheat			Coarse cereals			Total[a] food grains		
	A	P	Y	A	P	Y	A	P	Y	A	P	Y
NW Himalaya												
Himachal Pradesh	72	125	1728	330	646	1957	321	622	1936	755	1432	1896
J & K	276	517	1871	321	314	979	336	379	1131	957	1220	1275
Uttarakhand	262	604	2307	348	654	1881	222	314	1415	897	1626	1811
Total NW Himalaya	**610**	**1246**	**2042**	**999**	**1614**	**1615**	**879**	**1315**	**1496**	**2609**	**4278**	**1640**
Arunachal Pradesh	127	285	2241	3.8	7.5	1974	74	102	1378	261	409	1889
Assam	2495	5223	2093	24	29	1216	33	96	2895	2701	5459	2021
Manipur	224	334	1488	0	0	2602	26	59	2243	283	427	1509
Meghalaya	110	291	2703	0.4	0.8	1909	21	43	2075	140	354	2532
Mizoram	37	68	1643	0	0	0	5.7	9	1515	47	75	1607
Nagaland	195	454	2326	3.1	5.6	1824	79	147	1860	314	650	2065
Sikkim	11	20	1818	0.4	0.4	1077	47	76	1614	65	102	1579
Tripura	257	747	2903	0.1	0.3	2143	4.5	6	1303	274	761	2783
Total NE Himalaya	**3456**	**7422**	**2147**	**31.8**	**43.6**	**1371**	**290.2**	**538**	**1854**	**4085**	**8237**	**2016**
Total Indian Himalayas	4066	8668	2132	1031	1658	1608	1169	1853	1585	6694	12,515	1869

[a]Including pulses (A, P, and Y represent area, production, and yield, respectively).
Source Reserve Bank of India, 2016–17

in IHR as against 46% in India (Fig. 7.24). The area under pastures and grazing lands is 7% in IHR and 4% in India. The area, production, and average yield of major food crops (2014–15) in Indian Himalayan states are presented in Table 7.24. The average yield in IHR is relatively low than the All India average yield of various crops (Gupta 2006, Gupta et al. 2006) (Fig. 7.36).

Detailed land use under different categories of land utilization has been provided in Table 7.37 (GOI 2012–13)

7.16.2 Agriculture in Indian Himalayas

In the Indian Himalayan region, traditional agriculture is generally a low input, low production subsistence system. Its intensification without conservation involves high risk of natural resources degradation. Initially, cultivation evolved in hills was a closed cycle in harmony with nature. Disregard to ecosystem health, extreme variation in growing conditions, small holdings, remoteness, degraded watersheds, and poverty are challenges to hill agriculture. The agriculture in the Himalayas with poor accessibility, extreme resources diversity, fragility, and marginality of land pose significant challenges (Jodha 2004).

Jodha (2004) mentioned that the rate of exploitation of natural resources, efficiency of a system and compensatory mechanism, will influence the productivity and economic sustainability. The approaches and management followed in other ecosystems cannot be readily applicable to hill farming. One aspect is tillage intensive large field farming verses hand tool specific very small terrace hill farming (Jodha 2004). He elaborated that the terrain conditions, geological limitations, very low resilience and very difficult restoration in HKH region contribute to fragility and Ecosystem productivity, infrastructure, limitations, stress ecosystem, widening ratio of agriculture to forests and lack

Table 7.24 State-wise agricultural land use (000, ha) in Indian Himalayan Region states (2013)

State	Geographical area	Forest	Land not available for cultivation	Other un-cultivated lands excluding fallow	Fallow land	Net sown area	Total cropped area	Agri. land/cultivable land/arable land/cultivable land
Arunachal Pradesh	8374	5154	64	119	107	216	285	424
Assam[a]	1533	386	686	301	0	160	220	160
Himachal Pradesh	5567	1126	1131	1697	79	543	947	811
J & K	22,224	2025	573	314	126	745	1162	1070
Manipur	2233	1742	27	8	0	309	309	316
Meghalaya	2243	946	239	555	215	285	340	1056
Mizoram	2108	1585	95	53	244	116	116	408
Nagaland	1658	863	95	164	155	380	489	1285
Sikkim	710	584	11	11	9	77	140	98
Tripura	1039	629	141	20	4	256	360	277
Uttarakhand	5348	3485	449	897	136	706	1124	1547
West Bengal[a]	325	125	34	8	15	143	143	0.0
Total IHR	53,361	19,483 (36.5)	3545 (6.6)	4147 (7.8)	1090 (2.0)	3936 (7.4)	5495 (10.3)	7452 (13.9)
All India	328,726	70,007 (21.3)	43,738 (13.3)	25,976 (7.9)	26,283 (8.0)	139,932 (42.6)	194,390 (59.1)	181,950 (55.4)

[a]Karbi Anglong and N.C. Hills of Assam and, Darjeeling district of West Bengal
Sources Government of India 2015

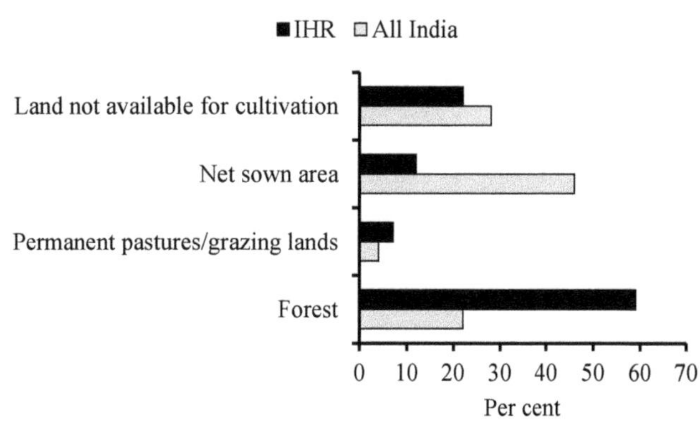

Fig. 7.36 Agricultural land use IHR (Indian Himalayan Region) and All India

of alternate livelihood to marginality. Variability and diversity in the region is found in climatic, resources and biodiversity, and farming system. Inaccessibility in the Himalayas is due to hostile terrain, high cost of infrastructure development, and limitations due to the mindset toward traditional means of transport (Jodha 2004). However, the region has some basic niche advantages of seasonality, rich biodiversity, largely non-chemical inputs, largest possible range of agroecosystems and eco-services. Specific features of mountain and hill agriculture productivity balance (Srivastva 2006) are given in Table 7.25 (Fig. 7.37).

Table 7.25 Hill agriculture productivity balance

Gains	Losses	Balance
Sunshine,	Rainfall and soil erosion,	Soil moisture regime,
Rainfall,	Evaporation,	Soil nutrient regime,
Soil, water and nutrient aggradation, Efficient crops and cropping patterns,	Loss of organic matter,	Energy balance,
Improved management,	Loss of nutrients,	SOM build-up,
Improved photosynthesis,	Loss of growing period,	Soil health,
On-site benefits	Loss of efficiency,	Soil productivity level,
Higher production	Poor management,	Environment impact,
	Loss of compensation mechanism for off-site benefits	Off-site benefits
	Declining production	Sustainability

Source Srivastva (2006)

Fig. 7.37 Agricultural land use in mid-Himalayas (Photos by Dr A. Pattanayak)

Table 7.26 Extent of soil erosion in the Himalayas vis-a-vis World

World/region	Total soil loss (million t yr^{-1})	Soil loss (t ha^{-1})	Reference/Remarks
NE Himalayas (India)	601	23.5	Sharma and Prasad (1995)
NW Himalayas (India)	403	19.2	Estimated by U. C. Sharma
India	5300	16.1	Dhruvanarayana and Rambabu (1983)
Hindu Kush Himalayas	7801	18.2	Estimated by U. C. Sharma
World	36,000	–	Brown (1984)

7.16.3 Management of Sustainability

Management of hill agriculture should aim at harmony of farm production sustainability and economic sustainability. The respective prioritization will depend on the criticality potential, opportunity, and environment. Increase in population beyond a threshold level and mismanagement of resources accelerate the threat of degradation and stress to fragile hill ecosystem (Lal 2001). Organic carbon in cultivated soils was reported to be 4.55%, 1.86%, 0.80%, and 0.60% in alpine region, temperate hills, subtemperate region, and sub-humid areas (Srivastva and Singh 2006). The N content was 0.45%, 0.19%, and 0.29% in alpine, temperate, and subhumid zones, respectively. The N, P, and K were estimated in sub-temperate region only in forest and agriculture ecosystems. The forest land had higher amount of these nutrients compared to agricultural land (Srivastva and Singh 2006).

7.16.4 Resources Degradation in Himalayas

Land degradation can be managed by caring, improving, and managing the natural resources. Degradation is a process of decline in productive capacity and quality of natural resources on a short or long-term basis (Srivastva 2006). The degradation may be due to (i) Encroachment of forest land, inappropriate management of steep marginal land and degrading farming systems, and (ii) Unsustainable intensification of farming with lack of suitable technology and over exploitation of natural resources. The extent of different form of degradation in Indian Himalayan Region is presented in Table 7.26. The extent of land affected by different land degradation processes is shown in Fig. 7.38.

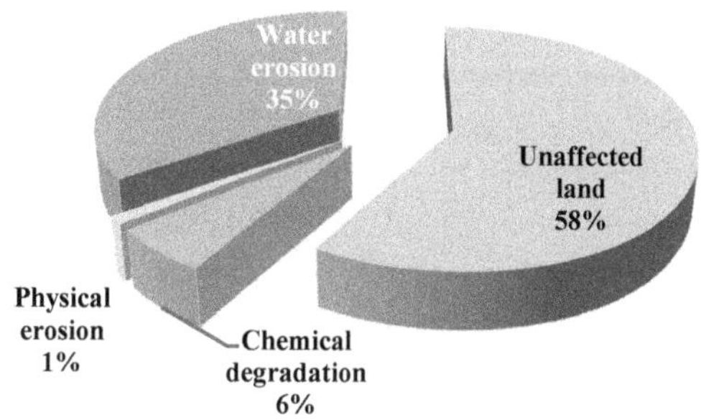

Fig. 7.38 Land degradation in Himalayas

7.16.5 Features of Indian Himalayas

Important features of Indian Himalayan Region are described below (Table 7.27).

7.16.6 Climate Change and Food Security in HKH

The Global Hunger Index (GHI) of the International Food Policy Research Institute (IFPRI) ranks India 63rd out of 78 countries, with the severity of India's food insecurity classified as "alarming" (von Grebmer et al. 2013). China, Nepal, Pakistan, and Bangladesh, in that order

Table 7.27 Some important features of Indian Himalayas (Samra et al. 1999)

Attributes	Northeastern Himalayas	Northwestern Himalayas
Geographical area (Mha)	20.22	33.49
Climate	Extremely wet	Dry and cold
Latitude	21° 5'–29° 3' N	29° 31'–36° 58' N
Longitude	85° 5'–97° 5' E	73° 26'–85° 30' E
Elevation (m, amsl)	200–8598	400–8611
Annual rainfall (mm)	2000–12,000	350–3000
Mean temperature (°C)	8–22	8–22
Physiography	Slopes are generally gentle	Slopes are generally steep
Soils	Deep to shallow, pH acidic to neutral	Shallow soils, pH acidic to neutral
Vegetation	Tropical forest	Temperate and drought resistant
Farming system	Shifting cultivation, settled	Settled
Livestock	Cattle, pig, mithun, Yaks, and poultry	Cow, buffalo, sheep, goats, and yak
Dominant land ownership	Community, private	Government and private

Source Samra et al. (1999)

are ranked 6th, 49th, 57th, and 58th. Food security is a serious challenge in India, particularly in the hilly regions, where rural communities consume 2098 kcal per capita per day, 50 units less than the national average (Giribabu 2013). The National Food Security Mission (NFSM) aims to increase rice, wheat, and pulse production. The government hopes to reach out to 75% of the rural population and 50% of the urban population by providing food grain subsidies (Gayrathri 2013).

The production of food grains in India increased at an ACGR of 2.33% between 1952–53 and 2015–16 (Table 7.28); however, the pace of growth slowed down during the last decade of the twentieth century. The highest ACGR of 3.12% was recorded between 1972–373 and 1992–93 which came down to 1.50% 1992–93 and 2015–16 against an increase in population at an ACGR of 2.18% and 1.76% during the corresponding periods, respectively. The reasons for slow growth are many but climate change and consequent droughts, floods, and inclement weather are certainly one of the major reasons (Kurvits et al. 2014). All the Himalayan states of India have, by and large, met their requirements of food grains from the central pool. Annual Compound Growth Rate (ACGR) of rice, wheat, pulses, and total food grain in India was 1.90, 1.81, 1.21, and 1.79%, respectively between

Table 7.28 Increase in population and foodgrains production during different periods

Year	Foodgrain production (Million t)	ACGR[a] (%)	Year	Population (Million)	ACGR (%)
1952–53	59.20		1952–53	36.81	
1962–63	80.15	3.07	1962–63	44.90	1.98
1972–73	97.03	1.93	1972–73	56.02	2.23
1992–93	179.48	3.12	1992–93	86.20	2.18
2015–16	253.16	1.50	2015–16	129.09	1.76
1952–53 to 2015–16		0.33	1952–53 to 2015–16		2.01

[a]Annual compound growth rate

1980–81 and 1999–2000. However, in the Himalayan states of India, the ACGR of food crops during this period remained at 0.87%.

7.16.7 Integrated Nutrient Management for Crop Production

India, with a geographical area extending over 329 million ha, having different types of climate, soils, flora and fauna, is a challenge for agricultural development and production. The country's resources are constantly threatened by the huge population (1342 million with a density of 451 persons km^{-2} in 2018), who need drastic measures of land management for ensuring food security. The 'green revolution', which was responsible for higher use of chemical fertilizers and pesticides, irrigation water, etc. boosted agricultural productivity, was also a cause for decline in soil fertility, imbalanced nutrient content of soils, and management problems. The major future challenge would be to keep the pace of production under situations arising out of decreasing per capita availability of arable land.

7.16.8 Gap Between Nutrient Demand and Supply

There is a big gap between the demand and supply of crop nutrients in India. At present (2019), a negative annual balance of more than 15 million tons exists (Fig. 7.39). It is expected that this gap is likely to increase to more than 20 million tons by 2025. Higher negative nutrient balance is a cause of concern as it poses a threat to soil quality and sustainable agriculture. This gap is required to be bridged to obtain optimum production of crops in future. The situation in Himalayan states of India is still more bleak. Though some of the Himalayan states of the country have switched to organic farming, they would continue to face food deficit in future too.

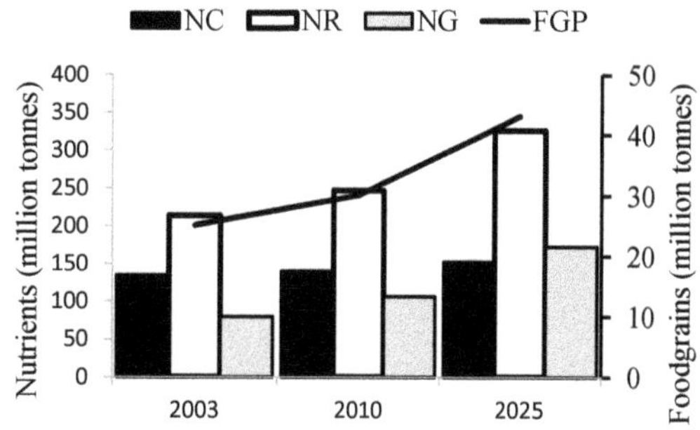

Fig. 7.39 Projected foodgrain production (FGP) in relation to nutrient (N-P2O5-K2O) consumption (NC), removal (NR), and Nutrient gap (NG). *Source* (Based on Rao and Reddy 2005)

7.16.9 Crop Nutrient Balance

Crop nutrient balance exercise involves the native soil nutrient content plus application through additional sources, that is, through organics and inorganic sources of nutrients, what the plants use during crop growth period and what amount is left in the soil (Fig. 7.40). It provides vital information on whether a soil's nutritional status is being maintained, built up, or decreased.

A declining trend in nutrients use efficiency has been observed in India over the years (Fig. 7.41). The figure shows the increase in food production in relation to increase in nutrient consumption and the nutrient response ratio from

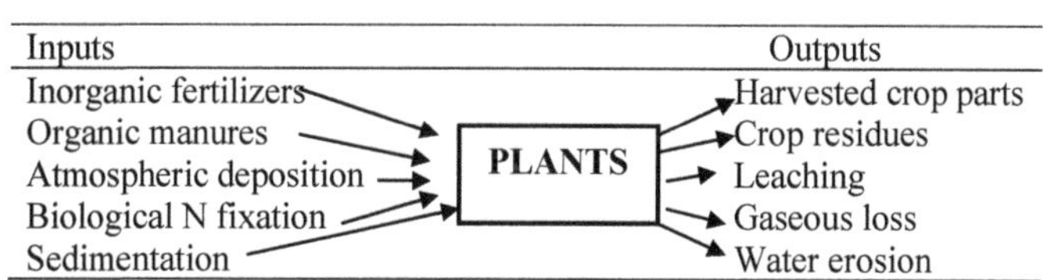

Fig. 7.40 Basics of nutrient balance approaches (Satyanarayana and Tewatia 2009)

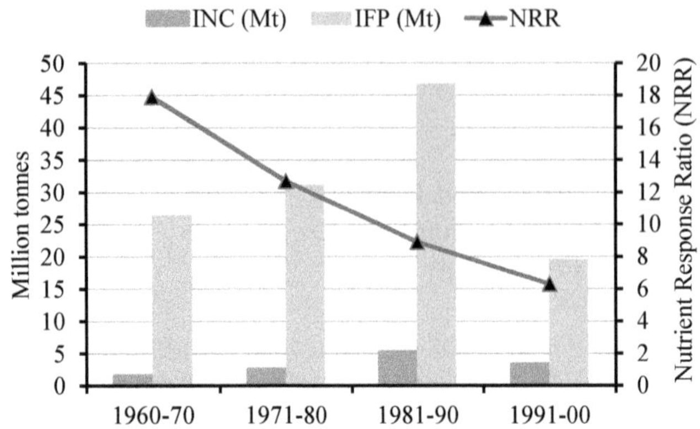

Fig. 7.41 Declining trend of nutrient use efficiency (INC, Increase in nutrient consumption; IFP, Increase in food production; NRR, Nutrient Response Ratio or kg grain per kg of nutrients applied)

1960 to 2000. The grains produced per kg of nutrient supply has decreased over the years.

Continuous nutrient mining owing to uneven nutrient usage is a primary contributor to deteriorating crop response, resulting in soil depletion of major, secondary, and micronutrients such as P, K, S, Zn, Mn, Fe, and B. Contrasting results have been observed in nutrients removal by the crop (Fig. 7.42a) and nutrients removed from the soil (Fig. 7.42b) (Satyanarayana and Tewatia 2009).

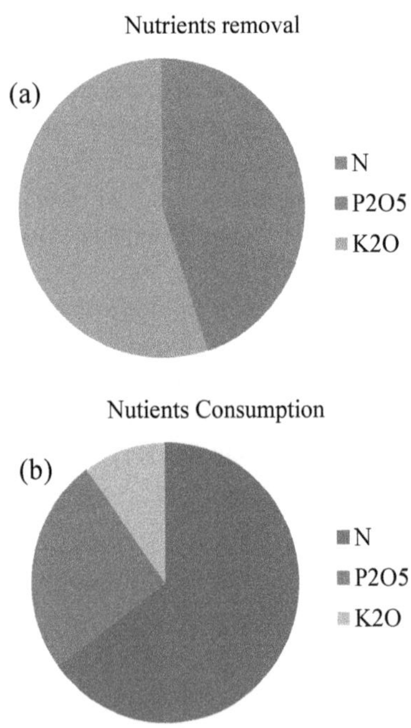

Fig. 7.42 Contrasting pattern of nutrient removal **a** and fertilizer consumption in Indian agriculture **b**

The balance sheet of main nutrients applied and withdrawn from Indian soils was presented by Satyanarayana and Tewatia (2009). They reported that in India, N, P_2O_5, and K_2O application annually was 10.923, 4.188, and 1.454 million tones, whereas removal was 9.613, 3.702, and 11.657 million tons with a gross balance of 1.310, 0.486, and −10.203 million tons, respectively. There was a slight build-up in case N and P but K had quite a large negative balance in the soil. It shows very low application of K and most of the requirement of crops for K was compensated by native soil K. Continuous nutrient mining owing to uneven nutrient usage is a primary contributor to deteriorating crop response, resulting in depletion of some of the major, secondary, and micronutrients like P, K, S, Zn, Mn, Fe, and B from the soil. Singh et al. (2002) reported about the advantage of applying green manure to rice–wheat cropping system and nutrients build-up in the soil. They mentioned that the yield of rice increased by 18.65 and 14.8% with the application of green manure through green gram and sesbania, respectively, over control (no application of green manure). A significant increase in OC and Olsen P in soil was found when green manuring was done.

7.16.10 Causes of Low Productivity in Indian Himalayas

National Academy of Agricultural Sciences (NAAS 2019) compiled the reasons for low crop productivity in mountain areas in India. Broadly it was concluded that the continuous use of fertilizer N alone with inadequate P and K application has led to nutrients imbalance in the soils. The use of nutrients in proper amounts is necessary to maintain a proper balance for higher crop productivity. Intensive cropping systems such as 'rice–wheat' with high-yielding varieties have depleted soils of secondary and micronutrients, particularly S, Zn, Mn, B, and Fe, even when using approved NPK. Generally, high-analysis fertilizers are recommended for crops and cropping systems and addition of organic manures has remained inadequate. This has caused widespread deficiencies of S and micronutrients. Moreover, wrong time and method of fertilizer application has also resulted in low crop productivity and poor soil health. The fertilizer application is done in absence of soil tests, which may lead to over or underuse of nutrients resulting in low productivity and imbalance. Aside from incorrect fertilizer use, nitrogen and other plant nutrients have been leached from rice–wheat cropping systems, sugarcane, and other intensively fertilized crops. There are also some operational issues, such as a lack of appropriate fertilizer availability, a lack of adequate and high-quality soil testing facilities, and non-availability of fertilizer recommendations, a lack of sufficient seeds of high-yielding crop varieties at the appropriate time, and a lack of more nutrient responsive crop genotypes.

Other natural and edaphic factors that contribute to low productivity include antagonistic reactions between some plant nutrients, low soil organic carbon levels in Indian soils, subsoil impedance to plant root systems that limits nutrient uptake, soil degradation due to high salinity/sodicity/acidity/waterlogging, and environmental degradation, which has a negative impact on agriculturally important microorganisms.

Fertilizer will continue to be the major input for attaining the country's expected foodgrain production objectives due to a lack of sufficient organic sources of crop nutrition. Because improved productivity per unit land is the only way to boost foodgrain output, there is a need to improve fertilizer usage efficiency. Balanced and adequate use of N, P, K, S, Zn, B, and Fe, or any other inadequate nutrient, are some of the suggested measures. To realize the full potential of various crops, an Integrated Plant Nutrient Supply System (IPNS), the development of high-quality soil-testing facilities at the district/block level, the timely availability of desired fertilizer materials, the availability of high-quality seed of recommended crop varieties, the implementation of recommended agronomic practices, and the availability of low-interest credit to farmers are all required. For main agricultural and farming systems in various agro-eco zones, customized soil and crop-specific fertilizer ingredients are required.

7.16.11 Performance of Agriculture in India

The annual average growth rate of GDP due to agriculture has remained low as compared to overall growth rate of GDP in India. Between 2004–05 and 2007–08, the agriculture sector grew at a pace of 5%, while between 2008–09 and 2013–14, it grew at a rate of 3%. The economy increased at an annual average of 9% and 7% during the same time periods, respectively (Fig. 7.43). The reliance on the vagaries of the monsoon is one of the key reasons for the large variation in GDP growth rate related to agriculture. Because of the large number of small and marginal holdings, policymakers are concerned about the high volatility.

The agriculture and farming communities have remained resilient even during unfavorable period for other sectors during the COVID-19 pandemic. In 2020–21, agriculture in India grew by 3.4% at constant prices, whereas other sectors nearly collapsed. Agriculture's contribution to GDP climbed to 19.9% in 2020–21, up from 17.8% in 2019–20. The agricultural industry contributes substantially more to the Indian economy than the global average of 6.4%. In India, the contribution of the industry and services sectors is lower than the global average of 30% for the industry sector and 63% for the

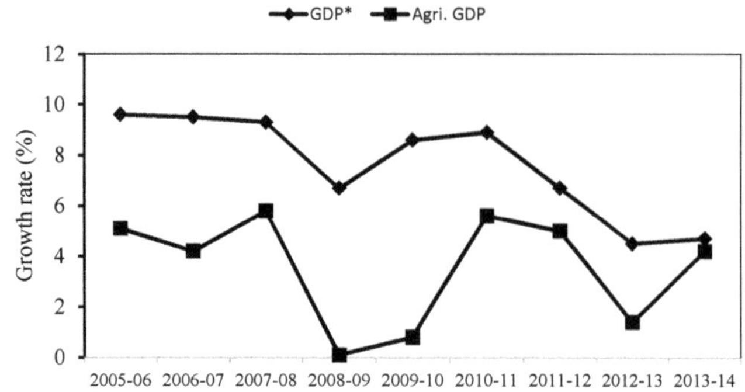

Fig. 7.43 Agricultural GDP Growth Rate at Constant (2004–05) Prices. *Overall GDP. *Source* CSO (2012)

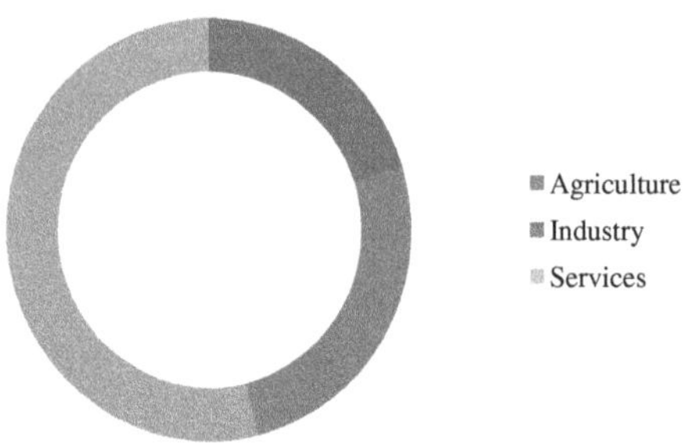

Fig. 7.44 Contribution of various sectors to GDP in India

services sector. Out of total contribution of agriculture and allied sectors to GDP, crops, livestock, forests, and fisheries and aquaculture accounted for 61.6, 25.8, 7.3, and 5.33%, respectively. During 2020–21, the agriculture contributed 20.2% as against 25.9% for industry and 53.9% for service sector (Fig. 7.44).

Ranging from 5.8% in 2005–06 to 0.4% in 2009–10 and −0.2% in 2014–15, agricultural growth has been quite variable during the past years, (Fig. 7.45). Looking into such a variation in growth and its declining trend, some of the decisions made by expert bodies from time to time include bringing in land leasing laws for agriculture, micro-irrigation techniques to improve efficiency of water use, availability of quality seeds by involving the private sector, and a national agricultural market for the trading of agricultural produce online (GOI 2015–16).

Change in GDP of agriculture and allied (livestock, forests, and fisheries) sectors in India as percent of total GDP at current (2020–21) and 2004–05 prices is given in Fig. 7.46. The contribution of agriculture and allied sectors towards GDP at current and 2004–05 prices was 51.8% in 1950–51 which declined to 16.4 and 20.2% in 2020–21. While the growth in manufacturing and service sectors improved from 1950–51 to 2015–16, it has steadily declined in case of agriculture (Fig. 7.47).

The production of various crops in India has been shown in Fig. 7.48 and Table 7.29.

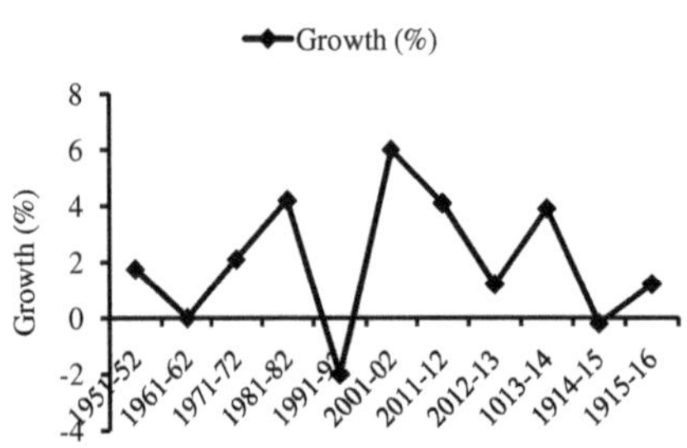

Fig. 7.45 Agricultural growth in India

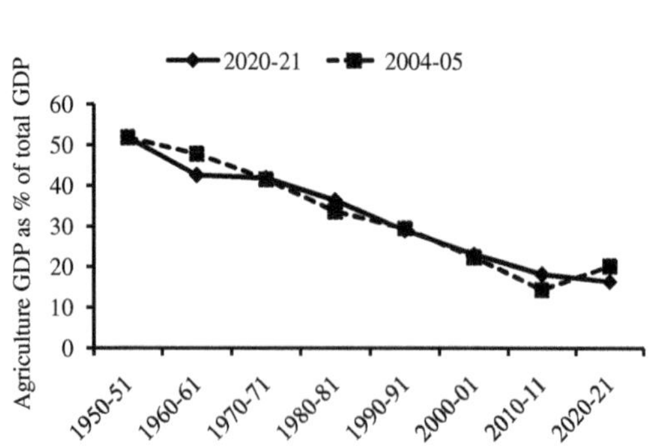

Fig. 7.46 Change in GDP of agriculture and allied (livestock, forests, and fisheries) sectors in India as percent of total GDP at current (2020–21) and (2004–05) prices

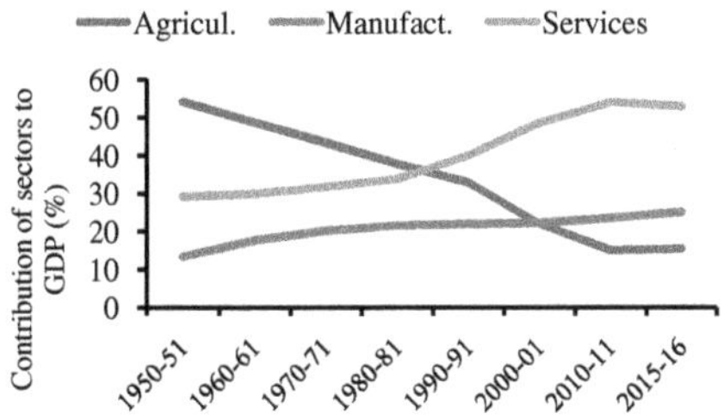

Fig. 7.47 Contribution to GDP of sectors (%). *Sources* Ministry of Statistics and Programme Implementation; PRS

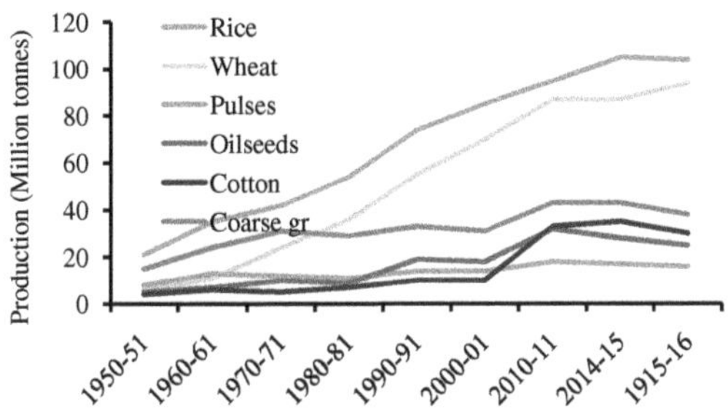

Fig. 7.48 Crop production in India

There has been a steady increase in the productivity of food grains in Indian Himalayan states (Table 7.30). The crop yields are comparatively lower in these hilly states than the plain areas and the states are, by and large, deficient in food grains.

7.16.12 Eastern India Himalayan Region

The Eastern India Himalayan region (EIHR) extends between 21°57' to 37° 5' N latitudes and 72° 40' to 97° 25' E longitudes. The area under cultivation is about 12% of total geographical area with per capita availability of 0.15 ha of land. Many ethnic communities are inhabiting the region. More than 171 of total 573 scheduled tribes of India are living in this region (Samal et al. 2000). Jodha (1992) mentioned that inaccessibility, fragility, marginality, diversity (heterogeneity), niche (natural suitability), and adaptability are major characteristics of the region. The groundwater development and % irrigation area in Indian Himalayan Region states are given in Table 7.31. The percent of irrigated area in Himalayan states is lower compared to All India percentage of 47.2 (CWC 2011).

7.16.13 Agricultural Land Use in EIHR

The elevation, climate, and mountainous terrain or topography are the major factors affecting land use pattern in EIHR. The usual land use pattern of EIHR is given in Table 7.32.

Table 7.29 Crop production in India (million tons)

	Rice	Wheat	Pulses	Oilseeds	Cotton	Coarse grains
1950–51	21	6	8	5	4	15
1960–61	35	11	13	7	6	24
1970–71	42	24	12	10	5	31
1980–81	54	36	11	9	7	29
1990–91	74	55	14	19	10	33
2000–01	85	70	14	18	10	31
2010–11	95	87	18	32	33	43
2014–15	105	87	17	28	35	43
1915–16	104	94	16	25	30	38

Table 7.30 State-wise yield of food grains in Indian Himalayan States (kg ha^{-1})

State	2004–05	2006–07	2008–09	2010–11	2012–13	2014–15
Arunachal Pradesh	1178	1216	1255	1673	1786	1796
Assam	1405	1286	1551	1763	1962	2012
Himachal Pradesh	1923	1714	1757	1787	1850	2011
Jammu & Kashmir	1686	1733	1851	1639	1962	1379
Manipur	2390	2241	2236	2244	1926	–
Meghalaya	1674	1800	1783	1803	1997	2387
Mizoram	1888	822	898	1246	1756	1506
Nagaland	1577	1482	1811	1958	2027	2018
Sikkim	1406	991	1351	1448	1608	1577
Tripura	2179	2610	2526	2587	2711	2680
Uttarakhand	1697	2057	1715	1841	1962	1824
All India	1652	1756	1909	1930	2129	2070

Sources Agricultural Statistics at a Glance 2015, Ministry of Agriculture; PRS

Table 7.31 Groundwater development and coverage of irrigation in Himalayan states of India

State	Groundwater Development (%)	Irrigated area (%)	State	Groundwater Development (%)	Irrigated area (%)
Arunachal Pradesh	0.0	13.0	Mizoram	4.0	9.5
Assam	14.0	5.4	Nagaland	6.0	7.0
Himachal Pradesh	71.0	11.4	Sikkim	137.0	15.9
Jammu & Kashmir	133.0	39.9	Tripura	7.0	21.4
Manipur	1.0	18.6	Uttarakhand	74.0	40.9
Meghalaya	0.0	20.9	All India	62.0	47.2

Sources Central Water Commission; Agricultural Census (2011); PRS

7.16.14 Agroforestry in EIHR

The shifting cultivation prevalent in EIHR presents a sort of primitive agroforestry system. Agroforestry is a long standing practice in this region, where many crops are being grown along with fruit and other trees, (mainly *Alnus nepalensis* and *Schima wallichii*) in the different agro-climatic zones. Waste and degraded land in EIHR of India is presented in Table 7.33. This land can be reclaimed and protected through agroforestry interventions.

7.16.15 Spatial Distribution of Degraded and Wastelands

Water, wind, physical, and chemical degradations have all been investigated in degraded and wasteland environments. In EIHR states, seventeen deterioration classes have been established, and area statistics for each class have been examined and are presented in Table 7.34.

Table 7.32 Land use pattern of N–E Himalayan region ('000 ha)

State	Geographical area	Forest	Land not available for cultivation	Other uncultivated lands excluding fallow	Fallow land	Net sown area	Total cropped area	Agri. Land/cultivable land/arable land/cultivable land
Arunachal Pradesh	8374	5154	64	119	107	216	285	424
Assam[a]	1533	386	686	301	0	160	220	160
Manipur	2233	1742	27	8	0	309	309	316
Meghalaya	2243	946	239	555	215	285	340	1056
Mizoram	2108	1585	95	53	244	116	116	408
Nagaland	1658	863	95	164	155	380	489	1285
Sikkim	710	584	11	11	9	77	140	98
Tripura	1039	629	141	20	4	256	360	277

[a]Hills of Assam

Table 7.33 Wasteland/degraded land in Indian Himalayan states (km^2) (Wasteland Atlas of India 2010)

State	Land with dense and open scrub	Waterlogged and marshy land	Shifting cultivation degraded land	Under-utilized degraded forests	Degraded pastures	Other wastelands (snow-covered, rocky)	Total wastelands	Wastelands as % of TGA
Arunachal Pradesh	3129.7	0	1531.5	20.5	186.1	876.0	5743.8	6.86
Assam	3592.8	1520.2	239.6	3433.3	0	0	8785.9	11.19
Manipur	4619.3	0	852.2	1555.8	0	0	7027.4	31.48
Meghalaya	3094.5	0	449.0	67.1	0	255.1	3865.7	17.24
Mizoram	36.3	0	2617.7	3367.2	0	0	6021.1	28.56
Nagaland	1983.5	0	2827.6	0	0	4.0	4815.1	29.04
Sikkim	6.5	0	0	60.9	0	3213.5	3280.9	46.24
Tripura	527.9	0.7	254.1	522.5	0	10.0	1315.2	12.54
J & K[a]	3897.9	75.5	0	4257.5	125.5	65,398.0	73,754.4	72.54
Himachal	3371.8	10.4	0	1290.4	164.3	17,633.1	22,470.0	46.36
Uttarakhand	1284.3	0	0	730.5	410.7	10,364.5	12,790.0	23.91
Total	25,544.5	1606.8	8771.7	15,305.7	886.6	97,754.2	149,869.5	31.7

[a]Jammu & Kashmir

7.16.16 Use of Vegetative and Other Barriers

The results from Central Soil and Water Conservation Research and Training Institute (CSWCR & TI Vision 2011), Dehradun (Uttarakhand state) showed that the Himalayan region indicates that vegetative barriers can decrease runoff by 18 to 21% and soil loss by 23 to 68% on slopes varying from 2 to 8%. When compared, the guinea grass barrier was found more effective over other barriers (khus khus,

Table 7.34 Land degradation due to water and soil acidity in Indian Himalayan states (000' ha)

State	Exclusively water erosion (>10 tons/ha/yr);	Water erosion under open forest	Exclusively acid soils (pH < 5.5)	Acid soils under water erosion	Acid soils under open forest	Other causes	Total Affected area	% of TGA
AP	165	215	300	501	968	5	2154	25.8
Assam	1,929	437	411	1319	265	210	4571	58.2
Manipur	36	114	115	86	1396	21	1768	79.1
Meghalaya	127	579	52	175	796	3	1732	77.4
Mizoram	0	0	150	0	1013	0	1163	55.2
Nagaland	1	30	17	45	1454	3	1550	93.5
Sikkim	2	0	2	43	13	0	60	8.4
Tripura	26	48	101	83	525	2	785	74.8
J & K	1,327	674	21	42	15	15	2094	20.7
Himachal	941	43	34	41	1	5	1065	19.1
Uttarakhand	829	180	13	189	198	26	1435	25.7
Total	5383	2320	1216	2524	6644	290	18,377	38.9

Notes Classes*: 1. Exclusively water erosion (>10 tons/ha/yr); 2. Water erosion under open forest; 3. Exclusively acid soils (pH < 5.5); 4. Acid soils under water erosion; 5. Acid soils under open forest; 6. Mining/Industrial waste; 7. Waterlogged area (Permanent)
Source NBSS & LUP (2004), Maji et al. (2010)

Bhabar) in arresting runoff and reducing soil loss through erosion. The grain yield of wheat and maize was much higher under the guinea grass barrier.

Despite having rich natural resources of soil, water, and vegetation, the states of the region have remained economically poor. Faulty agricultural practices have depleted the valuable resources of the region. Due to indiscriminate use of soil resources, there is a decline in soil fertility and a necessity to have sustainable, socially acceptable and economically viable land use systems is thus felt. There are quite a few potential indigenous farming systems prevalent for centuries. These systems of cultivation make use of locally available materials and organic source of nutrients thus making them sustainable and eco-friendly. Reduced cycle of shifting cultivation due to increase in population is insufficient for rejuvenation of the sharp decline in soil fertility. So the need of the hour is to adopt suitable land use systems in order to maintain high agricultural potential and soil health (Bhattacharyya et al. 2015).

7.16.17 Indigenous Farming Systems of Eastern Himalayas

Many indigenous farming techniques exist in the Northeastern Hill region, and they perform admirably in terms of resource conservation and ecological balance. These farming systems have been discussed in detail by many workers from time to time (Singh 1989; Sonowal et al. 1989; Prasad and Sharma 1994; Sharma 1997a, b, c, 1998a, b, c).

Zabo System

Zabo system is practised in Phek district of Nagaland and has a combination of forest, agriculture, livestock, and fisheries (Sonowal et al. 1989). Zabo means impounding of water. In this land use system, there is a forestland on the top of the hill followed by water harvesting system in the middle and livestock yard and paddy fields at the foothills. The slope of the catchment area is kept under natural vegetation. Water harvesting ponds are dug out with the formation of earthen embankment below the catchment area. The usual capacity of the pond varies from 300 to 600 cubic meters with depth varying between 1.5 and 2.5 m. The ponds are having overhead silt retention tanks and are compacted from the inner surface in order to arrest water loss. Below the pond, it is a common practice to make a cattle enclosure with wood and bamboo. Buffaloes, cows, and pigs are the common animals with the farmers. Cattle yards are washed with run-off water that is then diverted to the paddy fields for manuring (Sharma and Sharma 2003a, b). Rice is grown in the bench terraces constructed at the lower elevations. The area of paddy fields varies between 0.2 and 0.8 ha. Use of leaves and branches of Alnus nepalensis, Albizzia lebbeck, and Mekhonu for green manuring is common. Rice fields are thoroughly rammed at the puddling time to minimize percolation losses. Seepage loss through shoulder bunds is checked with the use of paddy husk on the upstream side. Farmers also practise paddy cum-fish culture.

Rice-Based Farming System of Apatani Plateau

The Apatani plateau occupying a stretch of 26 sq. km area in Subansiri district of Arunachal Pradesh is inhabited by "Apatani" tribe. The valley is called "The Rice Bowl" of the Apatanis who practise wet rice cultivation and have a good knowledge of land and water management (Sharma and Sharma 2005). Apatani farmers grow a number of tree and bamboo species in the hills surrounding the valley. Rice is cultivated on terraces prepared on mild slope in the valley. They have developed a remarkable system of water management to maintain a 10 cm water table in the plots. The plot-to-plot outlets in rice fields are made with wood or bamboo. Nutrient and fertility management of the terraces is done mainly through the recycling of agricultural wastes. All rice varieties grown are local. 'Emo' variety gives the highest yield (5.2 t ha^{-1}) followed by 'Pyaping' (4 t ha^{-1}). Paddy-cum-fish culture is also attempted. Finger millet is generally grown on the risers. This system is highly sustainable and efficient and helps in conserving natural resources, maintenance of ecological balance, and flow of streams.

Bamboo Drip Irrigation System

This Bamboo Drip Irrigation System (BDIS) is followed mainly in Hills of Meghalaya (Singh 1989). This irrigation system has been found to be extremely useful under water scarcity situations and/or soils have poor water-holding capacity, the topography is rocky and undulating and irrigation is required for such crops where water need is relatively low. The convenient technique would be to divert water from upper reaches of the hill top to lower reaches through gravitational flow. Different forms of bamboo pipes are used, and the rate of water flow varies from 800 to 1200 liters hr^{-1}. Bamboo drip is an excellent system to provide irrigation to specific crops such as beetle vines, areca nuts, and black pepper.

Agriculture with Alder (Alnus nepalensis)

Alder (Alnus nepalensis) is grown in Nagaland at altitudes varying between 1000 and 2000 m. It is a leguminous tree having the ability to fix 50 to 100 kg N ha^{-1}. Alder trees have multipurpose uses. It is used for firewood, furniture, and house construction. Alder trees become ready for pollarding after 6 to 8 years of planting. The alder is pollarded horizontally at a height of 2 to 2.5 m above ground level. Tree trunks are then stripped off sub-branches and leaves which are left in the field and burnt in heaps. Ash is spread uniformly in the field. From February to April, various crops, viz., Maize, job's tears, millet, potato, chillies, pumpkin, and barley are sown. Harvesting is done in October or November. By this time, alder tree sprouts new coppices. Stubbles of the previous crops and the new leaves are collected, burnt, and added to the soil to augment its fertility. Besides annual crops, alder tree is also useful for providing shade to tea, coffee, and cardamom (Fig. 7.49).

Rice Cultivation on Terraces (Panikheti)

This system of rice cultivation originated in Nagaland but is now prevalent in Manipur and Sikkim. Terrace construction on hill slopes becomes necessary to keep rainwater standing in

Fig. 7.49 *Alnus nepalensis* on pollarding

the terraces, and thus it is locally known as "Panikheti". The terraces are constructed and supported with strong bunds of width varying from 0.5 m to 1.2 m and height from 0.2 to 1.4 m. The bunds are rammed properly with sticks and local tools to make them firm. The terraces are leveled properly, and clayey soils are spread to check water percolation appreciably. The water coming down from the hill surrounding the terraced lands is diverted to the rice fields through channels and sub-channels. Local rice varieties are grown on the terraces and colocasia and some other crops are sown on risers.

The perusal of the data as presented in Table 7.35 indicates that the indigenous farming systems have a significantly large area under effective vegetative cover. Zabo system contains the maximum area under forest (46.5%) followed by Apatani (41.4%), Narrow valley cultivation (39.1%), Agriculture with Alder (33.6%), Panikheti (26.0%), and Shifting Cultivation (19.8%).

Homestead Agroforestry System

In every homestead, farmers of Assam and Tripura states grow number of tree species along with livestock, poultry, and fish mainly for the purpose of meeting their own needs. Jackfruit, coconut mango, areca nut are grown and in their interspaces pineapple, vegetables, and other crops are also grown. These traditional

Table 7.35 Effective land cover (ha) under different farming systems

Farming system	Forest (Open, dense, and scrubs)	Grazing lands	Trees and groves	Cropped area	Total
Zabo	0.465	0.029	0.022	0.054	0.570
Apatani	0.414	0.019	0.017	0.090	0.540
Agri-with alder	0.336	0.038	0.034	0.096	0.504
Panikhati	0.260	0.014	0.023	0.043	0.440
Valley cultivation	0.391	0.026	0.037	0.059	0.513
Shifting cultivation	0.198	0.034	0.022	0.050	0.304

agricultural land uses practised over centuries are well adapted to the local agro-climatic conditions thus providing food and employment to the rural masses as well as maintaining ecological balance. There is a need to bring about an improvement to the land uses for getting higher crop yields. This could be done through adoption of viable technology such as effective cropping sequence, new crop varieties, and integrated nutrient management, etc. Tea gardens in the Himalayas are good example of agroforestry systems (Figs. 7.48 and 7.49).

It is found that the EIHR has more than 13.06 M ha of land under forest accounting for 47.43% of the geographical area. The data also indicate that the area under forest has increased from 42.27% of the total geographical area in 1982 to 47.43% in 2006. According to the national forest policy, one-third of the geographical area of the country should be under forest and it should be 23% in the plain area and 66% in the hill areas. So, sub-zones I and II remain below the national norm. The area (Anonymous 2002) under recorded forest is 0.14 M ha, while the area under actual forest cover is 0.167 M ha, covering 53.4 and 63.6% of TGA in NE states, respectively. However, a decline of 15,800 ha of forest area has been indicated between 1995 and 1997 (Figs. 7.50 and 7.51).

It is generally reported that forest in Eastern India Himalayas remains in degraded state and the extent of degradation is so intense that forest

Fig. 7.51 Forest ecosystem in Assam

tree species having 10% crown density cover a large proportion of the reported area. Out of 74.88 million hectares forest (22.7 percent of TGA), India has only about 37.8 million ha of closed forests covered with crown density of over 40%, which is only 11.5% of total geographical area of the country. This shows the poor status of our forests despite the National Forest Policy of 1952 which envisaged to raise the forest area and its cover to 33% of geographical area of the nation. National Remote Sensing Agency reported loss of about 2.89% area of the country for forest cover just in a span of seven years between 1972 to 1975 and 1980 to 1982. The government is trying to increase the area under forest cover by planting trees on hill slopes (Fig. 7.52) and teak plantation in Himalayan states of Assam and Mizoram (Fig. 7.53).

Fig. 7.50 View of tea garden in Assam

Fig. 7.52 Forest in hill slopes of Nagaland

Fig. 7.53 Teak plantation in Mizoram

7.16.18 Agriculture in Indian Himalayan States

Besides forestry, agriculture which contributes about 18% of gross domestic product of India also becomes the mainstay of livelihood among the farming community of Eastern India Himalayan states. Area not available for cultivation for the region as a whole amounts to 21.03% of the reported area. Relatively higher percent of the area is reported to be used for non-agricultural purposes in Assam, precisely because of high population density and growing need for settlements, roads, and urban centers, etc. On the other hand, the highest proportion of barren and uncultivable land is found in Mizoram and Tripura states. Net area sown constitutes 18.75% of the geographical area of the region. A severe scarcity of cultivable land is generally noted in the hills owing to their physiographic situations. Assam state contains the maximum (34.59%) net area sown and the least (2.04%) area is under cultivation in Arunachal Pradesh.

Cultivators of Eastern Himalayan states of India have developed beautiful terraces for rice cultivation (Figs. 7.54 and 7.55) and cultivation of millets on slopy lands (Figs. 7.55 and 7.56).

7.16.19 Horticulture in Indian Himalayas

The agro-climatic situation prevailing in NE region is conducive for cultivation of wide variety of tropical and temperate fruits. The dominant horticultural crops of the region include pineapple, citrus, notable among them being mandarin oranges, and Assam lemon, banana, and areca nut. The EIHR has a total area of 23.78 thousand hectares under pineapple and 29.67 thousand hectares under citrus. Areca nut is an important horticultural crop of the region having a high commercial value. Total area under this crop is estimated at 56.94 thousand ha. A new crop which holds great promise for development is apple. Presently, 2.31 thousand hectares are put under this crop. Cultivation of pineapple, vegetables, coffee, rubber, etc. in Himalayan states has been depicted in Figs. 7.57, 7.58, 7.59, 7.60, 7.61 and 7.62.

Fig. 7.54 Terrace cultivation

Fig. 7.55 Rice cultivation in hill slopes of Mizoram, Valley under rice in Mizoram

Fig. 7.56 Crop cultivation in Tripura in Hilly slopes

Besides farming, rearing of milch cattle, goat and poultry keeping, piggery, sericulture, and weaving are some of the supplementary occupations for rural peoples of EIHR.

7.16.20 Cropping Systems

Several crops are grown in this region depending on the system of farming, food habits, and agro-climatic conditions. Rice, potato, and maize are the main crops of this region and rice-based cropping system is common (Table 7.36). But the yield is very low. Assam is by far a mono-crop state with 73% of the total cropped area under rice. The most profitable cropping pattern for Assam (Borthakur 1992; Sharma 2002) are as follows:

Cropping systems in Tripura are (i) tilla (upland) area: Rice-groundnut/green gram/black gram, sesamum—rice/cowpea. (ii) mid-high (chara land) are: Rice-rice, rice-potato, rice/sesamum-rice, rice-potato, (iii) lunga (low land) area: rice-rice-rice, rice-vegetables.

In Sikkim, maize, rice, finger millet, ginger, and potato are the prominent crops in the

Fig. 7.57 Banana cultivation in Mizoram

Fig. 7.58 Pineapple cultivation on hill slopes in EIHR

cropping systems. In maize-based cropping system, about 12 cropping patterns have been identified having maize as the base crop. Maize is rotated with wheat/barley/mustard/rabi maize or vegetable in rabi or buckwheat. Finger millet, pulses, potato, ginger, and vegetables in the Kharif are either the alternatives or intercrops. In rice-based cropping system, rice is rotated by rice/wheat/mustard/maize/potato/vegetables. In finger millet-based cropping system, finger millet is largely rotated with barley. Three cropping patterns are identifiable, viz. (i) Millet (pure), (ii) Maize + millet (relay cropping), (iii) Maize-millet (sequential cropping) in Sikkim. In potato-based cropping system, recommended cropping systems with potato as base crop are as follows:

Fig. 7.59 Cucurbit cultivation in Mizoram

Fig. 7.60 Areca nut cultivation in Assam

Fig. 7.61 Coffee plantation in Assam

Potato-maize-millet, Potato-maize-rice, Potato-pulses, and Potato-rice. In ginger-based cropping system, the recommended systems of cropping with ginger as base crop stand such as

Fig. 7.62 Rubber plantation in Tripura

Ginger–Tapioca (3 or 4 years rotation), Ginger-Sweet potato/yam/chilli, Ginger + Maize/ragi/vegetables/garlic/onion. Ginger is also grown under the shade of orange, lime, and lemon. Some of the forage-based systems found suitable for lower and mid hills are: Rice-Berseem—Fodder maize, Maize + Cowpea—Oat + Pea, Napier grass + Cowpea/Ricebean—Oat, Tall fescue + Ricebean, Perennial Ryegrass, Orchard grass + Maize/Cowpea, Fodder Triticale + Ricebean, Grassland + Rice/Ragi/Soybean. In plantation crops and fruit tree-based systems, crops under this category include large cardamom, orange, apple, other temperate fruits, guava, banana, and tea. Large cardamom is cultivated throughout the state. North district in Sikkim has more than 60% area followed by West district (17%), south (14%), and east (9%). Ginger, buckwheat, beans, and maize are the principal crops grown in orange orchard. Radish, cabbage, pea, and potato are the crops grown in apple orchard. Under other unconventional crops and cropping systems, floriculture and cultivation of medicinal plants hold great promise in Sikkim.

Potato-rice or monocropping of rice is common in central plateau of Khasi Hills in Meghalaya. It is also possible to grow wheat (irrigated) after rice in Meghalaya in mid and low altitudes. Under rainfed terraces, cultivation of linseed/mustard/pea after rice with life-saving irrigation will be an ideal cropping system (Patel 1983). Mandarin orange-based cropping system is also commonly found in Meghalaya. Wheat,

Table 7.36 Cropping systems for flood free and flood-affected areas of Himalayan state of India (Assam)

Region	Flood free	Flood affected
Upper Assam	Autumn rice—winter rice; Winter rice—pea; Autumn rice—mustard/potato/blackgram/green gram/vegetables Soybean-mustard; Rice-potato	Autumn rice—pea Autumn rice—mustard/pea
Central Assam	Autumn rice—winter rice; Winter rice—pea Summer rice—wheat; Autumn rice—rabi pulse/mustard/potato/vegetables; Rice-potato	Autumn rice—mustard/pea/lentil; Wheat; Rice-groundnut
Lower Assam	Autumn rice—winter rice Summer rice—wheat Autumn rice—rabi pulse/mustard/potato Autumn rice—winter rice Summer rice	Late winter rice Autumn rice Rice-rice Jute-rice-rice

oilseeds, pulses, potato, and other vegetables are being introduced in Manipur Valley after rice crop. In Arunachal Pradesh, temperate fruit-based cropping systems are present. Maize and rice-based cropping systems are also existing in the valleys. These crops are rotated by soybean/cowpea/groundnut/green gram. Sharma and Arora (1988) found increase in size and quality of potatoes by applying P fertilizers in potato deficient soils.

An Integrated Approach

In order to sustain the land productivity in accordance with the need of the people inhabiting the area, scientific land and water use are to be practised. It is pertinent to mention that shifting cultivation, the adverse effects of which are manifolds, is intermingled with cultural heritage of tribal population of NE states. Thus, Borthakur (1992) felt that improvement approaches to shifting cultivation such as provision of contour bunds or vegetative barriers, improved varieties of crops, growing of different crops in separate strips rather than as intimate mixture, dibbling of fertilizers along with seeds might be undertaken. An alternative approach to shifting cultivation gives birth to watershed-based land use systems. The different farming systems demonstrated at research stations are as follows:

Agriculture-Based Land Use

In this system, bench terraces, contour bunds, and flat land area, if available, are used for growing various arable crops to develop a suitable cropping system.

Livestock-Based Land Use

The entire watershed in this system is utilized for feed and fodder production, cultivating fodder crops and trees. The livestocks are also maintained in the watershed itself. Composite fish culture can also be implemented if water bodies are available.

Horticulture Land Use

In this watershed, various horticultural crops according to the suitability of the land and slopes are grown.

Agroforestry Land Use

The trees under this watershed provide fuelwood, timber, and animal fodder.

Agri-Horti-Silvipastoral Land Use

In this system, trees with pasture are planted at the top of the slope. The horticultural crops are grown in the middle portion and agricultural crops being grown on the lower portion of the slope. Animal husbandry and pisciculture can also be incorporated in the watershed.

7.16.21 Selection of Land Use System Hill Areas

Under the perspectives of various land use systems available in the region, the major factors for deciding a land use system in the hills are (i) soil depth (ii) slope, and (iii) extent of land degradation (Sharma 1998a, b, c). These factors are taken into consideration to delineate three categories as presented in Table 7.37.

The suitable land use systems (Sharma, 1998a, b, c) can be assigned taking into consideration the different land parameters. When the soils are too shallow on hill slopes, it is not advisable to grow annual crops which need frequent tillage. Only horticulture (H), forestry or other woody species and grasses or silvi-pastoral (SP) land uses have been assigned to D_1. All sorts of agricultural crops (A), horticulutre (H) and silvi-pastrol (SP), agri-horti-silvi pastoral system (AHSP) can be adopted under D_2 and D_3 soil depths.

In this way, land use systems like agriculture (A) to G_1, horti-silvi-pastoral (HSP) to G_2 and silvi-pastoral (SP) to G_3 categories of land degradation and agriculture (A) to S_1, horticulture (H) to S_2 and silvi-pastoral (SP) to S_3 soil depth are assigned. The suggested land use systems for eastern Himalayas are presented in Table 7.38 (Sharma 1990b; 1998a, b, c).

Two schematic diagrams to find a land use system for a particular location (Fig. 7.63) can be used as a ready reckoner to select a land use system under different situations of soil depth, slope, and degradation.

With the knowledge of slope of the land, depth of the soil, and the extent of land degradation, a suitable land use system can be ascertained through the diagrammatic presentation of the systems. Suppose a land has 1.20 m soil depth (D_2), 25% slope (S_1) and slightly degraded soil (G_1) or D_2 S_1 G_1, we can assign AHSP for D_2, A for S_1, again A for G_1or combinedly AHSP + A + A or A_3 HSP to D_2 S_1 G_1. Since agriculture (A) has been repeated thrice against once each for horticulture, silviculture, and pasture, agriculture would be the most suitable land use systems for

Table 7.37 Delineation of limits for different factors determining land use

Factors	**Low**	**Medium**	**High**
Soil depth	<0.5 m (D_1)	0.5 to 1.5 m (D_2)	>1.5 m (D_3)
Slope	<30% (S_1)	30–60% (S_2)	>60% (S_3)
Land degradation	Normal to sheet erosion (G_1)	Rill erosion (G_2)	Gully erosion (G_3)

Table 7.38 Land use systems for different land parameters

Land parameters	Land use systems
Soil depth	
D_1	Horti-silvi-pastoral (HSP)
D_2	Agri-horti-silvi pastoral (AHSP)
D_3	Agri-horti-silvi-pastoral (AHSP)
Land degradation	
G_1	Agriculture (A)
G_2	Horti-silvi-pastoral (HSP)
G_3	Silvi-pastoral (SP)
Soil slope	
S_1	Agriculture (A)
S_2	Horticulture (H)
S_3	Silvi-Pastoral (SP)

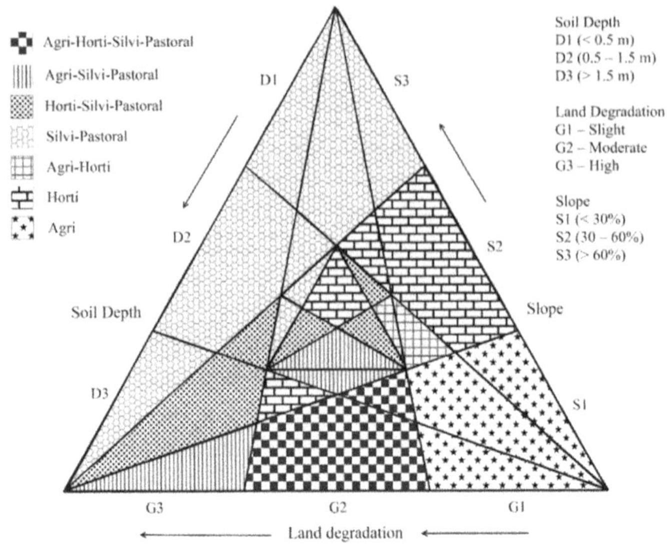

Fig. 7.63 Schematic diagram of land use systems

this area. Similarly, for land having 35 5 slope (S_2), 1.8 m soil depth (D_3), and moderately degraded (G_2), the appropriate land use system would be H + AHSP + HSP or $AH_3S_2P_2$. So horticulture being predominant is going to be preferred. The perusal of Fig. 5.2 indicates that on the left-hand side of the triangle, if we consider it a hillock, the agriculture is at the lower slope, horticulture in the mid slope and silvi-pastoral on the top slope of the hill. When the land is highly degraded (G_3), it is mainly silvi-pastoral land use system being more preferable. At the base, it is agriculture on the right, agri0 hort0 silvi-pastoral in the middle, and the silvi-pastoral on the left side according to land degradation. All other possible combinations of land configurations and relevant land use systems are shown in the triangle. Through permutation and combinations of different parameters, the following land use could be obtained (Table 7.39).

It is now possible that various farming systems can be practised depending on the slope gradient of the area, hydrological behavior of the watershed, soil depth, market availability, and need of the farmers. The need of the hour is to make a concerted approach to evaluate the sustainability and remuneration of the particular land use system to be adopted by the farming community. The Himalayas, being ecologically vulnerable, unplanned, reckless use of land, forest, and water resources may turn out to be disastrous. Land use planning, therefore, needs to be given the top priority.

The land is divided into *Rajs* or ruling units in the Khasi Hills district of Meghalaya, and each member of the unit can cultivate as much land as he pleases. But many times, the head of the clan or headman allots the land according to the number of members in a particular family. In Mizoram, earlier, there was a system of hereditary chieftainship but now the land belongs to the villages council. The Karbi tribe of Assam practises community ownership of land. Land tenure and small size of holdings are two major problems impeding the crop productivity and production due to the reason that farmers do not invest for soil and water conservation measures.

The productive activities related to agriculture, horticulture, plantation, animal husbandry, fisheries by the same households depend on the location, cultivation and nature of land available. For the plains of Assam, Tripura, and Manipur, the basic need is to expand assured irrigation facilities for the expansion of gross cropped area. A dependable system of supply of the inputs at reasonable prices and marketing of products matched by appropriate price policy needs to be built up. On the other hand, the hill farmers are unlikely to ensure adequate rice production as rice is the staple food of the hill people. Unless a

Table 7.39 Maximum land use systems are possible by using the figures above. (Sharma 1998a, b, c)

Code	De-coding	Suggested land use
D1G1S1	HSP + A + A = A_2HSP	Agriculture
D1G1S2	HSP + A + H = AH_2SP	Horticulture
D1G1S3	HSP + A + SP = AHS_2P_2	Silvi-pastoral
D1G2S1	HSP + HSP + A = AH2S2P2	Horticulture, silvi-pastoral
D1G2S2	HSP + HSP + H = H3S2P2	Horticulture
D1G2S3	HSP + HSP + SP = H2S3P3	Silvi-pastoral
D1G3S1	HSP + SP + A = AHS2P2	Silvi-pastoral
D1G3S2	HSP + SP + H = H2S2P2	Horticulture, Silvi-pastoral
D1G3S3	HSP + SP + SP = HS3P3	Silvi-pastoral
D2G1S1	AHSP + A + A = A3HSP	Agriculture
D2G1S2	AHSP + A + H = A2H2SP	Agri-horticulture
D2G1S3	AHSP + A + SP = A2HS2P2	Agri-silvi-pastoral /
D2G2S1	AHSP + HSP + A = A2H2S2P2	Agri-horti-silvi-pastoral
D2G2S2	AHSP + HSP + H = AH3S2P2	Horticulture
D2G2S3	AHSP + HSP + SP = AHS3P3	Silvi-pastoral
D2G3S1	AHSP + SP + A = A2HS2P2	Agri-silvi-pastoral
D2G3S2	AHSP + SP + H = AH2S2P2	Hort-silvi-pastoral
D2G3S3	AHSP + SP + SP = AHS2P2	Silvi-pastoral
D3G1S1	AHSP + A + A = A3HSP	Agriculture
D3G1S2	AHSP + A + H = A2H2SP	Agriculture, Horticulture
D3G1S3	AHSP + A + SP = A2HS2P2	Agri-silvi-pastoral
D3G2S1	AHSP + HSP + A = A2H2S2P2	Agri-horti-silvi-pastoral
D3G2S2	AHSP + HSP + H = AH3S2P2	Horticulture
D3G2S3	AHSP + HSP + SP = AH2S3P3	Silvi-pastoral
D3G3S1	AHSP + SP + A = A2HS2P2	Agri-silvi-pastoral
D3G3S2	AHSP + SP + H = AH2S2P2	Horti-silvi-pastoral
D3G3S3	AHSP + SP + SP = AHS2P2	Silvi-pastoral

strong public distribution system for meeting this basic need of the people is organized, it will be nearly impossible to motivate the people to adopt the suggested changes. Secondly, the requirement of land for agri-pastoral, horti-pastoral, or silvi-pastoral practices by the tribals after abandoning the practice of Jhuming cannot be met unless the tribal rights over the land where they have been traditionally practising shifting cultivation are respected. There is a trend of growing inequality in the distribution of land. In the tribal areas, an appropriate communal form of ownership is it the cooperative form, joint farming society form, village ownership form or any other system may be evolved. This is necessary to ensure an effective interface between the system of delivery of farming inputs and the systems of receipt and use of these inputs for production/disposal of the product on the market.

The model of using the land for mixed farming covering silviculture, horticulture, agriculture, animal husbandry, and pisciculture may be used to evolve the pertinent strategies. The achievement depends on an effective system of management of supply of inputs, credit, irrigation water on the one hand and implementation of price support policy/marketing of the products on the other.

7.16.22 Response of Crops to Nutrients

The Eastern Indian Himalayan Region (EIHR) is predominantly hilly and characterized by varied type of agro-climatic and geographical situations. Mainly dominated by forest, the region has about 15% area under cultivation (Sharma et al. 2006). In spite of concerted efforts made by the Government and the introduction of modern technologies for crop production, the region has remained deficient in food grains. One of the major causes of low crop productivity in the region is soil acidity. More than 95% of the soils are acidic in reaction requiring the use of amendments for the correction of acidity for optimum flow of crop nutrients. The other major causes of low crop productivity are (Venkatesh et al. 2006):

- Non-availability of seed material of recommended crop cultivars due to poor or no seed production programs,
- Very low application of recommended crop nutrients for exploitation of crop varietal potential,
- Poor extension mechanism for extending developed agricultural technologies,
- Absence of approach for commercial viability of non-conventional food crops,
- Low irrigation facility and low soil moisture during winters,
- Socio-economic issues such as prevailing land tenure system and stray grazing, and
- Subsidized availability of food grains from the central pool.

Response of various crops to nutrients in Indian Himalayas is mentioned below.

7.16.22.1 Rice

Rice is the major crop of the Eastern Himalayan Region. Productivity of the crop is low in this part of the World due to many reasons and low use of crop nutrients to exploit the full potential of high-yielding varieties is one of them. Phosphorus fixation in the soil due to high content of exchangeable Fe and Al is a major constraint.

Nitrogen

Nitrogen is one of the most important nutrients affecting the yield of crops in Eastern Himalayas of India. In Meghalaya area, the rice crop responded to 60–90 kg N ha^{-1} under upland conditions depending on the nature of the soil and crop variety. A linear response of rice up to 100 kg N ha^{-1} was found in Tripura. Response of rice to N in Sikkim has been reported to be 60–90 kg N ha^{-1} (Sharma and Singh 1996). In Nagaland area, 80 kg N + 10 t FYM ha^{-1} resulted in higher yield of 4.97 t ha^{-1}. Optimum requirement of N and P for rice variety RCM-8 was 90 and 26 kg ha^{-1}, respectively, in Manipur (Kailash Kumar and Rao, 1990).

Phosphorus and Potassium

Though the dose of P is low compared to N and K in Eastern Himalayas, its large doses are required to realize the full potential of rice crop. Data presented in Table 7.40 (Singh et al. 1993) revealed that maximum yield response was when 85 kg K + 45 kg P ha^{-1} were applied to rice crop. Differential response of rice to applied P has been reported from studies conducted under different soil and water management conditions. The economic dose of SSP, Mussoorie rock phosphate and Purlia rock phosphate has been found to be 60, 53, and 51 kg P_2O_5 ha^{-1}, respectively and uptake of P by the crop correlated with the yield of the crop (Sharma and Singh 1996). Laskar and Datta (1989) found that the mixture of SSP and RP (1:1) for increase in rice productivity in acid soils of Tripura. Phosphate-enriched poultry manure applied to rice-groundnut crop rotation was found to be highly suitable in acid soils (Datta and Laskar 1995). Among different P sources, DAP, MRP, and mixture of SSP + MRP (1:1) are equally effective, their relative efficacy being in the order DAP > MRP > MRP + SSP > SSP > PRP.

The response of rice to K varied between 30 and 80 kg K ha^{-1}. The adequacy limit of K and response of rice to applied K in rainfed direct seeded rice was 30 kg K ha^{-1}. A level of 213.5 kg K ha^{-1} was considered as critical level of K. However, Sharma and Singh (1996) reported that the economic dose of K for rice was

Table 7.40 Effect of sources of P (@ 26 kg P ha^{-1}) on the grain and straw yield of rice (t ha^{-1})

P source	Grain			Straw		
	Rainfed	Flooded	Mean	Rainfed	Flooded	Mean
Control	2.88	3.15	3.01	7.69	7.95	7.82
DAP	3.39	3.37	3.38	8.55	9.08	8.81
SSP	3.01	2.89	2.95	7.88	7.88	7.88
PRP	2.81	2.90	2.85	6.92	7.90	7.41
MRP	3.15	3.60	3.37	7.44	8.01	7.72
MRP + SSP (1:1)	3.03	3.58	3.30	7.80	7.80	7.80
Mean	3.04	3.24	3.14	7.71	8.07	7.89
C.D. (p = 0.05)						
Water management		NS			0.39	
P sources		0.18			0.41	
WM x P		0.25			NS	

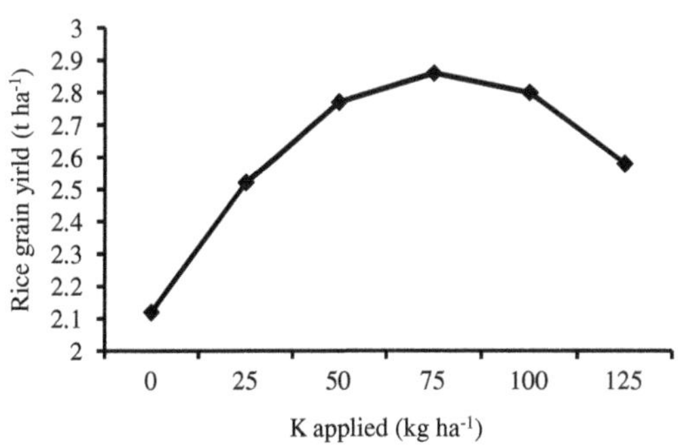

Fig. 7.64 Response of rice crop to potassium

69 kg K_2O ha^{-1} for Meghalaya soils. The regression equation developed in Meghalaya soils (Patiram and Prasad 1983) for response of rice to K was Y = 21.217 + 0.191x − 0.00123x^2 (r^2 = 0.887) (Fig. 7.64). Top dressing of K to rice at 10 days after transplanting and at the time of tiller initiation, tinkering, and maximum tillering stages was found superior to basal application (Patiram and Prasad 1989).

Micronutrients

With the intensification of agriculture and introduction of high-yielding varieties, deficiency of micronutrients has aggravated and rice crop has responded fairly to the application of various micronutrients. Zinc deficiency is found to be more common compared to other micronutrients. On a soil having available Zn of 0.79 ppm and B of 0.63 ppm, a dose of 3.0 kg Zn and 1.5 kg B ha^{-1} was found to increase the yield of rice significantly. Response of rice to higher levels of Zn (9 kg ha^{-1}) has also been reported under low land conditions (Singh et al. 1990). It has been observed that the requirement of Zn, B, and Cu varied between 5.0 to 12.5, 1.5 to 7.5, and 1.5 to 3.5 kg ha^{-1} for getting optimum yield of rice. In Tripura, 8.0 kg of $ZnSO_4$ and 1.4 kg of B ha^{-1} gave an additional yield of 0.7 t ha^{-1} and 0.64 t ha^{-1} over control. Datta and Laskar (1988) and Datta and Dhiman (2001) reported that 15.7, 30.8. 37.2, 21.2% increase in yield of rice could be obtained with the application of 20 kg $ZnSO_4$, 5 kg Boric acid, 1.5 kg ammonium molybdate or $CuSO_4$ under upland conditions and 19, 22.7, and 19.8% with Zn, B, and Mo, respectively, under flooded conditions (Table 7.41). No response to Cu was found under flooded condition.

Integrated Nutrient Management

With emphasis being given to sustainable agriculture, the use of organic sources of nutrients has become all the more important. The Eastern Himalayan Region is traditional user of organic manures for crop nutrition but vegetables and potatoes usually get preferential treatment. Though the organic manures such as FYM,

Table 7.41 Rice grain yield (t ha^{-1}) as influenced by micronutrients in Tripura

Micronutrient	Dose (kg ha^{-1})	Upland	Lowland	Micronutrient	Dose (kg ha^{-1})	Upland	Lowland
Control	0	1.17	2.71	H_3BO_3	2.0	1.36	s2.78
$ZnSO_4{\cdot}7H_2O$	10	1.29	–	H_3BO_3	5.0	1.53	3.32
$ZnSO_4{\cdot}7H_2O$	20	1.35	3.22	H_3BO_3	8.0	1.42	2.96
$ZnSO_4{\cdot}7H_2O$	30	1.28	3.17	$(NH_4)_6MO_7O_{24}$	0.5	1.43	2.62
$CuSO_4$	5	1.42	2.72	$(NH_4)_6MO_7O_{24}$	1.0	1.60	3.24
$CuSO_4$	10	1.20	2.86	$(NH_4)_6MO_7O_{24}$	1.5	1.30	2.95
$CuSO_4$	15	1.10	1.69	CD ($p = 0.05$)	–	0.07	0.35

Table 7.42 Integrated nitrogen management in lowland rice in two eastern Himalayan states of India

Treatment	Grain yield	(t ha^{-1})
	Meghalaya	Manipur
Control	2.29	2.01
Azolla (composted) + 50% N through urea	3.08	3.98
FYM + 50% N through urea	3.13	3.87
Azolla fresh (10 t ha^{-1}) + 50% N through urea	2.69	4.36
Urea @ 80 kg N ha^{-1} (three splits)	2.68	3.92
Green manure + 50% N through urea	3.82	3.75
Azolla inoculation (1 t ha^{-1}) + 50% N through urea	2.63	3.70
Azolla inoculation (fresh) @ 2 t ha^{-1}	2.63	–
Urea (40 kg N ha^{-1} (three splits)	2.63	–
CD ($p = 0.05$)	–	0.95

compost, and crop residues are short in supply, these are necessary to maintain soil fertility and soil physical properties. Patel and Singh (1984) reported that 50% of N each through FYM and urea gave 16.7% higher yield of rice over N applied through urea alone. Green manuring along with 40 kg N ha^{-1} as urea resulted in enhancing the rice productivity significantly (Table 7.42).

7.16.22.2 Maize

The studies conducted on response of maize crop to nitrogen have shown that a dose of 80 kg N ha^{-1} was optimum for getting significantly higher yield of the crop, though the results reported from various locations showed a variable response to N from 25 to 150 kg ha^{-1}. Application of N and P_2O_5 beyond 80 and 40 kg ha^{-1}, respectively did not produce profitable yield of maize in Assam and Meghalaya (Prasad et al. 1981). A mixture of SSP + RP (1:1) along with lime gave agronomic efficiency of 88% (Prasad et al. 1984). P use efficiency of inorganic P carriers could be increased by charging them with organic manures. Relative effectiveness of amended P carriers was in the order of poultry manure + inorganic P > pig manure + inorganic P > FYN + inorganic P for the first and second maize crop grown in succession.

7.16.22.3 Pulses

Important pulse crops of the Eastern Himalayan region are red gram, green gram, black gram, pea, rice bean, pigeon pea, and lentil. The productivity of pulse crops is very low in this area. The results obtained in All India Coordinated Agronomic Experiment Research Project

Table 7.43 Effect of P sources on the grain yield of field pea (t ha^{-1})

Cultivar	Control	SSP	MRP	PRP	Mean
DMR-4	0.52	1.00	0.71	0.71	0.74
DMR-7	0.62	0.98	0.83	0.64	0.77
Pant P-4	0.48	0.72	0.67	0.79	0.66
Rancha	0.93	1.12	0.83	0.75	0.91
T-163	0.83	0.97	1.10	1.10	1.00
Pant-9	0.96	1.02	1.07	0.91	0.99
DMR-13	0.77	0.82	0.85	0.87	0.83
KPMR-39	1.23	1.32	1.15	0.97	1.17
	0.79	0.96	0.90	0.84	
CD (0.05), Variety			0.40		
P sources			0.13		

(AICRP) showed that both lentil and pea gave significant increase in yield up to 25 kg N, 40 kg P_2O_5, and 20 kg K_2O ha^{-1}, respectively. The effect of P sources was studied (Datta and Ram 1993, Datta and Sharma 2006) on field pea and the results revealed that the trend response was SSP > MRP > PRP (Table 7.43). At other location in the region, application of 25 kg $ZnSO_4$ + 1 kg ammonium molybdate per ha along with NPK at recommended rates produced highest yield of pea.

7.16.22.4 Oilseeds

There is a considerable potential of oilseed crops in Eastern Himalayas. Oilseed crops like mustard, rapeseed, sesamum, linseed, groundnut, and soybean are the promising crops of the region. To ensure higher yields from these crops, judicious use of fertilizers and organic manures in proper amounts is imperative. Per capita availability of oilseeds in the region is around 7.24 kg as against the Indian average of 21.8 kg (Sharma 1995). The highest significant yield of 2.81 t ha^{-1} was obtained with 60 kg P_2O_5 and 30 kg K_2O ha^{-1} (Table 7.44). In another study conducted in valley soils of Manipur (Kailash Kumar and Rao 1990) showed that groundnut responded only up to 111 kg P ha^{-1} (Table 7.45). Soybean responded well to the application of P and Zn in acidic soils of Nagaland (Sharma, 1992b, 1993a, 1996; Sharma and Husain 1991). Application of FYM to the crop improved the efficiency of P utilization (Sharma 1992c). Sharma and Tripathy (1999) reported significant increase in the yields of rice and mustard in rice-mustard rotation when applied with P and FYM on acid hill soil.

Table 7.44 Effect of P and K application on the pod yield of groundnut (t ha^{-1})

P_2O_5 (kg ha^{-1})	K_2O (kg ha^{-1})				
	0	30	60	90	Mean
0	1.32	1.59	1.59	1.69	1.55
30	1.82	1.52	2.58	2.0.37	2.32
60	2.29	2.81	2.97	2.63	2.67
90	2.49	2.89	2.73	2.68	2.39
Mean	1.98	2.45	2.46	2.34	
CD (p = 0.05)	P = 0.43	K = 0.43	P x K = 0.86		

Table 7.45 Effect of P on yield, oil content, and P uptake by groundnut

P (kg ha^{-1})	Pod yield (kg ha^{-1})	P uptake (kg ha^{-1})	Oil content (%)	Shelling (%)
0	1939	15.0	49.6	64.6
11	2531	20.9	46.6	68.7
22	2607	22.4	49.8	71.7
33	2731	23.1	49.7	74.4
44	2689	23.7	49.9	74.8
CD (0.05)	509	–	NS	1.8

7.16.22.5 Vegetables

Proteins, vitamins, minerals, dietary fibers, micronutrients, antioxidants, and phytochemicals are all abundant in vegetables (Ganeshamurthy et al. 2016a). Vegetables play an important role in our daily nutrition. They include a wide range of possible phytochemicals, including anti-carcinogenic principles and antioxidants, in addition to nutrients. With its different agro-climates and distinct seasons, India allows for the cultivation of a vast range of vegetables. On 15.6% (9.2 million ha.) of the world's vegetable acreage, India produces 14% (162.19 million tons) of the world's veggies. Vegetable productivity in India (17.62 t ha^{-1}) is lower than the global average (19.7 t ha^{-1}). The principal vegetables, which account for 57.9% of total vegetable output in our country, are potato (28.0%), tomato (11.2%), onion (10.4%), and brinjal (8.3%). Cabbage (5.3%), cauliflower (4.9%), okra (3.9%), and peas are all prominent vegetables (2.4%). India is first in the world in okra output (73% of total production), and second in brinjal (27.1%), cabbage (12.2%), cauliflower and broccoli (35.6%), onion (20.2%), potato (12.4%), and tomato (11.2%) (Ganeshamurthy et al. 2016a).

A number of vegetables are grown in Eastern Himalayas, particularly cabbage, cauliflower, tomato and leafy vegetables in winter and potato, okra, cucumber, pumpkin, brinjal, and different gourds in summer are common. Some local vegetables are also grown in isolated pockets of the region. Potato is an excellent supplement for cereals and is of great significance in food security (Sharma 2000). People generally take potato in place of cereals, during fasting. In vegetable production, by and large, the use of inorganic and organic sources of nutrients is low. The experiments conducted at higher altitudes in Meghalaya soils have shown that potato responded up to a dose of 120, 66, and 50 kg ha^{-1} of N, P, and K (Sharma 1989a, 1990b). The crop also responds to micronutrients such as zinc, manganese, and copper (Sharma 1987a, b, c). It has been reported that with the use of FYM @ 10 to 15 t ha^{-1}, the recommended dose of major nutrients can be reduced by 20–25%, and the potato yields were significantly higher than NPK applied alone (Sharma, 1997a, b, c; Sharma and Arora 1995). Application of organic manures is particularly essential in soils having pH below 5.5 and P fixation is a problem, to get optimum yield of vegetables (Sharma 1990e). Application of gypsum-coated urea or rock phosphate-coated urea had the same effect on potato as that of ammonium sulfate and all were significantly superior to plain urea alone (Sharma and Patel 1991; Sharma and Arora 1993). Sharma (1992a, Sharma et al. 2014a, b, c) reported that the critical limit of available P for potato was 20.7 ppm in acidic soils of the hills. The application of FYM not only reduced the dose of P but also improved the efficiency of P utilization by 7.5 to 11.3%. Interestingly, Sharma and Grewal (1991) found that application of inorganic fertilizers 5 cm above the potato seed tubers in high rainfall areas of Eastern Himalayas gave a significantly higher yield than other methods fertilizer applications. Application of magnesium to potato @ kg ha^{-1} showed synergistic effect with P in increasing uptake of P and Mg (Singh and

Sharma 1987b, 1988). Leaching studies (Sharma 1989b, 1992a) showed that due to high rainfall in Eastern Himalayas, the leaching of nutrients, particularly N and K is very significant during summer (April–September). There is need for enhancing the dose of these nutrients by 15–20% to get potential yield of crops. Application of P and FYM to potato crop improved the quality and size of tubers (Sharma 1987b; Sharma and Arora 1988).

In potato-cauliflower cropping sequence, 75 kg K ha^{-1} to potato and 15 t ha^{-1} to cauliflower and in potato-radish cropping sequence, 50 kg K ha^{-1} to potato and 30 kg K ha^{-1} to radish, along with recommended dose of N and P for these crops, has been reported optimum (Sharma and Singh 1991). Tomato gave significantly the highest yield when 80 kg N and 90 kg P_2O_5 were applied to the crop. French bean, capsicum, and colocasia responded up to 60, 160, and 137 kg P_2O_5, respectively while 100 kg N was found optimum for all the three vegetable crops (Sharma 1997a, b, c, 1998a, b, c). Maximum yield of turmeric in Manipur was obtained with application of 20 kg N, 90 kg P_2O_5, and 60 kg K_2O ha^{-1}. Maize-autumn potato crop rotation in high rainfall hills was not found economical as reported by Singh and Sharma (1988, 1989). The major reason for low productivity of both the crops was high nutrient leaching beyond root access zone in case of maize during rainy season and low temperature during tuberization stage of spring potato. Establishment of critical N, P, and K ranges in potato petioles was helpful in applying optimum dose of these nutrients to the crop for getting potential yield (Sharma 1989a, b). In a long-term study with seven farming systems, Sharma and Sharma (2003b, 2009) observed positive plant-livestock-soil-hydrology interactions in highland agroecosystem.

Shifting cultivation (open burning) and bun (closed burning) methods of cultivation are predominant methods of food crop production in Eastern Indian Himalayas. In bun method, raised beds of about 5 m length, 1 m width, and 0.3 m height are prepared along the slope and dried vegetative biomass from adjacent forest area is burnt on beds (open burning) or underneath the soil (closed burning). After about a month, the crops like ginger, potato, chillies, okra, cabbage, cauliflower, etc. area raised on the beds, during first year and upland rice crop during second year. After 3 to 5 years, the land is left fallow for regeneration of fertility and the process is repated thereafter. Field experiment with four levels of biomass, that is 0, 40, 80, and 120 t ha^{-1} on air dry basis under open and closed burning was conducted (Munna Ram and Prasad 1992, Prasad and Ram 1985) in acid Hapludalf soil at Barapani, Meghalaya. In general, closed burning proved superior and gave 22.2% higher rhizome yield of ginger over open burning. Periodic monitoring of soil properties showed that burning increased the soil pH, Bray P_2-P and exchangeable K and. Lime requirement was significantly reduced.

7.16.23 Loss of Nitrogen

Loss of N occurs through leaching and runoff either by precipitation or irrigation. Precipitation causes heavy loss of N through leaching and runoff resulting in poor crop productivity (Sharma 1990a, 1999, 2000). Field experiments with potato grown by two methods of cultivation, terraced and bun, over a two-year period revealed that N loss through leaching and runoff increased with increased applied N and was significantly influenced by the nature of the soil, the amount of rainfall received, and the method of cultivation. In the terraced system, the mean loss was 14.2 kg, 16.5 kg, and 30.3 kg ha^{-1} through leaching and 5.8 kg, 5.5 kg, and 4.4 kg ha^{-1} through runoff; in the bun method of cultivation, the mean loss was 7.1 kg, 9.8 kg, and 19.9 kg ha^{-1} through leaching and 19.5, 19.4, and 33.7 kg in runoff in three soils of sandy clay loam, loam, and sandy loam in texture, respectively (Sharma 2000). (Tables 7.46 and 7.47).

Table 7.46 Nitrogen loss (kg ha^{-1}) through leaching and runoff for the terraced method of cultivation

Treatment	Leaching	Runoff	Total	% of applied N
First year				
Soils				
S1	16.3	6.7	23.0	10.2
S2	18.6	4.6	25.2	12.0
S3	32.8	4.7	37.5	21.8
CD (p = 0.05)	2.5	0.4	2.7	-
N applied (kg ha^{-1})				
0	7.6	2.6	10.2	-
100	21.0	5.0	26.0	15.8
200	29.3	7.8	37.1	13.4
200 + FYM (10 t ha^{-1})	32.2	8.2	40.4	15.1
CD (p = 0.05)	2.8	0.5	3.1	-
Second year				
Soils				
S1	12.2	5.0	17.2	7.8
S2	14.3	4.7	19.0	9.2
S3	27.8	3.4	31.2	19.0
CD (p = 0.05)	2.2	0.4	2.3	
N applied (kg ha^{-1})				
0	5.4	2.0	7.4	-
100	18.3	3.8	22.1	14.7
200	23.4	5.8	29.2	10.9
200 + FYM (10 t ha^{-1})	26.0	6.0	32.0	12.3
CD (p = 0.05)	2.5	0.5	2.6	-

Source Sharma (2000)

Table 7.47 N loss (kg ha^{-1}) through leaching and runoff for bun method of cultivation (mean of 2 years)

Treatment	Leaching	Runoff	Total	% of applied N
Soils				
S1	7.2	19.5	26.7	13.9
S2	9.8	20.0	29.8	16.5
S3	19.9	26.6	46.5	29.8
CD (p = 0.05)	1.6	2.9	3.7	-
N applied (kg ha^{-1})				
0	3.4	5.8	9.2	
100	10.4	18.6	29.0	19.8
200	16.6	31.0	47.6	19.2
200 + FYM (10 t ha^{-1})	18.6	32.7	51.3	26.1
CD (p = 0.05)	1.5	2.7	3.2	-

7.16.24 Nitrate–N Content of the Soil After Experimentation

The NO_3–N content of the soil varied significantly at different depths, both in terraced and bun methods of potato cultivation. It was found to be linked to the dose of allied N, nature of the soil, and time of application. Sharma (1999, 2000) and Sharma and Arora (1987a) reported that the NO_3–N content of soil was observed to be higher in surface (20 cm) compared to lower depth (20–40 cm and 40–60 cm) in both the methods of potato cultivation (Table 7.48). Application of FYM slightly increased leaching of N in relatively heavier soil while no significant effect was observed in light soil. More than 84% of the variability in N leaching was attributed to the amount of rainfall clay % and N content (available + applied) of the soil (Sharma 1991). Low content of NO_3–N at lower depths may be due to denitrification by facultative organisms in the absence of aerobic conditions and volatilization of N as nitrous oxide and elemental N. On an average, the NO_3–N content of the soil was higher by 22.3%, 32.7%, and 31.6% in the terraced than in the bun method at 0–20, 20–40, and 40–60 cm soil depths, respectively. The time of soil sampling after the planting of potato crop had a significant effect on the NO_3–N content in all the soils and at all the depths because it decreased with successive sampling dates at various depths in both the methods of cultivation.

7.16.25 Nitrogen Management

According to Prasad et al. (2014), nitrogen is required for plant protein production, which is a

Table 7.48 NO_3–N content of soils at different depths under terraced and bun methods of potato cultivation

Treatment	Terraced method			Bun method		
	0–20 cm	20–40 cm	40–60 cm	0–20 cm	20–40 cm	40–60 cm
Soils						
S1	32.1	26.6	20.0	25.8	19.7	16.1
S2	33.1	28.8	24.3	27.1	22.2	18.2
S3	35.2	31.2	25.6	29.1	23.2	19.0
C. D. (p = 0.05)	1.9	1.7	1.7	1.8	1.6	1.7
N applied (kg ha^{-1})						
0	14.4	13.9	11.2	13.2	12.0	10.6
100	28.9	23.8	19.8	24.3	19.6	15.7
200	41.9	35.5	29.9	34.3	25.0	21.1
200 + FYM (10 t ha^{-1})	48.7	42.1	32.4	37.4	30.2	23.7
C. D. (p = 0.05)	2.1	1.9	1.9	2.0	1.8	1.9
Time of sampling (DAP[a])						
30	40.6	35.7	29.4	33.8	28.1	23.3
45	40.5	33.7	27.0	31.2	25.3	20.3
60	33.1	28.6	23.9	27.3	22.6	18.4
75	28.1	24.8	20.0	23.3	18.5	14.8
90	25.1	21.3	16.4	20.9	14.0	11.9
C. D. (p = 0.05)	2.4	2.2	2.2	2.3	2.0	2.2

[a]Days after planting (Sharma 2000)

direct or indirect source of protein for animal and human nutrition. Proteins control all physiological processes in life, other than being part of the body tissue of animals and humans. To overcome protein malnutrition, more and more protein-rich food, feed, and fodder have to be produced. There is need to produce more than double global food production by 2050 to satisfy hunger. The globe will need to produce as much food annually as it did in the previous 8000 years. Only by properly applying fertilizer, particularly nitrogen, would this be viable. Furthermore, increased fertilizer nitrogen application is required to fully exploit the higher yield potential of traditional, high-yielding types as well as genetically modified (GM) crops (Roberts 2009). Native fixed ammonium in Indian soils may vary from 70 to 798 mg/kg soil and may account up to one-fifth of the total soil N (Grewal and Kanwar 1967).

7.16.26 Increasing Nitrogen-Use Efficiency

Since the crop yield (or any other economic produce) is the nominator for all the terms used for expressing NUE, adoption of Best Crop Management Practices (BCMP's) is the best way for increasing NUE. Components of BCMP's are

- Growing of high-yielding crop varieties recommended for the area.
- Use of recommended rate of quality seed.
- Timely sowing/transplanting.
- Use of organic manure.
- Site-specific nutrient management including application of balanced NPK an adequate amount of deficient secondary and micronutrient(s).
- Application of economic optimum dose of N.
- Placement of basal N, split application as per recommendation, foliar application if recommended.
- Use of neem-coated urea/urea supergranules.
- Efficient water management.
- Efficient weed management.
- Adequate pest management.

Nitrogen Uptake and Needs of Crops

Nitrogen is taken up by crop plants as nitrate (NO_3^-) or ammonium (NH_4^+) ions. Most crop plants will take nitrate or ammonium, though rice is reported to prefer ammonium (Prasad et al. 1983). Nevertheless, it is true that rice grows under anaerobic conditions, where ammonium ions dominate the soil solution. As regards crops grown under well-drained conditions, Joseph and Prasad (1993) reported that wheat preferred more ammonium at early growth stages and nitrate at later growth stages. Higher ammonium concentrations in soil solution could be toxic to plants for crops grown under well-drained conditions (Prasad 2007).

7.16.27 Environmental Implications of Fertilizer N Use

Prasad et al. (2014) reported that humans interferences have greatly altered the global N cycle, increasing both the availability and the mobility. This N burden has also disturbed the ecological N equilibrium. In order to revert back to original equilibrium, ammonium (NH_4^+) tends to be oxidized to nitrate (NO_3^-) by aerobic chemo-autotrophic bacteria ($NH_4^+ \rightarrow NO_2^- \rightarrow NO_3^-$). Again in anaerobic conditions (water or anoxic sediments), facultative anaerobic bacteria can utilize nitrite (NO_2^-) and nitrate (NO_3^-) resulting in formation of nitrous oxide (N_2O) and nitrogen (N_2) gases. All the above forms except gaseous N have individual potentials for adversely affecting the environment. In addition to natural sources, inorganic N can enter through point or non-point sources, mostly anthropogenic in origin. Major point sources could be livestock farming, aquaculture practices, sewage effluents, industrial wastewater, waste disposal, ore mining, and overflow of sewer. On the other hand, the most common non-point sources are agriculture (cultivation, fertilizer application), forest run-off, septic leachates, emissions, and biomass burning. Generally, non-point sources have greater relevance than point sources, since these are large and difficult to control (Prasad et al. 2014).

7.16.28 Plant Nutrient Requirements

Plants require water and nutrients from the soil, yet they are sensitive to excessive amounts of growth-inhibiting chemicals. Physical qualities (depth, texture, structure, pore space with water and air), physico-chemical properties (pH and exchange capacity), and chemical and biological properties (nutrient status, transformation, availability, and mobility) all influence the supply and uptake of nutrients from the soil (Roy et al. 2006). The best pH range for various crops as reported by them is 5.0–5.5, 5.1–7.5, 5.7–7.0, 5.5–6.5, 5.5–7.0, 5.5–7.5, 4.5–5.5, 6.0–7.0, 5.0–6.5, and 5.5–7.0 for potato, apples, cabbage, cotton, maize, rice, tea, spinach, beans, and wheat, respectively. Suitable pH range for different crops is depicted in Fig. 7.65.

7.16.29 Soil Conservation in Indian Himalayas

Sowing of crops on steep slopes in high rainfall areas without protection is generally faced with serious problems of resource degradation. Bench terraces, contour trenches, contour bunds, half-moon terraces were found to be effective soil conservation measures in the area (Mishra and Sharma 2001; Sharma and Sharma 2003b; Satapathy 2006).

Contour Bunds

Bunds are constructed along the contours (across the slope) (Fig. 7.66). The purpose is to divert excess runoff during rains and retain the soil within the bunds, which results in leveling of land between two bunds, finally taking the shape of a terrace in about 6 to 8 years. The vertical interval

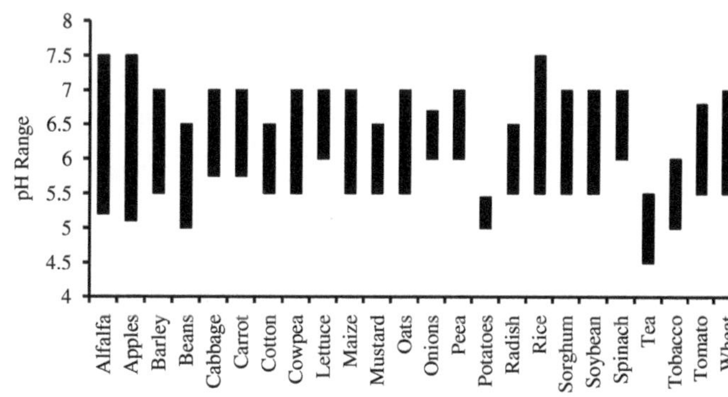

Fig. 7.65 Most favored soil pH for different crops. *Source* (Roy et al. 2006)

Fig. 7.66 A view of contour bunds Satpathy (2006). (*Source*)

Fig. 7.67 Citrus on halfmoon terrace. (*Source* Satapathy 2006)

of these bunds may vary from 0.5 m to 5 m depending on land use and soil depth. Maize, sweet potato, tapioca, and groundnut, etc. may be preferred to be grown during transformation of the bunds to bench terraces (Satapathy 2006).

Bench Terraces

Bench terracing consists of series of step like formation along contours (Fig. 7.64). Spaces between two terraces are leveled by cut and fill method. The vertical interval between two terraces should not be more than 1.0 m. Such measures can be adopted where soil depth is more than 1.0 m. Terracing of the entire hill slope is not necessary since trees and horticultural crops can be raised without terraces. Only the lower portion of the hill needs to be terraced for agricultural crops. The terrace risers can be utilized for growing perennial fodder grasses which not only help in conservation but also provides enough fodder. Some of the fodder crops, which can be successfully grown on terrace riser are stylo, guinea grass, hybrid/thin Napier, and lemon grass (Satapathy 2006).

Half-Moon Terraces

Where complete terracing is not desired or feasible, productive multiple use of steep slopes is possible by planting trees on half-moon terraces (Fig. 7.67). The beds are used mainly for planting and maintaining fruit trees like guava, citrus, etc. in horticultural land use.

Vegetative Bunds

Barrier hedges substantially reduce runoff and increase infiltration. Some runoff may cross the barrier, while the soil will be partly filtered out and deposited. Pineapple plantation has been successfully used (Anonymous 1976) as vegetative bunds in the hills. Datta et al. (2003) reported the use of hybrid Napier, broom grass, local perennial arhar, pineapple, subabul, and gliricidia as vegetative barrier to control the soil and nutrient loss in the hilly terrain of Tripura.

Grassed Water Ways and Grassland Management

These are open channels protected with suitable grasses, construction along the slope and act as a outlet for the terraces or bunds. As far as possible natural courses should be used without much disturbance for draining out the excess water (Satapathy 2006). Grassland development and management entails a set of practises based on ecological principles and scientific methods. The only area, which is available for pasture development, is classed under the category of wastelands. Closure of the area from biotic influences, removal of noxious vegetation, moisture conservation techniques, upgrading herbage quality and yield through HYVs and nutrient-rich grasses, legumes, and fodder trees forming a multitier system, proper grazing management practises, and forage conservation and carryover for lean season should all be included in the package. Native, cultivated, and silvipasture can be developed for low, high, and medium productivity of livestock, respectively. Improvement of native pasture should be the highest priority program out of the three. Cultivated pasture program needs labor intensive approach, where native species are replaced by cultivated pasture of legume and grasses, which has high nitrogen fixation and under cover ground production capacity. It is very difficult for goats and calves to graze in dense and high grass of native pasture. These pastures in comparison to cultivated pasture are faster in rainy season and very slow in dry season and very difficult to maintain a proper stocking density. Native pasture are less productive in winters for goats, but in the rainy season both pasture are equal for mutton production. For cattle grazing, cultivated pastures are always superior to native pastures.

Terracing Steep Slopes

Though the most prevalent agricultural practice in the hills is shifting cultivation, there are several pockets which have some kind of irrigation facilities where excellent terrace cultivation is practiced. Instances are available of destruction of upland terraces where cultivation under rainfed conditions is practiced. The high cost of terracing, the loss of land, the drop in crop output during the first years, and the lack of proper production technology for rainfed benches appear to be the key barriers to boosting terrace farming among tribal farmers. Studies in ICAR, however, have demonstrated that use of the latest technology with bench terrace as conservation base can provide a stable alternative to shifting cultivation.

Bench leveling of paddy lands results in reduction of surface area for cultivation under bunds, risers and channels. In EIHR, bench terrace cultivation exists in slopes up to 100% having terrace width as low as 3 m and situations on steep hills normally do not permit of longer length of terrace. Level beds with vertical interval of about 1 m along with 15 sq.m cross section shoulder bunds and 1/2:1 batter slope for terrace risers was found satisfactory for movement on the slope for performing various agricultural operations (Prasad et al. 1987).

Cultivation of steep slopes is hazardous due to serious soil erosion problem. Soil loss rate from slopes under shifting cultivation is 13 and 10 times the soil loss rates when the slope is fully terraced or only terraced at the bottom, respectively. Puertorican terraces and mixed forest along with habitation also caused very high rates of soil less. Surface runoff was found higher from slopes which reduced considerably (48.93%) in plots having various proportions under bench terracing with the highest reduction (93%) observed in the plots where 75% area was put to bench terracing. Studies on watershed-based farming systems coupled with mechanical soil conservation have demonstrated the scope of using steep slopes for crop production with no more soil losses than in similar forest area.

Geojute or Jute Agrotextile

Land slides in road risers, agricultural land, streamlines in hill areas are a very common and destructive phenomenon, especially during rainy season. Steep slopes, dwindling vegetation, unstable soil mass over a rock with heavy rainfall make the hill sides in the North East prone to soil erosion and landslides. Bringing such degraded areas under permanent vegetation cover is the only possible solution to this problem (Satapathy 2006). But development of permanent vegetative cover on such steep and unstable slopes by conventional plantation methods is not possible because of high runoff on such slopes. Mechanical measures, which also serve as first line of defence against flow of sediments and runoff from the catchment area, are invariably needed for protection of the slopes. Geojute matting is a natural product which may be used for erosion control and which decomposes

Table 7.49 *In situ* rainfall retention and soil loss under different land use systems

Land use	% Rainfall retained	Soil loss (t ha^{-1} yr^{-1})
Pine forestry	96.0	0.306
Agriculture (fodder)	99.7	0.160
Forestry (Fuel, fodder)	97.9	7.800
Agroforestry	98.0	0.190
Agriculture (food crops)	99.4	0.260
Agri-horti-silvi-pastoral	99.6	1.180
Horticulture	98.8	8.034
Natural vegetation	99.8	0.050
Timber forest	98.3	0.325

Source Satapathy (2006)

within the ecological cycle (Satpathy 2006). With normal thickness of about 5 mm, jute mesh provides an open area of about 65% through which planting can easily be done. It is bio-degradable and rots in about two years. The main purpose of using Geojute matting in revegetation over eroded slopes is to provide stability to the uppermost layer of soft soil and the seeds sown over the slopes till new vegetation is established over the area. The species such as *Arundinaria hirsute* (Sejnaka), *Ipomea carnea* (Besarma), *Thysenolaena maxima* (broom grass) can be planted for revegetating the area. Geojute material can prevent surface erosion for up to six months (one rainy season) if the material is spread out properly, the species are planted in a timely manner, and free grazing of animals is avoided. The material has been found suitable for use in low and medium slope group of areas where erosion is caused mainly by the surface runoff. The eroded areas can be further guarded with series of Gabion structures constructed with wire mesh and stone materials at the lower reaches of the slopes and across the waterways.

Geojute Technique

In Tripura state, open mesh woven jute agrotextile and non-woven jute agrotextile were used to control the weed growth, arrest the soil loss, and conserve soil moisture (Anonymous 2004–05).

Non-woven jute agrotextile (500 GSM) was placed in the soil between the double row of pineapple planted in the slopy areas of Tripura (Datta et al. 2005) and the material was used to suppress the weed growth. The weed density was 0.76 kg/m^{-2} in the area under the cover of jute but it was raised to 1.35 kg/m^{-2} in the area without jute.

In situ Retention of Rainwater

Crop cultivation in watersheds, having soil and water conservation measures, was found useful in in-situ retention of rainwater in the soil. Surface runoff from covered soil surface was low. The water retained in slopes of the watersheds can be used by the crop, recharge of streams, and groundwater (Satapathy 2006). During their first 6 years of operation, land uses such as forest, agroforestry, and pine afforestation (Table 7.49) yielded 72 to 93%, 76 to 92%, and 72 to 93% of total water output as base flow, respectively. Aside from rainfall management, the conservation technique proved to be highly effective in soil conservation, with losses being low in almost all watersheds, especially at Barapani, which got a lot of rain. Simple mathematical model has been developed by Sharma and Sharma (2003a, b) to predict soil erosion from catchments (Fig. 7.68).

7.16.30 Rainwater Harvesting Structures and Land Use in Eastern Himalayas

In some areas of eastern Himalayas, unique rainwater structures have been developed, around which specialized land use systems exist (Figs. 7.69 and 7.70). Typical examples are

Fig. 7.68 A view of Black Pepper

plantation crops (areca nut, coconut), agricultural crops (rice, fodders, beans, cucurbits, etc.), livestock (pisciculture, duckery, poultry, etc.). These systems give full food security to the farmers.

7.16.31 Alternate Land Use Model for Mountains

A multi-disciplinary long-term study with watershed approach was undertaken by ICAR Research Complex farm at Umiam in Meghalaya state of India to evaluate six farming system based on livestock, agroforestry, forestry, agriculture, Agri-horticulture, Agri-horti-silvi-pastoral vis-à-vis shifting cultivation (Fig. 7.71). It has been observed that all the studied farming systems are highly productive, resource and eco-friendly and economically viable in comparison to shifting cultivation. These farming systems are an appropriate alternative to replace shifting cultivation if desired or may also be used for the intensification of agriculture.

7.16.32 Nutrient Cycling Management

Nutrient management in mountain agriculture presents a complex scenario due to several factors. Due to the undulating terrain in mountain, the irrigation facilities are meager and majority of the areas are rainfed and, therefore, productivity is low. Introduction of short duration high-yielding crop varieties, appropriate inputs, and matching agro-techniques can result in higher productivity of rainfed crops. Soil fertility in these areas is largely supported by the nutrient inputs from the adjoining forest area (Bhattacharyya and Prakash 2006). With the intensification of agriculture and deforestation, this cycle of mutual sustenance has broken up. Thus, the external supplementation of nutrients is necessary to augment crop productivity. Magdoff

Fig. 7.69 Runoff water harvesting Structures in Hills

Fig. 7.70 Rainwater harvesting structures and land use

et al (1997) reported that the nutrient cycle is an example of a closed loop pattern of flow in which a particular atom ends up back in the same location from where it started. It is essentially movement of nutrients in plants and its environment. The nutrients pools and fluxes and the factors and processes affecting the same in a given condition are important components of efficient management of nutrient cycling, which is inherent in the functioning of an ecosystem.

In a long-term study (Bhattcharyya and Prakash 2006), maximum soybean and residual wheat yield was obtained when farmyard manure (FYM) was applied along with NPK fertilizers. In all the treatments, total soil C and N contents were increased after 30 years, while a negative balance was found in P and K in some treatments.

7.16.33 Crop Diversification and Requirements

Location specific crop diversification with acid-tolerant crops like potato, sweet potato, chow-chow, and Dioscoria spp, etc. should be followed (Ganeshamurthy et al. 2016a). Potatoes do well across a wide range of pH, but prefer slightly acidic soil pH of 5.3 to 6.0. The yield of potato increased when liming was combined with cattle manure (Sharma 1990e, Sharma and Arora

Fig. 7.71 Alternate land uses developed by Indian Council of Agricultural Research Complex for hill ecosystem in Eastern Himalayas. This photo belongs to Indian Council of Agricultural Research Complex for Northeastern Region of India

1987a, b, c, Sharma and Singh 1991a). The field experiments conducted under the ICAR scheme for maximization of phosphate potential showed that significant increase in yield of potato was recorded by the application of lime (3.3 t ha^{-1}) (Ganeshamurthy et al. 2016a).

7.16.34 Management of Nutrient Deficiencies in Fruit Crops

Phosphorus

P deficiency in fruit crops on acid soils is widespread. This is mainly because of poor availability of P in acid soils due to P fixation and P precipitation. A schematic diagram of how soil pH affect P availability is shown in Fig. 7.72. To make matters worse, phosphorus losses in horticulture systems might occur as a result of runoff water erosion. Furthermore, the removal of plant material (which might serve as a source of organic phosphorus) during harvest and annual pruning reduces phosphorus availability in perennial fruit crops. Although most hilly soils have a high P-fixing capacity, phosphorus leakage can occur if the soil approaches its maximum phosphorus holding capacity, particularly when P fertilizers are overapplied. Under extensive orchard management, such situations are common. Phosphorus leaching is particularly common in sandy soils. The pollution of groundwater reserves is a result of phosphorus leaching.

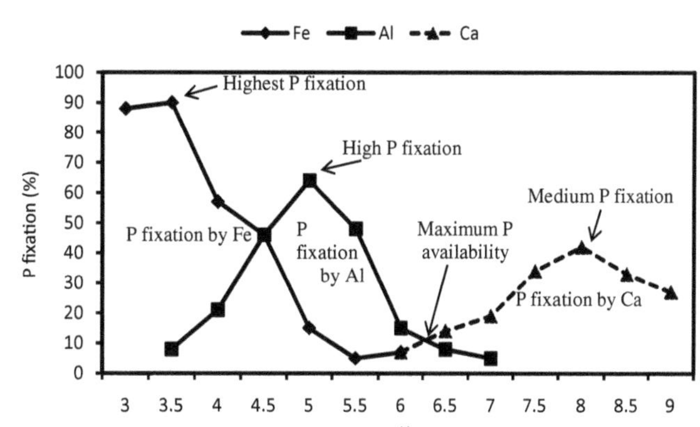

Fig. 7.72 P fixation in soil at different pH values by Fe, Al, and Ca

Potassium

Potassium deficiency occurs most commonly in fruit crops grown on sandy and acidic soils. Small fruit with low sweetness and acidity are the most common signs of potassium insufficiency. Banana plants that have been severely afflicted produce few and tiny fruits. The use of K becomes important in order to boost the size and quality of the fruit. Excess potassium intake, on the other hand, might lead to magnesium insufficiency. This is because a high potassium level can interfere with magnesium absorption. Fruits with a rough peel, such as pomegranate and mango, and high acidity, such as acid lime and mandarins, will suffer from a high potassium status.

Calcium and Magnesium

Deficiency of calcium and magnesium are common problem in fruit crops grown on acid soils. In mango, claims are made that spongy tissue in alphonso mango is associated with calcium deficiency. However, no direct evidence is available to prove this claim. In banana, calcium deficient orchards show severe lodging of leaf lamina. Magnesium deficiency in acid soils is a consequence of leaching of Mg from root zone. Deficiencies are also due to nutrient imbalances. In most cases, the deficiency is expressed in the middle-aged leaves or older leaves.

Zinc Deficiency

Zinc deficiency in fruit crops is very widespread in acid soils receiving heavy phosphorus application. Some fruits are more susceptible to this problem than others. For example, Poovan banana grown mostly on acid soils show Zn deficiency soon after heavy fertilization with phosphorus. However, in the same area mangoes fertilized with heavy dose of phosphorus do not exhibit any such symptoms. This is probably in mango the nutrients get locked up more in the woody parts. Similarly, papaya and pineapple receiving heavy phosphorus in most acid soils of India show Zn deficiency. Soils inherently low in available Zn in apple gardens of higher reaches of Himalayas, citrus gardens in northeastern states, Coorg mandarin, Kinnow mandarin, and Khasi mandarins show widespread Zn deficiency.

Sulfur Deficiency

Sulfur insufficiency is common on old, severely weathered land surfaces where the soils have been heavily leached over time. Banana, Guava, pineapple, apple, and papaya grown on acid soils are prone to sulfur deficiency than mangos, citrus, and other fruit crops. Sulfur deficiency in apple gardens of Himachal Pradesh is common wherever straight fertilizers are continuously used. Pale yellow leaves throughout the trees and poor growth and yield are typical symptoms associated with fruit crops in acid soils.

7.16.35 Farm Power to Harness Crop Yield Potential in Mountains

There are some important points necessary to raise crop yields in mountain ecosystem and the role of mechanization in realizing the potential is well established. The holdings of farmers in mountain areas being small, the mechanization in agriculture require specially designed equipment and power units suitable for hill agriculture and agro-processing. To improve efficiency in everyday work, implements should be chosen, tested, tweaked, and then implemented. Power units for horticulture crops, such as extra light power tillers, hill-side tractors, and efficient power-driven tools and implements, should be bought and examined for adoption. Subsoilers should be used on a regular basis for deep tillage to break up hardpan and eliminate perennial weeds. The need for proper equipment has been noted from the beginning of commercial farming in alpine locations. In order to eliminate drudgery in the threshing and cleaning of grains, pulses, and oilseeds, suitable power threshers should be widely used in the hills. Snow melt and rainwater harvesting, conservation, and storage should be heavily pushed, as should efficient conveyance equipment, sprinklers, and drip irrigation systems. In the region's foothills and 'terai', standard power-operated equipment utilized in the plains may also be useful. Serrated sickles, fruit pluckers, vertical conveyor reapers, and combine harvesters are examples of improved harvesting equipment that should be implemented (Alam).

7.16.36 Management of Soil Health in Hill Agro-Ecosystem

For optimum production of agricultural crops, a healthy soil base is very necessary. It has been found that biological degradation, increase in erodibility, acidity, and salinity, and ill-exploitation of soil with poor physico-chemical properties lead to decline in soil health and its potential to produce good crops. The extent of soil and nutrient transfer, causing environmental degradation in the Eastern India Himalayan states, has been estimated to be about 601 million tons of soil, and about 1.5 million tons of crop nutrients (Sharma and Sharma 2004a, b, c). There is a strong need to maintain equilibrium between resources and their use and to conserve the soil and its nutrients from losses due to run-off, leaching, and volatilization to have a stable ecosystem. Looking into the present level of resource degradation, Saha et al. (2012) advocated for applying resource conservation techniques like zero tillage/minimum tillage, hedge crop, mulching, and cover crops.

7.16.37 Multipurpose Trees (MPTs) and Soil Health

The MPTs help in improvement of soil health and other ecological conditions, besides providing fuel, fodder, timber, and other miscellaneous commodities. In a long-term study, Singh and Dhyani (1995) found a great variation in the soil properties such as the organic carbon, available P, and exchangeable cations contents caused due to different types of agroforestry system on the land. In all the agroforestry systems, significant increase in organic carbon was found as compared to initial status (Majumdar et al. 2004). The agroforestry systems studied were Agri-silviculture, Agri-silviculture, Assam lemon + crops, Alder + pineapple + fodder grass, and Alder + tea + black pepper + crops. Exchangeable Ca, Mg, K and available N, P, Fe, Mn, Cu, and Zn increased significantly over initial status of soil in all the agroforestry systems.

7.16.38 Role of Agroforestry in Soil Health Management

To improve soil health (quality) for a variety of purposes, such as agricultural productivity, rehabilitating degraded lands, and minimizing the greenhouse effect of shifting cultivation, a substantial paradigm shift is necessary. Agroforestry systems, rather than increasing crop production, offer a good option for buffering and preserving the production capacity of agricultural systems for sustainable mountain agriculture. It emphasizes the ecosystem's stability, sustainability, and equitability, as well as the diversity of land use systems' productivity (Patiram and Choudhury 2000). The soil productivity of the mountain environment has deteriorated due to the loss of vegetation cover on uncultivated fields. The need for animal food, fuel wood, and agricultural land has increased, putting pressure on the forest. The farmers have over generations evolved their own adaptation mechanisms. Anthropogenic influences in this fragile alpine ecosystem, combined with different tribes' existing land tenure systems, have resulted in serious degradation of land resources, decreased soil production, poverty, starvation, and economic regression. Over-exploitation of forest resources, shifting farming on hill slopes, and mining activities without effective land management have all contributed to land degradation (Patiram 2002).

7.16.39 Agroforestry and Hill Ecosystem

Toky and Ramakrishna (1983a, b) conducted research on changes in above-ground biomass production, nitrogen buildup in aerial shoots, ecology, and litterfall at lower elevations in the Eastern Himalayas. They discovered a consistent rise in biomass output, net primary production, nutrient buildup in aerial shoots, and litterfall. The biomass showed an increasing trend with increase in fallow period and it increased from 5.0 t ha^{-1} in the first year of fallow to 147.6 t ha^{-1} after 20 years of fallow. The N, P, K, Ca, and Mg content of living aerial shoots increased from 31 to 489 kg ha^{-1}, 5 to 64 kg ha^{-1}, 35 to 1379 kg ha^{-1}, 14 to 440 kg ha^{-1}, and 15 to 226 kg ha^{-1} with increase in the fallow period from 1 to 20 years (Toky and Ramakrishna 1983a, b).

7.16.40 Recycling of Nutrients

Alnus nepalensis is a non-leguminous tree fixing atmospheric N. Annual N fixation in Alnus nepalensis ranged from 117 to 155 kg ha^{-1}, depending on the tree population per unit of area and age of the tree, with the youngest stand having the highest annual N fixation (Sharma 1995). On the basis of nutrient usage efficiency, nutrient dynamics, and cycling in this agroforestry system, he indicated that replanting after 20 years for both Alnus and cardamom would be highly advantageous and sustainable. In comparison to other agroforestry systems, the agroforestry system with Alnus nepalensis maintains a higher level of soil organic matter content, total nitrogen, and N mineralization rate. In comparison to non-N_2-fixing species, N_2-fixing species store less nutrition and hence provide more of these elements in their litter, resulting in more cycling (Sharma 1997a, b; Sharma and Singh 1994; Sharma and Prasad 1995). This increases the production potential of companion crops in N fixer stands. Potato is grown under the pine tree (Pinus kesiya) at higher altitudes in Meghalaya, employing the litter fall, twigs, and branches under the bun system of agriculture. Nutrient build-up in various agroforestry systems is given in Table 7.50.

Table 7.50 Nutrients build up in topsoil (0–15 cm) in the sole tree stand and under agroforestry systems (Dhyani 1998)

Agroforestry system	Total N (%)	C/N	Exch	nutrients	(me/100 g)	Available nutrient
			Ca^{2+}	Mg^{2+}	K^{+}	P (ppm)
Tree only						
Alder	0.25	6.9	3.10	2.28	0.34	9.4
Albizia	0.22	7.6	3.26	4.58	0.35	11.1
Cherry	0.17	8.4	2.30	2.14	0.21	8.5
Mandarin	0.18	8.7	2.84	2.30	0.27	8.8
Tree + crop						
Alder	0.23	7.1	2.91	2.05	0.41	12.1
Albizia	0.20	8.0	3.02	4.46	0.39	18.1
Cherry	0.15	9.6	2.11	1.90	0.23	10.5
Mandarin	0.17	9.2	2.52	2.00	0.27	11.8
Sole crop	0.11	12.2	0.53	0.51	0.19	5.8
Mean	0.19	8.6	2.51	2.46	0.29	11.6
Sd (±)	0.04	1.61	0.83	1.28	0.08	3.39
CV (%)	23.2	18.63	33.23	51.90	27.02	31.8

7.16.41 Ameliorating Agents and Soil Properties

Application of ameliorating agents helped in maintaining favorable pH, augment phosphorus, calcium, magnesium content and decrease the toxic concentration of iron, aluminum and manganese contents in soils. Addition of 2.0 t lime ha^{-1} increased pH from 4.7 to 5.6 in Sikkim (Patiram and Prasad 1990b) and exchangeable Al^{3+} underwent a sharp fall from 2.0 to 0.1 c mole (p^{+}) kg^{-1}. Aluminum saturation was also tangibly reduced from 45.2 to 2.7%. Soil analysis indicated that exchangeable Al^{3+} disappeared completely after liming but again appeared on the exchange complex after three years of cropping in Meghalaya (Prasad et al. 1981). Besides pH, exchangeable Ca + Mg also increased from 0.85 to 11.52 c mole (p^{+}) kg^{-1} after liming in acid soils of Nagaland. Lime potential determining the activity of calcium and magnesium ions [pH − ½ p (Ca + Mg)] was also raised from 3.29 to 6.73. Free energy (F = ½ RT ln a_{Ca}/a_{H}) was decreased from—4597 to—9084 thereby indicating the enhanced availability of calcium ions in soils after liming. Similarly, application of press mud effectively increased the pH, exchangeable Ca + Mg and lime potential in soils. Data on the changes in physico-chemical properties of soils are reported in Table 7.51. In residual experiment, the doses of lime and press mud also increased the pH (4.20 to 6.30 and 4.25 to 6.80) and exchangeable Ca + Mg [0.56 to 2.80 and 0.60 to 2.76 c mole (p^{+}) kg^{-1}] respectively, thus evidencing the presence of residual action in soils. Available contents of Fe, Mn, Zn were also decreased (Datta et al. 1983a) after the addition of lime in acid soils of Nagaland.

7.16.42 Crop Nutrition

Concentration of nutrients in different crops is greatly influenced after amelioration of soil acidity. On the other hand, concentration of some other nutrients is also reduced. Leaf analysis in maize indicated a rise in contents of Ca and Mg from 0.46 to 1.09 and 0.32 to 0.60% after liming. There was slight increase in P content of maize from 0.28 to 0.30% and K content remained constant. But contents of Al underwent a fall from 136.7 to 71.3 mg kg^{-1} in ear leaf of maize grown in acid soils of Sikkim. As reported by Datta and Gupta (1984b), concentration of P was raised in plant tissues of wheat and maize followed by a decline after the incorporation of higher amounts of ameliorating agents. A progressive rise in calcium content was also observed. Data on nutrient uptake are presented in Table 7.52. Concentration of Zn and Fe in plant tissues was not markedly influenced but a sharp decline in Zn content in grain resulted with their increasing doses (Datta and Sharma 2006).

7.17 Myanmar

Myanmar covers a total size of 67.657 million hectares and has heights ranging from 915 to 2134 meters. The country has the distinction of being Southeast Asia's longest country (Win 2005). About 60% of the population relies on subsistence agriculture, primarily rice farming, as their major source of sustenance. Agricultural output variability is a major concern to the food production program, posing a threat to the country's economic stability and enhancing the population's living standards (FAO/WFP 2016). The statistics of the country show that during 2014, the agricultural output, service sector and industrial and manufacturing sectors contributed about accounted for 31, 22, and 29% of Gross Domestic Product (GDP) of Myanmar. Myanmar is ethnically diverse, with eight major ethnic groupings, 135 subgroups, and 108 ethnolinguistic groups residing there. About 30% of the population lives in cities, with the bulk of the population having a rural background. Myanmar has 17.38 million ha of arable land. Land under net sown area has increased over the years to produce more for the population which is being added to that of 0.8%, annually.

Table 7.51 Effect of ameliorating agents on soil properties in Nagaland

Crop/Dose (t ha^{-1})	pH	Lime potential	Free energy (-cal $mole^{-1}$)	Available P(kg ha^{-1})	Exchangeable cations {c mole(p^+) kg^{-1}}	
					Ca	**K**
Soybean						
Control	4.30	3.32	4182	6.27	0.53	0.21
2.0 pressmud	5.15	5.10	5640	8.96	1.22	0.16
2.5 lime	6.10	5.20	7115	9.41	1.98	0.16
Pea						
Control	4.60	3.80	4166	15.68	0.65	0.19
2.0 pressmud	5.70	5.55	5627	19.45	1.28	0.20
2.5 lime	5.90	5.78	5609	20.84	1.50	0.19
Maize						
Control	4.10	3.25	4190	6.80	0.84	0.64
2.0 pressmud	5.40	5.34	7060	16.58	2.28	0.23
2.5 lime	5.85	5.52	7148	14.34	2.34	0.19
Wheat						
Control	4.55	3.80	4568	9.05	1.05	0.75
2.0 pressmud	5.80	5.68	7295	18.12	2.35	0.22
2.5 lime	6.05	5.80	7450	18.49	2.49	0.26

Source Datta et al. (1983b)

Table 7.52 Effect of soil amelioration on nutrient uptake (kg ha^{-1}) in Nagaland

Doses (t ha^{-1})	P		K		Ca + Mg	
	Wheat	Maize	Wheat	Maize	Wheat	Maize
Control	1.50	2.91	4.75	4.06	0.40	0.85
2.0 (lime)	2.69	6.50	6.66	6.75	2.58	1.90
Control	2.97	5.02	5.69	10.12	0.74	0.96
2.0 (pressmud)	7.94	16.50	5.67	18.42	1.62	1.48

Source Datta et al. (1983b)

7.17.1 Ecosystems and Natural Resources of Myanmar

According to an assessment, the total area under cultivation is 79940 km^2 or about 12% of the total land area and forests cover 344,237 km^2, which is 51% of the total land area. Myanmar has a diverse ecosystem, ranging from the Himalayas to the tropical reefs along the Bay of Bengal. Changes in area under forest and agriculture during the years 1985–1993 are given in Table 7.53. The area under forest cover and water bodies decreased by 5.9% and 0.1%, while it increased under scrub land and agriculture by 3.7% and 3.6%, respectively.

The practice of shifting cultivation in Myanmar has affected 154,389 km^2 which is about 23.3% of the geographical area of the country. Shifting cultivation is the major factor of land degradation in hill ecosystem, resulting in accelerated soil erosion. The burning of biomass in shifting cultivation exposes the soils to different types of erosion and the inappropriate agronomic practices such as cultivation against the slopes cause rill erosion. It is why slight to severe soil erosion is found in almost all the hills and mountain ecosystem and has resulted in the slow but gradual decline of crop yields. This has encouraged the growers which in turn leads to more deforestation to produce enough food for their livelihood.

7.17.2 Agriculture and Crop Production

As of 2009, the net cropped area in Myanmar was 11.67 million ha; constituting about 18% of the total land area of the country. About with 5.7 million ha (8.4% of total area) area is still available to expand, if required (FAO/WFP 2009). According to the FAO/WFP (2009), the food crop subsector alone accounts for 80% of overall production value. The share of area under rice, among the food crops has declined over the years. Rice farming covered around 8.1 million hectares in 2006–07, accounting for about 40% of the total area sown. Maize is another important crop in Myanmar, and it is the preferred food crop after rice, especially in mountainous areas.

Table 7.53 Change in land use in Myanmar

Land cover category	1985–86		1992–93		% Change
	Area (km^2)	%	Area (km^2)	%	
Forest (including mangrove)	329,969.1	49.9	290,024.9	44.0	−5.9
Scrub land	193,816.3	29.3	220,676.3	33.0	3.7
Agriculture	114,051.6	17.3	138,252.2	20.9	3.6
Water bodies	1676.0	0.3	1036.4	0.2	−0.1
No data available (cloud covered)	21,407.9	3.2	10,908.1	1.9	−1.3

In 2007–08, maize was grown on 346,000 hectares in Sagaing Division, north and south Shan State, Chin State, and Magwe Division. Horticultural crops are especially important in the Sagaing and Shan States' mountainous regions, where higher precipitation levels allow for year-round production (UNDP Agriculture Sector Review 2004). Pulses are also becoming a popular export crop, with over 1 million tons shipped in the last few years (FAO/WFP 2009).

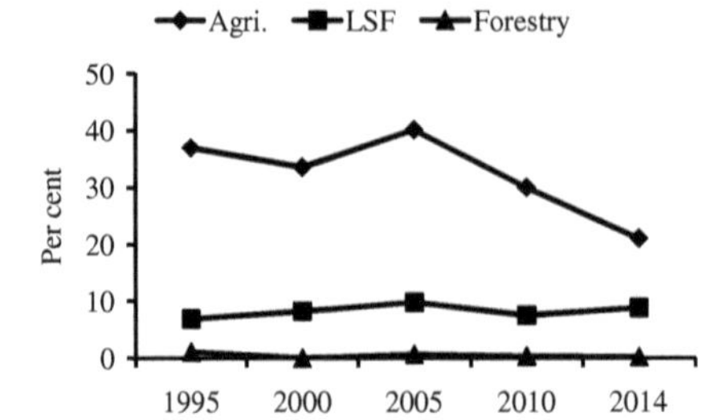

Fig. 7.73 Myanmar—Percentage share of GDP by agriculture, livestock, fishery, and forestry. *Source* Myanmar Statistical Information Service

7.17.3 Food Security and Nutrition

About 10% of population of Myanmar is below the official food poverty line, as revealed by Integrated Household Living Conditions Survey (UNDP 2007). However, there are significant differences between states, with Chin State (40%) having a very high poverty rate, northern Shan State (21%) having a high poverty rate, and eastern Shan State having a low poverty rate (20%). Food insecurity is primarily caused by a lack of year-round access to sufficient food. Access to irrigation, climate conditions, and household access to land and agriculture are all major correlates of food security, depending on the locality. In certain locations, such as northern Shan State, ethnic violence and the ensuing instability and development repercussions are also key drivers of food insecurity (Baroang 2013).

7.17.4 Agriculture

The contribution of agriculture to GDP of Myanmar decreased from 37% in 1995 to about 21% in 2014, in a span of about 20 years. At the same period, the livestock and fisheries sector's contribution of GDP climbed from 7 to 9%. (Fig. 7.73). Agricultural development and food security have been highlighted by the Myanmar government as crucial pillars supporting and enabling inclusive and sustained economic growth, as stated in various publications (FAO/WFP 2016).

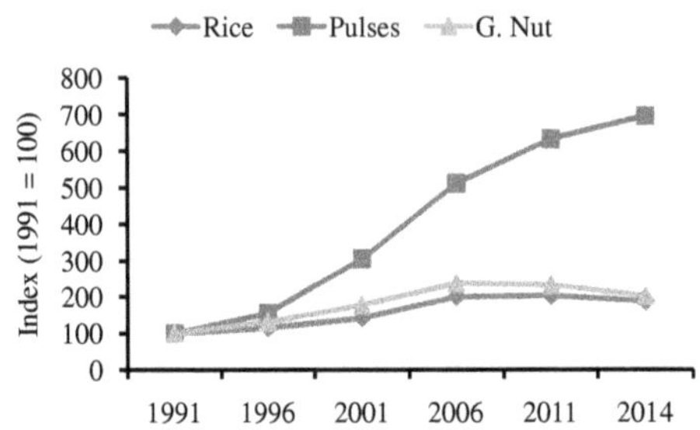

Fig. 7.74 Growth index of rice, pulses, and groundnut in Myanmar. *Source* Department of Planning, MoAI

7.17.5 Food Production

Between 1994 and 2014, rice-cultivated areas in Myanmar rose by more than 20% (5.9 to 7.2 million ha), while production increased by 37%, reaching 28.2 million tons in 2014. Rice production development was also aided by improved extension services and increased use of high-yielding varieties and fertilizers (FAO/WFP 2016). The growth rate index of rice, pulses, and groundnut is shown in Fig. 7.74.

7.17.6 Irrigation

The entire renewable water supply of the country is estimated to be 1168 km^3; however, only a small percentage of this is utilized for irrigation. In 2015, the Myanmar Ministry of Agriculture reported that dams and pumping stations watered 2.17 million ha (16.2% of the planted land). The rise in net sown area and irrigated area between 2001 and 2015 is depicted in Fig. 7.75.

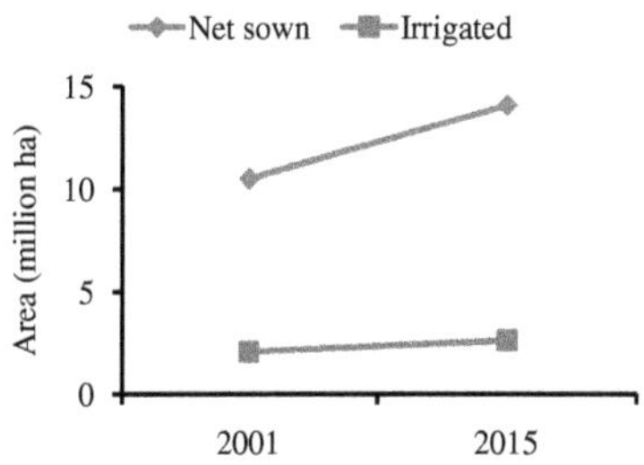

Fig. 7.75 Irrigated area in Myanmar

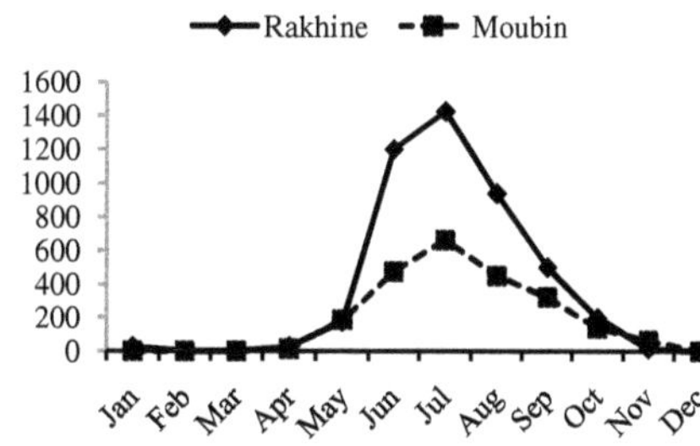

Fig. 7.76 Annual rainfall (mm) at two locations in Myanmar (Average of 2014 and 2015)

7.17.7 Crop Area and Production

The net sown area in the country has increased at an annual compound growth rate (ACGR) of 2.46%, from 10.14 million ha (14.98% of the GA) to 17.75 million ha (26.2% of GA). During the same period, the area under reserve forest has increased by 1.69% ACGR. About 30% of Myanmar's total land area is under permanent forest (CSO 2015). Myanmar Forest Resource Assessments show that the forest cover in 2010 was 31.773 million ha, approximately 46.96% of the total country area (MoECF 2014). Crop sown area and production are presented in Fig. 7.77.

About 10.18 million hectares of land in Myanmar (15%) was under shifting cultivation in 2003 (Htun 2007). Regarding forest areas under shifting cultivation, it has been estimated that 0.29 million hectares of land of reserved forests and protected forests of Myanmar has been brought under shifting cultivation (Schmidt-Vogt et al. 2009). Win (2005) gave a complete detail of area and production of various crops in Myanmar. Rice has been the leading crop in Myanmar's agriculture with a production of 26.27 million tons in 2019. Myanmar's farms are well diversified, and other than rice, beans, pulses, oilseeds, and maize are also grown during the cool and dry seasons.

The factors affecting the crop production in Myanmar are stated below.

Rainfall

Rainfall received during the cropping season is a major factor affecting the crop yields in Myanmar, as is true in other areas of the world. This is particularly important for those areas where irrigation facilities do not exist. The month-wise average (2014 and 2015) rainfall in two states of Myanmar, that is, Rakhine and Moubin is presented in Fig. 7.76. Kayah State and Mandalay Region experienced well below-average rains between May 2015 and November 2015.

Paddy Production

The paddy production in 2015 was 27.5 million tons in 2015, down 3% from the previous year's output of 28.2 million tons and 2% lower than the three-year average of 28.1 million tons (Fig. 7.77).

Maize

Maize is a small crop in the United States, but it is the main source of nutrition in certain rural places, such as parts of the Chin and Shan states. The majority of the country's maize, on the other hand, is utilized as livestock feed. Maize production in the country is predicted to be 1.267 million tons, with an average yield of 2.9 tons per hectare.

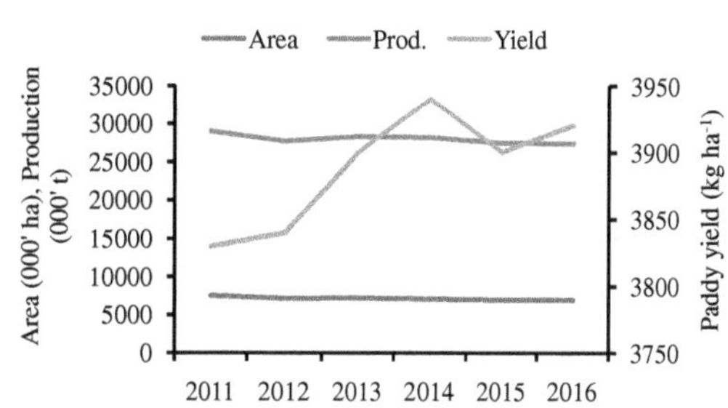

Fig. 7.77 Area, production, and yield of paddy in Myanmar between 2011 and 2016

7.17.8 Livestock, Fisheries, and Aquaculture

Like other countries of the HKH region, livestock forms an important component of the agricultural sector in Myanmar. The farmers and their families, particularly in the dry zone, besides getting milk, meat, etc. from the animals, have raised their income through sale of milk and milk products, and livestock trade. Official estimates show that most livestock numbers exhibited a very rapid increase between 2007 and 2014 (Fig. 7.78).

7.17.9 Water Resources and Irrigation

The country has abundant water resources. Irrigation and hydropower generation are done with groundwater and water from five major rivers. Only approximately ten percent of the country's total water resources are used, with irrigation accounting for 90% of that use (ADB 2013). Though water resources are plentiful, their availability varies greatly throughout time and geography. River flows are mostly determined by rainfall patterns, with 80% of flows occurring during the monsoon and barely 20% during the dry season. In places like the Dry Zone and Rakhine state, seasonal water scarcity is a major issue. While Rakhine has some of the highest rainfall in the world, water shortages are common outside of the monsoon season due to a lack of natural or manmade water storage capacity (Baroang 2013).

Between 1980 and 2000, the total area irrigated more than doubled, reaching nearly a quarter of all cultivated land by 2004, though it still accounts for less than 3% of total land area. According to UNDP, river-pumped systems served nearly 45% of all irrigated land in 2001/2; reservoir and river diversion systems served 32%; private, village-based systems served only about 12% of irrigated land; and groundwater production served about 5% (groundwater mainly used for domestic supplies (51.4%), with a share of 47% for irrigation and 1.6% for industrial uses) (2004). According to UNDP (2004), the Ayeyarwaddy delta is underlain by a succession of deep and shallow freshwater aquifers that might be used to irrigate a considerable section of land in the area, particularly during the dry season. According to FAO/WFP (2009), the percentage of irrigated land increased dramatically in the 1990s due to rising demand and support from government programs (from 12.6% in 1989 to 17.3% in 2002), but has remained stable at around 17% in recent years (even as net sown area increased by 25%). In general, the Delta has the biggest proportion of irrigated land. While different regions have a mix of rainfed and irrigated monsoon paddy (Ayeyarwaddy, Yangon, and others rainfed; Dry Zone parts irrigated), irrigation is required for all summer paddy crops in the country (FAO/WFP 2009). Access to irrigation, according to the World Food Programme (2011), is a critical aspect in ensuring food security in the Dry Zone.

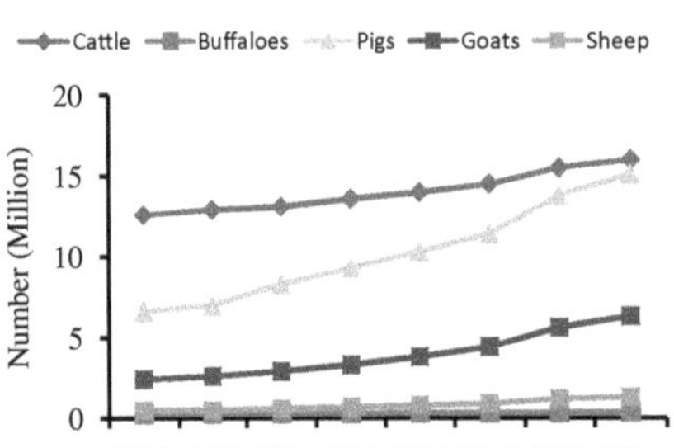

Fig. 7.78 Showing growth of livestock in Myanmar between 2007 and 2014. Livestock Statistics (2015). *Source* Livestock Breeding and Veterinary Department, MLFRD, Myanmar

7.17.10 Improving Plant Nutrient Management

Rice is the most important crop of Myanmar, covering an area of about 62% of the net sown area. Maize is next crop in importance. The deficiency of macro and micronutrients has been reported from the soils of different areas due to intensification of agriculture. The farmers mainly depend on indigenous sources for nutrient supply. Integrated use of crop nutrients and their efficient management is necessary to obtain optimum crop productivity. The country is mostly dominated by degraded soils (Win 2005) and, therefore, only judicious management of soil can ensure better yields of crops.

In the mountain ranges where the population is low and sufficient land availability is there, shifting cultivation is a common land use. This practice, however, has caused soil erosion and degradation and increase in nutrient deficiency. About five decades earlier, the government encouraged the use of high-yielding crop varieties and chemical fertilizers in order to increase agricultural production. The high yields, however, were not sustained because of lack of adequate supply of fertilizers as per crop requirement. The inability to apply balanced fertilizer nutrients because of unavailability and/or awareness is likely to result in nutrients deficiencies or toxicities in the soils. Increasing soil infertility is due to soil erosion by water and wind, soil acidity in high rainfall areas, soil degradation, and low use of organic sources of nutrients. Farmers in Myanmar are generally resource poor and so have limited resources to use adequate inputs for good crop production.

7.17.11 Nutrient Deficiency

Nitrogen deficiencies occur in all of the major rice growing regions of Myanmar. This is mainly due to losses through volatilization, denitrification, leaching, and immobilization by soil microbes. It has been found that the P deficiency rarely occurs in rice-producing areas except problematic soils such as acid soil, acid sulfate soil, saline and sodic and peat soils because of P fixation. Zinc deficiency has been a problem in intensively cropped soils. The deficiency of S occurs in dry zone areas, many rural areas, and irrigated rice-grown areas. Integrated Plant Nutrient Management has been encouraged by the Land Use Division and Extension Division of MOAI of Government of Myanmar. The government has undertaken long and short-term programs for prevention of soil and environmental degradation.

7.17.12 Nutrient Budget and Uptake

The results of the study of the Phosphate Potash Institute (PPI) on nutrient uptake in Myanmar are presented in Table 7.54.

Organic Fertilizer

Win (2005) reported that the raw materials for compost making were not enough because of the increasing multi-user demands for crop residues and weeds. The government of Myanmar has, however, encouraged to grow green manuring crops for incorporation of organic material into soils. In addition, the use of various kinds of biofertilizer like blue green algae, azolla, and mycorrhiza is being popularized. Rhizobium have gained wider acceptance with the farmers.

Table 7.54 Soil nutrient budget and nutrient uptake in Myanmar

Crop	Area sown (Mha)	Yield (Mt)	Nutrient uptake (000' t)				
			N	P_2O_5	K_2O	MgO	CaO
Edible crops[a]	16.1	18.2	451	140	618	79	87
Fruits	0.7	3.2	30	11	36	5	8
Vegetables	0.2	3.5	13	2	10	2	7
Perennial crops	0.5	0.2	3	1	7	1	1
Total	17.5	25.1	497	154	671	87	103

[a]Paddy, Maize, Soybean, Groundnut, Cassava, and Sweet potato. *Source* FAO (1994)

7.17.13 Nutrient Supply and Crop Production in Myanmar

The low productivity of cereal crops in Myanmar is due to inadequate application of nutrients, particularly nitrogen (Farquharson et al. 2017). In developing countries like Myanmar, there has been a shift from subsistence to semi-commercial farming. Actually, in general, the Green Revolution in Myanmar consisted of developing high-yielding crop varieties and applying high rates of fertilizer (de Janvry 2016). However, it has been found that new 'agricultural technologies' have been widely adopted in irrigated areas, while rainfed regions have, by and large, kept away from this change. According to Denning et al (2013), Htwe et al. (2016), and Aung Naing Oo et al. (2016), nitrogen was the most limiting nutrient for irrigated lowland rice, followed by phosphorus (P) and potassium (K). Myanmar's nitrogen (N) application rates are generally low. Increased usage of N fertilizer has the potential to improve rice output, which will increase regional demand for fertilizers and rice supply on the international market.

7.17.14 Drought-Prone Area in Myanmar

In Myanmar, some areas/states have been marked as drought prone where rains are either erratic or low. The agricultural productivity in these areas is low. The total drought-prone area in Mandalay, Sagaing, Magway, Rakhine, Bago East, and Bago West regions of Myanmar is 58482 ha, 45,741 ha, 44,868 ha, 39,854 ha, 35,985 ha, and 28,956 ha, respectively (Annual Report of Myanmar Agriculture Service 2003).

Hom et al. (2015) suggested the following strategy to be adopted to enhance agricultural production in Myanmar:

- Promoting high-yielding, stress-tolerant cultivars, as well as climate-smart management techniques
- Evaluating and supporting the growth of traditional rice types that are naturally tolerant of deep water, extended flooding, and drought
- Integrative farming systems are being evaluated and promoted
- Alternative seed flow technologies for obtaining high-quality seed are being developed
- Awareness campaigns and education programs (MOAI 2015)

7.18 Nepal

7.18.1 Sustainable Soil Fertility Management in Nepal

Whole of Nepal is located in the Himalaya range. It is bordered on the north by China, and on the east, west, and south by India. Nepal is a landlocked country with a population of 28 million people and an area of 14,7181 km^2. Despite the fact that agriculture is only suited for roughly 17% of Nepal's total land area, it remains a key economic activity in the country. The agriculture industry accounts for 38.8% of the country's total GDP. The population-to-arable-land ratio is among the highest in the world. Nepal has a wide range of climates, from humid subtropical in the foothills to alpine in the north, where the high mountains are located. Annual rainfall ranges from roughly 250 mm in rain-shadow places like the Mustang valley to more than 5200 mm in Pokhara area. June to September is the wettest time as more than 80% of rainfall is received during this period. In the summer, the highest temperature in the Terai is around 400 °C, whereas the highest temperature in the Hills is around 280 °C. Winter rains are also obtained in the western section of the country, albeit in small amounts. The country is separated into three ecological regions: the terai plains, the middle mountain range, and the high highlands. The Terai plains in the southern belt cover 23% of the country's total area, followed by hills in the middle belt (42%), and mountains in the northern belt (35%). As a result, land use practices in these zones varied significantly (FAO/GIEWS 2007). The Terai consists of flat and fertile alluvial land with a tropical and subtropical climate. Though the region has 17% of the total land area of Nepal, it accounts for about 46% of the gross cultivated area. The Terai region ends at the first range of foothills called the Shiwaliks. The Terai has a higher potential for productivity increase than the other regions since it can be irrigated and crops can be grown all year. Paddy, wheat, maize, legumes, sugarcane, jute, and vegetables may all be grown thanks to abundant water, fertile soil, and level terrain. The hill region lies between 700 and 4000 m above sea level, with steep slopes and a number of small narrow valleys. The climate of this region ranges from subtropical to moderate. With 68% of the land area, the region is the most populous. Rice, wheat, maize, fruits, and vegetables are among the crops grown on terraced fields. Rice—wheat/barley and maize—millet/legumes are the main cropping patterns. Misra (2011) reported that the region has good cultivation potential for fruits and vegetables.

7.18.2 Sustainable Land Use Management

Nepal is a mountainous country with a wide range of geography, climate, and plant life. The impact of both biophysical and human actions on the environment has been thoroughly documented (Maskey et al. 2003). On the basis of physiography, geology, and geomorphology, the terai, Shiwaliks, middle mountain, high mountain, and high Himalayas occupy 14, 14, 30, 19, and 23% of the total geographical area of the country, respectively. Resource deterioration and accompanying environmental repercussions are a natural element of the region's mass wasting process, which is a result of geological and climatic conditions (Ives 1987). Sustainable land use management would require soil and water conservation measures to be adopted with relevant cropping systems, appropriate nutrient application using proper methods and necessary tillage operations.

The mountains lie in the north and include the Himalayas and the high mountainous areas of the country. The region consists of large number of snow-covered areas and is mostly steep, rugged, and cold. Because of its geography and cold climatic conditions, it is sparsely populated. It has a 15% share of land area, only 2% of which is suitable for cultivation. Livestock rearing is the main occupation. Vegetables are cultivated as cash crops in a few areas. Administratively Nepal is divided into 75 districts.

7.18.3 Agriculture in Nepal

Agriculture is one of Nepal's most important contributors to the national economy, employing two-thirds of the people. Agriculture's growth is so critical for the country's economic development as well as food security. Approximately 80% of the overall population lives in rural areas and works in agriculture. Figure 7.79 depicts the area, production, and yield of food grains. Rice is the most important cereal crop, accounting for half of overall cereal production. The Terai region produces more than 70% of all rice. Over the last few years, rice production has been stable. The average production per hectare is around 2.8 t ha^{-1}. Maize is mostly farmed in rainfed environments on marginal soil with minimum fertilizer use. Soybeans, legumes, radish, potatoes, and upland rice are all important

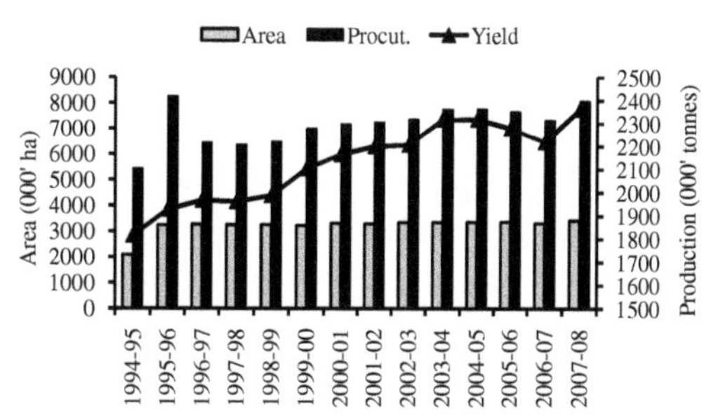

Fig. 7.79 Area, production, and yield of food grain crops in Nepal. *Source* FAO (2009)

Table 7.55 Important cropping patterns in two ecological regions of the Nepalese Mountains

Ecological region	Important cropping patterns	
Hills	Rainfed	Maize-fallow-fallow, Maize/finger millet-fallow, Maize/finger millet-wheat, Maize/soybean-mustard, Maize + soybean-mustard, Maize/upland rice-fallow
	Irrigated	Rice-wheat-fallow, Rice-wheat-maize, Rice-rice-maize
Mountains	Rainfed	Maize-wheat-finger millet (2-year pattern), Potato-wheat-finger millet (2-year pattern)
	Irrigated	Rice-wheat, Rice-barley, Buckwheat-barley, Potato-barley

maize-based cropping systems. Most of the yield increase is due to the introduction of hybrid varieties (Misra 2011). Yield levels remain relatively low at 2.2 t ha^{-1}. Wheat is the third most important cereal. Over 60% of wheat is produced in the Terai region. Average yields remain at 2.2 t ha^{-1}. Wheat production from the main winter crop in 2009 was estimated at 1.34 million tons, a decrease of about 14.5% from the preceding because of a severe drought. In recent years, the coverage of HYV wheat, rice, and maize has increased. However, irrigation coverage has remained stagnant. A high positive correlation between irrigation coverage, cropping intensity, use of HYV, and fertilizer use has been found. The synergetic effect of these factors has contributed towards increased crop yields for the major crops. Millets are usually planted with minimal inputs and is mostly grown on marginal lands. Production growth has virtually stagnated in the past several years (Misra 2011). Barley is another marginal cereal crop. Most barley production takes place in the mountains. The 2009 barley harvest was estimated at 23,000 tons. Pulses are mostly grown in wetland environments. In the western section of the country, lentil is a major pulse crop. Pulses and oilseeds production has grown at a relatively slow pace in recent years. Rapeseed, mustard, toria, groundnut, sesame, and sunflower are oilseed crops with a lot of promise, especially in the western half of the country. Oilseed yields have been modest, averaging around 740 kg per hectare. Between farmers' fields and research stations, and the yields that can be achieved, there is a significant. Misra (2011) attributed the reasons for low yield of crops to nutrient mining, soil erosion, and increasing cropping intensities without adequate fertilizer use.

7.18.4 Mountain Farming Systems

Important crop rotations in mountain areas are presented in Table 7.55. Cultivation is limited to lower slopes and relatively fertile patches in narrow-bottomed valleys along rivers where a number of crops are grown (Table 7.55).

Irrigation, recognized as an important factor in contributing to increased crop production, next only to fertilizer use, has remained stagnated at around 24% of the total cropped area for the past several years (Fig. 7.80).

The production of millets per capita in Nepal is depicted in Fig. 7.81. It has dropped since a

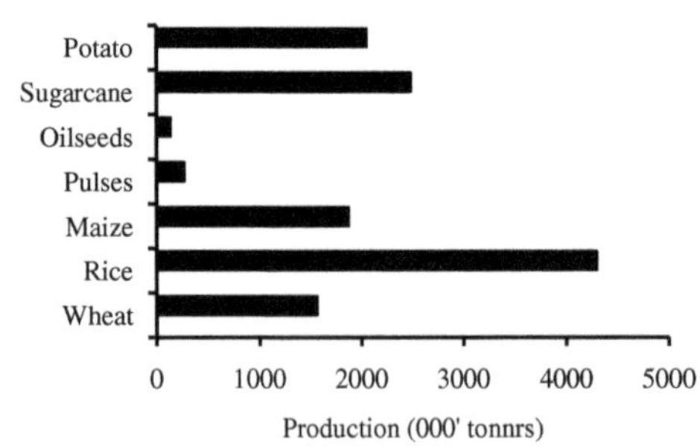

Fig. 7.80 Production of major crops in Nepal ('000 tons). *Source Ministry of Agriculture and Cooperatives*

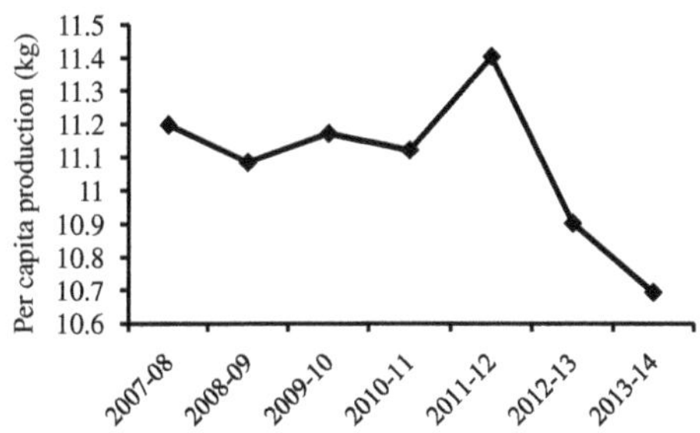

Fig. 7.81 Per capita production of millets in Nepal (Adhikari et al. 2017)

peak in 2010–11. Millets have a very low per capita productivity in Nepal when compared to wheat, rice, and maize. Farmers and policymakers do not choose coarse grain crops for local people's nutrition security and livelihoods, according to Adhikari et al. (2017). Millets, barley, buckwheat, black gram, and horse gram account for only 3.84% of total foodgrain production (GoN 2016).

7.18.5 Fertilizer and Soil Fertility Management And Food Production

Soil fertility, as elsewhere, is an important factor in agriculture productivity and production. Most agricultural soils in Nepal have a poor plant nutrient status. The Terai is the major agricultural zone for food crop production. Due to intensive cropping, the zone is showing signs of soil nutrient exhaustion and depletion with soil micronutrient deficiencies being increasingly observed at various locations. Use of organic manures is on the decline especially in the Terai because of an increased inclination towards mechanized farming. The hill and mountain soils are prone to erosion and therefore poor in soil fertility. Besides, the hills and mountains receive only a small fraction of the fertilizers available in the country. As a result, crop productivity is lower in these regions.

7.18.6 Role of Fertilizer and Soil Fertility Management in Crop Production

Limitations relating to suitable additional land that can be brought under cultivation in Nepal increase the critical role of soil fertility management and efficient fertilizer use in attaining agricultural production targets. If not managed properly, the sustainability of soil resources may be endangered, leading to economic adversity for farmers through lowered productivity and adverse impact on the environment. The decline in the soil resource base may be caused by depletion of nutrients or soil erosion as well as by acidity, salinity, alkalinity, waterlogging, nutrient toxicities, and decline of vegetative cover. Due to depletion of organic matter, decline in plant nutrient status, occurrence of micronutrient deficiencies, soil acidity, soil erosion and loss of vegetative cover, the soil resource base in Nepal has been adversely affected. Sometimes improper and imbalanced use of fertilizers has also led to ecological problems such as nitrates in water resources. It is important that sound fertilizer use and soil fertility management is crucial for providing stability to crop production and ensuring enduring food security in Nepal.

7.18.7 Fertilizer and Soil Fertility Management Policies

Assessment of existing national fertilizer management practices and policies and appropriate recommendations relating to sustainable soil fertility management and fertilizer use strategies and policy is a prerequisite in the present context of providing sustainability to the system (Misra 2011). As an immediate response to the APP (Agricultural Prospective Plan), increased fertilizer use and distribution, pricing, and technology policies have become critical issues. The less weight is given to subsidies, the more emphasis must be put on stepping up research, increasing profitability, providing timely and convenient supplies, and educating farmers on the proper use and profitability of fertilizer. In the long run, as the total volume of fertilizer use increases, both efficiency and environmental considerations will demand greater attention to optimal fertilizer use from both research and extension (APP 1995). The food grains production problem is complicated because of difficulties in transportation and a non-competitive market, especially in remote hilly areas.

7.18.8 Problem Soils

Soil acidity is a conspicuous problem. Eighty percent of the soil samples analyzed by the Department of Agriculture reveal acidic reaction requiring the application of lime. Adequate provision of lime is thus imperative. However, agricultural lime is not available despite there being ample deposits in various parts of the country.

The fertilizer consumption and food grains production is shown in Fig. 7.82.

7.18.9 Soil Health Management

Instances of organic matter depletion have been reported, especially in the Terai region, on account of intensive cropping and lower application of manures resulting from diminishing organic resources (Misra 2011). Besides the depletion of soil organic matter, the major factors contributing to decline in soil fertility are intensive cropping, inadequately supported with proper fertilization; loss of plant nutrients due to inappropriate soil management practices; soil erosion and lack of soil and water conservation measures. Intensively cropped areas like the Terai are showing increased signs of secondary and micronutrient deficiencies particularly of sulfur, zinc, boron, and molybdenum. It is now known that, apart from other reasons, the present levels of nutrient application are a major cause for a negative balance of various crop nutrients in the soil.

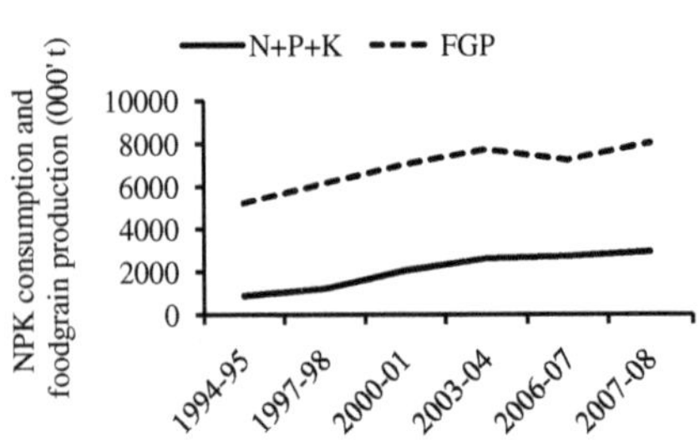

Fig. 7.82 Fertilizer nutrient consumption and foodgrain production in Nepal. *Source* FAO (2005)

Food security is highly variable throughout Nepal. Kurvits et al. (2014) reported that the country has a surplus of food, yet in 2013, 27 out of Nepla's 75 districts were food deficits. The western highlands have the most severe and chronic food shortages, with Humla, Dolpa, Darchula, Bajura, and Mugu suffering the worst. The food deficit districts are defined by (i) a preference for rice grains over traditional local crops such as maize, millet, buckwheat, and potato; (ii) outmigration of the agricultural labor force, particularly the rural youth population; (iii) weak public distribution systems and agricultural extension services; (iv) highly variable land productivity, depending on how agricultural practises are implemented throughout the seasons; and (v) rising food prices. Nepal imported 180 thousand tons of cereals in 2009, increasing to 753 thousand tons in 2012, demonstrating the severity of the country's food shortage (Shahi 2013). Demand is forecast to rise as the population grows from 28.8 million to 36 million by 2025. (Shahi 2013). Due to a late monsoon, rice production declined 11.03% in 2013, while imports soared 173% (Prasain 2013). Imported rice is increasingly becoming the most significant grain, both in terms of providing food help to food-insecure areas and meeting the demands of developing metropolitan centers.

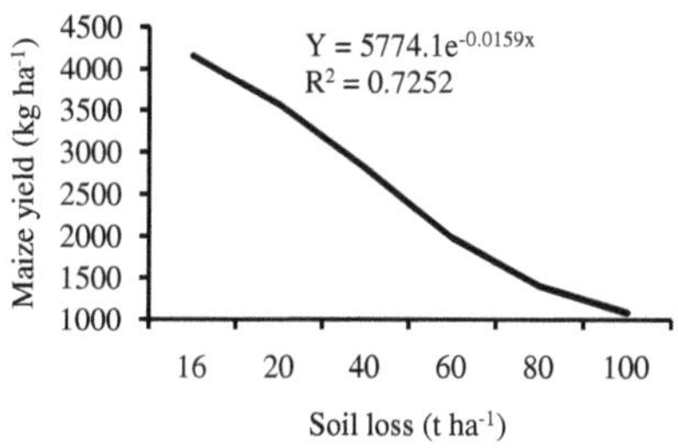

Fig. 7.83 Relationship between cumulative soil loss and the maize productivity. *Source* (Sherchan and Karki 2005)

7.18.10 Soil and Nutrient Losses from the Hill Slopes

There was a strong relationship between soil loss and maize crop productivity (Shercan and Karki 2005) (Fig. 7.83). This was mainly because of poor soil health and adequate availability of crop nutrients to sustain crop growth.

7.18.11 Biofertilizers, BNF, and Organic Manure

Bhattarai and Maskey (1987) reported that use of inoculums in the farmers' fields in Nepal, increased the productivity of the major legume crops (Fig. 7.84). They have standardized the application procedures for the effective Rhizobium strains of soybean, black gram, cowpea, lentil, chickpea, peanut, mungbean, and clover.

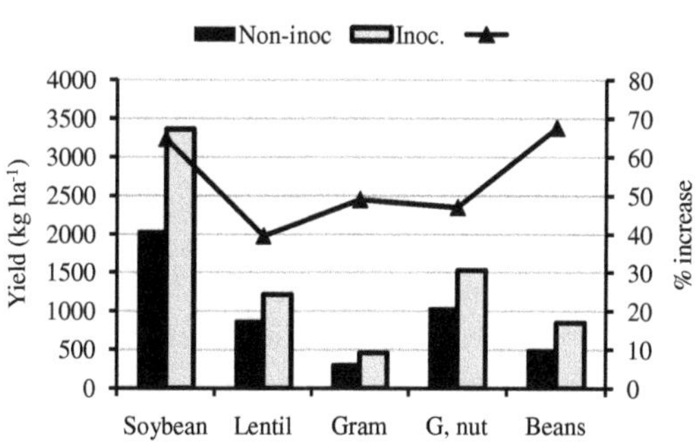

Fig. 7.84 Effect of the Rhizobium inoculum on the yield of leguminous crops

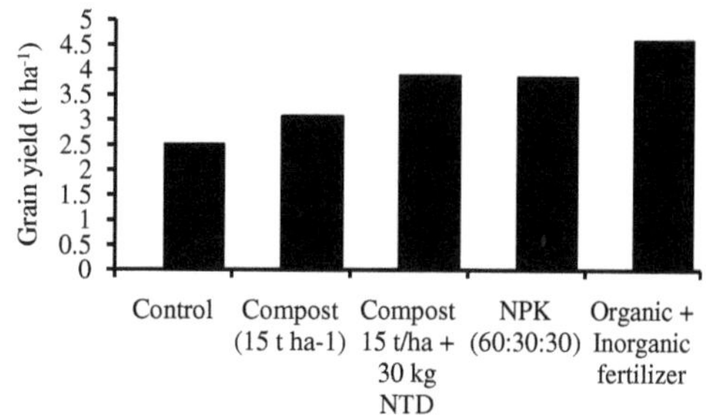

Fig. 7.85 Effect of application of compost on the grain yield of maize in Nepal. *Source* (Sherchan and Karki 2005)

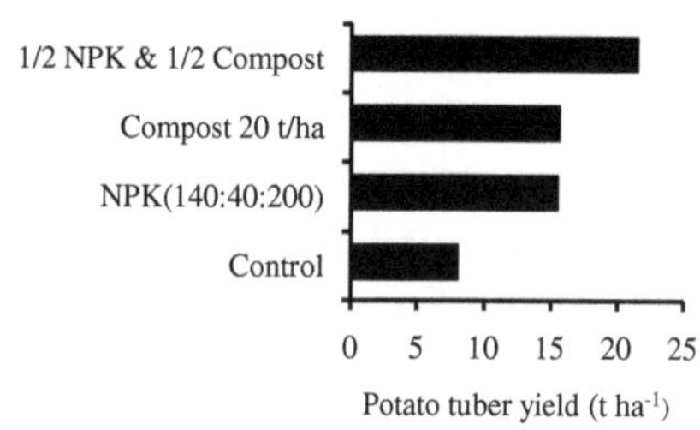

Fig. 7.86 Effect of compost application on the potato tuber yield in Nepal (Sherchan and Karki 2005)

The crop productivity of maize (Fig. 7.85) as well as potato (Fig. 7.86) has been found to increase significantly with the use of organic manures.

7.18.12 Fertilizers Recommendation

Recommendations on use of fertilizers for various crops in Nepal were standardized as per crop needs for optimum productivity (Joshi and Deo 1976). These recommendations have been revised from time to time according to the soil nutrient status and varietal requirements of newly introduced crop varieties. Though fertilizer recommendations for important crops are in place, the farmers use their own rate of fertilizer, possibly due to non-availability of desired fertilizer, lack of purchasing power or based on farmer's own experience. Shercchan and Karki (2005) identified integrated plant nutrient components in the Nepalese farming system. They mentioned about interrelationship among social environment (market and labor availability, social equity, food quality), soil health (soil nutrient content and availability, OM, leaching, erosion), nutrients use (FYM/compost, green manures, organic matter, biological fixation, chemical fertilizers, residue management), soil management (reduced erosion, balanced pH, reduced leaching, increased nutrient, availability) and crop management (yield, nutrient uptake, variety, intercropping, methodology, water, weeds, and pests management). Proper integration of all these factors can ensure potential crop productivity.

7.18.13 Micronutrient Problems in the Soil of Nepal

Food systems of the Himalayas are deficient in micronutrients mainly due to parent material with low micronutrients content, removal of topsoil rich in micronutrients with runoff, and increasing cropping intensity. According to studies, boron (B) deficiency was found in 80 to 90% of soil samples, zinc (Zn) deficiency was found in 20 to 50% of soil samples, and molybdenum deficiency was found in 10 to 15% of soil samples (Mo). The micronutrient deficiencies not only decrease agricultural productivity but also impact human nutrition directly or indirectly (Andersen 2007). Malnutrition caused by a lack of micronutrients in Nepalese soils, which then enter the food chain, poses a number of health risks to the Nepalese population (Anonymous 1998; Andersen et al 2005). Cropping intensity, according to Andersen (2007), can lead to nitrogen depletion in the soil (Andersen 2007). Deficits in zinc, molybdenum, and vitamin B limit agricultural yields and the availability of protein, iron, folate, and other minerals. The majority of the soil in the hill areas is made up of phyllite, schist, sandstones, quartzite, and granite, according to pedology (Carson 1992).

7.18.14 Impacts of Agricultural Intensification on the Environment

Large-scale mechanization of farming, the ever-increasing use of agrochemicals and improved hybrid varieties of crops have, though, led to increase in crop productivity but was also associated with environmental degradation. Earlier, the demands were low and traditional agricultural practices were done under those circumstances. The need for supply of nutrients from external sources was not felt due to low food demand, and the environment remained unaffected (Bajracharya et al. 2014). However, with the advent of high-yielding crop types and the advancement of chemical fertilizers, insecticides, and other agricultural products, a significant shift in production practices has occurred. Farmers have increased production with multiple and high-value crops due to changes in eating patterns, irrigation infrastructure, and expanding demands (Brown and Shrestha 2000; Dahal et al. 2009). According to Raut et al. (2012), high rice crop production in Nepal has resulted in soil acidification, increased denitrification rates, and the release of the strong greenhouse gas N_2O into the atmosphere.

7.18.15 Future Line of Work

Keeping in view the fact that the pressure on the soil resources will continue to increase to satisfy rising demands, it would be necessary to intensify agriculture (Schreier et al. 1994, Pandey et al. 1995). Schreier et al. (2000) and Bajarcharya (2002) mentioned that the demands for soil resources will be more in the Himalayas due to the huge populations and because of delicate balance in which ecosystems of this region. Water management is an important aspect of getting optimum productivity, so its efficient use is of paramount importance. Research should be done to improve the efficient use of inputs. There is need to timely arrange inputs such as good quality seed, fertilizers, and crop protection chemicals for the mountain areas having low accessibility and mobility.

7.19 Pakistan

Pakistan, with a total area of 79.61 million ha, is located in Southern Asia. The cultivated area in Pakistan was 21.28 million ha (26.73% of the total geographical area of Pakistan) during 2009, of which 20.09 million ha (94.4% of the cultivated area) for annual crops and 1.19 million ha (5.6%) for permanent crops. Total land use/land cover area of two mountain provinces of Pakistan has been shown in Table 7.56.

7.19.1 Natural Resources and Their Utilization

Pakistan's natural resource base is diverse, ranging from the Himalayas, Karakorams, and Hindu Kush, which include snow-capped peaks, to vast irrigated plains in the Indus basin that produce major and minor food crops, as well as deserts and undulating rocky plateaus in Balochistan. supplying rangelands and caring for a substantial amount of the country's livestock population (Mian and Mirza 1993a, b, c). The high Himalayas, as well as inter-mountain valleys in the north, provide water for irrigation and electricity generation, as well as growing fruits, vegetables, and other high-value products. The Indus basin features huge irrigated plains that produce major and minor food crops and are a substantial contributor to Pakistan's agricultural GDP. Deserts and undulating rocky plateaus in Balochistan's south-western corner provide rangelands and a substantial amount of the country's livestock population. Balochistan has 31.5 million hectares of rangeland, 7.97 million hectares of desert, 3.5 million hectares of trans-Himalayan grazing lands, 2.38 million hectares of Kohistan ranges, 1.68 million hectares of alpine pastures and Pothwar scrub ranges, 1.5 million hectares of Suleiman Mountain ranges, and 0.67 million hectares of Himalayan forest grazing lands.

7.19.2 Land and Soil Resources

Natural resource management has an impact on the degree of deterioration of land, water, and air, as well as the biological systems that support human life (Siddiqui 1997). Pakistan has a diverse range of soil types. It is necessary to

Table 7.56 Land use/land cover in the mountain provinces of Pakistan

S. No	Land use/land cover	NWFP	Province	Balochistan	Province
		Area (000' ha)	%	Area (000' ha)	%
1	Forest land	2311.9	16.5	508.1	1.5
2	Rangeland	3848.7	27.5	9255.8	26.7
3	Agriculture land	1174.1	8.4	822.2	2..4
4	Bare soil/current fallow	1290.5	9.2	4494.7	13.0
5	Exposed rocks	3451.1	24.7	16,425.1	47.3
6	Built-up area	26.4	0.2	6.7	0.2
7	Waterlogged/saline	0.2	0.0	15.2	0.4
8	Water bodies	57.1	0.4	1.8	0.001
9	Snow/glaciers	1829.6	12.1	–	–
10	Desert	–	–	3189.4	9.2
	Total	13,989.7	100	34,719.0	100

Source Ministry of Environment, Government of Pakistan (MEGoP) (2009)

protect important agricultural fields from overuse and to restore deteriorated soils (Farooq et al. 2007). About 70% of Pakistan's 79.6 million hectares of land area is classified as dry or semi-arid. Besides 50.88 million ha of rangelands, about 22.0 million ha (27.6%) is cultivated area. Soil erosion caused by runoff is one of Pakistan's greatest issues, with the Pothwar plateau seeing the most erosion of all. 5.328 million hectares of Pakistani soils are impacted by various types of salinity and sodicity, with half of them in Punjab, 40% in Sindh, and 9% in Balochistan province (Mian and Mirza 1993a, b, c). Waterlogging, like salinity and sodicity, is mostly the result of historic hydrological processes that are still active in much of the area with distinct geomorphological and physiographic characteristics.

Mian and Mirza (1993a, b, c) mentioned about the land capability classification of Himalayan Region of Pakistan. Maximum land (5.5778 million ha) is under Class VII and class VIII and is not suitable for cultivation. Land under class I, II, and III is most suitable for agriculture and covers an area of 1.3775 million ha in FATA, 1.1035 million ha in Balochistan and 0.338 million ha in Northern areas. The area under class IV land is 0.581, 0.699, and 0.046 million ha in histan and Northern areas, respectively.

Land Holdings

The demographic pressure has resulted in skewed distribution of landholdings in Pakistan. More than 56.7% of land holdings are below two ha and constitute only 15.5% of the total farm area; while 28.1% of land holdings vary from two to five ha, constituting 28.9% of the area. It is evident that about 85.7% of holdings are below 5 ha. Land holdings of more than 10 ha are 5.5% only but constitute 37.5% of the total area (GOP 2000, Pakistan Agricultural Census 2000).

7.19.3 Water Resources

Water has important role in the economy of a country, more so in Pakistan where most of the area comes under semi-arid to arid climatic conditions. In Pakistan, the agriculture sector uses about 88% of available water because agriculture is strongly reliant on water supply. Pakistan's per capita surface water availability has decreased from 5300 cubic meters in 1951 to 1100 cubic meters in 2006, and it is expected to continue to diminish in the coming years due to the country's rapid population growth. Groundwater irrigation accounts for over half of all irrigation needs. Water is being drawn from depths of over hundreds of meters in rainfed parts of Balochistan, according to the World Bank (2005), and depletion is now seen in all canal control areas in the Indus Basin. In terms of the balance between population and accessible water, Pakistan is one of the world's most water-stressed countries. In many locations, groundwater is over-exploited, and its quality is degrading. Furthermore, water productivity is low (World Bank 2005). In order to harness Pakistan's diminishing water resources, more investment is required. According to Qureshi et al. (2003), Punjab agriculture has grown largely reliant on groundwater, the quality of which varies widely from place to place. The quality of groundwater is affected by variations in its origin, source of recharge, and flow patterns in aquifers.

7.19.4 Natural Resource Utilization

The average annual growth in cultivated area, cropped area, and cropping intensity was 0.70%, 1.18%, and 0.48%, respectively, from 1947–48 to 2005–06 (Fig. 7.87).

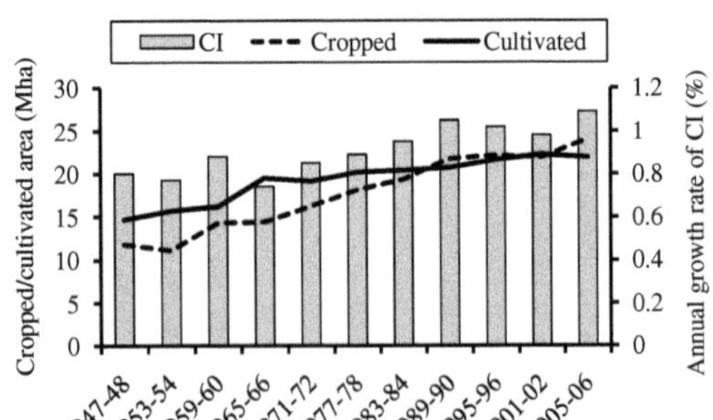

Fig. 7.87 Land Utilization Trends in Pakistan (CI, Cropping Intensity)

7.19.5 Land and Agriculture Resources in NWFP, Pakistan

The North Western Frontier Province of Pakistan almost completely comes under Hindukush-Karakoram-Himalayan ranges and so is typical representative of the Himalayan ecosystem. During the past few decades, the Agriculture sector of NWFP has been growing at a very satisfying rate (Atta-ur-Rahman 2003). More land has been brought under cultivation. Nonetheless, achieving self-sufficiency is a problem for the agricultural industry, which must compete with the province's rapidly rising population. As a result, increased adoption of high-yielding seed varieties and green revolution technology reversed agriculture's long-term stagnation (Atta-ur-Rahman 2003).

7.19.6 Land Use

According to Khan et al. (2012), around 23.39 million hectares of land were cultivated in 2006–07, compared to 14.99 million hectares in 1949–50. This accounts for around 29.38% of Pakistan's total land area of 79.61 million hectares. At a compound annual growth rate of 0.77%, the cultivated area rose by nearly 56.1% from 18.82% in 1949–50 to the year 2006–07. The average cropping intensity in Himalayan segment of Pakistan during 2016 was only 110% as against about 142% in plain provinces of the country. Of the total 8.345 million ha area in NWFP, 1.847 million ha (22.1%) were cultivated from 1999 to 2000. A total of 570,797 ha (30.1%) of the total farmed land was cultivated more than once (GoP 2000). The pace of development was so rapid that by the year 2000, NWF Province had very little land available for farming. Because of the minimal rainfall in the area, farming is heavily reliant on irrigation. The rainfed land, which is cultivated after the rains, is 923 948 ha. Still, from 1.15 million hectares in 1947–48 to 1.84 million ha in 2000–2001, the cultivated area (net area seeded plus current fallow) has increased. According to Atta-ur-Rahman (2003), out of a total area of 83,452 km^2 of NWFP, 18,479 km^2 (22%) is cultivated, 14,342 km^2 (17%) is the net sown area, and 4137 km^2 (5%) is current fallow. In FWFP, 9240 km^2 is the area under irrigation which forms 50% of the cultivated area. Land use classification is given in Table 7.57.

In NWFP agriculture is largely a two-season activity, that is, *Kharif* season (during summer) and *Rabi* season (during winter). Rabi season crops grown in NWFP include wheat, barley, mustard, sunflower, gram, *masoor,* tobacco, and sugar beet. Besides this, *Rabi* vegetables and fruits are also cultivated on large scale. *Kharif* or summer cropping season includes maize, rice, *bajra, jowar,* cotton, groundnut, sesamum, and soybean. The pulses are *mung, mash,* etc. and the major cash crop is sugarcane. Besides this, Kharif vegetables and fruits are also cultivated (GoNWFP 2000a).

Table 7.57 Land use classification, 1999–2000

Land use	Area (km^2)	%	Land use	Area (km^2)	%
Total area	83,452	100	Un-irrigated	9239	50
Cultivated area	18,479	22	Area sown more than once	5708	6.8
Net sown area	14,342	17	Uncultivated area	64,972	78
Current fallow	4137	5	Culturable waste	12,119	15
Irrigated area	9240	50	Forest area	13,608	16

Source Atta-ur-Rahman (2003)

7.19.7 Present Trends in Land Use

According to Mian and Mirza (1993a, b, c), current land use in the country is based on either the user's opportunity and economic standing or the user's socio-economic needs. A proper land use exercise is required to be done to use lands as per their capability. Much land on the steep slopes of mountains has been brought under various field crops to satisfy the food needs of the fast-growing population, which was otherwise suitable only for forestry or rangeland. Substantial area of excellent arable land in the Indus plains remains under forests or is under non-agricultural use. Because of climatic suitability, rice grows well in the area. This crop is also economically important because it gives a higher return. The requirement of water for rice is quite heavy. If the rice area is diverted to other potential crops requiring less water, more area can be brought under irrigation net. One of the reasons for the incorrect or irrational use of land is the lack of a coordination organization for agricultural research and extension activities (Mian and Mirza 1993a, b, c). The Pakistan Soil Study conducted a complete soil survey of all key agricultural research farms to aid in rational land use through more judicious use of agricultural research.

Wheat

Wheat is the most significant Rabi crop in NWFP, both in terms of value and planted area. It is mostly grown in areas with access to canal irrigation water (GoNWFP 1996a). Wheat is farmed in all of the districts and agencies along the border (Atta-ur-Rahman 2003). The canal irrigation system increased soil fertility by introducing a variety of organic and inorganic nutrients from the catchment area. Table 7.58 shows the acreage, production, and yield of key crops in the NWFP.

7.19.8 Livestock

Like other areas in HKH region, livestock forms an important component of agriculture in Pakistan. Livestock are also kept in the NWFP province as part of a traditional modest mixed farming system that includes crop production. About 87.2% of farms in the province are smaller than 5 hectares, with 61.2% of those being smaller than 2 hectares. The province has little area for crop production because it is largely made up of wet hilly terrain in the north and dry hilly terrain in the southwest. More than 85% of small farmers, including the landless, are completely reliant on animal production in this situation (Atta-ur-Rahman 2003). The livestock

Table 7.58 Area, production, and yield of major crops in NWFP Province (2000)

Crop	Area (000' ha)	Production (000' t)	Yield (kg ha^{-1})
Wheat	806.5	1067.8	1324
Tobacco	35.7	80.3	2250
Sugarcane	105.8	4803.5	45,401
Maize	533.9	8199.9	1536
Rice	67.1	129.2	1925
Fruits	47.1	541.1	11,488

Source GoNWFP (2000b)

Table 7.59 Livestock population in NWFP (1996)

Livestock	Number	Livestock	Number	Livestock	Number
Cattle	3,155,885	Goat	4,480,624	Mule	33,691
Buffaloes	1,261,109	Camel	41,568	Ass	395,789
Sheep	1,475,381	Horse	44,686	Poultry	16,020,638

population in NWFP province of Pakistan is shown in Table 7.59.

7.19.9 Nutritional Security

Nutrition security contributes to the achievement of many development goals related to agriculture, water, health, education, poverty alleviation, and gender development; investing in this sector will be highly beneficial to developing countries (Adhikari et al. 2017). Three Sustainable Development Goals of the United Nations such as Goals 2, 3, and 6 are directly related to nutrition security. A balanced diet is very necessary for individual health and health of a country. The nutritional value of neglected and underutilized food crops (millets like pearl, sorghum, finger, foxtail, proso baryard, and kodo) has been found to, by and large, more than normally used cereals such as rice, wheat, and maize. People are becoming more health conscious now and, it has been noticed that, of late, they have inclination towards underutilized food crops. Almost all millets are significantly more fibrous than normally used cereals. Generally, millets are highly rich than wheat, rice and maize in iron, calcium, zinc, riboflavin and folic acid.

7.19.10 Plant Nutrient Management

With a population of 200.8 million (2018), growing at an ACGR of 1.89% (between 2008 and 2018), Pakistan needs to make best efforts to boost and sustain agricultural production for food and economic security. Optimum soil health will be important in augumenting farm production. The use of fertilizer nutrients in Pakistan is high in the HKH region, only next to China. However, the nutrient use efficiency is low. For optimum crop productivity, more attention needs to be given to judicious and balanced use of fertilizer nutrients.

7.19.11 Fertilizer Consumption

Ahmed and Rashid (2003) reported that the cultivated soils in Pakistan are calcareous and alkaline with pH 7.0 to 8.5. These soils are generally deficient in N (almost all) and about 90% in P. The consumption of N, P and K in plains and mountains of Pakistan is presented in Table 7.60.

Crop use of fertilizers for different major crops in Pakistan is shown in Fig. 7.88.

This shows the comparative increase in the yield of major crops in relation to applied nutrients in Pakistan. Among cereal crops, wheat has responded well to the use of fertilizers followed by rice and maize (Fig. 7.89).

7.19.12 Use of Fertilizers in Pakistan

Wakeel (2015) reported that fertilizer consumption in Pakistan increased dramatically after the green revolution of the 1960s, with the introduction of high-yielding crop types that were fertilizer sensitive. Following the introduction of nitrogenous (N) fertilizer in the 1970s, phosphatic (P) fertilizer was introduced later in the decade, with levels gradually increasing.

Table 7.60 Comparison of N, P, and K in plain and mountain provinces of Pakistan (kg/ha)

Province	N	P	K	Total	N:P:K ratio
Plains					
Punjab	132.2	32.5	1.4	166.1	95:23:1
Sindh	156.2	42.7	1.4	200.3	111:30:1
Mountains					
KPK	111.4	21.8	0.9	134.1	124:24:1
Balochistan	88.7	14.1	1.0	103.8	89:14:1
Pakistan	**132.4**	**32.4**	**1.4**	**166.2**	95:23:1

Source National Fertilizer Development Centre

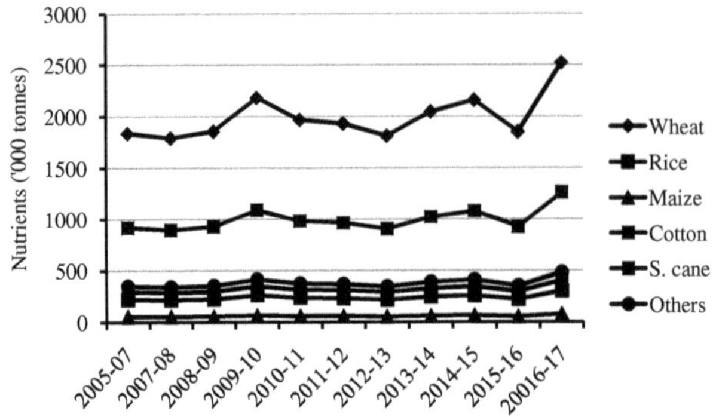

Fig. 7.88 Showing nutrients applied to various crops. *Source* National Fertilizer Development Centre (NFDC), Islamabad

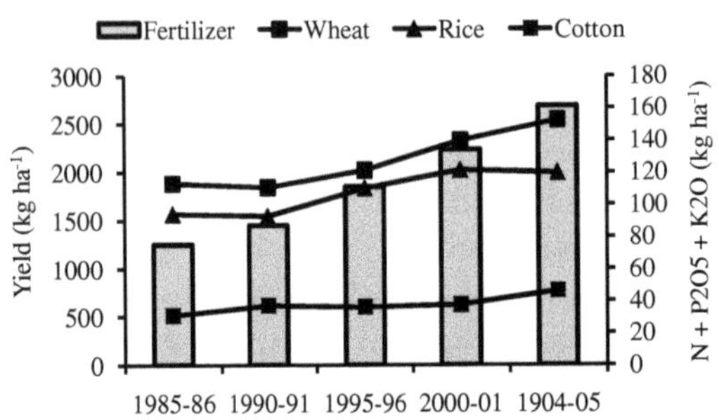

Fig. 7.89 Impact of fertilizer ($N + P_2O_5 + K_2O$) use on the crop yield over the years in Pakistan. (*Source* NFDC, 2004/05, 2014)

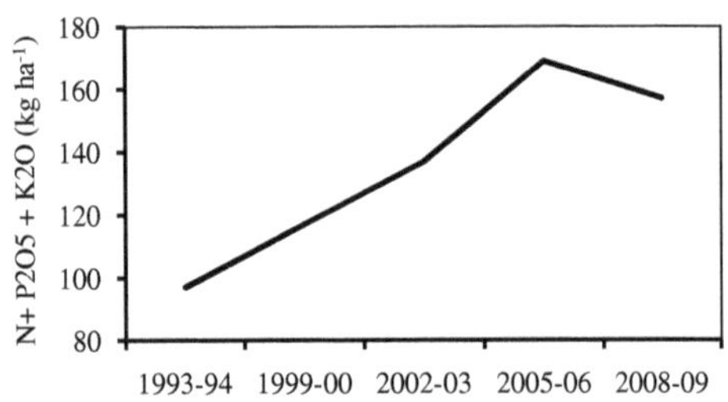

Fig. 7.90 Change in fertilizer nutrients ($N + P_2O_5 + K_2O$) applied per ha in Pakistan

Nitrogenous fertilizers were chosen, partially because they were readily available and partly because crops showed a visible reaction to nitrogen. The use of P fertilizer has continued to rise slowly and steadily. Slow increase in the use of P fertilizer was due to its availability at right time and amount. As the potassic fertilizers are imported, their use remained marginal till today. Restricted use of P and K has created imbalance of N, P, and K in the soils. Imbalance fertilization, particularly of macronutrients (NPK), degrades natural resources while also yielding low economic harvests (Wakeel 2015).

The use of fertilizer nutrients ($N + P_2O_5 + K_2O$) per ha increased substantially in Pakistan after 1960s to realize full potential of high-yielding varieties of crops. During 1993–94, the N, P_2O_5, and K_2O applied were at the rate of 97 kg ha^{-1} which increased to 157 kg ha^{-1} in 2008–09 (Fig. 7.90).

7.19.13 Impacts of Balanced Fertilizer Use

Solaiman and Ahmed (2005) reported that the present yield obtained by farmers is about 40 to 70% of that which can be obtained with recommended rates and balance use of fertilizer nutrients. Interestingly, the joint trials conducted by FAO, NFDC, IMPHOS and in different provinces of Pakistan for a period of three years showed that the yields of wheat (irrigated), wheat (rainfed), rice (IIRI), rice fine, cotton, Sugarcane, and maize increased by 59.4, 156.1, 88.4, 138.1, 37.7, 109.0, and 94.8% when balanced use of nutrients was made over the crop yields obtained under farmers' practice.

7.19.14 Organic Manures

Total livestock population in Pakistan is 138.37 million (excluding poultry). Total poultry is 357.18 million. This total population of livestock annually generates 555.2 and 274.4 million tons of dung (including poultry droppings) and urine, respectively. If this material is fully utilized for soil application to improve soil fertility, about 3.761, 1.213, and 3.049 million tons of N, P_2O_5, and K_2O, respectively, can be added to the soil (Table 7.61).

7.19.15 Limitations in the Adoption of IPNM

Though integrated plant nutrient management plays a significant role in realizing full yield potential of crops, there are some bottlenecks in its full application in Pakistan. The appropriate technologies are still not developed to suit different farming systems. More budgetary provisions are required to be made to develop state-of-the-art technology suitable under local specific conditions. Despite Pakistan's vast livestock population, the amount of animal dung accessible for usage in the field is insufficient to meet the needs of crops. The majority of cattle dung is burned as fuel. Because of the small size of the holdings and the strong demand for fodder, the farmer is unable to grow green manuring crops. The issue is further complicated by a lack of understanding of green manuring and legumes. There are limitations in the use of crop residues and biofertilizers. Many times the crop residues are used as fuel and also for feeding the livestock during winter. The use of biofertilizers is limited because of lack of application techniques and efficiency of strain, packing the product in marketable form, shelf life, and quality of the product.

7.19.16 Soil–Plant Micronutrient Dynamics

Khaliq et al. (2018) conducted a study on wheat-soybean cropping system with poultry manure (PM), wheat milling residues (WMR) and urea N (UN) and their integration. The integrated treatments improved wheat absorption of Cu, Fe, Mn, and Zn by 35.7 to 103%, 48.4 to 111.1%, 85.2 to 267.0%, and 33.8 to 128.2%, respectively, above control. In soybean, combined treatments increased Cu, Fe, Mn, and Zn micronutrient absorption by 18.3 to 60.3%, 27.5 to 87.4%, 14.1 to 54.6%, and 13.2 to 58.0%, respectively. The micronutrient content of the soil after harvest increased by 2 to 3 times, according to the post-harvest soil analysis. The findings revealed that using PM alone or in combination with WMR

Table 7.61 Livestock population, production of manure, and nutrient potential of manure in Pakistan

Livestock	Number (million)	Dropping (Mt y^{-1})	Urine (Mt y^{-1})	Dry matter (M t y^{-1})	N per year (000, t)	P_2O_5 per year (000, t)	K_2O per year (000, t)
Cattle	23.77	208.3	103.6	43.71	1159	477	1242
Buffaloes	26.15	267.0	134.2	56.1	1492	611	1600
Goat	59.15	33.1	21.5	11.6	595	204	512
Sheep	24.46	17.7	9.3	6.2	252	80	217
Poultry	357.18	8.1	0.0	3.7	120	68	31
Others	4.84	21.0	6.0	8.4	143	47	149
Total	495.55	555.2	274.6	129.7	3761	1213	3049

and UN can be an effective management technique for increasing soil micronutrient content (Khaliq et al. 2018).

7.19.17 Water Erosion

The Indus basin's soils are new and underdeveloped, while the surrounding mountains have high slopes (Khan 2012). Soil erosion is greatly exacerbated by heavy rains and melting snow in high mountains. Other elements that contribute to soil erosion include land use practises, vegetation cover, soil type, and structure, in addition to rain and snowmelt water. While water erosion is low in northern highland areas with steep slopes due to permanent forests, it is higher in places with arable crops on steep slopes. Water erosion affects around 11 million hectares of land. Each year, around 40 million tons of dirt are delivered into the Indus basin, reducing the life span of large reservoirs due to massive silt deposits, reducing their capacity for water storage and functional effectiveness. Sedimentation affects the efficiency of hydropower generation and reduces the effectiveness of irrigation systems in downstream areas.

7.19.18 Magnitude of the Problem

The province-wise extent of area affected by different degrees of water and wind erosion is given in Table 7.62.

Strategies for the Control of Water Erosion

The deforestation in large areas has enhanced the soil erosion in mountain of Pakistan. Forest replanting, conservation, and preservation are all needed in the country. Forest replanting, conservation, and preservation are all needed in the country. Cultivation on steep slopes must be stopped immediately. If necessary, bench terraces should be built and improved, field and terrace embankments should be strengthened, and contour-plowing and stripcropping techniques should be used on gently sloping land. To prevent deforestation, alternative fuel sources, food fodder, and other necessities must be provided to the local population. Proper vegetation cover has to be provided to reduce soil loss through erosion. Area affected by water and wind erosion in mountain provinces of Pakistan (000' ha) is given in Table 7.63.

7.19.19 Strategies for Sustainable Soil Fertility Management

Pakistan is trying to maintain daily per capita dietary intake at 2 400 kcal. Only cereals provide around half of the caloric demands, with livestock products accounting for 7 to 16% of the daily rural diet. Dietary intake increased from 1 754 cal per person per day in 1961 to 2 300 kcal/person/day by the end of the 1970s, but then plateaued in the 1980s and 1990s with no significant improvement. Inadequate food availability due to significant yearly population growth of 2.85% since 1947 is one of the key causes. In addition, the small and declining size of landholdings, 32% being less than 1 ha and another 24% of less than 2 ha each, is a constraint to productivity and agricultural income. Wheat is the country's principal staple food crop, and any shortage in production has an impact on both household food security and the national economy (Ahmed 2011).

Table 7.62 Area affected by water and wind erosion in mountain provinces of Pakistan (000' ha)

	NWFP and FATA	Balochistan	Northern areas
Water erosion	4,292.2	2634.6	2282.1
Wind erosion	36.5	280.5	–
Total	4328.7	2915.1	2282.1

Source Soil Survey of Pakistan. N.d. Reconnaissance Soil Survey Reports from 1965–88. Lahore

Table 7.63 Area affected by water[a] and wind[b] erosion in mountain provinces of Pakistan (000' ha)

Erosion degree	Water erosion				Wind erosion			
	NWFP & FATA	Balochistan	Northern areas	Pakistan	NWFP & FATA	Balochistan	Northern areas	Pakistan
Slight	156.3	–	180.5	398.0	13.1	36.0	–	2595.5
Moderate	853.8	1805.0	25.8	3581.4	3.8	143.6	–	496.7
Severe	1765.1	829.6	504.2	3745.9	19.6[c]	100.9c	–	1668.3[c]
V, severe	1517.0	–	1571.6	3446.5			–	
Total	4292.2	2634.6	2282.1	11,171.8		280.5	–	4760.5

[a] *Source* Soil Survey of Pakistan. 1988. Reconnaissance Soil Survey Reports from 1965–88. Lahore
[b] *Source* Reconnaissance Soil Survey Reports from 1965–88
[c] Including very severe

Other Nutrient Sources

Nutrient sources other than chemical fertilizers are organic materials such as farmyard manure (FYM), animal manures, crop residues, composts, and biofertilizers. Due to the high cost of fertilizers, deteriorating health of soils, and environmental concerns, the interest in the use of organic sources of nutrients has increased. Organic manures provide mainly nitrogen, phosphorus, and potassium as well as small quantities of the secondary and micronutrients. The addition of organic manures improves soil aeration and water retention. When used in combination with inorganic fertilizers, organic manures increase the efficiency of the inorganic fertilizers (Ahmed 2011).

Organic Fertilizers/Manure

This consists mainly of FYM, poultry manure, crop residues, filter cake stillage from the sugarcane industry, slaughterhouse waste, sewage and sludge, city refuse, and wastes from the food processing industries. The National Fertilizer Development Centre (NFDC) estimates that FYM can provide approximately 1.5 million tons of fertilizer to the country. Nitrogen accounts for 726 000 tons, P_2O_5 for 191 000 tons, and K_2O for approximately 617 000 tons. In Pakistan, nearly half of the dung is left uncollected, and about a quarter of it is used as fuel in the form of dry cake. The majority of crop wastes are consumed as animal feed or fuel, leaving only a little amount for recycling (Ahmed 2011).

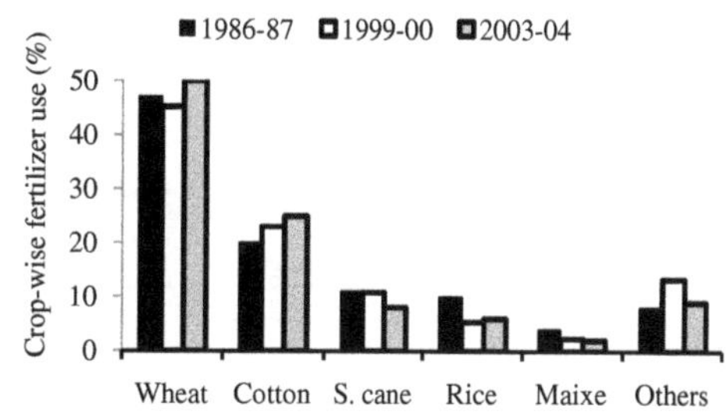

Fig. 7.91 Crop-wise fertilizer use in Pakistan. *Source* (Ahmed 2011)

Fertilizer Use by Crop

Λ number of field studies and surveys have been conducted to assess what share of total fertilizers is consumed by each crop. The results of three surveys conducted by NFDC, one during a regulatory regime (1986–87), and two after reforms (1999–2000; 2003–04) clearly indicate that wheat and cotton together consumed the lion's share of fertilizer nutrients (Fig. 7.91).

7.19.20 Fertilizer Use and Production of Food Crops

Ahmed and Rashid (2003) and Ahmed and Tila (1998) have reported about the use of fertilizers and crop response to nutrients in Pakistan. The population of Pakistan increased from 142.35 million in 2000–01 to 164.5 million in 2008–09, a growth rate of about 2.1% per annum. During the same period, the availability of cropped area

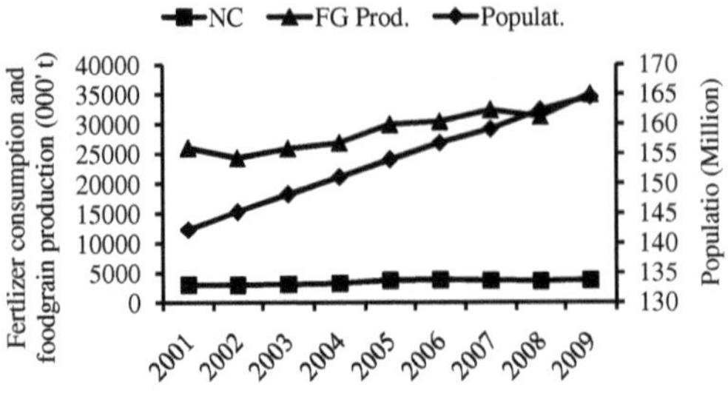

Fig. 7.92 Fertilizer consumption and foodgrain production in Pakistan (NC, Nutrient consumption' FG, Foodgrains)

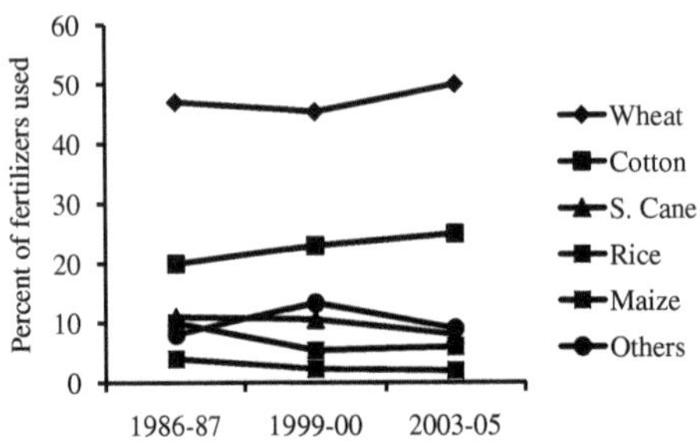

Fig. 7.93 Percent of fertilizers used for major crops in Pakistan. *Source* (Ahmed et al. 2011)

per capita decreased from 0.16 to 0.14 ha. The decreasing land per capita ratio has added to pressure to improve productivity per unit of land. The relationship between total fertilizer consumption from 1990 to 91 and production of food crops is presented in Fig. 7.92. Fertilizer consumption increased from 1.9 million tons in 1990–91 to 3.7 million tons in 2008–09. Correspondingly, the production of food crops increased from 19.6 million tons to 35 million tons. The production per capita increased from 182.5 kg in 2000–01 to about 212 kg in 2008–09. Per capita fertilizer use also increased. There appears to be a positive relation between fertilizer use and foodgrain production (Ahmed et al. 2011).

7.19.21 Trend in Fertilizers Use in Major Crops

Wheat claimed about half of the total fertilizers used in Pakistan, followed by cotton, sugarcane, and rice (Fig. 7.93). The share of fertilizer use has been increasing over time in case of wheat and cotton at the expense of other crops.

Pakistan was among the South Asian countries to benefit from the Green Revolution. Wheat and cotton yields in 2008–09 were triple those in 1966–67. The yields of other major crops such as rice and maize doubled. Despite this, the yields obtained are between 40 and 70% of economic yield potential. Fertilizer use and production of major crops in selected years with CAGR are given in Fig. 7.94. The rate per ha of applied nutrients has increased over time in Pakistan. The crop yields have increased considerably with increase in the use of fertilizer nutrients (Fig. 7.95).

7.19.22 Efficiency of Fertilizer Use

Despite impressive increases in fertilizer consumption in Pakistan, crop yields have not increased proportionately. It is estimated in studies of farm yields that between 20 and 50% of the N applied in fertilizer is recovered in the crop during the year of application. In most cases, between 15 and 20% of applied P fertilizer is utilized by the first crops, but part of the residual P remains available over a period of time to succeeding crops. Most of the fertilizer Zn is fixed by calcareous alkaline soils and only a small fraction remains available. Besides affecting the productivity of some crops such as rice, low Zn availability also affects the efficiency of N and P fertilizers. Several factors contribute to reduced fertilizer efficiency. They include poor seedbed preparation, poor quality of seed and improper seeding, delay in sowing, unsuitable crop variety, inadequate irrigation/drainage, weed infestation, pest and disease attacks and improper and unbalanced fertilizer application. The contribution of the various factors to reduced efficiency varies under different conditions (Ahmed 2011). Nutrient use efficiency generally decreases with increasing rates of application. In Pakistan, a steady decline in production response to nutrients has been observed. The period between 1985–86 and 2004–05 shows a continuing downward trend.

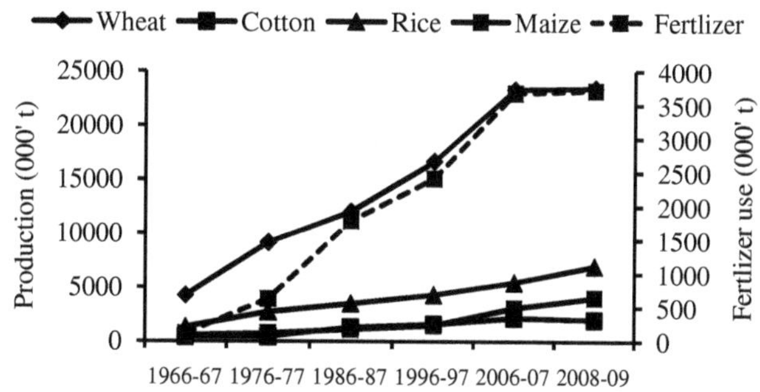

Fig. 7.94 Crop production and fertilizer use in Pakistan (Based on Ahmed et al. 2011)

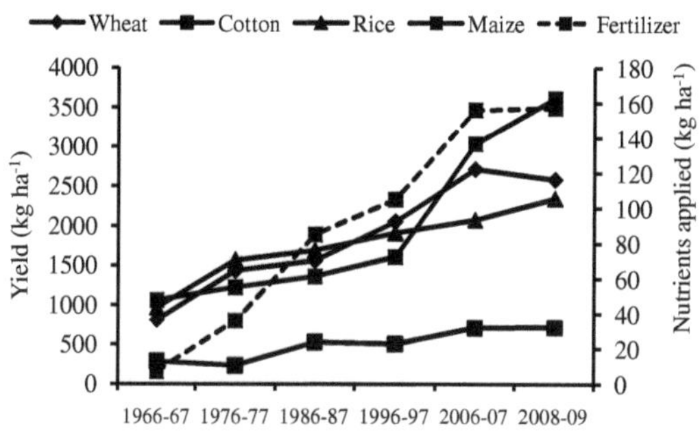

Fig. 7.95 Yield of different crops and nutrients used per ha in Pakistan (based on Ahmed et al. 2011)

7.19.23 Salinity and Sodicity

The problem of soil salinity and sodicity is severe in Pakistan restricting the pace of agricultural development. Canal irrigation is one of the factors responsible for this problem. However, the reconnaissance soil surveys in Pakistan have revealed that saline or saline-sodic soils occupy specific physiographic positions in the Indus plains because of long-term natural processes. The total area under different saline/sodic soils in Pakistan is about 550.6 thousand ha.

7.19.24 Food Security

Food security is a crucial issue in Pakistan's growth, which has been deteriorating since 2003 as a result of repeated flooding occurrences (Kurvits et al. 2014). In Pakistan, 80 out of 131 districts, or 48.6% of the country, lack adequate food (Guriro 2013). Food insecurity affects 68% of the Federally Administered Tribal Areas (FATA), 61% of Balochistan, and 56% of Khyber Pakhtunkhwa (KPK) residents in the mountain provinces. The mountains (ten districts in Balochistan, five in FATA, three in KPK, and one in Gilgit Baltistan) and the plains (one district in Gilgit Baltistan) account for 19 of the 20 most food insecure districts (Sindh). This demonstrates that food insecurity is widespread across the country, with the mountains having a higher rate of food insecurity than the plains. Food accounts for 46% of household income in Pakistan, compared to 35% in India, due to larger family sizes (Guriro 2013). Mountain farming systems are being restructured as a result of declining agricultural investments, labor outmigration, and a lack of young people interested in farming (Kurvits et al. 2014).

7.19.25 Flash Flood Risk Management in the Hindu Kush Himalayas

Flash floods are particularly harmful due to their unpredictability and the amount of water and debris they carry. Despite their strong effect on people, there is a general lack of policies to deal with flash floods in most of the countries of the HKH region (Shrestha and Pradhan 2015). The Hindu Kush Himalayan (HKH) region is prone to disasters. Percent share of different types of disasters occurring in HKH region is given in Fig. 7.96. About 40% of the disasters are caused

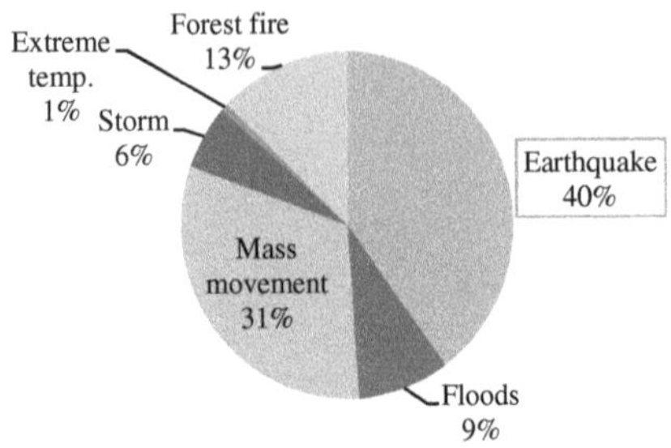

Fig. 7.96 Water-related disasters

by the earthquakes, followed by mass movement (31%), forest fires (13%), floods (9%), storms (6%), and extreme temperature (1%).

7.19.26 Water–Energy–Food Security Nexus

To secure growing demands for food, water, and energy, the HKH countries face a challenging task. This challenge becomes more complex with climate change scenarios. Adaptation to change requires that the natural resources may be used efficiently. The countries of the region generally deal with land, water, and energy on individual basis and do not consider the interconnections and interdependence among the three sectors. Despite the fact that the concept of a water–energy–food nexus is gaining traction and adaptation to climate change has become a pressing requirement, little research has been done to date on the ties between the nexus perspective and climate change adaptation (Rasul and Sharma 2016). Climate change adaptation is a global priority, and it is especially vital for impoverished countries (WRI 2011). The strategy has been to deal with the components of nexus partners, namely water, energy, and food, on an individual basis, with no regard for the overall effect. Sectoral adaptation techniques can increase vulnerability or erode net resilience (Barnett and O'Neill 2010; Lele, Klousia-Marquis and Goswami 2013; Urwin and Jordon 2008; Urwin and Jordon 2008). Multiple implications of climate change can be seen in the HKH region, particularly in terms of providing enough food, water, and energy for a growing population (Rasul and Sharma 2016). These difficulties are particularly acute in mountainous areas, where the effects of climate change are already evident and are projected to have major consequences for water, energy, and food supplies unless proper steps are implemented (Eriksson et al. 2009a, b; Rasul 2012; Shrestha and Aryal 2011).

7.19.27 Socio-Economic Constraints to Development of Highland Agriculture

Jodha (1992) explained that the decline of agriculture in fragile mountain agroecosystem is largely due to extreme poverty, inaccessibility, low technological absorption, lack of suitable farming systems, and, reluctant interest in agriculture. Proper harnessing of their niche and effective management of available resources would ensure higher level of productivity and financial returns.

Indicators of Unsustainability

On the basis of their actual or potential, the indicators of unsustainability have the following three categories (Jodha 1991).

Directly Visible Negative Changes

Increased landslides, drying of traditional irrigation channels, increased idle time of grinding mills or sawmills operated by natural water flows, prolonged fall in crop yields in mountains, reduced diversity of mountain agriculture, and increased migration of hill people are all examples of these changes.

Negative Changes Made Invisible

Substitution of shallow-rooted crops for deep-rooted crops, erosion of topsoil; substitution of cattle for small ruminants due to permanent degradation or reduced carrying capacity of grazing lands; substitution of cattle for small ruminants due to permanent degradation or reduced carrying capacity of grazing lands due to the loss of locally renewable resources, a public distribution system was established to address increasing inter-seasonal hunger gaps and greater external inputs in agriculture.

Development Initiatives with Potentially Negative Consequences

A number of measures adopted to increase production may jeopardize the ability of the system function normally with passage of time.

7.20 Crop Production Approaches in the HKH Region

7.20.1 Land Use

Land use and land use change, as well as changing ecosystem functions and services, are the primary causes of soil degradation, which influence ecosystems' ability to meet human demands for food, water, and energy. Integrated research on land use and land use change (LULUC) in the HKH region is generally scarce, according to a review of the literature. Rangelands occupy more than half of the entire area of around 3.4 million km^2. Despite being primarily mountainous, the HKH region has only 14% of its land covered with woods. Broad leaved woods, needle-leaved forests, and plantations are all included. Arable land makes up around a quarter of the territory, and wetlands make up about a tenth of the total area of HKH. Deforestation and degradation of grasslands at lower altitudes have generated the most substantial LULUC in recent decades.

7.20.2 Improving Food Security

The records show that the HKH region has made a visible progress in food production over the years, particularly after late sixties when high-yielding crop varieties were introduced. However, biophysical and socio-economic conditions in HKH play an important role in mountain agricultural systems as far as sustainability is concerned. In the olden days, a wide range of traditional crops like millets, sorghum, buckwheat, barley, etc. formed part of the food in mountain areas. China, India, Pakistan, and Bangladesh were chronically food deficit countries but transformed themselves to food-sufficient countries by making judicious use of land and water resources and appropriate use of inputs. In the late 1990s, all HKH countries except Afghanistan and Nepal exported some food grain. More than a tenth of the cereals consumed in the region are imported or provided as food aid (Rasul 2010; Rasul et al. 2017). Since the 1960s, India has achieved great progress in food production and was successful in ushering in the green, white, and blue revolutions. Significant agricultural success will be realized in the region if negative environmental impacts are mitigated. Sustainable agriculture encompasses all methods and activities that improve environmental and natural agricultural resource protection while also ensuring the production of adequate and high-quality goods at reasonable prices. Good soil health is a prerequisite for higher enhanced crop yields and quality of produce. Proper and appropriate additions of organic and inorganic inputs have to be made to compensate for the crop removal of nutrients and nutrient losses from the soil.

7.20.3 The Gap Between Land Use Policy and Adoption

The Hindu Kush Himalaya (HKH) region is profusely endowed with natural resources. It has an abundance of water, biodiversity and a wide range of crops are grown in different parts of the region due to variable climatic conditions. The rapid growth in agricultural productivity and the current global food system is a major driver of climate change, as well as increasingly vulnerable to it. The HKH Region has been benefitted up to a limited scale with the introduction of high-yielding crop varieties and improved management practices. The reason behind low crop productivity is marginality of the farmers and heterogeneity of resources. Climate change has direct impacts on food systems and food security. Women play a key role in food security in the HKH region and so, women empowerment in the region is important for sustainable agriculture. Land use changes have been found to be closely linked to hydrological processes. Household life is primarily reliant on agricultural production in the HKH region, where family farming practises are still common. In mountain regions, judicious land use techniques are one of the approaches for enhancing farmer livelihoods, safeguarding natural resources, and assuring long-term development. However, many of such farming practices have not reached the farming community, resulting in slow improvement in the Himalayan landscape as compared to plain areas. There is need to make the households as the units that make decisions, determines the objectives, and mobilizes resources to enhance crop yields. Based on the premise that appropriate land use policies will improve agricultural production, natural resource conservation has to be emphasized. Approach should be for integrated land use planning. Low crop yields are a result of conventional technology; however, such land limits can be solved by improved cultivation techniques. Land fragmentation may restrict the adoption of agricultural mechanization and a market-oriented agriculture on a wide scale. Tulachan (2001) found that, despite the fact that the area under foodgrain crops in HKH has not expanded, yields have not dropped, contrary to popular belief. Crop diversification in favor of horticultural and cash crops has been observed in the region, indicating that the cultivators are now showing interest in horticulture and cash crops. Crop diversification has not only added additional produce to the kitty of farmers but is also helping in cash earning and improvement in the quality of life of the people in HKH region. In terms of exploiting mountain niches and comparative advantages, these tendencies have encouraging implications for the future development of mountain agriculture.

7.20.4 Challenges and Opportunities for Crop Production

Challenges

Cropland Conversion: Change of forests, grasslands and rangelands to cropland has been aggravated due to economic and policy pressures. Many croplands have been found to be well managed for soil resource conservation, but conversion is frequently associated with severe soil and landscape cover disturbance in the beginning, resulting in rapid loss of soil carbon, physical compaction, and other changes that disrupt hydrologic functioning, soil nutrient cycles, and soil ecosystems, particularly the soil microbial community. These changes hamper crop production efforts.

Development activities: Developmental activities exert great pressure on soil resources and disrupt natural ecosystems. It has led to soil, water, and air pollution, and depletion of natural resources.

Threats to Soil Capacity: Forest soils are an important component of HKH landscapes. Although the forest occupies only about 30% of the region's land area, they are responsible for more than 75% of the surface freshwater. The forest soils are degrading at an alarming rate.

Land Cover Changes: Extreme weather events, such as droughts and floods, are well recognized to alter soil function. Extreme weather's impact on soil may continue to grow as a result of land use changes as their frequency rises as a result of climate change.

Opportunities

Research: In the absence of benchmarks, it is difficult to set targets for the severity of the effects of changes in land use due to a lack of full understanding of soil ecosystem services. Research activities are needed to improve fundamental understanding of soil ecosystem services and to develop metrics or indicators to evaluate and track soil function under changing land use conditions and its effects on crop yield and production.

Land Management: Land tenure issues can complicate programs related to the impact of climate change on land use and land cover.

Social Science: Improved understanding of the social forces underlying such changes that may affect soil function is needed to assess the effectiveness of existing incentives on properties managed under various ownerships and tenure arrangements.

Special focus is needed to create local food systems, strengthen the agricultural marketing system and infrastructure, manage water resources and other ecosystem services, and promote secondary sources of revenue for effective crop production.

7.20.5 Comparative Crop Production Approaches in HKH

Rice, wheat, and maize are the major cereal crops in HKH region for food security. This is in addition to a number of other traditional cereal, pulses, oilseeds, and cash crops, besides horticultural crops like fruits and vegetables. People in Afghanistan depend on consumption of cereals, particularly wheat, being staple food of majority of the population. The country has never been self-sufficient in food grains as the country could produce this commodity for a population varying from 53 to 96% (in 2009 when country benefitted from the most favorable weather) between the year 2000 and 2017. Wheat is the most important crop produced on irrigated land. About 60% of the area under irrigation is served through canals and/or river water. In 2012, 2.4 million tons of cereals (wheat, maize/sorghum, barley, rice, and millet) were produced. In addition to food crops, such as potatoes, fodder, and cotton are important production crops. Out of a total cereal production of 2.495 million tons in 2012, wheat, rice, maize/sorghum, barley, and millets accounted for 83.8, 7.5, 6.1, 2.4, and 0.1%, respectively.

Bangladesh has only 11.75% area under hills known as Chittagong Hill Tracts (CHT). At present, rice, wheat, jute, sugarcane, potato, pulses, and oilseed cover are about 79.4, 3.92, 3.71, 1.23, 1.11, 4.64, and 3.77% of cultivated land of the country. So, rice is the most predominant crop of Bangladesh, receiving maximum attention. This is, particularly, because of food security for a highly densely populated country but there is need to increase area under other crops to ensure proper nutrition also. Since area under other crops cannot be increased at the cost of area under rice for food security, the only option is to increase the productivity of rice with efficient use of inputs. In Bhutan, the national averages of the yields of various crops are still very low as compared to those of many other countries in the HKH region. Rice and maize are the most preferred crops in cereals, having 42.1% and 44.9% of area and contributing 48.1% and 46.4% to total production of cereals. Besides these two crops, wheat, buckwheat, oilseeds, and pulses are also grown.

Wheat is the major crop of Tibet, with an area of 51,000 ha (92.7% of rice, wheat, and maize put together), and production of 249,000 tons (93.2%). However, in Yunnan and Sichuan provinces of China, rice, wheat, and maize have comparable area, though rice and maize are leading crops in the former and rice and wheat in the latter are leading crops. The total area under these three crops is 55,000 ha, 25.9 million ha, and 71.5 million ha with a production of 267,000 tones, 10.52 million tones, and 36.19 million tones in Tibet, Yunnan and Sichuan provinces, respectively. Human interference and climate change have had an impact on these permafrost zones in recent decades, and they are now eroding and jeopardizing not only the unique species and nomadic way of life, but also the subsurface aquifers that recharge wetlands and great rivers. Ongoing permafrost degradation on the Tibetan Plateau is a matter of concern.

In the Indian Himalayan region, traditional agriculture is generally a low input, low production subsistence system. Its intensification involves high risk of natural resources degradation. Most of the food grains production has come from the irrigated plain lands while the mountain areas have remained deficit in food crops production for sustaining the population. Rice and maize are the major crops of mountain areas of India. A holistic approach, based on integrated fertilizer nutrient management, needs to be started to make the mountain region self-sufficient in food grains to ensure enduring food security. The average yield per ha varied from a low of 1379 kg to 2680 kg in Jammu & Kashmir and Tripura states of India, respectively. Rice is the major crop of the country with an area of about 7.0 million ha and production of 3.6

million tones. The other major crops are maize, sorghum, wheat, sesame, Black gram, green gram, pigeon pea, groundnut, and sunflower. Rice accounts for 91.2% of the area and 42.8% of production among cereals.

Agriculture is one of the major contributors to the national economy of Nepal. Growth in agriculture is, therefore, crucial for economic development of the country as well as food security. Among cereals, rice is the most important crop contributing to half of total cereal production. Over 70% of rice is grown in the Terai region. Average yield is in the order of 2.8 t ha^{-1}. In Nepal, maize is the second most important cereal, contributing about 25% of total cereal production in the country. Hills are more suitable for maize cultivation due to appropriate natural drainage and cover about 80% of the hill area. Grown under rainfed conditions on marginal land. Nutrients are applied at lower rates. Wheat is the third most important cereal and over 60% of wheat is produced in the Terai region. In recent years, the coverage of HYV wheat, rice and maize has increased. However, area under irrigation has remained stagnant. In mountain areas of Pakistan, subsistence agriculture, including livestock rearing, is the major source of livelihood. Maize and rice are the predominant crops in mountain landscape of Pakistan. Cash crops are also grown in some pockets. However, the per hectare yield of almost all crops has remained stagnant. Major constraints experienced in crop production in the HKH region are limited arable land, limited access to water, low crop productivity, lack of infrastructure and marketing facilities, and low value addition. A major limiting factor in crop productivity is soil degradation due to acidification, aluminum toxicity, and soil erosion.

7.20.6 Approaches to Produce Same Crop in Various HKH Countries

Food insecurity persists in the HKH countries, where low agricultural productivity is a result of the climate, rugged terrain, poor soils, and short growing seasons. The HKH's subsistence farming has always been excellent at utilizing the inherited mountain food systems. The state of food security varies widely between countries, and diversity has always been a key component of farming systems in the region's tough terrain. In the traditional subsistence systems, many farmers are interested in cash crop production. In Yunnan province, China, farmers have started growing tobacco, and to a lesser extent walnuts or fruits. This is supported by industry. The same is true in Assam, India, where the tea industry encourages small-scale farmers to cultivate tea on their land, which requires less labor and yields more profits. The rising market value of winter vegetables including cauliflower, potato, cabbage, carrot, radish, and medicinal herbs convinced an entire agricultural community in Nepal to switch from maize and buckwheat to more desirable cash crops. Production is threatened by poor soil management and water constraint. The impacts of agricultural intensification are visible, which pose a challenge to food security. Systems, not based on sustainability criteria, that were adopted to raise crop yields are, now, causing land degradation, deforestation, overgrazing, groundwater depletion, and water pollution.

With enhanced connectivity among different countries of the HKH region, production methodologies followed for growing a particular crop are almost similar, except those areas where still traditional methods are in place. The dissimilarities exist due to the following reasons:

1. Preferential treatment to a particular crop which has eating preference of the local population and is grown in a large area to meet the criteria of food security.
2. Access to inputs like fertilizers, pesticides, seeds, etc.
3. Mechanization of agriculture in mountains is a uphill task, however, limited farm machinery can be used in certain areas. Availability of suitable tools for tillage and inter-culturing, harvesting, etc.
4. Genetic make-up of a crop cultivar and its requirement for optimum production.
5. Local climatic conditions also determine the requirements of a crop/cultivar.

6. Length of growing period in a specific area/country for growing a crop needs specific methods for production.
7. Policies issues in a particular country of the HKH region.
8. Level of technology development for crop production in a particular country or access to technologies generated in other countries.

7.21 Conservation Agriculture

Conservation agriculture was developed in South American countries, such as Brazil and Argentina during the 1970s. In many countries of the world, this technique has been adopted to reduce the cost of production and enhance sustainability. The conservation programs help the cultivators to improve soil fertility, maintain water quality, and encourage biodiversity. Broadly speaking, the conservation programs assist in preventing deterioration of air and water quality in the ecosystem, preserving natural habitats of fauna, and biodiversity as well as maintaining ecosystem services. The cover crops and crop residues in conservation agriculture protect the soil from erosion and restrict weed proliferation. Though conservation agriculture has been adopted extensively in the Gangetic plains of India, there are some problems associated with adoption of CA in mountain areas in the Himalayas. Conservation agriculture (CA) is an agriculture system, primarily including no till or minimum disturbance of soil, maintaining of crop residues and follow crop diversification practices (Lumpkin and Sayre 2009). The concept was defined by the Food and Agriculture Organization (FAO) in Rome, Italy, as "resource-saving agricultural crop production that tries to produce acceptable profits while maintaining high and sustained production levels while simultaneously safeguarding the environment" (FAO 2007). Tillage operations should be kept to a bare minimum in California, as should outside inputs, to allow for improved biological processes in the soil. Conservation agriculture tries to mitigate and/or reverse the negative consequences of conventional farming, such as soil erosion, loss of organic matter, soil water loss, and soil physical degradation, among other things.

There is need to understand conservation agriculture in a larger context. In should be practiced in such a way that both soil and water conservation methods are mutually supportive rather than competitive. Sharma et al. (2014a, b, c) mentioned that although reduced tillage/no-tillage systems gave good output in many cases, one major constraint in arid and semi-arid regions is low availability of crop residue and its utilization for feed and fuel. It should be borne in mind that reduced tillage systems of developed countries cannot be implemented as such in mountain areas without modification as per the specific conditions prevailing in high altitudes. One has to consider the local soil, climatic, social, and economic conditions of the mountain ecosystems. Conservation agricultural practices need to be adopted holistically so that it helps in minimizing soil loss, conserving water, and controlling weeds. This is important for sustainable and successful crop production under hilly conditions.

The present procedures for sound mountain development have innumerable challenges. Globalization of economies, commercial agriculture, and edifice development have significantly changed old nature-community interactions, environmental matters, and agricultural strategic options in the mountains (Bhatta et al. 2019). Therefore, concerted efforts are required to help mountain communities to overcome these problems. Many transformations in mountain agriculture have taken place, such as climate change, demographic pressure, agriculture commercialization, food habits, and infrastructure development. There is paucity of information to capture these changes. Pradeep Kumar et al. (2021) reported that the agriculture sustainability not only ensures the food security but also holds promise for good net returns, environmental improvement, and socio-economic equity. Due to insufficient infrastructure, a lack of services, and frequent natural calamities, farmers in the Himalayas continue to practice subsistence agriculture despite enormous natural resources. Climate change's effects on agriculture, as well as social and economic security, are causing concern among planners. Mountain ecosystems represent over 22% of the world's land surface area, containing millions of people who rely mostly on agriculture for their livelihood security. Increased agricultural land abandonment has occurred in the Himalayan region as a result of flooding, landslides, drought, wildfire, and over-exploitation of resources (Tiwari and Joshi 2015). Major farming systems commonly found in the HKH region (IMI FAO Report 2019a, b) are:

1. Rainfed crop livestock mixed farming system (subsistence farming);
2. Horticulture-based system (commercial farming);
3. Fruit and vegetable-based farming system (commercial farming);
4. Shifting cultivation;
5. Pastoral and agro-pastoral system of the Himalayan Nomads;
6. The Himalayan people practice integrated farming, agroforestry, animal husbandry, and forestry.

Conservation Agriculture consists of a set of crop production operation, such as minimum soil disturbance, soil cover, diversified crop rotation, and weed control (Fig. 7.97).

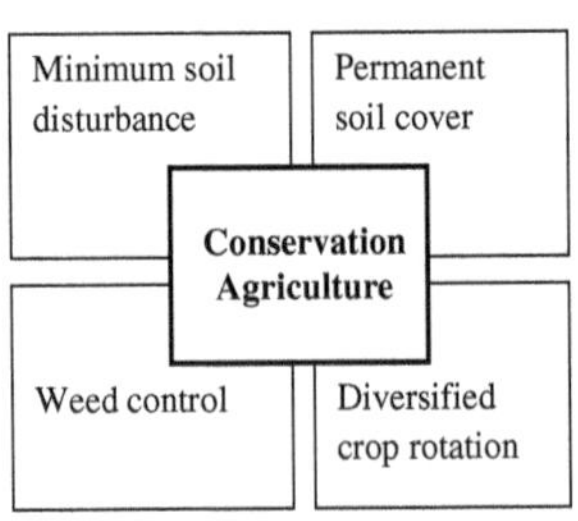

Fig. 7.97 Showing major features of conservation agriculture

7.21.1 Key Principles of CA

CA, according to the FAO, is a method of managing agroecosystems for greater productivity, earnings, and food security while protecting and improving the resource base and the environment. Conservation agriculture is epitomized by three connected principles, along with other associated best practices of growing crops and their judicious management (FAO 2014a, b). These are

- No-till or minimum possible tillage, which means no use of plow for land preparation. The sowing is done by direct seeding methods. No costs are involved on the labor and fuel for digging and plowing the fields. When no plowing is done, there is little or no soil disturbance as the seeds are either broadcasted or directly put in untilled soil. There is no soil disturbance due to cultural and harvesting operations as well as from farm traffic.
- Use of cover crops and mulches is recommended to avoid direct soil exposure and retention of soil moisture for longer periods. There should not be any burning and grazing of vegetation in the fields. The use of cover crops protects soil from (i) rainfall impact of rain drops, reducing runoff, increased water infiltration into soil profile and (ii) increased organic matter and improved structure of the soil. This will enhance soil water-holding capacity, and
- Ensure agricultural species diversification by using a crop rotation system that includes annuals and perennial crops, as well as a balanced mix of leguminous and non-leguminous crops.

7.21.2 Extent of CA in the World

The FAO has proposed a new paradigm of "sustainable production intensification" (FAO 2011a, b, c), which takes into account the requirement for productive agriculture that conserves and improves natural resource base and environmental quality. Above and below the ground surface, conservation agriculture has been shown to improve biodiversity and natural biological processes. External inputs are applied optimally and by techniques and quantities that do not mix or disarray the biological processes, and tillage is kept to a bare minimum or eliminated entirely (FAO 2014a, b). The highest area of 17 million ha under CA has been reported in Australia, followed by Canada, Russia, and China with an area of 13.48, 4.50, and 3.10 million ha, respectively. Total area covered under CA at global level during 2008–09 was about 106 M ha (7.5% of global cultivated area) (Kassam et al. 2009) (Fig. 7.92). During 2013–14 and 2015–16, it increased to about 157 M ha (11% of global cropland) and 180 M ha (12.5% of global cropland), respectively (Kassam et al. 2015). At present, CA is one of the most fast growing production system. The major reasons for this are (i) high factor productivity and field output, (ii) reduced production costs, (iii) improvement in net returns, (iv) more resilience to biotic and abiotic stresses, (v) low runoff and soil erosion and degradation, (vi) enhancement of soil fertility, and (vii) mitigation of climate change impacts (Kassam et al. 2017; Jat et al. 2014; Farooq and siddique 2014). CA was practised on only 2.8 million hectares worldwide in 1973–74 (Fig. 7.98), but it rose to 6.2 million hectares in 1983–84 and 38 million hectares in 1996–97. (Derpsch 1998). Between 1999 and 2013, California's farmland area grew at an annual average rate of around 8.3 million hectares per year (Kassam et al. 2015). From 2008–09 to 2015–16, the average annual growth rate was around 10.5 M ha, demonstrating farmers' strong interest in the CA land use system and demonstrating the system's adaptability. Initially, this accrual was essentially in North and South America, Australia, and New Zealand. However, presently, it is also happening in some countries of Asia, such as in Kazakhstan and China which have big farms, and in India and Pakistan, having small holdings. Wheat-based CA land use systems are on the rise in these countries. In the Indo-Gangetic Plains of India and Pakistan, wheat-rice cropping systems are being transformed into CA systems. These are also known as 'double no-till' rice–wheat systems. Conservation agriculture is also amplifying in small farms in Africa as well as in medium and large farms in Europe, Russia, and Ukraine

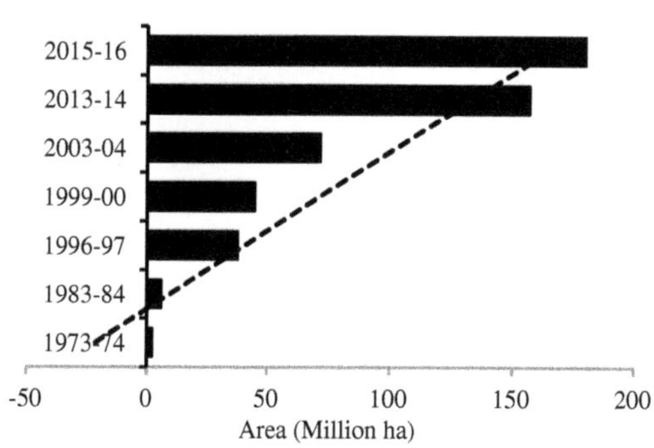

Fig. 7.98 Increasing trend of CA at global level

The extent of area under CA in HKH countries is nominal. From HKH countries, the area under CA has been reported from China, India, Pakistan, and Bangladesh, while no area has been reported from Afghanistan, Nepal, Bhutan, and Myanmar (FAO-AQUATSTAT 2014, 2016; Kassam et al. 2018). The area under CA increased from 1.33 million ha in 200–09 to 9.00 million ha in 2015–16. In 2015–16, the area under CA was 1.500, 0.600, and 0.0015 million ha in India, Pakistan, and Bangladesh, respectively. The area under CA in HKH region is non-existant, possibly due to the reason that CA principles are difficult to follow in cultivation on high mountain slopes, and where it is possible, the requisite techniques and information have not adequately percolated down among mountain communities to go for CA

7.21.3 Advantages of CA

Newly introduced technology must have sufficient advantages that may attract the cultivators. It is essential that farmers should know the difference between what they are doing and what they need. This will help wide adoption of a system. Conservation agriculture has the following advantages:

Economic Advantages

- It saves a lot of time required for land preparation and therefore reduced labor requirement. Tillage is one of the most "energy consuming" processes and if tillage is avoided as in the case of CA, major part of energy, in the shape of labor and machinery can be saved. The reduction in labor requirement is one of the major reasons for farmers in adoption of conservation agriculture, mainly those who fully depend on family labor.
- Reduction of costs on fuel, as well as machinery operating and maintenance costs.
- In general, the CA has higher benefit/cost ratio.

Agronomic Advantages

- Conservation agriculture improves soil productivity.
- Enhance organic matter content of soil.
- Helps in soil and water conservation.
- Improvement of soil structure, which in turn helps in proper root growth leading to good crop stand and crop productivity.

Crop leftovers are continuously incorporated into the soil, increasing the organic matter level. This is first limited to the top layer of the soil, but as time goes on, it will spread to deeper layers. Organic matter is crucial for fertilizer efficiency, water storage capacity, soil aggregation, rooting environment, and nutrient retention.

Environmental Advantages

- Conservation agriculture helps in arresting soil erosion. Crop residues on the soil surface lessen the impact of rains on the soil surface, and soil particles are not disturbed as much. As a result, there would be more infiltration and less runoff, resulting in less erosion. The residues also act as a physical barrier, slowing the flow of water across the surface.
- Improvement of water and air quality.
- CA is also responsible for increasing biodiversity and it helps in carbon sequestration. Monocropping encourages the loss of genetic variety, which can be mitigated by rotating crops and cover crops. There is more accumulation of carbon in the soil due to high crop residue left over soil surface in CA than the soils plowed frequently. The OM content of the soil is increased through the decomposition of roots during the first year of start of CA, and subsequently there is addition of leftover stubbles in the field which enhances organic matter content of the soil. Carbon is sequestered in the soil, transforming it into a carbon sink. This lowers the amount of greenhouse gases released into the environment.

7.21.4 Major Disadvantages of CA

- Starting of CA requires lot of financial involvement for the purchase of requisite machinery and equipment. It is why the practice could not be scaled up in HKH region as most of the farmers are marginal and their holdings are small. Even if the government policies encourage these farmers to avail loan facilities from the banks, they would be unable to pay back the loan by selling the produce from small holdings, over and above what is required for the family to sustain.
- There is lack of sufficient knowledge about the successful implementation of CA among the farmers.
- Low crop yields in some cases. With CA comes the idea of producing enough food for ever-increasing population. With this increase comes the responsibility for cultivators to increase food production using the same or less land than under cultivation at present. Complications would arise if CA lands do not produce as much as conventional farms, and this will leave the HKH region with less food for more mouths to feed.
- Many a times, the seed germination is poor in case of CA. This aspect has to be taken into cognizance is low germination will lead to poor crop stand.
- Stubbles left over in the field is irritating and sometimes injurious, the farmer has to use

proper gear (peripherals) to freely move in the plots.
- Increased risk of pests and diseases.
- Continuously keeping the soil surface undisturbed, without plowing for a long time, will harden upper soil.
- Many farmers believe in traditional way of cultivation.
- In developing countries, livestock is often an integral part of the agriculture, so it requires to be considered when introducing CA. Traditionally, the leftover crop residues were used for grazing livestock. As the application of CA requires a critical level of crop residues to be kept in the field, an alternate arrangement for feeding the domestic animals has to be made.

The CA, in larger perspective, give higher net returns to farmers, besides improving soil health and protecting the environment (Kumar et al. 2011). However, it is recommended that the CA practices should be adopted depending on rainfall and, soil and agro-ecological situations. Besides reduced tillage with mulching, other CA practices may include rainwater harvesting, soil–water conservation measures, integrated nutrient management, recycling of crop residues, green manuring, crop rotation with legumes.

7.21.5 Shifting Cultivation—CA by Default

Though the HKH Region has almost negligible area under CA, there are many other land use management practices that qualify the definition of CA in a broader context. For example, the prevalence of shifting cultivation in some HKH regions of some countries, such as India, Bhutan, Nepal, Bangladesh, and Myanmar also qualify for conservation agriculture. The shifting cultivation can said to be a sort of CA by default. The shifting cultivation fulfills two major principles of CA, that is, (i) it involves minimum tillage, as only hand hoe is used for land preparation and the sowing is done by a dibbler. The is absolutely minimum soil disturbance in this practice, (ii) there is great crop diversification in shifting cultivation because many complementary crops, including leguminous crops, are sown together. However, the shifting cultivation cannot be termed as perfect case representing CA because the shifting cultivators do not follow the best management practices (BMP) for raising the crops and, it has remained a subsistence type of farming practice. Consequently, the net returns are low as is expected from the perfect form of CA.

Another typical example of CA in HKH region is that of interventions through agroforestry systems. The agroforestry is now picking up in the region and the farmers are themselves coming forward to adopt the practice for economical and resource conservation reasons. CA can be described as a cogitation for crop production with sound resource conservation base that attempts to accomplish better gains along with high and perpetual production standards and environmental conservation. CA resembles agroforestry with its marginal soil disturbance and maintenance of sustained soil cover. The litter serves as soil surface cover. The erosion rates under CA, AF, and forest systems can be reduced to practically zero. Agroforestry diversifies and sustains production for increased advantages for farmers and the environment (Elevitch 2004). Agricultural areas cultivated in hilly and mountainous areas represent some region's historical heritages and providers of ecosystem services and so, need to be protected (Tarolli and Straffelini 2020).

7.21.6 Major Constraints in Adoption of CA in HKH Region

Major constraints to the adoption of CA in HKH Region are summarized below. These constraints include biophysical constraints (such as climate), knowledge constraints, land use constraints and government policy and land tenure systems constraints. It is rather arduous to deduce the comparative significance of these constraints due to the reason that requisite data is not available from the divergent landscapes.

Agro-Ecological Suitability

Conservation agriculture can be conducted in a variety of geographical regions, although it has been discovered that sub-humid zones have the most potential for CA, but arid and semi-arid environments are not ideal for CA due to low biomass output. In dry environments, reduced tillage and direct seeding can assist reduce water loss, promote early planting, and maximize use of sparse rainfall.

Farm Size Constraint

Small size of land holdings in the hills and mountains are not much suitable for CA. Though the cultivators themselves take care of their land to take maximum possible output, they are not able to keep proper soil cover during winter season under heavy snow. From small-sized holdings, the farmer tries to produce products of daily use, like food grain crops, vegetables, and fodder for the domestic animals.

Constraint of Land Tenure

In many areas of HKH countries, the land tenure is not secure, for example, in the Northeastern states of India, where the maximum land belongs to either village chief or whole community. The land is allotted to the cultivators as per their needs and they have no permanent rights over the land they cultivate. The farmers cannot take major decisions as far as land use is concerned.

Accessibility Constraints

Due to rugged terrain of the hills and mountains in HKH Region, it may not be possible for the cultivators to meet the requirements of CA, like paucity of necessary inputs and heterogeneity of available resources.

Socio-Cultural Constraints

Socio-culturally, the cultivators are tied to their traditions and taboos to follow the methods of cultivation of their ancestors because many festivals and rituals are associated with old methods of cultivation. Under such situations, these farmers are hesitant in going for CA. Conservation agriculture contradicts much of conventional farming traditions. Conventional agricultural practices are also strongly attached to local culture and ritual, making such practices even more difficult to be adopted.

Knowledge Constraints

Derpsch (2009) mentioned that adoption of CA needs many changes, not only in practices, but also in mindset of the farmers. There may be many cultivators who think that plows are essential part of agriculture and the optimum productivity cannot be achieved unless fields are properly plowed. For them, it is difficult to go for CA if they do not experience environmental or economic pressures to change.

Input Constraints

Adoption of CA is hampered by a lack of appropriate equipment, seeds, fertilizers, pesticides, and herbicides. Some of the equipment used in California agriculture differs from that used in conventional agriculture, and it is not always available. The most major distinctions are in the processes of land preparation and seeding.

Financial Constraints

Conservation agriculture is generally more remunerative in the long term than conventional farming. However, this needs more investment, which may prove risky for small farmers to undertake on their own. The small farmers of hills and mountains may not have the requisite resources (money, land with ownership rights, and water for irrigation) to invest and go for CA.

Policy Constraints

By and large, there is no proper policy of the Government to encourage adoption of CA in the HKH countries. Scaling up of CA in HKH region requires proper policy and structural support so as to make the farmers comfortable in transition from conventional agriculture to CA.

7.21.7 Conservation Agriculture and Carbon Sequestration

In many soil types and cropping systems, the CA has the ability to enrich soil organic matter storage. Datta et al. (2001) found that humic acid was greater in the 0–15 cm layer (4.23 g kg^{-1}) of Alfisol soil in zero tillage wheat-mung bean rotation than in farmer method of conventional tillage. In the 15–30 cm layer of the soil, the humic acid level was 2.08 g kg^{-1}. As a result, CA has been discovered to store carbon in the soil during conservation tillage rather than releasing it into the atmosphere as CO_2.

The potential impact of CA on C sequestration was explored by Govaerts et al. (2009). The reduction in soil tillage was found to have the greatest impact on lowering emissions from farming activities. The quantity and quality of organic matter input were found to alter soil C stores when crop rotation was changed. Conservation agriculture is a valuable method that improves soil processes, saves production costs, and minimizes soil erosion (Govaerts 2009). The level of growth in soil organic carbon through cropping systems is determined by the quality and quantity of organic material given to the soil, as well as its subsequent degradation by soil microbes. For soil microbes, organic matter is the primary source of carbon. When organic matter is mineralized, some of the carbon is needed for microbial development and maintenance, while the rest is respired as CO_2 and returned to the atmosphere (Lavelle et al. 1993). By conserving irrigation water, conservation agriculture can also help to minimize CO_2 emissions. Irrigation adds to CO_2 emissions because it requires energy to lift water, hence CA minimizes CO_2 emissions by conserving irrigation water (Schlesinger 2000). Rice–wheat systems are a type of rice–wheat system. Rice–wheat rotation is quite frequent in the Indo-Gangetic Plains, and zero tillage is said to conserve irrigation water by 20–35% in the wheat crop as compared to conventional tillage (Hobbs and Gupta 2003). Tillage practice can also impact the distribution of SOC in the profile, with zero tillage containing more SOM in the surface layers than conventional tillage. Cover crops increase the amount of plant wastes added to the soil and provide a vegetative cover during critical periods, which helps to increase SOC sequestration (Bowman et al. 1999).

7.22 Forces of Change

Understanding the evolution of Himalayan farming cultures is difficult due to the fact that the drivers of change and their dimensions alter throughout time and geography. There are variances in transformation among Himalayan states when it comes to local skills for managing change (Pratap 2020). A success story in one state, or in one region of a state, is difficult to replicate elsewhere. Organic Sikkim is not replicated in the same dimensions elsewhere, and altering the mountain economy through fruit cultivation and vegetable production in Himachal Pradesh may not be replicated elsewhere on the same scale. There are numerous local success stories and failures within the Himalayan area states from which others can learn what to avoid. Between the people and institutions of the Himalayan states, there is a dearth of lateral flow of knowledge, information, and methods for sharing experiences. The best term to describe it is "too little, too slow". There is a lot of room for developing lateral partnership structures among Himalayan farmers and institutions in these states.

What are the key factors of change in the HKH that affect mountain sustainability? Individual and cumulative impacts are represented at numerous regional and temporal scales, according to a detailed examination of the primary causes leading to changes in the HKH. The connections and consequences of these and other change factors are complicated, and they are frequently local and distinctive in the context of the dominating ecosystem socio-cultural and economic settings. The HKH Region provides a variety of services to many communities who live both within and beyond the region. Global and regional change, as well as a combination of natural and manmade forces, are all exerting immense stresses on the HKH region. Fast population and economic growth in most sections of the HKH has raised demand for natural resources, resulting in over-exploitation, considerable land use and land cover change, forest disintegration, and inequitable and uneconomic socio-economic actions in many cases. Fast change in the lifestyle of the people has altered consumption magnitude and forms, as well as investment pattern. Advances in agricultural technology may increase crop yields and people's food security, but they are fraught with risks.

7.22.1 Pedestals and Forces of Change

Individually or in combination, drivers of change have an impact on sustainability through environmental, societal, and economic factors. In addition, regional and global processes can influence local activities, and vice versa. The origin and influence of significant change drivers in the HKH can be seen at three different levels or scales: (1) local impacts of drivers on ecosystem, land and water, and their regulation, within the background of social and ecological systems; (2) regional influence of local forces, intervening largely through the provision of ecosystem services in relation to upland–lowland connection; and (3) regional and global impacts on the HKH through tele-coupled systems, where actions taken outside the region have prominent influence within the HKH. Multiple channels of impact are prevalent in each driver's widely interwoven areas of influence, including both direct and indirect relationships with mountain sustainability. There are three basic sustainability dimensions or pillars.

7.22.2 Environmental Forces

Land use and land cover change (LULCC) has a direct effect on the environment and causes socio-economic shifts in human groups. LULCC is the dominant force of change in soil resource, diminishing ecosystem activities and provisions, and thus ecosystems' ability to assist peoples' demands is seriously affected. In this way, LULCC assesses ecosystems' and human society's vulnerability and resilience to external disturbances such as policies of the government, climate change, and other facets of globalization. There is a general absence of integrated LULCC research across the HKH. The majority of published case studies have been conducted in China's Tibetan Plateau region, with a focus on the extensive grasslands. Because most accessible case studies use temporally and geographically varied satellite data, regional patterns of LULCC are difficult to discern (Harris 2010).

Land use and land cover are important because they are the primary source of anthropogenic CO_2 emissions and have regional and global implications. The demands of a changing climate, population expansion and migration, expanding urbanization, and market needs are all transforming land usage in mountain areas. Land use data is currently collected at the sub-basin level, and it is neither categorized nor widely available. The science of ecosystem services is poorly understood in the region, and user-friendly valuation methodologies are rarely mentioned in the context of the HKH countries. Natural calamities and disasters. The HKH region's physiographic setting and climatic characteristics render it vulnerable to both geological and hydrometeorological threats. Whether or not insurance is made available, disasters will have an impact on investments in agriculture through rural credit. The Hindu Kush Himalayan region contains a diverse agro-biodiversity, with land races of crops and cattle that have evolved over generations as adaptations to local micro-climates and environments, as well as farmer requirements (Partap and Sthapit 1998a, b). These species, subspecies, and land races constitute a vast genetic resource that could help researchers identify or develop crops that are better suited to changing climatic and soil conditions. However, these resources are being depleted at an alarming rate. Information, seeds, and other genetic material from as many different sources as possible should be collected and made available across the region.

7.22.3 Change in Agriculture

Various researches have been lacking in assistance to understand how changes take place in rural regions and how rural communities have adapted to the changes. Ghimire et al. (2022) highlighted the adoption of many new crop and vegetables species, which varied according to elevation and various communities living there, based on focused group interviews, climatic data analysis, and map data. Migration from the region and working in professions other than agriculture, infrastructure development, climate change, access to market, and changing dietary habits are the key drivers of change, according to a systematic review of many events and underlying processes. These forces began to act at different times and scales in different zones, and their intricate interactions eventually resulted in changes to the rural agricultural landscape.

Due to the interplay of environment and markets that are beyond farmers' control, Himalayan agriculture is predominantly subsistence, while slow progress has been noticed in market-oriented approach with the introduction of commercial crops in isolated areas (ICIMOD 2012). However, these factors have had a direct impact on farmers' decisions about which crops to plant and when to plant them (Martin et al. 2013). Importantly, recent climate change has resulted in unpredictably changing weather patterns and, as a result, cropping patterns. In the Himalayas, vertical resource distribution is prevalent (Guillet 1983). Agriculture has changed differently at different elevations in the Himalayas.

Agriculture provides a direct or indirect source of income for over 70% of the HKH's population and is a significant contribution to national earnings (Tiwari 2000a, b). Agriculture in this region is primarily rainfed, making it subject to variations in precipitation timing and frequency. Crop productivity is influenced by a variety of biophysical processes and factors, as well as their complex, nonlinear relationships (Challinor et al. 2009). Cereals alone have been ineffective in lifting Himalayan farmers out of poverty and food insecurity. Even with the use of new technological options, there is no reason to believe it will happen in the future. As a result, new cropping systems have emerged and continue to emerge, with fruit and vegetable cash cropping leading the way. Cash crops are producing much more jobs and assisting in the diversification of farm earnings. In the Himalayan states today, around 175 different types of vegetables are grown. This aspect is frequently overlooked due to a misunderstanding of the importance of preserving native agro-biodiversity (Pratap 2019). In recent decades, cropping patterns have shifted from agriculture to fruit crops due to the special suitability of hills and mountains for horticulture. These findings clearly show that throughout the Himalayan states, a process of change has begun that appears permanent, driving farmers to convert agricultural lands for commercial crop growing (Pratap 2019). Farmers have been able to better exploit local niches created by agro-climatic and socio-economic conditions as a result of this process. As a result, the shift in cropping patterns to cash crops appears to have pushed food grain farming in the Himalayas to a lesser extent, with future trends indicating a dominance of a wide range of cash crops, including vegetables, fruits, floriculture, medicinal, and aromatic plants, across Himalayan states. The ongoing total eradication of traditional indigenous crops from current cropping patterns is the single largest victim of the change of agriculture from subsistence to cash economy. The effects of rising temperatures are currently being seen in the form of challenges in continuing to grow some temperate fruit crops. Apples are one of these fruit crops that is being phased out in favor of plums, pomegranates, and persimmons. Warmer and often drier local weather conditions, as well as irregular rainfall patterns, have perplexed Himalayan farmers in recent decades, and they expect uncertain futures.

Climate change is also thought to have impacted the availability of irrigation water, which is required for cash crops. Intensive agriculture holds the key to making agriculture profitable on marginal and small hill farms. Cropping intensity has already exceeded 200% in some vegetable-growing areas of steep landscapes. The commercialization of Himalayan agriculture has increased farmers' reliance on outside inputs, such as seeds, fertilizers, and labor, including mechanization. One of the most essential inputs, labor, has emerged as one of the most crucial, accounting for around half of the overall variable cost of agriculture. As a result, migrant labor is increasingly being used not just by medium and large farmers, but also by small farmers, both on a seasonal and permanent basis. Farmers' interaction with market functions has greatly risen as farming has become more market-oriented. Commercial farmers are emerging and distancing themselves from the stereotype of subsistence farming. For price and market information, they rely more on media sources. New Age farmers are also familiar with the use of web marketing, particularly for the marketing of flowers, fruits, and vegetables. Cash transactions are being phased out in favor of internet banking choices.

Extensive research has revealed that grassland degradation has expanded across the whole high-altitude sections of the HKH during the last few decades (Wu et al. 2013a, b). High-altitude grassland in China's Tibetan Plateau is sensitive to disturbance, especially in the drier parts, where turf removal is permanent. Since the 1960s, several alpine meadows in the Tibetan Plateau's north-east region have been severely deteriorated (Wang et al. 2008). Warming changes on the Tibetan Plateau have a direct impact on the area of permafrost cover and snow melting. Over the previous half-century, human activities on the plateau have intensified dramatically, having a considerable impact on land use. Climate change and human activities are approximately equally blamed for desertification in the northern Tibetan Plateau, with climate change having a little edge over human activities (Yang et al. 2004). Estimating the magnitude of land cover changes under global warming is challenging due to problems with available climate models in this undulating and rugged terrain due to shortcomings in existing global climate models (Cui and Graf 2009). Change in forest composition due to afforestation and reforestation programs, expansion in agriculture, and developmental works has resulted in significant changes in land use and cultivation patterns in mountains (Xu et al. 2009; Sharma et al. 2010a, b). Because of Afghanistan's terrain and arid climate, wide swaths of land are prone to soil erosion (Saba 2001). The loss of vegetation and soil humus results in increasingly arid conditions. Desertification has been exacerbated by land abandonment, bad reclamation plans, overgrazing, and the removal or destruction of vegetation for fuel (Saba 2001), which has a severe influence on pastoral productivity and livelihoods in arid and semi-arid areas.

7.22.4 Forest Pattern Changes

Mountains are important ecosystems with an impact that extends well beyond their physical bounds. In the Himalayas, peoples' interferences have caused simultaneous cutting of forest vegetation and rehabilitation, and interposing to prevent forest loss and enhance forest regeneration require a comprehensive knowledge of the intricate and interconnected forces of forest change. Verma et al. (2021) employed a system dynamics technique paired with causal network analysis to locate, map, and explain the relationships between forest change causes, actors, and mechanisms across the Himalayan Mountain range from 1984 to 2020. There were 221 links discovered between five proximate drivers, 12 underlying drivers, two institutional factors, and five 'other' factors, according to the study. Forest change dynamics have been dominated by smallholder agriculture, extensive non-timber forest product extraction, widespread commercial and non-commercial wood exploitation, and high rates of agricultural abandonment. Population expansion, low crop yields, global assistance for expansion programs, and accomplished peoples' woods regulation systems are some of the underlying factors. A variety of contextual circumstances can result in contradictory links, which can have harmful consequences for conservation goals. Changes in governance, transnational infrastructure development initiatives, economic slowdowns, labor migrations, and climate change all have the potential to destabilize established dynamics and alter forest trajectories. Multiple avenues exist for the underlying and proximal forces to interact in order to achieve conservation goals. Based on our findings, we have identified five thematic target areas for reducing forest loss and promoting recovery: (1) reducing population pressure, (2) increasing agricultural productivity in a sustainable manner, (3) bolstering forest management institutions, (4) capitalizing on tourism growth and long-term infrastructure construction, and (5) converting to alternate fuels and devising fuelwood raising on deteriorated land. The widespread adoption of systems thinking, specifically a system dynamics approach and causal network analysis, will greatly improve policy rigor, aid in the design of site-specific interventions at multiple spatial scales that can respond to local and global changes, and guide deeper inquiry into driver-forest dynamics.

7.22.5 Variability in Climate Change

The temperature of the HKH has changed dramatically over time, and these variations are directly linked to the anthropogenic factors. Climate drivers like El Nio-Southern Oscillation (ENSO), North Atlantic Oscillation (NAO), Indian Ocean Dipole (IOD), Madden–Julian Oscillation (MJO), and Arctic Oscillation (AO) have influenced the region's weather and climate on multiple spatiotemporal scales, as evidenced by studies conducted in the region (Rajbhandari et al 2019). Climate change, as a major force of changes in mountain resilience today, interacts in complex ways with a number of other important variables, including globalization, population increase, urban swelling, and local change in land use, all of which can have substantial consequences. Climate change has environmental and social consequences for human populations across the HKH, including increased uncertainty in water supply and agricultural production. Temperature and precipitation changes could have major consequences for biodiversity and ecosystem goods and services (Chettri and Sharma 2016). Climate change is having evident repercussions in the HKH, with signs of phenological changes and lower agriculture production in some of the primary crops in some areas (Hart et al. 2014).

7.22.6 Environmental Effects

Climate change in the HKH (particularly increasing regional temperatures) has already had a direct and indirect impact on biological and ecological systems, in combination with other anthropogenic factors. Climate change also affects hydrological systems, causing shifts in carbon sequestration and nitrogen accumulation, as well as human foods and consumption (Singh et al. 2011). Hydrological changes can provide significant difficulties to human society and natural ecosystems.

7.22.7 Socio-Cultural Drivers of Change

Changing Demographic Situation

Population growth and decline, changes in age distribution and education, social and migration processes are all the outcome of environmental and economic forces, and they have an impact on both. In terms of the environment, rising population increases demand for and use of natural resources, as well as changes in land use, posing the risk of degradation due to overuse or underuse. Demographic shifts in mountain economies lead to the modification of social connections and structural changes. The shifting demographic situation has both quantitative and qualitative implications in either instance.

Changes wrought by a variety of causes in the HKH are posing both challenges and opportunities for long-term growth. A network of interactions among a variety of variables drives environmental and social changes in the HKH. Although some of the influences are global, such as climate change, globalization, and scientific and technological growth, others, such as pollution and migration, are regional. The interactions that cause change are more or less unique to each location and are linked to specific mountain ecosystems or trans-boundary landscapes. As a result, there is a considerable deal of regional variation in the drivers of change and their effects. The mountain definition of 'inaccessibility' will shift to some extent as access to global services improves. As a result, the demand for certain ecosystem services will alter and local communities' ability to adapt to global change will improve.

Despite its importance, little is known about the HKH, its ecosystems, and its inhabitants, particularly in the face of rapid change. More research has been done on the region in the previous half-century, but the information gained is typically dispersed, reaches a small audience, is sectoral or based on a single discipline, and, most critically, does not reach decision-makers in government, local communities, or the commercial sector. The reasons for this conclusion are numerous. The first is about expanding the knowledge base that is available. Bringing together people who are creating knowledge about the HKH to aggregate existing knowledge has been really beneficial. Furthermore, by collaborating across disciplines and countries, the assessment incorporates ideas from a variety of viewpoints on the mountains. Global evaluations and programs such as the Intergovernmental Panel on Climate Change (IPCC) now have access to a valuable knowledge base regarding this region, and the book will be useful in informing global debates and discourses. Then there's the importance of bringing together a network of people who can work across disciplinary and geographical borders in the future, which goes beyond the evaluation report itself.

Mountain resources, mainly water, satisfy about half of the world's population. Mountains have 25% of the world's terrestrial biodiversity and nearly half of the world's biodiversity 'hotspots' (Sharma et al. 2019). Out of twenty plant species that provide about 80% of the world's food (cereals, fruits, vegetables, etc.), 6 plant species originated in mountains (Fleury 1999). In humid places, mountains provide 30–60% of the world's fresh water, while in semi-arid and desert areas, 70–95% (Kapos et al. 2000; WCMC-UNEP 2002). Mountains make available different goods and services, hydroelectricity, timber, biodiversity and niche products, mineral resources, recreation, and flood management (Schild and Sharma 2011; Molden and Sharma 2013). Mountains have a more diverse ethnic and linguistic population.

7.22.8 Major Issues of the HKH

Even without human intervention, the HKH region is geologically unstable, with young and rising mountains that are prone to erosion and landslides. Stressors such as climate change and human conflicts, as well as causes like globalization, infrastructure development, migration, tourism, and urbanization, are causing rapid change in the region. The outcome of the interaction of these complex change factors is difficult to anticipate, but it will have far-reaching repercussions, not just in the region but internationally. It is necessary to examine the potential cost of these drivers to the HKH environment and human wellbeing, as well as the opportunities they may bring.

The Hindu Kush Himalayan Monitoring and Assessment Programme (HIMAP), strives to fill in the gaps in information and provide a path ahead. The following essential aspects have been considered by HIMAP:

- The most important change forces in the HKH, role of climate, and the possible consequences for biodiversity, ecosystem services, livelihoods, and water resources?
- Arrangements for improving community adaptation to climate change; how can they be augmented; and with what outcomes?
- Advantage of gender-equitable and inclusive approaches to sustainable mountain development.
- Outmigration trends in the region, its current impacts on livelihoods and the ecosystem, climate change-induced movement, could

migration be considered an adaptive strategy, and what are the choices for addressing migration.

- Cryosphere's current state, potential changes, and the potential consequences of these changes?
- Existing and expected future amount, variability, and quality of water in the HKH's ten major river basins; expected implications of change on water availability and measures to reduce negative impacts?
- Most effective ways to reduce the danger of flooding and drought, and how can they be implemented at different scales?
- Reasons for air pollution in the HKH increase, in which way air pollution from within and outside the HKH affecting the HKH, and what are solutions?
- HKH's energy demands and options, positive and negative aspects of hydropower development, and how can hydropower be developed in the region in the most sustainable way?
- How ecosystems should be managed to ensure soil and biodiversity conservation, as well as improved livelihoods?
- Ecological services provided by mountains, and how their management and supply be compensated?
- The best way for the HKH to establish a green economy? What technologies and approaches are best suited for sustainable mountain development in the region, and how can they be improved?

Perceptions

To enable HKH Region to a wealthy, self-reliant in food and energy, clean water; with climate impact and, floods and drought resilient mountain communities, the following measures need to be considered:

- Providing clean air, water, pollution control, and a healthy natural resource base ensures productive and dignified employment, as well as social, cultural, and environmental wellbeing.
- Equal access to opportunities and benefits from resources, regardless of gender and social standing.
- Food security—people who are healthy and have access to sufficient, and nutritious food.
- Without jeopardizing the current low-carbon status, access to a sufficient amount of inexpensive, non-polluting, and sustainable energy.
- Access to safe, inexpensive water as well as safety against harsh weather disasters.
- Bio-diverse ecosystem services to support culture and economies; ecosystem services must be protected and used wisely.
- Communities and countries that are climate and catastrophe resilient.
- Cooperation between countries at all levels for long-term and mutual benefit.

Natural ecosystems in the Himalayas have influenced environmental, social, and economic changes, resulting in the creation of traditional agro-ecosystems primarily in Pakistan, India, Nepal, Bhutan, and China. The Himalayan environment has suffered the effects of agricultural expansion, deforestation, and subsequent soil erosion as a result of large-scale population growth (Ma et al. 2012). The main causes of flooding and stagnant agricultural output are deforestation and land degradation. In the Himalayas, human and livestock population growth, as well as economic growth, are key causes of land use change and natural resource degradation. In HKH countries, agriculture continues to be the primary source of income for the bulk of the population. Anthropogenic disturbances are the main drivers of land use change in the form of increased agricultural operations and human settlements in a study area in India's Himalayan state of Uttarakhand. Natural causes, such as soil erosion and landslides, account for roughly 47% of land disturbances, whereas anthropogenic influences account for the balance (ADB/ICIMOD 2006). In the past, Nepal has shown a pattern in which the initial migration from the lowlands to the mountains has been supplanted by a faster migration from the hills to the plains. As a result of the influx of locals from hill and mountain areas to the Terai plains, many agricultural fields have been abandoned. Food shortages and a loss of livelihood possibilities are among the socio-economic implications of such abandonment, particularly among marginal and small farm communities (Hrabovszky and Miyan 1987). The trend of population movement in northern India's Himalayan region resembles Nepal's scenario. Young men are traveling to the plains and urban areas in search of work as agricultural output declines.

The topic of sustainable forest management in the Himalayas has grown in importance as it becomes a more difficult challenge to determine how effectively to use non-timber forest products to improve lifestyles. At the local level, proper management of existing forest resources, engaging people, government and non-government groups, as well as other research institutes, can prevent additional forest depletion and degradation. Local people's active participation, in particular, is a key role in minimizing deforestation. The grass-roots level "Chipko Andolan" (Hug the Tree Revolution) in the Uttarakhand Himalayan region became a world-renowned ecological movement, resulting in improved environmental consciousness among local people. The major objectives of this change were:

1. To maintain the unique ecosystem while also assisting local residents in improving their quality of life.
2. Finding a middle ground between ecological benefits, development, and conservation Increased livelihood possibilities are required to attain food security. Poor communities' lives will not improve unless ecosystem services other than food provisioning, such as water, land, forest, range land, and the natural environment, are maintained and both access and appropriate use are guaranteed.

It is critical to take a comprehensive approach to not only food provisioning services but also other ecosystem services (Saxena et al. 2001). Strengthening upstream and downstream interactions along a geo-gradient in a mountainous area, such as the Himalayan region, is part of a long-term food security strategy (Tiwari, 2000a, b). The productivity of an agricultural system is considered sustainable if it can be maintained eternally, the natural resources that drive the agricultural production process are protected, and production profitability, and hence financial income for farmers is guaranteed. To manage the inherent trade-offs between serving urgent human needs and sustaining natural resources for other ecosystem functions, a quantitative yet comprehensive approach to ecosystem reactions to land use is required. Singh (2006) indicated that the planning for sustainable development in the Himalayan region must dwell on trusted and extensive data on natural and socio-cultural aspects, as also the environment.

7.23 The Way Forward

Stakeholders' and others' growing concerns about the environment's deterioration during the previous six decades appear to be linked to massive cause-and-effect discussions on Himalayas and the northern plains, which are also rapidly approaching environmental and socio-economic collapse (Rao 1997). It's known as the 'Theory of Himalaya Environmental Degradation,' according to Ives and Messerli (1989). They said that the introduction of modern health care and medication, as well as the reduction of diseases, increased population pressure. Increased population in subsistence mountain societies has resulted in less land per family, extreme poverty, and significant deforestation, which will result in the total loss of all available forest cover in the Himalayan countries in the near future. While there are uncertainties, limitations in knowledge or a lack of access to information lead each researcher to reach different results.

7.23.1 Problems and Issues of the Himalayan Region

Natural resource exploitation trends are currently visible all across the world at rates far above those at which these resources are replenished. While a huge increase in per capita resource demand is highlighted as the primary driver of the Himalayan region's expanding gap between resource extraction and replenishment, population growth is believed to be the most powerful determinant of such trends. Environmental degradation in terms of deforestation, poor biological productivity and utility potential, soil erosion, hydrological imbalances, flood and other natural dangers, and socio-economic inequalities are all consequences of these trends. Lower rates of abiotic, biotic, and cultural exchanges, (ii) slower rate of growth, (iii) slower aging and late maturity, (iv) inferior reproductive efficiency, and (v) increased resistance characterize the systems as they climb in elevation. At the moment, the impact of environmental challenges is less obvious in these locations. Natural risks, on the other hand, have a greater influence as elevation rises. Several authors have previously reviewed several major regional issues (Swaminathan 1991; Khoshoo 1992; Ramakrishnan et al. 1994, 1996; Rao 1994; Qasim 1995; Joshi 1996; Valdiya 1997).

Natural forest deforestation and related challenges are a global issue (Guppy 1984; Sayer and Whitmore 1991). Since the invention of agriculture, humans have been destroying forests for millennia (William 1989). Deforestation is not a new occurrence in the Himalayas, either. It has a long history, dating back to at least the late eighteenth century (Mahat et al. 1986). In the Himalayas, forest degradation is a major issue that leads to a slew of other issues. Soil erosion, slope collapses, depletion of soil fertility, scarcity of fuelwood and feed, increased overland flows, reduced groundwater recharge, and loss of biological variety are all hastened by deforestation of slopes. The level of damage of many processes attributable to vegetal degradation, on the other hand, is dependent on a variety of other elements such as prior histories, the severity of natural vegetation loss, natural regeneration patterns, and/or other human interferences (Valdiya and Bartarya 1991; Singh et al. 1984; Gilmour et al. 1989; Ramakrishnan et al. 1992; Alford 1992). Understanding the interconnections between agriculture, animal husbandry, and forests is critical to understanding the significance of forests in the lives of Himalayan Mountain cultures. The increasing demands on the forests are strongly linked to population growth, increased area under subsistence crops, and an increase in livestock populations. Lopping for fodder/fuel is claimed to generate canopy openings, leading to an increase in invasive species, as demand for the same resources is competitive. Based on the information available, such programs have only been successful in a few regions (Rao and Swarup 1993). South-east Asian tropical plant species are more common in the north-eastern Himalaya due to its proximity to South-east Asia, and gradually reduce towards the north-west. Economic incentives and monetary returns to farmers were ensured through government subsidies and market demands, which helped the transition from conventional food crop farming to agro-horticultural systems succeed. Horticultural development in the central and north-eastern Himalayas has lagged behind that of the western Himalayas. Other than fruit trees, perennial cash crops play an important role in the economy of the eastern and northeastern Himalaya. In Sikkim, large cardamom is an important perennial cash crop growing as an understory crop in forest areas, whereas citrus, tea, rubber, and pineapple are the main cash crops in the north-eastern Himalaya.

Hill societies have long placed a premium on self-sufficiency in order to ensure their continued existence in remote and inaccessible places. In more remote and inaccessible regions, the element of self-reliance is likely to be more adaptive. Accelerating the pace of development while achieving the sustainability standards requires a right combination of self-reliance and reliance on external sources. Improving people's life quality with external inputs may gain credibility among those facing challenging circumstances, but it is unlikely to be sustained in the long run.

Indigenous development capacities must be leveraged and enhanced in development interventions. From the perspective of the people who live in the Himalayas and many more who live in the surrounding lowlands, current levels of acceptable technological knowledge and capacity to put that knowledge into effect require major increases. Environmental deterioration is more a matter of human interactions fighting for productive resources than of human-habitat relationships (Horowitz 1988). Mountain development priorities are justified in order to improve the quality of life for mountain people and to persuade plains people that the future of the mountains cannot be separated from their own (Eckholm 1979).

7.23.2 The HKHs' Challenges and Chances

With the exception of the Indus and other western basins, where monsoon precipitation contributions are minimal, annual river flows across the HKH will not change considerably this century because enhanced precipitation and runoff will tend to offset lower flow from glacial melt. Pre-monsoon flows, on the other hand, are predicted to decrease, affecting irrigation, hydropower, and ecosystem services. Data uncertainties are substantial, and regional and temporal heterogeneity is difficult to account for. Surface water variability is expected to grow, according to projections (Scott et al. 2019). Anecdotal information suggests that springs in the Himalayas' mid-hills are drying up. These factors contribute to lean season baseflow in local streams, which serve as the principal supply of drinking water in the HKH mid-hills before flowing into glacial melt-fed rivers. Due to population expansion and the concentration of economic and political power in cities and towns, growing urban demand for water will continue notwithstanding high proportions of water consumption in agriculture throughout the region. The water resources of HKH cities will be put under unprecedented strain in order to meet SDG targets for urban water supply, wastewater treatment, and sanitation. To fulfil expanding demand, larger cities in and around the HKH are expected to resort to long-distance water transfers from the HKH highlands in the future, potentially implementing appropriate payment arrangements for upstream populations for ecological services (Scott et al. 2019). Some of the larger cities are also likely to invest in functional wastewater-treatment facilities. Smaller growing towns, on the other hand, will face a true problem due to a lack of finances for infrastructure upgrades, particularly water supply and wastewater treatment, and a lack of appropriate governing structures. Hydropower is increasingly becoming the main source of electricity and revenue for Himalayan states, but shifting river flow regimes will necessitate the construction and management of hydropower projects to account for pre-existing irrigation water consumption. Improved benefit-sharing principles are needed to allow for the protection of agricultural water flows while allowing for significant, albeit moderated, hydropower development. The flow regimes of ecosystem flow in Himalayan rivers, and streams are substantially influenced by human water use. Based on numerous workshops and the opinions of the participants, Prabhakar et al. (2019) highlighted obstacles and possibilities for the HKH region. Disasters were considered as the most serious challenge in the region, followed by climate change and bad governance, poverty, and water scarcity. The most important opportunities were regional collaboration/integration, technology innovation, resource sharing, and harnessing of renewable energy.

The following issues and opportunities for the Himalayan Region have been identified based on our experience working in the eastern Indian Himalayas:

Challenges

Disasters, climate change, low crop productivity, poor governance, poverty, water scarcity, poor economic development, political instability, ecosystem degradation, pollution, population growth, out migration, unsustainable agriculture, new generation not interested in agriculture profession, lack of finance, poor connectivity, rugged terrain, non-availability of eco-friendly and high-yielding cropping systems, social transformations, etc.

Opportunities

Regional integration, technology innovation, resource sharing, renewable energy, eco-tourism, sustainable biodiversity and ecosystem services, non-farm sector development, mountain niche products, improved food security, cultural diversity, suitability for horticultural crops, great scope for improving productivity of agricultural crops, demographic pressure is still low compared to plains, joint family system still prevalent which solve a part of the labor problems, higher capability for hydropower, etc.

7.23.3 Natural Resources

The majority of HKH countries have some of the world's highest rates of hunger and malnutrition. As part of SDG2, hunger and malnutrition are expected to be completely eradicated for all people by 2030, particularly for the poor, babies, and individuals in vulnerable situations. Sustainable food production systems, the deployment of resilient agricultural methods, addressing the consequences of climate extreme events and disasters, land and soil quality management, and other productivity-enhancing characteristics will all play a key part in achieving this. Optimal land allocation and water usage, limiting land degradation, urbanization trends with increasing income, and changing population tastes are all major supply side concerns for agriculture. Strengthened local food systems and a consistent food supply from the plains also aim to increase food availability and accessibility. Reduced malnutrition rates will be difficult to achieve until food intake, people's awareness of nutrition and health, dietary diversity and micronutrient intake, health services and infrastructure, and the health environment all improve. Ecosystem degradation trends in the region will diverge by 2030, with countries such as Myanmar, Afghanistan, and Pakistan seeing higher degrees of degradation. Community-based efforts such as REDD+ in Nepal and Bhutan will play an important role in ecosystem protection and preserving ecosystem flows. In this context, nations like China, Myanmar, and Bhutan have pledged to mitigate climate change impacts by increasing forest-based carbon sink capacity to meet SDG 13 and 15 goals (Prabhakar et al. 2019).

7.23.4 Towards a Self-Reliant HKH

In the HKH Region, development pathways vary in terms of trade-offs, possibilities, and problems, but they ultimately lead to the same result. A decision maker's pathway is defined as a collection of actions and combinations of actions that he or she can take. Two broad paths are offered here to aid future visioning and decision-making in the direction of a thriving HKH.

1. The paths' actual efficacy will be determined by a variety of factors, including social acceptability, political agreements, and investment decisions. Using the HKH's naturally accessible resources, the prosperous scenario can be achieved by prioritizing centralized large-scale power, communication, infrastructure, and development initiatives. Water and energy will be crucial in assuring energy access, food security, and poverty eradication (Prabhakar et al. 2019). However, large-scale financial and professional human resource mobilization, as well as collaboration between state and non-state actors, will be required. There is a need for good connectivity and communication system, improved infrastructure, collaborations, and resource sharing.
2. It is also conceivable to achieve the prosperous scenario by prioritizing small-scale, decentralized, local or national developmental initiatives and projects that use the HKH's vital resources—energy and water. However, this would necessitate the participation of a significant number of national and subnational developmental and social players. In order to accommodate local variances, mobilize financial resources for a large

number of projects of varying scales, and track progress to ensure it is in line with the scenario goals, a layered governance structure will be required. There is no single best possible set of actions or a shortcut to minimize the challenges and seize the opportunities for the HKH in order to achieve the long-term goal of prosperity. Several acts must be carried out at the same time. Visioning and strategic planning must begin immediately in order for decision-makers to undertake measures beyond 2030. (Prabhakar et al. 2019). Prior analysis, comprehension, and identification of the HKH's unique and growing prospects and challenges are required in order to make a decision on which course to take. The HKH has significant problems, and overcoming them can help the HKH grow in the long run. At the same time, the region has a plethora of opportunities. Because the two alternative pathways are not fully mutually incompatible, some common acts between them are feasible. Activities were divided into three categories by Prabhakar et al. (2019): "Specific actions: how paths differ", "Common set of actions," and "Actions should be avoided". These exact activities are what distinguish the two paths.

7.23.5 Knowledge Gaps

There are several combinations of actions that can lead to wealth along the two paths outlined. Decisions will be strategic and made at regular intervals through various levels of cooperation. Otherwise, the prospects of obtaining prosperity will be questionable, as spiraling disputes can have a cumulative effect, diverging from long-term development. The choice of path, technology, and resource sharing will all influence the most crucial decision on the extent of collaboration. Because this is the first time a long-term scenario for the HKH has been developed, it is evident that a more extended approach involving many more stakeholders is required. More refined indicators developed through a continual process of participatory involvement in the region could be valuable in this context.

7.23.6 Energy Needs

The HKH has a wide range of contexts and, despite having abundant energy resources, is nonetheless energy poor and vulnerable. On the one hand, the immediate issue for addressing the HKH region's energy needs is to build on previous triumphs in order to exploit vast untapped renewable resource prospects. On the other hand, an enabling policy climate for clean energy investment in a sustainable way is required to promote the rapid diffusion and scale-up of appropriate, customized business models for off-grid renewable energy solutions (Palit et al. 2019).

For immediate solution to energy problems, we may set the following priority areas (Palit et al. 2019):

- Integrate mountain-specific energy policies and programs into the country's overall energy development strategy. The first step toward developing evidence-based policies for sustainable energy supply is to create a complete energy database, which includes information on supply and demand as well as resources.
- It is necessary to construct a mountain-specific multi-tier assessment system with adequate energy access metrics. This must enable for accurate measurement and tracking of progress on many aspects of renewable energy access, while also supporting integrated and holistic approaches to other SDGs.
- Support for energy planning, and infrastructure development is essential. The private sector is required for both large-scale energy generation and infrastructure projects as well as smaller-scale decentralized energy supply.
- Using a high-level, empowered regional mechanism, accelerate regional commerce and collaboration in sustainable energy. With the exception of a few bilateral hydropower exchanges, the region's cooperative attempts

have been largely unsuccessful. However, there is a huge chance to secure sustainable energy by leveraging the region's complementary energy resources. This can be accomplished through integrated regional energy planning, commerce, and the construction of institutional systems that regulate environmental externalities and provide revenue-sharing mechanisms for local communities.

7.23.7 Ecosystem Services

Many ecological services are provided by the HKH region, which help both the people who live in the mountains and those who dwell in the vicinity. The climate-related changes provide practical issues because they have an impact on ecosystems' ability to provide sustaining, supplying, regulating, and cultural functions to the region and beyond. Increased disasters will have an impact on agriculture, forestry, and all infrastructure (Singh et al. 2011).

The following areas should be the focus of monitoring initiatives (Singh et al. 2011):

(i) Recording historic coping methods of pertinent agro-biodiversity to document species' loss;
(ii) Recording current biodiversity adaptation responses;
(iii) Capturing indigenous community behaviors and coping methods;
(iv) Research into the consequences of increased CO_2 levels in the atmosphere on ecosystem productivity; and
(v) Investigations on impediments to the development of ecosystem due to peoples' interferences.

7.23.8 Supportable Mountain Development (SMD)

Climate change has created complex issues, but it has also created a number of opportunities for mountain countries. Increased warmth in the highlands, for example, may allow for the cultivation of a variety of novel crops, as well as the emergence of new grassland and forest species on higher slopes. It has also made room for new scientific studies to coexist with mountain people's livelihood pursuits. A complete reassessment of mountain development practises is required. One strategy is to build ecological infrastructures that can provide items for subsistence as well as services that can assist downstream populations (Karki et al. 2011). However, because the environment is a top priority, the focus must be on enhancing the resilience of existing programs, reducing poverty and unfairness, and improving property rights and governance. Other new potential for SMD include expanding green operations like ecotourism, high-value mountain products, and mountain forestry for carbon sequestration and biodiversity protection, all with built-in systems to address equitable and environmental sustainability concerns. In the context of SMD, some specific issues and questions pertaining to the green economic discussion include

(a) How do you gather documentation evidence of mountain ecological footprints' national, regional, and global influence or contribution?
(b) What specific functions may mountains play in countries' transitioning from a "brown" to a "green" economy, given that national policymakers' existing perceptions and attitudes are that they are marginal and negligible?

The following are the major obstacles to SMD's institutional strengthening. Information and knowledge based on research in order to bridge the knowledge gap, an appropriate institutional framework must incorporate the function of research organizations at both the national and regional levels. Water, biodiversity, markets, and quality standardization for green products are all trans-boundary challenges in the HKH. These countries must establish green accounting policies as well as green technology research and adaption institutions before they can transition to a green economy. Consumerism is an engine of change today, influencing people's beliefs and attitudes, but in the context of the green economy, customers must be prepared to pay for green products, which will necessitate rising awareness, which will rely heavily on data and information. Green economy will necessitate "green technologies", which will necessitate an appropriate structure and method to enable adaptation, retrofitting, and transfer to emerging mountain countries. Education, research, campaigning, and policy reform are all needed to promote and transition to a green economy. Similarly, incentives for producers of mountain-based ecosystem goods and services require appropriate policies and institutions. Women are more vulnerable to climate change than men, because they face more socio-economic obstructions (Nellemann et al. 2011).

7.23.9 Indigenous Technical Knowledge (ITK)

Indigenous knowledge has recently been observed as an important component of measures need to be undertaken for sustainable mountain development. Indigenous people consume less from the ecosystem and contribute more to it than modern-day city inhabitants. There is a delicate balance between humans and environment in mountain habitats. Traditional agricultural practises foster beneficial species, conserve forests, and maintain rangelands and waterways. Forests are chopped and regenerated, pastures are grazed, and lands are intensively tilled. These systems are entwined with cultural ideas and mores, which include taboos, language, technical practises, knowledge transfer, and customary social consent and governance institutions—collectively known as indigenous knowledge (IK) (Jianchu and Melick 2006). Some knowledge is sacred and only known by initiates, while others is shared among communities: it could be an ancient tradition or a new innovation. Rapid and drastic changes have occurred throughout Asia, with increased resource extraction, intensification of land use, migration, and infrastructure development eroding traditional values and indigenous knowledge, resulting in deteriorating ecosystems. How can policymakers and scientists collaborate to find and revive indigenous knowledge (IK) or coping techniques in a regional agenda to improve conservation and development outcomes? Bringing culture and nature together The four great ancient civilisations were China and India, as well as ancient Egypt and Babylon. In the Himalayas, upstream peoples are still blamed for environmental degradation and pay the price for over-exploitation, which is typically driven by downstream interests. Local communities are frequently denied access to their customary land and livelihoods as a result of extensive reforestation and watershed protection programs, as well as the formation of nature reserves.

Rediscovering ITK

Traditional Asian knowledge systems are widely known for their diversity. Interest in ITK has resurfaced in recent decades, resulting in several exciting research collaborations. ITK can aid in the discovery of useful compounds and the development of tourism potential, but more importantly, the reintroduction of traditional management values aids in the delivery of successful and long-term conservation. Indigenous peoples did not create biodiversity, but they did keep it alive. Because of weak science, competing goals, or a lack of support, current conservation and sustainable development programs in the Himalayan region are ineffective. Policies that promote reforestation at the expense of rangelands are paradoxically implemented in southwest China alongside efforts to expand livestock numbers in order to boost rural incomes. Moves to directly conserve places by establishing protected zones can lead to over-exploitation by residents who are afraid of being excluded and losing access. ITK techniques are being revived through community-based natural resource management and participatory technology development in a centralized society (Jianchu and Melick 2006).

7.23.10 Agricultural Sustainability

Sustainable agriculture is a comprehensive approach that considers how existing resources (soil, water, biodiversity, and so on) can be used to meet the needs of current and future generations. Agriculture sustainability ensures profitability, environmental health, and social and economic equality, in addition to global food security. Due to harsh weather conditions, habitat degradation, and fragmentation, the Indian Himalayan Region (IHR) is one of the most endangered ecosystems on the planet. Despite the IHR's vast natural resources, farmers still practise subsistence agriculture due to poor infrastructure, lack of facilities, and natural disaster risk (Kuma et al. 2021). Sustainable agriculture is the only way to meet the SDG 1 and SDG 2 goals of no poverty and zero hunger. Climate change has caused numerous disruptions in IHR, including flooding, landslides, drought, wildfires, and resource over-exploitation. It has resulted in an increase in the abandonment of agricultural land in the Himalayan region (Tiwari and Joshi 2015). Food instability and malnutrition are becoming big concerns. Around 50% of the population in the Hindu Kush Himalaya (HKH) region, including the IHR, has been found to be malnourished (Rasul et al. 2019). Overall, 38% of Indian Himalayan farmer families are experiencing transient/transitory food insecurity (IMI FAO Report 2019a, b).

Agriculture

The agriculture sector provides the direct or indirect source of livelihood for over 70% of the population residing in Hindu Kush Himalaya (HKH) including IHR (Wang et al. 2019). The agriculture system in entire Himalayan region has been based on locally available natural resources involving complex linkages among farm, forests, and livestock systems (Tiwari and Joshi 2015). Five types of farming systems commonly occur in IHR (IMI FAO Report 2019a, b). These are rainfed crop livestock mixed farming system, horticulture-based system (commercial farming), fruit and vegetable-based farming system (commercial farming), shifting cultivation and burn farming culture of north east Himalaya, and pastoral and Agro-pastoral system of the Himalayan Nomads. Food grain crops cover over 76% of the gross cultivated area in the Himalayan region (Partap 2011). Rice is the most common food crop, accounting for around 81% of all acreage. Mixed farming is popularly practiced with a mixture of diverse crops.

7.23.11 Alternatives to Constrained Land Resources of Himalayas

A farming family's access to appropriate crop land is a basic requirement and the foundation for economically viable farming in any location. Previously, lack of access to better production technologies and markets were the primary causes of food insecurity and poverty among mountain farmers. The main reason is and will continue to be the chronically insufficient crop land accessible to households for surviving solely on farming. Several attempts have been made in the past in certain Himalayan states to remedy this problem by consolidation of land holdings, but all have failed. Mountain farmers place a premium on the locational advantages and disadvantages of their crop land parcels, and this becomes a major roadblock (Pratap 2019).

Much of the Himalayan region's livelihoods were built on mixed crop livestock husbandry, with a significant forestry-farming relationship. The ties have weakened over the years for a variety of reasons. The biological degradation of available grazing areas in the Himalayan region has been one of the main ecological factors. Noxious weeds such as Lantana, Eupatorium, Ageratum, and congress grass have infested these support areas in large numbers. The backbone of farming-based lives has been these so-called waste lands around village communities, earning them the moniker "support fields". These poisonous weeds have seriously impacted the outer Himalayan region and mid Himalayan range's local grazing pastures, and their functional usefulness as support lands has been dramatically diminished.

Even though it had nutritional superiority over staple foods, it was once anticipated that Himalayan grain chenopods would become extinct (Partap and Kapoor 1987). The replacement of staple crops by high-yielding varieties has resulted in a decrease in their cultivation. Certain Himalayan Mountain locals eat Chenopodium grain, and the crop is well adapted to their mixedcropping dominated system since it produces high yields at low plant densities, which promotes the establishment of companion crops. Chenopodium will have a role in the future both as a healthy food crop, according to Partap (1986) and Pratap et al. (1998a, b), with a shift in concentration towards cultivation in poor soils.

7.23.12 Innovations for Resilience

Environmental dangers and changes, as well as a labor scarcity, drive individuals to innovate in the Central Himalayas (CH). Crops being devastated by wildlife, insufficient precipitation at certain periods of the year, and occasionally too much rainfall when it is not required for crop cultivation have all had a significant impact on production. The availability of agricultural labor has been limited due to migration and livelihood diversification. Economic and market needs, collaboration with scientists, and experimenting are the next most important motivations for innovation. People have been forced to reintroduce some crops that had vanished but are now in high demand due to economic pressures. Some farmers have learned new and improved agricultural practises in order to boost production while also improving soil quality and land productivity (Mukerjee et al. 2018). Due to outmigration to towns and cities, and the younger generation's loss of interest in agriculture due to better chances in non-farm activities, agricultural labor scarcity was a crucial innovation element.

Traditional agriculture practises have proven to be resilient in the Central and Eastern Himalayan (EH) regions over time. With many regions of the Himalayas in the midst of an agrarian crisis, fast declining productivity and soil health, and increasing droughts, some areas provide a richness of crop diversity and resilience methods and innovations to aid climate change adaptation. The crops and varieties farmed represent the high crop diversity of some locations, while other crops have declined and are now cultivated on a small scale. Farmers have adopted cash crops such as ginger, broomstick grass, and cardamom that are suited to their agro-ecological niches, resulting in a market-oriented agriculture in the EH. In the CH, diminishing agricultural returns have driven widespread migration, which has been exacerbated by wildlife crop destruction. Low acceptance of inventions has come from a weakening social fabric and waning interest in agriculture, and progress is limited by the need to preserve resources.

7.23.13 Soil Management

Soil is essential for growing food, as well as manufacturing feed, textiles, wood, and other commodities. It cleans our water, supports biodiversity, recycles nutrients, controls climate, and contributes to our landscapes and cultural heritage. As a result, it's critical to manage this finite resource responsibly so that farming communities can benefit from long-term output (Kumar et al. 2020). The land provides essential natural resources to Himalayan regions and supports a dynamic natural ecology, life, and human progress. More than 90% of the people in the HKH are small landowners, with below one hectare of land under cultivation (Partap 1999). Most agricultural tracts in this region are not only limited in their potential production, but they are also deteriorating in quality due to increased land degradation, decreased soil fertility, and the resulting impact on crop productivity. Most agricultural regions in this region are not only limited in their potential production, but they are also deteriorating in quality due to increased land degradation, decreased soil fertility, and the resulting impact on crop productivity. Furthermore, the geography in this region is diverse, with mild plains and steep slopes, deep valleys, gorges, saddles, terraces, lacustrine-deposited rivers, lakes, and snow-capped summits, among other features (Thapa and Paudel 2002). Increased urbanization, rising population, increased tourist traffic, and restricted agricultural land are gradually putting further strain on HKH agriculture. Land for agricultural growth is limited, and the majority of farming in the region is rainfed (86%). Furthermore, agricultural communities in this mountainous terrain are vulnerable due to complicated biophysical and marginal socio-economic conditions. Food shortages effect the poverty chain response for the majority of families in mountainous regions (Jodha and Shrestha 1993).

The scientific community declared 2015–2024 to be the International Decade of Soils at the conclusion of the year. The efforts initiated during the International Year of Soils will be continued in this decade. At the global level, a number of key efforts on soils are bringing science and policy together. One of these is the FAO-led Global Soil Partnership and its regional hubs. Soils play a key role in Agenda 2030 and the Sustainable Development Goals. Simultaneously, the European Commission is attempting to expand the evidence base for soil management through specialized research efforts such as the European Joint Program on Agricultural Soil Management.

Soil is one of the most significant natural resources in terrestrial ecosystems, providing necessary nutrients, water, oxygen, and support for plants, among other things. It absorbs all of the effects of human presence, both in terms of direct activity (heavy agriculture, irrigation, compaction, contamination, etc.) and in terms of decreasing its ability to react to other natural processes, such as water erosion. Soils play an important role in the carbon, nutrient, and hydrological cycles. By providing a biogeochemically active filtration and cleaning service that changes or retains materials or nutrients introduced on the ground surface, they can reduce flood danger and contribute to water purification. Soil qualities can change the way chemicals flow, protecting people and animals from the harmful consequences of pollution. Soils are home to a quarter of the world's biodiversity, and soil biodiversity is critical for soil health and ecosystem functions. Soil biodiversity ranges from microscopic bacteria, yeasts, fungus, molds, and protists, through visible nematodes, micro-arthropods, and insects, to larger animals such as earthworms and vertebrates that spend their whole lives in soil. It has an impact on aboveground biodiversity through regulating plant community composition and above-ground pests, plagues, and infections, as well as controlling plant abundance and invasiveness. The loss of soil organic matter (SOM) has an effect on soil structure and aggregate stability, water retention, soil biodiversity, and plant nutrient sourcing. Organic carbon (OC), the major ingredient of SOM, is frequently used as a proxy for SOM because it is easy to detect and may be linked to emissions from the land to the atmosphere. However, soil health in the Himalayan region is a matter of concern. Appropriate management of soil is necessary to make it highly productive without losing its original fertility status.

As the region is positioned in a climate change hotspot, rising temperatures, altering precipitation patterns, snowfall, monsoon onset, length, and intensity have all been documented (Kumar et al. 2008, Szabo et al. 2016). Climate change has the ability to harness the potential influence on food production and nutrition, particularly where appropriate indigenous and cash crops are concerned, therefore area-specific options should be investigated (Szabo et al. 2018). To alleviate food insecurity in the region, it is critical to assess the existing potential of land in terms of agriculture production. Agricultural intensification has resulted in considerable land conversion and soil deterioration in recent decades. Land holdings that are scattered and small remain uneconomic (Shah 1982). It also has an impact on terrain, resulting in soil nutrient loss and erosion (Shah 1982). Land suitability management focuses on promoting sustainable agriculture development and scientific land use planning (El Baroudy 2016). Land potentiality refers to the land's natural ability to harvest common crops in a region based on soil attributes and topographical characteristics (USDA 1961; FAO 1976), whereas land appropriateness is a mix of soil capability in relation to numerous social and economic conditions (FAO 1976). Land evaluation accuracy aids decision-makers in developing better agricultural production plans. The large number of effective environmental variables aids in the assessment of land use appropriateness in terms of preventing land degradation in the long-term management of sustainable usage (Bandyopadhyay et al. 2009). This helps land users to select the best site for agricultural production while avoiding soil erosion and other problems. The assessment not only identifies the appropriate land and environmental boundary, but also reflects conservation and management methods for a variety of natural resources. Finally, agricultural expansion will benefit not just the rural economy, but will also boost rural tourism and diversification among poor farmers, preventing poor people from migrating to the plains and supporting the mountainous region's long-term development (Boori et al. 2015). For the classification and administration of land units, the most effective environmental elements are required (FAO 2006). Human involvement, on the other hand, has severely harmed land resources in the mountainous region due to demographic pressure, migration to cities in search of jobs and developmental works (Sharma et al. 2010a, b). Forecasting the niche of farmland for long-term production is crucial in Himalayan regions where poor drainage, rugged topography, and less productive mountainous soils are the major limiting factors. Establishing agriculture's adaptability is a critical stage in agricultural management and planning (FAO 1993).

Multi-criteria analysis has been used to anticipate and construct land assessment models in recent decades (Bozdag et al. 2016). Furthermore, the analytic hierarchy process (AHP) and frequency ratio (FR) are standardized methods for finding and deriving weights based on the relative significance of various characteristics through pair-wise comparisons. It is commonly used in multi-criterion decision-making (MCDM) to assess land suitability for different applications (Miller et al. 1998). AHP, FR, and MCDM approaches are frequently used to solve a wide range of problems involving complex variables at multiple levels where interaction between parameters is critical (Tiwari et al. 1999). In addition, a number of recent studies have discovered that geographic information systems (GIS) and remotely sensed data offer a great deal of potential for improving land suitability evaluation (Zolekar and Bhagat 2015; Baroudy 2016; Montgomery et al. 2016). As a result of fast population growth and urban expansion, land has become a relatively scarce commodity for agricultural and rangeland purposes, and the demand for efficient land use is greater than ever. There is a growing requirement to match land capabilities and uses as efficiently as feasible (Zarkesh et al. 2010). Sustainable agricultural development is one of the HKH Region's major concerns. The overall purpose of sustainable agriculture is to balance the natural land resource with agricultural requirements, with a special emphasis on resource optimization to ensure long-term productivity (Joshua et al. 2013). As the world's population grows, so does the demand on land resources, necessitating the use of technology like GIS to assist in determining the most ideal sites for sustainable agricultural production for food supply based on environmental potential. In the context of land suitability analysis (LSA), selecting the optimal place for agricultural production is a complex process encompassing not only technical criteria, but also physical, economic, social, and environmental considerations that may conflict. Due to the complications, multiple decision support tools must be used at the same time (Joshua et al. 2013). This application of agricultural land suitability analysis based on RS, GIS, and AHP is useful in directing agricultural activities to areas with good physical and environmental conditions (Rajasekhar et al. 2021).

7.23.14 Suitability of Soil

The most essential aspect in land suitability for agricultural activities is the kind of soil. Soil behavior can be used to estimate soil performance for agricultural purposes. Thus, knowing the soil types is crucial when determining the suitability of land for agricultural development (Akinci et al. 2013). Alluvial, brown soil, reddish-brown, hydromorphic, colluvial, siorezem, and regosol soil are among the soil types studied. Alluvial soils are high in organic matter and supply abundant nutrients for crops (Brady 1974). Alluvial soils are therefore classified as "very appropriate" for agriculture. Climate elements such as precipitation and temperature have an impact on crop yield and plant growth (Aldababseh et al. 2018). The climate of the region ranges from subtropical to subarctic (Kumar 2019). Hill farming has traditionally been thought of as subsidence agriculture, but it is now done with technological inputs to yield products that require speedy transit to market to sell. Road connectivity is critical to making agriculture more productive and cost-effective. Furthermore, as the distance from perennial rivers and streams grows, the amount of moisture in soils decreases. It took into account the inhabitants' primary occupation (animal husbandry), as well as their proximity to roads classified as economically viable for cultivation.

7.23.15 Capacity Building

The purpose of enhancing the HKH's capability is to meet the growing need for skilled people in for training programs, and business opportunities. Private sector companies such as NGOs, research and educational institutes, and business and technology centers will need to be involved at both the national and state levels to generate capabilities to run and expand mountain services (Niti Aayog 2018). Capacity development program on adaptation planning and implementation to increase officials' capacities to respond to climate change issues at various levels of government in the HKH region. Institutional capacity building is used to provide relevant recommendations, including participation from both the public and private sectors. The construction of local biogas systems was supported by this synthesis of operations from various sectors and their partners, which was utilized to determine the biogas production potential (Lindfors et al. 2020).

References

Aayog N (2018) Niti Aayog annual report. Niti Aayog, New Delhi, India

ADB (2013) Myanmar: agriculture, natural resources, and environment initial sector assessment, strategy, and road map. ISBN 978–92–9254–353–2

Adhikari L, Hussain A and Raul G (2017) Tapping the potential of neglected and underutilized food crops for sustainable nutrition security in the mountains of Pakistan and Nepal. Sustainability 9:291. https://doi.org/10.3390/su9020291

Ahmad I, Ahmad MSA, Hussain M, Ashraf M, Ashraf MY (2011) Spatio-temporal variations in soil characteristics and nutrient availability of an open scrub type rangeland in the sub mountainous Himalayan tract of Pakistan. Pak J Bot 34:565–571

Ahmed N and Rashid M (2003) Fertilizer use in Pakistan, NFDC. Planning and Development division, Islamabad, Pakistan

Ahmed NA, Tila M (1998) Fertilizer, plant nutrition management and self-reliance in Agriculture. Pak Dev Rev 37:217–233

Akıncı HO, Zalp AY, Turgut B (2013) Agricultural land use suitability analysis using GIS and AHP technique. Comput Electron Agric 97:1–87

Aldababseh A, Temimi M, Maghelal P, Branch O, Wulfmeyer V (2018) MultiCriteria evaluation of irrigated agriculture suitability to achieve food security in an arid environment. Sustainability 10:803

Alden Wily L (2003) Land rights in crisis: restoring tenure security in Afghanistan. Afghanistan Research and Analysis Unit, Prime Minister's Compound, Kabul, Afghanistan

Alford D (1992) Streamflow and sediment transport from mountain watersheds of the Chao Phraya basin, Northern Thailand: a reconnaissance study. Mt Res Dev 12(257–268):35

Ali A and Shaoliang Y (2013) Highland rangelands of Afghanistan: significance, management issues, and strategies. In: Wu N, Rawat GS, Joshi S, Ismail M, Sharma E (eds) High-altitude rangelands and their interfaces in the Hindu Kush Himalayas. ICIMOD, Kathmanu, Nepal

Alvarez-Jubete L, Arendt EK, Gallagher E (2010) Nutritive value of pseudocereals and their increasing use as functional gluten-free ingredients. Trends Food Sci Technol 21(2):106–113

Andersen P, Tuladhar JK, Karki KB, Maskey SL (2005) Micronutrients in South and South East Asia. Proceedings of an international workshop, Kathmandu,

Nepal, 8–11 September 2004. http://books.icimod.org/index.php/search/publication/62

Andersen P (2007) A review of micronutrient problems in the cultivated soil of Nepal. In: Mountain research and development 27(4):331–335. https://doi.org/10.1659/mrd.0915 https://doi.org/10.1659/mrd.0915

Anonymous (1976) Development of suitable planting techniques of pineapple on hill slopes. Annual Report, ICAR Research Complex for Northeastern Hill Region, Umiam, Meghalaya, India, pp 122

Anonymous (1998) Nepal micronutrient status survey. Kathmandu, Nepal: Ministry of Health, Child Health Division, His Majesty's Government of Nepal, New ERA, MI Nepal, UNICEF, WHO

Anonymous (2002) Basic statistics of North Eastern region. NEC Secretariat, Shillong, Meghalaya, India

Anonymous (2004–05) In: Annual report, ICAR Research Complex for Northeastern Region of India. Umiam, Shillong, Meghalaya, India

APP (1995) Agriculture perspective plan (1995–2015). NPC, Government of Nepal, Kathmandu

Asian Development Bank/ICIMOD (2006) Environment assessment of Nepal: emerging issues and challenges, Kathmandu

Aubrecht E, Biacs PA (2001) Characterization of buckwheat grain proteins and its products. Acta Aliment 28:261–268

Bajaj M (2014) Thimphu's growing pains: challenges of implementing the city plan. Centre for Bhutan Studies and GNH Research

Bajarcharya RM (2002) Fertility and productivity parameters for soil from five mid-hill districts of Central Nepal. Proceedings International Seminar on Mountains, Royal Nepal Academy of Sciences and Technology, Kathmandu, Nepal, pp 141–150

Bajracharya RM, Dahal BM (2012) Agricultural intensification. In: Craig RK, Nagle JC, Pardy B, Schmitz OJ, Smith WK (eds) The Berkshire encyclopedia of sustainability, vol 5. Ecosystem Management and Sustainability. Berkshire, New York, pp 7–10

Bajracharya RM, Sherchan DP, Dahal MB and Raut N (2014) Soil management for sustainable agricultural intensification in the Himalayan region. In: Soil management of smallholder agriculture. Taylor and Francis Publication. https://doi.org/10.1201/b17747-7

Baligar VC and Fageria NK (1997) Nutriet use efficiency in acid soils: nutrient management and plant use efficiency. In: Moniz AC, Furlani AMC, Schaffert RE, Fageria NK, Rosolem CA and Cantarella H (eds) Plant-soil interactions at low pH: sustainable agriculture and forestry production. Brazilian Soil Science Society, Brazil

Baligar VC, Bennett OL (1986) Outlook on fertilizer use efficiency in the tropics. Fertil Res 10:83–96

Bandyopadhyay S, Jaiswal RK, Hegde VS, Jayaraman V (2009) Assessment of land suitability potentials for agriculture using a remote sensing and GIS based approach. Int J Remote Sens 30(4):879–895

Banskota K (2006) New opportunities for mountain economies. In: Securing sustainable livelihoods in the Hindu Kush-Himalayas: directions for future research. Development and Cooperation, ICIMOD, Kathmandu, Nepal

Barnett J, O'Neill S (2010) Maladaptation. Glob Environ Chang 20(2):211–213

Baroang K (2013) Myanmar Bio-Physical Characterization: Summary Findings and Issues to Explore, Background Paper No.1. Center on Globalization and Sustainable Development, Earth Institute at Columbia University

BBS (1996) Statistical year book. Bangladesh Bureau of Statistics, Dhaka, Bangladesh

BBS (2006) Bangladesh economic review. Bangladesh Bureau of Statistics, (HIES), DAE and Food

BBS (2007) Bangladesh economic review. Bangladesh Bureau of Statistics, (HIES), DAE and Food

BBS (2009) Bangladesh economic review. Bangladesh Bureau of Statistics, (HIES), DAE and Food.

Bhatta LD, Shrestha A, Neupane N, Jodha NS and Wu N (2019) Shifting dynamics of nature, society and agriculture in the Hindu Kush Himalayas: perspectives for future mountain development. J Mt Sci 16(5):1133–1157. https://doi.org/10.1007/s11629-018-5146-4

Bhattacharyya R and Prakash V (2006) Nutrient cycling management. In: Gupta HS, Srivastva AK and Bhatt JC Sustainable (eds) Production agricultural watersheds in Northwest Himalayas, pp 311–327, Vivekananda Parvatiya Krishi Anusandhan Sansthan, Almora, Uttrakhand, India, p 615

Bhattacharyya R, Ghosh BN, Mishra PK, Mandal B, Rao CS, Sarkar, Das K, Anil KS, Lalitha M, Hati KM, Franzluebbers AJ (2015) Soil degradation in India: challenges and potential solutions. Sustainability 7:3528–3570. https://doi.org/10.3390/su7043528

Bhattarai S, Maskey SL (1987) Use and promotion of bacterial fertilizer for legumes in Nepal. Proceedings of the first review and working group meeting on biofertilizers, pp 199–205

Blas J, Dyer G (2009) China sows seeds of food self-sufficiency . The Financial Times, April 16, 2009

Bondeau A, Smith PC, Zaehle S, Schaphoff S, Lucht W, Cramer W, Gerten D, Lotze-Campen H, Mueller C, Reichstein M, Smith B (2007) Modelling the role of agriculture for the 20th century global terrestrial carbon balance. Glob Change Biol 13:679–706

Boori MS, Voz˘enílek V, Choudhary K (2015) Land use/cover disturbance due to tourism in Jeseníky Mountain, Czech Republic: a remote sensing and GIS based approach. Egypt J Remote Sens Sp Sci 18:17–26. https://doi.org/10.1016/j.ejrs.2014.12.002

Borthakur N (1992) Agriculture of the North Eastern region with special reference to hill agriculture, p 265. Bee Cee Prakashan, Guwahati, India

Bowman D, Minehart D, Rabin M (1999) Loss aversion in a consumption-saving model. Journal of Economic Behaviour and Organization, 38: 155–178.

Bozdag A, Yavuz F, Gunay AS (2016) AHP and GIS based land suitability analysis for Cihanbeyli (Turkey) County. Environ Earth Sci 75(9):813–827. https://doi.org/10.1007/s12665-016-5558-9

Brady NC (1974) The nature and properties of soils. McMillian Publishing Company, London

Brown LR (1984) State of the World. World Watch Institute, New York

Brown S, Shrestha B (2000) Market-driven land use dynamics in the middle mountains of Nepal. J Environ Manage 59:217–225

Carson B (1992) The land, the farmer, and the future. A soil fertility management strategy for Nepal. ICIMOD Occasional Paper 21. Kathmandu, Nepal

Central Water Commission (CWC) (2011) Agricultural census 2011, Government of India, PRS

Challinor AJ et al (2009) Crops and climate change: progress, trends, and challenges in simulating impacts and informing adaptation. J Exp Bot 60(10):2775–2789

Chang DHS (1983) The Tibetan Plateau in relation to the vegetation of China. Ann Mo Bot Gard 70:564–570

Chettri N, Sharma E (2016) Reconciling the mountain biodiversity conservation and human wellbeing: drivers of biodiversity loss and new approaches in the Hindu-Kush Himalayas. Proc Indian Natl Sci Acad 82(1):150–158

Clark RB (1982) Plant response to mineral element toxicity and deficiency. In: Christiansen MN and Lewis CF (eds) Breeding plants for less favourable environment. John Wiley & Sons, New York, NY, pp 71–73

CSO (Central Statistics Organization) (2012) Afghanistan living conditions survey 2016–17. CSO, Kabul, Afghanistan

CSO (Central Statistics Organization) (2015) Workshop on strategic planning for agriculture and rural statistics, Myanmar

CSWCR&TI (2011) Vision 2030 of the central soil and water conservation research and training Institute, pp 1–46. Allied publisher, Dehradun, India

Cui X, Graf HF (2009) Recent land cover changes on the Tibetan Plateau: a review. Clim Change 94:47–61

Dahal BM, Nyborg I, Sitaula BK, Bajracharya RM (2009) Agricultural intensification: food insecurity to income security in a mid-hill watershed of Nepal. Int J Agric Sustain 7(4):249–260

Datta M, Bhattacharyya BK, Saikh H (2001) Soil fertility - a case study of shifting cultivation sites in Tripura. Journal of Indian Society of Soil Science, 49: 104–109

Datta M, Dhiman KR (2001) Effect of some multipurpose trees on soil properties and crop productivity in Tripura. J Indian Soc Soil Sci 49:511–515

Datta M, Dhiman KR, Singh NP (2003) Success story of National Watershed Development Project for rainfed agriculture. Publication No. 21, ICAR Research Complex for NEH Region, Lembuchhera Centre, Agartala, Tripura, India

Datta M, Gupta RK (1984) Studies on P, Ca, Zn and Fe contents in soil and plant tissues as influenced by pressmud and lime. J Indian Soc Soil Sci 32:450–458

Datta M, Laskar S (1995) Influence of phosphate enriched poultry manure on rice-groundnut rotation in an acid soil of Tripura. Indian J Hill Res 8:125–127

Datta M, Gupta RK, Prasad RN (1983a) a) Response of wheat and maize to lime in acid soils of Nagaland. J Indian Soc Soil Sci 31:236–240

Datta M, Gupta RK, Prasad RN (1983b) b) Bulletin, Amelioration and phosphate management of acid soils in Nagaland ICAR Research Complex for NEH Region, 22. Nagaland, India

Datta M and Lashkar S (1988) In: Annual reports. ICAR Research Complex for NEH Region, Umiam, Meghalaya, India

Datta M, Ram M (1993) Status of micronutrients in some soils of Tripura. Journal of Indian Society of Soil Science, 41:776–777

Datta M and Sharma UC (2006) Soil acidity and amelioration. In: Sharma UC, Datta M and Samra JS (eds) Soils and their management in Northeast India, p 502. Indian Council of Agricultural Research Complex for Northeastern Hills Region, Meghalaya, India

Datta M, Singh NP, Choudhury PK and Mitra S (2005) Jute agrotextile—its uses in agriculture, Publication no 21. ICAR Research Complex for NEH Region, Tripura, India

de Janvry A, Kyle E, Dar SE, Manzoor (2016) The agricultural technology adoption puzzle: what can we learn from field experiments?, FERDI Working paper 178

Demaine H (1998) A livelihood systems approach to the sustainable development of upland farming systems. In: Chapman EC, Bouahom B and Hansen PK (eds) Upland farming systems in LAO DPR, problems and opportunities for livestock. ACIAR Proceeding No. 87. Australian Center for International Agricultural Research, Canberra

Dendup T (2018) Agricultural transformation in Bhutan: from peasants to entrepreneurial farmers. Asian J Agric Ext, Econ Sociol 23(3):1–8. https://doi.org/10.9734/AJAEES/2018/40289.

Denning G, Kye BT, Sandar MIN (2013) Background Paper No. 2 rice productivity improvement in Myanmar. Michigan State University, US, p 33

Derpsch R, Friedrich T (2009) Development and current status of no-till adoption in the world. Proceedings on CD of the 18th Triennial Conference of the International Soil Tillage Research Organization (ISTRO), June 15–19, 2009, Izmir, Turkey

Derpsch R (1998) Historical review of no-tillage cultivation of crops. Proceedings of the 1st JIRCAS Seminar on Soybean Research on No-Tillage Culture & Future Research Needs, pp 1–18 March 5–6, 1998. Iguassu Falls, Brazil. Working Report No. 13, JIRCAS

Devi PB, Vijayabharathi R, Sathyabama, S, Malleshi NG and Priyadarisini VB (2011) Health benefits of finger millet (*Eleusine coracana* L.) polyphenols and dietary fiber: a review. J Food Sci Technol. https://doi.org/10.1007/s13197-011-0584-9

Dhanai R, Negi RS, Singh S, Parmar MK (2015) Fuelwood consumption by villagers in different altitudinal gradient: a case of Takoligad watershed of Garhwal Himalaya, India. Int J Curr Eng Technol 5:72–80.

Dhruvanarayan VV, Rambabu(1983) Estimation of soil erosion in India. Journal of Irrigation and Drainage Engineering, 109: 419–434

Dhyani SK (1998) An analysis of agroforestry systems on crop productivity and soil characteristics. Ph. D. Thesis, Department of Botany, North Eastern Hill University, Shillong, Meghalaya, India

Dorji KD and Dorji T (2015) The status of national soil resources of Bhutan. Asian Soil Partnership Consultation Workshop on Sustainable Management and Protection of Soil Resources, Bangkok, Thailand

Dorji KD (2008) Agriculture and soil fertility management in Bhutan: an overview. Paper presented in the meeting of Asia-Pacific Net on Integrated Plant Nutrient Management & International Workshop on Sustainable Nutrient Management: Technology and Policy, Shijiazhuang, Hebei, China, May 27–31, 2008

Dorji T (2015) Bhutan: organic by 2020. National Soil Services Centre, Dept. of Agriculture, MoAF Thimphu-Bhutan. Paper presented at the conference "The Next Phase of Organic Development—Visions, Trends and Innovations", Korea

Dulal R (1976) Inventory of the major soils of the World with special reference to mineral stress hazards. In: Wright MJ (ed) Plant adaptation to mineral stress in problem soils. Cornell University Press, Ithaca, NY, pp 3–13

Eckholm E (1979) Planting for the future: forestry for human needs. World Watch Paper 26, World Watch Institute, Washington, DC

El Baroudy AA (2016) Mapping and evaluating land suitability using a GIS-based model. CATENA 140:96–104

Elevitch CR (ed) (2004) The overstory book: cultivating connections with trees, 2nd edn. Permanent Agricultural Resources, Holualoa, Hawai'i, p 526

Eriksson M, Xu J, Shrestha A, Hydr R, Nepal S, Sandstrom K (2009a) The changing Himalayas: impact of climate change on water resources and livelihoods in the greater Himalayas. International Centre for Integrated Mountain Development, Kathmandu, Nepal

Eriksson M, Xu J, Shrestha AB, Baidya RA, Nepal S, Sandstrom K (2009b) The changing Himalayas: impact of climate change on water resources and livelihoods in the Greater Himalayas. Nepal, Kathmandu

FAO (1976) A framework for land evaluation, soils bulletin 32. Food and Agriculture Organization, Rome

FAO (1993) Guidelines for land-use planning. FAO Development Series I, FAO, Rome

FAO (1994) FAOSTAT, Agricultural data. http://apps.fao.org/default.htm

FAO (2002) FAOSTAT, Agricultural data. http://apps.fao.org/default.htm

FAO (2003a) Agriculture and food production in post-war Afghanistan: a report on the winter agricultural survey. Food, Agriculture and Animal Husbandry Information Management and Policy Unit Food and Agriculture Organization of the United Nations, Kabul, Afghanistan

FAO (2003b) FAOSTAT, Agricultural data. http://apps.fao.org/default.htm

FAO (2005) Current world fertilizer trends and outlook to 2009/2010. Italy, Rome

FAO (2006) Guidelines for soil profile description. Food and Agriculture Organization of the United Nations, Rome, Italy

FAO (2007) Conservation agriculture, agriculture and protection department. Food and Agriculture Organization, Rome, Italy. http://www.fao.org/ag/ca/

FAO (2009) Review of evidence on drylands pastoral systems in climate change. Implications and opportunities for mitigation and adaptation. In: Neely C, Bunning S, Wilkes A (eds) Land and water, discussion paper 8

FAO (2010) Global forest resources assessment 2010. Rome, Italy. www.fao.org/docrep/013/i1757e/i1757e.pdf

FAO (2011a) Case studies on policies and strategies for sustainable soil fertility and fertilizer management in South Asia. Food and Agriculture Organization of the United Nations Regional Office for Asia and the Pacific, Bangkok

FAO (2011b) Current world fertilizer trends and outlook 2015. Food and Agriculture Organization, Rome

FAO (2011c) The state of the world's land and water resources for food and agriculture (SOLAW)—managing systems at risk. Food and Agriculture Organization of the United Nations, Rome and Earthscan, London

FAO (2012) Country programmming framework (CPF) for 2012–2015 for Afghanistan. Food and Agriculture Organization of the United Nations. Rome, Italy

FAO (2014a) CA adoption worldwide, AQUASTAT: FAO-CA. http://www.fao.org/

FAO (2014b) What is conservation agriculture? FAO CA. http://www.fao.org/ag/ca/1a.html. Accessed 15 May 2018

FAO (2015a) FAOSTAT database: agriculture production. Food and Agriculture Organization of the United Nations, Rome, Italy

FAO (2015b) Understanding mountain soils. In: Romeo R, Vita A, Manuelli S, Zanini E, Freppaz M and Stanchi S (eds) A contribution from mountain areas to the international year of soils 2015b, Rome, Italy

FAO (2016) Food and Agriculture Organization of the United Nations, conservation agriculture

FAO/GIEWS (2007) Special report, FAO/WFP food security assessment mission to Nepal, 25 July 2007, GIEWS country briefs—Nepal, 7 October 2009. Rome, Italy

FAO/UNEP (1982) In: Lanly JP (ed) Tropical forest resources, FAO Forestry Paper 30. Rome, Italy

FAO/WFP (2009) Special report—crop and food security assessment mission to Myanmar. http://documents.wfp.org/stellent/groups/public/documents/ena/wfp196368.pdfM10.1080/07900627.2015.1030494

FAO/WFP (2016) Crop and food security assessment mission to Myanmar. Food and Agriculture Organization of the United Nations World Food Programme, Rome, Italy

FAOSTAT (2013) FAOSTAT-agriculture, production, crops. http://faostat3.fao.org/

Farid ATM, Ahmed JU, Anwar MN, Islam MS (1994) Integrated fertilizer management in rice-wheat cropping system. Bangladesh J Agril Res 19(1):60–65

Farid ATM, Hossain SM (1988) Diagnosis of farming practices and their impact on soil resources loss and economic loss in the hill tracts areas of Bangladesh. Gazipur, Bangladesh Agricultural Research Institute, Bangladesh

Farid ATM, Miah MAM, Karim Z (1998b) Innovative residue management in a typical rice soil of Bangladesh. Thai J Agril Sci 31(3):360–368

Farooq M, Siddique K (2014) Conservation agriculture: concept, brief history and impacts on agricultural systems. In: Farooq M, Siddique KHM (eds) Conservation Agriculture, 1st edn. Springer International Publishing, pp 3–17. https://doi.org/10.1007/978-3-319-11620-4_1

Farooq U, Ahmed M, Jasra AW (2007) Natural resource conservation, poverty alleviation, and farmer partnership. Pak Dev Rev 46(4 Part II):1023–1049

Farquharson B, Ramilan T, Thar SP, Than SM, Aung NM (2017) Nitrogen for smallholders and cereal crops in Myanmar: economics and social dimensions for fertility decisions. Paper presented at 61st Australian agricultural and resource economics society, annual conference, Brisbane, Australia

Feng S, Hu Q (2004) Changes in agro-meteorological indicators in the contiguous United States: 1951–2000. Theor Appl Clim 78:247–264

Fiedler JL (2014) Food crop production, nutrient availability, and nutrient intakes in Bangladesh: exploring the agriculture–nutrition nexus with household income and expenditure survey, food and nutrition bulletin, vol 35(no. 4) The Nevin Scrimshaw International Nutrition Foundation

Fleury JM (1999) Mountain biodiversity at risk. IDRC briefing 2:1–6

Foy CD (1984) Physiological effects of hydrogen, aluminium and manganese tocicities in acid soils. In: Adams F (ed) Soil acidity and liming, ASA-CSSA-SSSA, Madison, pp 57–97

FRMD (2017) Land use and land cover of Bhutan. Maps and statistics, forest resources management division department offorests & park services ministry of agriculture and forests Thimphu, Bhutan. ISBN: 978-99936-743-2-0

Fuhrer J (2003) Agroecosystern responses to combinations of elevated CO_2, ozone, and global climate change. Agric Ecosyst Environ 97:1–20

Gamel TH, Linssen JP, Alink GM, Mosallem AS, Shekib LA (2004) Nutritional study of raw and popped seed proteins of *Amaranthus caudatus* L and *Amaranthus cruentus* L. J Sci Food Agric 84(10):1153–1158

Ganeshamurthy AN, Kalaivanan D, Satisha GC (2016a) Management of vegetable crops in acid soils of India. In: Peter KV (ed) Innovations in horticultural sciences. New India Publishing Agency, New Delhi, India, pp 559–584

Ganeshamurthy AN, Kalaivanan D, Satisha GC (2016b) Management of fruit crops in acid soils of India. In: Peter KV (ed) Innovations in horticultural sciences. New India Publishing Agency, New Delhi, India, pp 539–558

Gayathri A (2013) India's ambitious $19.5 billion national food security bill wins lower house approval. Int Bus Times. http://www.ibtimes.com/indiasambitious-195-billion-national-food-security-bill-wins-lowerhouse-approval-1400125, 27 August 2013

Ghalley DR (1990) Agriculture in Bhutan. In: Mountain agriculture and crop genetic resources. International workshop on mountain agriculture and crop genetic resources, Kathmandu, Nepal. International Development Research Centre, Oxford & IBH Publishing Co. Pvt. Ltd., New Delhi, pp 69–74

Ghimire P, Chapagain PS, Yan J, Cheng X, Liu L, Zhang Y (2022) Drivers of changing crops and cropping patterns in the different elevation zones of middle mountain Nepal. Geogr J Nepal 15:1–22. https://doi.org/10.3126/gjn.v15i01.42883

Gilmour DA, Bowell M, Cassells DS (1989) The effects of forestation on soil hydraulic properties in the middle hills of Nepal. Mt Res Dev 7:239–249

Giribabu M (2013) Food and nutritional security in North East India: some contemporary issues. Int J Dev Res 3 (5):1–8

Giuliani A (2007) Developing markets for agro biodiversity: securing livelihood in dryland areas. Earthscan Publications, London

Godfray HCJ, Beddington JR, Crute IR, Haddad L, Lawrence D, Muir JF, Pretty J, Robinson S, Thomas SM, Toulmin C (2010) Food security: the challenge of feeding 9 billion people. Science 327:812–818

Goedert WJ, Lobato E and Lourenco S (1997) Nutrient use efficiency in Brezilian acid soils: nutrient management and plant efficiency. In: Moniz AC, Furlani AMC, Schaffert RE, Fageria NK, Rosolem CA, Cantarella H (eds) Plant-soil interactions at low pH: sustainable agriculture and forestry production. Brazilian Soil Science Society, Brazil

GoI (2013) State-wise land use statistics. Directorate of economics and statistics. Department of Agriculture, Cooperation and Farmers Welfare. Ministry of Agriculture and Farmers Welfare, Government of India

GoNWFP (1996a) SarlzadProvincial conservation strategy. Prepared by the Govt. of NWFP in collaboration with IUCN-the world conservation Union, p 248

GoNWFP (2000a) Important district-wise socio-economic indicators of NWFP. Bureau of Statistics, planning, environment and development department, p 105

GoNWFP (2000b) Agricultural statistics, NWFP for 1999–2000b. Agriculture statistics wing, Agriculture Department, Peshawar, p 144

GOP (2009) Pakistan economics survey 2008–09. Finance Division, Government of Pakistan, Islamabad

Gopalan C, Rama Sastri BV, Balasubramanian SC (2009) Nutritive value of Indian foods. National Institute of Nutrition, Indian Council of Medical Research. Hyderabad, India

Gopalan C, Rama Sastri BV, Balasubramanian SC (1989) Nutritive value of Indian foods. National Institute of Nutrition, Indian Council of Medical Research, Hyderabad, India (Re-printed 2004)

Govaerts B, Verhulst N, Castellanos-Navarrete A, Sayre KD, Dixon IJ, Dendooven L (2009) Conservation agriculture and soil carbon sequestration: between myth and farmer reality. CritAl Rev Plant Sci 28:97–122. https://doi.org/10.1080/07352680902776358

Government of India (2015–16) Ministry of Agriculture & Farmers Welfare Department of Agriculture, Cooperation & Farmers Welfare, Directorate of Economics and Statistics, New Delhi, India

Government of Nepal (GoN) (2016) Annual household survey 2014–2015. Central Bureau of Statistics, National Planning Commission Secretariat, Kathmandu, Nepal

Government of Pakistan (GoP) (2000) 1998 provincial census report of NWFP. Population Census Organization, Islamabad, Pakistan, p 280

Grewal JS, Kanwar JS (1967) Ammonium fixation in Punjab soils. J Res, Punjab Agric Univ, Ludhiana, India 4:1–6

Gruère GP, Giuliani A, Smale M (2006) Marketing underutilized plant species for the benefit of the poor: a conceptual framework. IFPRI Environmental and Protection Technology (EPT) Discussion Paper 154

Guillet D (1983) Toward a cultural ecology of mountains: the Central Andes and the Himalayas compared. Curr Antropol 561–574

Guo JH, Liu XJ, Zhang Y, Shen JL, Han WX, Zhang WF, Christie P, Goulding KWT, Vitousek PM, Zhang FS (2010) Significant acidification in major Chinese croplands. Science 327:1008–1010

Guppy N (1984) Tropical deforestation: a global review. Foreign Aff 62407:928–965

Gupta A, Sood S , Agrawal PK, Bhatt JC (2013) Underutilised food crops of Himalaya regions: utilization and prospective. In: Behera KK (ed) Newer approaches to biotechnology. Narendra Publishing House, India, pp 101–120

Gupta HS (2006) Status of research and production potential. In: Gupta HS, Srivastva AK, Bhatt JC (eds) Sustainable production agricultural watersheds in Northwest Himalayas. Vivekananda Parvatiya Krishi Anusandhan Sansthan, Almora, Uttrakhand, India, p 615

Guriro A (2013) 48.6% Pakistanis suffer food insecurity. Daily Times. 7 June 2013. http://archives.dailytimes.com.pk/national/07-Jun-2013/48-6-percent-pakistanis-suffer-foodinsecurity

Han JG, Zhang YJ, WangCJ BWM, Wang YR, Han GD, Li LH (2008) Rangeland degradation and restoration management in China. Rangel J 30:233–239

Harris RB (2010) Rangeland degradation on the Qinghai-Tibetan Plateau: a review of the evidence of its magnitude and causes. J Arid Environ 74:1–12

Hart R, Salick J, Ranjitkar S, Xu J (2014) Herbarium specimens show contrasting phenological responses to Himalayan climate. Proc Natl Acad Sci USA 111 (29):10615

Hasler K, Olfs HW, Omta O, Bröring S (2017) Drivers for the adoption of different eco-innovation types in the fertilizer sector: a review. Sustainability 9:2216

Hauck R (1985) Slow-release and bio-inhibitor-amended nitrogen fertilizers. In: Fertilizer Technology and Use, (eds) Engelstad OP, pp 293-322, Soil Science Society of America, Madison, WI , USA

Hobbs PR, Gupta RK (2003) Resources conserving technologies in wheat in rice-wheat system. In: Improving the Productivity and Sustainability of Rice-Wheat System: Issues and Impact. Vol. 65, (ed.) Ladha JK, Hill J, Gupta RK, Duxbury J and Buresh RJ, paper 7, ASSA Special publication, pp 149–171, Madison, WI: ASA

Hom NH, Htwe NM, Hein Y, Than SM, Kywe M, Htut T (2015) Myanmar climate-smart agriculture strategy. Ministry of Agriculture and Irrigation (MOAI). Naypyitaw, Myanmar: CGIAR Research Program on Climate Change, Agriculture and Food Security (CCAFS), International Rice Research Institute (IRRI). https://www.researchgate.net/publication/315384337

Hong C (1990) Agricultural systems of Xizang. In: Mountain agriculture and crop genetic resources. International Workshop on Mountain Agriculture and Crop Genetic Resources, Kathmandu, Nepal. International Development Research Centre, Oxford & IBH Publishing Co. Pvt. Ltd., New Delhi, pp 43–50

Horowitz MM (1988) Anthropology and the new development agendas. Dev Anthropol Netw 6:1–4

Hrabovszky JP, Mitan L (1987) Population growth and land use in Nepal: "the great turnabout." Mt Res Dev 7:264–270

Htun AA (2007) Local communities and environmental conservation Paunglong watershed area. Ph. D. dissertation, Department of Botany, University of Yangon, Yangon, Myanmar

Htwe TN (2016) Soil types and general soil information in Myanmar. The 2nd meeting of national soil and nutrient management expert group for ASEAN guidelines on soil and nutrient management, Yangon, Myanmar

ICARDA (2002) Needs assessment on soil and water in Afghanistan. International Center for Agricultural Research in the Dry Areas (ICARDA), Aleppo, Syria (Sponsored by the United States Agency for International Development)

ICIMOD (2008) Food security in the Hindu Kush Himalayan region: a position paper. Nepal, Kathmandu

ICIMOD (2012) Policy and institutions in adaptation to climate change: case study on tree crop diversity in China, Nepal and Pakistan. International Centre for Integrated Mountain Development, Kathmandu

ICIMOD (2006) In: Annual Report 2005. Int'l. center for integrated mountain development, Kathmandu, Nepal, pp 92

ICRISAT (2011) Production and productivity trends of ICRISAT mandate crops

IFDC (2002) Afghanistan fertilizer market: 2002 and beyond. Internal working document, IFDC, Muscle Shoals, AL 35662

IMI FAO Report (2019a) State of the Himalayan farmers and farming. A part of the project "Strengthening Institutional Capacities for Sustainable Mountain Development in the Indian Himalayan Region" awarded by FAO under the TCP/IND/3601/C1 to IMI

IMI FAO Report (2019b) State of the Himalayan armers and arming. A part of the project "Strengthening Institutional Capacities for Sustainable Mountain Development in the Indian Himalayan Region" awarded by FAO under the TCP/IND/3601/C1 to IMI. IPCC 2014

IPCC (2007a) Impacts, adaptation and vulnerability, contribution of working group II to the fourth assessment report of the intergovernmental panel on climate change. Cambridge University Press, UK

IPCC (2007b) Climate change mitigation. In: Contribution of working group III to the fourth assessment report of the intergovernmental panel on climate change. Cambridge University Press, Cambridge, UK and USA, p 841

Ives Jack D (1987) The theory of Himalayan environmental degradation: its validity and application challenged by recent research. Mt Res Dev 7(3):189–199. University of California Press, USA

Ives JD, Messerli B (1989) The Himalayan dilemma—reconciling development and conservation. Routledge, London

Jat RA, Sahrawat KL, Kassam AH (eds) (2014) Conservation agriculture: global prospects and challenges. CABI, Wallingford

Jianchu X, Melick D (2006) Emancipating indigenous knowledge: can traditional cultures assist Himalayan sustainable development? In: Sustainable mountain development No. 50. International Center for Integrated Mountain Development, Kathmandu, Nepal, pp 31–35

Jodha NS (1991) Mountain farming systems: coping with unsustainability prospects. Paper presented at the 11th FAO session of FAO farm management commission for Asia and Pacific, Kathmandu, Nepal, December 4–6, 1991

Jodha NS (1992) Mountain perspective and its utility: a framework for development strategies. Paper presented in the workshop at Manali on approaches to sustainable development of Indian Himalayas, Manali, Himachal Pradesh, India, 1–4 August

Jodha NS (1993) Perspective on property generating process in mountain areas. In: Proceedings of international forum on development of poor mountain areas, Beijing, China, 22–27 March 1993, pp 38–46

Jodha NS (1995) Enhancing in the warming and more croded world: factors and processes in fragile zones. In: Downing TE (ed) Climate change and world food security, NATO, Asia series 1, global environmental change 37. Springer Verlag, NY, pp 381–420

Jodha NS (2004) Highland low and linkages in the globalized world. In: Bhadra B, Khanal NR, Richter J (eds) Jodha NS. Poverty alleviation in mountain area of China, ICIMOD, Kathmandu, Nepal pp, pp 167–183

Jodha NS, Shrestha S (1993) Sustainable and more productive mountain agriculture: problems and prospects. Mount Environ Develop 3:1–65

Jodha NS, Shreshtha S (1994) Towards sustainable and more productive mountain farming. In: Proceedings of international symposium of mountain environment and development: constraints and opportunities. ICIMOD, Kathmandu, Nepal

Joseph PA, Prasad R (1993) Correlation studies on ammonium/nitrate concentration in soil and growth and yield of wheat. J Agron Crop Sci 171:26–30

Joshi D, Deo GP (1976) Fertilizers recommendations for major crops of Nepal. Division of Soil Science and Agricultural Chemistry, Department of Agriculture, HMG/Nepal

Joshi L, Shrestha RM, Jasra AH, Joshi S, Gilani H, Ismail M (2013) Rangeland ecosystem services in the Hindu Kush Himalayan Region. In: Wu N, Rawat GS, Joshi S, Ismail M, Sharma E (eds) High-altitude rangelands and their interfaces in the Hindu Kush Himalayas, ICIMOD GPO Box 3226, Kathmandu, Nepal

Joshi SK (1996) Some issues related to the sustainable development of the Himalayan Region. 6th G.B. Pant Memorial Lecture. G.B. Pant Institute of Himalayan Environment and Development, Kosi-Katarmal, Almora

Joshua JK, Anyanwu NC, Ahmed AJ (2013) Land suitability analysis for agricultural planning using GIS and multi criteria decision analysis approach in Greater Karu Urban Area, Nasarawa State. Nigeria. Afr J Agric Sci Technol 1(1):14–23

Kailash Kumar, Rao KVP (1990) Fertility status of soils of Manipur. India Journal of Hill Farming, 3: 27–31

Kalindi Roy RC (2000) Lang rights of the indigenous peoples of Chittagong Hills Tract, Bangladesh. IWGIA Document No. 99, Copenhagen. ISBN: 87-90730-29-1

Kapos V, Rhind J, Edwards M, Price MF, Ravilious C (2000) Developing a map of the world's mountain forests. In: Price MF, Butt N (eds) Forests in sustainable mountain development: a state-of-knowledge report for 2000. CAB International, Wallingford, pp 4–9

Karim AJMSK, Egashira Y, Yamada HJ, Nahar K (1995) Long-term application of organic residues to improve soil properties and to increase crop yield in terrace soil of Bangladesh. J Fat Agr Kyushu Uni 39:149–165

Karki M, Sharma S, Mahat TJ, Tuladhar A, Aksha S (2012a) Sustainable mountain development in the Hindu Kush—Himalaya International Centre for Integrated Mountain Development (ICIMOD). Nepal, Kathmandu

Karki M, Sharma S, Mahat TJ, Tuladhar A, Aksha S (2012b) Sustainable mountain development 1992, 2012b, and beyond—Rio+20 assessment report for the Hindu Kush Himalayan Region

Karki MB, Shrestha AB, Winiger M (2011) Enhancing knowledge management and adaptation capacity for integrated management of water resources in the Indus river basin. Mountain Research and Development, 31: 242-251

Kassam AH, Friedrich T, Shaxson F, Pretty J (2009) The spread of conservation agriculture: justification, sustainability and uptake. Int J Agric Sustain 7(4):292–320. https://doi.org/10.3763/ijas.2009.0477

Kassam AH, Mkomwa S, Friedrich T (eds) (2017) Conservation agriculture for Africa. CABI, Wallingford

Kassam AH, Friedrich T, Derpsch R, Kienzle J (2015) Overview of the worldwide spread of conservation agriculture. Field actions science reports 8, https://journals.openedition.org/factsreports/3966

Kassam A, Friedrich T, Derpsch R (2018) Global spread of conservation agriculture. Int J Environ Stud. https://doi.org/10.1080/00207233.2018.1494927, http://www.tandfonline.com/loi/genv20

Keift H (1993) The potential of low input agriculture in sub-Saharan Africa. In: van Realer H, Prins WH (eds) The role of plant nutrition for sustainable food crop production in sub-Saharan Africa. Dutch Association of Fertlizer Producers, AK Leidschendam, The Netherlands, pp 127–140

Khaliq A, Zafar M, Abbasi MK, Hussain I (2018) Soil-plant micronutrients dynamics in response to integrated fertilization under wheat–soybean cropping system at Rawalakot. Pak J Arch Agron Soil Sci 64(5):640–653. https://doi.org/10.1080/03650340.2017.1376245

Khan MA, Ahmed M, Hashmi HS (2012) Review of available knowledge on land degradation in Pakistan. International Center for Agricultural Research in the Dry Areas (ICARDA)

Khoshoo TN (1992) Plant diversity in the Himalaya: conservation and utilization. G.B. Pant Memorial Lecture: III. G.B. Pant Institute of Himalayan Environment and Development, Kosi-Katarmal, Almora

Kira T, Shidei T (1967) Primary production and tumover of organic matter in different forest ecosystems of Western Pacific. Japan, I Eeol 17:70–91

Kobayashi M, Chhetri R, Fukamachi K (2015) Transition of agriculture towards organic farming in Bhutan. Himalayan Study Monographs 16:66–72

Koerner C (2003) Alpine plant life, 2nd edn. Springer, Berlin, p 337

Kuma P, Sharma PK, Kumar P, Sharma M, Butail NP (2021) Agricultural sustainability in Indian Himalayan region: constraints and potentials. Indian J Ecol 48 (3):649–662

Kumar A, Pramanik M, Chaudhary S et al (2020) Land evaluation for sustainable development of Himalayan agriculture using RS-GIS in conjunction with analytic hierarchy process and frequency ratio. J Saudi Soc Agric Sci. https://doi.org/10.1016/j.jssas.2020.10.001

Kumar K, Rao KVP (1992) N and P requirement of upland rice in Manipur, Oryza 29:306–309

Kumar SV, Reichle RH, Peters-Lidard CD, Koster RD, Zhan X et al. (2008). A land surface data assimilation framework using the land information system: Description and Applications. Advances in Water Resources, 31: 1419–1432

Kumar S, Sharma KL, Kareemulla K, Chary GR, Rama Rao CA, Rao SC, Venkateswerlu B (2011) Techno-economic feasibility of conservation agriculture in rainfed regions of India. Curr Sci 101(9):1171–1182. https://www.researchgate.net/publication/259334946. Accessed on November 10, 2021

Kumar P, Sharma PK, Kumar P, Sharma M, Butail NP (2021) Agricultural sustainability in Indian Himalayan region: constraints and potentials. Indian J Ecol 48 (3):649–662

Kumar A (2019) A study of land and water resource for the sustainable development in Rudraprayag District (A mico-level analysis). Doctoral dissertation, H.N.B. Garhwal University. Unpublished

Kurvits T, Kaltenborn B, Nischalke S, Karky B, Jurek M, Aase TH (2014) The last straw: food security in the Hindu Kush Himalayas and the additional burden of climate change. ICIMOD, Kathmandu, Nepal. ISBN: 978-82-7701-129-5

Lal R (2001) Managing world soils for food security and environmental quality. Adv Agron 75:155–192

Laskar S, Datta M (1989) Response of Rice to phosphate fertilizers in soils of Tripura. J Indian Soc Soil Sci 37:836–840

Leduc B, Choudhury D (2012) Agricultural transformations in shifting cultivation areas of North-east India, implications for land management, gender, and institutions, social exclusion and adverse inclusion. In: Dev N, Xaxa V (eds) Development and deprivation of adivasis in India, pp 237–258

Lele U, Klousia-Marquis M, Goswami S (2013) Good governance for food, water and energy security. Aquatic Procedia 1:44–63

Li SQ, Zhang QH (2001) Advances in the development of functional foods from buckwheat. Crit Rev Food Sci Nutr 41(6):451–464

Liang S, Yang G, Ma Y (2010) Chemical characteristics and fatty acid profile of of foxtail millet bran oil. J Aer Oil Chem Soc 87:63–67

Lindfors A, Gustafsson M, Anderberg S, Eklund M, Mirata M (2020) Developing biogas systems in Norrköping, Sweden: an industrial symbiosis intervention. J Clean Prod 277:122822. https://doi.org/10.1016/j.jclepro.2020.122822

Lumpkin TA, Sayre K (2009) Enhancing resource productivity and efficiency through conservation agriculture. In: Compendium of lead papers, 4th world

congress on conservation agriculture, New Delhi, 4–7 February 2009, pp 3–9
Ma M, Singh RB, Hitala R (2012) Human driving forces for ecosystem services in the Himalayan region. Environ Econ 3(1):52–57
Magdoff F, Lanyon L, Liebhardt B (1997) Nutrient cycling, transformations and flows: implications for more sustainable agriculture. Adv Agron 60:1–66
Mahat TBS, Griffin DM, Shepherd KR (1986) Human impact on some forest of the middle hills of Nepal. I. Forestry in the context of the traditional resources of the State. Mt Res Dev 6:223–232
Maji AK, Obi Reddy GP, Sarkar D (2010) Degraded and wastelands of India—status and spatial distribution. Indian Council of Agricultural Research and National Academy of Agricultural Sciences. New Delhi
Majumdar M, Kumar K, Venkatesh MS, Patiram, Bhatt BP (2004) Effect of different agroforestry systems on soil properties in acid Alfisols of Meghalaya. J Hill Res 17(1):1–5
Martin G, Martin-Clouaire R, Duru M (2013) Farming system design to feed the changing world. A review. Agron Sustain Dev 33:131–149
Maskey RB, Binod P, Sharma BP, Madhav Joshi M (2003) Human dimensions in sustainable land use management in degraded land areas of Nepal. Paper presented at the open meeting of the global environmental change research community, Montreal, Canada
Mathou T (1999) Bhutan: political reform in a Buddhist monarchy. J Bhutan Stud, pp 114–145
Mian A, Mirza MYJ (1993a) Pakistan's soil resources. A joint publication of IUCN and the environment and urban affairs division of the Government of Pakistan. (Pakistan National Conservation Strategy Sector Paper No. 4)
Mian A, Mirza MYJ (1993b) Pakistan's soil resources, environment & urban affairs division. Government of Pakistan
Mian M, Mirza MYJ (1993c) Pakistan's soil resources: a Pakistan national conservation strategy sector paper 4. IUCN—The World Conservation Union, Pakistan 1 Bath Island Road, Karachi, p 75530
Michalk DL, Zhi-Kai H (1992) Response of subterranean clover (Trifolium Subteraneum) to lime, magnesium, and boron on acid infertile soil in subtropical China. Fertil Res 32:249–257
Miller W, Collins MG, Steiner FR, Cook E (1998) An approach for greenway suitability analysis. Landscape Urban Plann 42(2–4):91–105. https://doi.org/10.1016/S0169-2046(98)00080-2
Miller DJ (1996) Pastoral development in the HKH: organising rangeland and livestock research for the twenty-first century. In: Chaudhry MA, Safdar A, Muhammad IA (eds) Proceedings of the seminar on farming systems research in the context of food security. University of Arid Agriculture, Rawalpindi, Pakistan, 4–6 August 1996
Ministry of Environment, Government of Pakistan (2009) Land use Atlas of Pakistan.
Ministry of Environmental Conservation and Forestry (MoACF) (2014) Facts about environmental conservation and forestry in Myanmar. Ministry of Environmental Conservation and Forestry, Myanmar.
Ministry of Livestock, Fisheries, Myanmar (2015) Livestock breeding and veterinary department, MLFRD, Myanmar Livestock Statistics, Myanmar
Mishra BK, Ramakrishnan PS (1983) Secondary succession subsequent to slash and burn agriculture at higher elevations of north-east India. Species diversity, biomass and litter production. Acta Oecologica-Oecology Appl 4:95–107
Mishra AK, Sharma UC (2001) Traditional wisdom in range management for resource and environment conservation in northeastern region of India. ENVIS Bull Himal Ecol 9(1):263–269
Misra RV (2011) Bangladesh. In: Case studies on policies and strategies for sustainable soil fertility and fertilizer management in South Asia. Food and Agriculture Organization of the United Nations Regional Office for Asia and the Pacific, Bangkok
MoAF (2014) Policy objectives. Ministry of Agriculture and Forests, Royal Government of Bhutan, Thimphu. http://www.moaf.gov.bt/aboutus/policy-objectives
MoAF (2015) Bhutan RNR statistics. Ministry of Agriculture and Forests, Royal Government of Bhutan, Thimphu
MoAI (Ministry of Agriculture and Irrigation) (2015) Myanmar agriculture in brief, pp 81
Molden D, Sharma E (2013) ICIMOD's strategy for delivering high-quality research and achieving impact for sustainable mountain development. Mt Res Dev 33 (2):179–183
Montgomery B, Dragićević S, Dujmović J, Schmidt M (2016) A GIS-based logic scoring of preference method for evaluation of land capability and suitability for agriculture. Comput Electron Agric 124:340–353. https://doi.org/10.1016/j.compag.2016.04.013
Moody PW, Cong PT, Legrand J, Chon NQ (2008) A decision support framework for identifying the soil constraints to the agricultural productivity for upland soil. Soil Use Manag 24:148–155
Mukerjee P, Sogani R, Gurung N, Rastogi A, Swiderska K (2018) Smallholder farming systems in the Indian Himalayas Key trends and innovations for resilience. IIED country report. IIED, London. http://pubs.iied.org/17618IIED
NAAS (2019) Agricultural transformations: a road map to new India. National Academy of Agricltural Sciences, New Delhi, India
Naing OA, Khaung T, Thein SS (2016) Nutrient management for rice production in Myanmar. Practice-oriented Area Studies Series No.9. Kyoto University, Japan
NBSSLUP (2004) Soil resource management reports. National Bureau of Soil Survey and Land Use Planning, Nagpur, India
Nellemann C, Verma R, Hislop L (eds) (2011) Women at the frontline of climate change: gender risks and hopes. A rapid response assessment. United Nations

Environment Programme, GRID-Arendal. ISBN: 978-82-7701-099-1

NFDC (National Fertilizer Development Center) (2005) Fertilizer use survey (2002–03). A micro-level assessment of situation in Pakistan, Publication No. 2/2005 Islamabad, Pakistan

NFDC (2014) Fertilizer related statistics. Islamabad, Pakistan

Norbu C, Floyd C (2001) Changing soil fertility management in Bhutan: effect on practices, nutrient status and sustainability. National Soil Services Centre, Ministry of Agriculture, Thimphu, Bhutan

Norbu T (2011) The significance of the Tibetan Plateau, River waters: perspectives and challenges for Asia IIC, New Delhi, India, 18–20 Nov 2011

NSB (2017) Statistical yearbook of Bhutan. Royal Government of Bhutan

NRVA (National Risk and Vulnerability Assessment) (2011-12) Afghanistan living condition survey, 2011-12, Kabul, Afghanistan

Obilana AB, Manyasa E (2002) Millets. In: Belton PS, Taylor JRN (eds) Pseudo cereals and less common cereals: grain properties and utilization potential. Springer-Verlag, New York, pp 177–217

Palit D, Mainali B, Nepal R, Purohit P, Goswami A, Malikyar GM, Wakhley KB (2019) Meeting future energy needs in the Hindu Kush Himalaya. In: Wester P, Mishra A, Mukherji A, Shrestha AB (eds) The Hindu Kush Himalaya assessment—mountains, climate change, sustainability and People Springer Nature Switzerland AG, Cham

Pandey SP, Tamang DB and Baidya SN (1995) Soil fertility manaagent and agriculture production issues with reference to middle mountain regions of Nepal. In: Scheier H, Shah PB and Brown S (eds) Challenges in mountain resources management in Nepal: processes, trends and dynamics in Middle Mountain Watersheds. Proc. Workshop, ICIMOD, Kathmandu, Nepal, April 10–12 1995, pp 41–49

Partap T (2011) Hill agriculture: challenges and opportunities. J Agric Econ 66(33):52

Partap T, Kapoor P (1987) The Himalayan grain chenopods III: an underexploited food plant with promising potential 19(1):71–79

Partap T, Sthapit B (eds) (1998a) The challenges of managing agrobiodiversity in HKH region—an overview of issues. In: Mt Res Dev 20(4):1–30. ICIMOD, Kathmandu, Nepal

Partap T, Joshi BD, Galwey N (1998b) Chenopods, Chenopodium spp. Monograph. IPGRI Rome Italy, p 67

Partap T (1986) Agrobiodiversity of Himachal Pradesh: crops and cropping patterns: a field survey of the state in 1985–86

Partap T (1999) Sustainable land management in marginal mountain areas of the Himalayan region. Mt Res Dev 251–260

Patel CS, Singh J (1984) In: Annual reports. ICAR Research Complex for NEH Region, Umiam, Meghalaya, India

Patel CS (1983) In: Annual report. ICAR Res Complex for NEH Region, Shillong, Meghalaya, India

Patiram, Choudhury BU (2000) Role of agroforestry in soil health management. ICAR Research Complex for NEH Region, Umiam, Meghalaya, India

Patiram, Prasad RN (1983) Yield responses to fertilizer potassium and balance sheet of potassium in potato-rice sequence. J Indian Soc Soil Sci 31:502–504

Patiram, Prasad RN (1989) Effect of top dressing of potassium on rice grown in valley soils of Meghalaya. Indian J Hill Farming 2:21–24

Patiram RRN, Prasad RN (1990) Efffect of liming on aluminium and yield of wheat on acidic soils. J Indian Soc Soil Sci 38:719–722

Patiram (2002) Environmentally sustainable hill agriculture for soil and water conservation in north-eastern hills region. Indian J Hill Agric 15:1–9

Paustian K, Lehmasnn J, Ogle S, Reay D, Robertson GP, Smith P (2016) Climate-smart soils. Nature 532:49–57

Prabhakar SVRK, Khattak BN, Shi P, Huggel C, Chitale V (2019) Exploring futures of the Hindu Kush Himalaya: scenarios and pathways. In: Wester P, Mishra A, Mukherji A, Shrestha AB (eds) The Hindu Kush Himalaya assessment—mountains, climate change, sustainability and People Springer Nature Switzerland AG, Cham

Prasad RN, Ram M (1985) Soils of north-eastern hill region and their management. Soils of India and their management. The Fertilizer Association of India. New Delhi, India, pp 266–283

Prasad R, Thomas J, Gupta VVSR (1983) Ammoniphillic plants for reducing water pollution. Environ Conserv 10:260–261

Prasad RN, Sharma UC (1994) Potential indigenous farming systems of North Eastern Hill Region. ICAR Research Complex for NEH Region, Barapani, Meghalaya, India, pp 48

Prasad RN, Patiram, Ram M (1984) Annual Report ICAR Research Complex for NEH Region, Umiam, Meghalaya, India

Prasad RN, Patiram Barooah RC, Ram M (1981) Soil fertility management in North Eastern Hills Region. Research Bulletin No. 9. ICAR Research Complex for NEH Region, Shillong, India

Prasad RN, Patiram, Ram M (1984) Annual Report ICAR Research Complex for NEH Region, Umiam, Meghalaya, India.Prasad RN, Patiram and Ram M (1984). Annual Report ICAR Research Complex for NEH Region, Umiam, Meghalaya, India

Prasad RN, Singh A, Arun V (1987) Application of research findings for management of land and water resources in Eastern Himalayas region. Bulletin ICAR Research Complex, Shillong, India, pp 1–50

Prasad R, Pathak H, Patra AK, Shivay YS (2014) Nitrogen management. In: Textbook of plant nutrient management, p 407. https://www.researchgate.net/publication/280716753

Prasad R (2007) Nitrogen in Indian agriculture. In: Abrol YP, Raghuram R, Sachdev MS (eds) Agricultural nitrogen use and its environmental implications. IK

International Publishing House Pvt. Ltd, New Delhi, India
Prasain S (2013) Nepal to face rice deficit of 900k this year. Kathmandu Post, 10 June 2013. http://www.ekantipur.com/thekathmandu-post/2013/06/10/money/nepal-to-face-rice-deficit-of-900k-tonnes-this-year/249829.html
Pratap T (2001) Mountain agriculture, marginal lands and sustainable livelihoods: challenges and opportunities. Intl symposium on mountain agriculture in HKH region. ICIMOD, Kathmandu, Nepal, 21–24 May 2001
Pratap T (2003) Mountain agriculture, marginal lands, and sustainable livelihoods: challenges and opportunities. In: Ya T, Tulachan P (eds) Mountain agriculture in the Hindu Kush-Himalayan Region. ICIMOD, Kathmandu, Nepal
Pratap T (2019) State of the Himalayan farmers and farming—a study about social, economical and ecological factors changing the mountain agricultural scenario. FAO, Rome, Italy
Pratap T (2020) Pandit Gobind Ballabh Pant memorial lecture. G. B. Pant National Institute of Himalayan Environment, Kosi-Katarmal, Almora, Uttrakhand, India
Qasim SZ (1995) Some major socio-economic and environmental issues of the Himalayan Region. G. B. Pant Memorial Lecture: V. G. B. Pant Institute of Himalayan Environment and Development, Kosi-Katarmal, Almora
Qureshi AST, Shah, Akhtar M (2003) The groundwater economy of Pakistan. Pakistan country series No. 19. International Water Management Institute, Lahore, Pakistan. http://www.iwmi.cgiar.org/pubs/working/WOR64.pdf. (IWMI Working Papcr 64)
Rahman KMA, Zhang D (2018) Effects of fertilizer broadcasting on the excessive use of inorganic fertilizers and environmental sustainability. Sustainability 2018(10):759–773. https://doi.org/10.3390/su10030759
Rajasekhar M, Sudarsana RG, Nanabhau Kudnar N, Ramachandra M, Badapalli PK (2021) Multi-criteria land suitability analysis for agriculture in semi-arid region of Kadapa District, Southern India: geospatial approaches. Remote Sens Land 5:59–72. https://doi.org/10.21523/gcj1.2021050201
Rajbhandari R, Saeed S, Sanjay J, Syed MA, Vellore R, Xu Y, You Q, Ren Y (2019) Unravelling climate change in the Hindu Kush Himalaya: rapid warming in the mountains and increasing extremes. In: Wester P, Mishra A, Mukherji A, Shrestha AB (eds) The Hindu Kush Himalaya assessment—mountains, climate change, sustainability and People Springer Nature Switzerland AG, Cham
Ram M, Prasad RN (1992) In: Annual reports. ICAR Research Complex for NEH Region, Umiam, Meghalaya, India
Rao S, Reddy KS (2005). Integrated nutrient management vis-a-vis crop production/productivity/nutrient balance. Indian Journal of Fertilizers, 1: 77–80.
Ramakrishnan PS (1992) Shifting agriculture and sustainable development: an interdisciplinary study from North-Eastern India. UNESCO, Paris and Oxford University Press, Delhi/Bomay/Madras
Ramakrishnan PS, Toky OP (1981) Soil nutrient status of hill agro-ecosystems and recovery pattern after slash and burn agriculture (Jhum) in north-eastern India. Plant Soil 60:41–64
Ramakrishnan PS, Rao KS, Kothyari BP, Maikhuri RK, Saxena KG (1992) Deforestation. In: Singh JS (ed) Restoration of degraded land: concepts and strategies. Rastogi Publications, Meerut
Ramakrishnan PS, Purohit AN, Saxena KG, Rao KS (1994) Himalayan environment and sustainable development. Indian National Science Academy, New Delhi
Ramakrishnan Ps, Purohit AN, Saxena KG, Rao KS, Maikhuri RK (1996) Conservation and management of biological resources in Himalaya. Oxford & IBH Publication Ltd., New Delhi
Rao KS, Swarup R (1993) Energy efficient technology in Indian Himalaya: an assessment. Paper submitted to science and technology in third world development—second triennial conference, 5–7 April 1993. University of Strathclyde, Glasgow. ENVIS Monograph 1
Rao UR (1994) Space for sustainable development with special emphasis on Himalayan Region. G. B. Pant Memorial Lecture: IV. G. B. Pant Institute of Himalayan Environment and Development, Kosi-Katarmal, Almora
Rao KS (1997) Natural resource management and development in Himalaya—a recourse to issues and strategies. In: Rao KS (ed) ENVIS monograph 1, G.B. Pant Institute of Himalayan Environment and Development Kosi-Katarmal, Almora, 263 643, India
Rasul G (2010) The role of the Himalayan mountain systems in food security and agricultural sustainability in South Asia. Int J Rural Manag 6(1):95–116
Rasul G (2012) Contribution of Himalayan ecosystems to water, energy and food security in South Asia: a nexus approach. International Centre for Integrated Mountain Development, Kathmandu
Rasul G, Karki M (2007) Pro-poor policy agenda for sustainable agriculture development in the Hindu Kush-Himalayan region: ICOMOD talking points 2 (7). Nepal, Kathmandu
Rasul G, Joshi L (2011) Climate change and food security in the Himalayas. Regional workshop on climate change and food security in ASEAN+3 countries, Beijing, China. International Centre for Integrated Mountain Development Kathmandu, Nepal, 29–31 March 2011
Rasul G, Kollmair M (2008) Sustainable livelihood promotion through agricultural development in the hills of South Asia. ICIMOD, Nepal
Rasul G, Sharma B (2016) The nexus approach to water–energy–food security: an option for adaptation to climate change. Climate Policy 16(6):682–702. https://doi.org/10.1080/14693062.2015.1029865

Rasul G, Hussain A, Sutter A, Dangol N, Sharma E (2016) Towards an integrated approach to nutrition security in the Hindu Kush Himalayan Region. ICIMOD Working Paper 2016/7 Kathmandu, Nepal

Rasul G, Hussain A, Mahapatra B, Dangol N (2017) Food and nutrition security in the Hindu Kush-Himalayan region. J Sci Food Agric. https://www.researchgate.net/publication/319165543. https://doi.org/10.1002/jsfa.8530

Rasul G, Saboor A, Tiwari PC, Hussain A, Ghosh N, Chettri GB (2019) Food and nutrition security in the Hindu Kush Himalaya: unique challenges and niche opportunities. In: Wester P, Mishra A, Mukherji A, Shrestha A (eds) The Hindu Kush Himalaya assessment. Springer, Cham, pp 301–338. https://doi.org/10.1007/978-3-319-92288-1_9

Raut NP, Dorsch BK, Sitaula, Bakken LR (2012) Soil acidification by intensified crop production in south Asia results in higher $N_2O/(N_2 + N_2O)$ product ratios of denitrification. Soil Biol Biochem 55:104–112

Reserve Bank of India (2017) Hand book of statistics of Indian States. Mumbai, India

Roberts TL (2009) The role of fertilizer in growing the world's food. Better Crops 93(2):12–15

Roder W (1990) Traditional use of nutrient inputs. ILEIA Newsletter 6(3):3–4

Roder W, Calvert O, Dorji Y (1993) Effect of burning on selected soil parameters in a grass fallow shifting cultivation system in Bhutan. Plant Soil 149:51–58

Roder W, Dorji K, Gratzer G (2003) Nutrient flow from the forest life source for traditional Bhutanese Agriculture. 120. Jahrgang, Heft 1:65–72

Romeo R, Vita A, Manuelli S, Zanini E, Freppaz M, Stanchi S (2015) Understanding mountain soils: a contribution from mountain areas to the international year of soils. FAO, Rome, Italy

Roy RD (1995) Land rights, land use and indigenous people in the Chittagong Hill Tracts. Bangladesh land forest and forest people. Society for Environment and Human Development (SEHD), Gain, Dhaka, Bangladesh, pp 53–118

Roy RN, Farid ATM (2011) Bangladesh. In: Case studies on policies and strategies for sustainable soil fertility and fertilizer management in South Asia. Food and Agriculture Organization of the United Nations Regional Office for Asia and the Pacific, Bangkok

Roy RN, Finck A, Blair GJ, Tandon HLS (2006) Plant nutrition for food security—a guide for integrated nutrient management. FAO Fertilizer and Plant Nutrition Bulletin No. 16, Rome, Italy

Roy AH (2003) The rebuilding of Afghanistan's agriculture: the IFDC solution. International Center for Soil Fertility and Agricultural Development P.O. Box 2040 Muscle Shoals, Alabama 35662, USA

Royal Government of Bhutan (RoGB) (2016) Bhutan RNR statistics. Policy and Planning Division, Ministry of Agriculture and Forests, Royal Government of Bhutan, Thimpu, Bhutan

Saba DS (2001) Afghanistan: environmental degradation in a fragile ecological setting. Int J Sust Dev World 8:279–289

Saha R, Chaudhary RS, Somasundaram J (2012) Soil health management under hill agro-ecosystem of North East India. Appl Environ Soil Sci 2012 (2012):696174. Hindawi Publishing Corporation. https://doi.org/10.1155/2012/696174

Salick J, Yongping Y, Amend A (2005) Tibetan land use and change near Khawa Karpo, Eastern Himalayas. Econ Bot 59:312–325

Samal PK, Fernando R, Rawat DS (2000) Influences of economy and culture in development among mountain tribes of Indian central Himalaya. Int J Sust Dev World 7:41–49

Samra JS, Dhyani BL, Sharma AR (1999) Problem and propects of natural resources management in India Himalayas—a base paper. Hill and Mountain Agroecosystem Directorate, Dehradun, India

Sanchez P (1976) Properties and management of soils in the tropics. John Wiley & Sons, New York, USA, p 618p

Sanchez PA (1994) Tropical soil research: towards the second paradigm. Transanctions 15th world congress of soil Science. Acapulco, Maxico 1:65–88

Sanchez PA (1997) Changing tropical soil fertility paradigms: from Brazil to Africa and back. In: Moniz AC, Furlani AMC, Schaffert RE, Fageria NK, Rosolem CA and Cantarella H (eds) Plant–soil interactions at low pH: sustainable agriculture and forestry production. Brazilian Soil Science Society, Brazil

Satapathy KK (2006) Soil conservation and management strategies. In: Sharma UC, Datta M, Samra JS (eds) Soils and their management in Northeast India. Indian Council of Agricultural Research Complex for North-eastern Hills Region, Meghalaya, India, pp 372–388

Satyanarayana T, Tewatia RK (2009) State wise approaches to crop nutrient balances in India. International Plant Nutrition Institute, IPI-OUAT-IPNI Intl Symp, Nov. 5–7, 2009

Saxena KG, Rao KS, Sen KK, Maikhuri RK, Semwal RL (2001) Integrated natural resource management: approaches and lessons from the Himalaya. Conserv Ecol 5:14. http://www.consecol.org/vol15/iss2/art14

Sayer JA, Whitmore TC (1991) Tropical moist forests: destruction and species extinction. Biol Cons 55:119–124

Scheier H, Shah PB, Lavkulich LM, Brown S (1994) Maintaining soil fertility under increasing land use pressures in the middle mountains of Nepal. Soil Use Manag 10:137–142

Schlesinger WH (2000) Soil respiration and the global carbon cycle. Biogeochemistry, 48: 7–20.

Schild A, Sharma E (2011) Sustainable mountain development revisited. Mt Res Dev 31(3):237–241

Schmidt-Vogt D, Leisz S, Mertz O, Heinimann A, Thiha T, Messerli P, Epprecht M, Cu PV, Chi VK, Hardinon

M, Dao T (2009) An assessment of trends in the extent of swidden in Southeast Asia. Human Ecol 37(3):269–280. Retrieved from http://www.jstor.org/stable/40343971

Schoenlechner R, Drausinger J, Ottenschlaeger V (2010) Functional properties of gluten-free pasta produced from Amaranth, Quinoa and Buckwheat. Plant Foods Hum Nutr 65(4):339–349

Schreier H, Shah PB (2000) Soil fertility status and dynamics in the Jhikhu and Yarsha khola watershed. In: Richard A, Hans S, Brown S, Shah PB (eds) The people and resource dynamics project: the first three years (1996–1999). ICIMOD, Kathmandu, Nepal

Schreier H, Shah PB, Brown S (1995) Challenges in mountain resource management in Nepal. Processes, trends, and dynamics in middle mountain catchments. Proceedings of a workshop held in ICIMOD, Nepal 10–12 April 1995, pp 163–170

Scott CA, Zhang F, Mukherji A, Immerzeel W, Mustafa D, Bharati L (2019) Water in the Hindu Kush Himalaya. In: Wester P, Mishra A, Mukherji A, Shrestha AB (eds) The Hindu Kush Himalaya assessment—mountains, climate change, sustainability and People Springer Nature Switzerland AG, Cham

Shah SL (1982) Ecological degradation and future of agriculture in the Himalayas. Indian J Agri Economics 37:1–22

Shahi P (2013) Rice dependency causing food deficit. Kathmandu Post, Nepal, 12 March 2013

Sharma KP (1993a) Role of melt-water in major river systems of Nepal'. In: Young GJ (ed) International symposium on snow and glacier hydrology, Kathmandu. International Association of Hydrological Sciences, IAHS Publication No. 218. Wallingford, UK, pp 113–122

Sharma UC (1987) Potato yield as affected by zinc, iron and their interactions with phosphorus. J Indian Pot Assoc 14:21–25

Sharma UC (1989a) Effect of levels of N P and K and their interactions on the yield and nutrient uptake of potato on acis soils. J Indian Pot Assoc 19:77–80

Sharma UC (1989b) Influence of soil texture on leaching of potassium and its recovery by potato. Indian J Agric Sc 59:713–717

Sharma UC (1990a) Effect of sources and methods nitrogen application on the yield and N uptake of potato in Meghalaya. Indian J Agric Sci 60:119–122

Sharma UC (1990b) Evaluation of two sources of phosphorus on acidic soils of Meghalaya with potato as test crop. J Indian Soc Soil Sci 38:46–50

Sharma UC (1990c) Effect of FYM on potato tuber yield and N uptake in Meghalaya. Indian J Hill Farming 3:11–14

Sharma UC (1990d) Studies on nitrogen leaching due to precipitation in Alfisols of Meghalaya. J Indian Soc Soil Sci 38:218–223

Sharma UC (1990e). Effect of sources and methods of nitrogen application on the yield and N uptake of potato in Meghalaya. Indian Journal of Agricultural Sciences, 60: 119-122

Sharma UC (1991) Release pattern of nitrate-N from nitrogenous fertilizers under field condition in Meghalaya and its effect on the yield of potato. Indian J Agric Sci 61:561–566

Sharma UC (1992a) Effect of soil texture and precipitation on phosphorus leaching in Alfisols of Meghalaya. J Indian Soc Soil Sci 40:413–416

Sharma UC (1992b) Critical limit of P for potato in Alfisols of Meghalaya. J Indian Soc Soil Sci 40:211–212

Sharma UC (1992c) Response of rainfed soybean to phosphorus and farmyard manure on acid soils of Nagaland. Indian Journal of Hill Farming, 5: 75–76.

Sharma UC (1993b) Response of soybean to micronutrients in Nagaland. Indian J Hill Farming 6:33–36

Sharma UC (1995) Nutrient management in oilseed crops in northeastern states for higher productivity. Fert News 40:29–36

Sharma UC (1996) Use of half-life technique to diagnose nitrate-N status in the petioles of potato during the growing season. Indian J Agric Sci 66:1–7

Sharma UC (1997a) Balanced fertilization in improving productivity in the north-eastern region. Fert News 42:71–77

Sharma UC (1997b) Some indigenous farming systems of water management. Asian Agri-History 1:267–273

Sharma UC (1997c) Constraints in sustainable agricultural production in NE states under different agro-climatic situations. Paper presented at International seminar on 'Sustainable Productivity in Energy, Food and Industry, held in Braunschweig, Germany, June 22–28

Sharma UC (1998a) Methods of selecting suitable land use system with reference to shifting cultivation in NEH region. Indian J Soil Conserv 26:234–238

Sharma UC (1998b) Nutrient management in potato and potato based cropping systems for higher productivity in north-eastern region of India. Scene Annual, Commercial Publication, Bombay, India, Indian Fert

Sharma UC (1998c) Water management by the tribes of northeastern region: the traditional wisdom and future strategies. Fertilizer News 43:43–49

Sharma UC (1999) Loss of N through leaching and runoff from two-potato land use systems on different soils. In: Heathwaite L (ed) Impact of land use change on nutrient loads from diffuse sources. Birmingham, IAHS Publ. No. 257. IAHS Press, Wallingford, UK, pp 27–32

Sharma UC (2000) Loss of N through leaching and run off from two potato based land use systems. In: Healthwaite L (ed) Impact of land use change on nutient load from diffuse sources. IAHS Press, Wallingford, UK, pp 27–32

Sharma UC (2002) Land use options for the northeastern region of India. In: Dyer FJ, Thomas MC, Olley JM (eds) The structure, function and management of Fluvial sedimentary system. IAHS Publ. no. 276, Alice Springs, Australia, pp 239–242

Sharma UC (2004a) Integrated catchment management in NE region of India to control fluvial sediment

transport. In: Webb B et al (ed) Hydrology—science and practice for the 21st Century, vol II, London, UK, pp 474–479
Sharma UC, Arora BR (1987a) The effect of nitrogen, phosphorus, potassium and FYM on the yield of potato. J Agric Sci, Cambridge 108:321–329
Sharma UC, Arora BR (1987b) Effect of phosphorus application on the micronutrients uptake of potato. J Micronutr Anal (England) (merged in Food Chemistry in 1990) 3:137–145
Sharma UC, Arora BR (1987c) The effect of nitrogen, phosphorus, potassium and FYM on the yield of potato. J Agric Sci (camb.) 108:321–329
Sharma UC, Arora BR (1988) Effect of nutrient application to potato on the starch, sugars and protein content of potato tubers. J Food Chem (England) 30:313–317
Sharma UC, Arora BR (1993) Uptake of zinc, manganese, iron and copper by potato as affected by applied K. Indian J Hill Farming 6:5–8
Sharma UC, Arora BR (1995) Effect of N application on the yield and N, P and K uptake of potato. Indian J Hill Farming 8:207–208
Sharma UC, Grewal JS (1991) Effect of levels and methods of P application on the yield and P uptake of potato on acid hill soil. Indian J Agric Sci 59:231–235
Sharma UC, Hussain SM (1991) Relative performance of soybean cultivars in acidic soils of Nagaland. Indian J. Hill Farming 4:13–18
Sharma UC, Patel CS (1991) Relative efficiency of N sources and amendments on acid soils of Meghalaya with potato as test crop. J Indian Soc of Soil Sci 36:128–132
Sharma UC, Prasad RN (1995) Socio-economic aspects of acid soil management and alternate land use systems for northeastern states of India. In: Date RA, Grundon NJ, Rayment GE, Probert ME (eds) Plant-soil internations at Low pH. Principles and Management, Kluwer Academic Press, The Netherlands, pp 689–696
Sharma UC, Sharma V (2009) Plant-livestock-soil-hydrology interactions in the north-eastern region of India. Biologia 64(3):460–464
Sharma UC, Singh K (1991) Integrated management of phosphorus and FYM in potato-radish cropping sequence in acidic soils of Meghalaya. J Indian Soc Soil Sci 39:468–471
Sharma UC, Singh RP (1996) Nutrient management in rice in north-eastern states of India. J Hill Res 9:177–191
Sharma UC, Tripathy AK (1999) Integrated phosphate management in rice-mustard cropping sequence in Nagaland. J Indian Soc Soil Sci 47:1–7
Sharma E, Chettri N, Oli K (2010a) Mountain biodiversity conservation and management: a paradigm shift in policies and practices in the Hindu Kush-Himalayas. Ecol Res 25:909–923
Sharma E, Chettri N, Oli KP (2010b) Mountain biodiversity conservation and management: a paradigm shift in policies and practices in the Hindu KushHimalayas. Ecol Res 25(5):909–923
Sharma UC, Datta M, Sharma V (2014b) Soil fertility, erosion, runoff and crop productivity as affected by different farming systems. Ecopersia 2:629–650
Sharma LN, Vetaas OR, Chaudhary RP, Maren IE (2014c) Ecological consequences of land use change: Forest structure and regeneration across the forest-grassland Ecotone in Mountain Pastures in Nepal. J Mt Sci 11:838–849
Sharma UC, Bujarbaruah KM (2016) Issues and challenges in shifting cultivation in the present context. Policy paper No. 83, Naational Academy of Agricultural Sciences, New Delhi, India
Sharma UC, Bujarbaruah KM (2017) Issues and challenges in shifting cultivation and its relevence in the present context. National Academy of Agricultural Sciences, Policy Paper No. 83
Sharma UC, Datta M, Samra JS (2006) Soils and their management in Northeast India Indian Council of Agricultural Research Complex for Northeastern Hills Region, pp 505
Sharma UC, Sharma V (2003a) Mathematical model for predicting soil erosion by flowing water in ungauged watersheds. Paper presented at International Union of Geodesy and Geophysics, IUGG. In: Boer DH, Froehlich W, Mizuyama T, Pietroniro A (eds) Erosion prediction in ungauged basins: integrating methods and techniques (Supporo, Japan) IAHS Publ. 279. IAHS Press, Wallingford, UK, pp 79–83
Sharma UC, Sharma V (2003b) The 'Zabo' soil and water management and conservation system in northeast India: tribal beliefs in the development of water resources and their impact on society—an historical account of a success story. In: Rodda JC, Ubertini L (eds) The basis of civilization, water science. Rome, Italy, IAHS Publ. 286. IAHS Press, Wallingford, U.K, pp 184–191
Sharma UC, Sharma V (2004c) Implications of nutrient and soil transfer with runoff in the northeastern reghion of India. In: Golosov V, Belyaev V, Walling DE (eds) Sediment transfer through the fluvial system. IAHS Publ. No. 288. (Moscow, Russia), IAHS Press, Wallingford, UK, pp 488–493
Sharma UC, Sharma V (2005) Traditional wisdom of tribal people in integrated water resources management and environmental quality. In: Takara K, Tachikawa Y, Nawarathna NMNSB (eds) Monitoring, prediction of water related disasters, Kyoto University, Japan, pp 709–712
Sharma UC, Singh KA (1994) Agriculture with alder (*Alanus nepalences*): a potential farming system of Nagaland. Paper presented at International seminar on management of soils for higher productivity, New Delhi, India, Nov 29–Dec 2
Sharma UR, Yonzon PB, Steiner A (2005) People and protected areas in South Asia. IUCN World Commission on Protected Areas (WCPA), Resources Himalaya Foundation, NP, p 177 ISBN:
Sharma NK, Ghosh BN, Mandal D, Singh RJ, Mishra PK (2014a) Conservation agriculture for resource conservation in North-Western Himalayan Region. In:

Somasundaram J, Chaudhary RS, Subba Rao A, Hati KM, Sinha NK, Coumar V (ed) New India Publishing Agency, New Delhi, p 528
Sharma E, Molden D, Rahman A, Khatiwada YR, Zhang L, Singh SP, Yao T, Wester P (2019) Introduction to the Hindu Kush Himalaya assessment. In: Wester P, Mishra A, Mukherji A, Shrestha AB (eds) The Hindu Kush Himalaya assessment—mountains, climate change, sustainability and People Springer Nature Switzerland AG, Cham
Sherchan DP, Karki KB (2005) Pland nutrient management for improving crop productivity in Nepal. In: Improving plant nutrient management for better farmer livelihood, Food Security and Environmental Sustainability. Proc. Regional Workshop, Beijing, China, pp 41–57, 12–16 December
Shobana S, Malleshi NG (2007) Preparation and functional properties of decorticated finger millet (*Eleusine coracana*). J Food Eng 79:529–538
Shoemaker HE, McLean EO, Pratt PF (1961) Buffer methods for determining lime requirement of soils with appreciable amounts of extractable aluminum. Soil Sci Soc Am J 25(10):2136. https://doi.org/10.2136/sssaj1961.03615995002500040014x
Shrestha AB, Pradhan NS (2015) Strengthening flash flood risk management in the Hindu Kush Himalayas and better interfaces with local institutions. ICIMOD 7th World Water Forum, Daegu and Gyeongbuk Republic of Korea, 12–17 April 2015
Shrestha A, Aryal R (2011) Climate change in Nepal and its impact on Himalayan glaciers. Reg Environ Change 11(1):565–577
Shrestha UB, Bawa KS (2015) Harvesters' perceptions of population status and conservation of Chinese caterpillar fungus in the Dolpa region of Nepal. Reg Environ Change 15(8):1731–1741
Siddiqui K (1997) Land management in South Asia: a comparative study. Oxford University Press, New York, Karachi
Singh JS (2006) Sustainable development of the Indian Himalayan region: linking ecological and economic concerns. Curr Sci 90:784–788
Singh K, Sharma UC (1987) Response of potato to magnesium on acidic hill soil. J Hill Res 1:88–95
Singh K, Sharma UC (1988) Response of potato cultivars to phosphorus in acidic soil of Meghalaya. Indian J Agric Sci 58:600–602
Singh K, Sharma UC (1989) Nutrients needs of maize-potato cropping sequence in acidic soils of Meghalaya. Indian J Agric Sci 59:159–161
Singh JS, Pandey U, Tiwari AK (1984) Man and forests: a central Himalayan case study. Ambio 13:80–87
Singh BP, Das M, Prasad RN (1990) Evaluation of available Cu status in high altitude wetland rice soils. J Indian Soc Soil Sci 38:464–468
Singh S, Singh RN, Prasad J, Kumar B (2002) Effect of green manuring, FYM and biofertilizer in relation to fertilizer nitrogen on yield and major nutrient uptake by upland rice. J Indian Soc Soil Sci 50:313–314
Singh SP, Bassignana-Khadka I, Karky BS, Sharma E (2011) Climate change in the Hindu Kush-Himalayas: the state of current knowledge. ICIMOD, Kathmandu
Singh BP, Das M, Ram M, Dwivedi BS, Prasad RN (1993) Ann Agric Res 14:100–103
Singh BP, Dhyani SK (1995) Significance of jackfruit in restoration of soil fertility. In: Annual report. ICAR Research Complex, Umiam, Meghalaya, India
Singh A (1989) Bamboo drip irrigation. Res Bull. Indian Council of Agriculture Research Complex for North-eastern Hills Region of India, Shillong, Meghalaya, India
Siqueira JO, Moreira FMS (1997) Microbial population and activities in highly weathered acidic soils: highlights of the Brazillian research. In: Moniz AC, Furlani AMC, Schffert RE, Fageria NK, Rosolem CA, Cantarella H (eds) Plant–soil interactions at low pH: sustainable agriculture and forestry production. Brazilian Soil Science Society, Brazil, pp 139–156
Siquiera JO, Moreira FMS, Grisi BM, Hungria M, Araujo RS (1994) Microrganisms processes biologicos do solo perspective ambiental. EMBRAPA, Brazil, p 142
Soil Survey of Pakistan (1988) Reconnaissance soil survey reports from 1965–88. Lahore
Solaiman S, Ahmed N (2005) Plant nutrition management for better farmer livelihood, food security and environment. In: Improving plant nutrient management for better farmer livelihoods, food security and environmental sustainability. Proceedings of a regional workshop, Beijing, China, FAO, Regional Office for Asia and the Pacific, Bangkok. 12–16 December 2005
Sonowal DK, Tripathi AK, Krirekha V (1989) Zabo, an indigenous system of Nagaland. Indian J Hill Res 2:1–8
Sousa DMG, Carvalho LICB, Miranda LN (1986) Correcao de acidez do solo. In: Goedert WJ (ed) Solos dos Cerrados technologias e Estrategias de Manejo Livraria Nobel, Sao Paulo, Brazil, pp 99–127
Srivastava AK, Singh S (2006) Biochemical markers and nutrient constraints diagnosis in citrus: a perspective. J Plant Nutr 29(5):827–855
Srivastva AK (2006) Sustainability of hill agriculture ecosystems and watersheds. In: Gupta HS, Srivastva AK and Bhatt JC (eds) Sustainable production agricultural watersheds in Northwest Himalayas. Vivekananda Parvatiya Krishi Anusandhan Sansthan, Almora, Uttrakhand, India, p 615
Statastical Year Book of Tibet, Yunnan and Sichuan (1997) The state of mountain agriculture in China. Tibet, China
Stoddart LA, Smith AD, Box TW (1975) Range management. McGraw Hill, New York, USA
Swaminathan MS (1991) Pandit Govind Ballabh Pant Memorial Lecture: I. G. B. Pant Institute of Himalayan Environment and Development, Kosi-Katarmal, Almora

Swarup A, Purakayastha TJ (2008) Conservation and management of soil and water resources. Division of Soil Science and Agricultural Chemistry Indian Agricultural Research Institute New Delhi, India

Szabo S, Nicholls RJ, Neumann B, Renaud FG, Matthews Z, Sebesvari Z, AghaKouchak A, Bales R, Ruktanonchai CW, Kloos J, Foufoula-Georgiou E, Wester P, New M, Rhyner J, Hutton C (2016) Making SDGs work for climate change hotspots. Environ Sci Policy Sustain Develop 58(6):24–33

Szabo S, Hossain MS, Renaud F, Traore D, Hussain A, Matczak P, Ahmad S, Singh DR, Neumann B, Matthews Z (2018) Accelerating progress toward the zero hunger goal in cross-boundary climate change hotspots. Environ Sci Policy Sustain Develop 60 (3):18–27

Tarolli P, Straffelini E (2020) Agriculture in hilly and mountainous landscapes: threats, monitoring and sustainable management. Geogr Sustain 1:70–76

Thapa GB, Paudel GS (2002) Farmland degradation in the mountains of Nepal: a study of watersheds? with? and? without? external intervention. Land Degrad Dev 13(6):479–493

Theng BKG (1991) Soil science in the tropics—the next 75 years. Soil Sci 151:76–91

Tiwari AK (2000a) Infrastructure and economic development in Himachal Pradesh. Indus Publishing, New Delhi

Tiwari PC (2000b) Land-use changes in Himalaya and their impact on the plains ecosystem: need for sustainable land use. Land Use Policy 17(2):101–111

Tiwari PC, Joshi B (2015) Climate change and rural out-migration in Himalaya. Chang Adapt Socio-Ecol Syst 2(1):8–25

Tiwari DN, Loof R, Paudyal GN (1999) Environmental–economic decisionmaking in lowland irrigated agriculture using multi-criteria analysis techniques. Agric Syst 60(2):99–112. https://doi.org/10.1016/S0308-521X(99)00021-9

Toky OP, Ramakrishnan PS (1982) A comparative study of the energy budget of hill agro-ecosystems with emphasis on the slash and burn system (Jhum) at lower elevations of north-eastern India. Agric Syst 9:143–154

Toky OP, Ramakrishnan PS (1983a) Secondry succession following slash and burn agriculture in Northest India. I—Biomass, litterfall and productivity. J Ecol 71 (3):735–745

Toky OP, Ramakrishnan PS (1983b) Secondry succession following slash and burn agriculture in Northest India II—nutrient cycling. J Ecol 71(3):747–757

Tsering K, Sharma E, Chettri N, Shrestha A (eds) (2010) Climate change vulnerability of mountain ecosystems in the Eastern Himalayas—synthesis report. International Centre for Integrated Mountain Development (ICIMOD), Nepal, Kathmandu

Tubiello N, Soussana JF, Howden SM (2007) Crop and pasture response to climate change. Proc Natl Acad Sci USA 104:19686–19690

Tulachan PM (2001) Mountain agriculture in the Hindu Kush-Himalaya: a regional comparative analysis. In: Mountain research and development. ICMOD, Kathmandu, Nepal 21:260–267

UNDP (2009) Socio-economic baseline survey of Chittagong Hill Tracts. Chittagong Hill Tracts Development Facility (CHTDF), Dhaka, Bangladesh

United Nations Development Programme (UNDP) (2007) Integrated household living conditions survey in Myanmar. Poverty profile, Myanmar

United Nations Development Programme (2004) Myanmar—agricultural sector review investment strategy. Volumes 1 and 2. Sector review. Accessed on 23 September 2018

ur-Rahman A (2003) Land and agriculture resources: a case of NWFP. Pak Geogr Rev 58(1 & 2):1–94

Urwin K, Jordan A (2008) Does public policy support or undermine climate change adaptation? Exploring policy interplay across different scales of governance. Glob Environ Chang 18:180–191

USDA (1961) The agriculture handbook No. 210: land-capability classification. United States Department of Agriculture. United States Government Printing Office, Washington, DC

Valdiya KS, Bartarya SK (1991) Hydrogeological studies of springs in the catchment of the Gaula River, Kumaon Lesser Himalaya India. Mt Res Dev 11:239–258

Valdiya KS (1997) Developing a paradise in peril. G. B. Pant Memorial Lecture: VII. G. B. Pant Institute of Himalayan Environment and Development, Kosi-Katarmal, Almora

Venkatesh MS, Kumar K, Raychaudhuri M, Datta M, Sharma UC (2006) Crop production and crop response to nutrients. In: Sharma UC, Datta M, Samra JS (eds) Soils and their management in Northeast India Indian Council of Agricultural Research Complex for North-eastern Hills Region, India, pp 505

Verma A, Schmidt-Vogt D, De Alban JD, Lim CL (2021) Drivers and mechanisms of forest change in the Himalayas. Glob Environ Chang. https://doi.org/10.1016/j.gloenvcha.2021.102244

von Grebmer KD, Headey T, Olofinbiyi D et al (2013) The challenge of hunger: building resilience to achieve food and nutrition security. 2013 Global Hunger Index. International Food Policy Research Institute (IFPRI), Welthungerhilfe, and Concern Worldwide

von Uexkull HR, Bosshart RP (1989) Management of acid upland soils in Asia. In: Craswell ET, Pushparajah E (eds) Management of acid soils in the humid tropics of Asia. ACIAR Monograph No. 13 (IBSRAM Monograph No.1), x + p 118

Wakeel A (2015) Balanced use of fertilizer in Pakistan: status and perspective. International Potash Institute, Switzerland, Faislabad, Pakistan

Wang H, Zhou X, Wan C, Fu H, Zhang F, Ren J (2008) Eco-environmental degradation in the northeastern margin of the Qinghai-Tibetan Plateau and comprehensive ecological protection planning. Environ Geol 55(5):1135–1147

Wang Y, Wu N, Kunze C, Long R, Perlik M (2019) Drivers of change to mountain sustainability in the Hindu Kush Himalaya. In: Wester P, Mishra A, Mukherji A and Shrestha A (eds) The Hindu Kush Himalaya assessment. Springer, Cham, pp 17–56. https://doi.org/10.1007/978-3-319-92288-1_2

Wangchuk P, Tashi K (2004) Shifting cultivation in Bhutan. Policy and Planning Division, Ministry of Agriculture, Thimphu, Bhutan

Wasteland Atlas of India (2010) Government of India. New Delhi, India

WCMC-UNEP (2002) Mountain watch: environmental change and sustainable development in mountains. UNEP—World Conservation Monitoring Centre, Cambridge, UK

Wilkes A (2008) Towards mainstreaming climate change in grassland management: policies and practices on the Tibetan Plateau. Working Paper No. 67. World Agroforestry Centre, ICRAF, China, p 43

William M (1989) Deforestation: past and present. In: Progress in Human Geography, pp 176–208

Win SH (2005) Improving plant nutrient management for better farmer-livelihood, food security and environment. In: Improving plant nutrient management for better farmer livelihoods, food security and environmental sustainability. Proceedings of a Regional Workshop, Beijing, China. FAO, Regional Office for Asia and the Pacific, Bangkok, 12–16 December 2005

Woomer PL, Swift MJ (1994) The biological management of tropical soil fertlity. John Wiley & Sons, New York, USA, p 243

World Bank (2015) Agriculture for development: fertilizer use in the World. World Bank, Washington

World Bank (2005) Pakistan: country water resources assistance strategy, water economy running dry. Agriculture and Rural Development Unit, South Asia Region, Report No. 34081-PK, World Bank

WRI (2011) Decision making in a changing climate—adaptation challenges and choices, World resources 2010–2011

Wu N, Rawat GS, Joshi S, Ismail M, Sharma E (2013a) High-altitude rangelands and their interfaces in the Hindu Kush Himalayas. ICIMOD, Kathmandu, Nepal

Wu N, Rawat GS, Joshi S, Ismail M, Sharma E (2013b) High-altitude rangelands and their interfaces in the Hindu-Kush Himalayas. ICIMOD, Kathmandu, pp 1–189

Xia GM, Tao Y, Zheng LL, Chu FY (2013) Effects of alternate irrigation on the water infiltration and utilization efficiency of the peanut. J Irrig Drain 33:113–115

Xu J, Badola R, Chaudhary RP, Zomer R, Pokhrel B, Hussain SA, Pradhan S, Pradhan R (2019) Sustaining biodiversity and ecosystem services in the Hindu Kush Himalaya. In: Wester P, Mishra A, Mukherji A, Shrestha AB (ed) The Hindu Kush Himalaya assessment: mountains, climate change, sustainability and people. Basel, Switzerland: Springer Nature, pp 127–166. https://doi.org/10.1007/978-3-319-92288-1

Xu B, Xin X, Tang H, Zhou Q, Chen Y (1999) The influence and strategy of global climate change to agricultural geographical distribution. Prog Geogr 18:316–321

Xu J, Grumbine RE, Shrestha A, Eriksson M, Yang X, Wang YUN et al (2009) The melting Himalayas: cascading effects of climate change on water, biodiversity, and livelihoods. Conserv Biol 23(3):520–530

Yang M, Wang S, Yao T, Gou X, Lu A, Guo X (2004) Desertification and its relationship with permafrost degradation in Qinghai-Xizang (Tibet) plateau. Cold Reg Sci Technol 39:47–53

Yi S, Wu N, Luo P, Wang Q, Shi F, Sun G, Ma J (2007) Changes in livestock migration patterns in a Tibetan-style agropastoral system. Mt Res Dev 27:138–145

Yi S, Sharma E (2009) Climate change and the Hindu Kush–Himalayan rangelands. ICIMOD, Kathmandu, Nepal

Zapata F, Hera C (1995) Enhancing nutrient management through use of isotopic techniques. In: Nuclear techniques in soil-plant studies for sustainable agriculture environmental preservation. Int Atomic Energy Agency, Vienna, Austria, pp 83–105

Zarkesh MMK, Ghoddusi J, Zaredar N, Soltani MJ, Jafari S, Ghadirpour A (2010) Application of spatial analytical hierarchy process model in land use planning. J Food Agric Environ 8(2):970–975

Zhang XS (1983) The Tibetan Plateau in relation to the vegetation of China. Ann Mo Bot Gard 70:564–570

Zhang G, Dong J, Zhou C, Xu X, Wang M, Ouyang H, Xiao X (2013) Increasing cropping intensity in response to climate warming in Tibetan Plateau. China, Field Crops Research 142:36–46

Zhang G, Shen R (2013) Impact of high intensity land uses on soil and environment in China. In: Toth G, Li X 2013 (eds) Threats to the soil resource base of food security in China and Europe. A report from the Sino-EU panel on land and soil. Report EUR 25632 EN

Zhao J, Michalk DL, Guozhen Du, Yifei W, Kemp DR, Helen N (2007) Effect of fertilization on pasture in acid soil in Yunnan Province in China. N Z J Agric Res 50(4):523–535. https://doi.org/10.1080/00288230709510320

Zheng D (1996) The system of physico-geographical regions of the Qinghai-Tibet (Xizang) Plateau. Science in China (series d) 39:410–417

Zhou ZH, Du SH, Liu GC (2009) Acidification of surface soil in croplands in the semiarid middle Tibet Plateau, China. In: Guerrero JE (ed) International conference on environmental science and information. Application Technology, CPS Publishing, California, USA, pp 209–212

Zhou Z, Zhang G, Song X, Du S, Liu G (2010) Soil acidification with various planting patterns in Lhasa, Tibet, China. Paper presented at the 19th world congress of soil science, soil solutions for a changing world, Brisbane, Australia, 1–6 August 2010

Zolekar RB, Bhagat VS (2015) Multi-criteria land suitability analysis for agriculture in hilly zone: remote sensing and GIS approach. Comput Electron Agric 118:300–321

Index

U. C. Sharma et al., *Soils in the Hindu Kush Himalayas*, Geography of the Physical Environment,
https://doi.org/10.1007/978-3-031-11458-8

R

S

GPSR Compliance
The European Union's (EU) General Product Safety Regulation (GPSR) is a set of rules that requires consumer products to be safe and our obligations to ensure this.

If you have any concerns about our products, you can contact us on

ProductSafety@springernature.com

In case Publisher is established outside the EU, the EU authorized representative is:

Springer Nature Customer Service Center GmbH
Europaplatz 3
69115 Heidelberg, Germany

www.ingramcontent.com/pod-product-compliance
Ingram Content Group UK Ltd.
Pitfield, Milton Keynes, MK11 3LW, UK
UKHW051640031026
15588UKWH00007B/1250

* 9 7 8 3 0 3 1 1 1 4 5 9 5 *